Donkey and Mule Medicine: Foundations, Breeds, and Management

Amy K. McLean
Francisco Javier Navas González
Erin L. Goodrich
Editors

Donkey and Mule Medicine: Foundations, Breeds, and Management

A Practical Approach to the Treatment and Care of Long Ears
Volume 1

Springer

Editors
Amy K. McLean
Department of Animal Science
University of California
Davis, CA, USA

Francisco Javier Navas González
Department of Genetics
University of Córdoba
Córdoba, Spain

Erin L. Goodrich
Department of Population Medicine
and Diagnostic Sciences
Cornell University College of Veterinary
Ithaca, NY, USA

ISBN 978-3-032-12725-9 ISBN 978-3-032-12726-6 (eBook)
https://doi.org/10.1007/978-3-032-12726-6

Cover image: Photo credit to Cassidy Cooper (pictured from left to right, a mule, hinny and donkey).

This Springer imprint is published by the registered company Springer Nature Switzerland AG
The registered company address is: Gewerbestrasse 11, 6330 Cham, Switzerland

Preface – A Long-Awaited Resource for Long-Eared Equids

Volume 1: Foundations, Breeds, and Management

Donkeys, mules, and hinnies have served humanity for millennia—yet their origins, breeds, and management practices have long been overlooked or misunderstood. This first volume of *A Long-Awaited Resource for Long-Eared Equids* offers the foundational knowledge necessary to appreciate these remarkable animals, tracing their journey from domestication to their essential roles in agriculture, companionship, and industry worldwide.

Written by leading veterinarians, researchers, and equid specialists from across the globe, this volume explores the roots of donkey diversity and the genetic, environmental, and cultural forces that have shaped their populations. It brings attention to urgent challenges in conservation, welfare, and sustainable breeding while celebrating the resilience and adaptability of these long-eared equids.

Readers will discover not only the historical and biological foundations of donkey and mule populations but also their continued importance to communities and economies today. The chapters emphasize practical management, nutrition, welfare, and behavioral understanding—tools essential for improving the daily lives of donkeys and mules everywhere.

This volume represents the groundwork upon which the second, clinical volume builds.

Together, they complete a comprehensive resource dedicated to the understanding and care of donkeys, mules, and hinnies. We hope this first volume deepens your respect for these animals and inspires continued efforts to preserve their genetic heritage and ensure their welfare across the world.

Davis, CA, USA — Amy K. McLean
Córdoba, Spain — Francisco Javier Navas González
Ithaca, NY, USA — Erin L. Goodrich

Introduction – Volume 1: Foundations, Breeds, and Management

For thousands of years, donkeys—and their hybrid offspring, mules and hinnies—have faithfully served humankind. From ancient trade routes to modern farms, these animals have carried goods, plowed fields, and sustained livelihoods in some of the world's most challenging environments. Despite their endurance and intelligence, donkeys and mules have long remained in the periphery of scientific inquiry and veterinary education. This neglect stems, in part, from the widespread assumption that they can be treated simply as small horses with long ears. Yet, their evolutionary path, physiological adaptations, and behavioral characteristics make them distinct species that require equally distinct management and medical approaches.

The purpose of this first volume, *Foundations, Breeds, and Management*, is to establish a comprehensive scientific and practical framework for understanding donkeys and mules within their historical, biological, and cultural contexts. It is structured in three main parts, each addressing critical dimensions of long-eared equid stewardship and conservation.

Part A: Origins and Conservation traces the domestication and global dispersal of donkeys, revealing how mitochondrial haplogroups and genomic analyses illuminate their evolutionary journey from African wild asses to today's diverse breeds. This section emphasizes the importance of conservation genetics and sustainable breeding, presenting case studies on threatened populations, breed preservation programs, and policies that balance cultural heritage with scientific advancement. It also explores the sociocultural dimensions of donkey conservation, recognizing that these animals' survival is often linked to human communities facing their own economic and environmental challenges.

Part B: Roles and Industry explores the expanding roles of donkeys and mules in modern economies. Working equids remain vital to agricultural and transport systems in low-income regions, while new industries—such as dairy donkey farming and gelatin production—have created both opportunities and welfare concerns. The section offers insight into production systems, welfare impacts, and ethical considerations surrounding the global skin trade, industrial use, and the growing market for donkey milk and meat. By examining these transformations, the volume contextualizes the species' economic significance within global development and trade networks.

Part C: Husbandry and Functional Aspects provides the applied knowledge needed to manage donkeys, mules, and hinnies effectively and humanely. Readers will find detailed discussions on nutrition, anatomy, parasitology, dentistry, biomechanics, orthopedics, and hoof care, as well as practical approaches to behavior, harnessing, and saddle fit. The section integrates physiological understanding with clinical practice, emphasizing that welfare begins with informed daily management.

Together, these parts present a foundation for the next generation of equid science—one that unites molecular biology, field observation, and traditional husbandry into a holistic understanding of long-eared equids. By acknowledging their evolutionary uniqueness, cultural significance, and biological complexity, this volume seeks to elevate the scientific and ethical treatment of donkeys and mules worldwide.

It is our hope that this work empowers veterinarians, students, and owners to not only enhance animal health and welfare but also preserve the genetic and cultural heritage that these species represent. The long-eared equids, once underestimated, now stand at the center of a growing field of research and compassion—a testament to their enduring partnership with humankind.

University of California, Davis, CA, USA — Amy K. McLean
University of Córdoba, Córdoba, Spain — Francisco Javier Navas González
Cornell University, Ithaca, NY, USA — Erin L. Goodrich

Contents

Part A

Origins and Conservation

Companion and Breed Considerations: Donkey Breeds and Populations Worldwide—Origins, Challenges, Welfare and Opportunities

Miguel Quaresma and Francisco Javier Navas González

Abstract

The donkey was historically one of the main working animals of Africa, Asia and South Europe. Not fit for battle, capable of surviving with relatively low costs and stoic, the donkey, and its hybrid with the horse, the mule, were fundamental for the development of agriculture and commercial routes. They were particularly important for the poor population, and this may be the reason why today they are not the main focus of governmental programs, social welfare concerns and scientific funding.

Domesticated in Africa, this equid spread worldwide. Susceptible to cold conditions, but very resistant to dry, hot climates, the donkey stayed mainly in the regions with these climates. Further adaptation to local work needs, climate and food availability led to the development of local breeds or phenotypic differentiation. However, with the loss of the need for working donkeys due to industrialization, this breed diversity is being lost. The species is not endangered and is still useful, but most breeds are in a critical situation, due to low numbers of individuals and high levels of inbreeding. Although solid information is missing about the real situation of most donkey breeds, sparse literature points out that of the around 200 phenotypically distinct of donkeys, most are extinct, or at least

M. Quaresma (✉)
Department of Veterinary Sciences, School of Agrarian and Veterinary Sciences, University of Trás-os-Montes and Alto Douro, Vila Real, Portugal

CECAV, Animal and Veterinary Research Center, University of Trás-os-Montes and Alto Douro, Vila Real, Portugal

AL4AnimalS-Associate Laboratory for Animal and Veterinary Sciences, Vila Real, Portugal
e-mail: miguelq@utad.pt

F. J. Navas González
Genetics Department, Veterinary Sciences, University of Córdoba, Rabanales University Campus, Cordoba, Spain

A. K. McLean et al. (eds.), *Donkey and Mule Medicine: Foundations, Breeds, and Management*, https://doi.org/10.1007/978-3-032-12726-6_1

vulnerable to extinction, with just a few thriving. Officially, just around 30 breeds of donkeys were recognized. Furthermore, some populations that are named donkey breeds by some authors should be just considered donkey populations as they lack the technical criteria of genetic and phenotypic homogeneity to be considered true animal breeds.

This chapter cannot summarize the situation of pure donkey breeds worldwide, due to a scarcity of reliable sources with data about this subject. It will, however, describe the current situation of donkey breeds in different parts of the world, based on peer-reviewed published papers and, when these are lacking, official data published online. Programs, run by civil society and governments, are in place to save genetic diversity and promote the donkey. Also pivotal to a positive future for this marvelous animal is the rise in awareness of the welfare problems donkeys face globally.

Keywords

Donkey breeds, welfare, non-governmental organizations · Working animals · Human-animal relations · Non-industrialized countries

1 Introduction

Approximately 59 million donkeys are used worldwide today with most of them in economically underdeveloped countries, in which they are essential for the subsistence of millions of people (Kim et al. 2021). Developing countries are home to 96% of the donkeys, where they are a vital component of rural development for pastoral communities. They enable farmers to access distant markets in which they sell their products, which would be inaccessible if it were not for their donkeys being able to carry goods and people inexpensively over long distances. Despite these important roles played by donkeys, not much value has been given to them in most parts of the world (Pritchard et al. 2005).

The domestic donkey is integrated into the Equidae family, that includes a single genus Equus, which contains four subgenera with eight species. The subgenus Asinus includes *Equus asinus africanus* (donkeys), *Equus hemionus* (Onagers and Asian wild assess), and *Equus kiang* (Kiangs). The horse is in the subgenus Equus, with which the donkey can interbreed, producing mules and hinnies (Nowak and Paradiso 1983).

Donkeys are amongst the least studied and most neglected domestic species (Blench and MacDonald 2000). Around the World, the survival of most of the donkey breeds is threatened by the decline of their populations (Ivankovic et al. 2002; Ciampolini et al. 2007; Bordonaro et al. 2012; Quaresma et al. 2014). In Fig. 1 the most representative breeds of some countries can be seen. Despite the still large numbers of donkeys worldwide, there is a real risk of genetic conservation failure, not only of endangered donkey breeds, but of this domestic species worldwide, due to the disparities in conservation and improvement measures taken by different

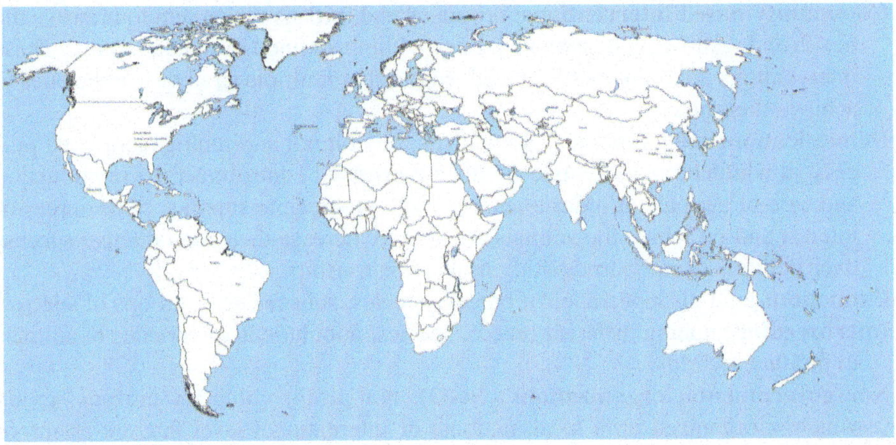

Fig. 1 Officially recognized breeds of donkeys worldwide. Although there are almost 200 donkey populations that could be potentially considered as donkey breeds, there are just around 30 officially recognized pure donkey breeds, with the necessary closed Studbook and genealogy registry

countries (Kugler et al. 2008). The donkey breeds are a non-renewable genetic resource (Mariante et al. 2002) resulting from long-term ecological conditions, genetic drift and natural and human selection (Seyiti and Kelimu 2021). Hence, there are clear benefits in describing and defending the genetic diversity of the numerous donkey breeds and populations (Zhang et al. 2023). Measures must be taken by official authorities and local breeders to conserve donkey populations, with the establishment of robust breeding programs (FAO 2015), which also rely on identification of donkey breeds and breeding populations, with a correct phenotypic and genetic characterization, which is still incomplete in most cases (Kugler et al. 2008).

As transportation and farming mechanization becomes more common, the number of donkeys may decrease. However, this process has also led to a diversification of the ways donkeys are employed by humans (Xie 1987; Xu et al. 2012). With the development of agricultural mechanization, donkeys gradually have lost their traditional role as working animals (Li et al. 2022a, b) and become multi-purpose farm animals (Chai et al. 2022). However, the donkey is still an important domestic animal that provides a reliable source of work and other benefits for many human populations (Han et al. 2014).

2 Key Concepts

Breeds: specific group of domestic animals with a homogeneous phenotype, behavior and other characteristics that distinguish it from other individuals of the same species, officially recognized and with registry in a Studbook.

Community-based interventions: programs and initiatives that aim to improve the health and well-being of specific groups within a defined local community. These interventions are often multi-faceted, employing multiple strategies and tactics to achieve their goals.

Domestication: coevolutionary, mutualistic, biocultural and multigenerational process, in which humans assume significant levels of control over the reproduction and care of animals in order to ensure more predictable supplies of resources of interest and by which the animals are able to increase their reproductive success over individuals not participating in this relationship.

Extinction: total disappearance of breeds, species, subspecies or groups of species.

Interbreeding: mating different breeds, species, populations or varieties of animals to produce hybrids.

Non-governmental organizations (NGO): non-profit, voluntary citizens' group which is organized on a local, national or international level that are nonprofit entities independent of governmental influence (although they may receive government funding).

Non-industrialized countries: countries that exhibit poor economic growth and have the lowest Human Development Index (HDI) measures.

Populations: community of animals among whose members interbreeding occurs.

Studbook: an official record of the pedigree of purebred animals, belonging to a breed.

Welfare: well-being of non-human animals.

Working animals: an animal that is kept by humans and trained to perform tasks.

3 Historical Framework

3.1 Donkey Domestication and Breed Development

In ancient periods, the donkey played a pivotal role in agriculture and transportation, remarkably in the more arid areas of the globe (Smith and Pearson 2005). Particularly in Eurasia and Africa, the donkey has been broadly used for work for close to six millennia (Clutton-Brock 1992; Blench 2004; Smith and Pearson 2005; Vilá et al. 2006; Rossel et al. 2008).

Donkey breeds worldwide frequently have an uncertain genetic background (Navas et al. 2017). Donkey remains are scarce in the archaeological record and sometimes mistaken with other equid remains, if using only morphological criteria (Geigl and Grange 2012; Han et al. 2014). Tracing the origin of the different breeds is not easy, as there is a great lack of historical or trade records, as well as limitations in genetic studies, as the origin of some genetic markers, mtDNA haplotypes (group of genes within an organism that was inherited together from a single parent) for example, is not always clear (Stanisic et al. 2017).

To better trace the origins of the domestic donkey, their spread around the world and the origin of the different breeds, mitochondrial DNA (mtDNA) has been of paramount importance, despite some caveats. It can be used for species identification and in the study of genetic diversification of domestic species into different breeds with better results than nuclear DNA because of characteristic such as the presence of more copies, well-known gene structure, lack of introns, high mutation rate, and an absence of recombination events (Aranguren-Mendez et al. 2004; Cai et al. 2011; Sun et al. 2016; Han et al. 2014). Usually, the portion tested is in the mtDNA D-loop, which is an extremely variable region, characterized by fast evolution and a non-coding function, with the largest mutation rate in the whole mtDNA (Sbisa et al. 1997), which makes it particularly useful to assess the genetic variation, evolutionary relationships, and matrilineal genealogy, populations and breeds (Xia et al. 2023). Besides studying the mtDNA in current populations, testing ancient remains is important to provide insights into the domestication and early spread around the world of the donkey (Orlando et al. 2009).

Several studies on donkey mtDNA have assessed evolutionary relationship, variation, and genetic diversity among several Asian, European, and African breeds (Lei et al. 2007; Han et al. 2014, 2014; Perez-Pardal et al. 2014; Ozkan Unal et al. 2020; Ma et al. 2020). These studies of donkey mitochondrial DNA proved that there were two distinguished maternal genetic lineages (Clade I and Clade II) during donkey domestication, which seem to represent the two wild ancestral subspecies that are in the origin of the domestic donkey. The survey of domestic donkey mtDNA D-loop diversity identified two highly divergent mtDNA lineages, estimated to have diverged over 0.303 to 0.910 million years ago. This divergence was found to be much earlier than the domestication of the donkey, suggesting that two separate maternal origins of the domestic donkey were from two distinct wild populations (Beja-Pereira et al. 2004; Chen et al. 2006; Kimura et al. 2011). The data obtained so far highlighted the ancient Nubian wild donkey was an ancestor of Clade I donkeys, while the ancestors of Clade II donkeys are still unknown because the findings trace their origin to a relative of the Somali wild donkeys (*Equus africanus somaliensis*), which is already extinct (Beja-Pereira et al. 2004; Kimura et al. 2011). This molecular evidence of two distinct lineages also supports two or more independent domestication events, both in Africa (Beja-Pereira et al. 2004) and both contributing to the genetic background of most live donkeys. There used to be an old differentiation between large, dark skin donkeys and the smaller, grey or white donkeys, naming the first European type and the second African type (Fig. 2). These were not confirmed by genetic testing, with all live donkeys being of African origin and with genetic contributions that are in most cases more indicative of convergent selection of phenotypic characteristics than of a different genetic origin.

The origin of Clade 2 donkeys may be much more complex than previously thought (Cozzi et al. 2017). New molecular evidence questions the hypothesis of Beja-Pereira et al. (2004) and Kimura et al. (2011) that extinct Somali wild donkeys were the ancestors of the Clade II donkeys (Kefena et al. 2014; Rosenbom et al. 2015; Stanisic et al. 2017; Xia et al. 2019) suggesting alternative origins of donkeys belonging to this lineage (Rosenbom et al. 2015). A morphological feature, black stripes on legs, present in Banat and Somali wild ass present populations, but also in Amiata donkeys, an Italian breed, suggests that the origin of Clade 2 (Somali lineage) donkeys may be much more complex than previously thought (Stanisic

Fig. 2 Two Portuguese breeds of donkeys, the (**a**) Burro de Miranda with a large frame (over 130 cm) and dark color and the (**b**) Burro Anão da Graciosa smaller (less than 110 cm) and typically grey (picture) or white (not represented). Although the phenotypic characteristics of these donkeys represent the two classical different morphotypes of donkeys, the "European" and the "African", no clear genetic distinct origin can back up this classification

et al. 2020). However, due to the complex origin of genetic characteristics convergent evolution or phenotypic similarities of different genetic origin cannot be ruled out. A specific genetic haplotype is found in Anatolian and donkeys from Cyprus (Cinar-Kul et al. 2016). Therefore, there is data supporting a close genealogical relationship of donkeys from all three south European Peninsulas, Turkey/Cyprus and Africa; and furthermore, reveal that haplotypes found in several Italian donkeys occupy an intermediate position at the backbone of the donkey genetic network between Clade 1 and Clade 2 haplotypes (Stanisic et al. 2017).

Morphologically, body length, carpal circumference, chest depth, hip height and tibia circumference, are the most common morphological traits used for donkey breed characterization (Folch and Jordana 1997; Sargentini et al. 2009, 2018; Kefena et al. 2011; Kosuková et al. 2015; Labbaci et al. 2018; Ayad et al. 2019).

3.2 Historical Roles of Donkeys and Mules

Probably in response to desertification of Northeastern Africa around 6000 years ago, humans domesticated the donkey (Beja-Pereira et al. 2004). Since then, until the last century, the donkey has played an important role as a working animal in human development, transportation and trading activities across Africa and Eurasia (Blench and MacDonald 2000). Since its domestication, the donkey has been used for the transportation of people and goods, as well as for working the land, in many cultures. Its domestication, along with the horse, marked cultural modification from mainly isolated agrarian lifestyles towards trade, both economic and cultural (Beja-Pereira et al. 2004). Aside from the use as a working equid or meat producing animal, historically, and still, donkey milk has been regarded as a good source of nutrition for treating conditions such as psoriasis and dry skin; joint pains, asthma, stomach ulcer, sore throat, liver problems, infectious diseases, fever, edema, poisoning, and wounds (Alessandri and Mari 2007; Tesse et al. 2009; Gastaldi et al. 2010; Fantuz et al. 2012; Karabagias 2017).

Closely related to the use of the donkey was the rearing and usage of mules. However, it has been very hard to document the evolution of mules' history, due to lack of both historical documents and fossil identification. When a robust historical registry is missing, it is hard to know mule production history as mule remains are difficult to detect archaeologically, because their remains are almost impossible to reliably distinguish from horses or donkeys (Blench and MacDonald 2000).

3.3 African Populations and Breeds

Although Africa is the place of origin of the donkey, the economic constraints that scientific investigation suffers in most of its countries do not allow it to produce much scientific knowledge on the genetic structure of the populations and breeds of the continent. The genetic population structure of native Ethiopian donkeys distinguished Abyssinian, Afar and Haraji donkeys from Ogaden, Omo and Sinnar donkeys, but did not separate Ethiopian donkeys into any genetically distinct group (Kefena et al. 2021).

3.4 European Populations and Breeds

Some haplotypes found in individuals belonging to several donkey breeds from Italy are positioned within the same central node of Clade 2, as Spanish ATI-1 haplotype, indicating a close genealogical relationship of donkeys from Iberian and Italian Peninsulas donkeys (Cozzi et al. 2017). The genetic distance of the Spanish donkey breeds studied by Aranguren-Méndez et al. (2001) is lower than the one observed inside the Dezhou donkey (Wang et al. 2023).

The Iberian Peninsula donkey population also derives from the two formerly identified ancestral sources, the well-known mtDNA Clades I and II. Furthermore, there has been historically a continuous transfer of animals from Africa, through the

Strait of Gibraltar (Epstein 1984), shown in the Spanish Catalana, Mallorquina and Zamorano-Leonessa breeds, with highly similar extant maternal markers to the populations observed in African donkey populations, both from Morocco and Zimbabwe. The mtDNA haplotypes ATI-1 and ATI-3, common to three Spanish donkey breeds and two African donkey populations, from Morocco and Zimbabwe, in similar proportions, embody the most ancestral mtDNA types (Aranguren-Mendez et al. 2004).

Donkeys were introduced around the Mediterranean Basin by the Greeks, brought from their African colonies in the second millennium BC (Vilá et al. 2006). In the case of the Balkan Peninsula, these introductions ensued in various waves, as donkeys fitting to Clade 2 appeared in Greece prior to those from Clade 1, but they expanded throughout the Balkans and Europe later than Clade 1 donkeys. According to local tradition, in the XVIII century the Habsburg Queen Maria Theresa transferred a herd of Spanish donkeys to the north-eastern part of Serbia, the Banat region, to work in local vineyards. These donkeys were stronger, taller and thinner than the local animals at the time, which made them more able to pass between the rows of the vineyards. This story was confirmed, once it was shown that the Banat donkey mitochondrial haplotypes, analyzed genealogically together with those reported in ancient and modern individuals from Spain, Italy, Turkey, Cyprus and Africa, are shared with three Spanish breeds and individuals belonging to Amiata and some other Italian breeds (Stanisic et al. 2020).

3.5 Asian Populations and Breeds

Chinese historical records uncovered that the donkey and mule were treasures rare in the northern and western regions of China 4000 years ago (Zheng 1985). In China, historically, donkeys were extensively used by the small farmers for work on agriculture and transport of people and goods (Zhang et al. 2010). Donkey husbandry in China is closely related to a cultural, economic, social and ecological heritage, with a historical relevance in alimentation, mule production, and transportation (Han et al. 2014; Seyiti and Kelimu 2021). Both historical records and genetic studies of donkey archeological remains showed that domestic donkeys spread into western and northern China before the emergence of the Han dynasty, (about 2000 yr BP), before the Silk Road opened, perhaps in conjunction with ethnic migrations. Initially the population of Chinese domestic donkeys increased to meet the expansion of trade, used as commodities or for shipping goods along the Silk Road, and later during the Tang Dynasty (Han et al. 2014). The small geographic and body size structure differences in Chinese donkeys, related with a high genetic diversity, seems to imply that there was a high gene flow, associated with humans bringing new donkeys in subsequent migrations in Chinese history (Zhang et al. 2010).

There is a rich genetic diversity in Chinese domestic donkeys. Chinese domestic donkey sequences suggest that the maternal ancestor of Chinese domestic donkeys, as all the other donkey populations in the world (Beja-Pereira et al. 2004) is the

Somali and Nubian of African wild ass (Chen et al. 2006; Lu et al. 2008; Zhang et al. 2010; Han et al. 2014; Sun et al. 2016). The Clade II (Somali lineage) is prevalent in Chinese domestic donkey breeds, although the presence of Clade I (Nubian lineage) is also present (Lei et al. 2007; Xia et al. 2023). The Asian wild ass can be excluded as an ancestor of the domestic donkeys in China (Xia et al. 2023).

3.6 American Populations and Breeds

The donkey was later brought to the American continent, just in the post-Colombian era, but has been important in the exploration and livelihood of that continent since then. In America, the establishment of important trade routes coincides with the spread of donkeys throughout the Caribbean and Latin regions. (Carneiro et al. 2018). Spanish donkey breeds, such as the Catalonian donkey have been frequently used throughout the last centuries to improve the New World American breeds (Jordana and Folch 1996; Jordana et al. 2016). The genetic origin and genetic diversity of the Mexican Creole donkey, studied by mtDNA D-loop evolution, showed, as in all other breeds and donkey populations, an African origin (Lopez et al. 2005).

4 Worldwide Context

4.1 Donkey Roles and Challenges in the XXI Century

In 1991, over 95% of all donkeys and mules were found in developing countries, being used for the most part for work (Fielding 1991). Even today, in several countries, the sustenance of hundreds of millions of humans greatly benefits from the donkey. This species provides valuable work aid, as well as food products such as meat and milk (Ali et al. 2014). Donkeys are good vehicles and will probably remain one of the main means of transport in several developing countries, where roads for motor vehicles are unavailable and where households face economic scarcity (Molla et al. 2017). In richer countries, other products besides work, such as donkey milk, are of relevance, being used in the food, medical, assisted therapy and cosmetic industries (Li et al. 2022a, b; Martini et al. 2021).

In the last three decades of last century, there was a continued increase in the global donkey and mule population, especially in Africa, South Asia, and Latin America (Starkey and Starkey 2000). According to the United Nations Food and Agricultural Organization, population estimates of donkeys around the world have increased from 38 million in 1968 to 50 million in 2018 (FAOSTAT 2017). While the global donkey population was showing an upward trend, there were large regional differences with most of the donkeys found in Sub-Saharan Africa, the Northern region of the Indian Subcontinent, and Latin America (Starkey and Starkey 2000).

When compared to other domestic species, the inclusion of donkeys in agricultural or livestock health and nutrition policy, education, and research sectors is negligible, although these animals are vital in sustaining the livelihoods of some of the most underprivileged communities, where donkeys can help to empower women and give them a way to work and make a living. Donkeys and mules are not given enough attention in terms of developing strategies regarding their role in reducing poverty. There is a dearth of data on their impact on different economic activities, factors contributing to the donkey population dropping, socioeconomic status of the dependent communities and related animal and human welfare issues (Ravichandran et al. 2023).

4.2 Welfare Challenges Faced by Donkeys

Millions of donkeys and mules live their entire life in very poor welfare conditions and even in their final moments, instead of being euthanized, are left to die from exhaustion or famine. For example, in the Egypt Equine Aid clinic (Cairo, Egypt), many of the attended animals suffered from wounds, orthopedic problems, colic, and infectious diseases, that sometimes, given a correct clinical evaluation, renders euthanasia as the most human solution (Benedetti et al. 2024). In poor, underdeveloped areas, non-governmental organizations, sanctuaries, and charity hospitals are the main, or only, entities providing health and welfare care (Kay et al. 2019). As work is sometimes the only productive output of a donkey, the care given to donkeys that are unable to work properly is on most occasions unsatisfactory. Owners should realize that animals in good condition live longer, more productive lives (Yalew et al. 2024).

Donkeys' increased cognitive abilities have been mistaken for stubbornness. The refusal to participate in new situations may be related to a higher capacity to evaluate hypothetically unsafe states (Navas González et al. 2018). These traits often lead to frustration by the owners and are addressed with violence. The human-equid relationship is critical to the care of these animals and has profound implications for their psychological status (Ali et al. 2016). Inferior welfare can result in donkey dullness, which in turn can be interpreted by the handler as laziness, making these donkeys more likely to be beaten and suffer from chronic fear, in a vicious circle (Burn et al. 2010). The preservation of a positive human–donkey relationship is crucial to avoid accidents and improve equine welfare. The evaluation of donkeys' reactions to unfamiliar people, using appropriate standardized tests, is an effective tool in assessing the quality of their past and ongoing relationships with human-beings (Dalla Costa et al. 2015).

Indicators of general health that can be used for the assessment of donkey and mule welfare conditions include the presence of wounds, lameness or poor body condition (Pritchard et al. 2005).In a study carried out in Afghanistan, Egypt, India, Jordan and Pakistan, the assessment of the welfare of working horses, mules and donkeys used for draught, packing and riding, showed that, although mules and donkeys had better body condition scores than horses (maybe due to their greater capacity of coping with feeding of lower nutritional quality), they presented with a high incidence of several problems like limb deformities and abnormalities of gait, abnormal mucous membrane color, ectoparasites or poor coat condition (Pritchard et al. 2005).

Overloading, beating, improper harnessing with fabric straps and/or hobbling, and overworking with load-bearing packs are frequent causes of wounds and lameness both in cart pulling and pack donkeys worldwide (Alujia and Lopez 1991; Pearson et al. 2001; Sells et al. 2010; Henok et al. 2017), which originates from poor awareness of the welfare of the donkeys (Yalew et al. 2024). Due to their stoic nature, donkeys are assumed to be hardy and are often forced to work for long hours without proper management, severely affecting their health and welfare. Moreover, in several countries, donkeys are neglected due to low social status (Tesfaye and Curran 2005). This prevailing mistreatment and lack of veterinary care for donkeys have contributed to early death and work life expectancy of sometimes just 4 to 6 years, compared to the up to 30 years in countries where animal welfare is in practice (Fred and Pascal 2006).

The type of work they are used in affects the response to humans in donkeys, as well as these responses are different than in mules and/or horses used for the same work in the same geographical areas (Pritchard et al. 2005). These observations can be due to species specific responsive behaviors, but most likely can be attributed to different handling and by caretakers. A proper owner education and timely medical assistance in case of sick donkeys is fundamental in giving working donkeys a life worth living (Benedetti et al. 2024). The use of animal-based welfare assessment methods demonstrated that non-governmental organizations had made a positive impact on donkey welfare through owner education (Leeb et al. 2013). To maximize the advantages of community education and extension programs for the users of equids, the knowledge transfer method must be properly designed, adapted to the target group (Stringer et al. 2011).

A good welfare assessment of donkeys and mule welfare status should, as in other animals, describe significant aspects that matters from the point of view of the animals; are possible to repeat in successive valuations, evaluating changes in welfare overtime; be dependent of the decisions taken by the caretaker or owner and be measurable in a cheap and easy way (Sørensen et al. 2001). The results of welfare studies on mules and donkeys may not necessarily represent the welfare status of all the population in those countries, as circumstances vary within and between countries and animals raised in more isolated areas are seldom evaluated (Pritchard et al. 2005). Tests validated for the quality of the relation of donkeys with humans can be the avoidance distance and walking downside, at least in a typical Western Countries farm environment, they can be performed on-farm, in 60 to 90 seconds per animal. However, considering the subjective nature of the evaluation process, training of the evaluators should always be considered (Pritchard et al. 2005; Burn et al. 2009; Dalla Costa et al. 2015). Another test used in donkeys can be the observation of tucking in the tail in hindquarters (tail tuck') in response to an observer walking down their side, a behavioral response not commonly observed in mules (Pritchard et al. 2005).

Although working donkeys and mules provide transport in a lot of developing countries, most work in harsh conditions, raising concerns about their welfare. In some countries, donkeys are more likely than other domestic equids to demonstrate avoidance or aggressive behavior towards humans, which might be related to their husbandry (Pritchard et al. 2005). The preservation of a positive human–donkey relationship is crucial to avoid accidents and improve equine welfare. The evaluation of donkeys' reactions to unfamiliar people, using appropriate standardized tests, is an effective tool in assessing the quality of their past and ongoing relationships with human-beings (Dalla Costa et al. 2015).

Proper working management of working equids is a requirement to determine good welfare standards (Yalew et al. 2024). Even though the cases of mistreatment were historically common and still exist, the donkey is being more and more valued in the context of Assisted Interventions (Lima et al. 2021). Worldwide, several non-governmental organizations and breeders' associations work to improve donkey and mule living conditions. Several non-governmental charitable organizations provide veterinary treatment for equids working in developing countries and advise their owners on good management practices (Pritchard et al. 2005).

4.3 Breed Conservation Programs

Major concerns on the productive sustainability of most donkey breeds *in situ* conservation are raising international demand, increasing feeding costs and a decrease in government subventions (Navas et al. 2017). The poor knowledge of ancestry of most donkey breeds results in the underestimation of inbreeding level, as individuals with unknown parents and their offspring are assigned a no-inbreeding coefficient even when they are related (Lutaaya et al. 1999; Quaresma et al. 2014; Navas et al. 2017).

Most of the world's "pure" donkey breeds are now in an endangered situation. Several factors contribute to this situation, including mainly the abandonment of traditional agriculture, ageing of the owners and inadequate breeding policies (Quaresma et al. 2014, 2015). In donkeys, intraherd breeding policies and the excessive contribution of few ancestors of the gene pool could lead to narrower pedigree bottlenecks. Traditional breeding practices may have led to an inbreeding increase, an effective population size reduction and a consequent genetic diversity loss and genetic bottlenecks (Navas et al. 2017). Although these genetic bottlenecks do not necessarily lead a population towards extinction, diversity loss, once fixed in the population, may lead it to extinction in the long term, as it results in a lower adaptation capacity to an ever-changing environment (Frankham et al. 2010). In donkey conservation programs, studbook information, genetic diversity status, population structure and the assessment of breeding practices are crucial tools for the development of conservation plans (Santana and Bignardi 2015).

Microsatellite markers have been widely applied to analyze genetic diversity and variability within and among different donkey breeds, in several countries (Aranguren-Méndez et al. 2001; Bordonaro et al. 2012; Zhang et al. 2016; Kefena et al. 2021; Kim et al. 2021; Wang et al. 2023), providing a scientific basis for the conservation of biodiversity and genetic resources in donkey populations (Wang et al. 2023). One expected finding in these studies was that populations with lower numbers of individuals present with lower genetic variation (Aranguren-Méndez et al. 2001; Bordonaro et al. 2012; Wang et al. 2023), making it challenging to maintain these populations.

Due to their long gestation period of around 12 months, donkeys are expensive to breed as a pure production animal. In China, to circumvent this, favorable polices, such as subsidizing donkey rearing and creation of breeding centers, were adopted by the Government to encourage donkey breeding (Bennet and Pfuderer 2020).

4.4 Regional Challenges and Adaptations of Donkeys

Donkeys are one of the most adaptable domestic species, capable of performing as an amazing working animal or being a great companion. Their worldwide spread is a direct result of this malleability and capacity to endure severe working conditions, even with limited resources (Davis 2019). In the different continents and countries, donkeys were able to adapt to quite different conditions, both in terms of natural environment and socio-economic context. Due to this same worldwide spread of the species, they suffer very different challenges and opportunities, in terms of welfare, preservation and use.

Working equids in poor countries are used for carting goods and people, carrying packs, bricks and other construction materials, riding, tillage, weeding, water-carrying, rubbish collection, tourism and for ceremonial purposes such as weddings and festivals. These animals are often neglected concerning food, shelter and appropriate equipment, because of their ownership by people that face poverty and very difficult living conditions themselves (Pritchard et al. 2005). Due to strenuous work donkeys are forced into and the lack of care they receive in developing countries, these animals face several health and welfare problems, including wounds, lameness, colic, parasitosis, and various infectious diseases (Getachew et al. 2002; Pritchard et al. 2005; Stringer et al. 2017). These and other issues affecting the welfare of working equids can be improved through creating community awareness (Whay et al. 2015; Yalew et al. 2024).

Donkeys naturally save energy, an adaptation to the harsh environment where they evolved as a species (Delgado et al. 2014). For that reason, they tend to decrease their activity under stressful weather conditions (Zakari Friday et al. 2015). For example, extreme high wind speed and rainfall has been reported to be a welfare distorting factor for donkeys (Ayo et al. 2008). Also, high temperatures associated with high humidity have been reported to be a stressful factor from a welfare perspective that alters donkey behavior (Zakari Friday et al. 2015; Gebresenbet et al. 2016).

4.4.1 Donkey Breeds, Farms and Uses in Africa

In the Continent of its domestication is probably where the donkey still retains most of its socioeconomic importance. However, with many countries facing generalized poverty, famine and war, the challenges that donkeys and their owners must face are sometimes almost overwhelming. However, there, in rural and peri-urban areas, each equid may support the daily survival needs of up to 20 people (The Brooke 2015). Donkeys play an important role through support of income-generating activities in areas such as agriculture, transportation, tourism and construction industry; alleviating poverty, all as described in the Sustainable Development Goals (SGD), set by the United Nations and designed to create a more sustainable future for humanity (Zarins and Hauser 2014). For example, before the introduction in Uganda's highlands of donkeys for transportation, post-harvest food losses were high, due to a lack of farm transport. The initiation of donkey transportation to these areas had positive impacts, providing inexpensive and sustainable transport. This is highly applicable to women, as they are the ones who are associated with harvest and post-harvest management-related farm activities in these areas (Ayo-odongo et al. 1999).

The donkey is a very important asset for an African farmer. But most of these farmers are poor and vulnerable to making short-term economic decisions, agreeing to give away their animals for much needed money. For example, the increased exports of donkey ejiao are hampering the livelihoods of South African donkey-owning communities for whom donkeys are the primary source of income (Binda 2019). In Ethiopia, several works have been made by NGO, like the Donkey Sanctuary, SPANA (Society for the Protection of Animals Abroad), and Brooke, to improve the management by donkey owners related to the health and welfare of working donkeys. These community-based interventions increased the health and welfare of working animals. Based on these good results, widespread and constant equid health and welfare campaigns through community-based interventions should be conceived also in other countries, where similar conditions exist (Yalew et al. 2024).

Poor infrastructure in many parts of rural Ethiopia, as in other developing countries, have made motorized transportation difficult. In these areas, people must still depend on equids to transport crops, wood, water, and even themselves. The donkey is one of the most important draught animals, playing a key role in the agricultural livelihood of these countries (Admasu and Shiferaw 2011). There are around 10.80 million donkeys in Ethiopia, Africa's largest population and the second largest in the world (CSA 2021). In Ethiopia, in 2011, approximately 56% of donkeys were kept mainly for pack services, 26% for cart use, 14% for pack use but exclusively for homestead use, and 4% exclusively for renting, breeding, or petty trade (Admasu and Shiferaw 2011). The donkey populations in Ethiopia are dominated by male indigenous donkeys purchased from different markets in the region (Yalew et al. 2024). Males tend to be more aggressive, which can have deleterious outcomes to these animals.

In a study recently carried out in central Ethiopia, the selected donkeys were managed by the owners (78%) or by daily laborers hired to work for the owners (22%). Most of these individuals were aged between 15 and 30 years old, with an elementary school education, married and with four or more family members who were dependent on their donkey (Yalew et al. 2024). Most donkeys are fed wheat bran, chopped sugar cane tops, and household food waste during working time and graze on communal lands or forage through trash on roadsides (Haddy et al. 2021). A cross-sectional study conducted to assess donkeys' health and welfare in community-based intervention areas vs. non-intervention areas in Southern Ethiopia, found that the prevalence and severity of lameness and wounds on donkeys managed in non-intervention areas were higher than those observed in community-based intervention areas. Likewise, donkeys in the areas intervened had better BCS and were alert and tolerant to people (Yalew et al. 2024).

Africa poses, along with some areas of Asia, major challenges in terms of animal welfare. In Ethiopia, The Donkey Sanctuary, SPANA, and Brooke, have been intensively working since 1993 to improve the health and welfare of equines through outreach treatment centers and community-based intervention approaches. The outreach treatment approach directly provides health and veterinary services, whereas the community-based approach develops the capacity of the community for keeping their donkeys healthy, wound and lameness free, through education and training, supporting veterinary services, and training farriers and pack saddle and harness makers. Community-based interventions also include discouraging practices like beating and overloading through the training of children, teachers, and volunteers for equine welfare (Yalew et al. 2024).

Brooke Ethiopia and The Donkey Sanctuary initially focused primarily on working donkeys through the provision of treatment for sick donkeys and different harness materials, better carts, and implements for resource-poor owners. However, both NGOs have now adopted community-based approaches/interventions whereby change agents (model equine owner/users) and school children receive tailored training and education to work jointly with these NGOs as an advocacy group and concomitantly transfer their skills and knowledge to peers (Demissie and Desalegn 2014; Haddy et al. 2021).

In regions affected by war, animals also suffer and are sometimes killed along with their human caretakers. As an example, in the armed conflict that went on in Tigray, Ethiopia from 2020 to 2022, an estimated 610,976 donkeys died, both killed directly in warfare and due to lack of proper care. War causes an immense impact on animal welfare and a collaborative effort between governmental and non-governmental organizations is necessary to recover afterwards. This study also highlighted how the war (Tedla et al. 2023). Campaigns to hold the forces on the ground during war accountable for deliberate and malicious harm to animals, as to humans, must be considered a war crime and treated as such (International Fund for Animal Welfare 2022).

4.4.2 Donkey Breeds, Farms and Uses in Europe

In the industrialized world, the donkey populations, mainly present around the mediterranean basin, started a steep decline with the fading of the causes of its breeding and selection: agricultural work, transport and mule production. This process occurred halfway through the nineteenth century (Navas et al. 2017) and had its apex after World War II.

In the countries where they are no longer used in traditional roles, or are seldom used, donkeys are acquiring the role of a relational subject, appreciated for their special connection with humans. The donkey is one of the most evolved species in Animal-Assisted Interventions (AAIs), which, because it involves physical, mental and emotional engagement with the managers of the animal, should be regarded as work, and not as a companion role (Sobrero et al. 2024). Donkey-Assisted Interventions represent one variation of the designated Social Farming or Greencare, with the objective of enhancing the quality of life of people with or without vulnerabilities, through contact and care of donkeys (Borioni et al. 2012; Galardi et al. 2022). Distinct from horses, ground-based relational work is more common with donkeys (Portaro et al. 2020).

The welfare of donkeys enrolled in AAI should not only be regarded as a way of safeguarding the animal, but also as a step towards the delivery of patient care services, embodying the concept of One Welfare, an interconnection between human well-being, animal welfare and the environment (Tarazona et al. 2020; García Pinillos 2021; Fine and Griffin 2022). This implies that the user's welfare cannot be considered more important than the animal's, a challenge in a species like the donkey that manifests conditions of discomfort quite subtly. Despite this, there are currently no specific welfare indicators regarding donkeys involved in educational farms or in Animal-Assisted Interventions (Sobrero et al. 2024).

In a study conducted in Italy in 2024, that analyzed six facilities that consistently conduct Donkey-Assisted Interventions, just 17% were recognized as a Specialized Center in Animal-Assisted Education and Therapy. Staff wise, in 67% of the facilities, personnel were trained for AAIs, while 33% held either a degree in Animal Husbandry and Animal Welfare and were professional veterinarians. Half of the facilities did not have any veterinarians trained for AAIs. In 83% of the institutions, the feeding and cleaning operations were carried out by the same individuals who organized the AAIs. It seems that establishing a multidisciplinary team is not practical due to the cost. However, it would be beneficial to have a higher level of integration of expertise (Sobrero et al. 2024).

The estimated number of donkeys in Serbia is between 500 and 1000 (FAO 2015). A small number of these donkeys are kept by individual households, mostly in rural regions. Most sexually mature individuals are kept at three locations where large donkey herds are currently maintained in the Country (Stanisic et al. 2020). In the Balkan Peninsula, the donkey breed present today is the Balkan donkey, usually regarded as an unselected, unstructured, traditionally managed donkey breed (Kugler et al. 2008). However, genetically distinct sub-populations of the Balkan donkey, may be present in the region, as well as eventual new breeds that may remain uncharacterized and close to the brink of extinction, such as the Istrian and North-Adriatic donkeys in Croatia (Ivankovic et al. 2002; Stanisic et al. 2017).

The most common donkey type in the Balkans is the mid-sized Balkan donkey, with a grey to chocolate coat color. However, in Serbia, there are a few larger individuals of a lighter coat color, named Banat donkeys. Although severely depopulated, the genetically distinct Banat donkey is not severely affected by the loss of genetic diversity and inbreeding, but actions to preserve the Banat donkey, a

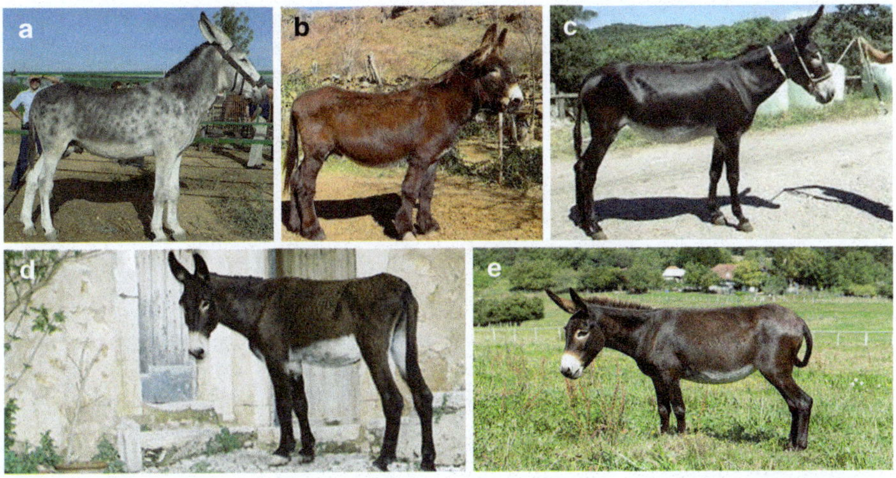

Fig. 3 Current officially recognized Spanish breeds of donkeys: (**a**) Andaluz, (**b**) Zamorano-Leonês, (**c**) Catalan, (**d**) Mallorquino, (**e**) Encartaciones. The average height of the Spanish donkey breeds varies from 150 cm (59 inches) in the Andaluz to 120 cm (47 inches) in the Encartaciones

valuable but critically endangered genetic resource, with a population of less than 100 individuals, are urgent (Stanisic et al. 2020). In Serbia, the Banat donkey type is morphologically distinct from the Balkan donkey by its larger body, different coat color and pigmentation scheme, with a cross on the back and black stripes on legs resembling those typical for Somali wild donkeys (Groves 1986; Moehlman 2002). Such pigmentation represents primitive markings of the species (Johnson and Johnson 2008) typical also for the Italian Amiata donkey (Sargentini et al. 2009, 2018). The Banat donkey is also characterized by a unique nuclear gene pool, predominantly harboring Clade 2 mtDNA haplotypes, that are nowadays less abundant in Ethiopian donkeys (Kefena et al. 2014; Stanisic et al. 2017), and more common in those from south-west Asia and Europe, including several Spanish (Fig. 3) (Aranguren-Mendez et al. 2004) and Italian donkey breeds (Fig. 4) (Cozzi et al. 2017). Even with these indications of morphological and genetic individuality, the severely depopulated Banat donkey is not formally acknowledged as a distinct donkey breed and is critically endangered (Stanisic et al. 2020).

4.4.3 Donkey Breeds, Farms and Uses in Asia

With countries in very different stages of industrialization and very diverse cultures and economies, for the donkeys, Asia is a place of big contrasts and different challenges, although, at the same time, a continent where the donkeys have a very important role in quite different areas, from work to ejiao and meat production.

China, a country where the donkey was introduced nearly 4000 years ago, has one of the largest populations in the world (Lei et al. 2005), but statistics from China's Statistical Yearbook showed that the number of donkeys dropped from

Fig. 4 Current officially recognized Italian breeds of donkeys: (**a**) Sardo, (**b**) Martina- Franca, (**c**) Amiata, (**d**) Ragusano, (**e**) Asinara. The relative size of the donkeys varies from 135 cm (53 inches) in the Martina-Franca to 105 cm (41 inches) in the Asinara donkey

9.444 million in 1994 to 2.324 million in 2020 (Wang et al. 2023; Zhang et al. 2023). The cause for this substantial drop in the last few decades was agricultural modernization and the depopulation of rural districts in China (Tian et al. 2020). Donkeys are reared in 17 provinces of China; primarily in the dry, arid, semi-arid, and warm climates of western China around the Huanghe Valley, but they also are present in the cold climate of three provinces of northeastern China (Xie 1987). China officially hosts 24 registered indigenous donkey breeds (Zhang et al. 2023). This genetic diversity in Chinese donkey breeds helps preservation of these genetic resources. However, small-sized breeds, such as Cullen donkeys, Qingyang donkeys, Xiji donkeys and Huaibei have been slaughtered indiscriminately, due to their lower production of meat and hide, and may face the risk of extinction (Lei et al. 2007). The Guanzhong breed is also an endangered breed, with a low number of individuals (Zhang et al. 2023).

Currently, Chinese citizens appear to be more open to environmental and biodiversity conservation, which can allow for a new role for the donkeys in a more sustainable agricultural setting (Seyiti and Kelimu 2021). There has been an increase in raising donkeys in China, as donkey meat and milk have been perceived to have high nutritional value and the Asini Corii Collas (hide) is increasingly used as a component of "traditional Chinese medicine" (Lei et al. 2007; Kumeta et al. 2014). Still today, donkeys are common in the rural areas of China where they continue to be an important contributor to the socioeconomic progress for millions of households (Seyiti and Kelimu 2021; Ma et al. 2003). The use of donkeys is changing from draft to meat, milk, hide, "traditional medicine", cosmetics and other bioproducts (Cong et al. 2018), with the development of a livestock industry around donkeys rather than the means of support of farmers who live in rural areas (Nolan et al. 2008). Nevertheless, in China still today over 70% of donkeys are raised in traditional smallholder farms (Deng et al. 2021), but other donkey products, like donkey meat, milk and hide production are gaining prominence, becoming an important industry in recent years (Zhang et al. 2019; Li et al. 2020; Gao et al. 2022).

When China started its industrialization in 1978, donkey populations rapidly declined, particularly in economically developing regions. Nowadays, conservation status protects donkey breeds, such as the famous Guanzhong breed, preserving their genetic resources. Also, despite China's industrialization, in less developed regions, donkeys still serve as important domestic animals, used for transportation, meat, and sources of products for traditional Chinese medicine. In recent years, the demand for donkey meat has increased rapidly and some farmers are raising donkeys for this purpose (Zhang et al. 2010). The number of the large-scale donkey breeding farms in China has increased dramatically with 52% of them keeping between 100 and 500 donkeys with a predominance of the Dezhou donkey (Zhang et al. 2023), although there are nearly 30 donkey breeds in China, (Seyiti and Kelimu 2021). At least three donkey farms possess more than 2000 donkeys (Zhang et al. 2023). China currently keeps a donkey reserve system based on national, provincial and private donkey original breeding farms (Zhang et al. 2023). Donkeys raised on large scale farms account for 13% of the total donkey population in the Country (Bennet and Pfuderer 2020), having also an important role in the genetic conservation and germplasm resource utilization of local donkey breeds. Currently, Chinese authorities implemented a donkey reserve system based around national, provincial and non-governmental donkey original breeding farms (Zhang et al. 2023).

The Dezhou donkey breed, named after its distribution center in Dezhou City, is the most common donkey breed in China, known for its flat back, round belly, strong limbs, round ribs and sturdy hooves. It is mainly divided into two strains: Sanfen and Wutou, with differences in coat color between the two; the former shows light points around the eyes, muzzle and under the belly, while the latter is fully black. A study on the genetic transmission of these traits was done, identifying the genes and variants responsible for the phenotypic differences observed (Liu et al. 2024).

In China large-scale farms, artificial insemination has gained popularity, but in most donkey farms breeding by in-hand natural mating is the predominant breeding system. Reproductive efficiency and donkey productivity are quite different among farms, pointing to diversity in management and breeding practices. In larger government farms, the selection of animals is associated with larger animals and higher production indices, in milk, meat and hide. In the large-scale donkey breeding farms of China, trained donkey breeders, milkers and breeding staff are employed. These modern donkey farms have sheltered stables with pastures, nutritional feed supplements and free access to water, also providing facilities and management conditions for improving welfare of donkeys such as proper ventilation, lighting and bedding materials. In these farms, according to the answers on a voluntary survey of farm employees, the highest education level of more than half of the donkey farm managers was a college or university degree (Zhang et al. 2023). This high level of formal education may have improved the management and health care of the donkeys compared to smallholder farm owners in this area (Deng et al. 2021). In other locations around the world assisted reproduction technology has helped, and continues to aid, in saving, preserving and developing donkey breeds.

The Dezhou donkey is the main native donkey breed in China, reared essentially in the Northeast region of the country. The average height at the withers is usually between 135 and 140 cm (53–55 inches) and the coat color is present in two types; one where parts around the nose, eyes, and abdomen are white with the other parts black and another where the entire coat is black. Another Chinese breed, the Yunnan donkey, belongs to small varieties in southwest China, with a height around 100 cm (39 inches) The coat color of this breed is mainly grey, and most Yunnan donkeys have a dorsal stripe, shoulder stripes and leg striping (Zhang 2011). The Dezhou donkey breed of China is generally divided into the Sanfen and the Wutou varieties (Fig. 5) (Zhang et al. 2022). While both strains have dark coats, the Sanfen donkey also has a cream to gray-white colored pattern around the muzzle, the eyes and the belly (Shi et al. 2020). In addition, other contrasts between these two types are that the Wutou donkey has thick limbs, large hooves, a wide and deep chest and a heavy body, while the Sanfen donkey presents thinner limbs, high and small hooves, a thin and short-haired coat, and is lightweight. Because of these features, in terms of slaughter traits, the Wutou donkeys have a higher net skin percentage, making it more widely used to provide gelatin products, while the Sanfen donkeys have a higher net meat percentage, being suitable for meat production (Gao et al. 2021). The Huaibei grey donkey is an endangered native breed from the Anhui Province, in China. The most probable origin of this breed is the Somali lineage, with the median-joining network and phylogenetic tree indicating two possible maternal lineages (Xia et al. 2023). It is deemed an important breed in this country, due to its disease resistance, coarse feed tolerance, and good adaptability (Xia et al. 2023).

In South Korea, a rich developed country, with a small remaining population of around one thousand donkeys, these are used mainly for research, to pull tourist carriages, or for meat production. A study was developed to evaluate the diversity of donkey populations, also comparing it with the diversity of horse breeds, local and imported, mainly from China, in the country, with the intention of identifying breeding backgrounds to contribute to management and conservation of donkeys. This study showed that the genetic relationship among donkeys was generally consistent with pedigree records and showed that genetic diversity is decreasing

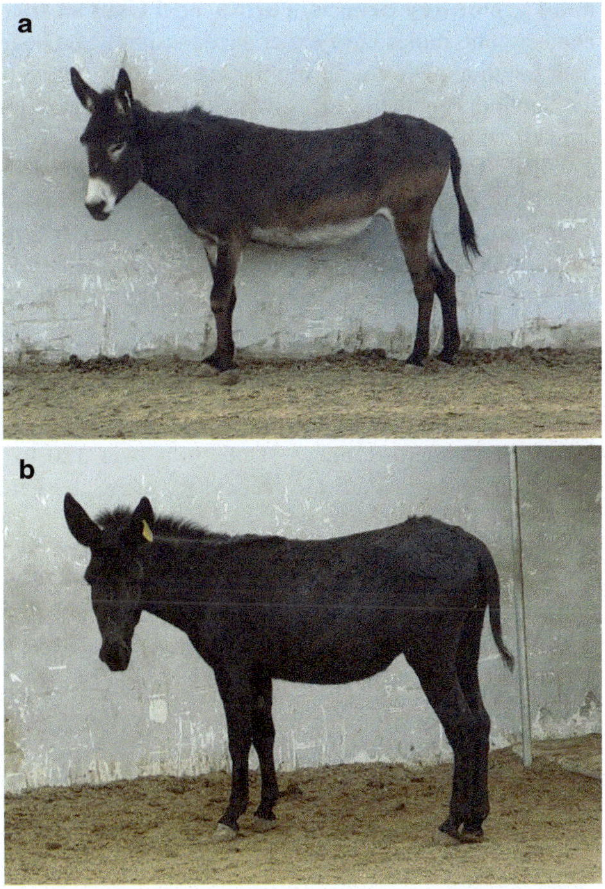

Fig. 5 Dezhou Chinese donkey varieties. Both strains have dark coats, (**a**) Sanfen donkey has a cream to gray-white colored pattern around the muzzle, eyes and belly, presents thinner limbs, high and small hooves, a thin and short-haired skin, and is lightweight (**b**) Wutou donkey has thick limbs, large hooves, a wide and deep chest and a heavy body. (Adapted from Zhang et al. 2022)

due to inbreeding. To overcome this, it is suggested that a breeding program is needed to minimize inbreeding within farms and more stallions should be used (Kim et al. 2021).

In another part of Asia, in the subcontinent of India, donkeys' population has decreased significantly, mainly due to mechanization of labor previously executed by donkeys. According to the 20th livestock census, the donkey population has declined from 0.32 million in 2012 to 0.12 million in 2019 (DAHD 2022). In northern parts of India, donkey ownership is mostly dictated by caste and ethnicity, linked to impoverished families with low status and limited access to resources. Such systems are oppressive, both to the people and to the animals managed by these communities. In the brick kiln industries, both humans and their donkeys are exploited and experience poor welfare (Watson et al. 2020). In India, donkeys are divided into three breeds: the Spiti, Halari, and Kachchhi (Yadav et al. 2024).

4.4.4 Donkey Breeds, Farms, and Uses in the Americas

Donkeys and mules have experienced a significant decrease in numbers in America, both South and North, during the last century, with a more pronounced decrease in the South during the last few decades (Norris et al. 2021) For example, in Brazil the donkey numbers are declining, mainly because of agricultural and transportation mechanization that has led to the abandonment and the consequent indiscriminate slaughter of these animals (Alves et al. 2022).

In North America, it is more common to classify donkeys by their size rather than as traditional breeds, like in Europe. Mediterranean Sicilian miniature donkeys and the American Mammoth Jack are among the most popular breeds in North America. Both the Canadian Donkey and Mule Association and the American Donkey and Mule Society register donkeys according to their size, as miniature, standard or mammoth donkeys (Fig. 6). Adult miniature donkeys stand 91 cm (36 inches) or less at the withers. Small standard donkeys stand from over 91 to 122 cm (36–48 inches) and large standard donkeys stand from over 120 to 140 cm (48–54 inches) for jennies, or over 140 cm (56 inches) for jacks and geldings. Mammoth jennies are taller than 140 cm (54 inches), and mammoth jacks and geldings are over 145 cm (56 inches) (The Donkey Government of Alberta 1990; The American Donkey and Mule Society 2011; Canadian Donkey and Mule Association 2012). Any donkey with at least two spots behind the throat latch and above the legs can be registered with the American Council of Spotted Asses under the trademarked name (American Council of Spotted Asses 2009).

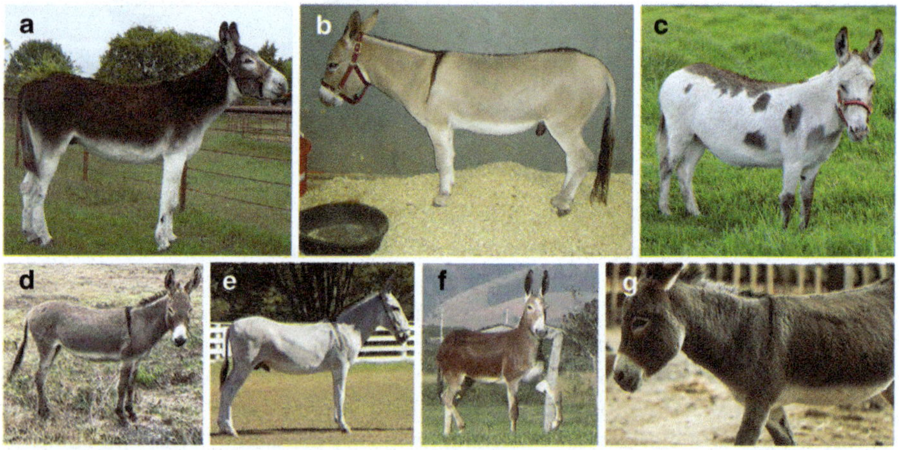

Fig. 6 American breeds of donkeys (**a**) American Mammoth Jack (**b**) Standard Donkey, (**c**) Miniature Spotted Donkey, (**d**) Northeast (Nordestino Ecotype), (**e**) Brazilian or Paulista, (**f**) Pêga, (**g**) Mexican Creole donkey. Most breeds in both South and North America have open Studbooks, that allow for registration of animals of unknown ancestry, only by phenotypic features, which allows for a great diversity in genetical characteristics, although it is well established that most American donkeys are direct descendants of donkeys brought from Iberian Peninsula by Portuguese and Spanish, as they settled in their colonies, having suffered a divergent selection afterwards, such is the case of the American Miniature Spotted Donkey, a phenotype not common in Europe, but present in American donkeys that have the same ancestry as the other American breeds

With the largest herd of these animals in South America, Brazil hosts three types of donkeys (Fig. 6), brought from the Old World: the Northeast (Ecotype) Donkey, a class of highly rustic animals adapted to the adverse conditions of the Brazilian semi-arid regions; the Brazilian Donkey, brought from Italy, with a close genetic relationship to the small donkey still seen in Africa; and the Pêga Donkey, of Iberian origin, a special type of Donkey developed mainly in the state of Minas Gerais for use in the mining industry (Carneiro et al. 2018). The body indices for Brazilian donkeys show that they vary from medium (Paulista) to small (Northeastern and Pêga) animals with an aptitude for strength (Pimentel et al. 2023).

5 Conclusion

The situation of donkey populations and donkey breeds varies worldwide. In the industrialized world, where industrialization deprived donkeys of their traditional role as a working animal, breeds are registered and there are some efforts for conservation, although donkey populations have significantly decreased. On the other hand, in countries where donkeys and mules are still common as working animals, breed registration and animal welfare are, to say the least, very conflicting and unreliable. China, with the new uses for donkeys and a strong economy and reliable official services has a particular kind of industrial use of the donkey, with care for preservation and breeding, but with some concerns being raised on animal welfare.

Study Questions

1. What is the origin of all donkey breeds and populations seen worldwide, geographically and genetically?
2. In which areas of the world are donkeys more common?
3. What is the difference between a donkey breed and a donkey population?
4. What are the main menaces to donkey breed preservation worldwide?
5. What are the main measures and resources that can contribute to donkey breeds preservation?

Answers

1. The origin of all domestic donkey breeds and populations worldwide traces back to Northeast Africa, particularly the Horn of Africa and surrounding arid regions, where wild ancestors of the modern donkey—the African wild ass (*Equus africanus*)—were first domesticated. Genetic studies have identified two main maternal lineages (clades) of domestic donkeys, both tracing back to different populations of African wild asses:Nubian lineage (Clade I)—linked to the Nubian wild ass (*Equus africanus africanus*). Somali lineage (Clade II)—associated with the Somali wild ass (*Equus africanus somaliensis*). These clades reflect multiple domestication events, likely occurring between 5000–6000 years ago, driven by

human need for pack animals in desert trade routes. Archaeological and mitochondrial DNA evidence supports early domestication in Egypt, Sudan, and the Horn of Africa, spreading into the Middle East, Europe, Asia, and eventually the Americas.
2. Donkeys are most common in developing regions, particularly in Africa, Asia, and parts of Latin America, where they are essential for transport, agriculture, and livelihood support.
3. The difference between a donkey breed and a donkey population lies in how each is defined in terms of genetics, origin, and standardization. In short, breeds are standardized and often conserved, while populations are more loosely defined and may vary widely even within regions. Many donkey populations around the world remain unclassified due to limited research and lack of structured breeding programs.
4. The main threats to donkey breed preservation worldwide are: 1. Crossbreeding and Genetic Erosion, 2. Decline in Working Roles, 3. Donkey Skin Trade (Ejiao Market), 4. Lack of Institutional Support, 5. Poor Public Perception and Low Status, 6. Loss of Indigenous Knowledge
5. The main measures and resources that contribute to donkey breed preservation include:Breed Identification and Documentation, Community-Based Conservation Programs, Governmental and Institutional Support, Ex Situ Conservation, Education and Awareness, Sustainable Use and Marketing

References

Admasu B, Shiferaw Y (2011) Donkeys, horses and mules – their contribution to people's livelihoods in Ethiopia. The Brooke, Addis Ababa, p 72

Alessandri C, Mari A (2007) Efficacy of donkey's milk in treating cow's milk allergic children: major concerns. Pediatr Allergy Immunol 18:625–626

Ali M, Baber M, Hussain T, Awan F, Nadeem A (2014) The Contribution of Donkeys to Human Health. Equine Vet J 46:766–767

Ali ABA, El Sayed MA, Matoock MY, Fouad MA, Heleski CR (2016) A welfare assessment scoring system for working equids – a method for identifying at risk populations and for monitoring progress of welfare enhancement strategies (trialed in Egypt). Appl Anim Behav Sci 176:52–62. https://doi.org/10.1016/j.applanim.2015.12.001

Alujia AS, Lopez F (1991) Donkeys in Mexico. In: Fielding D, Pearson RA (eds) Donkeys, mules and horses in tropical agricultural development. CTVM, Edinburgh, pp 1–7

Alves JS, de Araujo Oliveira CA, Escodro PB, Pinto LFB, Costa RB, deCamargo GMF (2022) Genetic origin of donkeys in Brazil. Trop Anim Health Prod 54:291. https://doi.org/10.1007/s11250-022-03280-x

American Council of Spotted Asses (2009) Accessed March 2012

Aranguren-Méndez J et al (2001) Genetic diversity in Spanish donkey breeds using microsatellite DNA markers. Genet Sel Evol 33:433–442

Aranguren-Mendez J, Beja-Pereira A, Avellanet R, Dzama K, Jordana J (2004) Mitochondrial DNA variation and genetic relationships in Spanish donkey breeds (Equus asinus). J Anim Breed Genet 121:319–330. https://doi.org/10.1111/j.1439-0388.2004.00464.x

Ayad A, Aissanou S, Amis K, Latreche A, Iguer-Ouada M (2019) Morphological characteristics of donkeys (Equus asinus) in Kabylie area, Algeria. Slovak J Anim Sci 52:53–62

Ayo JO, Dzenda T, Zakari FO (2008) Individual and diurnal variations in rectal temperature, respiration, and heart rate of pack donkeys during the early rainy season. J Equine Vet Sci 28: 281–288

Ayo-odongo J, Mutyaba C, Kalunda P (1999) Improving on-farm transport using animal draught power in two hilly districts of Western Uganda. In: Empowering Farmers with Animal Traction. In Proceedings of the Workshop of the Animal Traction Network for Eastern and Southern Africa (ATNESA), Mpumalanga, South Africa, 20–24 September 1999, pp 210–212, ISBN 0-907146-10-4

Beja-Pereira A, England PR, Ferrand N, Jordan S, Bakhiet AO, Abdalla MA et al (2004) African origins of the domestic donkey. Science 304:1781. https://doi.org/10.1126/science.1096008

Benedetti B, Freccero F, Barton J, Elmallah F, Refat S, Padalino B (2024) A retrospective study on the status of working equids admitted to an Equine Clinic in Cairo: Disease Prevalence and Associations between Physical Parameters and Outcome. Animals 14:817. https://doi.org/10.3390/ani14050817

Bennet R, Pfuderer S (2020) The potential for new donkey farming systems to supply the growing demand for hides. Animals 10:718. https://doi.org/10.3390/ani10040718

Binda K (2019) A Donkey's worth in South Africa: domestic laborer or export product; socioeconomic impacts of China's skin trade on South African Donkey Owners. Diploma Thesis. Department of Political Science, Faculty of Social, Charles University, Prague, Czechia

Blench RM (2004) The History and Spread of Donkeys in Africa. Wageningen, ACP-EU Technical Center for Agriculture and Rural Cooperation

Blench RM, MacDonald KC (2000) The origins and development of African livestock: archaeology, genetics, and ethnography. University College London Press, London, pp 191–220

Bordonaro S et al (2012) Genetic diversity and variability in endangered Pantesco and two other Sicilian donkey breeds assessed by microsatellite markers. Sci World J 2012:648427

Borioni N, Marinaro P, Celestini S, Sole FD, Magro R, Zoppi D, Mattei F, Armi VD, Mazzarella F, Cesario A et al (2012) Effect of Equestrian Therapy and Onotherapy in Physical and Psycho-Social Performances of Adults with Intellectual Disability: A Preliminary Study of Evaluation Tools Based on the ICF Classification. Disabil Rehabil 34:279–287

Burn CC, Pritchard JC, Whay HR (2009) Observer reliability for working equine welfare assessment: problems with high prevalences of certain results. Anim Welf 18:177–187

Burn CC, Dennison TL, Whay HR (2010) Environmental and demographic risk factors for poor welfare in working horses, donkeys and mules in developing countries. Vet J 186:385–392. https://doi.org/10.1016/j.tvjl.2009.09.016

Cai D, Tang Z, Yu H, Han L, Ren X, Zhao X, Zhu H, Zhou H (2011) Early history of Chinese domestic sheep indicated by ancient DNA analysis of Bronze Age individuals. J Archaeol Sci 38(4):896–902

Canadian Donkey and Mule Association. Archived 16 January 2012

Carneiro GF, Cavalcante Lucena JE, de Oliveira Barros L (2018) The Current Situation and Trend of the Donkey Industry in South America. J Equine Vet Sci 65:106–110

Central Statistical Agency [CSA] (2021) Statistical report on livestock and livestock characteristics (private peasant holdings). Agricultural Sample Survey 2020/2021 (2013 E.C.), vol. II. Central Statistical Agency, Federal Democratic Republic of Ethiopia, Addis Ababa. Statistical Bulletin 589

Chai W et al (2022) Differential proteomic analysis to identify potential biomarkers associated with quality traits of Dezhou donkey meat using a data-independent acquisition (DIA) strategy. LWT-Food Sci Technol 166:113792

Chen S, Zhou F, Xiao H, Sha T, Wu S, Zhang Y (2006) Mitochondrial DNA diversity and population structure of four Chinese donkey breeds. Anim Genet 37:422–431

Ciampolini R, Cecchi F, Mazzanti E, Ciani E, Tancredi M, De Sanctis B (2007) The genetic variability analysis of the Amiata donkey breed by molecular data. Ital J Anim Sci 6(supp 1): 78–80. https://doi.org/10.4081/ijas.2007.1s.78

Cinar-Kul BC, Bilgen N, Akyuz B, Ertugrul O (2016) Molecular phylogeny of Anatolian and Cypriot donkey populations based on mitochondrial DNA and Y-chromosomal STRs. Ank Üniv Vet Fak Derg 63:143–149

Clutton-Brock J (1992) Horse-Power: a history of the horse and the donkey in human societies. Harvard University Press, Cambridge, MA

Cong W, Chen L, Shan XF, Qian AD, Meng QF (2018) First genetic characterization of Toxoplasma gondii infection in donkey meat slaughtered for human consumption in Shandong province, eastern China. Infect Genet Evol 61:1–3. https://doi.org/10.1016/j.meegid.2018.03.008

Cozzi MC, Valiati P, Cherchi R, Gorla E, Prinsen RTMM, Longeri M, Bagnato A, Strillacci MG (2017) Mitochondrial DNA genetic diversity in six Italian donkey breeds (Equus asinus). Mitochondr DNA Part A 29:409–418. https://doi.org/10.1080/24701394.2017.1292505

Dalla Costa E, Dai F, Murray LAM, Guazzetti S, Canali E, Minero M (2015) A study on validity and reliability of on-farm tests to measure human–animal relationship in horses and donkeys. Appl Anim Behav Sci 163:110–121

Davis E (2019) Donkey and Mule Welfare. Vet Clin North Am Equine Pract 35(3):481–491. https://doi.org/10.1016/j.cveq.2019.08.005

Delgado JV, Navas FJ, Miranda JC, Miró M, Arando A, Pizarro MG (2014) Preliminary body weight estimation methodology and its applicaction to the Andalusian donkey breed as an energetic producer. Actas Iberoam Conserv Anim 4:207–209

Demissie TD, Desalegn T (2014) Including the excluded: use of governmentextension services to improve equine welfare. Proceedings of the 7th International Colloquium on Working Equids. Royal Holloway, University of London, pp. 87–89

Deng L, Shi S, Li J, Tang C, Han Y, Xie PA et al (2021) survey of smallholder farms regarding demographics, health care, and management factors of donkeys in northeastern China. Front Vet Sci 8:626622. https://doi.org/10.3389/fvets.2021.626622

Department of Animal Husbandry and Dairying (DAHD), Ministry of Fisheries, Animal Husbandry and Dairying Government of India—Annual Report 2022–2023. https://dahd.nic.in/sites/default/filess/ FINALREPORT2023ENGLISH.pdf

Epstein H (1984) Ass, mule and onager. In: Mason IL (ed) Evolution of domesticated animals. Longman, London and New York, pp 174–184

Fantuz F, Ferraro S, Todini L, Piloni R, Mariani P (2012) Donkey milk concentration of calcium, phosphorus, potassium, sodium and magnesium. Int Dairy J 24:143–145

FAO (2015) In: Scherf BD, Pilling D (eds) The second report on the state of the world's animal genetic resources for food and agriculture. FAO Commission on Genetic Resources for Food and Agriculture Assessments, Rome. Available at http://www.fao.org/ 3/ i4787e/ i4787e.pdf

FAOSTAT (2017) Food and Agriculture Organization of the United Nations

Fielding D (1991) The number and distribution of equines in the world. In: Proceedings of the colloquium on donkeys, mules and horses in tropical agricultural development, Edinburgh, 3–-6 September, pp 62–66

Fine AH, Griffin TC (2022) Protecting animal welfare in animal-assisted intervention: our ethical obligation. Semin Speech Lang 43:008–023

Folch P, Jordana J (1997) Characterization, reference ranges and the influence of gender on endangered Catalonian donkey. J Equine Vet Sci 17:102–111. https://doi.org/10.1016/S0737-0806(97)80347-4

Frankham R, Ballou JD, Briscoe DA (2010) Introduction to conservation genetics. Cambridge University Press, Cambridge, UK

Fred O, Pascal K (2006) Extension approaches to improving the welfare of workingequines. Kenya Network for Dissemination of Agricultural Technologies (KENDAT), Nairobi, Kenya, pp. 1–28

Galardi M, Filugelli L, Moruzzo R, Riccioli F, Mutinelli F, Diaz SE, Contalbrigo L (2022) Challenges and perspectives of social farming in North-Eastern Italy: the farmers' view. Sustainability 14:8390

Gao Q, Wang JP, Li YH, Sun Y, Yang CH, Li HJ et al (2021) Preliminary study on production performance of Dezhou donkey. Acta Ecol Anim Domastici 42:56–61

Gao R, Shi L, Guo W, Xu Y, Jin X, Yan S et al (2022) Effects of housing and management systems on the growth, immunity, antioxidation, and related physiological and biochemical indicators of donkeys in cold weather. Animals 12:2405. https://doi.org/10.3390/ani12182405

García Pinillos R (2021) One welfare impacts of COVID-19—a summary of key highlights within the one welfare framework. Appl Anim Behav Sci 236:105262. [CrossRef]

Gastaldi D, Bertino E, Monti G, Baro C, Fabris C, Lezo A, Medana C, Baiocchi C, Mussap M, Galvano F et al (2010) Donkey's milk detailed lipid composition. Front Biosci 2:537–546

Gebresenbet G, Aradom S, Kaumbutho PG (2016) Performance and welfare status of working donkeys. Changes 12:15

Geigl E-M, Grange T (2012) Eurasian wild asses in time and space: Morphological versus genetic diversity. Ann Anat 194(1):88–102

Getachew M, Feseha G, Andrew F, Boja F, Fanta A, Bayissa A (2002) Some common clinical cases and interventions of the Donkeys Health and Welfare project, Donkey sanctuary Debre Zeit Ethiopia. In: Pearson A, Fielding D, Tabba D (eds) Fourth international colloquium on working equines. University Hama, Syria, pp 96–109

Groves CP (1986) The taxonomy, distribution and adaptations of recent equids. In: Meadow RH, Uerpmann P (eds) Equids in the ancient world. Ludwig Reichert Verlag, Wiesbaden, pp 11–65

Haddy E, Burden F, Fernando-Martínez JA, Legaria-Ramírez D, Raw Z, Brown J et al (2021) Evaluation of long-term welfare initiatives on workingequid welfare and social transmission of knowledge in Mexico. PLoS One 16:e0251002. https://doi.org/10.1371/journal.pone.0251002

Han L, Zhu S, Ning C, Cai D, Wang K, Chen Q, Hu S, Yang J, Shao J, Zhu H, Zhou H (2014) Ancient DNA provides new insight into the maternal lineages and domestication of Chinese donkeys. BMC Evol Biol 14:1–10. https://doi.org/10.1186/s12862-014-0246-4

Henok G, Gebertsadik Y, Nejash A, Ibrahim N (2017) Prevalence of wounds and associated risk factors in working equines in Jimma town of Oromia region, South-Western Ethiopia. Acad J Anim Dis 6:23–29. https://doi.org/10.5829/idosi.ajad.2017.23.29

IFAW (International Fund for Animal Welfare), Animals, People and War: the Impact of Conflict (2022) Available at: https://www.ifaw.org/press-releases/animals-people-war-impact-conflict

Ivankovic A, Kavar T, Caput P, Mioc B, Pavic V, Dovc P (2002) Genetic diversity of three donkey populations in the Croatian coastal region. Anim Genet 33:169177. https://doi.org/10.1046/j.1365-2052.2002.00879.x

Johnson S, Johnson D (2008) Horse breeds. Voyageur Press, Minneapolis

Jordana J, Folch P (1996) The endangered Catalonian donkey breed: the main ancestor of the American ass or mammoth. J Equine Vet Sci 16:436–441. https://doi.org/10.1016/S0737-0806(96)80209-7

Jordana J, Ferrando A, Miró J, Goyache F, Loarca A, Martínez López OR, Canelón JL, Stemmer A, Aguirre L, Lara MAC, Álvarez LA, Llambí S, Gómez N, Gama LT, Nóvoa MF, Martínez RD, Pérez E, Sierra S, Contreras MA, Guastella AM, Marletta D, Arsenos G, Curik I, Landi V, Martínez A, Delgado JV (2016) Genetic relationships among American donkey populations: insights into the process of colonization. J Anim Breed Genet 133:155–164. https://doi.org/10.1111/jbg.12180

Karabagias IK (2017) Donkey milk exploitation: health benefits, potential applications and prospective invigoration of the hellenic economy. Agric Res Technol Open Access J 12:10–12

Kay G, Tligui N, Semmate N, Azrib R, González FJN, Brizgys L, McLean A (2019) Determining factors and interspecific modeling for serum amyloid a concentrations in working horses, donkeys, and mules. Res Vet Sci 125:256–265

Kefena E, Beja-Pereira A, Han JL, Haile A, Mohammed YK, Dessie T (2011) Eco-geographical structuring andmorphological diversity of Ethiopian donkey populations. Livest Sci 141:232–241

Kefena E, Dessie T, Tegegne A, Beja-Pereira A, Kurtu MY, Rosenbom S, Han JL (2014) Genetic diversity andmatrilineal genetic signature of native Ethiopian donkeys inferred from mitochondrial DNA sequence polymorphism. Livest Sci 167:73–79

Kefena E et al (2021) Genetic diversity and population genetic structure in native Ethiopian donkeys (Equus asinus) inferred from equine microsatellite markers. Trop Anim Health Prod 53:334

Kim SM, Yun SW, Cho GJ (2021) Assessment of genetic diversity using microsatellite markers to compare donkeys (Equus asinus) with horses (Equus caballus). Anim Biosci 34:1460–1465

Kimura B, Marshall FB, Chen S, Rosenbom S, Moehlman PD, Tuross N, Sabin RC, Peters J, Barich B, Yohannes H, Kebede F, Teclai R, Beja-Pereira A, Mulligan CJ (2011) Ancient DNA from Nubian and Somali wild ass provides insights into donkey ancestry and domestication. Proc R Soc B Biol Sci 278:50–57. https://doi.org/10.1098/rspb.2010.0708

Kosuková M, Cernohorská H, Bihuncová I, Oravcová I, Sobotková E, Jiskrová I (2015) Characteristics of morphological parameters of donkeys in the Czech Republic. Acta Universitatis Agriculturae et Silviculturae Mendelianae Brunensis 63:419–424. https://doi.org/10.11118/actaun201563020419

Kugler W, Grunenfelder HP, Broxham E (2008) Donkey breeds in Europe. St. Gallen. Monitoring Institute for Rare Breeds and Seeds in Europe, St. Gallen, Switzerland, Switzerland

Kumeta Y, Maruyama T, Asama H, Yamamoto Y, Hakamatsuka T, Goda Y (2014) Species identification of Asini Corii Collas (donkey glue) by PCR amplification of cytochrome b gene. J Nat Med 68(1):181–185

Labbaci M, Djaout A, Benyarou M, Ameur Ameur A, Gaouar SBS (2018) Morphometric characterization and typology of donkey farming (Equus asinus) in the wilaya of Tlemcen. Genet Biodivers J 2:6072

Leeb C, Henstridge C, Dewhurst K (2013) Welfare assessment of working donkeys: assessment of the impact of an animal healthcare project in West Kenya. Anim Welf 12:689–694. https://doi.org/10.1017/S0962728600026385

Lei CZ, Chen H, Yang GS, Bin Sun W, Lei XQ, Ge QL et al (2005) Study on mitochondrial DNA D-loop polymorphism in Chinese donkeys. Acta Genet Sin 32:481–486

Lei CZ, Ge QL, Zhang H, Liu RY, Zhang W, Jiang YQ et al (2007) African maternal origin and genetic diversity of Chinese domestic donkeys. Asian-Austral J Anim Sci 20:645–652. https://doi.org/10.5713/ajas.2007.645

Li W, Li M, Cao X, Han H, Kong F, Yue X et al (2020) Comparative analysis of whey proteins in donkey colostrum and mature milk using quantitative proteomics. Food Res Int 127:108741. https://doi.org/10.1016/j.foodres.2019.108741

Li M et al (2022a) Comparing the profiles of raw and cooked donkey meat by metabonomics and lipidomics assessment. Front Nutr 9:851761

Li Y, Ma Q, Liu G, Wang C (2022b) Effects of donkey milk on oxidative stress and inflammatory response. J Food Biochem 46:e13935

Lima YF, Tatemoto P, Santurtun E, Reeves E, Raw Z (2021) The human-animal relationship and its influence in our culture: the case of donkeys. Braz J Vet Res Anim Sci 58:e174255

Liu S, Su J, Yang Q, Sun M, Wang Z, Yu J, Jafari H, Lei C, Sun Y, Dang R (2024) Genome-wide analyses based on a novel donkey 40K liquid chip reveal the gene responsible for coat color diversity in Chinese Dezhou donkey. Anim Genet 55:140–146. https://doi.org/10.1111/age.13379

Lopez CL, Alonso R, de Aluja AS (2005) Study of the genetic origin of theMexican Creole donkey (Equus asinus) by analysis of the mitochondrial DNA D-loop region. Trop Anim Health Prod,37 (Suppl 1):173–188. https://doi.org/10.1007/s11250-005-9001-6

Lu C-J, Xie W-M, Su R, Ge Q-L, Chen H, Shen S-Y, Lei C-Z (2008) African origin of Chinese domestic donkeys. Yi Chuan 30:324–328. https://doi.org/10.3724/SP.J.1005.2008.00324

Lutaaya E, Misztal I, Bertrand JK, Mabry JW (1999) Inbreeding in populations with incomplete pedigrees. J Anim Breed Genet 116:475–480

Ma Y, Xu G, Wang D, Liu H, Yang Y (2003) Study on dynamic information of animal genetic resources in China. Agric Sci China 2:80–84

Ma XY, Ning T, Adeola AC, Li J, Esmailizadeh A, Lichoti JK, Agwanda BR, Isakova J, Aldashev AA, Wu SF et al (2020) Potential dual expansion of domesticated donkeys revealed by worldwide analysis on mitochondrial sequences. Zool Res 41:51–60

Mariante A, Da S, Egito AA (2002) Animal genetic resources in Brazil: result of five centuries of natural selection. Theriogenology 57:223–235. https://doi.org/10.1016/S0093-691X(01)00668-9

Martini M, Altomonte I, Tricò D, Lapenta R, Salari F (2021) Current knowledge on functionality and potential therapeutic uses of donkey milk. Animals 11:1382

Moehlman PD (2002) Equids: zebras, asses and horses. Status survey and conservation action plan. IUCN/SSC equid specialist group. IUCN, Gland, Switzerland/Cambridge, UK

Molla B, Dembela S, Megersa B, Mekuria W (2017) The welfare, watering, housing, feeding and working management of working donkeys in and around Hawassa City, Southern Ethiopia. J Vet Res Anim Husb 2:106

Navas FJ, Jordana J, León JM, Barba C, Delgado JV (2017) A model to infer the demographic structure evolution of endangered donkey populations. Animal 11(12):2129–2138. https://doi.org/10.1017/S1751731117000969

Navas González FJ, Jordana Vidal J, León Jurado JM, McLean AK,Pizarro Inostroza G, Delgado Bermejo JV (2018) Genetic parameter estimation and implementation of genetic evaluationfor gaits in a donkey-assisted therapy breeding program. Vet Res Commun 42:101–110. https://doi.org/10.1007/s11259-018-9712-6

Nolan S, Unkovich M, Shen Y, Bellotti L, Li W (2008) Farming systems of the Loess Plateau, Gansu Province, China. Agric Ecosyst Environ 124:13–23. https://doi.org/10.1016/j.agee.2007.08.009

Norris SL, Little HA, Ryding J, Raw Z (2021) Global donkey and mule populations: figures and trends. PLoS One 16(2):e0247830. https://doi.org/10.1371/journal.pone.0247830

Nowak RM, Paradiso JL (1983) Walker's mammals of the world, 4th edn. The Johns Hopkins University Press, Baltimore, MD

Orlando L, Metcalf JL, Alberdi MT, Telles-Antunes M, Bonjean D, Otte M, Martin F, Eisenmann V, Mashkour M, Morello F (2009) Revising the recent evolutionary history of equids using ancient DNA. Proc Natl Acad Sci 106(51):21754–21759

Ozkan Unal E, Ozdil F, Kaplan S, Gurcan EK, Genc S, Arat S, Soysal MI (2020) Phylogenetic relationships of Turkish indigenous donkey populations determined by mitochondrial DNA D-loop region. Animals 10:1970

Pearson RA, Mengistu A, Agaije T, Allen EF, Smith DG, Mesfin A (2001) Use and management of donkey in peri-urban areas of Ethiopia. Draught Animal Power Technical Report. Centre for Tropical Veterinary Medicine, University of Edinburgh, Scotland

Perez-Pardal L, Grizelj J, Traore A, Cubric-Curik V, Arsenos G, Dovenski T, Markovic B, Fernandez I, Cuervo M, Alvarez I et al (2014) Lack of mitochondrial DNA structure in Balkan donkey is consistent with a quick spread of the species after domestication. Anim Genet 45:144–147. [CrossRef]

Pimentel F, Paiva S, Dias LT, Pimentel D, McManus C (2023) Donkeys in Brazil: bibliometric mapping and breed information. Int J Equine Sci 2(1):34–46

Portaro S, Maresca G, Raffa A, Gemelli G, Aliberti B, Calabrò RS (2020) Donkey therapy and hippotherapy: two faces of the same coin? Innov Clin Neurosci 17:20

Pritchard JC, Lindberg AC, Main DCJ, Whay HR (2005) Assessment of the welfare of working horses, mules and donkeys, using health and behaviour parameters. Prev Vet Med 69(3–4):265–283. https://doi.org/10.1016/j.prevetmed.2005.02.002

Quaresma M, Martins AM, Rodrigues JB, Colaço J, Payan-Carreira R (2014) Pedigree and herd characterization of a donkey breed vulnerable to extinction. Animal 8(3):354–359. https://doi.org/10.1017/S1751731113002218

Quaresma M, Martins AMF, Rodrigues JB, Colaço J, Payan-Carreira R (2015) Viability analyses of an endangered donkey breed: the case of the Asinina de Miranda (Equus asinus). Anim Prod Sci 55:1184–1191

Ravichandran T, Perumal RK, Vijayalakshmy K, Raw Z, Cooke F, Baltenweck I, Rahman H (2023) Means of livelihood, clean environment to women empowerment: the multi-faceted role of donkeys. Animals 13:1927. https://doi.org/10.3390/ani13121927

Rosenbom S, Costa V, Al-Araimi N, Kefena E, Abdel-Moneim AS, Abdala MA, Bakhiet A, Beja-Pereira A (2015) Genetic diversity of donkey populations from the putative centers of domestication. Anim Genet 46:30–36. https://doi.org/10.1111/age.12256

Rossel S, Marshall F, Peters J, Pilgram T, Adams MD, O'Connor D (2008) Domestication of the donkey: timing, processes, and indicators. Proc Natl Acad Sci USA 105(10):3715–3720

Santana ML, Bignardi AB (2015) Status of the genetic diversity and population structure of the Pêga donkey. Trop Anim Health Prod 47:1573–1580

Sargentini C, Tocci R, Lorenzini G, Gianangeli B, Martini A, Gallai S, Giorgetti A (2009) Morphological characteristics of Amiata donkey reared in Tuscany. Ital J Anim Sci 8:721–723. https://doi.org/10.4081/ijas.2009.s2.721

Sargentini C, Tocci R, Martini A, Bozzi R (2018) Morphological characterization of Amiata donkey through multivariate analyses. Rev Bras Zootec 47:e20170310. https://doi.org/10.1590/rbz4720170310

Sbisa E, Tanzariello F, Reyes A, Pesole G, Saccone C (1997) Mammalian mitochondrial D-loop region structural analysis: identification of new conserved sequences and their functional and evolutionary implications. Gene 205:125–140

Sells PD, Pinchbeck G, Mezzane H, Ibourki J, Crane M (2010) Pack wounds of donkeys and mules in the Northern high atlas and lowlands of Morocco. Equine Vet J 42:219–226. https://doi.org/10.2746/042516409X478532

Seyiti S, Kelimu A (2021) Donkey industry in China: current aspects, suggestions and future challenges. J Equine Vet Sci 102:103642. https://doi.org/10.1016/j.jevs.2021.103642

Shi T, Hu W, Hou H, Zhao Z, Shang M, Zhang L (2020) Identification and comparative analysis of long non-coding RNA in the skeletal muscle of two Dezhou donkey strains. Genes 11(5):508

Smith DG, Pearson RA (2005) A review of the factors affecting the survival of donkeys in semi-arid regions of sub-Saharan Africa. Trop Anim Health Prod 37(Suppl 1):1–19

Sobrero L, Dalla Costa E, Minero M (2024) Management of donkeys in assisted interventions: a snapshot. Animals 14:670. https://doi.org/10.3390/ani14050670

Sørensen JT, Sandøe P, Halberg N (2001) Animal welfare as one among several values to be considered at farm level: the idea of an ethical account for livestock farming. Acta Agr Scand A: An Suppl 30:11–16

Stanisic LJ, Aleksic JM, Dimitrijevic V, Simeunovic P, Glavinic U, Stevanovic J, Stanimirovic Z (2017) New insights into the origin and the genetic status of the Balkan donkey from Serbia. Anim Genet 48:580–590. https://doi.org/10.1111/age.12589

Stanisic L, Aleksi¢ JM, Dimitrijevic V, Kova£evi¢ B, Stevanovic J, Stanimirovic Z (2020) Banat donkey, a neglected donkey breed from the central Balkans (Serbia). PeerJ 8:e8598. https://doi.org/10.7717/peerj.8598

Starkey P, Starkey M (eds) (2000) Regional and world trends in donkey populations, pp 10–21

Stringer AP, Bell CE, Christley RM, Gebreab F, Tefera G, Reed K et al (2011) A cluster-randomised controlled trial to compare the effectiveness of different knowledge-transfer interventions for rural working equid users in Ethiopia. Prev Vet Med 100:90–99. https://doi.org/10.1016/j.prevetmed.2011.02.001

Stringer AP, Christley RM, Bell CE, Gebreab F, Tefera G, Reed K et al (2017) Owner reported diseases of working equids in Central Ethiopia. Equine Vet J 49:501–506. https://doi.org/10.1111/evj.12633

Sun Y, Jiang Q, Yang C et al (2016) Characterization of the complete mitochondrial genome of Dezhou donkey (Equus asinus) and evolutionaryanalysis. Curr Genet 62:383–390. https://doi.org/10.1007/s00294-015-0531-9

Tarazona AM, Ceballos MC, Broom DM (2020) Human relationships with domestic and other animals: one health, one welfare, one biology. Animals 10:43

Tedla MG, Berhe KF, Grmay KM (2023) The impact of armed conflict on animal well-being and welfare, and analyzing damage assessment on the veterinary sector: the case of Ethiopia's Tigray region. Heliyon 9(12):e22681. https://doi.org/10.1016/j.heliyon.2023.e22681. PMID: 38076045; PMCID: PMC10709505.

Tesfaye A, Curran MM (2005) A longitudinal survey of market donkeys in Ethiopia. Trop Anim Health Prod 37:87–100. https://doi.org/10.1007/s11250-005-9010-5

Tesse R, Paglialunga C, Braccio S, Armenio L (2009) Adequacy and tolerance to ass's milk in an Italian cohort of children with cow's milk allergy. Ital J Pediatr 35:19

The American Donkey and Mule Society. Archived 14 August 2011

The Brooke (2015) Invisible workers: the economic contributions of working donkeys, horses and mules to livelihoods. The Brooke, London, UK. Available online: https://www.thebrooke.org/research-evidence/invisible-workers-economic-contribution-workingequids-livelihoods

The Donkey (1990) Government of Alberta: Agriculture and Rural Development. Archived 15 August 2011

Tian F, Wang JP, Li YH, Yang CH, Zhang RT, Wang XG et al (2020) Integrated analysis of mRNA and miRNA in testis and cauda epididymidis reveals candidate molecular markers associated with reproduction in Dezhou donkey. Livest Sci 234:103885. https://doi.org/10.1016/j.livsci.2019.103885

Vilá C, Leonard JA, Beja-Pereira A (2006) Genetic documentation of horse and donkey domestication. In: Zeder MA, Bradley DG, Emshwiller E, Smith BD (eds) Documenting domestication, pp 342–353

Wang T, Liu Z, Shi X et al (2023) An investigation of genetic diversity in three Dezhou donkey original breeding farms. Sci Rep 13:11203. https://doi.org/10.1038/s41598-023-38219-1

Watson TL, Kubasiewicz LM, Chamberlain N, Nye C, Raw Z, Burden FA (2020) Cultural "Blind Spots," Social influence and the welfare of working donkeys in Brick Kilns in Northern India. Front Vet Sci 7:214

Whay HR, Dikshit AK, Hockenhull J, Parker RMA, Banerjee A, Hughes SI et al (2015) Evaluation of changes in equine care and limb-related abnormalities in working horses in Jaipur, India, as part of a two year participatory intervention study. PLoS One 10:e0126160. https://doi.org/10.1371/journal.pone.0126160

Xia X, Yu J, Zhao X, Yao Y, Zeng L, Ahmed Z, Shen S, Dang R, Lei C (2019) Genetic diversity and maternal origin of Northeast African and South American donkey populations. Anim Genet 50:266–270. https://doi.org/10.1111/age.12774

Xia J, Chang L, Xu D, Jia Y, Ding Y, Cao C, Geng Z, Jin S (2023) Next-generation sequencing of the complete Huaibei Grey Donkey Mitogenome and Mitogenomic Phylogeny of the Equidae Family. Animals 13:531. https://doi.org/10.3390/ani13030531

Xie C (1987) Horse and Ass Breeds in China. Shanghai Scientific and Technical Publishing House, Shanghai

Xu ZJ, Liao XX, Mao YJ, Yang ZP (2012) Analysis on present genetic resources of Chinese Ass. Chin Herbivorous Anim Sci 32(4):70–73

Yadav RK, Singh N, Dilawar M, Archana FA (2024) Donkey farming in India. SR Publication

Yalew A, Darge D, Melake BM (2024) Assessment of community-based intervention approaches to improve the health and welfare of working donkeys in selected areas of Sidama region, Southern Ethiopia. Front Vet Sci 10:1253448. https://doi.org/10.3389/fvets.2023.1253448

Zakari Friday O, Ayo Joseph O, Kawu Mohammed U, Rekwot I (2015) The effect of season and meteorological stress factors on behavioural responses and activities of donkeys (Equus asinus)—A review. Ann Anim Sci 15:307–321

Zarins J, Hauser R (2014) The Domestication of Equidae in Third-Millennium BCE Mesopotamia, vol 24. Cornell University Studies in Assyriology and Sumerology (CUSAS), University Park, PA, p XI + 432. ISBN 978-1934309-513

Zhang BW (2011) China National Commission of Animal Genetic Resources. Animal genetic resources in China: Horses, Donkeys, Camels. China Agricultural Press, China

Zhang YS, Yang XY, Wang XB, Zhang CM, Qin F, Zhou ZH et al (2010) Cytochrome b genetic diversity and maternal origin of Chinese domestic donkey. Biochem Genet 48:636–646. https://doi.org/10.1007/s10528-010-9345-0

Zhang R et al (2016) High polymorphism at microsatellite loci in the Chinese donkey. Genet Mol Res 15(2):gmr.5028291

Zhang X, Li H, Yu J, Zhou X, Ji C, Wu S et al (2019) Label-free based comparative proteomic analysis of whey proteins between different milk yields of Dezhou donkey. Biochem Biophys Res Commun 508:237–242. https://doi.org/10.1016/j.bbrc.2018.11.130

Zhang Z, Gao X, Faheem M, Wang Y, Wang T, Shi X et al (2022) Comparative analysis of growth and development characteristics of two Dezhou donkey strains. Livestock Sci 263:105024. https://doi.org/10.1016/j.livsci.2022.105024

Zhang Z, Huang B, Wang Y, Zhu M, Liu G, Wang C (2023) A survey report on the donkey original breeding farms in China: current aspects and future prospective. Front Vet Sci 10:1126138. https://doi.org/10.3389/fvets.2023.1126138

Zheng PL (1985) Livestock Breeds in China and their ecological characteristics. Agriculture Press, Beijing

Challenges in Donkey Conservation and Breeding: Historical, Genetic, and Sociocultural Perspectives

Francisco Javier Navas González and Miguel Quaresma

Abstract

This chapter offers an integrative examination of the historical, genetic, and sociocultural dimensions shaping donkey conservation and breeding worldwide. It begins by tracing the domestication and global dispersal of donkeys from their African wild ass ancestors during the mid-Holocene to their contemporary roles across diverse ecological and cultural landscapes. The historical narrative highlights donkeys' centrality to trade, agriculture, and colonization, their subsequent decline under industrialization, and recent pressures driven by global demand for donkey-derived products. The second section addresses genetic challenges, including high coancestry, inbreeding, and the erosion of purebred lineages, alongside opportunities for sustainable genetic management. Topics such as incomplete pedigrees, introgression, feral population undervaluation, long generation intervals, and assisted reproduction inefficiencies underscore the complexity of maintaining genetic diversity. Special attention is given to the ambiguous contributions of wild ass populations to modern breeds and the implications of selective breeding for health and adaptability. Finally, the chapter explores anthropological and sociocultural factors influencing donkey populations, including owner-driven inbreeding practices, foreign breed introgression, shifting breed standards, and paradoxical conservation outcomes—endangered in native regions yet invasive abroad. It also discusses economic, cultural, and ethical contexts shaping conservation policies and breeding

F. J. Navas González (✉)
Department of Genetics, University of Cordoba, Cordoba, Spain

Worldwide Donkey Breeds Project, Cordoba, Spain
e-mail: fjnavas@uco.es

M. Quaresma
Departamento de Ciências Veterinárias, Universidade de Trás-os-Montes e Alto Douro, Vila Real, Portugal

A. K. McLean et al. (eds.), *Donkey and Mule Medicine: Foundations, Breeds, and Management*, https://doi.org/10.1007/978-3-032-12726-6_2

practices. By integrating archaeological evidence, genetic insights, and sociocultural analysis, the chapter underscores the urgent need for multidimensional strategies combining scientific tools, policy frameworks, and community engagement to secure the future of donkey genetic resources.

Keywords

Donkey domestication · Genetic diversity · Inbreeding · Conservation strategies · Breed introgression · Assisted reproduction · Feral populations · Historical livestock dispersal · Sociocultural factors · Donkey welfare · Sustainable breeding · Ejiao trade · Pedigree management · Selective breeding impacts · Anthropological perspectives · Endangered breeds

1 Introduction

Donkeys (*Equus asinus*) have accompanied human societies for over 7000 years, shaping the development of civilizations across arid and semi-arid landscapes through their indispensable roles in transport, agriculture, and trade (Kimura et al. 2022). Originating from African wild ass lineages, their domestication during the mid-Holocene marked a pivotal adaptation to the challenges of increasing aridity and expanding human mobility (CNRS 2022). Over millennia, donkeys dispersed globally through complex networks of migration, commerce, and colonization, becoming deeply embedded in economic systems and cultural traditions (Zhang et al. 2024). Despite their historical significance and resilience, donkeys have often been marginalized in scientific research and agricultural policy, overshadowed by other livestock species such as horses and cattle.

Today, donkey populations face a paradoxical reality. While they remain vital to subsistence economies in many developing regions (Biffa and Woldemariam 2020), they have suffered sharp declines in industrialized nations, driven by mechanization and rural depopulation (Gebreab 2020). More recently, global demand for donkey hides for ejiao production has created new demographic and welfare crises, exacerbating population pressures and threatening genetic diversity (Pinna et al. 2025). These challenges intersect with longstanding issues such as inbreeding, incomplete pedigree documentation, and insufficient institutional frameworks for conservation.

Addressing these concerns requires an interdisciplinary approach that integrates historical perspectives, genetic science, and sociocultural understanding. This chapter explores the trajectory of donkey domestication and global diffusion, the genetic vulnerabilities and management opportunities that shape their current status, and the anthropological and cultural factors influencing breeding practices and conservation outcomes. By combining archaeological evidence, molecular insights, and social dimensions, the chapter seeks to provide a holistic foundation for strategies aimed at safeguarding donkey populations—both as a genetic resource and as a cultural and economic asset for future generations.

2 Overview of the History of Donkeys Since Domestication to Our Days

Sociocultural, anthropological, archaeological and genetic data point to a long evolutionary history for the donkey. African wild asses *(Equus africanus)* split into deeply divergent maternal lineages long before humans began to domesticate them (Wang et al. 2022) involving Nubian, Somali and Atlas or Algerian Wild asses, parallel to zooarchaeological evidence placing the earliest donkey domestication in northeast Africa during the mid-Holocene. Donkeys were first tamed by mobile pastoralists adapting to aridification, roughly between 5500 and 4500 BCE (Wang et al. 2022). Morphological changes in early domestic donkeys include modest reductions in body size and more robust limbs—adaptations for load-carrying—compared to contemporaneous wild asses (Kimura et al. 2011). Zooarchaeological finds at predynastic Egyptian sites (El Omari in Upper Egypt, Maadi and Helwan in the Delta, Hierakonpolis, among others) dated ~4600–3500 BCE show the first securely domesticated riding or draft donkeys: standard-bodied individuals with homogeneous age profiles and dental wear suggestive of bit use or tethering (Mitchell 2017). By the end of the fourth millennium BCE (predynastic to Early Dynastic Egypt), complete donkey burials and skeletal remains (e.g., from Abydos and nearby funerary sites) exhibit degenerative lesions on vertebrae and joints diagnostic of heavy pack work (Mitchell 2017). Contemporary iconography and texts (e.g., the Libyan Palette, ~3200–3000 BCE) portray donkeys in ritual and transport roles, and later Old Kingdom inscriptions routinely record donkeys in the state economy—hauling stone for pyramid building, moving commodities, collecting taxes, and managing caravan logistics (Mitchell 2017).

Donkeys had spread beyond Egypt by the early Bronze Age. They appear by ~3000 BCE in the Levant and Mesopotamia, both archaeologically and textually. Egyptian reliefs and cuneiform accounts indicate that donkeys were being imported or bred in the Near East by the late fourth millennium BCE (Todd et al. 2022) populating territories of the Arabian Peninsula and Sokotra Island. Art suggests the existence of an already perceived variability within the population and their differential functional application (Fig. 1).

The renowned "kungas" of Bronze-Age Syria (~2500 BCE)—elite draft and warfare animals depicted in Sumerian and Akkadian art—have recently been confirmed by ancient DNA to be first-generation hybrids of female domestic donkeys and male Syrian wild asses (hemiones) (Bennett et al. 2022) (Fig. 2). This finding proves that by ~4500 years ago donkeys were fully domesticated in Mesopotamia and were even being crossbred with wild equids to produce superior beasts of burden.

During the 3rd–2nd millennia BCE, domestic donkeys dispersed widely along trade and pastoral networks. Westward from Egypt they appear in Anatolian and Aegean contexts; eastward they moved through Mesopotamia into Iran and the Indus Valley as supported by remains found at Tell es-Sâfi sites in Gath dating back from the Early Bronze Age III (c. 2800–2600 BCE). Chinese and Eurasian donkey populations carry African donkey ancestry both Nubian and Somali (Todd et al.

Fig. 1 Egyptian donkey population variability depicted through art

2022), consistent with historical spread along Persian, Silk Road, and maritime routes. In Africa, donkeys proliferated beyond Egypt after ~3000 BCE: they become common in Nubian and Saharan contexts (Kerma in Sudan, Jebel Uweinat and Libyan oases) by the second millennium BCE and are first documented in East African pastoral sites (Kenya/Tanzania) around 1000 BCE alongside cattle and caprines (Mitchell 2017).

In India, donkeys are believed to have arrived by the third millennium BCE (ca. 2500–2000 BCE), likely through overland trade corridors traversing the Iranian plateau and via maritime exchanges along the Persian Gulf and Arabian Sea.

Fig. 2 Depiction of Kungas and their uses in the Standard of Ur. Peace Panel (upper left first row) and War Panel (upper left second row); Suckling donkey foal and male in two herds of donkeys Scene is in a wall relief block, at Saqqara and the funerary chapel from the mastaba of Neferirtenef, respectively (upper right and central row); Herd of donkeys on their way to Omo Valley in Ethiopia (lower row)

Archaeological findings at major Indus Valley Civilization sites such as Harappa and Mohenjo-Daro include osteological remains consistent with *Equus asinus*, attesting to their early integration into protohistoric economies. These animals were primarily valued for their capacity to endure arid environments, and their diffusion across the subcontinent was likely facilitated by their use in transportation, agricultural labour, and regional trade. Their presence coincides with broader cultural and economic networks linking Mesopotamia, the Iranian highlands, and the northwest of the Indian subcontinent.

In Iran and the adjacent highlands, the role of donkeys and their hybrids—especially mules—appears prominently in both linguistic and historical records. Notably, the Tocharian B word for 'mule' is derived from Old Iranian (*atswa-* 'horse'), suggesting that mules—rather than donkeys—were likely the first equids to traverse the rugged passageways of the Pamirs into the Tarim Basin. This implies a sophisticated use of equine hybrids well adapted to mountain travel, and perhaps a staged introduction of donkey parentage through subsequent trade and migration flows. Iran thus served not only as a conduit but potentially as an early adopter of equid management techniques central to donkey diffusion eastward.

In China, domestic donkeys appear to have been introduced before the Hàn period (202 BCE–220 CE), although clear archaeological and textual evidence of their widespread use only becomes prominent by the second century BCE. This period marks their strategic deployment by the imperial court for crossing the Taklamakan Desert—an endeavor crucial to the logistics of long-distance diplomacy and Silk Road trade. Genetic studies indicate that Chinese donkeys are not descended from local or Asian wild equids but from African ancestors, confirming their ultimate derivation from the Nubian and Somali wild asses. Early breeding centers are believed to have been located in the region of modern-day Xīnjiāng, from which domestic donkeys dispersed along three principal axes: (1) eastward via Níngxià and Gānsù into the Guānzhōng Plains (關中) and Shǎnxī; (2) northeastward into Inner Mongolia and southwestward into Yúnnán; and (3) from the Guānzhōng Plains into the wider regions of central and southern China. Despite their late attestation in the historical record, donkeys came to occupy a vital role in imperial transport, agriculture, and regional economies, mirroring their earlier trajectories in South Asia and the Near East.

Although the dispersal of donkey breeds is largely linked to human movements, their southward progression into the southern regions of Africa encountered a notable terminus. Arabic accounts for West-Central Africa mention donkeys several times (Levtzion and Hopkins 1981). Al-Bakr¯i noted the use of donkeys to carry salt in the Kingdom of Ghana (a use which they will recover later on, as donkeys primary use in some parts of the Caribbean will be to work alongside black slaves in salt mines), and Al-Um¯ar¯i commented on their small size in the Empire of Mali.

Textual records indicate large herds of donkeys; many used for portage. Under Pharaoh Pepi II (c. 2270 BCE), trading expeditions to Punt (Ethiopia) consisted of caravans with pack donkeys (Kitchen 1993). The extent to which the domestic donkey departs from its wild relative can be tracked through Egyptian wall paintings: the dark shoulder stripe of the ass gradually disappears as the Old

Kingdom transitions to the Middle Kingdom (Brewer et al. 1994: 100). Donkeys also occur archaeologically at Shaqadud in the Butana grasslands of Sudan during the second millennium BCE (Peters 1991). The historical and archaeological evidence for domestic donkeys in the Maghreb is reviewed by Camps (1988), and faunal assemblages at Carthage during the Roman period (first–fourth centuries AD) include donkey remains (Levine 1994). In Morocco, Kaache (1996) discusses possible ass bones from the Neolithic sites of Dar-es-Soltane and Tangier, though no certain rock-art representations exist.

Despite these early and widespread occurrences, research on livestock dispersal in Sub-Saharan Africa suggests that infectious diseases—trypanosomiasis, equine babesiosis (piroplasmosis), and African horse sickness—limited the distribution of domestic animals such as cattle (Gifford-Gonzalez 2000, 2017) and dogs (Mitchell 2015). Studies confirm that donkeys were absent from southern Africa before European arrival; the combined threat of these diseases likely halted their southward spread.

The most plausible explanation for donkeys' arrival in southern Africa is their importation by the Dutch East India Company (VOC) between 1652 and 1656, most likely from Europe (though Persia or Indonesia have also been suggested). The VOC's settlement at the Cape of Good Hope (1652) served initially as a provisioning station for ships to the East Indies. Livestock—including donkeys and mules—were introduced to trade with the Khoikhoi and supply fresh meat to passing vessels, later to support the settler population. Mules became increasingly common through the eighteenth century, hauling wagons inland. A rock painting at Stompiesfontein in the Koue Bokkeveld, likely dating to the late 1700s, depicts mules yoked to transport wagons (working farms as early as 1728).

Although mules and donkeys pulled wagons and ploughed wheat farms in the Western Cape into the early 1900s—and even worked on sugar plantations in KwaZulu-Natal—neither equid found widespread employment among European farmers. Trypanosomiasis in the more humid eastern regions constrained their use. Correspondingly, archaeological finds are rare: a pre-1870 (likely nineteenth-century) specimen at a Bushman–occupied rockshelter in Haaskraal (eastern Karoo); a possible example at a British cavalry outpost (Steinaecker's Horse) during the South African War (1899–1902) in northern Kruger National Park; and a potentially intrusive donkey at Ondini, capital of King Cetshwayo (1872–1879).

Artistic and literary references to livestock among the first Boer colonists often mention cattle, horses, and mules, but donkeys remain scarce and ambiguous. Yet donkeys actively participated in developing the Lower Orange River Valley—from Lesotho into South Africa and Namibia—during the Great Trek (beginning in 1836) and the Pioneer Column under Cecil Rhodes (1890), which led to Rhodesia's establishment. Bridle paths between South Africa and Lesotho saw goods carried on donkeys, mules, and oxen (Mitchell 2017).

Some colonial-era references to donkeys come from indigenous groups like the Sonquas (Hottentots) (Wright 1996). Spread across Botswana, Namibia, Angola, Zambia, Zimbabwe, Lesotho, and South Africa, the Sonquas neither owned cattle nor kept livestock but hunted large game, including wild horses and mules—suggesting interactions with feral equids.

In the Swartruggens region of South Africa's Western Cape, rock shelters preserve a distinctive colonial-period art corpus. Painted between the 1870s and 1885 rock paintings depict wagons, horses, donkeys and/or mules (with varied ear lengths), European figures, and symbolic motifs such as handprints and dots. Hall and Mazel interpret these works not as passive records but as satirical, "hidden transcripts" of laborers and drovers responding to colonial expansion and the diamond rush. One painting shows four human figures, nine equids (donkeys, mules, and horses), three wagons, and a right-hand red handprint—underscoring donkeys' presence in everyday life and subtle resistance narratives.

One tragic episode illustrating donkeys' complex social position occurred in 1983 in Bophuthatswana. Framed by colonial legacies and apartheid's separate-development policies, authorities decreed that households could own no more than four donkeys and ordered the slaughter of all others. Donkey owners were initially reluctant to disclose the number of donkeys they owned. This hesitation stemmed from concerns that the research data might be used for further culling or harm, rather than for aiding. Police and army units enforced the policy brutally, shooting donkeys—sometimes repeatedly—near their owners. Rom 20,000 to 60,000 donkeys were killed. The massacre traumatized rural communities, particularly women reliant on donkeys for transport and subsistence. Grassroots resistance erupted, notably when an elderly man killed a soldier in retaliation, sparking broader rural rebellions. The cull inflicted ecological damage too, as carcasses polluted water sources and created markets for water monopolized by white farmers.

Despite scant official records, donkey populations in Bophuthatswana plunged, followed by declines in cattle—undermining ecological justifications. The massacre revealed the coercive power of an undemocratic state and the use of environmental narratives to suppress dissent. In subsequent decades, local communities around Kuruman reframed donkeys as symbols of resilience, dignity, and justice, integrating the massacre's memory into broader anti-apartheid struggles and popular protests.

Alongside donkeys, quaggas—a now-extinct subspecies of zebra native to southern Africa—occasionally featured in human equid economies. Though never fully domesticated, quaggas were sometimes tamed, especially when captured young. Early observers such as Barrow noted their remarkable strength: one farmer claimed a single quagga could pull more than six horses combined, though their gait was largely limited to walking or slow cantering, and they fatigued when driven faster than a walk.

A striking example of habituation is the quagga photographed by Gustav Theodor Fritsch on April 8, 1864 at Andrew Hudson Bain's farm, Quaggafontein. The animal's calmness during stereophotography—requiring stillness—demonstrates its tractability, even if it was not part of a formal breeding program or employed in work.

To overcome quaggas' gait limitations and disease carriage, breeders experimented with hybrids—zebroids (zorses, zebdonks) produced by crossing zebras or quaggas with horses or donkeys. These hybrids aimed to combine domestic equids' tractability with quaggas' resistance to trypanosomiasis and African horse sickness. In southern Africa, zebdonks were used alongside mules to pull stagecoaches, illustrating practical utility (Fig. 3).

Fig. 3 "Royal Gift". Only known depiction showing the long characteristic long coat of the Zamorano-Leonés donkey (upper left). "Royal Gift" was a jack given as a present from Spanish King Charles III to President George Washington (upper middle); Decree CXLI. Repeal of all laws, ordinances, etc., related to the breeding of mules and horses: abolition of sub-delegations, inspectorates, etc., concerning the breeding of mares (upper right); Prospector Standing in Front of Building with donkey Loaded with Gold Mining Equipment (North American Gold Rushes between 1848 and 1855) (Central right); South African Boer farmer kids riding donkeys or using them in caravans (central left and middle) Boer Wars (1899 to 1905) caricature of the Battle of Elandslaagte in which Boer Soldiers flighted against British soldier who were depicted as hinnies or mules in this postcard, in a mocking reference to the white donkey given as a present from Egypt to the queen of England (central left); Painted rock depicting long ear equines in Swartruggens Mountains (South Africa); Tamed Quagga (Central Middle) and Quagga mare x Male Donkey hybrid in Doncaster Museum

One notable hybrid was the Doncaster foal (Owston Park, England, 1830), born to a quagga mare and donkey sire. Though it lived only 3 weeks, taxidermist Hugh Reid preserved the foal, making it the sole confirmed hybrid among 23 known quagga skins worldwide. Historical breeding experiments included painting an ass stallion with stripes (1773) to attract a zebra mare and maintaining domesticated quaggas among horse herds in the Cape Colony. In nineteenth-century Britain, a quagga stallion owned by the Earl of Morton sired a filly with a high-Arabian mare, fueling debates on heredity and telegony—the belief a female's first mate influences all future offspring. Beyond breeding, quaggas and their hybrids captured public fascination through exhibitions, artwork, and training. In southern African folklore, oral traditions link quaggas to donkeys, encoding these equids into cosmologies—such as the Khoisan interpretation of the Milky Way as a donkey figure—highlighting their enduring cultural resonance.

According to Navas-González et al. (2016), over the first millennium BCE and into the Classical period, donkeys became central to long-distance trade. They were the workhorses of the trans-Saharan salt and gold caravans that linked West Africa with Mediterranean markets. Here they recovered their herd animal conditions while gaining individual relevance for riding, transport and draft labours. Greek and Roman writers (Xenophon, Pliny, among others) describe advanced donkey husbandry—bit-fitting, pack-saddle design, and nutritional regimes—reflecting the importance of the species to commerce and military logistics. The Phoenicians (c. 1200–600 BCE) and Ptolemaic Egyptians (c. 305–30 BCE) introduced donkeys to Mediterranean islands (Cyprus, Crete) and beyond.

Remains from donkeys have been found at Iron I levels in Spanish archaeological sites of Sansol (Muru-Astrain) and La Hoya dating back to 600–501 BCE, which may support the fact that donkeys may have anecdotally arrived in Europe crossing the Mediterranean, primarily they did it all the way around across Asia and Eastern Europe. The Roman era (27 BCE–fall of the Western Roman Empire in 476 CE) saw selective breeding to produce larger, stronger donkeys for heavy tasks (quarrying, mining, milling, mule breeding, or even amusing people in circuses) and also spreaded donkeys through northern European territories like Britain (mid-first century CE (after the Emperor Claudius Roman invasion in 43 CE; it is presumed that there were no donkeys in the territory prior to this event) and from here to Ireland (first evidence in 1642 capture of Maynooth Castle). Imperial veterinary texts (e.g., Nero's writings) detail special diets and stud programs for "mammoth" jackstock, and estates routinely bred donkeys with horses to make mules, a major economic enterprise throughout the Empire (Todd et al. 2022).

In the medieval period, donkeys continued to spread with expanding trade networks. Islamic Umayyad caliphates, who would standout the role of large donkeys with particular white coat phaneroptic features as exceptional for riding among high classes, brought African donkey lineages across the Maghreb into Iberia after the eighth century, where Andalusi agricultural manuals describe pack-donkey care. Indeed, is this caliphates era during when the first protection against exportations and banning against horse crossbreeding regulations were formulated. This in turn would become the background reasons for the first historical donkey population bottlenecks experienced worldwide. To the east, donkeys accompanied camel caravans along the Silk Road and into Tang China; archaeozoological finds in Han-dynasty tombs (second century BCE) confirm their early arrival in East Asia (Wang et al. 2022). By 1500 CE, donkeys were well established as indispensable transport animals from the Mediterranean to Central Asia.

The Age of Exploration (fifteenth to seventeenth centuries) marked the global dispersal of donkeys. Iberian explorers deliberately stocked Atlantic islands with asses, either directly from the Iberian Peninsula but also donkeys that have arrived to Macaronesia (Madeira, Canary Islands and Cape Verde) from Western Africa along with slave trade by the Dutch: Portuguese and Spanish ships carried jacks and jennies from Macaronesia to support colonial agriculture and mining. Columbus's second voyage (1493–1496) introduced donkeys to Hispaniola, and subsequent Spanish (Alonso de Ojeda among others) and Portuguese expeditions brought them throughout the Caribbean, South America, and North America by the fifteenth and mid-sixteenth centuries.

Donkeys proved essential in colonial salt, gold and silver mines and plantations, which led to feral herds establishment on remote islands (e.g., the Virgin Islands, Bonaire, among others) as a legacy of these introductions and gradually moved northward with Spanish expeditions, reaching Mexico and the American Southwest by the late 1500s.

In the Eighteenth Century, not only working standard donkeys were being shipped, but also taller ones with the intentional role to breed for mule production. Exportation of large jacks belonging to mainly Andalusian, Zamorano-Leonés and Catalonian breeds was still forbidden by law in some parts of Spain such as Andalusia, Murcia and Extremadura by King Ferdinand VII of Spain until 1822 (prohibition lasted from XV to XIX century). A donkey named "Royal Gift", a gift from the King Charles III of Spain to George Washington, arrived in America by boat in 1785. Specifically, it arrived in Gloucester, Massachusetts, and was then transported to Mount Vernon. This Spanish Zamorano-Leonés jack was crucial for Washington's mule breeding program, as mules were highly prized for their strength and work ethic in agriculture and marked the start of the interest in European donkeys and successful mule industry in North America (Fig. 3).

The utility of donkeys as pack animals in mining regions made them indispensable during the California and Colorado gold rushes of the nineteenth century, where they were favoured for their endurance in rugged terrain. Indeed, after their release they became useful and valued animals among North American Indian Tribes (like Navajo). In the Southern Hemisphere, the first African donkeys reached the Cape of Good Hope in 1656 with Dutch settlers; they spread slowly through eastern Africa, thereafter, documented in colonial farm records for labour tasks related to wheat harvesting, water supply, draft and diamond and asbestos mining (Mitchell 2017).

On the other side of the world a parallel process occurred as donkeys were introduced to Australia during the nineteenth-century gold rushes. Donkeys arrived in Australia in 1793, having been introduced from India—the destination in their global dispersal. It may be important to highlight the crucial role of donkeys in haulage in New South Wales during the 1840s and their later importance in facilitating colonial expansion into the interior regions of South Australia (1850s–1860s) and Western Australia (1890s) together with British and Afghan camel caravans and became widespread in arid interior region. The very traits that had made donkeys so valuable in the Americas—remarkable stamina, resilience in arid climates, ability to thrive on sparse forage, and resistance to ailments like red water disease, walkabout disease, and cattle ticks—also made them preferable to horses or oxen for draught work across the vast and challenging landscapes of inland Australia. On remote sheep stations, they were often the only feasible transport solution. In some areas, notably the wagon routes from Carnarvon on the far western coast to the Kimberley region over 1000 km away, donkeys and mules were essential. Oral histories describe wagons drawn by teams of 20 to 30 animals, with donkeys imported from diverse locations such as Mauritius, Pakistan, Spain, Mexico, and Indonesia.

Aboriginal Australians played an indispensable role in the successful colonization of Australia's interior and the establishment of commercial livestock enterprises. Their intimate knowledge of the land and its resources, combined with their contribution of both skilled and unskilled labour, proved vital to the operation of sheep and cattle stations. Where donkeys were among the animals managed—such as in the Murchison Ranges near the center of the Northern Territory—Aboriginal workers became adept at handling and caring for them. In some cases, particularly in smaller-scale pastoralist contexts, these relationships fostered more reciprocal dynamics, occasionally including intermarriage. Archaeological investigations, supported by oral testimony, have identified various sites where livestock—including donkeys—were housed, watered, and utilized in local haulage systems that sometimes employed up to 100 donkeys to pull a single cart. In the Pilbara region of Western Australia, equids also entered the realm of Aboriginal rock art. While horses and cattle predominate in these depictions, some equids are shown being ridden or standing, signifying their integration into local visual culture. Today, Australia hosts the world's largest population of feral donkeys—an enduring, if unintended, legacy of the Columbian Exchange. Indeed, current feral donkey populations, descended from these working animals, and persist in parts of the American West and the Australian Outback maintaining a rich diverse gene pool from those colonial donkeys.

With the Industrial Revolution, mechanization undercut donkey use in Europe and North America. Railways, steam engines, and motor vehicles largely supplanted equid traction. Nonetheless, formal breed registries emerged in the nineteenth century to preserve working breeds: e.g., various European studbooks conserved the largest types for mule production and draught work (c. 1880s onwards), 1884 Baudet du Poitou or Poitou Donkey breed and the American Mammoth Donkey Association (est. 1888). In the twentieth century, many autochthonous donkey breeds in Europe (with Italy, France and Spain as the most diverse countries in terms of donkey breeds) declined to near extinction amid agricultural mechanization and urbanization. By contrast, in the developing world, donkeys remained vital to millions of rural households, used extensively for water transport, agriculture, and livelihood activities. However, such populations often suffer from malnutrition, overwork, scant veterinary care and lack of attention by researchers and authorities (Wang et al. 2022).

A new crisis emerged in the early twenty-first century, global demand for donkey hides (for production of the traditional Chinese medicine ejiao) surged. China's donkey population collapsed from over 11 million in 1990 to 2–3 million by 2019, driving intensive donkey-hide exports from Africa and Latin America. It is estimated that several million skins are needed annually to meet ejiao production, far outstripping the renewal rate of donkey populations, therefore their supply. Reports warn that excessive slaughter and trade at a way faster rate than they are bred for have threatened the existence of donkeys in some regions (Matlhola and Chen 2020). This crisis has prompted emerging conservation and management efforts: NGOs, researcher centers and governments are now developing donkey welfare standards, livelihood-based conservation programs, and genetic management plans to preserve unique lineages (in Ethiopia, North Africa, China, among others).

Despite their significant genetic and ecological value—as reservoirs of the gene pool of the first "New World" introduced donkeys and as effective water diggers in extremely arid conditions when feralized, thereby promoting biodiversity around water wells—feral donkey populations have increasingly been regarded as a threat in the regions where they have become relegated and established, such as the Bonaire Islands, North America's Death Valley, and central and northern Australia. Furthermore, NGOs lack of interest on the genetic background of individuals in their quest to ensure welfare castrate individuals promoting the historical reservoirs of allelic richness that feral donkeys represent to be lost.

To counteract this, genetic-based management plans implemented by both public but mostly private associations underpin regional breed preservation and international organizations have begun recognizing donkeys' role in mediated therapy, climate-resilient agriculture and women's empowerment. By 2025, global donkey numbers are still on the order of 40–50 million, concentrated in sub-Saharan Africa, South Asia, and parts of Latin America, where they remain essential to subsistence transport and small-scale farming (FAOSTAT 2019).

A new perspective arises for the future as selective breeding programs (for example, Brazil's Pêga donkey initiative) are reviving local breeds by emphasizing traits like temperament, endurance, and multi-purpose productive utility, demonstrating that focused breeding can improve productivity and animal welfare simultaneously. New economic roles are emerging for donkeys worldwide—from eco-tourism and therapeutic use to gourmet niche products (cheese from donkey milk, traditional crafts). Throughout their history, the donkey has been a flexible, innovative partner to people in nearly every environment; current efforts in genetic preservation, sustainable breeding, and welfare reflect a broader appreciation of the donkey's continued importance.

In this context, there is a significant lack of reliable knowledge and data regarding global donkey population figures, both for standardized, officially recognized breeds and for non-standardized or indigenous types. Despite the critical roles donkeys play in agriculture, transportation, and livelihoods—particularly in developing regions—comprehensive and up-to-date records are often missing, outdated, or inconsistent. Many countries do not maintain centralized registries for donkey populations, and where data exists, it frequently excluds non-pedigreed animals, which make up the vast majority. This gap hinders effective conservation, breed management, and welfare efforts, and reflects a broader neglect in recognizing the economic and cultural value of donkeys. Without improved systems for documentation and monitoring, the true status of donkey populations remains largely obscured, which in turn may be even riskier than populations being endangered as, such obscure background prevents in situ and ex situ conservation actions from being implemented.

For these reasons, this chapter aims to present integrative approaches to donkey breed conservation and breeding, drawing on historical evidence, genetic management practices, and forward-looking policy perspectives. Given the profound gaps in data and recognition regarding both officially standardized and non-standardized donkey populations, there is a pressing need for multidimensional strategies that combine traditional knowledge of the situation that donkey populations face with modern science. The historical context of donkey domestication and breed development is explored, highlighting how cultural and regional factors have shaped genetic diversity. Then, current practices in genetic monitoring, breeding programs, and population management are examined, emphasizing the importance of inclusive frameworks that consider all donkey types—not just those formally recognized. Finally, this chapter advocates for cohesive policy actions that promote sustainable conservation, ensure animal welfare, and acknowledge the cultural and socio-economic value donkeys provide, especially in marginalized rural communities. By integrating these diverse perspectives, the chapter offers a comprehensive foundation for future efforts in safeguarding donkey genetic resources.

3 Genetic Challenges and Opportunities in Donkey Breeds and Populations

While robust institutional frameworks and standardization systems are essential for preserving the genetic integrity of donkey breeds, these structural foundations alone are not sufficient. They must be complemented by a thorough understanding of the genetic challenges that donkey populations—both feral and domestic, standardized and unstandardized—have historically faced and continue to endure. Across the globe, donkeys have experienced similar pressures, including demographic bottlenecks, elevated inbreeding levels, and the erosion of purebred lineages. Without addressing these genetic vulnerabilities, conservation and protection efforts risk falling short of their objectives. It is therefore crucial to move beyond administrative and organizational measures and adopt a genetics-focused approach that confronts the biological consequences of these long-standing issues. While many of these challenges are shared across donkey populations, their capacity to respond to them varies considerably depending on management context, population structure, and institutional support. The following section delves into these genetic constraints and opportunities, highlighting the urgent need to integrate institutional protection with strategic genetic management to ensure the long-term sustainability of donkey populations.

4 Historical Bottlenecks

Throughout history, donkey breeds have experienced several bottlenecks that have significantly impacted their genetic diversity and overall population health. One of the most prominent bottlenecks occurred during the Industrial Revolution, when mechanization began to replace traditional animal labor. As machines took over tasks previously performed by donkeys, the demand for these animals plummeted, leading to a drastic reduction in their populations. This decline was not uniform across all regions; in some areas, certain breeds nearly vanished, resulting in a loss of genetic diversity that continues to affect these populations today. The reduced number of breeding individuals during this period caused a genetic bottleneck, wherein the gene pool became limited and more susceptible to the effects of inbreeding and genetic drift (Colli et al. 2013; Ivanković et al. 2022; Quaresma et al. 2014; Shen et al. 2021).

Another significant bottleneck occurred during periods of war and economic hardship. During these times, donkeys, like many other livestock, were often requisitioned for military use or as sources of meat and leather, leading to a sharp decline in their numbers. Post-war reconstruction efforts and economic instability further exacerbated these declines, as resources for maintaining and breeding donkeys were limited. In some cases, specific breeds were pushed to the brink of extinction, with only a handful of individuals surviving. These events not only reduced the population size but also led to the fragmentation of breeding populations, making it difficult to maintain genetic diversity and manage effective breeding programs (Navas et al. 2016).

More recently, a new bottleneck has emerged in association with the development of novel production niches, such as donkey meat and eijao (donkey-hide gelatin). These markets have grown rapidly in certain regions, often without accompanying measures to support sustainable breeding practices. In many cases, these activities occur outside the bounds of legal regulation and oversight, resulting in the removal of large numbers of animals from local populations without ensuring their replacement through structured reproduction. This unregulated exploitation further depletes breeding stocks, threatens the continuity of traditional breeds, and exacerbates the erosion of genetic diversity already affected by earlier historical pressures. The historical and ongoing bottlenecks experienced by donkey breeds underscore the importance of continuous and proactive conservation efforts to ensure their survival and genetic health (Bennett and Pfuderer 2020; Hassan et al. 2022; Matlhola and Chen 2020; Seyiti and Kelimu 2021).

4.1 High Coancestry and Inbreeding

High rates of coancestry and inbreeding, surpassing the critical threshold of 10%, pose a severe threat to the health and viability of donkey breeds. When the genetic contribution of common ancestors becomes too pronounced, it leads to a significant increase in the expression of deleterious traits. This can manifest in various health issues such as reduced fertility, increased susceptibility to diseases, and the prevalence of genetic disorders. These negative outcomes undermine the overall fitness and longevity of the breed, making it more challenging to sustain healthy populations (Ciampolini et al. 2007; Colli et al. 2013; Gutiérrez et al. 2005; Quaresma et al. 2014; Quaresma 2015).

Inbreeding levels in donkey populations significantly contribute to severe immunological and reproductive inefficiencies. While jennies may not face challenges in becoming pregnant, the likelihood of successful deliveries and the survival of newborn foals dramatically decreases due to the adverse effects of inbreeding. This genetic issue manifests in a concerning percentage of terminal abortions and newborn deaths, with rates exceeding 30% in some populations. Such high mortality rates not only threaten the sustainability of donkey breeds but also highlight the urgent need for effective breeding management practices that prioritize genetic diversity and health (Chaouat et al. 2019; Miraglia 2012).

The implications of inbreeding extend beyond reproductive outcomes; they also compromise the overall health of donkey populations. Increased susceptibility to diseases and genetic disorders often accompanies low genetic variability, resulting in a population that is less resilient and more vulnerable to environmental stresses. Addressing inbreeding and its consequences requires a multifaceted approach, including the implementation of breeding programs that promote genetic diversity and minimize the risks associated with close genetic relationships. By focusing on improving the reproductive health and immune function of donkey populations, stakeholders can work towards ensuring their viability and sustainability in the long term (McAnulty et al. 2011; McGregor 1990; Mogil and Wegmann 2020).

As the basis for all of the aforementioned, excessive inbreeding diminishes genetic diversity, which is crucial for adapting to changing environmental conditions and emerging health threats. A limited gene pool restricts the ability of the breed to evolve and cope with new challenges, potentially leading to a decline in population numbers and even extinction. To mitigate these risks, it is essential to implement strategic breeding programs that prioritize genetic diversity and monitor coancestry rates closely. By doing so, breeders can ensure the long-term health and sustainability of donkey breeds, preserving their unique characteristics and contributions to agricultural and cultural heritage (Mazzatenta et al. 2021; Quaresma et al. 2014; Rosenbom et al. 2015; Shen et al. 2021; Todd et al. 2022).

4.2 Purebred Individual Decline

The lack of pure individuals in standardized pedigrees is a critical issue affecting the sustainability of many donkey breeds. The inclussion of new individuals, even if particular policies for parentage origins are implemented when protection structures such as conservation programmes exist, is performed by solely considering the phenotype of the animals, which often results in the consideration of crossbred individuals as pure. In numerous cases, the number of completely purebred individuals has diminished to half of the existing population, which poses a significant threat to the genetic integrity of these breeds. This decline not only hampers efforts to maintain breed standards but also increases the risk of inbreeding, which can lead to the emergence of genetic disorders and other health issues. As a result, the population of purebred donkeys often dwindles to just a few hundred individuals, further compounding the challenges faced in conservation and breeding efforts (Ciampolini et al. 2007; Colli et al. 2013; Criscione et al. 2024; Mazzatenta et al. 2021; Rosenbom et al. 2015; Wang et al. 2020).

The reduction in purebred individuals significantly impacts the ability to implement effective breeding programs aimed at preserving unique traits and characteristics inherent to each breed. With such a limited gene pool, breeders may struggle to make informed decisions about pairings and may inadvertently perpetuate genetic weaknesses. This situation underscores the urgency of establishing comprehensive breeding strategies that prioritize the conservation of purebred individuals while also promoting genetic diversity. By developing initiatives that support the maintenance of pure standardized pedigrees, stakeholders can help ensure the long-term viability of donkey breeds and protect them from the threat of extinction (Legrand et al. 2014; Sponenberg 2020).

4.3 Breed Adscription and Introgression

The lack of genomic tools to determine breed ascription, percentage of introgression, and levels of inbreeding or coancestry presents a significant obstacle to the effective management and conservation of donkey breeds. In the context of strongly incomplete pedigrees, these genomic tools are essential for accurately assessing genetic diversity and understanding the genetic relationships between individuals within a breed. Without access to advanced genomic technologies, breeders and conservationists are left with limited information about the genetic makeup of their populations, making it challenging to identify issues related to inbreeding and genetic health. This gap in knowledge can result in poorly informed breeding decisions that may inadvertently exacerbate existing problems, such as reduced fertility and increased susceptibility to diseases.

Moreover, the inability to accurately measure introgression from other breeds complicates efforts to maintain the purity and distinctiveness of standardized breeds. Understanding the extent of genetic influence from external breeds is crucial for implementing targeted breeding strategies that preserve the unique characteristics of donkey breeds while mitigating the risks associated with inbreeding. To address these challenges, there is an urgent need for the development and deployment of genomic tools tailored to the needs of donkey breeding programs. By investing in genomic research and technologies, stakeholders can gain valuable insights into the genetic health of donkey populations, allowing for more effective conservation strategies and the preservation of these vital breeds for future generations (Huang et al. 2023; Renaud et al. 2018; Todd et al. 2022; Wang et al. 2020).

4.4 Definition of the Exact Role of Diverse Wild Ass Populations in Donkey Breed Configuration and Evolution

Despite major advances in genomic technologies and archaeological methodologies, a significant gap persists in our understanding of the full spectrum of wild ass contributions to the genetic foundation of domestic donkey breeds. Much of the current scholarship and genetic modelling has focused predominantly on two extinct African wild ass subspecies—the Nubian wild ass (*Equus africanus africanus*) and the Atlas wild ass (*Equus africanus atlanticus*). These two lineages are closely associated with the two deeply divergent mitochondrial haplogroups, A and B, which are found across modern donkey populations (Kimura et al. 2011).

Haplogroup A, generally attributed to the Nubian wild ass, exhibits a wide geographic distribution and is thought to reflect a primary lineage involved in the earliest stages of domestication. Haplogroup B, in contrast, remains more enigmatic. While it has sometimes been linked to the now-extinct Atlas wild ass, its precise origins remain unresolved due to the lack of genomic data from this and other potential ancestral populations (Kimura et al. 2011). The asymmetrical focus on these two haplogroups—and the African wild asses associated with them—has overshadowed the possible influence of other extant and extinct wild ass populations, including the Somali wild ass (*Equus africanus somaliensis*) and several Asiatic species such as the kiang (*Equus kiang*) and the onager (*Equus hemionus onager*) as also supported by historical, cultural and archaeological evidence.

This narrow emphasis in both sampling and interpretation not only limits our understanding of donkey domestication but also obscures the diverse evolutionary pathways that may have shaped regional breed characteristics. For instance, some historical sources and breed-specific traits—such as the long-haired coat phenotype found in the Poitou, Ragusano, Mirandese, and Zamorano-Leonés donkeys—suggest possible introgression from Somali wild asses, whose adaptive features may have influenced coat length, texture, or thermoregulatory traits. Yet, such hypotheses remain largely unexplored due to the absence of comprehensive comparative genomic studies involving a broader range of extant and extinct wild ass populations.

A promising avenue to address this gap is to test the genetic basis of long coat expression in the aforementioned donkey breeds. Genetic proximity among such breeds (except for Poitou, which was not included) was molecularly addressed by the study by Jordana et al. (2016). Recent genomic investigations have linked this trait to a mutation in the fibroblast growth factor 5 (FGF5) gene, which is known to regulate hair cycle termination across a range of mammalian species. A mutation in this gene likely inhibits normal hair shedding, resulting in a prolonged growth phase and the expression of a long or shaggy coat. While the FGF5 mutation has been characterized in some domestic donkey populations (Legrand et al. 2014), its evolutionary origin remains unclear.

One hypothesis posits that the source of this mutation may lie in ancestral traits retained from African wild ass populations, particularly the Somali wild ass. Like all equids, Somali wild asses grow long winter coats to cope with colder seasonal temperatures and shed them in the spring. This ability to regulate coat growth seasonally suggests the presence of genetic mechanisms that, if disrupted or altered, could result in the perpetually long coats observed in some domestic breeds. The mutation may therefore represent a modification of an existing ancestral trait rather than a completely novel adaptation.

Alternatively, a second hypothesis proposes that the mutation could reflect introgression from Asiatic wild asses, such as the kiang (Equus kiang), native to the high-altitude, cold environments of the Tibetan Plateau. These equids exhibit pronounced winter coat growth as an adaptation to extreme cold and may have contributed relevant genetic material through ancient admixture with domestic donkey populations along trans-Eurasian trade or migration routes. Recent genomic data revealing complex admixture signals in donkey lineages—some of which do not correspond cleanly with haplogroups A or B—support the plausibility of such scenarios (Todd et al. 2022; Wang et al. 2023). Determining the source of the FGF5 mutation—and whether it was selected independently, inherited from a wild ancestor, or introduced via hybridization—could provide crucial insights into how environmental pressures and human selection shaped phenotypic diversity in donkeys.

The issue is further compounded by methodological challenges in archaeological and paleogenomic studies. Donkeys have often been difficult to distinguish from wild equids in the archaeological record due to overlapping morphological traits, and ancient DNA studies have been constrained by poor preservation conditions in key regions such as the Sahara and the Horn of Africa. Moreover, cultural and economic undervaluation of donkeys relative to horses has led to their underrepresentation in artistic, textual, and burial contexts, further obscuring the historical role of diverse wild ass populations in their evolution.

Addressing this knowledge gap is crucial for several reasons. First, a more inclusive approach that considers all candidate wild populations—African and Asian alike—can yield a more accurate picture of the genetic diversity and population structure of early domestic donkeys. This, in turn, could clarify ambiguities surrounding mitochondrial divergence, maternal lineage mixing, and regional breed differentiation. Second, recognizing the potential contributions of under-studied wild ass populations can deepen our understanding of local adaptations in domestic breeds, including traits related to climate resilience, disease resistance, and physical characteristics like size, gait, or coat morphology.

Third, the conservation implications are substantial. Many wild ass populations, including the Somali and Asiatic wild asses, are endangered and under increasing threat from habitat loss, climate change, and human conflict. Understanding their historical and genetic linkages to domestic donkeys can elevate their importance in conservation planning and may even justify new efforts to preserve wild genetic lineages as reservoirs of adaptive traits.

Finally, the revaluation of these neglected lineages has implications for broader models of animal domestication. The traditional binary of wild versus domestic is increasingly understood as a spectrum involving recurrent admixture, gradual morphological change, and complex socio-ecological interactions. The presence of diverse mitochondrial and nuclear DNA lineages in domestic donkeys—including signals of admixture with other equids such as the Asiatic kiang—suggests that donkey domestication, like that of many animals, was not a singular event but a diffuse, dynamic process.

In summary, the role of lesser-studied wild ass populations in the evolution of domestic donkey breeds deserves urgent and focused research attention. Expanding the geographic and taxonomic scope of genetic and archaeological investigations will not only refine our models of donkey domestication but also contribute to the conservation of threatened equid lineages. A more inclusive and multi-source perspective is essential for capturing the full complexity of donkey evolution and for informing sustainable management of both domestic and wild populations in a rapidly changing world.

4.5 Feral Donkey Populations Gene Pool Misconsideration

The lack of comprehensive studies on the genetics behind feral donkey populations poses significant challenges to effective management and conservation strategies. Feral asses can truly be reservoirs for ancestral genetic material of those individuals which eventually became released into the wild. Often, feral donkeys are subjected to policies involving castration and relocation to foster homes, aimed at controlling their numbers and mitigating their impact on local ecosystems. However, these approaches frequently overlook the genetic diversity present within feral populations, which can result in a loss of valuable gene pool which regulates the expression of genetically based traits because of the inability to understand the broader implications of massive castration interventions. Without an in-depth understanding of the genetic makeup of feral donkey populations, it becomes difficult to assess how such policies may affect their long-term viability and adaptability (Blake et al. 1981).

By neglecting the genetic aspects of feral donkey populations, management practices risk undermining the potential benefits these animals could provide if preserved in their natural environments. Research into the genetics of feral donkeys can reveal important insights into their adaptability, health, and contributions to local ecosystems. Understanding the genetic diversity within these populations can also inform strategies that balance population control with conservation efforts, ensuring that valuable genetic traits are not lost in the process. To develop more effective management policies, there is an urgent need for targeted research that focuses on the genetics of feral donkey populations, paving the way for practices that promote both ecological health and the preservation of genetic resources (Pitt 2021; Rodríguez-Rodríguez et al. 2022).

4.6 Nodal Ancestors and Breeding Loops

In donkey breeding, the phenomenon of certain individuals becoming nodal ancestors can significantly shape the genetic landscape of populations. These individuals often display desirable traits, leading them to be favoured for breeding.

As a result, they are frequently included in the pedigree of a substantial fraction of the population. This high representativity can create a situation where the genetic contribution of these nodal ancestors is disproportionately large, particularly evident in popular jacks whose widespread use amplifies their influence on the gene pool. Additionally, the presence of breeding loops—where the same ancestor appears multiple times in the pedigree of a specific animal—can further enhance the impact of these nodal ancestors. These loops can be beneficial for concentrating certain desirable traits but also increase the risk of inbreeding by reinforcing genetic similarities within the lineage.

While the presence of nodal ancestors and breeding loops can initially benefit breeding programs by concentrating desirable traits, they raise concerns about the risks of inbreeding and loss of genetic variability. Over-reliance on a small number of influential individuals can lead to a genetic bottleneck, where the gene pool becomes limited, making the population more susceptible to diseases and environmental changes (Smith 2008). The repeated presence of the same ancestor in breeding loops can exacerbate this situation by intensifying the genetic similarities among offspring. To mitigate these risks, it is crucial for breeders to adopt strategies that promote genetic diversity while still capitalizing on the strengths of successful nodal ancestors. By carefully managing breeding practices and incorporating a broader range of genetic backgrounds, conservationists and breeders can ensure the long-term health and viability of donkey populations, safeguarding them against the challenges posed by concentrated genetic lines and inbreeding effects (Chen et al. 2010; Santana and Bignardi 2015).

4.7 Long Generation Intervals

Long generation intervals present both challenges and opportunities for conservation and breeding programs involving donkeys. On one hand, the lengthy time it takes to produce new foals—typically around 12 months—can complicate breeding efforts, especially in situations where rapid population growth is needed to recover endangered breeds or enhance genetic diversity. This delay can lead to potential setbacks in achieving conservation goals, as the ability to assess the effectiveness of breeding strategies is hampered by the time it takes to produce and evaluate offspring. Additionally, managing inbreeding and coancestry becomes increasingly complex, as breeders must carefully monitor genetic relationships over longer periods to avoid detrimental effects on the population's health and viability (Navas et al. 2017).

On the other hand, longer generation intervals can be advantageous for conservation programs focused on preserving genetic diversity. With extended timeframes, there is a greater opportunity to implement careful breeding strategies that emphasize genetic management, thereby reducing the risks associated with inbreeding. Breeders can take the time needed to select the best individuals for mating, considering their genetic backgrounds to foster a more resilient population. This approach allows for a more thorough evaluation of the health, behavior, and adaptability of foals, ultimately enhancing the long-term success of breeding programs. By balancing the challenges posed by long generation intervals with the opportunities for thoughtful genetic management, conservationists can work toward maintaining and revitalizing donkey populations effectively (Ivanković et al. 2022; Navas González 2019; Navas et al. 2019a, b; Rizzi et al. 2011).

4.8 Pedigree Completeness Levels

Monitoring inbreeding and coancestry is essential for the success of donkey breeding programs, particularly in the context of the Pedigree Completeness Index (PCI). The PCI acts as a critical measure of the reliability of pedigrees, offering insights into the genetic relationships among individuals and their potential for inbreeding. By evaluating the completeness of an animal's pedigree, breeders can identify the degree of relatedness between potential mating pairs. This knowledge allows for informed decisions that help mitigate the risks associated with inbreeding depression and related matings, which can negatively impact reproductive fitness and the overall health of the population. Maintaining a high PCI is vital to ensuring that breeding practices contribute to genetic diversity rather than diminish it, thereby enhancing the resilience and adaptability of the donkey population.

Employing the PCI as a strategic tool for planning breeding pairs is crucial for counteracting the adverse effects of inbreeding and fostering a more robust genetic framework. By carefully managing breeding practices and selecting mates based on comprehensive pedigree information, breeders can promote healthier offspring and minimize the likelihood of inbreeding-related issues, such as reduced fertility, increased susceptibility to diseases, and overall decreased vitality. Additionally, the use of the PCI encourages the inclusion of diverse genetic backgrounds in breeding programs, which is essential for sustaining the long-term health and viability of donkey populations. Through diligent monitoring and strategic pairings informed by the PCI, breeders can safeguard the genetic integrity of the breed while preserving its unique traits and enhancing its potential for future generations (Fava 2024; Navas et al. 2017; Quaresma et al. 2014).

4.9 Inbreeding-Based Term Abortions and Neonatal Deaths in Donkey Foals

Inbreeding within small or isolated populations of donkeys, particularly in endangered or rare breeds, is a significant concern for the overall health and survival of the population. Deleterious effects derived from inbreeding have been suggested to become apparent at concerning rates when generalized inbreeding coeffient in a population surpasses 10%. The limited genetic pool in such populations leads to a higher probability of genetic mutations being passed down through generations, which in turn increases the occurrence of reproductive failures and immunological depression. These failures, including term abortions and neonatal deaths, are often the result of the accumulation of harmful recessive genes that may not be expressed in more genetically diverse populations (Gutiérrez et al. 2006).

The genetic bottleneck created by inbreeding reduces the immune system's ability to combat diseases, leaving foals more vulnerable to infections. Moreover, developmental problems associated with inbreeding can result in serious congenital defects, such as heart or kidney malformations, which can lead to stillbirths or early neonatal deaths (Quaresma et al. 2014). The lack of genetic variation also weakens the population's ability to adapt to environmental stressors or changing conditions, which compounds the overall risk of species extinction (Norris et al. 2021).

To mitigate these risks, careful genetic management strategies are essential. Genetic monitoring programs can help identify individuals with valuable genetic traits and track the diversity within a population. Controlled outcrossing, in which genetically diverse individuals from different populations are introduced, can reduce the harmful effects of inbreeding. This introduction of new genetic material promotes genetic variation, which can improve the overall health, fertility, and resilience of the population (Gutiérrez et al. 2006).

In addition, advanced reproductive technologies such as embryo transfer and artificial insemination offer alternative ways to manage genetic diversity. These techniques allow the introduction of genetically valuable, unrelated donkeys into the breeding pool without the immediate risks associated with direct mating. By carefully selecting breeding pairs and monitoring genetic health, these interventions can prevent the buildup of harmful recessive genes and decrease the incidence of term abortions and neonatal deaths (Quaresma et al. 2014).

4.10 Assisted Reproduction Compromises

Reproductive techniques play a crucial role in breed conservation, yet they present significant challenges. One of the primary methods, artificial insemination (AI) using frozen semen, has proven to be inefficient for many breeds, particularly in jennies (female donkeys). The freezing and thawing process can damage sperm cells, resulting in reduced fertility rates. This inefficiency hampers the ability to maintain genetic diversity and preserve endangered breeds, as fewer successful pregnancies limit the potential to expand and sustain populations. Compounding this issue is the prevalence of endometritis in jennies, an inflammatory condition of the uterus that may occur after insemination. Endometritis further reduces reproductive efficiency by affecting embryo implantation and overall pregnancy rates, making it even more difficult to achieve successful outcomes with frozen-thawed semen (Allen et al. 1985; Segabinazzi et al. 2025; Vargas et al. 2025; Zanardi et al. 2025).

Another important factor influencing donkey semen quality is the expression of spermadhesins—surface proteins that play key roles in sperm capacitation, interaction with the female reproductive tract, and oocyte binding. Recent studies have revealed variations in the presence and functionality of these glycoproteins in donkey semen, particularly after cryopreservation. Alterations in spermadhesin integrity may partly explain the low fertility rates observed. Investigating their expression and stability could be key to improving cryopreservation protocols and enhancing AI efficiency in this species.

Cryoprotectants, such as glycerol or dimethyl sulfoxide (DMSO), are used in semen freezing to protect sperm cells from ice crystal damage. However, in donkeys, these agents often present cytotoxic effects and can destabilize sperm membrane integrity. Donkey spermatozoa appear particularly sensitive to cryoprotectant-induced osmotic stress, oxidative damage, and functional impairment, which compromises post-thaw viability and motility. Selecting appropriate cryoprotectant types, concentrations, and exposure protocols is therefore essential to improve outcomes.

In this context, the use of donkey seminal plasma has emerged as a promising strategy to mitigate cryoinjury and support sperm function after thawing. Seminal plasma contains a unique composition of proteins, antioxidants, and signalling molecules that help stabilize sperm membranes, neutralize reactive oxygen species, and maintain motility. Adding donkey seminal plasma—either as a component of freezing extenders or during post-thaw processing—has been shown to enhance sperm recovery, improve acrosomal integrity, and potentially increase fertility rates. Further research is needed to optimize the use of homologous plasma and identify key bioactive factors responsible for its protective effect.

To address these challenges, innovative techniques such as Intracytoplasmic Sperm Injection (ICSI) and embryo preservation through vitrification offer promising alternatives. ICSI, which involves injecting a single sperm directly into an egg, has shown higher fertilization rates compared to traditional AI. However, it requires specialized equipment and expertise, limiting its widespread application. Vitrification, a rapid freezing method, has emerged as a viable option for preserving donkey embryos. This technique minimizes ice crystal formation that could damage embryos, leading to higher survival rates upon thawing. Research indicates that vitrified embryos have better implantation and development potential, thereby improving reproductive outcomes in conservation efforts.

Embryo transfer (ET) in donkeys is a valuable assisted reproductive technology, especially for conserving endangered breeds and optimizing breeding in genetically superior jennies. While ET is routinely used in equine species, transferring donkey embryos into mares (female horses) or even mules as surrogate mothers has become a promising strategy to overcome the limited reproductive efficiency and availability of jenny recipients.

Hybrids such as mules (offspring of a male donkey and a mare) and hinnies (offspring of a jenny and a stallion) offer a unique model to study interspecific maternal inheritance, chromosomal origin, and gestational compatibility in equids. Although both hybrids possess 63 chromosomes ($2n = 63$)—an unbalanced number resulting from the union of a donkey ($2n = 62$) and a horse ($2n = 64$)—the direction of the cross significantly affects mitochondrial DNA, the maternal uterine environment, and epigenetic regulation of foetal development. In mules, the maternal contribution comes from the horse, providing a uterine environment more aligned with equine gestation, which may explain the higher viability and reported success of mules as surrogate mothers for donkey embryos. Conversely, hinnies inherit a maternal environment from donkeys and may face more significant developmental challenges when gestating embryos with a horse-derived paternal genome. These maternal-foetal mismatches, along with karyotypic incompatibilities such as asynchronous chromosome pairing, centromeric origin differences, and disrupted meiotic processes, can impair implantation, placentation, or gestational progression. Despite their sterility, mules have been shown to support pregnancies from donkey embryos, making them a promising—albeit unconventional—resource for assisted reproduction and conservation programs. However, the feasibility of using hinnies as embryo recipients remains uncertain and under-researched. Given incongruent gestation timing, potential chromosomal conflicts, and hormonal differences, further research is necessary to evaluate their reproductive utility. Understanding how chromosome number and parental origin shape maternal compatibility is crucial for advancing cross-species embryo transfer strategies in equids.

Successful pregnancies have been achieved with donkey embryos gestated in horse mares, although differences in gestation length and placental development between species can pose challenges. Recent studies have also shown that mules, despite being typically sterile hybrids, can serve as effective embryo recipients due to their equine uterine physiology and maternal behaviour. However, implantation rates and pregnancy maintenance may vary depending on genetic compatibility and immunological tolerance between donor and recipient. As research progresses, cross-species embryo transfer could become a practical tool not only for enhancing reproductive output in donkeys but also for preserving rare genetic lines and extending the use of valuable breeding females without removing them from productive roles.

Addressing the inefficiencies associated with frozen semen, including functional disruptions like spermadhesin modification, cryoprotectant toxicity, and complications such as endometritis, is critical for effective breed conservation. Continued research and development in reproductive technologies are necessary to enhance their reliability and accessibility. By improving these methods and incorporating techniques such as embryo vitrification and seminal plasma supplementation, conservationists can better manage and preserve the genetic diversity of endangered breeds, ensuring their survival for future generations. This comprehensive approach to reproductive management not only improves the likelihood of successful breeding outcomes but also supports the overall health and welfare of jennies and their offspring (Bragulat et al. 2023; Dobbie et al. 2024).

4.11 Implications of Breed Selection on Health and Genetic Vulnerability

Selective breeding in donkeys, while often aimed at enhancing desirable traits such as size, temperament, or coat color, carries significant implications for the overall health and genetic resilience of various breeds. As breeders prioritize specific characteristics, the genetic diversity within populations can become restricted, inadvertently amplifying the risk of hereditary conditions and diminishing the animals' ability to resist disease or adapt to environmental stressors (Frankham et al. 2010).

One of the most pressing concerns is inbreeding, which tends to rise in closed or narrowly selected breeding populations. This genetic bottleneck can lead to the expression of deleterious recessive genes, resulting in compromised immune function and increased prevalence of congenital disorders (Frankham et al. 2010). A reduced genetic pool can impair the development of robust immune systems, leaving individuals more vulnerable to infectious diseases and reducing the population's overall capacity to adapt to changing environments or emerging pathogens (Frankham et al. 2010).

Miniature donkey breeds are a clear example of how extreme selection for size can lead to developmental complications. The intentional reduction of body size may contribute to an increased incidence of skeletal malformations and congenital anomalies such as dwarfism, which, beyond its physical manifestations, can be associated with respiratory, cardiovascular, and orthopedic complications. Similarly, other genetic anomalies, including reproductive disorders, may become more common in breeds subjected to intense selective pressures without concurrent health screening protocols.

In large donkey breeds, such as those bred for strength or work, disproportionate growth and increased body mass can predispose individuals to orthopedic conditions such as limb deformities or joint diseases. The prioritization of traits like stature and muscularity may not always be matched with corresponding skeletal support or joint health, resulting in an increased likelihood of mobility issues that compromise welfare and utility.

Coat colour selection also introduces health considerations. Breeds with lighter skin pigmentation may have a reduced capacity to withstand ultraviolet (UV) radiation, making them more prone to dermatological issues and skin cancers. In such cases, aesthetic preferences intersect with environmental vulnerabilities, necessitating preventive care and environmental modifications to safeguard health.

Ultimately, breed selection, while beneficial for preserving phenotypic diversity and cultural heritage, must be approached with a critical awareness of its biological trade-offs. Without careful genetic management, selective breeding can inadvertently compromise the very vitality and sustainability it seeks to preserve. Responsible breeding practices must balance the desire for specific traits with proactive health screening, genetic monitoring, and an emphasis on long-term population resilience. By integrating genetic health assessments into breeding decisions, stakeholders can ensure that breed preservation does not come at the cost of animal welfare or genetic robustness.

5 Antropological and Sociocultural Factors Affecting Donkey Breeds and Populations

While genetic tools and pedigree-based strategies offer critical insights into the conservation of donkey breeds, they represent only one facet of a complex challenge. The issues of inbreeding, declining effective population sizes, and incomplete pedigree information underscore the need for urgent genetic management (Pierini et al. 2025; Gambini et al. 2025). Yet, these efforts often overlook the human dimensions that shape breeding outcomes on the ground. Donkey conservation is not conducted in isolation from cultural, economic, and social realities; rather, it is embedded within them. Breeding decisions are largely influenced by owner beliefs, regional preferences, and traditional knowledge systems, which may not align with formal conservation goals (Wassie et al. 2023; Maswana et al. 2022). As Hall (2019) notes, livestock breeds are deeply interwoven with human identities and livelihoods. Therefore, an integrative approach to conservation must move beyond genetics alone to consider how sociocultural practices affect the use, selection, and preservation of donkey populations. The following section explores these anthropological and sociocultural dimensions, emphasizing how local perceptions, community norms, and informal breeding traditions can either support or undermine long-term genetic sustainability.

5.1 Owner Misbeliefs for Trait Consolidation and Own-Built Bloodlines

A recurring issue in the management of local donkey populations—particularly in regions where formal breeding programs are lacking—is the widespread misconception among breeders and owners that inbreeding is a useful tool to “fix” desirable traits or even to be of help when following an specific purpose or defining “owner’s breed lineages or bloodlines”. Many donkey owners, especially in rural or economically marginalized areas, operate under the mistaken belief that mating related animals, such as siblings or parents with offspring, will reinforce valuable characteristics such as strength, endurance, or temperament. While selective breeding does aim to enhance specific traits, this approach is often confused with inbreeding, leading to unintended and harmful consequences (de Boer et al. 2021).

In reality, inbreeding—which could practically be defined as the mating of closely related individuals—dramatically increases the risk of expressing deleterious recessive genes. Over time, this can result in a range of negative outcomes, including reduced fertility, higher susceptibility to disease, structural deformities, poor growth rates, and compromised immune function (de Boer et al. 2021; Huang et al. 2023). In donkeys, which already have a relatively narrow genetic base in many regions, inbreeding further limits genetic diversity, accelerating the decline of population viability and resilience (Jere et al. 2025).

The belief in the supposed benefits of inbreeding is often reinforced by anecdotal successes. For example, if an owner mates two related animals and produces a foal that appears healthy or exhibits a desired trait, this may be interpreted as confirmation of the method’s value. However, such outcomes are typically exceptions rather than the rule. The cumulative genetic damage from repeated inbreeding often only becomes visible after several generations, by which point the negative effects may be widespread and difficult to reverse (Maswana et al. 2022).

Recovery from the negative effects of inbreeding in donkeys following endogamous crosses is a prolonged and complex process, particularly due to the species' long reproductive cycle. Donkey pregnancies typically last between 12 and 13 months, meaning that even under ideal conditions, only one generation can be produced per year. As a result, genetic recovery from inbreeding depression—which may include reduced fertility, poor growth, structural deformities, and compromised immune function—can take a decade or more. Although some initial benefits such as increased vigor may be observed after the first outcrossing with unrelated individuals, full restoration of genetic diversity and fitness usually requires at least five to ten generations.

In practice, this equates to 12 to 15 years of carefully managed breeding. The pace of recovery is influenced by the severity of prior inbreeding, the size and genetic variability of the breeding population, and the availability of unrelated or distantly related males for outcrossing. In many rural or under-resourced donkey-rearing communities, these conditions are rarely met, either it is intentionally (due to genealogical strategical interest, economical or administrative reasons mainly) or not. Poor recordkeeping continued use of related sires, and lack of access to veterinary or genetic guidance often perpetuate inbreeding cycles. Therefore, reversing the genetic damage requires more than just time—it demands structured intervention, including the introduction of new genetic lines, establishment of community breeding programs, and, where possible, use of molecular tools to monitor genetic diversity. Without such measures, even with long-term efforts, the population may continue to exhibit the cumulative effects of inbreeding depression across generations.

Moreover, when genealogical information exists, the lack of access to breeding records or genetic assessment tools by breeders, who are the main conductors and protectors of the genetic health of donkey breed populations, exacerbates the problem. In many cases, owners are unaware of the exact relationships among animals within a herd, or they do not recognize the genetic risks associated with continuous use of the same few breeding males. Without guidance from veterinarians, geneticists, or conservation programs, inbreeding becomes a default strategy—not out of malice or negligence, but due to misinformation and lack of alternatives (Jere et al. 2025).

To combat this issue, there is a critical need for targeted educational campaigns that inform donkey owners about the risks of inbreeding and the importance of genetic diversity, which optimally should be provided from reliable sources and organized by the associations of breeders relaying on experts from different contexts (universities, research center, among others). Outreach should emphasize the long-term health and sustainability benefits of outcrossing—breeding between unrelated individuals—as a way to improve overall herd performance. In parallel, establishing community breeding cooperatives, mobile genetic monitoring programs, or registries for tracking lineage could provide practical support to owners and help prevent the silent erosion of donkey breed viability caused by misguided breeding practices.

5.2 Foreign Breed Introgression into Local Standardized Breeds

Strong introgression of foreign breeds into already standardized donkey breeds poses a significant threat to the integrity and distinctiveness of these genetic lines. Introgression in donkey breeds has been documented historically, frequently lead either by territorial proximity or phenotypic similarity. For example, Castejón y Martínez de Arizala (1945) would speak about the proximity of Morrocan Moruno (Moorish) and Saharian donkey breeds to resemble, at least in regards their coat, Andalusian donkey individuals (Figs. 4 and 5). Such proximity is reflected today in the wide variation of coats which, though not being accepted by the Andalusian standard, appear in the across the population to this date.

When foreign genetic material is introduced—whether deliberately or unintentionally—it can dilute or even override the specific traits that define a breed, leading to the progressive loss of features that have been carefully selected and preserved over generations. This not only compromises breed purity but also undermines their historical, cultural, and functional significance. Traits such as body

Fig. 4 Andalusian (left) and Saharian (right and lower) Donkey Breeds

Fig. 5 Moorish or Moruno Donkey Population. It is not an Standardized breed

Fig. 6 Spanish Ase Balear Donkey Breed

conformation, coat type and color, temperament, and adaptation to regional environments may be progressively eroded, reducing the breed's distinctiveness and long-term viability (Hoffmann and Scherf 2006; Kong et al. 2025).

A key factor exacerbating this threat is phenotypic proximity between certain breeds. Donkeys from geographically or morphologically related populations—such as the Miranda and Zamorano-Leonés, Andalusian and Moruno (Moorish), Catalonian and Balearic (Fig. 6), or even the Mammoth donkey—may appear strikingly similar to the untrained eye. This visual resemblance often leads to misidentification, relaxed breeding practices, or intentional crossings, especially when formal registration systems are lacking or inconsistently applied. In some cases, breeders or owners may not be aware that they are facilitating introgression, believing the animals belong to the same or closely related breeds (FAO 2010; Ginja et al. 2013).

Several causes contribute to this problem beyond phenotypic similarity. Geographical proximity of breeding areas—especially in rural zones where different breeds co-exist—greatly increases the risk of interbreeding. Poor or informal breeding practices, including a lack of controlled mating, absence of pedigree verification, and inadequate knowledge of breed standards, often result in unrecorded or inappropriate crosses. Moreover, owner preferences driven by subjective tastes, such as selecting animals based on size or color rather than genetic background, may lead to deliberate crossbreeding that compromises breed identity. Accidental escapes or inadequate fencing can also result in unintended matings, particularly in extensive or semi-feral management systems (Kong et al. 2025; FAO 2010).

The introduction of foreign or unverified genes into standardized breeds disrupts structured breeding and conservation programs, which depend on maintaining genetically coherent and traceable bloodlines. This complication makes it difficult to monitor genetic health, assess inbreeding risks, and ensure the preservation of unique breed attributes. Over time, such disruptions may lead to reduced fertility, increased genetic disorders, and diminished resilience to environmental stressors or disease (Hoffmann and Scherf 2006; Ginja et al. 2013).

To safeguard the future of standardized donkey breeds, stringent measures must be adopted. These include mandatory genetic registration, controlled mating programs, and increased use of molecular tools to verify parentage and genetic compatibility. Educational efforts targeting breeders, owners, and local authorities are equally crucial to foster awareness of the risks associated with uncontrolled introgression. By addressing not only the genetic and management challenges but also the role of phenotypic confusion in breeding decisions, we can protect the distinct heritage of each donkey breed and ensure their survival as genetically and culturally valuable livestock resources (Navas González et al. 2017).

5.3 Exclusion of Phenotypic Diversity Due to Changing Breed Standards

For the seek of breeds identity, breed registries often enforce narrow standards that can unintentionally erode a breed's genetic integrity. For example, the official standard for the Andalusian donkey (Asno Andaluz) only permits a gray/roan coat ("capa torda") (ASNOPRA n.d.) and does not say anyhting about the pointed ears feature that appears in animals with gray coats belonging to the Egyptian Hassawi linked group of breeds (Andalusian, Grigio Viterbese and Pêga, among others, Fig. 7). Any other coat color is explicitly listed as a defect (ASNOPRA n.d.). However, melaos or reddish individuals have traditionally allegedly been registered and there are even references explaining that although the only recognized coat color is 'rucia'—a traditional Spanish term encompassing various shades of diluted gray—ranging from dark rucio to silvery white. Within this spectrum, there exists a diversity of tones that do not receive specific names. The most frequent shades include dark rucio, 'apardado' (a grayish-brown or dull-brindled tone), bluish, light, and 'melado' (a golden or honey-colored chestnut dilution). It is particularly

Fig. 7 Egyptian donkey foal (left) and Andalusian Jack displaying the pointed ears feature typical of the breeds from the same Hassawi descending trunk

noteworthy that animals with the melado coat tend to transmit this coloration as alazan consistently to their offspring when crossed with mares, preserving the melao rucia phenotype in their donkey descendants. The appearance of dark chestnut coats (*castaño oscuro*) in some Andalusian jacks, as well as black coats in various shades, is uncommon. The same author would say that when such individuals do appear, they are invariably the result of crossbreeding with the Ausetan or Ausetano Catalonian donkey breed, which is characterized by its solid black coat and larger build (Manjón-Cabeza 1933; Ruiz 1950). However, black Andalusian are still anecdotally born from studbook officially registered individuals.

Similarly, the Normand donkey standard recognizes only bay (including "chocolate") or pangaré black coats, stating that "any other coat is excluded" (Association Nationale de l'Âne Normand n.d.). This has also been contemplated in the Pantesco Donkey Breed, known for its ambling gait, from Italy and the Catalonian Donkey Breed from Spain (Figs. 8 and 9). These rigid criteria have the effect of excluding dark or dilute-colored donkeys from the breeding population. In practice, this has an even stronger repercussion in Standardized breeds and leads to very limited founder stock: e.g. a pedigree analysis of the Italian Martina Franca (Pugliese donkey Breed) donkey found only ~16 founders remained in the registry, yielding a critically low effective population size (Ne $\approx$ 3) and rising inbreeding (Pierini et al. 2025). Such findings underscore how strict selection of a few "ideal" individuals (by color or other traits) concentrates ancestry and produces genetic bottlenecks (Pierini et al. 2025).

Despite these narrow standards, historical and genetic evidence shows greater coat diversity within these breeds. Genomic studies have identified two major coat-color patterns in donkeys: the ancestral Dun type (diluted, gray or light chestnut) and the non-Dun type (undiluted, yielding black or chestnut coats) (Lara-Chamarro and Jordana 2020). In other words, true black and dark chestnut coats are part of the donkey's natural variation. Among French donkeys, for instance, seven recognized

Fig. 8 No Light Points (excluded from registries) and Light Points (included in registries) Catalonian Donkeys

Fig. 9 Light points and No light points Normand Donkey Breed Individuals

breeds display black, bay, or gray coats (Lara-Chamarro et al. 2020). The Normand bay donkey (Fig. 9) only rarely produces red/chestnut foals, but such occurrences have been documented (Lara-Chamarro et al. 2020). Likewise, although the Andalusian standard demands a gray coat officially (ASNOPRA n.d.), administrations have been lenient with Andalusian-type donkeys with dark or golden ("melado or melao", Fig. 10) coats, and except for those drastically dark animals which clearly departed fromt he standard, these where registered in the studbook. In sum, colors like black, dun, or melado may reflect genuine heritage phenotypes rather than aberrations—they stem from the breed's ancestral gene pool that once included a variety of coat colors.

Historically, breed organizations have misinterpreted these normal variants as faults. The Andalusian registry explicitly labels any disallowed coat as a disqualifying defect (ASNOPRA). The Normand rules likewise ban all but the prescribed colors (Association Nationale de l'Âne Normand n.d.). As Marín Navas et al. (2022) note in horses, "bans against certain coat colours and patterns may have prevented an appropriate registration of genealogical information," effectively treating unusual colors as undesirable. In donkeys, this means that animals carrying

Fig. 10 Melao or Melado Andalusian Donkey

historically valuable color genes are often culled or excluded by the studbook. Rather than seeing a black or dun coat as a sign of richness in the breed's genetic history, registries have viewed it as a deviation from a rigid "ideal."

Removing these animals from breeding programs has predictable genetic consequences. Excluding even a small number of variant-colored donkeys shrinks the effective gene pool. For example, pedigree analyses warn that when "excessive contribution of few ancestors to the gene pool" occurs, it leads to "narrower pedigree bottlenecks" (Pierini et al. 2025). In practical terms, each culled donkey reduces founder diversity. Studies of endangered European donkeys confirm this: the Martina Franca breed's effective population has fallen far below the FAO-recommended minimum ($Ne > 50$), with only ~16 founders contributing most diversity (Pierini et al. 2025). This mirrors patterns seen in other equines, where color-based culling elongated generation intervals and raised inbreeding (Marín Navas et al. 2022). Over time, such culling of "non-standard" donkeys erodes genetic variation and flexibility. Conservationists warn that "risks of genetic erosion and inbreeding" grow unless breed bases are broadened (Pierini et al. 2025).

Recognizing and conserving such variation requires a multi-pronged approach. The FAO's Global Plan of Action emphasizes that breed characterization must combine phenotypes, genetic data, and historical records (FAO 2010). In practice, this means using molecular tools (microsatellites, SNPs, mitochondrial DNA) alongside surveys of coat, morphology, and archival breed descriptions. For instance, genomic sequencing of donkeys has not only confirmed distinct lineages—showing the Andalusian and Canary-Island Majorera donkeys share ancestry distinct from other Iberian breeds (Lara-Chamarro and Jordana 2020)—but also revealed the genes underlying coat patterns (e.g., a TBX3 mutation for non-dun color) (Lara-Chamarro and Jordana 2020). Combining such data with knowledge of traditional breeding practices can identify valuable alleles currently hidden in "outlier" animals. In short, maintaining donkey breed diversity will depend on valuing phenotypic outliers (not discarding them) and using genetics to inform rather than override historical phenotypes (FAO 2010; Lara-Chamarro and Jordana 2020). Only by integrating all available information can registries promote true genetic integrity and long-term conservation of these donkey breeds.

5.4 The Conservation Paradox: Endangered at Home, Invasive Abroad

The conservation paradox in donkey populations highlights a complex and often contradictory situation where certain breeds or populations are considered extinct or at high risk of endangerment in their regions of origin while simultaneously posing a threat in other parts of the world.

The first examples of this situation appear possibly as early as the time of Columbus himself, and certainly during the sixteenth century, donkeys were introduced to Fuerteventura in the Canary Islands. Their establishment proved highly successful for the donkeys but problematic for the Spanish settlers, as the animals caused significant damage to crops. This led to deliberate efforts to exterminate the donkeys. Similar challenges and dilemmas have arisen in most other regions of the New World where donkeys were introduced after 1492, with the subsequent establishment of trade between Europe and the New World facilitated the spread of donkeys throughout the continent.

The introduction of new donkeys became increasingly frequent by Portuguese and Spanish colonizers during the sixteenth century. In Brazil, the first donkeys were brought from the Madeira and Canary Islands by Martim Afonso de Sousa in 1534, arriving in the town of São Vicente (São Paulo). Later, in 1549, Tomé de Souza introduced donkeys from Cape Verde to Bahia aboard the caravel *Golf*. Meanwhile, the Spaniards introduced the species through two main routes: (i) the Bay of Panama, which distributed animals to the Antilles and northern South America (Colombia, Venezuela, Ecuador, Peru, and northern Brazil), and (ii) the Rio de la Plata region, facilitating introduction into the southern cone of the continent. These introductions allowed donkeys to spread across Brazil's southeastern and northeastern regions during the fifteenth and sixteenth centuries, forming the basis for their use in transport and agriculture. During the colonial and imperial periods, mules—produced by crossing donkeys with mares—became the principal means of transport in inland Brazil, especially along routes such as the Caminho das Tropas, which linked the southern and central-southern regions to supply coffee and sugarcane plantations. At the end of the nineteenth and early twentieth centuries, additional imports of Italian and Spanish donkeys, promoted by immigrants and the Ministry of Agriculture, further enriched the genetic base.

From these historical introductions and subsequent adaptations, distinct donkey breeds emerged in Brazil through both artificial and natural selection (Fig. 11). The Pêga donkey, of Iberian origin, developed mainly in Minas Gerais, initially for mining and mule production. Selective breeding emphasized traits such as smooth gait, strength, and agility, leading to official recognition in 1947 under the Brazilian Association of Pêga Donkey Breed Owners (MAPA). Animals excluded from Pêga standards, often due to coat color, were grouped as the Paulista breed in São Paulo, similar in type but never formally recognized. In contrast, the Nordestino donkey arose through natural selection in the semi-arid Northeast, adapting to harsh conditions with traits like small size, rusticity, and endurance, rather than through controlled genetic improvement. These processes illustrate how donkeys introduced

Fig. 11 Nordestino donkey (upper left), Pêga donkey (upper right) and Brazilian Paulista lineage donkeys (lower row)

by Iberian colonizers diversified in Brazil through a combination of historical trade, human-driven selection, and environmental adaptation, resulting in breeds with distinct morphological and functional traits suited to different regions (Messias et al. 2021).

Island ecosystems, due to their small size and often delicate and specialized environments, are especially vulnerable to invasive species without natural predators. For example, in the Galápagos Islands of South America, feral donkeys, along with goats, pose a significant threat to several native species on Santiago Island and are believed to have contributed to the extinction of the endemic plant *Blutaparon rigidum*, last observed in 1906. More broadly, donkeys negatively affect local land iguana and tortoise populations throughout the Galápagos by trampling nests, destroying eggs, consuming native plants, and possibly competing directly with indigenous species for food. Removing donkeys and other invasive animals through aerial hunting may be the only viable strategy to enable native flora and fauna to recover and survive.

A striking current example of this phenomenon is observed with the Bonaire feral donkeys (Fig. 12). In Bonaire, these donkeys have become a significant concern for local ecosystems, where they are claimed to impact native flora and fauna due to overgrazing and coral habitat degradation. Despite their classification as feral and problematic in this context, Bonaire donkeys have been appointed as the last remnants of Nubian Wild Asses which faced severe decline and became presumably

Fig. 12 Already extinct Nubian Wild Ass at the Fiumicino Zoological Garden in Rome (left), Sokotran Wild Asses (upper right) and Bonaire Feral Donkeys (lower right)

extinct in their native habitat since the 1950s–70s (relationship with extinct nubians has also been suggested for Sokotran Wild Asses from the island of Sokotra in Yemen).

This paradox underscores the challenges faced in conservation strategies, as the status of donkey populations can vary dramatically depending on geographical and ecological contexts. In their regions of origin, efforts may focus on breeding programs and habitat preservation to sustain dwindling populations if it is not too late and population have run extinct. In contrast, in areas where they have become invasive, management strategies may prioritize removal or control of these donkeys to protect local biodiversity. Addressing this conservation paradox requires a nuanced understanding of both the ecological impact of donkey populations in different regions and the cultural significance of these animals in their native environments. Developing tailored strategies that consider both conservation needs and ecological impacts is crucial for ensuring the long-term sustainability of donkey breeds globally.

Yet, the assumption that feral donkeys are uniformly destructive does not hold up under ecological scrutiny and may potentially highlight economical or administrative reasons underlying. While overgrazing can occur in confined or sensitive habitats, recent studies have shown that donkeys can act as ecosystem engineers in arid environments. In regions such as the American Southwest and Australia, feral donkeys have been observed digging wells—deep holes that tap into underground

Fig. 13 Kulan or Mongolian Wild Ass (left) and Australian Feral Donkey (right) digging wáter wells

water sources during dry seasons. These wells significantly boost local biodiversity by providing critical water access to native species and supporting vegetation growth around the wells (Lundgren et al. 2021; Main 2021). In some cases, species richness near these wells was markedly higher than in surrounding dry zones, demonstrating that donkey behavior can enhance, rather than harm, certain ecosystems (Fig. 13).

Furthermore, claims about donkeys damaging coral reefs in places like Bonaire may be overstated or misattributed. Donkeys do not interact directly with marine ecosystems such as reefs. The supposed link between donkey presence and reef degradation likely stems from broader land-based disturbances, such as sediment runoff from degraded vegetation near coastal zones. However, such runoff is a complex process influenced by multiple factors including land use, deforestation, and infrastructure development. Placing blame solely on donkeys oversimplifies the ecological dynamics involved and risks diverting attention from more impactful anthropogenic sources of reef stress.

In addition to their ecological role, feral donkey populations may serve as critical genetic reservoirs. Genetic studies suggest that some of these populations, including those on Bonaire, may preserve lineage from the Nubian wild ass (*Equus africanus africanus*), one of the primary ancestors of domestic donkeys (Clancy et al. 2021; Heckadon 2020). In such cases, animals dismissed as invasive may actually carry irreplaceable genetic heritage. Uninformed eradication campaigns risk losing this genetic material forever, undermining conservation objectives in the regions where these animals originally evolved.

Recognizing these complexities calls for integrated management approaches. Rather than defaulting to eradication, conservation planning should evaluate both the ecological function and genetic value of feral donkey populations. Controlled grazing, genetic monitoring, and habitat impact assessments can help determine appropriate local strategies. In some regions, maintaining low-density donkey populations may sustain ecosystem services like water provisioning while minimizing overgrazing (Lundgren et al. 2021). In others, targeted preservation of genetically valuable individuals may be warranted. This multifaceted approach—

rooted in context-specific evidence—offers the best path forward in resolving the donkey conservation paradox while respecting both biodiversity goals and cultural heritage.

5.5 The Role of Human Preference in Shaping Donkey Breeds

Over the millennia of global trade and migration, humans have continuously bred donkeys for particular traits, so that today a rich variety of local breeds exists. Genome studies confirm that early domestication entailed strong human-driven selection: domestic donkeys today commonly exhibit non-diluted black or chestnut coat colors, whereas their wild ancestors uniformly bore gray ("dun") coats (Universitat Autònoma de Barcelona 2020; Wang et al. 2020). In other words, people deliberately favored intense, non-dun coat hues, though researchers note the exact motives—be it visibility, symbolism or simple aesthetics—remain debated. Such selection for appearance and utility underlies much of donkey breed diversification. For example, in the case of large donkeys in fashion in the North American south, muleteers adscribed better physical qualities to dark animals with light points, hence the basis for their preference.

Color has long shaped breed standards. As noted above, the ancestral wild ass (*Equus africanus*) was gray dun, but nearly all domestic breeds display rich brown or black coats (Wang et al. 2020). University researchers explicitly attribute this to human preference for strong pigmentations: a European study found that both horses and donkeys underwent selection at the same pigmentation gene (TBX3) during domestication, likely driven by "preference for intense coat colours" (Wang et al. 2020; Universitat Autònoma de Barcelona 2020). Historical breed records bear this out. For example, the American Mammoth Jackstock (a U.S. breed developed from imported European, mainly Spanish stock) was for generations registered only in black. Black jacks bred to gray Belgian mares produced uniformly dark mules, which teams of mules-makers favored for uniformity; hence "black [was] considered the only acceptable color" for this giant donkey (Livestock Conservancy 2024). Only in the late twentieth century did market fashions shift toward paler ("sorrel") draft mules, and the Mammoth Jackstock herd correspondingly became predominantly sorrel (Livestock Conservancy 2024). Similarly in France, the endangered Poitou donkey has long been defined by a dark coat standard—"black or brown with a grey underbelly"—and the traditional breed standard even forbids the natural shoulder-back "cross" marking (Oklahoma State University 2025). In short, breeders historically fixed coat color ideals (often dark shades) as part of breed identity, whether for mule-production needs or regional aesthetic taste. (In contrast, evidence of any superstition—e.g. beliefs that a white or colored donkey is lucky—is much less documented in scientific literature.).

This human driven process has occurred from Egyptian times on since donkey were domesticated. Historical sources from the Islamic Caliphate period suggest that in certain regions, including eastern Arabia, specific aesthetic and symbolic preferences governed the selection of donkeys used in public and ceremonial roles. The Hassawi donkeys—native to the Al-Ahsa region—were reportedly required to be white, reflecting both a visual ideal and religious-cultural symbolism associated with purity, humility, and prestige. White donkeys were often favored in Islamic traditions, sometimes linked to Hadith accounts describing the Prophet Muhammad's own white mount, and this likely influenced the preference for white-coated animals in administrative or urban contexts.

Prophet Muhammad's was linked to two main donkey based figures. The first one, Burāq is described as "a white (although roan and blueish gray depictions are available) animal, half-mule, half-donkey, with wings on its sides...," Burāq was originally introduced into the story of Muḥammad's night journey (*isrāʾ*) from Mecca to Jerusalem and back, thus explaining how the journey between the cities could have been completed in a single night. As the tale of the night journey (*isrāʾ*) became connected with that of Muḥammad's ascension to heaven (*miʿrāj*), Burāq replaced the ladder as Muḥammad's means of access into heaven. The fact that it was considered half donkey half mule may stem from the fact that Hassawi donkeys indeed where rather aesthethical donkey breeds with finer bones and characteristic white coats. Specifically, An excerpt from a translation of Sahih al-Bukhari describes Buraq as a white animal which was smaller than a mule and bigger than a donkey, to later highlight the quality of its gait by saying that the animal's step (was so wide that it) reached the farthest point within the reach of the animal's sight, as it was often noted for those Egyptian Hassawi white riding donkeys (Fig. 14).

The second one, Yaʿfūr (also spelled Yaʿfoor, Yaʿfour, ʿUfayr, ʿOfayr, among others, meaning "Deer" in Arabic), was a donkey described as blue—which can refer to a gray, roan, or bluish-gray coat—that the Islamic prophet Muhammad frequently rode, often without a harness. Many stories surround this donkey, but the most common recount that it was a gift from the Byzantine governor of Egypt sometime between 628 and 632 AD (8–11 AH). According to a Shi'ite Islamic tradition, Yaʿfūr possessed the ability to speak and told Muhammad it was the last in a line of donkeys ridden by prophets, descending from the donkey Jesus rode during his triumphant entry into Jerusalem, which was also named Yaʿfūr.

For this, in the case of the Alexandrian or Egyptian White Hassawi donkeys or even the Saharian donkey breed described by Castejón y Martinez de Arizala (1945) (Fig. 15), their color were not merely incidental but a deliberate selection criterion shaped by both spiritual association and social status during the Caliphate's governance of the region. This emphasis on coat color illustrates how sociocultural norms and religious symbolism played a direct role in configuring the traits of regional donkey populations.

Body size and conformation have likewise been guided by human use. Large donkeys have been prized where heavy work or mule production was important. Early French mule breeders deliberately chose jackasses with extreme stature and specific features: Poitou breeders selected jacks with "large ears, head and leg joints," believing these would yield "exceptionally large and strong mules" (Oklahoma State University 2025; Livestock Conservancy 2023). Their Poitou breed grew to heights of 1.35–1.50 m (and often much larger) with massive frames. Likewise, the American Mammoth Jackstock was defined by strict height

Fig. 14 Ya'fur (upper left) and Burāq (upper right and lower)

Fig. 15 Alexandrian or Egyptian Hassawi white riding donkeys

standards—historically at least 14.2 hands (about 1.47 m) for males and 14.0 for females—to maximize size and strength (Livestock Conservancy 2024). Most Mammoth jacks still weigh 900–1200 lb and possess very long ears (up to 33″) as a by-product of this selection (Livestock Conservancy 2024). These huge donkeys were bred expressly to produce hefty draft mules for carriage or farm work. Even in subsistence contexts, size matters: a recent survey of Ethiopian highland farmers found that growth performance and body size were the top criteria for choosing breeding donkeys (Wassie et al. 2023). In short, wherever greater draft power or load-carrying was needed, human preferences favored breeding the largest, most robust donkey types.

Personality and behavior have also influenced selection, though often secondary to physical traits. Across cultures, docility and tractability are generally valued in pack animals. The American Mammoth Jackstock is noted for its "good temperament"—the breed is described as docile, intelligent and easy to handle—and similarly in Ethiopia, farmers did consider temperament (together with fertility and health) when culling stock (Livestock Conservancy 2024; Wassie et al. 2023). However, empirical studies show temperament typically ranks below size or productivity in importance. For example, in the South Omo Zone of Ethiopia, herders prioritized growth, body size and disease resistance as breeding criteria, and only included color or temperament lower down on the list (Wassie et al. 2023). Nevertheless, these secondary traits were still recognized. Temperamental stability helps ensure a donkey will work faithfully; thus, even aesthetic appreciation of a "gentle" donkey (with a quiet expression or calm demeanor) may subtly shape breeders' choices, if only at the level of keeping favored animals. Overall, though, the main evolution of donkey breeds has been driven by utilitarian demands (strength, fertility) with temperament held as a desirable but lesser factor.

5.6 Complexities of the Production Systems of Donkey-Derived Products

Donkeys have played a vital role in agriculture and cultural practices across the globe for centuries, providing essential resources such as milk, meat, and hides. Historically, donkey-derived products have been valued for medicinal and nutritional properties; for example, in China donkey milk and hide gelatin (ejiao) have long been used in traditional medicine (The Donkey Sanctuary 2017; Gao and Clark 2020). However, the increasing global demand for donkey-derived products poses significant challenges to the sustainability and welfare of donkey populations. This rising demand, especially for ejiao, has driven up prices of donkey hides dramatically and led to widespread illegal trade and theft of donkeys (Waldie 2017; The Donkey Sanctuary 2017). Many specific breeds are at risk due to overexploitation and inadequate management practices. In particular, the global donkey population has been declining sharply in many countries as traditional uses (like transport and draft work) wane (Svoboda 2020). Economic development, mechanization, and now the ejiao industry have all contributed to dramatic population declines; for example, China's donkey population fell by 76% from 1992 to 2020 (Gao and Clark 2020). This situation jeopardizes not only the donkeys themselves but also the cultural and agricultural heritage they represent.

The global demand for donkey-derived products, such as meat, milk, and hides, poses significant challenges to the sustainability and welfare of donkey populations worldwide. Many regions rely on specific donkey breeds for these products, yet a widespread misunderstanding of the species' physiological and psychological needs often leads to inadequate care and management practices. This lack of awareness can result in systems that do not accommodate the welfare requirements of donkeys, leaving them vulnerable to neglect and abuse (Burn et al. 2020). For example, the surge in demand for ejiao has driven prices of donkey hides so high that it has incited theft and illegal slaughter of donkeys in source countries (Waldie 2017). Governments in donkey-exporting regions (e.g., Burkina Faso, Uganda) have had to ban donkey exports in response to widespread thefts linked to the skin trade (The Donkey Sanctuary 2017). These losses not only threaten local donkey breeds but also harm rural livelihoods, as donkeys are crucial for transport and agriculture in many communities (Brooke 2019, 2020).

Ejiao production in China relies on donkey hides, predominantly from Chinese breeds like the Jiangsu and Hebei donkeys, which are favored for their larger body size and higher hide yield. Donkey skin is the main raw material for ejiao, a gelatin used in traditional Chinese medicine to "tonify the blood" and improve wellness (Gao and Clark 2020). The skyrocketing popularity of ejiao has led to extreme pressure on donkey populations: current estimates suggest that around 4.8 million donkey hides are needed annually for ejiao production (Gao and Clark 2020). This demand has caused donkey populations to plummet and has forced China to import hides from abroad. The impact is felt globally—African countries with large donkey populations have become key sources of skins, prompting bans and international concern (The Donkey Sanctuary 2017).

In Italy, the Martina Franca and Ragusano donkey breeds (Fig. 16) are valued for their milk. The Martina Franca donkey of Puglia is the largest Italian breed and has gained a new role producing rich milk and meat in recent decades (Pilla et al. 2012). Its milk is high in protein and nutrients, traditionally used to make cheeses and other

Fig. 16 Martina-Franca (Pugliese) (left) and Sicilian Ragusano (right) Donkey Breed

dairy products. Similarly, the Ragusano donkey from Sicily yields milk prized for its quality and unique flavor (Polidori and Vincenzetti 2013). Donkey milk as a niche dairy has gained attention as a hypoallergenic alternative to cow's milk, being suitable for infants or individuals with cow's milk allergies (D'Alessandro et al. 2022).

The Zamorano–Leonés donkey of northwest Spain is another breed famed for its milk. This robust, hardy donkey is adapted to rugged terrain and produces nutrient-rich milk often used in artisanal cheeses. Donkey milk, in general, is low in fat and high in lysozyme, and is increasingly marketed for its nutritional and health-promoting qualities (D'Alessandro et al. 2022). As with the Italian breeds, the Zamorano–Leonés helps sustain rural livelihoods through its dairy products, and there is growing recognition of donkey milk's value as an alternative for those intolerant to cow's milk.

Meanwhile, donkey meat remains a traditional food in some regions but is generally overlooked in global discussions of sustainable protein. Certain breeds are prized for meat quality: the Encartaciones or Catalonian donkey breeds (Fig. 17) of Spain have traditionally supplied tender meat as a regional delicacy (Deng et al. 2020). In China, the Dezhou donkey (notably its SanFen and Wutou strains, Fig. 18) is a large local breed historically bred for meat, hides, and even milk. In Italy, the Amiata and Martina Franca donkeys are also appreciated for meat, used in traditional stews and cured specialties.

There is a strong case for promoting effective breeding and conservation programs to offset these risks. Stakeholders can create productive systems that respect donkeys' needs while meeting market demands by implementing controlled breeding to maintain genetic diversity and prevent over-exploitation of any single breed (Pilla et al. 2012). Education and awareness are key: by understanding donkeys' natural behaviors, communities can develop farming methods that enhance welfare. Policies that enforce welfare standards (such as EU animal welfare regulations) can help ensure that donkey farms provide a life worth living for these animals (Burn et al. 2020).

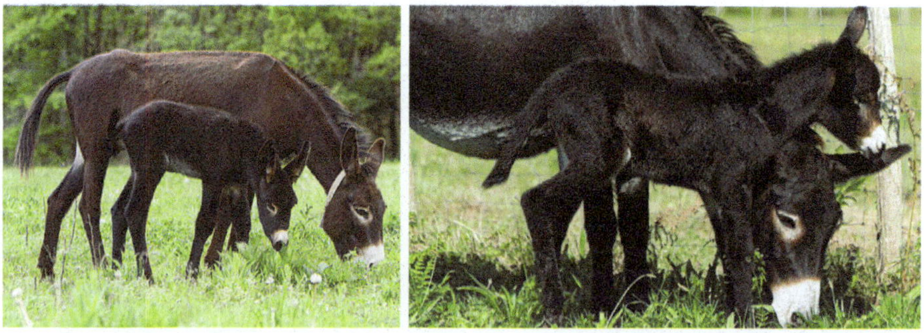

Fig. 17 Catalonian or Ruc Catalá (left) and de las Encartaciones (right) Donkey breed

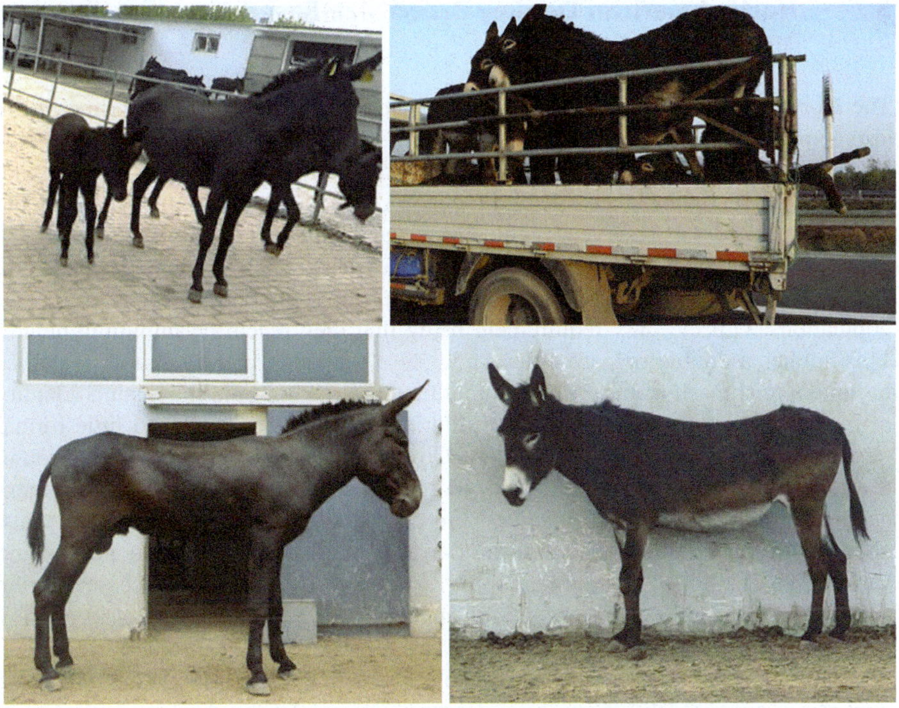

Fig. 18 Chinese Dezhou donkey breed Wutou (left) and SanFen (right) strains

Crucially, donkey product systems often overlook donkeys' psychological needs. Donkeys are intelligent, social creatures with complex behaviors and emotions, yet these are frequently neglected. Studies in equids show that abrupt weaning causes large spikes in stress hormones and behavioural signs of distress in young foals (Houpt et al. 2001). If a donkey foal is suddenly separated from its dam without gradual habituation or continued contact, it can suffer anxiety and its cognitive development may be impaired (Burn et al. 2020). Donkeys also tend to hide pain and discomfort, meaning that unless caretakers are attentive to subtle signs, a donkey's emotional distress might go unnoticed.

In conclusion, while the production of donkey-derived products—whether ejiao, milk, or meat—has the potential to be sustainable and economically viable, it must be managed under proper breeding and welfare programs (Gao and Clark 2020). Current practices often ignore the species' social structure and behavioural needs, focusing narrowly on output. Without addressing donkeys' unique physiology and psychology, we risk compounding welfare issues and driving endangered breeds toward extinction. A balanced approach is urgently needed: one that balances market demand with the ethical imperative to give donkeys the care, environmental enrichment, and social contact they require.

5.7 North American Donkey Breed Size-Based Classification and Registration

The North American practice of classifying and registering donkeys primarily by size—grouping them into categories like miniature, standard, and mammoth—has led to significant issues with breed dilution and loss of distinct breed lines. Unlike systems that maintain breed-specific registries, this size-based approach lacks the structure needed to preserve genetic lineage and unique breed characteristics, as there is often no historical or genetic registry beyond size distinctions (Beja-Pereira et al. 2004; Rossel et al. 2008). As a result, breeds may absorb individuals from other populations with the only requirement being that they reach the minimum height. This implies a double risk: American miniature, standard, and mammoth registries become a mishmash of unrelated stock, ultimately risking both the distinct identity of native donkey populations and the dilution of unique traits due to little formal record-keeping of bloodlines (Kavar and Dovč 2008). There may be an exception in the case of the Mammoth Jackstock, which has a dedicated registry; its stringent height requirements can—in practice—limit registration of non-lineage individuals, although exceptionally large European donkeys may still qualify (Oldenbroek and van der Waaij 2015). The broad grouping of donkeys by size fails to account for the particular heritage and characteristics of smaller, distinct breeds, which risks eroding genetic diversity (Kavar and Dovč 2008).

As previously introduced, in the United States, the registration of mammoth donkeys is governed by the rules and procedures set by the American Mammoth Jackstock Registry (AMJR) and the Standard Jackstock Jackass Registry (SJJR). The registration process can be done through two main methods: by pedigree and by measurement (Oldenbroek and van der Waaij 2015).

Jacks, jennets, and geldings (which cannot therefore breed) of any age may be accepted into the registry by pedigree as long as their sire and dam are registered with the AMJR/SJJR. However, the registry reserves the right to visually inspect an animal at any time to ensure it meets the standards, and may require certification from a licensed veterinarian verifying the height, size, color, and age of the animal. When applying by pedigree, the breeder must provide details including the animal's name, color, sex, date of foaling, and the name and address of the breeder and owner, along with the registration numbers of the sire and dam, if available (Oldenbroek and van der Waaij 2015).

For animals to be registered by measurement, certain physical criteria must be met. Jacks must measure at least 14.2 hands in height, 61 inches around the heart girth, and have a cannon bone circumference of at least 8 inches. Jennets and geldings must meet similar height and girth requirements, with slightly smaller measurements for the cannon bone. Applicants must submit three clear photos of the animal—front, rear, and side views—along with the application. The registry retains the right to inspect the animal visually and may require certification from a licensed veterinarian. It also reserves the right to deny registration for animals of questionable pedigree or breed characteristics, such as burro crosses or spotted coats, which do not conform to the breed standard (Oldenbroek and van der Waaij 2015).

If an animal is registered before maturity, it is issued a temporary certificate without measurements. The animal must be updated upon reaching maturity (typically around 5 years of age) for a permanent certificate to be issued. This ensures that the animal meets the final registry requirements after it has fully developed (Oldenbroek and van der Waaij 2015). The AMJR/SJJR maintains strict guidelines to prevent fraudulent entries, and any individual caught submitting false pedigree or measurement information will be permanently banned from the registry. This emphasizes the importance of maintaining the integrity of the breed (Oldenbroek and van der Waaij 2015). The registration process comes with a range of fees, including individual applications for jacks, jennets, and geldings, as well as discounted herd rates for larger groups. Additional fees apply for duplicate or corrected certificates and the transfer of ownership. International registrations incur an additional postage fee (Oldenbroek and van der Waaij 2015).

While these detailed rules ensure the consistency and integrity of the breed, they also highlight the challenges that the system poses in terms of limiting genetic diversity. The strict standards on size and conformation can encourage introgression from outside breeds in an attempt to meet these criteria. This focus on "purity" and the avoidance of certain traits (such as burro crosses or spotted coats) may inadvertently reduce the overall genetic pool of the breed, potentially weakening its long-term health and adaptability (Kavar and Dovč 2008; Rossel et al. 2008).

Ultimately, while the AMJR/SJJR's registry system provides a framework for maintaining a standardized breed, the process also underscores the tension between the preservation of pure physical characteristics and the need to maintain a healthy, genetically diverse population. The continued focus on conformation and purity may limit the breed's ability to adapt to environmental changes or withstand emerging health challenges, making the balance between standardization and genetic diversity a critical issue for the future of the mammoth donkey breed (Beja-Pereira et al. 2004; Rossel et al. 2008).

Registering animals solely based on measurements, as it was the case for the mammoth donkey breed in the United States, can lead to several significant problems that affect both the genetic diversity and overall health of the breed. These issues arise from the narrow focus on physical criteria, often at the expense of broader ecological and genetic factors that are essential for long-term sustainability (Kavar and Dovč 2008).

When registration is based exclusively on specific measurements, such as height, heart girth, and cannon bone circumference, breeders are often incentivized to prioritize physical conformation over genetic variability. This practice can result in a limited gene pool, known as genetic bottlenecking, which occurs when only animals that meet these strict physical criteria are bred. Over time, this reduces genetic diversity, making the breed more susceptible to inherited diseases, genetic disorders, and inbreeding depression (Beja-Pereira et al. 2004). With a smaller genetic pool, there is also a higher risk of losing important adaptive traits that would help the breed thrive in changing environmental conditions (Rossel et al. 2008).

To meet the specific measurement requirements set by the registry, some breeders may introduce animals from other breeds that possess the desired size or physical traits. This "introgression" can result in the mixing of genetic material from outside the breed, diluting the original genetic identity of the breed. While this may temporarily help meet the physical standards, it compromises the integrity of the breed's genetic lineage. Introgressed animals may not exhibit the same ecological and functional adaptations as the original breed, which can undermine the breed's resilience and long-term viability (Kavar and Dovč 2008).

The emphasis on size and conformation based on specific measurements often disregards other important characteristics, such as temperament (Navas González et al. 2019), disease resistance (Navas González et al. 2018), and the animal's ability to thrive in diverse environments. For example, donkeys that meet the height and cannon bone standards may still have underlying health issues, poor fertility, or less desirable working characteristics. Without considering the broader picture of genetic health and adaptability, the breed risks developing individuals that may not be well-suited to the challenges of modern agricultural and ecological systems (Beja-Pereira et al. 2004).

The reliance on measurements can exclude animals that may possess valuable genetic traits but fail to meet the strict physical standards. For example, a donkey might have exceptional work ethic, endurance, or disease resistance but may not measure up in terms of height or girth. By excluding these animals from the registry, the breed misses out on valuable genetic material that could contribute to the breed's overall health, resilience, and adaptability (Kavar and Dovč 2008).

Donkeys that are selected only for specific measurements may be less adaptable to changing environmental conditions, such as shifts in climate, land use, or farming practices. As climate change and other environmental factors become more significant, breeds that have not been selected for genetic diversity and environmental adaptability may struggle to survive. The narrow genetic base created by focusing on physical measurements can make the breed more vulnerable to environmental stressors, disease outbreaks, and market shifts that require flexibility and resilience (Rossel et al. 2008).

The emphasis on achieving certain measurements can sometimes lead to the overbreeding of animals with exaggerated traits, which could result in physical issues or health complications. For instance, a donkey bred solely for its size may suffer from musculoskeletal issues or other health problems due to the pressure to meet size standards. This focus on measurements rather than overall well-being can compromise the animal's welfare and result in long-term health problems for the breed (Kavar and Dovč 2008).

By prioritizing measurements over broader breeding objectives, the registry may inadvertently fragment the breeding goals of the donkey community. Some breeders may focus exclusively on producing animals that meet physical standards, while others may aim to breed for functional traits, such as endurance or working ability. This divergence in breeding goals can lead to inconsistencies in the breed's characteristics, making it harder to maintain a unified breed standard and potentially reducing the breed's overall utility in agricultural or working contexts (Oldenbroek and van der Waaij 2015).

5.8 Coat Colour Based Registries: American Council of Spotted Asses

In animal breeding, registration systems play a pivotal role in shaping breeding practices, maintaining genetic standards, and preserving breed characteristics. However, when registries prioritize aesthetic traits—such as coat color—over functional attributes like health, genetic lineage, and working ability, serious long-term consequences can arise. This issue is particularly evident in the case of donkeys, where certain organizations have adopted registration criteria based primarily on visual appearance, especially spotted coat patterns. While these practices may promote visually distinctive animals, they raise major concerns regarding genetic diversity, breed integrity, animal welfare, and adaptability to different environments.

The most prominent example is the American Council of Spotted Asses (ACOSA) and its registry, the American Spotted Donkey Registry (ASDR). ACOSA is an independent organization, founded in Billings, Montana, in 1966 by David Parker (American Council of Spotted Asses 2025). The registry is open—no pedigree documentation is required, although a two-generation history is encouraged (American Council of Spotted Asses 2025). Donkeys of any type—miniature, standard, or mammoth—as well as mules and hinnies, may be registered as long as they exhibit at least two visible spots located behind the throatlatch and above the legs in their photographs (American Council of Spotted Asses 2025). In contrast, "Masked Spotted Donkeys," which feature minimal spotting (e.g., a blaze and light belly), do not qualify for ACOSA registration but may be recorded under the American Donkey and Mule Society's "Masked Spotted Factor (MSF*)" option (American Donkey and Mule Society n.d.). ACOSA also sponsors its own shows, requiring registration for entry into spotted classes.

However, the core issue with aesthetic-based registries like ASDR is their tendency to contribute to genetic homogenization. The registry's emphasis on coat color over lineage, health, or utility traits can lead to a blending of genetic lines and a loss of distinctive breed attributes (Huang et al. 2023; The Livestock Conservancy 2025). Over time, this undermines the ability to trace purebred lines, identify valuable breed traits, and support conservation efforts for rare or regionally adapted donkey breeds.

ASDR is fundamentally a color-based registry, gathering donkeys that exhibit spotted coat patterns, rather than evaluating individuals based on conformation, performance, or environmental adaptability (American Council of Spotted Asses 2025). While some demographic details like age, sex, and ancestry may be recorded, the dominant registration criterion is coat spotting.

This approach raises significant genetic concerns. By allowing only spotted donkeys into the registry, ACOSA inadvertently encourages breeding decisions based on appearance alone. Such selection pressure can reduce the effective gene pool, potentially leading to inbreeding and a loss of valuable genetic traits such as fertility, disease resistance, and adaptability (Huang et al. 2023). As genetic diversity narrows, the long-term viability and resilience of the donkey population diminish (The Livestock Conservancy 2025).

Moreover, donkeys that do not meet the aesthetic standard—even if they are physically robust, hardy, or possess excellent working qualities—are excluded from the registry. As a result, such animals are marginalized from breeding programs that could improve overall population fitness (Smith and Pearson 2005). This exclusion not only removes important functional traits from circulation but also undermines the breed's ability to adapt to different ecological and agricultural contexts.

In regions where donkeys are vital for tasks like draft work, transportation, and farming, this emphasis on visual traits may render animals less suited to real-world labor demands. Breeding for color rather than capacity can lead to donkeys that lack stamina, strength, or environmental resilience (Smith and Pearson 2005). The Livestock Conservancy (2025) stresses that even miniature donkeys, often bred for appearance, retain work-capable characteristics that should not be neglected in favor of novelty.

The risk of crossbreeding to achieve specific visual traits further complicates breed conservation. In efforts to produce the desired coat pattern, breeders may resort to introgression—mixing in genes from other donkey types or equid species. While this can temporarily yield animals that "fit the look," it often introduces undesirable or unadapted traits, weakening the integrity of the breed (Huang et al. 2023; The Livestock Conservancy 2025).

Another major concern is the ethical dimension. Breeding based solely on appearance raises welfare issues. The Canadian Code of Practice for the Care and Handling of Equines explicitly recommends that breeders prioritize animal health, soundness, and freedom from genetic defects—criteria often overlooked in appearance-based registries (National Farm Animal Care Council 2025). Donkeys bred only for specific spotting patterns may suffer from inherited conditions or structural flaws that go unchecked because visual appeal is the primary goal.

Finally, donkeys bred strictly for aesthetic traits may lack the adaptive capacity needed to survive in certain climates or perform agricultural tasks. Without selection pressure for heat tolerance, digestive efficiency, or disease resistance, these animals become increasingly fragile and less useful across varied environmental settings (Smith and Pearson 2005; Huang et al. 2023).

In summary, the American Spotted Donkey Registry (ASDR) represents a case study in how registries centered on aesthetic novelty can compromise the genetic integrity, health, functionality, and adaptability of animal populations. While preserving visual traits like spots can have cultural or commercial value, registries must adopt balanced breeding frameworks that include health, conformation, performance, and adaptability as core registration criteria. Without this balance, registries risk creating populations that are visually distinct but genetically and functionally compromised—a dangerous trade-off for both animal welfare and long-term conservation (Huang et al. 2023; National Farm Animal Care Council 2025; The Livestock Conservancy 2025).

5.9 Owner Reluctance to Information Disclosure in Conservation Efforts

In conservation and breeding efforts, owner reliability can sometimes be a significant challenge, particularly when dealing with mistrustful owners who are very protective of their animals' information. These owners may fear judgment or consequences from conservation authorities or other entities, leading them to withhold critical details about their animals' habitats, behaviors, or health (Norris et al. 2021). This protective nature can stem from past negative experiences or a general mistrust of external organizations. As a result, conservationists may face difficulties in obtaining accurate and comprehensive information necessary for developing effective conservation strategies (The Brooke Hospital for Animals 2022). The lack of transparency can hinder efforts to monitor species, understand ecological impacts, and implement protective measures, potentially compromising the success of conservation initiatives (The Donkey Sanctuary 2017).

Deliberate omission of information by owners can further complicate the situation, as vital details about the animals' environments or interactions with humans and other species might be undisclosed. Additionally, it can be challenging for owners to follow mating decisions focused on reducing inbreeding, which is crucial for maintaining genetic diversity and the long-term viability of species (Gutiérrez et al. 2006). This intentional withholding can be motivated by various factors, including fear of regulatory consequences, loss of land use rights, or a desire to manage the situation independently. However, without complete and accurate information, conservationists are at a disadvantage, unable to make fully informed decisions. Building trust and fostering open communication between owners and conservation professionals is essential to overcoming these challenges (Norris et al. 2021). Encouraging a non-judgmental, supportive environment where owners feel comfortable sharing all relevant details can significantly improve the effectiveness of conservation efforts and the protection of endangered species (The Brooke Hospital for Animals 2022).

6 Conclusions

The donkey's journey from its African wild ancestors to a global companion of human societies reflects a unique interplay between environmental adaptation, human mobility, and cultural evolution. Historically essential for transport, agriculture, and trade, donkeys have demonstrated remarkable resilience in arid and marginal landscapes. Yet, this species now faces critical challenges stemming from demographic decline, genetic erosion, and sociocultural neglect. Mechanization, rural abandonment, and the growing demand for donkey-derived products—particularly hides for *ejiao*—have intensified pressures on already vulnerable populations. These trends risk accelerating the loss of irreplaceable genetic diversity and undermining the cultural heritage embedded in local breeds. The genetic dimension of donkey conservation presents specific challenges, including high coancestry, inbreeding depression, incomplete pedigrees, and the mismanagement of feral populations as mere nuisances rather than reservoirs of valuable alleles. Assisted reproduction technologies, while promising, remain constrained by biological and technical limitations, highlighting the need for continued research and protocol optimization. Furthermore, the ambiguous role of wild ass lineages in donkey evolution emphasizes the importance of genomic studies for clarifying ancestry and informing breeding programs aimed at preserving adaptive traits. Equally critical are the anthropological and sociocultural factors influencing donkey management. Owner-driven practices—such as intentional inbreeding, unregulated introgression of foreign breeds, and exclusion of phenotypic diversity—often conflict with formal conservation objectives. At the same time, shifting socio-economic contexts reveal paradoxes: donkeys are endangered in their regions of origin yet classified as invasive in others. Addressing these issues requires not only technical solutions but also participatory approaches that integrate local knowledge, cultural values, and livelihood needs into conservation strategies. Future efforts must therefore adopt a multidimensional framework that combines genetic monitoring, sustainable breeding programs, and culturally sensitive policies. This includes implementing molecular tools for breed verification, promoting community-based breeding cooperatives, and fostering international collaboration to regulate trade and protect endangered lineages. By acknowledging both the biological and cultural dimensions of donkey conservation, we can ensure that these resilient animals continue to serve as a vital resource for human societies while maintaining their genetic integrity and ecological role. The donkey's survival in the twenty-first century will depend not only on science but on our collective willingness to value and protect a species that has long sustained human progress.

Chapter Supplement: Timeline of Donkey Domestication, Global Dispersal, Decline and Revival

The history of the domestic donkey (*Equus asinus*) is a story that spans hundreds of thousands of years, intersecting with climatic shifts, human migrations, empire-building, and industrial decline. This timeline traces the evolutionary, cultural, and economic journey of donkeys from their early divergence as African wild asses through their initial domestication in northeastern Africa, their spread across Eurasia and beyond, and their eventual marginalization in the modern era. Each phase reveals how donkeys adapted to new environments and roles—from critical trade companions in Bronze Age Mesopotamia to keystone laborers in Roman logistics and colonial agriculture. Yet this same historical arc also underscores the fragility of donkey populations in the face of mechanization, neglect, and recent exploitative trends like the ejiao trade. By mapping out these chronological milestones, this supplement offers a contextual framework for understanding how donkeys became indispensable to human societies—and why their conservation in the twenty-first century is both a biological imperative and a matter of cultural responsibility.

1. **Deep Pre-Domestication (~900,000–300,000 years ago)**
   - **Lineage divergence**: Mitochondrial DNA analyses indicate two highly divergent haplogroups (A and B) within the African wild ass (*Equus africanus*), which split somewhere between ~900 ka and 300 ka. These two maternal clades persisted in separate wild-ass populations across northeast Africa long before humans intervened (Fig. 19).

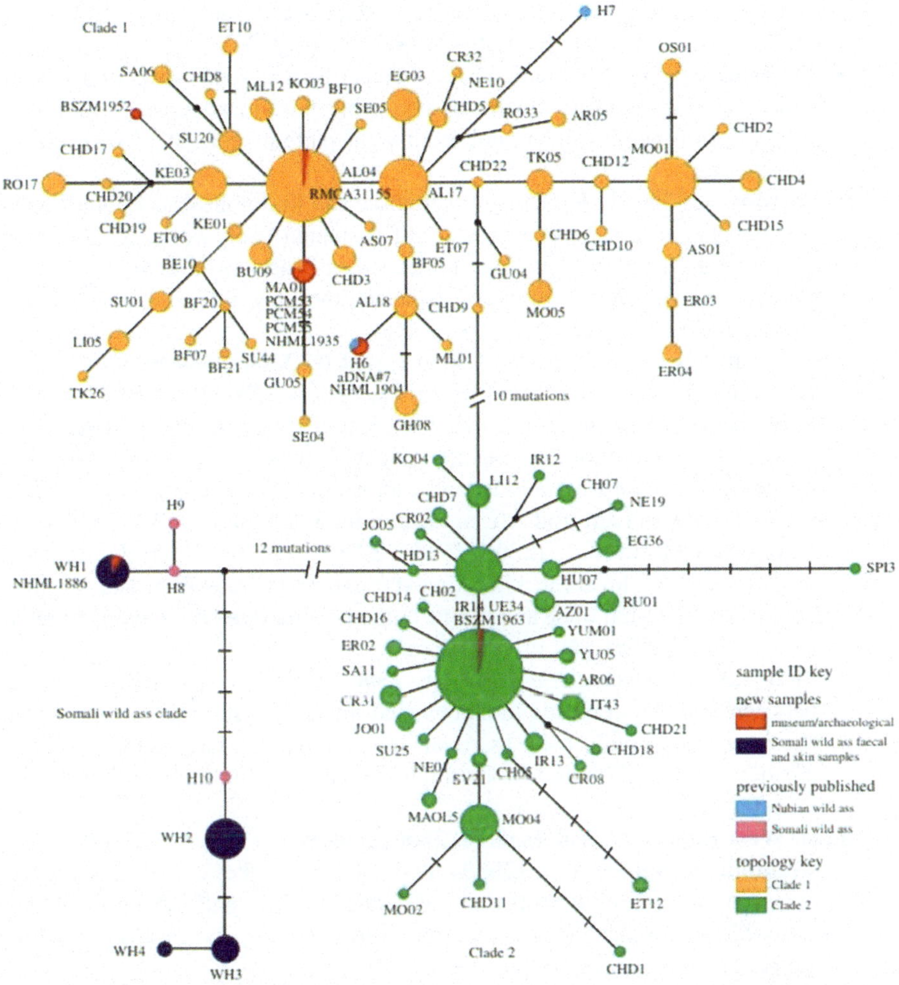

Fig. 19 Haplogroup (Clade) Divergence. Reduced median-joining network of 108 mtDNA haplotype sequences from domestic and wild asses. Coloured circles represent sampled haplotypes, while black dots represent hypothesized, unsampled haplotypes. Size of the circle is proportional to haplotype frequency and branch length is proportional to number of mutations

2. **Emergence of the Domestic Donkey (c. 5500–4500 BCE)** (Fig. 20).
    - **Initial domestication zone**: Zooarchaeological and contextual data from northeastern Africa—especially along the Nile and the Red Sea Hills—point to the first stages of donkey domestication between roughly 5500 and 4500 BCE, likely spurred by mid-Holocene aridification that made a hardy pack animal invaluable for mobile pastoralists.
    - **Morphometric changes**: Early domestic donkeys show modest but measurable reductions in size compared with contemporaneous wild asses, as well as changes in limb robustness indicative of load-bearing selection.
3. **Early Egyptian Contexts (c. 4600–3000 BCE)** (Fig. 22).
    - **Predynastic sites** (El Omari, Maadi, Hierakonpolis; 4600–3500 BCE): Skeletal remains exhibit small body size, herd-management indicators (e.g. uniform age-profiles), and tooth-wear patterns consistent with bit-use or tethering.
    - **Abydos (c. 3000 BCE)**: Donkey vertebrae and limb joints bear compressive and degenerative lesions diagnostic of sustained pack work. By this time, they already functioned as fully domesticated beasts of burden.
    - Population variability already depicted through art (hieroglyphs and etchings) (Fig. 1).
4. **From Symbolic to Widespread Use in Egypt (c. 3200–2500 BCE)**
    - **Iconography & texts**: The Libyan Palette (c. 3200–3000 BCE) depicts donkeys in processional and symbolic roles. By 2800–2500 BCE, hieroglyphic records routinely mention donkeys in state economy—taxation, transport of stone for pyramids, and caravan organization (Fig. 21).
5. **Spread into the Levant and Mesopotamia (c. 2800–2000 BCE)**
    - **Levant/Mesopotamia (2800–2500 BCE)**: Archaeological evidence (skeletons, dung layers in kiln contexts) and texts from Ebla and Ur attest donkeys as essential trade and tribute animals. Supported by remains found in sites such as Tell es-Sâfi in Gath (Israel)
    - **"Kungas" of Bronze-Age Syria (c. 2500 BCE)**: First-generation hybrids between female domestic donkeys and male Syrian wild asses (*Equus hemionus hemippus*), used as royal draft animals. Ancient-DNA analyses (Bennett et al. 2022) confirm their hybrid status and highlight early selective breeding (Fig. 22).
6. **East–West Eurasian Dispersals (c. 2600–1000 BCE)**
    - **Anatolia & South Asia (2600–2000 BCE)**: Donkeys moved westward across Anatolian trade routes and eastward via Iran and the Indus Valley. Early Mesopotamian caravan records mention donkey-led caravans reaching the Iranian plateau (Fig. 25).
    - **Central Sahara & East Africa (2000–1000 BCE)**: In Sudan and central Saharan oases, donkeys appear in settlement refuse and rock-art scenes. By ~1000 BCE, they reach pastoralist sites in Kenya and Tanzania alongside cattle and goats.

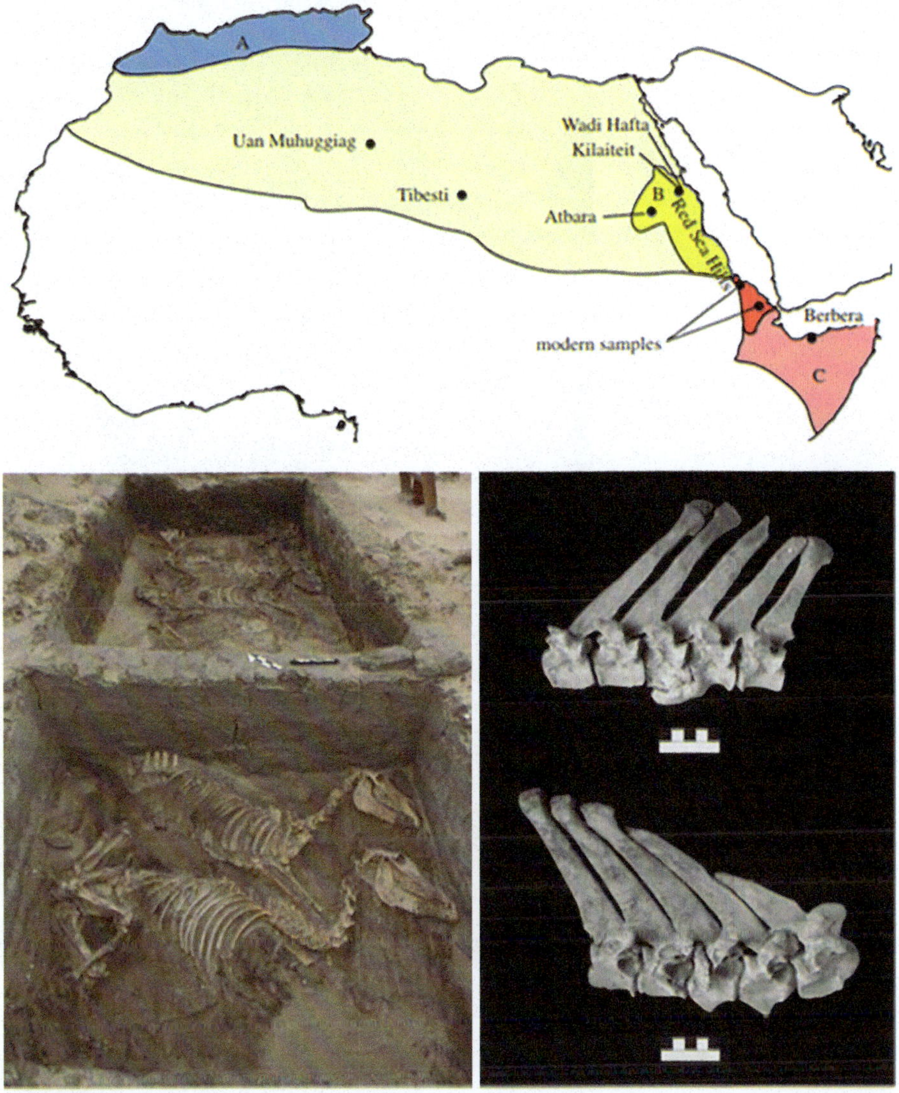

Fig. 20 Composite illustration combining archaeological and biogeographic data on African asses. (Lower Left) Abydos donkeys in situ within brick tombs. (Lower right) Vertebral arthropathies in Abydos donkey skeletons, showing osteophyte formation (upper) and compression-related spinal remodeling (lower). (Upper) Map illustrating the distribution of: (a) the ancient Atlas wild ass, (b) the historic Nubian wild ass, and (c) the historic Somali wild ass, with corresponding images or illustrations of each animal shown below. The hypothesized broader range of the ancient African wild ass across North Africa is shaded in light yellow. Locations of ancient and historic populations are marked in blue (Atlas), yellow (Nubian), and pink (Somali). The current distribution of the modern Somali wild ass is indicated in red. All sites where genetic samples were successfully analyzed are also shown

Fig. 21 The Libyan Palette (upper) and Donkey Panel Mastaba of Mereruka

Fig. 22 A 12th Dynasty (1981–1802 B.C.) fresco from the tomb of Khnumhotep III at Beni Hassan depicts a group of Semitic individuals—possibly Canaanite merchants—arriving in Egypt. They are believed to share cultural ties with the later dynasty known as the Hyksos (Left) and Standard of Ur (Right)

- **Arrival in China (2000 BCE)**
- Misconceptions and bad connotations towards donkeys likely start due to anthropological, sociocultural and religious reasons.

7. **Trans-Saharan and Mediterranean Integration (first millennium BCE)**
   - **Saharan trade**: Donkeys become pivotal for trans-Saharan salt and gold caravans, enabling long-distance trade between West Africa and Mediterranean emporia.
   - **Classical world**: Greek and Roman authors (e.g., Xenophon, Pliny) describe donkey husbandry, bit-fitting, and pack-saddle design. Coastal Phoenician and later Ptolemaic traders carry donkeys to Cyprus and Crete (Fig. 23).
   - **Parallel arrival in Europe via South and East**. Supported by archaeological sites finding in Mediterranean coasts, France and northern Spain (Fig. 24).
8. **Roman Breeding and Logistics (c. 200 BCE—500 CE)**
   - **Selective breeding**: From ~200 BCE onward, Roman stud farms—particularly in Campania and Sicily—produce larger, more muscular donkeys for mining, quarrying, and grain-mill work. Imperial records (e.g., Nero's veterinary treatises) describe dietary regimes and veterinary care.
   - **Mule production**: Roman estates routinely cross donkeys with mares to produce mules; this becomes a major economic enterprise.
   - **China breeds development**.
9. **Medieval to Early Modern Europe and Asia (500–1500 CE)**
   - **Islamic Umayyad caliphate (661–750 CE)**: Donkeys spread westward across the Maghreb into Iberia (Córdoba, Spain) after 711 CE, where Andalusi agricultural manuals detail pack-donkey husbandry. Eastward, Silk Road trading caravans carry donkeys into Central Asia and into Tang-dynasty China; zooarchaeological finds in second-century BCE Han tombs confirm their early introduction (Fig. 25).
10. **Age of Exploration and Colonial Era (fifteenth to seventeenth centuries)**
    - **Macaronesia & Atlantic crossings**: Portuguese and Spanish voyages deliberately stock domestic asses on Madeira, the Canaries, and Cape Verde from the 1420s onward.
    - **New World introductions**:
      - Columbus's second voyage (1493–96) brings donkeys to Hispaniola.
      - Tomé de Sousa's 1549 Bahia expedition establishes them in Brazil.
      - By mid-1500s, they support silver mining in Peru and Mexico; feral populations arise in Caribbean isles (U.S. Virgin Islands, Jamaica, Socotra).
    - **Southern Africa**: The first donkeys land at the Cape of Good Hope in 1656 with Dutch colonists; their slow southward spread is recorded in colonial farm logs (Fig. 26).

Fig. 23 Relief from the lion hunt of king Ashurbanipal. North Palace of Nineveh. Bas-reliefs represent the process of domestication from hunting onagers to their killing or separation for breeding purposes and their uses after domestication (first and second row);Transaharan donkey routes (third row left) and Desultor (athlete who jumped from one horse to another during race) riding a tamed Atlas Wild Ass backwards and holding up kantharos (winner's cup), mosaic in House of Desultor in Roman city of Volubilis (third row right) and Large sized donkey pushing a Roman harvester from a relief at Trier, Germany (fourth row)

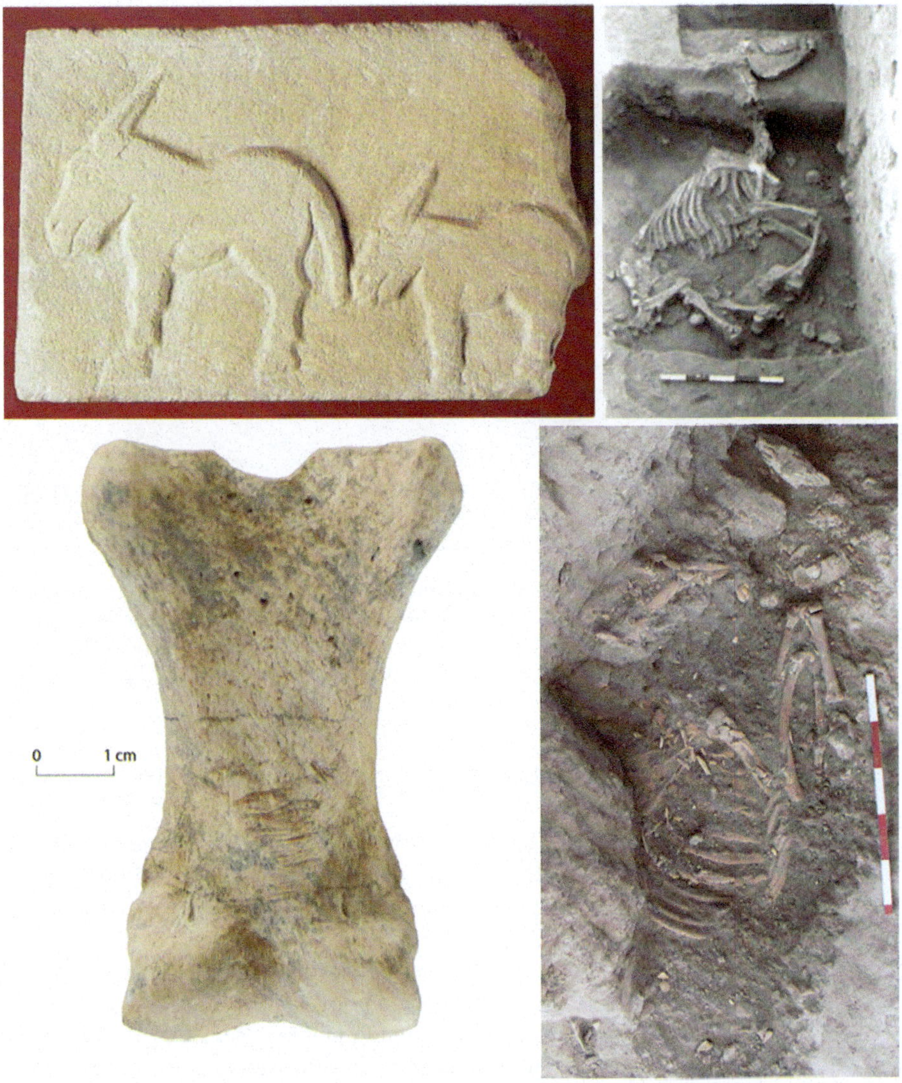

Fig. 24 Donkey and foal representation in el Cigarralejo archaeological site (Spain) (upper left) and Donkey remains of the Early Bronze age in the southern levant at Tell eş-⍰âfi/Gath (upper and lower right); Plantar face of a donkey *(Equus asinus)* third phalange bone with butchery slicing marks from Tell eş-⍰âfi/Gath

11. **Industrial Decline and Eighteenth and Nineteenth-Century Persistence**
    - **Mechanization**: Railways and steam power marginalize donkey traction in Europe and North America. Still they were used in cities for minor traction labours (Fig. 27).

Fig. 25 Onager Hunt in Qusayr'Amra (Jordan) (left) and Maqamat of al-Hariri, fol. 19r. Al-Wasiti depicting a war donkey among horses (right) and Royal hunt relief showing king Ashurbanipal (668-c.627 BCE) hunting onagers on horseback from the Nineveh Palace, kept at the British Museum, UK (lower row), which is one of the earliest depictions of these Near East wild assess hunts

- **The Great Bophuthatswana Donkey Massacre (1983)** 20,000 to 60,000 donkeys killed.
- **Small Donkey Imports**. Documentation of breeding prior to the Revolutionary War is lacking in detail, however, one account indicates the first jacks to be imported were from the Cape de Verde Islands, tracing back to the Portuguese.
- **Large donkeys Imports from Europe mainly Spain**: "Royal Gift" to President George Washington in 1785, among others.
- **Breed registries**: The American Mammoth Donkey Registry (est. 1888) and a handful of European studbooks (1880s onwards) attempt to conserve large working types, but most national programs remain informal (Fig. 28).

12. **Twentieth Century: Marginalization and Endangerment**
    - **Europe**: Native breeds in Italy (e.g. Amiata), France (e.g. Poitou), and the U.K. plunge to near-extinction by mid-twentieth century.
    - **Developing world**: Donkeys remain indispensable for rural poor—especially for water transport, women's labor, and micro-enterprises—but suffer from malnutrition, overwork, and lack of veterinary infrastructure.
13. **Twenty-First-Century Crisis and Conservation**
    - **Skin-trade emergency**: The global ejiao market—requiring millions of donkey skins annually—has driven precipitous declines across Africa, Asia, and Latin America, threatening unique genetic lineages in Ethiopia, North Africa, and China (Fig. 29).

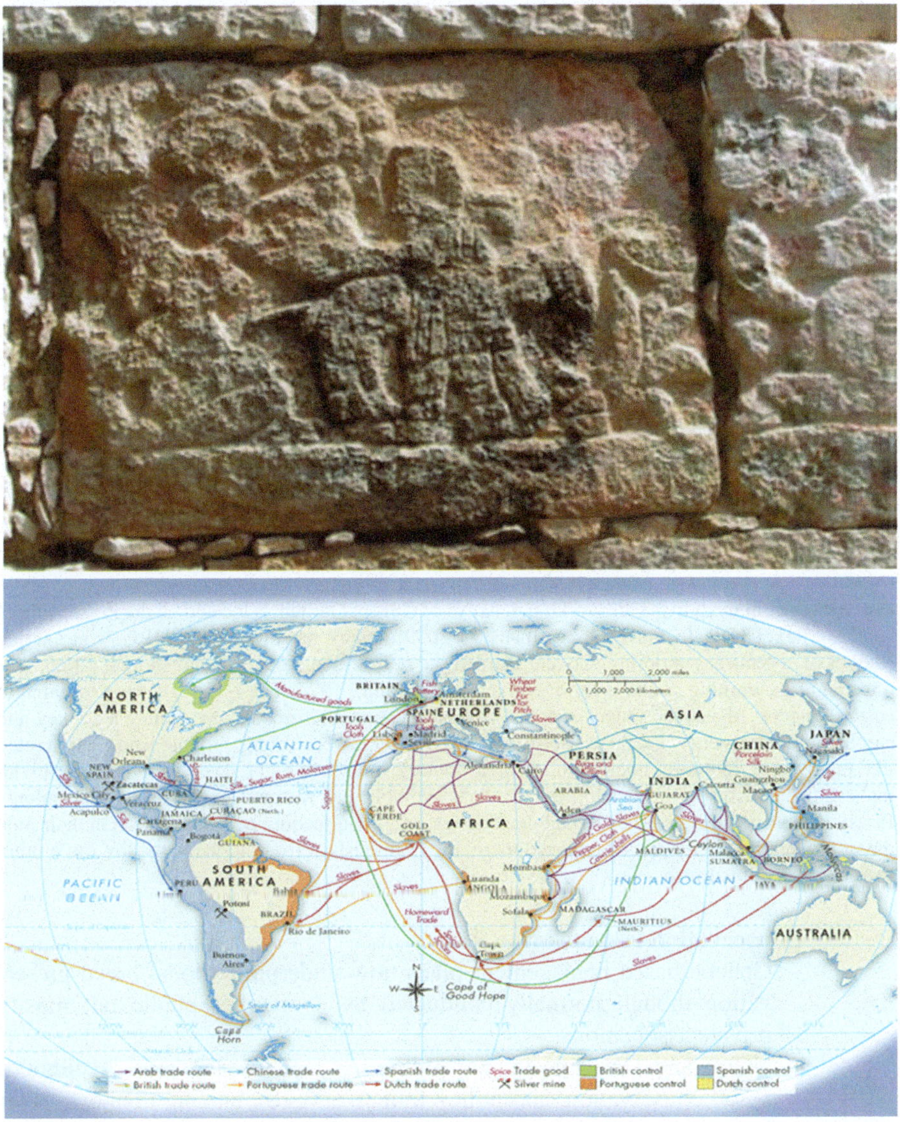

Fig. 26 Chichen Itza (Mexico) Donkey Rider depicted on Stone carving (Upper) and Map of Seaborne Trading Empires in the Sixteenth and Seventeenth Centuries

- **Feral Populations being considered a threat**: Gene pool lost risk (Fig. 30).
- **NGOs lack of care after genetic background of individuals**: Castration without testing.

Fig. 27 One, in memory of the working donkeys of Covent Garden, is located on the Piazza side of Jubilee Market Hall. In the 1860s as many as 2000 donkey barrows would fill Covent Garden Market on a Saturday morning. Covent Garden, now an upscale cultural and shopping hub, was once home to working donkeys for over five centuries until 1974. These animals were essential to costermongers—the street traders of fruits, vegetables, and other goods—as owning a donkey and cart signified prosperity. Donkeys, despite their small size, were strong, reliable, and often treated like family pets, returning home through their owners' houses at night. They were decorated with makeshift harnesses or brass-adorned gear and symbolically featured in the attire of London's Pearly Kings and Queens. They would demand rest and good feeding between loads. On their final day at the market, when trucks replaced them, they famously refused to leave by any route other than their traditional path

- **Conservation responses**:
  - Genetic-based management plans now underpin regional breed preservation though normally conducted by sometimes public but mostly private associations.
  - NGOs and UN agencies are promoting welfare standards, livelihood-based conservation, and policy frameworks recognizing donkeys' roles in climate resilience and women's empowerment.
- **Current population**: Despite localized collapses, global donkey numbers are still estimated at 40–50 million, heavily skewed toward sub-Saharan Africa and South Asia, where they remain vital to millions of households.

14. **Future Prospects (Mid-Twenty-First Century and Beyond)**
    - **Selective breeding for multifunctionality**: Programs like the Brazilian **Pêga donkey** initiative are revitalizing local breeds by selecting for temperament, endurance, and adaptive traits—demonstrating how focused genetic management can enhance both productivity and welfare in smallholder contexts (Fig. 31).

Fig. 28 American Mammoth Jackstock Association logo and Historical Prominent Mammoth Jacks born from 1846 to 1910. Black with light points was the preferred coat standard, though Andalusian types also anecdotally appeared

- **Emerging commercial niches:** Sustainability through the development of functions developed by the donkey along its history worldwide (Fig. 32).

Self-Evaluation Questionnaire

1. **What is a significant genetic challenge faced by donkey populations worldwide?**
   A. Excessive hybrid vigor due to crossbreeding
   B. High levels of coancestry and inbreeding
   C. Overproduction of purebred individuals
   D. Short generation intervals limiting selection
   Answer: B. High levels of coancestry and inbreeding

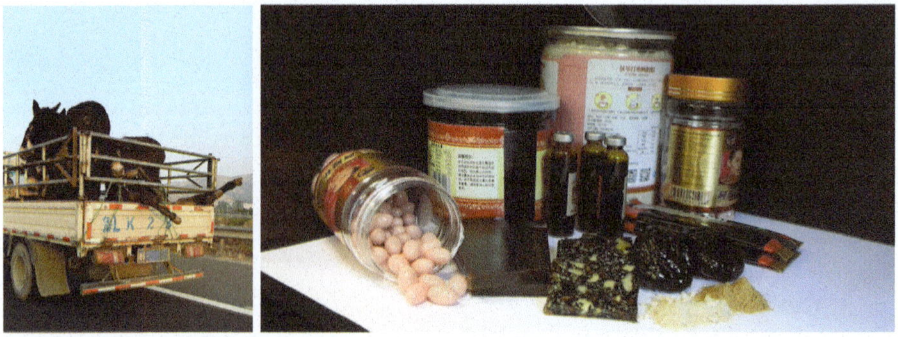

Fig. 29 Dezhou SanFen strain donkeys being transported to farms and factories (left) and ejiao products (right)

Fig. 30 Gebel Elba National Park (Egypt–Sudan border), potentially Nubian Wild Asses remnants. Claims exist of wild ass-like animals in this protected region, possibly representing pure or near-pure Nubian wild asses to date, however, no genetic studies have confirmed their lineage (left) and Bonaire ferals (right)

2. **Why do feral donkey populations present a conservation paradox?**
   A. They have no genetic value for domestic breeds
   B. Their population growth outpaces conservation breeding
   C. They are often culled despite harboring valuable genetic diversity
   D. They thrive only in regions with minimal human presence
   Answer: C. They are often culled despite harboring valuable genetic diversity
3. **What is one of the main risks of foreign breed introgression into local donkey breeds?**
   A. Creation of more robust hybrid vigor
   B. Improvement of local adaptability and disease resistance
   C. Erosion of genetic integrity and loss of unique breed traits
   D. Increased reproductive efficiency and shorter gestation
   Answer: C. Erosion of genetic integrity and loss of unique breed traits

Fig. 31 Pêga Donkey Breed and Bazilian Cup of Donkeys Competition Poster. Event supported and organized by ABCJPÊGA, the association of Pêga donkey breeders

4. **Which of the following is a primary inefficiency in using artificial insemination for donkeys?**
   A. Jennies' inability to ovulate during AI procedures
   B. High success rate leading to overpopulation
   C. Cryopreservation damage reducing semen viability
   D. Limited demand for AI in donkey breeding programs
   Answer: C. Cryopreservation damage reducing semen viability
5. **What major sociocultural factor contributes to inbreeding in donkey populations?**
   A. Widespread knowledge of genomic selection
   B. Owner misconceptions about reinforcing desirable traits
   C. Strict adherence to government breeding programs
   D. Overdependence on advanced reproductive technologies
   Answer: B. Owner misconceptions about reinforcing desirable traits
6. **Why do long generation intervals complicate conservation efforts in donkeys?**
   A. They reduce the need for genetic monitoring
   B. They make it difficult to rapidly recover endangered breeds

Fig. 32 Donkey milk, cheese, meat, assisted therapy and herd protection as emerging commercial niches for the species

C. They allow more careful evaluation of breeding outcomes
D. They minimize the risk of inbreeding
Answer: B. They make it difficult to rapidly recover endangered breeds

References

Allen W, Kydd J, Boyle M, Antczak D (1985) Between-species transfer of horse and donkey embryos: a valuable research tool. Equine Vet J 17(S3):53–62

American Council of Spotted Asses (2025) About us. ACOSA website. Retrieved Month Day, Year, from https://spottedass.com/pages/welcome:contentReference[oaicite:26]{index=26}

American Donkey and Mule Society (n.d.) Longear Lingo. ADMS website. Retrieved Month Day, Year, from https://www.lovelongears.com/longearlingo.html:contentReference[oaicite:27]{index=27}

ASNOPRA (Asociación Nacional de Criadores de Ganado Asnal) (n.d.) ASNOPRA – Asociación Nacional de Criadores de Ganado Asnal. https://www.asnopra.es

Beja-Pereira A, England PR, Ferrand N, Jordan S, Bakhiet AO, Abdalla MA et al (2004) African origins of the domestic donkey. Science 304(5678):1781–1781

Bennett R, Pfuderer S (2020) The potential for new donkey farming systems to supply the growing demand for hides. Animals 10(4):718

Bennett EA, Weber J, Bendhafer W et al (2022) The genetic identity of the earliest human-made hybrid animals, the kungas of Syro-Mesopotamia. Sci Adv 8(2):eabm0218

Biffa D, Woldemariam F (2020) Understanding the attitudes of communities to the social, economic, and cultural importance of working donkeys. Front Vet Sci 7:60

Blake JG, Douglas CL, Thompson LF (1981) Spatial variation in transferrin allele frequencies among herds of feral donkeys in Death Valley National Monument, California. J Mammal 62(1):58–63

Bragulat APF, Ortiz I, Catalán J, Dorado J, Hidalgo M, Losinno L, Dordas-Perpinyà M, Yánez-Ortiz I, Martínez-Rodero I, Miró J (2023) Time-lapse imaging and developmental competence of donkey eggs after ICSI: effect of preovulatory follicular fluid during oocyte in vitro maturation. Theriogenology 195:199–208

Brewer DJ, Redford DB, Redford S (1994) Domestic plants and animals: The Egyptian origins. Warminster: Aris and Phillips

Brooke (2019) Invisible workers: donkeys, mules and the people who depend on them. https://www.thebrooke.org

Burn CC, Dennison TL, Whay HR (2020) Donkeys and their welfare. In: Appleby MC, Olsson IAS, Galindo F (eds) Animal welfare, 3rd edn. CABI, pp 131–147

Camps G (1988) Âne. Encyclopédie berbère. Cahier 14:647–654

Castejón y Martínez de Arizala R (1945) Razas de ganados de Marruecos. Zootecnia VI(11–12): 69–120

Chaouat G, Menu E, Djian V, Delage G, Dang DC, Meliani AA, Martal J, Ropert S (2019) Immune control of pregnancy. In: Control of uterine contractility, pp 311–354

Chen J, Sun Y, Manglai D, Min L, Pan Q (2010) Maternal genetic diversity and population structure of four Chinese donkey breeds. Livest Sci 131(2–3):272–280

Ciampolini R, Cecchi F, Mazzanti E, Ciani E, Tancredi M, De Sanctis B (2007) The genetic variability analysis of the Amiata donkey breed by molecular data. Ital J Anim Sci 6(supl): 78–80

Clancy CL, Kubasiewicz LM, Raw Z, Cooke F (2021) Science and knowledge of free-roaming donkeys—A critical review. J Wildl Manag 85(7):1–14

CNRS (2022) September 9. The origins of donkey domestication, CNRS News

Colli L, Perrotta G, Negrini R, Bomba L, Bigi D, Zambonelli P, Verini Supplizi A, Liotta L, Ajmone-Marsan P (2013) Detecting population structure and recent demographic history in endangered livestock breeds: the case of the Italian autochthonous donkeys. Anim Genet 44(1): 69–78

Criscione A, Chessari G, Cesarani A, Ablondi M, Asti V, Bigi D, Bordonaro S, Ciampolini R, Congiu M, De Palo P, Ablondi M (2024) Analysis of ddRAD-seq data provides new insights into the genomic structure and patterns of diversity in Italian donkey populations. J Anim Sci 102(Supplement 3):skae165

D'Alessandro AG, Palazzo M, Martemucci G (2022) Donkey milk production and qualitative traits. Animals 12(9):1146

de Boer RA, Vega-Trejo R, Kotrschal A, Fitzpatrick JL (2021) Meta-analytic evidence that animals rarely avoid inbreeding. Nat Ecol Evol 5(7):949–964

Deng K, Zhang H, Wang Y, Wei Y, Zhang W, Huang H (2020) Nutritional composition and quality characteristics of donkey meat: a review. Food Rev Intl 36(5):537–547

Dobbie T, Felix MR, Gleason K, Hinrichs K (2024) Transvaginal aspiration, oocyte maturation and production of blastocysts and a pregnancy via ICSI in an endangered breed: Baudet du Poitou donkey. Theriogenology 224:34–40

FAO (2010) Molecular characterization of animal genetic resources: FAO guidelines. Food and Agriculture Organization of the United Nations

FAOSTAT (2019) Food and Agriculture Organization of the United Nations statistical database. Retrieved from http://www.fao.org/faostat/en/

Fava M (2024) Caratterizzazione della diversità genetica tramite pedigree e dato genomico di due razze asinine italiane Università di Parma, Dipartimento di Scienze Medico-Veterinarie

Frankham R, Ballou JD, Briscoe DA (2010) Introduction to conservation genetics, 2nd edn. Cambridge University Press

Gambini A, Smith JM, Gurkin RJ, Palacios PD (2025) Current and emerging advanced techniques for breeding donkeys and mules. Animals 15(7):990

Gao Y, Clark B (2020) The global trade in donkey skins: a review of the current situation and future implications. Animals 10(10):1868

Gebreab F (2020) Donkeys' economic and social contribution is urgently undervalued in Ethiopia. Africa at LSE

Gifford-Gonzalez D (2000) Animal disease challenges to the emergence of pastoralism in sub-Saharan Africa. Afr Archaeol Rev 17(3):95–139

Gifford-Gonzalez D (2017) Pastoralism in sub-Saharan Africa: emergence and ramifications. In Umberto Albarella, and others (eds), The Oxford Handbook of Zooarchaeology, Oxford Handbooks. Oxford Academic. https://doi.org/10.1093/oxfordhb/9780199686476.013.27

Ginja C, Gama LT, Cortés O, Martin-Burriel I, Vega-Pla JL, Penedo MCT et al (2013) Analysis of STR markers reveals high genetic structure in Iberian native equine breeds. Anim Genet 44(6): 711–721

Gutiérrez JP, Marmi J, Goyache F, Jordana J (2005) Pedigree information reveals moderate to high levels of inbreeding and a weak population structure in the endangered Catalonian donkey breed. J Anim Breed Genet 122(6):378–386

Gutiérrez JP, Marmi J, Goyache F, Jordana J (2006) Pedigree information reveals moderate to high levels of inbreeding and a weak population structure in the endangered Catalonian donkey breed. J Anim Breed Genet 122(6):378–386

Hall SJG (2019) Livestock biodiversity as an interface between people, landscapes and nature. People Nat, 1(1):284–290

Hassan ZM, Manyelo TG, Nemukondeni N, Sebola AN, Selaledi L, Mabelebele M (2022) The possibility of including donkey meat and Milk in the food chain: a southern African scenario. Animals 12(9):1073

Heckadon W (2020) Can Bonaire's donkeys be saved? Bonaire Reporter

Hoffmann I, Scherf BD (2006) Animal genetic resources—time to worry? Anim Genet Resour Inf 38:27–39

Houpt KA, Hintz HF, Wagner D (2001) Behavioural and physiological responses of foals to early weaning. J Anim Sci 79(5):1133–1139

Huang B, Khan MZ, Chai W, Ullah Q, Wang C (2023) Exploring genetic markers: mitochondrial DNA and genomic screening for biodiversity and production traits in donkeys. Animals 13(17): 2725

Ivanković A, Bittante G, Šubara G, Šuran E, Ivkić Z, Pećina M, Konjačić M, Kos I, Kelava Ugarković N, Ramljak J (2022) Genetic and population structure of Croatian local donkey breeds. Diversity 14(5):322

Jere ZM, Zidana R, Chatambalala J, Mwafulirwa S, Nyirenda Z (2025) Surviving amidst neglect: management and marketing practices of donkeys (Equus asinus) in the Central Region of Malawi. BMC Agric 1:3. https://bmcagriculture.biomedcentral.com/articles/10.1186/s44399-025-00003-z

Jordana J, Ferrando A, Miró J, Goyache F, Loarca A, Martínez López OR, Canelón JL, Stemmer A, Aguirre L, Lara MAC, Álvarez LA, Llambí S, Gómez N, Gama LT, Nóvoa MF, Martínez RD, Pérez E, Sierra A, Contreras MA, Guastella AM, Marletta D, Arsenos G, Curik I, Landi V, Martínez A, Delgado JV (2016) Genetic relationships among American donkey populations: insights into the process of colonization. J Anim Breed Genet, 133:155–164. https://doi.org/10.1111/jbg.12180

Kaache B (1996) L'Origine des animaux domestiques au Maroc: état des connaissances. Préhistoire et Anthropologie Méditerranéennes 5:85–92

Kavar T, Dovč P (2008) Domestication of the horse: Genetic relationships between domestic and wild horses. Livest Sci 116(1-3):1–14

Kitchen KA (1993) The land of Punt. In: Shaw T, Sinclair P, Andah B, Okpoko A (eds) The archaeology of Africa: food, metals and towns. Routledge, London, pp. 587–608

Kimura BT, Marshall FB, Chen S, Rosenbom S, Moehlman PD, Tuross N et al (2011) Ancient DNA from Nubian and Somali wild ass provides insights into donkey ancestry and domestication. Proc R Soc Lond B Biol Sci 278:50–57

Kimura B, Librado P, Fages A, Ishida Y, Nakamura K, Orlando L (2022) The genomic history and global expansion of domestic donkeys. Science 377(6611):1172–1176

Kong Y, Wen Z, Cai X, Tan L, Liu Z, Zhao Y (2025) Genetic traceability, conservation effectiveness, and selection signatures analysis based on ancestral information: a case study of Beijing-You chicken. BMC Genomics 26:402

Lara-Chamarro JA, Jordana J (2020) New data on the evolutionary and history of donkey coloration. Universitat Autònoma de Barcelona. http://uab.cat/news/donkey-coloration

Legrand R, Tiret L, Abitbol M (2014) Two recessive mutations in FGF5 are associated with the long-hair phenotype in donkeys. Genet Sel Evol 46:1–7

Levine Marsha A (1994) The analysis of Mammal and bird remains. In Excavations at Carthage, the British Mission. Vol II(I) Hurst HR (ed), 314–317. Oxford: Oxford University Press for the British Academy

Levtzion N, Hopkins JF (eds) (1981) Corpus of early Arabic sources for West African history. Cambridge University Press, Cambridge, UK, 484 pp. ISBN 978-0521224222

Livestock Conservancy (2023 May 18) Count me in: the Livestock Conservancy launches first American Poitou census. https://livestockconservancy.org/2020/09/24/count-me-in-the-livestock-conservancy-launches-first-american-poitou-census/

Livestock Conservancy (2024) American mammoth Jackstock donkey. Heritage Breeds. https://livestockconservancy.org/heritage-breeds/american-mammoth-jackstock

Lundgren EJ, Ramp D, Stromberg JC, Wu J, Nieto NC, Sluk M, Moeller KT, Wallach AD (2021) Equids engineer desert water availability. Science 372(6541):491–495

Main D (2021, April 29) Wild horses and donkeys dig wells in the desert, providing water for wildlife. National Geographic

Manjón-Cabeza L (1933) El Garañón andaluz y sus características de raza. Andalucía Ganadera 10: 230

Marín Navas C, Delgado Bermejo JV, McLean AK, León Jurado JM, Rodríguez de la Borbolla A, Navas González FJ (2022) One hundred years of coat colour influences on genetic diversity in the process of development of a composite horse breed. Vet Sci 9(2):68

Maswana M, Mugwabana TJ, Tyasi TL (2022) Evaluation of breeding practices and morphological characterization of donkeys in Blouberg Local Municipality, Limpopo province: implication for the design of community-based breeding programme. PLoS One 17(12):e0278400

Matlhola DM, Chen R (2020) Telecoupling of the trade of donkey-hides between Botswana and China: challenges and opportunities. Sustainability 12(5):1730

Mazzatenta A, Vignoli M, Caputo M, Vignola G, Tamburro R, De Sanctis F, Mirò Roig J, Bucci R, Robbe D, Carluccio A (2021) Maternal phylogenetic relationships and genetic variation among rare, phenotypically similar donkey breeds. Genes 12(8):1109

McAnulty PA, Dayan AD, Ganderup N-C, Hastings KL (2011) The minipig in biomedical research. CRC press

McGregor DD (1990) James A. Baker Institute for Animal Health Annual Report 1990: the first forty years

Messias TBON, Araújo EDOM, Sant'Ana AMDS, Lucena JEC, Pacheco MTB, Queiroga RDCRDE (2021) Challenges and perspectives for exploiting donkey milk in the Brazilian Northeast. Cienc Rural 52:e20210058

Miraglia N (2012) Equids contribution to sustainable development in rural areas: a new challenge for the third millennium. In: Forages and grazing in horse nutrition. Wageningen Academic, pp 437–452

Mitchell RW (2015) Creativity in the interaction: the case of dog–human play. In Animal creativity and innovation (pp. 31–44). Academic Press

Mitchell PD (2017) Why the donkey did not go south: disease as a constraint on the spread of Equus asinus into southern Africa. Afr Archaeol Rev 34(1):21–41

Mogil RJ, Wegmann TG (2020) Maternal-fetal immunogenetics. In: Human Immunogenetics, pp 69–89

National Farm Animal Care Council (2025) Code of Practice for the Care and Handling of Equines (Donkeys and Mules). Retrieved Month Day, Year, from https://www.nfacc.ca/codes-of-practice/equine-code:contentReference[oaicite:29]{index=29}

Navas González FJ (2019) Structural organization of genetic diversity and ethofunctional characterization in the Andalusian donkey breed. Universidad de Córdoba, UCOPress, Córdoba, Spain

Navas González FJ, Delgado Bermejo JV, Vargas Burgos JC (2016) Current donkey production and functionality: relationship with humans [Libro 1]. UCOPress, University of Córdoba

Navas González FJ, Jordana J, León Jurado JM, Barba C, Delgado Bermejo JV (2017) A model to infer the demographic structure evolution of endangered donkey populations. Animal 11(12): 2129–2138

Navas González FJ, Jordana Vidal J, León Jurado JM, McLean AK, Pizarro Inostroza G, Delgado Bermejo JV (2018) Genetic parameter estimation and implementation of the genetic evaluation for gaits in a breeding program for assisted-therapy in donkeys. Vet Res Commun 42:101–110

Navas F, Delgado J, Vargas J (2016) Current donkey production and functionality: relationships with humans. Book, 1, Córdoba

Navas FJ, Jordana J, León JM, Arando A, Pizarro G, McLean AK, Delgado JV (2017) Measuring and modeling for the assessment of the genetic background behind cognitive processes in donkeys. Res Vet Sci 113:105–114

Navas F, Jordana J, León J, McLean AK, Delgado J (2019a) Dumb or smart asses? Donkey's (Equus asinus) cognitive capabilities share the heritability and variation patterns of human's (Homo sapiens) cognitive capabilities. J Vet Behav 33:63–74

Navas F, Jordana J, McLean A, León J, Barba C, Arando A, Delgado J (2019b) Modelling for the inheritance of multiple births and fertility in endangered equids: determining risk factors and genetic parameters in donkeys (Equus asinus). Res Vet Sci 126:213–226

Norris SL, Little HA, Ryding J, Raw Z (2021) Global donkey and mule populations: figures and trends. PLoS One 16(2):e0247830

Oklahoma State University Department of Animal and Food Sciences (2025) Poitou donkey. Breeds of Livestock. https://breeds.okstate.edu/other-breeds-of-livestock/donkeys/poitou-donkeys.html

Oldenbroek JK, van der Waaij LH (2015) Textbook animal breeding and genetics for BSc students, Centre for Genetic Resources and Animal Breeding and Genomics Centre, Wageningen UR, viewed 21 May 2025, https://www.wur.nl/en/Research-Results/Chair-groups/Animal-Sciences/Animal-Breeding-and-Genomics/Education/Textbook-Animal-Breeding-and-Genetics-for-BSc-students.htm

Peters J (1991) The faunal remains from Shaqadud. In: Marks AE, Mohammed-Ali AS (eds) The archaeology of Shaqadud: A prehistoric site in eastern Sudan. Southern Methodist University Press, Dallas, TX, pp.197–235

Pierini C, Landi V, Ciani E, Maggiolino A, Campanile D, Gomez Carpio M, De Palo P (2025) Conservation and genetic analysis of the endangered Martina Franca donkey using pedigree data. Front Anim Sci 6:1588467

Pilla F, Moioli B, Napolitano F (2012) The Martina Franca donkey: recovery and perspectives. Ital J Anim Sci 11(4):e56

Pinna W, Manca E, Mura L (2025) Conservation and genetic analysis of the endangered Martina Franca donkey. Front Anim Sci 6:1588467

Pitt D (2021) Unravelling local adaptation in the genome of livestock. Cardiff University

Polidori P, Vincenzetti S (2013) Donkey milk production: state of the art. Anim Sci Paper Rep 31(2):103–110

Quaresma MNP (2015) The Asinina de Miranda donkey breed (Equus asinus): demographic analysis and characterization of the reproductive cycles. Universidade de Tras-os-Montes e Alto Douro (Portugal)

Quaresma M, Martins AMF, Rodrigues JB, Colaço J, Payan-Carreira R (2014) Pedigree and herd characterization of a donkey breed vulnerable to extinction. Animal 8(3):354–359

Renaud G, Petersen B, Seguin-Orlando A, Bertelsen MF, Waller A, Newton R, Paillot R, Bryant N, Vaudin M, Librado P (2018) Improved de novo genomic assembly for the domestic donkey. Sci Adv 4(4):eaaq0392

Rizzi R, Tullo E, Cito A, Caroli A, Pieragostini E (2011) Monitoring of genetic diversity in the endangered Martina Franca donkey population. J Anim Sci 89(5):1304–1311

Rodríguez-Rodríguez EJ, Gil-Morión J, Negro JJ (2022) Feral animal populations: separating threats from opportunities. Land 11(8):1370

Rosenbom S, Costa V, Al-Araimi N, Kefena E, Abdel-Moneim A, Abdalla MA, Bakhiet A, Beja-Pereira A (2015) Genetic diversity of donkey populations from the putative centers of domestication. Anim Genet 46(1):30–36

Rossel S, Marshall FB, Peters J, Pilgram T, Adams MD, O'Connor D (2008) Domestication of the donkey: timing, processes, and indicators. Proc Natl Acad Sci 105(10):3715–3720

Ruiz H (1950) Bol Zootec 55:71

Santana ML, Bignardi AB (2015) Status of the genetic diversity and population structure of the Pêga donkey. Trop Anim Health Prod 47:1573–1580

Segabinazzi L, Oliveira P, Freitas-Dell'Aqua C, Papa F (2025) Alternatives for donkey cryopreserved semen. J Equine Vet Sci 145:105323

Seyiti S, Kelimu A (2021) Donkey industry in China: current aspects, suggestions and future challenges. J Equine Vet Sci 102:103642

Shen J, Yu J, Dai X, Li M, Wang G, Chen N, Chen H, Lei C, Dang R (2021) Genomic analyses reveal distinct genetic architectures and selective pressures in Chinese donkeys. J Genet Genomics 48(8):737–745

Smith DC (2008) Book of mules: selecting, breeding, and caring for equine hybrids. Rowman & Littlefield

Smith DG, Pearson RA (2005) A review of the factors affecting the survival of donkeys in semi-arid regions of sub-Saharan Africa. Trop Anim Health Prod 37(Suppl 1):1–19

Sponenberg DP (2020) Conservation of criollo livestock in the United States: challenges and solutions. Arch Latinoam Prod Anim 28(3–4):103–109

Svoboda M (2020) The hideous trade in donkey skins. The Ecologist. https://theecologist.org/2020/feb/04/hideous-trade-donkey-skins

The Brooke (2020) Invisible workers: the economic contributions of working donkeys, horses and mules to livelihoods. https://www.thebrooke.org/research/invisible-workers

The Brooke Hospital for Animals (2022 February 3) Gaps in population data put working horses and donkeys at threat. https://www.thebrooke.org/news/gaps-population-data-put-working-horses-and-donkeys-threat

The Donkey Sanctuary (2017) Under the skin: the emerging trade in donkey skins and its impacts on donkey welfare and livelihoods. https://www.thedonkeysanctuary.org.uk/resources/under-the-skin

The Livestock Conservancy (2025) Miniature donkey. The Livestock Conservancy. Retrieved Month Day, Year, from https://livestockconservancy.org/breed-profiles/miniature-donkey/:contentReference[oaicite:31]{index=31}

Todd ET, Tonasso-Calvière L, Chauvey L, Schiavinato S, Fages A, Seguin-Orlando A, Clavel P, Khan N, Pérez-Pardal L, Patterson-Rosa L, Librado P, Ringbauer H, Verdugo M, Southon J, Aury J-M, Perdereau A, Vila E, Marzullo M, Prato O et al (2022) The genomic history and global expansion of domestic donkeys. Science 377(6611):1172–1180

Universitat Autònoma de Barcelona (2020 November 18) New data on the evolutionary history of donkey coloration [Press release]. UAB Newsroom. https://www.uab.cat/web/newsroom/news-detail/new-data-on-the-evolutionary-and-history-of-donkey-coloration-1345668003610.html

Vargas JH, Zarco L, Salinas EM, Urías-Castro C, Boeta M (2025) Effects of seminal plasma on uterine polymorphonuclear dynamics and fertility of jennies inseminated with glycerol-free, frozen-thawed donkey semen. J Equine Vet Sci 149:105571

Waldie D (2017 July 26) Donkeys in distress: the rising global demand for ejiao. National Geographic. https://www.nationalgeographic.com/animals/article/donkeys-africa-china-skin-trade

Wang C, Li H, Guo Y, Huang J, Sun Y, Min J, Wang J, Fang X, Zhao Z, Wang S (2020) Donkey genomes provide new insights into domestication and selection for coat color. Nat Commun 11(1):6014

Wang Y, Hua X, Shi X, Wang C (2022) Origin, evolution, and research development of donkeys. Genes (Basel) 13(11):1945

Wang J, Li W, Yan Y, Liu S, Zhang L, Zhang X et al (2023) Donkey genomes provide new insights into domestication and selection for coat color. Nat Commun 11:6014

Wassie AM, Getachew TB, Kassa AH, Megersa AG, Aseged TA (2023) Donkey production systems and breeding practices in selected districts of South Omo Zone, Southern Ethiopia. Rangeland J 45(3):97–108

Wright J (1996) Sonqua, Bosjesmans, Bushmen, abaThwa: comments and queries on pre-modern identifications. S Afr Hist J 35(1):16–29

Zanardi C, Carvalho A, Freitas-Dell'Aqua C, Alvarenga M, Papa F, Monteiro G, Segabinazzi L (2025) Characterization of sperm traits of donkey semen cryopreserved with different permeable cryoprotectants. J Equine Vet Sci 145:105336

Zhang Y, Wang X, Li J (2024) Unraveling the genetic and physiological potential of donkeys. Mamm Genome 35(1):45–58

Strategies for Donkey Conservation and Breeding: Policy Implementation, Research, and Case Studies

Francisco Javier Navas González and Miguel Quaresma

Abstract

This chapter examines the critical strategies required for the conservation and sustainable breeding of donkey populations worldwide, emphasizing the intersection of policy frameworks, genetic management, and socio-economic considerations. It highlights the urgent need for institutional protection structures, including standardized breed definitions, comprehensive studbooks, and accurate census systems, to mitigate genetic erosion and safeguard endangered breeds. The discussion explores governmental policy gaps, the role of NGOs, and the necessity of consolidating unstandardized populations into recognized breeds while acknowledging cultural and functional diversity. It further addresses research challenges, advocating for integrated approaches that combine quantitative and molecular genetics to enhance breed resilience. Through case studies—such as Brazil's Pêga donkey and France's Baudet du Poitou—the chapter illustrates successful and emerging models for conservation, including advanced reproductive technologies and transnational collaborations. Ultimately, the work underscores that effective donkey conservation demands coordinated efforts across scientific, institutional, and community domains to ensure genetic integrity, biodiversity preservation, and the continued socio-economic relevance of these animals.

F. J. Navas González (✉)
Department of Genetics, University of Cordoba, Cordoba, Spain

Worldwide Donkey Breeds Project, Cordoba, Spain
e-mail: fjnavas@uco.es

M. Quaresma
Departamento de Ciências Veterinárias, Universidade de Trás-os-Montes e Alto Douro, Vila Real, Portugal

A. K. McLean et al. (eds.), *Donkey and Mule Medicine: Foundations, Breeds, and Management*, https://doi.org/10.1007/978-3-032-12726-6_3

Keywords

Donkey conservation · Genetic diversity · Breed standardization · Census systems · Pedigree management · Policy frameworks · Molecular genetics · Sustainable breeding · Endangered breeds · Reproductive technologies · Case studies

1 Introduction

Donkeys have played an indispensable role in human societies for millennia, serving as resilient working animals across diverse agro-ecological and cultural landscapes. Despite their historical and socio-economic importance, donkey populations face an accelerating decline worldwide due to mechanization, demographic shifts, and neglect in both policy and research. Unlike more commercially valued livestock, donkeys are often marginalized in agricultural development agendas and excluded from formal conservation strategies. This oversight has led to alarming gaps in genetic data, incomplete pedigrees, and the erosion of unique local lineages that hold cultural and adaptive significance.

Effective conservation of donkey breeds requires more than isolated welfare interventions; it demands a holistic framework integrating institutional protection, standardized breed recognition, accurate census systems, and scientifically informed breeding programs. However, these efforts are complicated by fragmented ownership patterns, parallel registries, limited governmental support, and the absence of robust research funding. Additionally, NGOs often prioritize immediate welfare outcomes over long-term genetic preservation, further widening the gap between animal rescue and sustainable population management.

This chapter explores these multifaceted challenges and proposes strategies for building coordinated conservation programs grounded in both scientific evidence and socio-cultural realities. It examines the role of policy, institutional frameworks, and research in safeguarding genetic diversity, presents innovative tools for breeding management, and illustrates best practices through case studies such as the Brazilian Pêga donkey and the Baudet du Poitou in France. By addressing these issues, the chapter seeks to establish a roadmap for ensuring the long-term viability, genetic integrity, and cultural heritage of donkey populations worldwide.

2 Safeguarding Genetic Heritage: Building Institutional Protection Frameworks and Strengthening Census Systems for Donkey Breeds and Populations Conservation

2.1 The Urgent Need for Institutional Protection Structures

The absence of protection structures like breed standards, breeder associations, studbooks, and breeding or conservation programs significantly jeopardizes the integrity and survival of donkey breeds (Navas et al. 2017a, b). Without breed standards, there is no formal guideline to preserve the unique characteristics and qualities of each breed, leading to the risk of crossbreeding and dilution of genetic traits derived from the loss of alleles present in the gene pool of breeds (Bunevski et al. 2018). Breeder associations, which typically advocate for the interests of specific breeds, are vital for promoting responsible breeding practices and raising awareness about breed-specific needs. The lack of such associations means there is no organized body to champion the causes of these breeds, potentially leading to their neglect and decline (Quaresma et al. 2014).

Additionally, the absence of studbooks and breeding programs hampers efforts to track and maintain the lineage and health of donkey breeds. Studbooks (with first donkey studbooks or preferably official registries dating back to 1880s–1940s) are essential for recording the pedigree and ensuring the purity of the breed, while conservation and breeding programs focus on preserving and enhancing breed characteristics, respectively (Navas González 2019).

Without these tools, it becomes challenging to manage genetic diversity and prevent inbreeding and the deterioration of the genetic health of populations, which can lead to health problems concerning reproduction and immune system compromises and the eventual extinction of rare breeds, strains or varieties. Conservation programs and breed contests also play a crucial role in recognizing and rewarding breeders and owners' efforts to maintain breed standards, thus encouraging more people to participate in preservation activities. Without these protective measures, many donkey breeds face an uncertain future, with their unique traits and historical significance at risk of being lost forever.

2.2 Inaccurate or Incomplete Information from Official Sources

The lack of up-to-date or accurate information provided by official sources, such as the Food and Agriculture Organization (FAO) of the United Nations, poses significant challenges for the effective management, conservation, and sustainable use of donkey breeds worldwide (Norris et al. 2021). Reliable and current data are essential for understanding population dynamics, assessing genetic diversity, and monitoring the overall health and viability of donkey populations (Kindsvater et al. 2018). These data underpin scientific research, inform policy development, and guide conservation strategies. When official sources fail to deliver accurate or timely information, it severely undermines efforts to protect and manage donkey populations effectively (Kindsvater et al. 2018).

Inaccurate, outdated, or incomplete data can result in misguided policies and ineffective management strategies that do not reflect the actual conditions or needs of donkey populations. This misalignment can lead to inappropriate allocation of resources, insufficient protection of endangered or vulnerable breeds, and the neglect of critical areas in need of intervention. For example, areas experiencing rapid declines in donkey populations may go unnoticed, while breeds facing genetic erosion may not receive the conservation attention they urgently require.

Moreover, this information gap directly impacts the ability of key stakeholders—including breeders, researchers, animal welfare organizations, and policymakers—to make informed decisions. Without a solid, evidence-based understanding of donkey demographics and distribution, it becomes nearly impossible to implement targeted breeding programs, prioritize conservation efforts, or support rural communities that depend on donkeys for their livelihoods. The absence of robust data impairs efforts to track trends over time, assess the effectiveness of current interventions, or anticipate future challenges, especially in the face of climate change and socio-economic pressures (Norris et al. 2021).

Encouragingly, the FAO has recently begun to recognize these challenges and is taking steps to address the data gap. Through initiatives such as the Domestic Animal Diversity Information System (DAD-IS), the FAO is working to collect, consolidate, and make accessible comprehensive data on donkey breeds at both national and global levels. These efforts are gradually improving the availability and quality of information regarding donkey populations, including breed status, geographical distribution, and population trends (Norris et al. 2021; Kindsvater et al. 2018). The FAO is also increasingly engaging with national governments, research institutions, and local communities to enhance the accuracy and coverage of reporting.

While these steps mark important progress, there remains a pressing need for the FAO and its partners to accelerate and expand their data collection efforts. This includes standardizing methodologies, ensuring regular updates, incorporating local knowledge, and promoting transparency and open access to data. Only through such coordinated and sustained efforts can the global community ensure that decision-making around donkey conservation and management is grounded in reliable evidence (Kindsvater et al. 2018).

2.3 The Role of Breed Standardization in Effective Conservation

Standardizing the process of transitioning from undefined diverse donkey populations to recognized standardized breeds, lines, or ecotypes is a critical step in improving breed management and conservation efforts (Kugler et al. 2008). This standardization process involves the multilevel characterization of donkey populations (socioculturally, zoometrically, genetically, phenotypically, ethologically, reproductively, and functionally among others) establishing clear criteria for particular breed characteristics, such as zoometric traits (Cecchi et al. 2007; De Berardinis et al. 2024; Labbaci et al. 2018; Ljubodrag et al. 2015; Mustefa et al. 2020; Susana Lopes et al. 2023; Urosevic et al. 2019), sports, work and performance metrics (Aroua et al. 2024; Ivanković et al. 2023; Navas González et al. 2018b; Polidori et al. 2009a, b; Yılmaz et al. 2012), behavioral traits (Canisso and McDonnell 2010; De Santis et al. 2021; Mazzatenta et al. 2019; Navas et al. 2019a, b, 2017a, b), resistance and hypersensitivity features (Navas González et al.

2018a, b) or even reproductive traits (Mazzatenta et al. 2019; Navas et al. 2018; Quaresma 2015; Tosi et al. 2013) which can help facilitate more effective breeding programs. By creating standardized classifications, stakeholders can enhance the understanding of genetic diversity, promote responsible breeding practices, and ensure the preservation of unique traits within specific populations. Furthermore, this process can aid in the development of breeding guidelines that prioritize the health and welfare of donkeys, ultimately contributing to the sustainability of these breeds.

Implementing a standardized approach also allows for better tracking and monitoring of donkey populations, which is essential for assessing their health and viability (Bunevski et al. 2018). It can provide a framework for data collection, enabling researchers and conservationists to analyse population dynamics and genetic diversity more effectively (Delgado et al. 2018). This standardized classification can enhance collaboration among breeders, researchers, and policymakers, fostering a unified approach to the management and conservation of donkey breeds. By establishing a clear pathway from diverse populations to standardized breeds, stakeholders can work towards ensuring the long-term sustainability and viability of donkey populations in an ever-changing world (Delgado Bermejo et al. 2019).

2.4 The Critical Importance of Comprehensive Census and Distribution Data

Much of the information about donkey breed censuses and their distribution remains incomplete or non-existent, posing significant challenges to conservation and breeding efforts (Kugler et al. 2008). The lack of comprehensive data makes it difficult to assess the population size, genetic diversity, and geographical distribution of different breeds accurately. Without this critical information, it is nearly impossible to develop effective conservation strategies or monitor the health and viability of donkey populations. The absence of reliable data also hinders the ability to identify and protect endangered breeds, leaving them vulnerable to extinction (Carneiro et al. 2018; Navas et al. 2017a, b; Norris et al. 2021; Starkey and Starkey 2000).

Furthermore, the lack of thorough documentation and research on donkey breeds limits the potential for scientific advancements in breeding and genetics (McLean and Navas Gonzalez 2018). Detailed censuses and distribution records are essential for understanding the unique characteristics and needs of each breed, which in turn informs targeted breeding programs and conservation initiatives. The current gaps in knowledge reflect a broader issue of neglect and underinvestment in donkey research compared to other livestock species. Addressing these gaps requires coordinated efforts to conduct comprehensive surveys and establish centralized databases, ensuring that all stakeholders have access to accurate and up-to-date information on donkey breeds worldwide (Almejnah et al. 2023).

2.5 Multiple Parallel Extra Official Associations and Registries

Parallel associations that operate alongside the officially recognized bodies for the management of donkey breeds can significantly impact conservation and breeding efforts. Administratively, there should always be only one official breeder association in charge of the management of the studbook and breeding programs, this means in charge of the control of the individuals in the population. However, these parallel associations often arise due to differing interests and goals, which can lead to several challenges and complications in effectively preserving and improving donkey breeds. One of the most notable consequences is the fragmentation and misconnection of owners and breeders which in turn ends up hindering conservation and breeding efforts, where resources, including funding, expertise, public attention and manpower, are divided between the official association and the parallel ones (Akin et al. 2018). This division can dilute the overall effectiveness of conservation programs, as each association may struggle with limited resources. Additionally, different associations might adopt varying strategies and goals for conservation and breeding, leading to inconsistent practices and hindering the establishment of a unified and effective approach necessary for the long-term preservation of donkey breeds (Demirbük and Kızılaslan 2019).

Alternatively, parallel registries can be handled by the same association to respond to the anthropological, sociocultural, and administrative situations that may arise. For instance, the studbook of the Baudet du Poitou was created in 1884 and managed by La SABAUD, whose main objective was to safeguard the Poitou donkey. The book comprised two registries: Book A, where pure-bred (thoroughbred) Poitou donkeys were registered, and Book B, which gathered those not fully pure-bred—offspring of one Poitou donkey crossed with a donkey of a non-Poitou ancestry. Because the Poitou morphology dominates, these cross-bred animals usually resemble the pure-bred type. Since 2000, Poitou donkeys born have been subject to paternity testing using genetic markers to ensure the integrity of genealogical information. SABAUD's work concluded in 2019, after which its herd of 25 donkeys was donated to Races Mulassières du Poitou. This new association now represents and promotes several rare breeds—"Mulassières," "Poitou Donkeys," and "Mules of Poitou"—with the dual aim of safeguarding the Poitou donkey and the "trait mulassier" (a French cob–type draft horse), as well as fostering the cross-breeding that produces the "mule poitevine." (Fig. 1).

According to ASNOPRA association conservation programme (ASNOPRA n. d.), in the Andalusian donkey breed, the functioning of the studbook follows a structure designed to support conservation efforts for this critically endangered population. Animals are identified using electronic microchips in accordance with national regulations, and the studbook is divided into an Annex Section—for animals with unknown or partially known genealogy—and a Main Section, which includes candidates, approved reproducers, evaluated reproducers, and elite reproducers. Parentage verification is mandatory through genetic testing, and all breeding animals must meet morphological evaluation thresholds. In some cases, offspring from the Annex Section may be promoted to the Main Section to help

Fig. 1 French Baudet du Poitou (upper) and Poitevin mule (lower)

increase population numbers. Due to the breed's endangered status, the association reserves the right to restrict the use of individual animals or reproductive material to preserve genetic diversity. While this structure is tailored to the specific needs of the Andalusian donkey, it broadly follows the general model established for the management of endangered native breeds in the European Union. Most official studbooks are organized into similar sections (foundational or annex, main, and elite), with mandatory animal identification, parentage verification—often by DNA—and selection based on performance or morphological evaluations. However, the specific criteria and thresholds, as well as the flexibility for promotion between sections, are often adjusted based on the demographic and genetic situation of each breed.

Among other examples of parallel registries, the American Donkey/Jackstock Registry (ADR), established in 1967 by Paul and Betsy Hutchins, registers any Miniature donkey whose sire or dam—or both—are not already recorded in the Miniature Donkey Registry (MDR), provided the animal meets the breed standard. First-generation jacks and jennets require inspection by a qualified inspector or veterinarian, whereas geldings are exempt. Donkeys in the ADR may carry either a documented pedigree or an "untraced" pedigree designation. There is one pathway for ADR-registered donkeys to upgrade into the MDR: by successive breeding to MDR stock. An ADR donkey bred to an MDR donkey produces offspring that, if each subsequent generation continues to be bred back to MDR animals, will—by the fourth generation overall (third generation from ADR × MDR matings)—become eligible for MDR registration. For instance, MDR × ADR; ADR × MDR; ADR × MDR; and ADR offspring qualify for MDR upgrade.

At National Miniature Donkey Association (NMDA)–sanctioned shows, both MDR and ADR donkeys are eligible to compete, provided they hold current registration in one of these two books. Unregistered donkeys may still appear in open shows but are not permitted in NMDA-sanctioned classes. Proper genetic management is crucial for maintaining the health and resilience of donkey breeds. However, when parallel associations do not coordinate their efforts, there is an increased risk of inbreeding within the smaller groups managed by each association, reducing genetic diversity and increasing susceptibility to disease (Sponenberg and Bixby 2007). Accurate record-keeping is vital for tracking lineage and genetic health, but separate registries may maintain disconnected records, leading to inconsistencies and gaps in data. The lack of data sharing further complicates efforts to maintain comprehensive genetic and health records, which are essential for informed breeding decisions. Moreover, multiple associations can adopt divergent breeding and care standards, resulting in uneven enforcement of policies and market confusion among buyers and breeders uncertain which certifications to trust. Paradoxically, some purebred individuals may remain unregistered while others that fail to meet rigorous criteria are admitted, due to discrepancies among associations' requirements, strategical reasons or even personally biased interests. This fragmentation hampers the creation of a cohesive breeding population, making it difficult to manage and sustain donkey populations efficiently. To address these challenges, collaboration and communication between associations are crucial, as is the establishment of umbrella organizations to coordinate efforts and promote data-sharing agreements. However, the effectiveness of additional organizations may be limited by their lack of official status (Demirbük and Kızılaslan 2019).

2.6 Ensuring Pedigree Completeness for Genetic Integrity

Remarkably incomplete pedigrees in standardized donkey breeds present significant obstacles to effective breeding and conservation efforts (Navas et al. 2017a, b). Pedigrees are essential for tracking the lineage and genetic history of individual donkeys, ensuring that breeding practices maintain the purity and desirable traits of the breed. When pedigrees are incomplete or missing, it becomes difficult to make informed breeding decisions, making the reliability of the estimation of conservation parameters such as inbreeding or coancestry increase steadily and the response to conservation strategies to become sharply inefficient. This in turn leads to misinformed, frequently inappropriate, breeding decisions which instead of counteracting the effects of inbreeding promote it and with it the incidence of reproductive or immune linked genetic disorders, strongly affecting overall genetic health of donkey populations at a large scale in a relatively reduced amount of time, but whose effects can be patent in the medium to long term decades after. For this, the lack of comprehensive pedigree information undermines the integrity of the breed, potentially leading to a decline in health, but also of productive performance, and overall breed standards which can no longer possibly be met (Gutiérrez et al. 2005; Quaresma et al. 2014; Santana and Bignardi 2015).

Furthermore, incomplete pedigrees impede the ability to implement and monitor conservation programs effectively. Accurate pedigrees are crucial for identifying and preserving genetic diversity within standardized breeds, which is essential for their long-term viability (Baumung and Sölkner 2003). Without reliable lineage information, it is challenging to manage breeding populations and protect against the loss of unique genetic traits. To address these issues, there is a pressing need for rigorous documentation and maintenance of pedigrees within standardized donkey breeds. Establishing robust record-keeping systems and encouraging breeders to meticulously document and share pedigree information will help safeguard the genetic heritage and future of these valuable breeds (Baumung and Sölkner 2003; Gutiérrez et al. 2005; Navas et al. 2017a, b; Toro et al. 2011).

2.7 The Need for Distinct Census Data on Donkeys, Mules, and Hinnies

The practice of not providing separate official figures for donkey and mule censuses poses a significant challenge. Accurate data on their populations is crucial for effective management, conservation, and utilization strategies. Without clear distinctions in these figures, it becomes difficult to address the specific needs and health concerns of each group. Moreover, the absence of data on hinnies, another hybrid equine, further complicates the understanding of the broader equine population dynamics (Rowan).

Addressing these gaps in data collection would enhance the ability to support and sustain the vital roles that donkeys and mules play in various communities around the world. Donkeys are often used for transportation and agricultural work, especially in developing regions. Mules and hinnies, known for their strength and endurance, are also essential in similar roles and in more challenging terrains where other species might not perform as well (McLean et al. 2019).

Collecting and analysing specific data for donkeys, mules, and hinnies would enable better-targeted health interventions, improved breeding programs, and more effective resource allocation. For instance, understanding the population size and distribution of each group can help in planning vaccination campaigns, disease control measures, and nutritional support. Additionally, accurate data can aid in the development of policies and programs aimed at conserving these animals, many of which face threats from overwork, neglect, and environmental changes (Smith 2008).

2.8 Atomization of Donkey Breeder and Ownership and Its Genetic Risks

The issue of population atomization in donkey breeds, particularly those under conservation and breeding efforts, presents a significant challenge. Typically, donkey ownership is characterized by owners handling one or two animals. This widespread distribution helps maintain genetic diversity and reduces the risks associated with inbreeding. However, when only a few breeders accumulate large numbers of animals, the population becomes highly centralized. This concentration can lead to a bottleneck effect, where the genetic pool is narrowed, increasing the likelihood of hereditary health problems and reducing the overall resilience of the breed. Additionally, the centralization of the population in the hands of a few breeders can result in less genetic variation, making the population more vulnerable to diseases and environmental changes (Colli et al. 2013; Legha et al. 2023; Navas et al. 2017a, b).

Moreover, the concentration of donkeys within a limited number of breeding operations can undermine broader conservation efforts. When most animals are held by a few breeders, there is a risk that the genetic traits represented may not fully capture the diversity of the breed. This can lead to a homogenization of the population, where unique and valuable genetic traits are lost. The atomization of the population thus exacerbates the challenges of implementing effective conservation strategies. It is crucial to encourage a more equitable distribution of animals among a larger number of breeders to maintain genetic diversity and ensure the long-term sustainability of donkey breeds under conservation. This approach not only supports genetic health but also fosters a more resilient and adaptable population (Colli et al. 2013; Navas et al. 2017a, b).

3 Governmental Management and Policy Considerations

Donkey breeding and conservation practices are shaped as much by cultural context as by biological imperatives. Local communities often prioritize traits like resilience, endurance, and temperament based on traditional knowledge, environmental demands, and social valuation, rather than standardized genetic criteria (Wassie et al. 2023; Maswana et al. 2022). These informal systems of animal husbandry, while highly adaptive, frequently operate beyond the scope of institutional oversight. Moreover, donkeys are often viewed as culturally marginal animals, which further influences the low prioritization of their conservation in both public perception and policy agendas (Hall 2019). As emphasized in global livestock policy frameworks, meaningful conservation efforts must connect local practices with institutional support mechanisms that include coordinated breed characterization, state recognition, and inclusive stakeholder governance (FAO 2007). Bridging grassroots breeding knowledge with formal policy frameworks is thus essential to developing sustainable and context-sensitive conservation strategies that ensure the long-term viability of donkey genetic resources.

3.1 Census Inaccuracies Involving Donkeys, Mules, and Hinnies

Accurate census data for donkeys, mules and hinnies are frequently hindered by the tendency to group them together under a single category, overlooking the distinctions between the two species and entirely omitting hinnies. This combined categorization creates significant issues for understanding population dynamics, health needs, and the economic roles of each species. Donkeys and mules, while related, have distinct characteristics, uses, and care requirements that are essential to recognize for effective management and welfare practices. The exclusion of hinnies, a less common hybrid, further skews census results, leading to an incomplete view of the hybrid population and underrepresenting their specific needs and contributions (Allan 2022; McLean et al. 2019).

In practice, many countries do not count equids in detail—governments often list them only generically or exclude them altogether (Allan 2022). Hybrids like hinnies are particularly vulnerable to being overlooked in national surveys. Population statistics that fail to distinguish between donkeys and mules can distort veterinary research, nutritional programs, and breeding practices, as the two species have different vulnerabilities and uses. Moreover, the absence of comprehensive data on hinnies means that any unique health or welfare considerations for this hybrid are neglected, leaving gaps in animal care knowledge (The Brooke 2020).

This lack of differentiation and undercounting leads to broader misrepresentations that affect funding, research priorities, and the resources allocated to these animals. As Allan (2022) notes, "horses, donkeys and mules are often excluded from livestock censuses," creating severe data gaps. Crucially, the poor data feeds into policy: working equids are often excluded from animal health strategies, meaning that veterinary services and welfare programs do not account for them appropriately (Bonsi and Anderson 2023). Without accurate census data, policymakers cannot identify needs or allocate resources effectively for donkeys, mules, and hinnies.

To address these challenges, more precise census practices are needed to distinguish among donkeys, mules, and hinnies, enabling tailored approaches to care, management, and resource allocation (Allan 2022; Bonsi and Anderson 2023).

3.2 Global Need for Consolidation of Unstandardized Populations into Standardized Breeds

Donkey "breeds" are seldom formalized the way cattle or horses are; instead, local populations are usually known by region or appearance. Over centuries of crossing and trade, distinct donkey types have blended, so that few standardized breeds exist outside of Europe. Genetic surveys confirm this: donkey populations have been declining for decades and "the diversity contributed by autochthonous gene pools has been eroded," meaning unique local lineages are being lost (Rambaldi Migliore et al. 2024). For instance, high genetic diversity in some domestic donkeys is undermined by introgression from wild asses and uncontrolled breeding. Rambaldi

Migliore et al. (2024) emphasize that this erosion of genetic distinctiveness is global, and they call for identifying hotspots of variation to conserve local donkey lineages.

The vast majority of the global donkey population remains unstandardized, largely shaped by geographical, political, and cultural influences. This lack of formal breed recognition fosters an environment where introgression—crossbreeding between different types of donkeys—is widespread. In the absence of established breed standards, the distinct traits that define individual breeds are not systematically preserved, leading to the erosion of genetic purity and breed identity (Rambaldi Migliore et al. 2024). This issue is particularly pronounced in regions such as Central Africa and parts of Central and South America, where local breeding practices are often driven by immediate economic or practical needs rather than long-term conservation goals. In these areas, political instability can further hinder the development and enforcement of structured breeding programs. The fragmented nature of donkey breeding in such regions not only limits opportunities for breed development but also poses a significant threat to the preservation of genetic diversity (Rambaldi Migliore et al. 2024).

Compounding the problem is the tendency to name donkey populations after the countries in which they are found, with only anecdotal references to ecotypes or subtypes. This practice reflects the absence of systematic classification and fails to account for the subtle but significant variations that may exist within national populations. In some cases, heterogeneous donkey populations are even named after the human communities or occupational groups that traditionally manage them—such as the Zabbaleen donkeys in Egypt managed by Zabbaleen Coptic Christians in Cairo (Egypt)—further underscoring the lack of formal taxonomy. While such naming conventions offer cultural and historical insights, they rarely reflect clear genetic or morphological criteria. Without formal recognition of local adaptations or subtypes, valuable genetic and phenotypic diversity risks being overlooked or lost, further complicating conservation efforts (Rambaldi Migliore et al. 2024).

To mitigate these challenges, there is a pressing need for international collaboration aimed at establishing comprehensive breeding and conservation strategies. These efforts must be culturally sensitive and adaptable to local conditions while emphasizing the importance of maintaining the unique characteristics of distinct donkey breeds (Rambaldi Migliore et al. 2024).

3.3 Economical Support of Endangered or Non-Productive Breeds

Donkeys are often economically invisible in policy. In most subsidy schemes they count only as generic livestock units if at all, and rarely produce marketable goods, so their value goes unrecognized. Indeed, one study notes that working equids are "excluded by governments and international organisations from animal health policies and interventions", which means subsidies and welfare support ignore them (Bonsi and Anderson 2023). Similarly, research in Ethiopia found that donkeys remain “absent from the government’s nonhuman animal health and welfare policies,” overlooked in development goals and undervalued by society (Geiger 2024). This lack of economic benchmarking—no formal metrics for a donkey’s contribution—prevents policymakers from including donkeys in conservation or subsidy programs. Advocates argue that policy frameworks must be reformed: government schemes should explicitly recognize donkeys’ roles and establish support payments or conservation funds for endangered local breeds (Geiger 2024; Bonsi and Anderson 2023).

A major challenge in securing governmental support for donkey breeds—especially endangered or functionally misallocated ones—stems from how agricultural subsidies are typically calculated. Most government subsidy systems are based on productivity comparisons. That is, they assess the income potential of a highly productive breed (such as a high-yield dairy cow or fast-growing meat breed) and compare it to that of a less productive or endangered breed. The income gap is then partially or fully offset through subsidies, helping to keep the lower-yielding breed economically viable (Bonsi and Anderson 2023) following the evaluation of what it is called *loss of profit* calculation.

However, this approach becomes problematic when it comes to donkeys, particularly breeds that are either endangered or no longer used in traditional roles (such as transport or agriculture). These functionally misallocated animals no longer generate measurable economic output in a modern industrial context. As a result, governments find it difficult to calculate meaningful comparisons or justify financial support, since there is no baseline of “productive income” to measure against. Without a clear economic benchmark, the normal subsidy mechanisms break down (Geiger 2024).

In these cases, either no subsidies are granted, or the support is minimal—often just enough to cover basic feed costs, if that. This insufficient assistance makes it nearly impossible for keepers of these breeds to maintain them, leading to further decline in already vulnerable populations.

The absence of a standardized framework for recognizing the non-economic value of donkey breeds—such as their cultural, ecological, or historical importance—further exacerbates the problem. As a result, many governments overlook the broader significance of these animals and focus instead on livestock that offer direct economic returns (Geiger 2024).

To address this gap, there is an urgent need for policy reform. Governments must develop new evaluation frameworks that account for the multifaceted contributions of donkey breeds—beyond just productivity. These should include their roles in cultural heritage, landscape management, biodiversity, and community resilience. By adopting a more holistic approach, policymakers can create conservation strategies and financial support systems that sustain endangered donkey populations and preserve their value for future generations (Bonsi and Anderson 2023; Geiger 2024).

3.4 The Importance of Function Determination

The lack of function determination in donkey breeds poses a significant challenge to their conservation and utilization, as even authoritative organizations like the Food and Agriculture Organization (FAO) do not officially recognize the various roles that these animals play, such as in work, sports, meat production, or milk production in databases such as DAD-is because they are not officially recognised by national authorities from the country of origin of donkey breeds (Norris et al. 2021). This oversight undermines the potential value of donkey breeds and their contributions to agriculture and society. Donkeys are versatile animals, known for their strength and endurance, making them suitable for a variety of tasks, including riding, packing, and agricultural work (The Brooke Hospital for Animals 2022). However, without formal recognition of their specific functions, the unique characteristics and advantages of different breeds may be overlooked, leading to a decline in interest and investment in their conservation (Gutiérrez et al. 2006).

The absence of clearly defined roles for donkey breeds can also hinder efforts to promote sustainable breeding practices and develop targeted breeding programs. By not acknowledging the specific functions that different breeds fulfill, organizations and breeders may struggle to implement effective conservation strategies that enhance genetic diversity and improve breed performance (Quaresma et al. 2014). Furthermore, this lack of recognition can limit research funding and resources, making it more difficult to advance knowledge about the potential uses and benefits of donkeys in various contexts (The Donkey Sanctuary n.d.). To address these issues, it is crucial for stakeholders to advocate for a more comprehensive understanding of donkey breeds and their functions. By recognizing and promoting the diverse roles that donkeys can play in agriculture, sports, and other sectors, it becomes possible to bolster their conservation and ensure their continued contributions to society (The Brooke Hospital for Animals 2022).

3.5 Comprehensive Multilevel Breed Characterization

The implementation of policies aimed at the zoometric, functional, socio-economic, anthropological, reproductive, and organizational characterization of donkey breeds should be approached progressively. Such a comprehensive policy framework is essential for ensuring that the diverse roles and characteristics of donkeys are effectively understood and managed. By gradually introducing these policies, stakeholders can assess their impacts, adapt to changing circumstances, and refine their approaches based on ongoing research and feedback from the community (Norris et al. 2021). This stepwise implementation is particularly important given the complexity of donkey breeds, which can vary significantly in size, function, and cultural significance across different regions (Maswana et al. 2023).

Progressive policy implementation allows for the integration of scientific knowledge and local practices, ensuring that the unique needs of donkey populations and the communities that rely on them are met. For instance, understanding the reproductive traits of specific breeds can inform breeding programs, while socio-economic analyses can highlight the value of donkeys in local economies (Quaresma et al. 2014). Furthermore, incorporating anthropological perspectives can help preserve the cultural significance of donkeys within communities. By fostering collaboration among researchers, policymakers, and local stakeholders, these policies can promote the sustainable management of donkey breeds, ultimately enhancing their welfare and contributing to their conservation for future generations.

3.6 NGOs' Lack of Attention to Donkey Genetic Resources Conservation

Non-governmental organizations (NGOs) play a critical role in improving donkey welfare around the world, especially in regions where mistreatment and neglect are rampant. Their efforts in rescue, rehabilitation, castration, and rehoming have substantially enhanced the lives of countless animals, particularly in low-income countries where donkeys are integral to agriculture, transportation, and daily life under harsh and demanding conditions (The Donkey Sanctuary n.d.; The Brooke Hospital for Animals 2022). However, despite their undeniable contributions to immediate welfare, many NGOs fall short in addressing a crucial aspect of long-term donkey conservation: the preservation and responsible management of genetic resources (Gutiérrez et al. 2006).

A significant gap in current NGO practices is their general exclusion of breeding of purebred animals from their core activities. In high- and medium-income countries, many NGOs focus on removing donkeys from neglectful environments and routinely castrate males to prevent unwanted reproduction. This strategy is often driven by a desire to control overpopulation and mitigate further animal suffering, which, from a welfare perspective, is entirely understandable (The Donkey Sanctuary n.d.). Yet, this approach is frequently executed without any genetic evaluation, overlooking the broader consequences of removing healthy donkeys from the gene pool. Particularly in cases involving rare or endangered donkey breeds, this could result in the permanent loss of valuable genetic lines that are essential for breed survival and resilience (Gutiérrez et al. 2006).

The absence of genetic oversight in these programs has serious long-term implications for the sustainability of purebred donkey populations. When purebred donkeys are removed from the breeding pool without genetic evaluation, the population's overall genetic diversity diminishes (Norris et al. 2021). This reduced diversity weakens the resilience of the population, making it more vulnerable to diseases, environmental changes, and other threats. Furthermore, diminished genetic diversity increases the risk of inherited disorders and limits the potential for future genetic improvement (Gutiérrez et al. 2006). Such outcomes are especially concerning for endangered or lesser-known breeds, where every individual may carry unique or critical genetic traits that could play a key role in the breed's survival (Quaresma et al. 2014).

Moreover, the way many NGOs define their role in animal welfare further compounds this issue. Several organizations explicitly exclude breeding from their mandates, even when they rescue pregnant jennies or animals that could contribute positively to conservation efforts. This exclusion often arises from limited resources, a lack of expertise in reproductive management, or a belief that breeding is a distraction from the immediate task of rescuing and rehabilitating animals (The Donkey Sanctuary n.d.). However, this well-intentioned focus on short-term welfare often undermines long-term conservation goals. By disregarding reproductive potential, NGOs inadvertently hinder the recovery and stabilization of endangered donkey populations, preventing the development of sustainable breeding programs that could help secure their future (Gutiérrez et al. 2006).

The approach taken by some organizations, which prohibits all breeding and foaling unless jennies are already pregnant, adds another layer of complexity to this challenge. While this policy may be intended to control population growth and prevent overpopulation, it inadvertently hinders efforts to preserve endangered donkey breeds. For breeds with small, isolated populations, the restrictions on breeding drastically reduce the opportunities to produce offspring, accelerating the genetic erosion of these already vulnerable groups (Quaresma et al. 2014). The lack of reproduction directly impacts the ability to maintain essential breed traits, increasing the risk of extinction for rare donkeys (Norris et al. 2021).

To address these challenges, a more holistic and integrated approach to donkey welfare and conservation is needed. NGOs must recognize that while immediate welfare is a priority, the long-term survival of donkey breeds hinges on the preservation of genetic diversity. This shift in perspective does not require NGOs to engage in large-scale breeding operations. Rather, they should embrace the idea of responsible genetic management and recognize the importance of genetic evaluation as part of their rescue and rehabilitation efforts (The Donkey Sanctuary n.d.). By collaborating with equine geneticists, veterinarians, and experienced conservation breeders, NGOs can make informed decisions about which animals should be retained for breeding programs and which should be excluded from the gene pool, based on their genetic value (Gutiérrez et al. 2006).

Such partnerships could also facilitate the creation of centralized genetic databases or studbooks, which would track the lineage of individual donkeys and ensure the prevention of inbreeding or the loss of valuable bloodlines (Quaresma et al. 2014). By combining genetic management with their welfare initiatives, NGOs can help stabilize the population of endangered breeds, making informed decisions that protect genetic diversity without compromising animal welfare (Norris et al. 2021).

A more tailored and nuanced approach to breeding is critical for endangered donkey populations. Rather than a blanket ban on breeding, NGOs could implement controlled, responsible breeding programs specifically designed to protect endangered breeds. These programs would aim to strike a balance between population control and genetic preservation (Gutiérrez et al. 2006). By ensuring that breeding is conducted in a way that maintains genetic diversity and protects unique breed characteristics, NGOs can play a key role in safeguarding the future of these vulnerable animals (Quaresma et al. 2014).

Encouraging controlled breeding among jennies in endangered breed populations would not only ensure the continuation of these breeds but also help to restore genetic health, prevent inbreeding, and support biodiversity goals for donkeys worldwide. This strategy could be integrated with careful management protocols, including veterinary oversight and genetic counseling, to ensure that the overall welfare of the animals is maintained while contributing to their long-term conservation (The Brooke Hospital for Animals 2022).

4 Research and Development Framework

Government policy failures in donkey conservation—including outdated or incomplete livestock censuses, weak legal recognition of donkeys, and minimal financial support—leave critical information and incentive gaps that only targeted research can fill. In many low-income countries, reliable equid population data are simply not collected: Brooke (2022) reports that most nations lack recent livestock censuses, so that governments cannot spot worrying trends and threats to donkey numbers. At the same time, donkeys are seldom prioritized in policy, with planners giving them little formal status in livestock plans. For example, working donkeys frequently face poor welfare conditions often due to lack of recognition in policy frameworks like the national livestock master plan (Kithuka et al. 2025). This oversight extends to subsistence offtake—subsidies or incentives for donkey breeders are virtually nonexistent outside a few exceptions (some Chinese provinces have only recently offered breeding subsidies)—and nongovernmental groups alone lack the resources to substitute for official programs. Indeed, Brooke (2022) notes that working animals are not prioritised enough and that there is minimal funding for much-needed research into equid livelihoods. These policy deficiencies have already prompted experts to call for an evidence-based response and livestock authorities now explicitly urge new research on donkey breed status and breeding programs to halt steep population declines.

4.1 Overcoming Academic and Researchers Neglect

Donkeys are often undervalued relative to other livestock, despite their crucial roles in agriculture and transport. As one review notes, "donkeys, often overlooked in comparison to other domesticated animals like horses or cattle, have played a crucial role in human societies" (Rowan 2022). In practice, donkey welfare and management frequently receive little institutional attention. For example, a recent study in Malawi found that the "fate of donkeys ... has been overshadowed by larger and more glamorous livestock," and that donkey husbandry has "received limited attention" in policy and statistics (Jere et al. 2025). This neglect translates into minimal funding and scarce literature: Jere et al. (2025) report a "dearth of literature on donkey production and marketing practices" in Malawi. Without more research highlighting donkeys' economic and cultural importance, public awareness remains low and conservation advocacy is hampered (Aroua et al. 2025; Jere et al. 2025). Increasing academic focus on donkey genetics, biology, and management is essential to elevate awareness and attract the resources needed for their conservation (Geiger et al. 2020a, b; Bordonaro et al. 2012).

Although, dependent on the area, the situation is progressively improving, research groups often pay only anecdotal attention to the study of donkey populations, primarily due to a perceived lack of value and interest in these animals (Geiger et al. 2020a, b; Bordonaro et al. 2012). This neglect can lead to significant gaps in knowledge regarding the biology, genetics, and conservation of donkey breeds. While donkeys have played essential roles in agriculture, transportation, and cultural heritage for centuries, their contributions are frequently overshadowed by more economically prominent livestock species (Rossel et al. 2008; Geiger et al. 2020a, b). Consequently, institutional funding for research on donkey populations is often minimal or entirely absent, as these animals are considered less relevant than other livestock species. This lack of institutional support forces researchers to rely on particular or private funding sources, which may not always be sufficient to cover comprehensive studies and conservation efforts (Geiger et al. 2020a, b; Snedecor and Cochran 1989).

The perception that donkeys lack value not only affects academic interest but also influences public awareness and appreciation for these animals. Without robust research to highlight the importance of donkey breeds, it becomes challenging to advocate for their conservation and demonstrate their ecological and economic significance (Geiger et al. 2020a, b). This situation underscores the need for increased awareness and a paradigm shift in how research communities view donkeys. By promoting the study of donkey populations and emphasizing their unique attributes and contributions, researchers can foster a greater appreciation for these breeds and help secure the resources necessary for their conservation and management.

4.2 Compulsory Use of Combination of Quantitative and Molecular Genetics

The conservation of threatened taxa, including critically endangered domestic animals such as donkey breeds, requires integrated management approaches that address both genetic integrity and demographic stability. The overarching goal is to preserve genetic diversity, maintain reproductive viability, and enhance the adaptive potential of populations, whether *in situ* or *ex situ*. There has always existed a controversy on how rather old quantitative techniques may have lost their relevance in favour or rather recent genomic tolos available today. In this context, pedigree analysis and the strategic use of genomic tools are both essential, and their better suitability for informed conservation planning and sustainable population management may depend on the characteristics of each donkey population, standardized or not.

Fig. 2 Asino Pantesco or "di Pantelleria" Donkey Breed

Pedigree analysis serves as a cornerstone of genetic management, particularly for domesticated taxa with existing breeding records (Ballou and Lacy 1995). By reconstructing genealogical relationships, conservation managers can estimate important metrics such as inbreeding coefficients, mean kinship values, and founder contributions. This facilitates the identification of genetically valuable individuals, helps prevent the mating of close relatives, and allows for the strategic inclusion of underrepresented lineages in breeding programs (Ballou and Lacy 1995). For small and isolated populations, such as standardized donkey breeds, pedigree data can reveal the disproportionate genetic influence of popular sires or dams, enabling efforts to rebalance founder representation and minimize the risks associated with genetic bottlenecks (Ballou and Lacy 1995).

A pertinent example is provided by the Pantesco donkey, a critically endangered breed native to the island of Pantelleria (Italy) (Fig. 2), with fewer than 100 remaining individuals (Bordonaro et al. 2012). In such a limited gene pool, pedigree management is vital to prevent the loss of rare alleles and to maintain reproductive performance without increasing inbreeding levels (Bordonaro et al. 2012). Through carefully planned mating strategies, the conservation of this breed aims to stabilize genetic diversity and improve long-term population viability.

Genetic management must be closely aligned with demographic strategies to ensure not only diversity but also sustainable population growth. Pedigree-based data support demographic goals by enabling breeding decisions that promote equal founder contributions, facilitate genetic rescue through translocations, and prioritize individuals with high reproductive success and low kinship values (Frankham et al. 2010). Additionally, effective population management must address demographic variables such as age structure, sex ratio, and reproductive lifespan to avoid stochastic threats (Frankham et al. 2010). For example, maintaining a balanced sex ratio and avoiding overuse of a few breeding males can help prevent rapid increases in inbreeding as the breeding population ages.

Despite the utility of pedigree analysis, its effectiveness is limited by the accuracy and completeness of the data. Incomplete records, parentage errors, or undocumented founder origins can lead to misleading estimates of relatedness and genetic diversity (Criscione et al. 2024). These issues are especially pronounced in local or landrace donkey populations, where historical breeding information is often sparse or inconsistently maintained. To address these limitations, conservation programs increasingly incorporate molecular tools to validate and supplement pedigree data. Microsatellites and single nucleotide polymorphisms (SNPs) can confirm parentage, resolve inconsistencies in genealogical records, and provide empirical estimates of genetic relatedness. This hybrid approach—combining pedigree and molecular data—is particularly effective in populations with uncertain or incomplete histories and allows for more robust and evidence-based conservation strategies (Criscione et al. 2024).

Financial and technical constraints, however, limit the widespread application of advanced genomic tools in many conservation programs, particularly those focused on domestic breeds such as donkeys. High-throughput sequencing and large-scale SNP genotyping, while precise, may be prohibitively expensive for routine use. Moreover, the relatively low genetic variability observed in many donkey breeds reduces the utility of certain molecular markers. In these contexts, pedigree strategies become not only cost-effective but also essential for ongoing population monitoring. Genomic tools are often reserved for specific high-priority tasks, such as identifying founders, validating critical mating pairs, or investigating genetic traits relevant to health and reproduction. In the Ragusano donkey, for instance, conservationists have combined pedigree records with targeted SNP analysis to detect bottlenecks and guide breeding strategies that enhance genetic diversity (Criscione et al. 2024). This selective integration of genomic and pedigree tools exemplifies a pragmatic approach to breed conservation under budgetary constraints.

An important conceptual distinction in genetic management is that between identity by descent (IBD) and identity by state (IBS). IBD refers to alleles inherited from a common ancestor, while IBS refers to alleles that are identical in sequence but may not originate from the same ancestor. Genomic methods typically capture IBS, which can lead to overestimates of relatedness, particularly when the population's history is poorly understood (Balloux et al. 2004). In contrast, pedigrees offer direct estimates of IBD, though their accuracy depends on the completeness of the recorded ancestry. By combining pedigree-based IBD estimates with genomic IBS data, conservationists can derive a more accurate understanding of population structure, which is critical in making effective breeding and management decisions. This integrated approach is particularly important for endangered donkey breeds, where reliance on a single method may lead to misinterpretations and suboptimal outcomes (Balloux et al. 2004).

Practical implementation of these principles requires a coordinated and comprehensive conservation framework. This includes the establishment of centralized, digitized pedigree management systems that enable collaboration among breeders, zoos, and conservation organizations. Genomic support, even if limited to key individuals or specific traits, can strengthen the validity of kinship assessments and breeding choices. Translocations between isolated or genetically depauperate subpopulations can function as a form of genetic rescue, reintroducing lost alleles and enhancing overall diversity. Moreover, consistent monitoring of key indicators such as inbreeding coefficients, effective population size, and genetic load provides a feedback mechanism to adapt strategies over time.

In summary, the integration of genetic and demographic management tools forms the foundation for the effective conservation of endangered donkey breeds. While genomic technologies offer a high level of precision, the accessibility and utility of pedigree-based strategies make them indispensable, especially in resource-limited settings. When combined thoughtfully, these approaches provide a flexible and powerful toolkit to preserve the viability, functionality, and heritage of donkey populations at risk of extinction due to genetic erosion, demographic instability, and environmental change.

4.3 Limited Scientific Interest and Difficulty to Publish in Impact Journal

Research on donkey breeds faces significant publication challenges, largely due to the perceived lack of scientific interest compared to more prominent species like horses. Donkeys are often overshadowed in equine research, as horses hold higher economic, sporting, and cultural value in many regions. This leads to a prioritization of horse-related studies, resulting in less funding, fewer resources, and limited attention from the scientific community for donkey research (Rossel et al. 2008). Consequently, donkey health, genetics, and conservation often remain underrepresented in scientific literature, leaving critical aspects of their welfare and management insufficiently addressed (Miragaya et al. 2008).

Moreover, studies focused on donkeys frequently struggle to find placement in high-impact journals, as they are often published in lower-quartile journals with narrower audiences and reduced visibility. This lack of high-impact publication outlets limits the reach and influence of donkey research, making it difficult for findings to inform wider conservation and welfare policies. To improve the visibility and support for donkey-related research, targeted funding, advocacy, and the creation of specialized journals or sections dedicated to donkey studies are necessary. Elevating the profile of donkey research would not only support better welfare and conservation outcomes but also enhance the understanding of this historically significant and versatile species (Smith and Pearson 2005).

Publishing research on donkey breeding in high-quartile journals presents additional challenges that hinder the advancement of knowledge and conservation efforts. One primary issue is the limited availability of comprehensive data on donkey populations and breeding practices. Many studies lack the depth required to meet the rigorous standards of high-impact journals, resulting in a shortage of high-quality research on the unique reproductive needs and challenges faced by different donkey breeds (Norris et al. 2021). This data gap limits the ability to draw meaningful conclusions and inform effective conservation strategies.

Furthermore, donkeys are often seen as "less significant" compared to more economically valuable species, such as horses or cattle. Research on these more popular species tends to receive more attention and funding, creating a cycle where the lack of published donkey studies results in fewer researchers focusing on this species, perpetuating the knowledge gap (Kindsvater et al. 2018; Rossel et al. 2008). As a result, vital issues like genetic diversity, reproductive health, and welfare challenges of donkey breeds remain underrepresented in scientific literature.

To overcome these barriers, researchers must advocate for the importance of studying donkey breeds and their unique challenges. Collaborations between veterinarians, geneticists, and conservationists can help generate high-quality data that meets the standards of high-impact journals. Funding agencies and academic institutions should recognize the value of donkey breeding research and support initiatives that address existing knowledge gaps. By tackling these challenges, the scientific community can raise the visibility of donkey breeding issues, ultimately contributing to more effective conservation strategies and improved welfare for these important animals.

Importantly, the risks associated with selective breeding can be significantly reduced through strategic outcrossing using different family lines, or the intentional mating of genetically unrelated individuals. This approach helps maintain or restore genetic diversity within a population, reducing the likelihood of inbreeding depression and its associated health problems. By introducing new genetic material, breeders can dilute the expression of harmful recessive genes while still promoting the inheritance of favorable traits. When carefully planned, outcrossing can preserve or even enhance characteristics such as temperament, coat quality, or physical strength, without the detrimental effects commonly linked to narrow genetic pools. This method supports both breed improvement and long-term population health, making it a critical tool in sustainable donkey breeding and conservation programs-

4.4 Research Feedback and Research/Owner Mistrust

A significant challenge in donkey-related research is the lack of trust between owners and researchers, which often hinders collaborative efforts aimed at improving donkey health, welfare, and breed preservation. Many donkey owners feel that researchers, typically focusing on broader equine studies, lack an understanding of the unique qualities and needs of donkeys as distinct from horses. Additionally, some owners feel researchers make use of donkeys, that they daily handle and care after, in search for self-success instead of providing valuable feedback from research which may improve their lives or that of their donkeys. This perception is compounded by the fact that research findings or recommendations may sometimes seem disconnected from the realities of caring for donkeys in day-to-day contexts. As a result, owners may view research efforts with scepticism or even resistance, which can prevent valuable insights from reaching those who work most closely with these animals (Geiger et al. 2020a, b).

By contrast, some owners seek the obtention of economic benefits that surpass the possibilities that research funds can obtain, which hinders the obtention of reliable results, and in turn conditions the eventual development of studies.

This trust gap has implications for data collection, the application of research findings, and even funding for donkey-specific studies. Owners may be reluctant to participate in studies, fearing that the research will not address their practical concerns or that findings could be misapplied in ways that do not benefit their animals. Bridging this gap requires researchers to actively engage with donkey owners, respecting their firsthand experience and incorporating their feedback into research design and communication. Building mutual understanding and involving owners in the research process can help foster a cooperative environment, leading to more accurate data, meaningful studies, and ultimately, outcomes that benefit both the scientific community and donkey caretakers (Geiger 2024).

A further complication is that research outputs frequently are not transmitted back to the communities and individuals who contributed data—particularly donkey owners. In many participatory studies, the knowledge generated remains within academic or institutional circles rather than being communicated in accessible formats to the participants (Geiger 2024). For instance, in Ethiopia, participatory workshops and interviews captured rich owner knowledge and priorities, yet findings were often not shared with those same communities afterward, limiting opportunities for on-the-ground improvement and undermining trust (Geiger 2024; Turn0search0). Ensuring that study results are fed back—in local languages, via workshops, summary briefs, or practical guides—is essential. Without this, owners may feel exploited, reducing future participation and hindering the very goals of welfare and conservation research.

5 Case Studies and Models

Effective conservation of donkey breeds hinges not only on theoretical frameworks and policy reforms but also on tangible, real world examples that illustrate both successful strategies and the challenges of preserving genetic diversity. Case studies serve as critical tools for translating abstract recommendations into actionable practices. This section examines two contrasting models that offer valuable insights into breed conservation. The first spotlights the Brazilian Pêga donkey, a success story in structured breeding and economic integration that demonstrates the potential of systematic, well-supported programs to elevate a breed's status and utility. The second focuses on the Baudet du Poitou of France, an endangered breed whose recovery depends on transnational collaboration, genetic monitoring, and cutting-edge reproductive technologies. Together, these case studies reveal how diverse strategies—ranging from grassroots breeder networks to high-tech interventions—can be tailored to breed-specific contexts. They also underscore the necessity of sustained institutional support, genetic planning, and international cooperation in safeguarding the future of donkey breeds worldwide.

5.1 The Success of Brazilian Pêga Donkeys: A Model for Conservation Efforts Turning into Highly Standardized Breeding Programs

The Pêga donkeys from Brazil (Fig. 3) stand out as a prime example of a highly standardized breed benefiting from advanced breeding structures and a comprehensive census. The Brazilian Association of Pêga Donkey Breeders (ABCJ Pêga) exemplifies how organized breeding programs and robust support systems can significantly enhance the conservation and development of a donkey breed (Canisso and McDonnell 2010). With around 2000 breeder members across Brazil and approximately 20,000 mules and donkeys registered, the ABCJ Pêga maintains meticulous records and upholds stringent breed standards. This level of technification ensures that the unique characteristics and genetic integrity of Pêga donkeys are preserved and promoted.

The success of the Pêga donkeys highlights the importance of having well-established breeding structures, including breed standards, breeder associations, and comprehensive studbooks. These elements are crucial for managing and documenting the genetic lineage of the breed, facilitating informed breeding decisions, and promoting overall breed health and diversity. The remarkable census maintained by the ABCJ Pêga provides valuable data for ongoing research and conservation efforts, demonstrating the significant benefits of organized and systematic breeding programs. The Pêga donkeys serve as a model for other breeds, showcasing the potential for success when proper structures and support systems are in place.

Fig. 3 Brazilian Pêga Donkey Breed

Pêga donkeys possess substantial economic value, contributing significantly to Brazil's agricultural and rural economies. Renowned for their strength, endurance, and versatility, these donkeys are commonly used for various agricultural tasks, such as ploughing fields, transporting goods, and assisting in livestock management but also due to their smooth gait. Their ability to navigate challenging terrains and perform heavy labour makes them invaluable assets for smallholder farmers, who rely on them for efficient and sustainable farming practices (Canisso and McDonnell 2010).

Moreover, Pêga donkeys are often bred for their mules, which are highly sought after due to their superior strength, smooth gait and work capabilities. The mule production not only generates income for breeders but also enhances agricultural productivity. Additionally, the organized breeding programs facilitated by the Brazilian Association of Pêga Donkey Breeders (ABCJ Pêga) further support the economic viability of this breed, as they promote high-quality breeding stock and improve overall herd health. The economic contributions of Pêga donkeys underscore the importance of preserving and promoting this breed for the benefit of local communities and the broader agricultural sector in Brazil.

The production of mules for Brazilian beef farms typically occurs on the farms themselves or at specialized stud farms located away from major beef cattle production areas. Beef farmers usually maintain a small herd of crossbred mares, typically ranging from 80 to 200 broodmares, which are generally a mix of Quarter Horses and Brazilian saddle-type breeds. To sire the mules, these farms typically use 1 to 5 donkey jacks. The Pêga breed or Pêga-type donkey jacks are the predominant choice for both specialized breeding operations and beef farm mule production (Canisso and McDonnell 2010).

Most of these donkey jacks are produced in the Southwest region of Brazil, particularly in the states of Minas Gerais and São Paulo, where small stud farms with 20 to 150 donkey dams operate. At around 1.5–3 years of age, the donkeys are sold to serve as sires on beef cattle farms or specialized mule breeding farms. Pêga breed donkey jacks carry significant monetary value, ranging from \$4000 to \$50,000, allowing them to be utilized for breeding well into their old age; in fact, it is common for them to be bred beyond the age of 20 in Brazil (Canisso and McDonnell 2010).

5.2 Baudet du Poitou Breed of France: French/North American Transnational Fight Against Extinction

Many donkey breeds around the world face a high risk of complete extinction, with the Poitou donkey breed from France serving as a stark example of this alarming trend (Fig. 4). Once a prominent breed known for its unique characteristics and contributions to agriculture, the Poitou donkey has seen its population dramatically decline over the years. The breed is characterized by its large size, long ears, and distinctive coat, which can be a variety of colours, including grey and dark brown. Historically, Poitou donkeys were valued for their strength and endurance, making them essential for agricultural tasks such as ploughing and transportation. However,

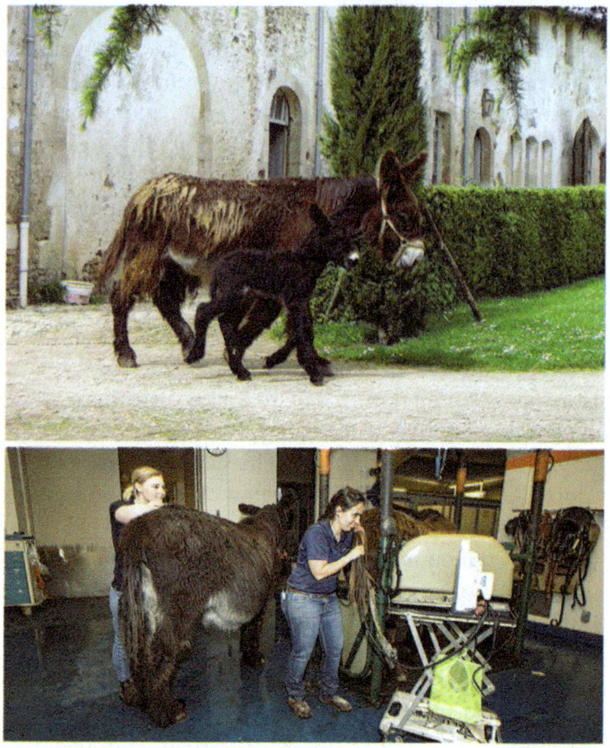

Fig. 4 French (upper) and American (lower) Baudet du Poitou Donkey Breed

changes in farming practices, urbanization, and the shift towards mechanization have led to a steep decrease in demand for this breed, pushing it closer to extinction (Livestock Conservancy 2025).

Conservation efforts for the Poitou donkey breed highlight the importance of targeted initiatives to preserve genetic diversity and maintain the breed's unique traits. Notably, the Mirandese and Zamorano-Leonés donkey breeds have been integral to these efforts. These breeds are known for their resilience and adaptability, and they have been used in carefully managed crossbreeding programs to enhance the genetic pool of the Poitou donkey. By introducing compatible genetic material, breeders aim to bolster the Poitou's population while preserving its distinctive morphological and functional characteristics (Yates 2022).

In addition to traditional breeding programs, recent advances in assisted reproductive technologies (ART) have played a growing role in conservation strategies for endangered donkey breeds like the Poitou. Techniques such as artificial insemination (AI), embryo transfer (ET), and in vitro fertilization (IVF) are being explored and applied in limited pilot initiatives. Of note is the development of intracytoplasmic sperm injection (ICSI), a sophisticated form of IVF that allows for the fertilization of an egg using a single sperm cell, significantly enhancing the efficiency of embryo production, even when viable gametes are scarce (Dobbie et al. 2024). ICSI-generated embryos can be cryopreserved and implanted into surrogate jennets, offering new possibilities for managing small, fragmented populations and ensuring the survival of valuable genetic lines (Dobbie et al. 2024).

These technologies are still in early stages of application for donkeys, and the procedures require specific adaptations due to the unique reproductive physiology of the species. Nonetheless, collaborations between veterinary reproductive specialists, geneticists, and conservation organizations are making progress. The use of ICSI embryos represents a promising frontier for augmenting the reproductive potential of aging or low-fertility individuals within critically endangered populations (Dobbie et al. 2024).

Alongside these technological interventions, organizations and breeders have launched public outreach initiatives to raise awareness about the Poitou donkey's plight and promote its preservation. These efforts include establishing breeding centers, developing educational resources, organizing breed showcases, and encouraging involvement from hobby farmers and conservation-minded owners. By fostering a renewed appreciation for the Poitou donkey and integrating both traditional and cutting-edge conservation methods, these initiatives aim to secure a sustainable future for the breed (Livestock Conservancy 2025).

The situation of the Poitou donkey serves as a powerful reminder of the broader challenges many donkey breeds face worldwide and underscores the urgent need for comprehensive and innovative conservation strategies. Integrating assisted reproduction into these efforts not only broadens the toolkit available to conservationists but also enhances the chances of successfully preserving endangered donkey breeds for future generations (Yates 2022).

In both the United States and France, concerted efforts are underway to recover and protect the endangered Poitou donkey through the revival and management of imported lineages. In France—the Poitou donkey's country of origin—government-backed breeding programs, breed registries, and specialized conservation centers have taken a central role in preserving the remaining purebred animals. The National Studs and local associations have implemented strict pedigree verification protocols to ensure genetic authenticity and to avoid dilution of the already limited gene pool. France also supports genetic monitoring and the use of assisted reproduction techniques, including semen preservation and embryo transfer, to facilitate controlled breeding and maintain lineage diversity (Dobbie et al. 2024; Yates 2022).

In the United States, a small but dedicated network of breeders and conservationists has focused on recovering and stabilizing imported Poitou lineages that arrived during the twentieth century (Fig. 4). Because the number of original imported individuals was limited, and few were purebred, American efforts have required careful coordination with French registries and experts to verify pedigrees and assess genetic compatibility. U.S. breeders often rely on collaboration with French institutions to obtain semen from genetically valuable jacks in France for artificial insemination, aiming to expand the genetic base of the North American Poitou population without compromising breed integrity (Livestock Conservancy 2025; Yates 2022).

Both countries are increasingly turning to international cooperation to share reproductive materials, coordinate breeding plans, and establish a viable global population. Cryopreservation of semen and embryos—especially through ICSI and embryo transfer—is viewed as a critical tool for securing genetic material from rare and aging individuals, both in France and abroad. These combined efforts reflect a shared recognition of the Poitou donkey's historical, cultural, and genetic value, and the urgency of saving this iconic breed from extinction through responsible, science-based conservation (Dobbie et al. 2024).

6 Conclusions

The conservation and sustainable breeding of donkey populations represent a complex challenge at the intersection of genetics, policy, culture, and economics. Unlike more commercialized livestock species, donkeys remain undervalued, resulting in fragmented institutional frameworks, incomplete census data, and insufficient research investment. These gaps have accelerated genetic erosion, undermining the long-term viability of local breeds and their capacity to adapt to environmental and socio-economic changes. Addressing these challenges requires a multi-layered approach. At the foundation lies the establishment of robust institutional structures—standardized breed definitions, centralized studbooks, and accurate, species-specific census systems. These measures enable effective monitoring of genetic diversity and informed breeding decisions, reducing the risk of inbreeding and loss of unique traits. Complementing institutional action, governmental policies must provide economic incentives and incorporate donkey genetic resources into national and international livestock development plans, while NGOs need to expand their focus beyond welfare to include genetic conservation objectives. Equally critical is the role of research. Advancing the integration of quantitative genetics with molecular tools can significantly enhance conservation outcomes, particularly when applied to small or endangered populations. However, these scientific strategies must be supported by strong community engagement and respect for local knowledge, ensuring that conservation efforts remain socially and culturally relevant. Case studies, such as the Brazilian Pêga donkey and the Baudet du Poitou, demonstrate that both structured breeding programs and innovative reproductive technologies can reverse population declines when implemented collaboratively and strategically. These examples underscore that success depends on coordinated, transnational efforts that combine scientific rigor with practical breeding management and policy support. Ultimately, donkey conservation is not solely about preserving an animal species; it is about safeguarding a vital genetic, cultural, and economic heritage. Only through integrated, inclusive, and evidence-based strategies can we ensure that donkey populations remain resilient, genetically diverse, and relevant for generations to come.

Self-Evaluation Questionnaire

1. What is one key function of breed standards in donkey conservation?
   A. Determining donkey feed requirements
   B. Regulating transport of donkeys
   C. Preserving unique breed characteristics and preventing genetic dilution
   D. Training donkeys for tourism
   Answer: C
2. Why is genetic diversity critical in donkey breeding programs?
   A. It ensures faster training for agricultural work
   B. It minimizes the risk of inbreeding and genetic disorders
   C. It helps in reducing feed costs
   D. It promotes early maturity in donkeys
   Answer: B
3. What is one recommended strategy for improving donkey welfare in conservation programs?
   A. Restricting natural mating completely
   B. Encouraging community-based breeding and farmer participation
   C. Limiting donkey population through selective culling
   D. Focusing only on artificial insemination techniques
   Answer: B
4. How do pedigree records support donkey conservation efforts?
   A. By determining the best diet plans for donkeys
   B. By identifying and maintaining genetic variability within breeds
   C. By classifying donkeys based on their working capacity
   D. By improving the speed of donkey training programs
   Answer: B
5. Which modern technology is increasingly used to monitor genetic diversity in donkey populations?
   A. GPS tracking
   B. Pedigree analysis only
   C. Genomic tools like SNP and ddRAD sequencing
   D. Hormonal therapy
   Answer: C
6. What is one major challenge in implementing donkey conservation programs in developing countries?
   A. Lack of interest in donkey meat production
   B. Limited infrastructure and funding for genetic resource management
   C. Overpopulation of donkey breeds
   D. High cost of donkey tourism activities
   Answer: B

References

Akin S, Kara A, Tutkun M (2018) Factors affecting membership to breeders association and contributions of the associations to the farmers training. J Agric Sci Technol 20(5):883–894

Allan FK (2022) What do we know about working equid populations? Livestock Data for Decisions. https://www.livestockdata.org/member-spotlight/what-do-we-know-about-working-equid-populations

Almejnah O, Saed HA, Marzok M, Almubarak A, Kandeel M, Shosha S, El-khodery S, Farag A (2023) Current profiling of research on donkeys and its implications in global studies based on bibliometric analysis. J Adv Vet Res 13(9):1835–1841

Aroua M, Haj Koubaier H, Rekik C, Fatica A, Ben Said S, Malek A, Mahouachi M, Salimei E (2024) Comparative study of carcass characteristics and meat quality of local mediterranean donkey breeds. Foods 13(6):942

Aroua M, Ben Said S, Mahouachi M (2025) Donkeys: unsung heroes of agriculture, health, and sustainability in a changing world. J Arts Human Soc Sci 2(1)

ASNOPRA (Asociación Nacional de Criadores de Ganado Asnal). (n.d.). Programa de conservación del ganado asnal en España. https://www.asnopra.es

Ballou JD, Lacy RC (1995) Identifying genetically important individuals for management of genetic variation in pedigreed populations. In: Ballou JD, Gilpin M, Foose TJ (eds) Population management for survival and recovery. Columbia University Press, pp 76–111

Balloux F, Amos W, Coulson T (2004) Does heterozygosity estimate inbreeding in real populations? Mol Ecol 13:3021–3031

Baumung R, Sölkner J (2003) Pedigree and marker information requirements to monitor genetic variability. Genet Sel Evol 35(4):369–383

Bonsi M, Anderson NE (2023) The socioeconomic impact of diseases of working equids in low and middle-income countries: a critical review. Animals 13(24):3865

Bordonaro S, Guastella AM, Criscione A, Zuccaro A, Marletta D (2012) Genetic diversity and variability in endangered Pantesco and two other Sicilian donkey breeds assessed by microsatellite markers. Sci World J 2012:648427

Bunevski G, Nikitovic J, Janzekovic M, Mergedush A, Prishenk J, Trajkovski B, Antić M (2018) Defining breed standards and breeding goals for domestic donkey. Acta Agric Serbica 23(46): 213–221

Canisso I, McDonnell S (2010) Donkey breeding behavior with an emphasis on the Pêga breed. Vet Care Donkeys Int Vet Inf Serv 20:0310

Carneiro GF, Lucena JEC, de Oliveira Barros L (2018) The current situation and trend of the donkey industry in South America. J Equine Vet Sci 65:106–110

Cecchi F, Ciampolini R, Ciani E, Mazzanti E, Tancredi M, Presciuttini S (2007) Morphological characterization of the Amiata donkey breed through the data reported in the Anagraphic register. Ital J Anim Sci 6(sup1):70–70

Colli L, Perrotta G, Negrini R, Bomba L, Bigi D, Zambonelli P, Verini Supplizi A, Liotta L, Ajmone-Marsan P (2013) Detecting population structure and recent demographic history in endangered livestock breeds: the case of the Italian autochthonous donkeys. Anim Genet 44(1): 69–78

Criscione A, Chessari G, Cesarani A, Ablondi M, Asti V, Bigi D, Bordonaro S, Ciampolini R, Congiu M, De Palo P, Ablondi M (2024) Analysis of ddRAD-seq data provides new insights into the genomic structure and patterns of diversity in Italian donkey populations. J Anim Sci 102(Supplement 3):skae165

De Berardinis A, Bucci R, De Amicis I, Del Signore F, Parrillo S, Massirio I, Vignoli M, Carluccio A (2024) Phenotypic characterization of the Martina Franca. Animals (Basel) 14(13):1950

De Santis M, Seganfreddo S, Galardi M, Mutinelli F, Normando S, Contalbrigo L (2021) Donkey behaviour and cognition: a literature review. Appl Anim Behav Sci 244:105485

Delgado Bermejo JV, Martínez Martínez MA, Rodríguez Galván G, Stemmer A, Navas González FJ, Camacho Vallejo ME (2019) Organization and management of conservation programs and research in domestic animal genetic resources. Diversity 11(12):235

Delgado J, Camacho M, Benavente M, Consorcio C (2018) FCPT (FAO/CONBIAND Census Projection Tool) Una herramienta para estimar los censos raciales nacionales. Acta Iberoam Conserv Anim 12:179–184

Demirbük M, Kızılaslan N (2019) Socio-economic analysis of breeders, breeders' associations and member relations, a case study of Sivas

Dobbie T, Felix MR, Gleason K, Hinrichs K (2024) Transvaginal aspiration, oocyte maturation and production of blastocysts and a pregnancy via ICSI in an endangered breed: Baudet du Poitou donkey. Theriogenology 224:34–40

FAO (2007) The state of the world's animal genetic resources for food and agriculture. Food and Agriculture Organization of the United Nations, Rome

Frankham R, Ballou JD, Briscoe DA (2010) Introduction to conservation genetics, 2nd edn. Cambridge University Press

Geiger MR (2024) Being with donkeys: insights into the valuing and wellbeing of donkeys in Central Ethiopia. Soc Animals 32(5–6):602–620

Geiger MR, Hockenhull J, Buller H, Engida GT, Getachew M, Burden FA, Whay HR (2020a) Understanding the attitudes of communities to the social, economic, and cultural importance of working donkeys in rural, peri-urban, and urban areas of Ethiopia. Front Vet Sci 7:60

Geiger M, Whay HR, Engida GT (2020b) Understanding the attitudes of communities to the social, economic, and cultural importance of working donkeys in Ethiopia. Front Vet Sci 7:60

Gutiérrez JP, Marmi J, Goyache F, Jordana J (2005) Pedigree information reveals moderate to high levels of inbreeding and a weak population structure in the endangered Catalonian donkey breed. J Anim Breed Genet 122(6):378–386

Gutiérrez JP, Marmi J, Goyache F, Jordana J (2006) Pedigree information reveals moderate to high levels of inbreeding and a weak population structure in the endangered Catalonian donkey breed. J Anim Breed Genet 122(6):378–386

Hall SJG (2019) Livestock biodiversity as an interface between people, landscapes and nature. People Nat 1(1):284–290

Ivanković A, Šubara G, Bittante G, Šuran E, Amalfitano N, Aladrović J, Kelava Ugarković N, Pađen L, Pećina M, Konjačić M (2023) Potential of endangered local donkey breeds in meat and milk production. Animals 13(13):2146

Jere ZM, Zidana R, Chatambalala L, Mwafulirwa S, Nyirenda Z (2025) Surviving amidst neglect: management and marketing practices of donkeys (Equus asinus) in the Central Region of Malawi. BMC Agric 1:3. https://bmcagriculture.biomedcentral.com/articles/10.1186/s44399-025-00003-z

Kindsvater HK, Dulvy NK, Horswill C, Juan Jordá M-J, Mangel M, Matthiopoulos J (2018) Overcoming the data crisis in biodiversity conservation. Trends Ecol Evol 33(9):676–688

Kithuka JM, Wachira TM, Onono JO, Gichure MN (2025) Assessing environmental factors and human practices on the welfare of working donkeys in Kenya. Discover Animals 2(1):1–16

Kugler W, Grunenfelder H-P, Broxham E (2008) Donkey breeds in Europe. Save Foundation, St. Gallen

Labbaci M, Djaout A, Benyarou M, Ameur AA, Gaouar SBS (2018) Morphometric characterization and typology of donkey farming (Equus asinus) in the wilaya of Tlemcen. GABJ 2(1): 56–68

Legha R, Talluri T, Bhardwaj A, Pal Y (2023) Spatial and temporal distribution of donkey and mule population in India. J Agric Ecol 15:45–48

Livestock Conservancy (2025) Poitou donkey. Retrieved from https://livestockconservancy.org/poitou-donkey/

Ljubodrag S, Vladimir D, Predrag S, Nada L, Ivica R, Ante I, Jevrosima S, Zoran S (2015) Morphological, biochemical and hematological characterization of endangered Balkan donkey breed. Acta Vet Brno 65(1):125–136

Maswana M, Mugwabana TJ, Tyasi TL (2022) Evaluation of breeding practices and morphological characterization of donkeys in Blouberg local municipality, Limpopo province: implication for the design of community-based breeding programme. PLoS One 17(12):e0278400

Maswana M, Mugwabana TJ, Tyasi TL (2023) Morphological differentiation among donkey breeds: a review. Pak J Zool 55(1):413–418

Mazzatenta A, Veronesi MC, Vignola G, Ponzio P, Carluccio A, De Amicis I (2019) Behavior of Martina Franca donkey breed jenny-and-foal dyad in the neonatal period. J Vet Behav 33:81–89

McLean AK, Navas Gonzalez FJ (2018) Can scientists influence donkey welfare? Historical perspective and a contemporary view. J Equine Vet Sci 65:25–32

McLean AK, Varnum A, Ali A, Heleski CR, Navas González FJ (2019) Comparing and contrasting knowledge on mules and hinnies as a tool to comprehend their behavior and improve their welfare. Animals 9(8):488

Miragaya MH, Neild DM, Losinno LN (2008) Donkey reproduction: current status and future prospects. Anim Reprod Sci 105(3–4):201–211

Mustefa A, Assefa A, Misganaw M, Getachew F, Abegaz S, Hailu A, Emshaw Y (2020) Phenotypic characterization of donkeys in benishangul Gumuz national regional state. Online J Anim Feed Res 10(1):25–35

Navas FJ, Jordana J, McLean AK, León JM, Barba CJ, Arando A, Delgado JV (2019) Modelling for the inheritance of multiple births and fertility in endangered equids: Determining risk factors and genetic parameters in donkeys (Equus asinus). Res Vet Sci 126:213–226. https://doi.org/10.1016/10.1016/j.rvsc.2019.09.007. Epub 2019 Sep 12. PMID: 31610472.

Navas González FJ (2019). Structural organization of genetic diversity and ethofunctional characterization in the Andalusian donkey breed. PhD thesis, Universidad de Córdoba, UCOPress.

Navas González FJ, Delgado Bermejo JV, Vargas Burgos JC (2016) Current donkey production and functionality: relationship with humans [Book1]. UCOPress, University of Córdoba

Navas González FJ, Jordana J, León Jurado JM, Barba C, Delgado Bermejo JV (2017a) A model to infer the demographic structure evolution of endangered donkey populations. Animal 11(12): 2129–2138

Navas FJ, Jordana J, León JM, Arando A, Pizarro G, McLean AK, Delgado JV (2017b) Measuring and modeling for the assessment of the genetic background behind cognitive processes in donkeys. Res Vet Sci 113:105–114

Navas González FJ, Jordana Vidal J, Camacho Vallejo ME, León Jurado JM, de la Haba Giraldo MR, Barba Capote CJ, Delgado Bermejo JV (2018a) Risk factor meta-analysis and Bayesian estimation of genetic parameters and breeding values for hypersensibility to cutaneous habronematidosis in donkeys. Vet Parasitol 252:9–16

Navas González FJ, Jordana Vidal J, León Jurado JM, McLean AK, Pizarro Inostroza G, Delgado Bermejo JV (2018b) Genetic parameter estimation and implementation of the genetic evaluation for gaits in a breeding program for assisted-therapy in donkeys. Vet Res Commun 42:101–110

Navas F, Jordana J, León J, McLean AK, Delgado J (2019a) Dumb or smart asses? Donkey's (Equus asinus) cognitive capabilities share the heritability and variation patterns of human's (Homo sapiens) cognitive capabilities. J Vet Behav 33:63–74

Navas F, Jordana J, McLean A, León J, Barba C, Arando A, Delgado J (2019b) Modelling for the inheritance of multiple births and fertility in endangered equids: determining risk factors and genetic parameters in donkeys (Equus asinus). Res Vet Sci 126:213–226

Norris SL, Little HA, Ryding J, Raw Z (2021) Global donkey and mule populations: figures and trends. PLoS One 16(2):e0247830

Polidori P, Beghelli D, Mariani P, Vincenzetti S (2009a) Donkey milk production: state of the art. Ital J Anim Sci 8(sup2):677–683

Polidori P, Cavallucci C, Beghelli D, Vincenzetti S (2009b) Physical and chemical characteristics of donkey meat from Martina Franca breed. Meat Sci 82(4):469–471

Quaresma MNP (2015) The Asinina de Miranda donkey breed (Equus asinus): demographic analysis and characterization of the reproductive cycles Universidade de Tras-os-Montes e Alto Douro (Portugal)

Quaresma M, Martins AMF, Rodrigues JB, Colaço J, Payan-Carreira R (2014) Pedigree and herd characterization of a donkey breed vulnerable to extinction. Animal 8(3):354–359

Rambaldi Migliore N, Bigi D, Milanesi M, Zambonelli P, Negrini R, Morabito S et al (2024) Mitochondrial DNA control-region and coding-region data highlight geographically structured diversity and post-domestication population dynamics in worldwide donkeys. PLoS One 19(8): e0307511

Rossel S, Marshall FB, Peters J, Pilgram T, Adams MD, O'Connor D (2008) Domestication of the donkey: timing, processes, and indicators. Proc Natl Acad Sci 105(10):3715–3720

Rowan A (2022) The donkey: an unsung hero of human civilization. WellBeing International. Retrieved from search engine

Santana ML, Bignardi AB (2015) Status of the genetic diversity and population structure of the Pêga donkey. Trop Anim Health Prod 47:1573–1580

Smith DC (2008) Book of mules: selecting, breeding, and caring for equine hybrids. Rowman & Littlefield

Smith DG, Pearson RA (2005) A review of the factors affecting the survival of donkeys in semi-arid regions of sub-Saharan Africa. Trop Anim Health Prod 37(Suppl 1):1–19

Snedecor GW, Cochran WG (1989) Statistical methods, 8th edn. Iowa State University Press

Sponenberg DP, Bixby DE (2007) Managing breeds for a secure future: strategies for breeders and breed associations. Albc

Starkey P, Starkey M (2000) In: Starkey P, Fielding D (eds) Regional and world trends in donkey populations, pp 10–21

Susana Lopes M, Azevedo AR, Mendonça D, Rojer H, Cabral V, Ceraolo F, Canto Brum C, Mendes B, da Câmara Machado A (2023) Morphological and genetic characterization of the Graciosa donkey breed. J Appl Anim Res 51(1):166–173

The Brooke (2022). Data on working equids. Retrieved from https://www.thebrooke.org/our-work/data-workingequids

The Brooke (2020) Invisible workers: the economic contributions of working donkeys, horses and mules to livelihoods. https://www.thebrooke.org/research/invisible-workers

The Brooke Hospital for Animals (2022) Gaps in population data put working horses and donkeys at threat. https://www.thebrooke.org/news/gaps-population-data-put-working-horses-and-donkeys-threat

The Donkey Sanctuary (n.d.) Who we are. https://www.thedonkeysanctuary.org.uk/what-we-do/who we are

Toro M, Fernández J, Shaat I, Mäki-Tanila A (2011) Assessing the genetic diversity in small farm animal populations. Animal 5(11):1669–1683

Tosi U, Bernabò N, Verni F, Valbonetti L, Muttini A, Mattioli M, Barboni B (2013) Postpartum reproductive activities and gestation length in Martina Franca jennies, an endangered Italian donkey breed. Theriogenology 80(2):120–124

Urosevic M, Nemecek M, Drobnjak D, Gangl A, Dameski P, Stojić P, Stanisic G (2019) Morphological characterization of the white baroque donkey. Anim Health Prod Hygiene 8(1)

Wassie AM, Getachew TB, Kassa AH, Megersa AG, Aseged TA (2023) Donkey production systems and breeding practices in selected districts of South Omo Zone, Southern Ethiopia. Rangel J 45(3):97–108

Yates D (2022) Saving an endangered breed of donkey. University of Illinois News. Retrieved from https://news.illinois.edu/view/6367/170984365

Yılmaz O, Boztepe S, Ertuğrul M (2012) The domesticated donkey: II-types and breeds. Can J Appl Sci 2:267–286

Donkey Evolution and Global Dispersal: From Domestication Mitochondrial Haplogroups to Breed Diversification Molecular Signals

Francisco Javier Navas González

Abstract

Donkey domestication represents a pivotal milestone in human history, driving the expansion of trade networks and adaptation to arid environments. This chapter synthesizes archaeological, genomic, and ecological evidence to reconstruct the origins, global dispersal, and breed diversification of donkeys (Equus asinus). Comprehensive analyses of mitochondrial haplogroups, nuclear genome variation, and Y-chromosome patterns reveal a single domestication event in Northeast Africa within a genetically diverse wild ass population, giving rise to Haplogroups A and B. Subsequent dispersals across Africa, Asia, and Europe were shaped by environmental constraints, hybridization events, and selective breeding. Genome sequencing highlights functional adaptations in energy metabolism, immunity, and coat color genetics, while molecular differentiation studies illuminate breed-specific traits and phenotypic diversity, such as long-hair phenotypes and color variants linked to TBX3, ASIP, and KIT genes. Conservation concerns emerge from ongoing population declines, driven by mechanization, market pressures (e.g., Ejiao industry), and neglect of traditional breeds. Integrating molecular insights with historical context underscores the importance of genomic tools for safeguarding donkey genetic diversity, informing conservation strategies, and unraveling the evolutionary dynamics of one of humanity's oldest domesticated species.

F. J. Navas González (✉)
Department of Genetics, University of Cordoba, Cordoba, Spain

Worldwide Donkey Breeds Project, Cordoba, Spain
e-mail: fjnavas@uco.es

A. K. McLean et al. (eds.), *Donkey and Mule Medicine: Foundations, Breeds, and Management*, https://doi.org/10.1007/978-3-032-12726-6_4

Keywords

Donkey domestication · Mitochondrial haplogroups · Genomic diversity · Phylogeography · Coat color genetics · TBX3 · ASIP · KIT gene · Breed differentiation · Conservation genomics · Equus asinus evolution

1 Introduction

The domestication of the donkey (*Equus asinus*) marks one of the most pivotal transitions in the evolution of human–animal relationships, particularly in arid and semi-arid landscapes. Donkeys enabled long-distance trade and transportation, supported mobile pastoralism, and contributed to the formation of early states and trade networks from Africa to Asia (Clutton-Brock 1992; Marshall 2007). However, compared to horses and cattle, donkeys have received relatively little attention in archaeological and genetic studies. Much of their history remains ambiguous due to limited archaeological remains, challenges in distinguishing them morphologically from wild equids, and underrepresentation in textual or artistic sources (Rossel et al. 2008). The evidence that does exist is complicated by overlapping wild and domestic traits, multiple mitochondrial lineages, regional dispersals, and inconsistent visibility in the archaeological record (Marshall 2007).

The archaeological evidence for donkey domestication is strongest in early Egyptian contexts, with the most definitive material found at Abydos, dating to approximately 3000 BCE (Rossel et al. 2008). It was in Abydos, where ten well-preserved donkey skeletons were discovered within a royal mortuary complex, all exhibiting a suite of osteopathological indicators consistent with long-term load carrying. These include spinal lesions, vertebral deformities, and limb compression, strongly suggesting they were used as pack animals (Rossel et al. 2008). Despite these signs of functional use, the animals retained morphometric features characteristic of wild African asses. This dichotomy—between pathological evidence of domestication and morphometric similarity to wild forms—reflects a domestication process that was gradual and incomplete, particularly in its early stages (Marshall 2007). Morphological changes lagged behind behavioral and functional integration into human societies. This pattern supports a broader model in which animal domestication is not a singular moment, but rather a continuum of interactions shaped by human management, environmental pressures, and ongoing genetic exchange with wild populations (Zeder 2012).

Mitochondrial DNA studies have revolutionized our understanding of donkey domestication by identifying two deeply divergent maternal lineages present in domestic donkeys worldwide. These lineages, historically termed Clade 1 and Clade 2 and now referred to as Haplogroup A and Haplogroup B, diverged from one another between roughly 300,000 and 900,000 years ago—long before domestication occurred (Beja-Pereira et al. 2004; Kimura et al. 2011). Haplogroup A is strongly linked to the Nubian wild ass (*Equus africanus africanus*), whose range once spanned parts of Northeast Africa, while Haplogroup B is presumed to have originated from the now-extinct Atlas wild ass (*Equus africanus atlanticus*), which likely inhabited the Maghreb region (Beja-Pereira et al. 2004), although at a later stage mitochondrial DNA studies may have identified the implications of Somali wild ass (*Equus africanus somaliensis*) in the conformation of for example Chines Donkey Breeds though the evaluation of Cytochrome b genetic diversity (Zhang et al. 2010). This deep divergence initially led some researchers to propose two independent domestication events (Beja-Pereira et al. 2004). However, both lineages are globally distributed across modern donkey populations, often co-occurring in individual regions, suggesting that they were both present in the ancestral population at the time of domestication (Kimura et al. 2011).

Whole-genome sequencing has added further insights, revealing a strong phylogeographic structure with three major nuclear DNA clusters corresponding to African, Asian, and European populations (Rosenbom et al. 2023). Reduced Y-chromosome diversity points to sex-biased breeding, possibly due to the selective use of a few male lineages during historical breeding practices (Rosenbom et al. 2023). Moreover, evidence of admixture with other equids, such as *Equus kiang* (the kiang) in Asia, and hybridization with wild African asses, suggests that the genomic makeup of domestic donkeys has been shaped not only by initial domestication but also by centuries of interbreeding and human-mediated selection (Rosenbom et al. 2023; Orlando et al. 2013).

The existence of two major mitochondrial lineages initially gave rise to the hypothesis of dual domestication events, potentially occurring in different regions of Africa (Beja-Pereira et al. 2004). This theory was supported by the significant genetic divergence between the clades, which appeared too large to have emerged within a single domestic population over a short timescale. However, this interpretation has been increasingly challenged by more recent genomic studies, which propose that both clades were already present within a single, genetically diverse population of African wild asses prior to domestication (Kimura et al. 2011). Under this scenario, the presence of both clades in the domestic population reflects ancestral polymorphism rather than separate domestication episodes. Whole-genome data from both ancient and modern donkeys have failed to identify clear signs of independent domestication centers, and the archaeological record does not support the presence of parallel, regionally distinct domestication events (Rosenbom et al. 2023). Consequently, the current consensus is shifting towards a model of a single domestication in Northeast Africa, likely in the 4th millennium BCE, involving a source population that already harbored the two divergent maternal lineages (Kimura et al. 2011; Rosenbom et al. 2023).

A prominent model linking donkey domestication to environmental change is the pastoralist hypothesis. According to this view, the domestication of the donkey was a strategic response by pastoralist societies to the increasing aridity of the Sahara during the mid-Holocene (approximately 7000–5000 BP) (Marshall 2007). As formerly verdant regions became dry and water sources retreated, people were forced to adopt a more mobile way of life. Donkeys, with their high endurance, low water requirements, and ability to consume coarse vegetation, were ideally suited to this shift. Their use allowed pastoralists to transport water, food, and household goods across increasingly inhospitable terrain (Marshall 2007). Unlike cattle, which require regular watering and rich forage, donkeys could move further and longer, opening up new trade and migration routes. In this view, domestication was not simply about taming animals for food or ritual, but about adapting to climatic instability. The shift toward donkey-based transport facilitated long-distance pastoral mobility and likely underpinned the development of expansive trade systems linking the Nile Valley with the Red Sea, the Levant, and beyond (Clutton-Brock 1992; Marshall 2007).

After their domestication in Northeast Africa, donkeys dispersed across large parts of the Old World, spreading into the Levant, Mesopotamia, Iran, South Asia, and eventually Europe by 2000 BCE. Archaeological evidence, such as pack-animal burials and artistic representations, attests to their increasing economic and symbolic importance in early complex societies (Clutton-Brock 1992). Genetic evidence mirrors this spread. Distinct mtDNA haplotype distributions and nuclear DNA clusters reflect early dispersals and regional differentiation (Rosenbom et al. 2023). For example, modern donkeys in China possess genetic signals of introgression from the kiang, indicating contact between domestic donkeys and local wild equids (Rosenbom et al. 2023). African and Arabian populations retain traces of admixture with wild African asses, highlighting a long-term dynamic of local introgression and adaptation (Kimura et al. 2011). The presence of a previously unrecognized donkey lineage in the Levant around 200 BCE, uncovered through ancient DNA studies, suggests that donkey management during the Roman period included selective breeding for size and strength, making these animals crucial to military logistics and urban economies (Rosenbom et al. 2023).

One of the most intriguing puzzles in donkey domestication is their failure to spread into southern Africa in pre-colonial times, unlike cattle and sheep, which had reached the region centuries earlier. This has been attributed not to cultural or technological limitations, but to ecological and disease-related constraints (Spinage 2012). Three major epizootic diseases likely formed a biological barrier: trypanosomiasis, equine piroplasmosis, and African horse sickness. Trypanosomiasis, in particular, is spread by tsetse flies, which infest the humid savannas and woodlands of central and southern Africa. Donkeys are especially vulnerable to this disease, which causes severe anemia and often death. Unlike cattle, which have some trypanotolerant breeds, and dogs, which were periodically introduced and reintroduced, donkeys lacked sufficient resistance (Spinage 2012). This made it practically impossible for them to survive or reproduce in affected regions. Additionally, their utility as burden animals may have been diminished in environments where riverine or forested terrain, combined with tsetse presence, made overland transport difficult. These constraints thus explain the archaeological and historical absence of donkeys from southern Africa until the colonial period, when disease control and veterinary care allowed for their introduction and survival (Spinage 2012).

Despite their importance in many ancient societies, donkeys are underrepresented in the archaeological and historical record. In part, this reflects a cultural bias: donkeys were often viewed as low-status animals, used by commoners, peasants, and traders, unlike the horse, which was associated with warfare, nobility, and ritual (Clutton-Brock 1992). This perception may have limited their inclusion in funerary art, elite burials, and written records. Additionally, donkeys were rarely consumed for food, making their bones less likely to be found in domestic refuse sites. Their remains may also be underreported due to the difficulty of distinguishing donkey bones from those of horses or onagers without detailed morphometric or molecular analyses (Rossel et al. 2008). In regions like the Sahara or Arabian Peninsula (Sokotra Island too), the two potential directions followed by donkeys in their process of dispersion, poor preservation conditions and the ephemeral nature of nomadic settlements further complicate recovery. However, advances in techniques such as Zooarchaeology by Mass Spectrometry (ZooMS) and ancient DNA analysis are beginning to fill these gaps, enabling more accurate identification of donkey remains and reconstruction of their roles in past societies (Rossel et al. 2008).

In the modern world, donkeys face a paradox. Although they remain essential to rural economies in many parts of the Global South, their overall population is in decline, or they are even considered an environmetal threat (Bonaire ferals connection to extinct Nubian Wild Asses, but considered a pest), which leads their genetic diversity to be increasingly threatened. Mechanization in agriculture and transport has reduced demand in industrialized countries, leading to the neglect or extinction of many traditional breeds (Starkey and Starkey 2004). At the same time, a surge in the commercial demand for donkey skins—particularly for the Chinese ejiao market—has placed unsustainable pressure on donkey populations across Africa and Asia (Norris and Redmond 2019). Genetic studies reveal that some populations, such as those in Ethiopia and South America, retain high levels of mitochondrial and nuclear diversity, representing important reservoirs of adaptive traits. These include genes related to disease resistance, fertility, and climatic tolerance (Rosenbom et al. 2023). Recent genomic work has also uncovered mutations associated with coat color (e.g., a TBX3-linked deletion responsible for the non-dun phenotype), which may have implications for thermoregulation and camouflage (Rosenbom et al. 2023). Effective conservation will require not only protecting these genetic resources but also recognizing and supporting the socio-economic roles that donkeys continue to play for millions of smallholder farmers, herders, and transporters around the world (Starkey and Starkey 2004).

The synthesis of archaeological, genetic, and ecological evidence supports a unified model of donkey domestication. Donkeys were likely domesticated only once, in the grasslands of Northeast Africa, by mobile pastoralists responding to Holocene climatic instability (Marshall 2007). This domestication occurred within a genetically diverse wild ass population, giving rise to two major maternal lineages (Haplogroups A and B) that have persisted in modern donkeys (Kimura et al. 2011). Following domestication, donkeys spread rapidly across Africa and Eurasia, undergoing local adaptation and admixture with wild equids (Rosenbom et al. 2023). Their movement was shaped by trade, empire, and environmental constraints, especially disease barriers that prevented their expansion into southern Africa (Spinage 2012). Although often marginalized in both ancient records and modern policy, donkeys have played a foundational role in human history. Ongoing research—particularly in ancient DNA, high-throughput genomics, and integrative zooarchaeology—holds the potential to resolve remaining uncertainties and to inform future strategies for the conservation, sustainable use, and cultural recognition of this remarkable species (Rosenbom et al. 2023).

2 Genomic Tools in the Study of Donkey Domestication, Expansion, Evolution, and Breed Consolidation

Genomic technologies have profoundly transformed the study of donkey (*Equus asinus*) biology, providing unprecedented clarity on their domestication, global dispersal, adaptive evolution, and breed structure. The integration of ancient and modern DNA, high-resolution whole-genome sequencing, and mitochondrial haplotyping has not only resolved historical ambiguities but also opened new avenues for conservation and sustainable breeding strategies (Todd et al. 2022; Wang et al. 2020).

Comprehensive genome-wide analyses strongly support a single African origin of donkey domestication around 7000–5000 BCE. Whole-genome sequencing studies, including the high-coverage assembly of the Dezhou donkey genome, have identified the Nubian wild ass (*Equus africanus africanus*) as the principal ancestor, with mitochondrial and autosomal evidence indicating the domestication likely occurred in the Horn of Africa or the Nile Valley in response to increasing aridity in the Sahara (Todd et al. 2022; Ma et al. 2019). Interestingly, mitochondrial DNA studies reveal the presence of two major haplogroups (now standardized as Haplogroups A and B), both traceable to African wild ass populations. This diversity within a single source population suggests that both haplogroups coexisted before domestication, rather than being products of independent domestication events (Ma et al. 2019).

Genome-wide analyses have also uncovered discordant paternal and maternal histories, with drastically reduced Y chromosome variability contrasting with richer mitochondrial lineage diversity. This discrepancy likely reflects male-biased management practices—such as the preferential use of a few elite sires—while females were maintained in broader genetic pools. This finding is consistent with patterns observed in other domesticated species with polygynous mating systems (Todd et al. 2022; Wang et al. 2020).

Following domestication, donkeys underwent a stepwise expansion into Eurasia, with genomic data documenting their presence across Europe and Central Asia by the mid-third millennium BCE. Ancient DNA confirms that as donkeys moved into new regions, they diverged into geographically distinct populations, leaving clear genetic signatures. Principal component analyses and population structure models distinguish modern African, Eurasian, and Australian donkey populations and highlight post-domestication introgression events, including gene flow from now-extinct wild lineages and interspecies hybridization with kiangs (*Equus kiang*) and onagers (*Equus hemionus*) (Todd et al. 2022).

Genomic studies have further shown that while effective population sizes declined after domestication, a demographic rebound occurred around 7000 years ago, possibly linked to increased human reliance on donkeys for transport and agriculture. This demographic expansion is reflected in mitochondrial Bayesian skyline plots and contrasts with the relatively steady levels of nuclear genetic diversity, which remain moderate but stable across both ancient and modern populations (Ma et al. 2019; Todd et al. 2022).

At the breed level, genomics has illuminated the historical consolidation of donkey lineages, especially in association with Roman and later Islamic and colonial breeding systems. Selection for traits such as body size, coat color, and hybrid vigor in mule production is detectable in the genome. For example, the non-dun phenotype—a domestication-related coat color variant absent in wild asses—is caused by a single base-pair deletion downstream of the *TBX3* gene, leading to increased pigmentation. This mutation likely underwent positive selection in early domesticated populations due to human preference for distinct coat colors (Wang et al. 2020).

Modern genomic tools are also critical for identifying endangered or under-characterized donkey populations and for managing breed conservation programs. Mitochondrial and nuclear markers reveal that some regions, such as Eastern and Northern Africa, remain hotspots of genetic diversity, underscoring the importance of these areas for global conservation efforts (Ma et al. 2019). Furthermore, the genomic distinctiveness of breeds highlights the urgent need for breed-specific management in regions where donkey populations have sharply declined (Wang et al. 2020).

Notable examples of the application of genomic tools to under-characterized populations is, on the one hand, provided by a multi-country study of Ibero-American donkeys. Jordana et al. (2012) analyzed 21 populations from 14 countries using microsatellite markers, revealing a strong genetic differentiation between European and New World donkey populations. Despite originating from Iberian introductions during the colonial era, most American donkeys did not cluster genetically with current European breeds, instead forming two distinct groups: a Caribbean cluster (including Cuba and parts of Colombia) and an Andean/Southern cluster (including countries such as Bolivia, Peru, and Argentina).

Prior to this, Lopez et al. (2005) demonstrated through molecular genetic studies of the mitochondrial DNA D-loop region that the Mexican Creole donkey (Equus asinus)—a population of major cultural and economic significance, not only for its role in rural livelihoods but also for its historical importance in the spread of donkeys from the Old World to the New, and subsequently across Central America toward both the northern and southern regions of the continent—harbors a rich and largely underexplored genetic heritage. These authors revealed ten haplotypes in Mexican populations, seven of them previously undescribed. This finding underscores not only the high degree of genetic diversity present but also the survival of unique maternal lineages that are absent from European reference breeds. Such diversity is remarkable given the limited attention historically devoted to donkey populations in Mexico. The molecular comparisons demonstrated that three of the haplotypes (designated HBurro1, HBurro2, and HBurro4) showed close similarity to haplotypes previously reported in Spanish and African breeds, while the seven novel haplotypes (HBMex1–HBMex7) were unique to Mexico. Phylogenetic analyses consistently clustered the Mexican haplotypes with Spanish lineages, particularly those of the Andaluza, Zamorano-Leonesa, and Majorera breeds, all of which themselves display African genetic influence. Importantly, haplotypes ATI1 and ATI2, characteristic of Spanish breeds, were also the most frequent in Mexican donkeys, strongly supporting an Iberian origin introduced during the colonial period. The presence of additional haplotypes, however, indicates genetic differentiation within Mexico and the persistence of lineages that may have vanished from the Iberian Peninsula.

This differentiation is attributed to long-term geographic isolation, founder effects, and minimal gene flow between continents following the initial introduction as first donkeys may have arrived to the New World through Christopher Columbus expeditions via Madeira and the Canary islands with his 2nd Voyage and via Cape Verde in his 3rd Voyage from 1493 to 1496 and 1498 to 1500, followed by a consecutive process led by Spanish Alonso de Ojeda's first and second voyages (1499 and 1502, respectively) via Cape Verde, as well (Jordana et al. 2012). These colonization donkeys reach the trade routes after crossing Africa from East to West to later be transported alongside with black slaves (who would have herded them as it can still be seen in African countries such as Ethiopia) for their assistance in the production of marine salt. Within the American metapopulation, the genetic structure analysis detected significant population substructure, with eight of the eleven populations showing strong membership (>0.85) to one of the two main clusters. The observed patterns of genetic drift and allelic richness—particularly the elevated diversity found in Cuba and Brazil—suggest that these populations have developed distinct genetic identities over time, likely due to their roles as key introduction points and genetic crossroads (Jordana et al. 2012), but also because animals introduced were at relatively early stages of domestication, frequently resembling Nubian wild asses. This may also constitute the milestone from which donkeys expanded through North America and the origin for feral populations of donkeys in California or Texas, after those donkeys intended for mining during the Gold Rush in the 1840s–1850s were released back into the wild. A paralel process occurred in Australia, which would also lead to the consolidation of donkey feral populations across the territory.

In practical terms, these findings emphasize the need to recognize Ibero-American donkeys as distinct genetic resources, potentially warranting formal breed designation, but also which represent a gene pool reservoir generated during the pivotal moment of expansion of donkeys worldwide. For instance, molecular results confirm that the even if Creole donkey population groups such as the Mexican Creole donkey may be of a predominantly Iberian maternal origin they also harbor novel genetic variants that distinguish it as a unique population. The combination of shared and exclusive haplotypes provides evidence of both historical introductions and subsequent local differentiation. This high level of genetic diversity enhances the adaptive potential of such undefined populations in the challenging rural environments where they remain indispensable for transport and agriculture.

Jordana et al. (2012) argue that developing country-specific or region-specific conservation strategies could help preserve these unique lineages. Such strategies would benefit from integrating genomic data into decisions about population management, conservation prioritization, and even the development of localized breeding programs aimed at genetic recovery or hybrid vigor.

On the other hand, another example is the set of mitochondrial DNA studies have revealed that the maternal genetic background of Chinese domestic donkeys derives from two ancestral African lineages (Zhang et al. 2010): the Nubian wild ass (*Equus africanus africanus*) and the Somalian wild ass (*Equus africanus somaliensis*). Through complete sequencing of the mitochondrial cytochrome b (Cytb) gene in 244 individuals across 13 native breeds, researchers identified 58 haplotypes and documented a clear bifurcation between these two African maternal lines.

Crucially, the study found that both the Nubian and Somalian lineages contributed to every single Chinese donkey breed examined, revealing a panmictic maternal structure without any clear geographic or morphological pattern. For instance, breeds such as Guanzhong, Taihang, and Biyang displayed a higher proportion of Somalian haplotypes (e.g., 14 out of 18 Guanzhong donkeys), whereas others like Xiji and Huaibei showed either a predominance or balance favoring Nubian lineages. The Gunsha breed had an equal representation from both maternal origins, while breeds like Liangzhou, Dezhou, and Xinjiang displayed a more intermediate distribution.

Specifically, the overall distribution of the 244 individuals showed 56.15% belonging to the Somalian lineage (137 individuals) and 43.85% to the Nubian lineage (107 individuals). The presence of both maternal lineages across all 13 breeds—including Guanzhong (GZ), Dezhou (DZ), Mongolia (MG), Xiji (GY), Qingyang (QY), Jiami (JM), Biyang (BY), Huaibei (HUB), Liangzhou (LZ), Taihang (TH), Xinjiang (XJ), Gunsha (GS), and Qinghai (QH)—suggests a historical pattern of admixture likely driven by transregional trade and human-mediated livestock exchange across Central Asia and the Silk Road corridors.

This widespread admixture underscores that the Chinese donkey breeds were not formed through isolated domestication events, but rather shaped by continuous influxes of African maternal genetic material. The detection of five diagnostic SNPs in the Cytb gene further enabled the clear delineation of Nubian and Somalian haplotypes, reinforcing the dual-African maternal origin model and expanding previous insights obtained through mtDNA D-loop analyses.

Therefore, the genetic conformation of Chinese donkeys represents a unique synthesis of two ancient African lineages. This not only emphasizes the continent's central role in donkey domestication but also reveals the complexity of genetic transmission over long distances and through extended historical timelines. The co-occurrence of Nubian and Somalian mitochondrial lineages in all examined Chinese breeds is a compelling testament to the mobility of both animals and people across continents in antiquity.

This case exemplifies the transformative power of genomic tools when applied to historically under-characterized livestock populations. By leveraging complete mitochondrial cytochrome b sequencing, researchers have illuminated the complex biogeographic origins and evolutionary trajectories of Chinese domestic donkeys—animals once overlooked in global livestock genomic assessments. The identification of both Nubian and Somalian maternal lineages within all surveyed breeds not only underscores the depth of genetic diversity retained in these populations but also challenges assumptions of geographic isolation in domestication processes. It highlights how ancient networks of trade and mobility shaped the genomic landscape of domestic animals as profoundly as they did human populations. In doing so, this research elevates Chinese donkey breeds as a key model for studying admixture, domestication, and gene flow across time and space, reinforcing the value of including traditionally neglected populations in global genomic narratives.

3 Donkey Genome Sequencing and Annotation

The first high-quality donkey (*Equus asinus*) genome was generated by Huang et al. (2015) using deep shotgun sequencing (Huang et al. 2015). The de novo assembly spans ~2.36 Gb in 2166 scaffolds (N50 ~ 3.8 Mb), outperforming earlier drafts (Huang et al. 2015). Comprehensive annotation was achieved with transcriptome data: RNA-seq from eight tissues (heart, liver, spleen, lung, kidney, brain, spinal cord, muscle) produced ~1.39 million reads (avg 522 bp) to guide gene models (Huang et al. 2015). In total 23,214 protein-coding genes were predicted (mean CDS ≈ 1.28 kb), of which 15,648 had supporting RNA evidence (Huang et al. 2015). This rich gene catalog—validated by extensive transcriptome support—provides the foundation for studying donkey biology and genetics.

Comparative analysis reveals donkey-specific expansions in several functional gene categories. A total of 283 gene families are significantly expanded in donkey versus only 206 in horse (Huang et al. 2015). Donkey-expanded families are enriched for olfactory transduction and nutrient metabolism (e.g. protein digestion pathways), whereas horses show expansions in immune and stress-response gene sets (Huang et al. 2015). Beyond gene families, many individual donkey genes are rapidly evolving in energy and immune functions. In donkey versus horse, fast-evolving genes are significantly associated with aerobic respiration, the tricarboxylic acid cycle, and lymphocyte differentiation (Huang et al. 2015). These shifts likely underlie donkeys' characteristic physiology: the genome reflects more effective energy metabolism and enhanced immunity compared to horses (Huang et al. 2015). Together, the expansions and rapid evolution in donkey point to selection on metabolic and immune pathways during domestication and adaptation.

Genomic data clarify the donkey's place in *Equus* evolution and its population history. Phylogenomic dating (using 5665 single-copy orthologs) places the split between donkey and horse at ~7.7–15.4 million years ago, and the divergence of domestic donkey from the Asiatic wild ass at ~1.5–3.3 Ma (Huang et al. 2015). Demographic reconstructions (PSMC) reveal stark contrasts: the horse lineage (and similarly the Asiatic wild ass) experienced multiple Pleistocene bottlenecks, whereas the donkey population size remained remarkably stable over the last million years (Huang et al. 2015). In fact, unlike northern equids impacted by Ice Age climate swings, the African wild ass ancestors of donkeys seem to have avoided severe contractions (possibly due to more stable tropical/subtropical environments) (Huang et al. 2015). These findings indicate that, despite ancient bottlenecks in horses, donkey ancestry maintained a steady effective population size through glacial cycles.

Equids exhibit extraordinary karyotypic plasticity, and the donkey genome highlights this dynamic evolution. At least 13 centromere repositioning events have been documented in the donkey lineage: at least 7 shifts between donkey and horse chromosomes plus 6 more unique to donkey (Huang et al. 2015). CENP-A ChIP-seq and FISH experiments show that 16 of the donkey's 31 chromosomes now carry satellite-free neocentromeres (Piras et al. 2010). In contrast, the horse karyotype ($2n = 64$) has only one known satellite-free centromere (ECA11) (Piras et al. 2010). Thus, most donkey centromeres were "born" by repositioning rather than inheritance of ancestral satellite arrays. Consistent with this, Huang et al. (2015) found abundant classical satellite repeats only at inactive (formerly centric) horse/donkey loci, and no satellite DNA at any active donkey neocentromere (Huang et al. 2015). Genome-wide, repeats account for ~42% of donkey DNA, but canonical alpha-satellite sequences are virtually absent (~0.05% of the genome) compared to ~1.6% in horse (Huang et al. 2015). Remarkably, many donkey neocentromeres contain ribosomal DNA (rDNA) repeats (Huang et al. 2015)—a novel centromeric feature. In short, donkey karyotypes are shaped by frequent centromere shifts producing satellite-free centromeres, in contrast to the satellite-based centromeres of most mammals.

The donkey genome also revealed an expanded repertoire of regulatory RNAs that may influence chromosome biology. Huang et al. (2015) annotated 1198 miRNAs (versus 881 in horse), along with 512 snoRNAs, 530 snRNAs and 189 lncRNAs. Several miRNA families are significantly expanded in donkey, and their predicted targets are enriched in cell-cycle regulation, cancer, and oocyte meiosis pathways (Huang et al. 2015). Notably, 40 novel donkey-specific miRNAs were discovered, five of which target genes directly involved in meiosis (Huang et al. 2015). Integration of miRNA target data with analyses of rapidly evolving genes highlighted the Anaphase-Promoting Complex (APC/C)—a master regulator of sister chromatid segregation, cytokinesis and the G1 cell cycle phase—as a central node under accelerated evolution in donkey (Huang et al. 2015). These patterns suggest that miRNA-mediated regulation of meiosis and the cell cycle may have contributed to the donkey's fast karyotype turnover.

4 Breed Molecular Differentiation Signals

Molecular differentiation among donkey breeds reflects both their evolutionary history and the impact of human-mediated selection, drift, and geographic isolation. Advances in molecular genetics—particularly the use of mitochondrial DNA, microsatellites, and, more recently, SNP genotyping—have enabled the identification of distinct genetic signatures that differentiate donkey populations across regions. These signals are essential not only for tracing domestication pathways and phylogeographic structure, but also for informing conservation priorities and breed management. In the case of either standardized or under-characterized breeds, either it is genetically, phenotipically, reproductively, functionally or socioculturally, understanding molecular differentiation helps clarify relationships between local populations and reveals reservoirs of unique allelic diversity. When interpreted alongside estimates of heritability and other quantitative parameters for conservational or functional traits, these molecular patterns provide a foundation for identifying candidate loci under selection and designing breed-specific genomic strategies. As such, molecular differentiation is not merely a descriptive tool, but a critical entry point into the functional and adaptive genomics of donkey populations.

4.1 Coat and Phaneroptics

4.1.1 Albinism in Asinara and Austro-Hungarian Donkeys and Ivory Miniature Donkeys

Albinism in donkeys is an uncommon autosomal recessive condition characterized by the complete absence of melanin pigmentation, resulting in white coat coloration, pink skin, and light blue or gray-blue eyes. This phenotype is notably present in two distinct populations: the Asinara donkey, a semi-feral breed from the island of Asinara (Sardinia, Italy), and the white Austro-Hungarian donkey, historically distributed in Central Europe (Fig. 1). In the Asinara population, the albino phenotype has been traced to a specific missense mutation in the tyrosinase (TYR) gene (c.604C > G; p.His202Asp), which affects the copper-binding site critical for tyrosinase enzymatic activity and thus melanogenesis (Utzeri et al. 2016). All Asinara albino donkeys are homozygous for this mutation, while gray individuals carry the wild-type allele. This mutation, inherited in a recessive Mendelian manner, confirms TYR as the causal gene for Asinara albinism. Despite their lack of melanin, Asinara donkeys display remarkable tolerance to solar radiation, likely due to behavioral adaptations such as photophobic shade-seeking and potentially metabolic adjustments. In fact, research has documented elevated levels of fat-soluble antioxidants like retinol (vitamin A) and tocopherol (vitamin E) in their blood during summer months, which may serve as internal photoprotective agents in the absence of melanin (Cappai et al. 2024).

The Austro-Hungarian white donkey population exhibits a strikingly similar phenotype—white coat, pink unpigmented skin, and blue eyes—but genetic analysis has demonstrated that these animals do not carry the TYR mutation found in the Asinara breed. Furthermore, no known depigmentation variants at the KIT gene locus have been identified in the Austro-Hungarian population (Grilz-Seger et al. 2020). Despite their outward similarities, comparative genetic studies indicate that the two groups are not closely related, and no common ancestor for the albino trait can be assumed. This suggests that the phenotype in the Austro-Hungarian population likely arises from alternative, currently unidentified mutations. Both populations are of high conservation value. The Asinara donkey, with a population of

Fig. 1 Albino Asinara Feral (upper right), Albino Austrian Hungarian White Baroque (upper left) and (below) Ivory American Miniature Donkey Breeds

approximately 100–120 individuals, has undergone fixation of the TYR allele due to geographic and reproductive isolation. Conservation efforts must therefore balance the protection of this unique allele with the management of genetic diversity to avoid inbreeding depression. Similarly, the Austro-Hungarian white donkey, recently recognized as a distinct breed, should be preserved as a separate genetic lineage. Ongoing and future studies aimed at identifying the unknown mutations responsible for albinism in this population will further aid in the development of targeted conservation strategies.

By contrast with the aforementioned, the term *ivory* (also referred to as BEW or *blue-eyed white*) describes a donkey with a cream to pale cream coat, pink skin, and icy blue eyes. Ivory is anecdotally observed and even selectively sought after by breeders through directed matings in American Miniature Donkeys. Although often mistaken for albino or pure white, ivory donkeys are not albino—they retain normal pigmentation in ocular and dermal tissues. Foals are typically born with a slightly deeper cream tone and may display a visible dorsal cross, which often fades with age. As they mature, the coat tends to lighten further, and small pigmented freckles may emerge on the pink skin surrounding the eyes, muzzle, and genitals.

The ivory trait follows a fully autosomal recessive mode of inheritance, requiring two copies of the mutant allele (homozygous *i/i*) to express the phenotype. Breeding two ivory donkeys consistently results in 100% ivory offspring. If one parent is ivory and the other is a carrier (*I/i*), approximately 50% of the foals may be ivory. Because heterozygotes are phenotypically indistinguishable from wild-type (*I/I*) individuals, foals can occasionally be born ivory from two seemingly non-ivory parents. This sets ivory apart from the cream gene in horses (*SLC45A2*), which is incompletely dominant: one copy results in partial dilution (e.g., palomino, buckskin), while two copies yield nearly white phenotypes (e.g., cremello, perlino). In contrast, the ivory gene shows no observable phenotype in heterozygotes, and does not produce any intermediate coat colors.

Other equine dilution genes—including champagne (*SLC36A1*), silver (*PMEL17*), and dun (*TBX3*)—typically act in a dominant or incomplete dominant fashion, producing visible effects in heterozygotes and frequently combining in complex ways. The ivory gene, by contrast, behaves like a classic recessive mutation, functionally more akin to the red coat color (*ee*) in horses. However, unlike most pigment-related genes in equids, the molecular identity of the ivory allele remains unknown, though its phenotypic consistency and inheritance pattern suggest a loss-of-function mutation.

Given its lack of interaction with eumelanin or pheomelanin pathways in heterozygotes, the ivory gene likely involves a novel genetic mechanism. Unlike dilution genes that affect melanin synthesis, the ivory mutation may disrupt melanosome maturation, pigment cell migration, or pigment deposition, only when present in homozygous form. Importantly, ivory donkeys do not exhibit neurological or auditory impairments, unlike certain blue-eyed or white-pigmented animals in other species (e.g., Dalmatians or white cats), suggesting the mutation is not pleiotropic in affecting non-pigment systems. That said, ivory individuals may show increased photosensitivity, occasionally lowering their heads or squinting in bright light, and are more prone to sunburn on exposed pink areas such as the muzzle or ears. These effects are manageable with proper care, including shade access and topical protection.

Because of its rarity, the ivory trait raises concerns regarding inbreeding and genetic bottlenecks. Breeding strategies that include mating ivory donkeys to unrelated non-ivory carriers can help maintain diversity while preserving the trait. However, at present, no commercial genetic test exists to detect the ivory allele, unlike those available for cream, dun, silver, or champagne genes in horses. Thus, identification relies entirely on phenotypic observation and pedigree analysis, which poses challenges for large-scale management. Collaborative efforts between breed registries and genetic laboratories are ongoing, with the aim of sequencing and characterizing the mutation to enable molecular diagnostics.

In summary, the ivory phenotype in miniature donkeys is a rare and recessively inherited trait producing a cream coat, pink skin, and blue eyes. It is genetically distinct from known horse dilution genes, with no visible heterozygous expression, no known pleiotropic effects, and likely involves a unique mechanism not yet molecularly characterized. Its conservation requires strategic breeding, genetic monitoring, and future research to identify the underlying mutation and ensure responsible propagation of this visually striking and genetically intriguing trait.

4.1.2 Long-Hair Phenotype in Poitou, Ragusano, Zamorano-Leonés, and Miranda Donkeys

The distinctive long, shaggy coat of certain European donkey breeds (notably the French Poitou, Italian Ragusano, Spanish Zamorano-Leonés, and Portuguese Miranda, Fig. 2) has been traced to recessive mutations in the *FGF5* gene (fibroblast growth factor 5), which encodes a key regulator of the hair growth cycle in mammals. Some sources address Somalian Wild Ass or even Asiatic Wild Asses as the source for the mutation, though this has not been genetically contrasted. In a comprehensive genetic study of Poitou donkeys, researchers sequenced *FGF5* and identified two independent loss-of-function mutations strongly associated with the long-hair phenotype. The potential connection between these breeds has been suggested to radish upon the Zamorano-Leonés donkey breed, which enjoyed protection of the Crown of Castile and León. Indeed, it was forbidden to export animals belonging to this breed until the eighteenth Century, when King Charles III (maybe earlier with his father Phillip V) would remove the prohibition. Despite the ban, individuals of the breed managed to cross the Portuguese – Spanish border which has been the suggested connection between Mirandese and Zamorano-Leonés donkeys. It would be then when Zamorano Leonese donkeys would eventually be exported to France to contribute to the formation of the Baudet du Poitou breed.

The connections between Ragusano and Spanish donkeys rather suggest the links with Catalonian donkeys. Catalonian donkey has a history of being exported to various countries, including Italy. In Tunisia, for instance, the French protectorate introduced Catalan donkeys to improve local breeding stock, starting in 1938. This indicates a precedent for the exportation of Catalan donkeys to regions outside Spain, including Sicily. However, the interest and value given to Zamorano-Leonés donkeys by Charles III of Spain, son of Philip V has been historically acknowledged. Indeed, before becoming monarch of Spain, Charles III reigned in Naples and Sicily, so it may be likely that the monarch acted as a potential connection and that Zamorano-Leonés were imported, which was supported by molecular genetic evidence (Jordana et al. 2016).

One is a 2-base-pair frameshift deletion (c.433_434delAT) and the other is a nonsense substitution (c.245G > A). These mutations introduce premature stop codons, producing truncated proteins that lack essential receptor-binding domains. Animals homozygous or compound heterozygous for these variants develop the characteristic long, matted coat of the Poitou breed, while donkeys with at least one normal allele show a short-haired phenotype (Legrand et al. 2014).

FGF5 normally acts as a molecular signal that terminates the anagen (growth) phase of the hair cycle and induces catagen (regression). This regulatory role has been conserved across mammals and was first identified in mice carrying natural *Fgf5* mutations, which grow abnormally long hair due to delayed catagen induction. In donkeys, the identified *FGF5* mutations likely abolish this signaling function, prolonging the anagen phase and resulting in excessive hair growth (Legrand et al. 2014).

Interestingly, the same long-hair *FGF5* mutations have been found in low frequencies in Zamorano-Leonés and Miranda donkeys. This suggests a shared genetic background, possibly due to historical gene flow between the Poitou

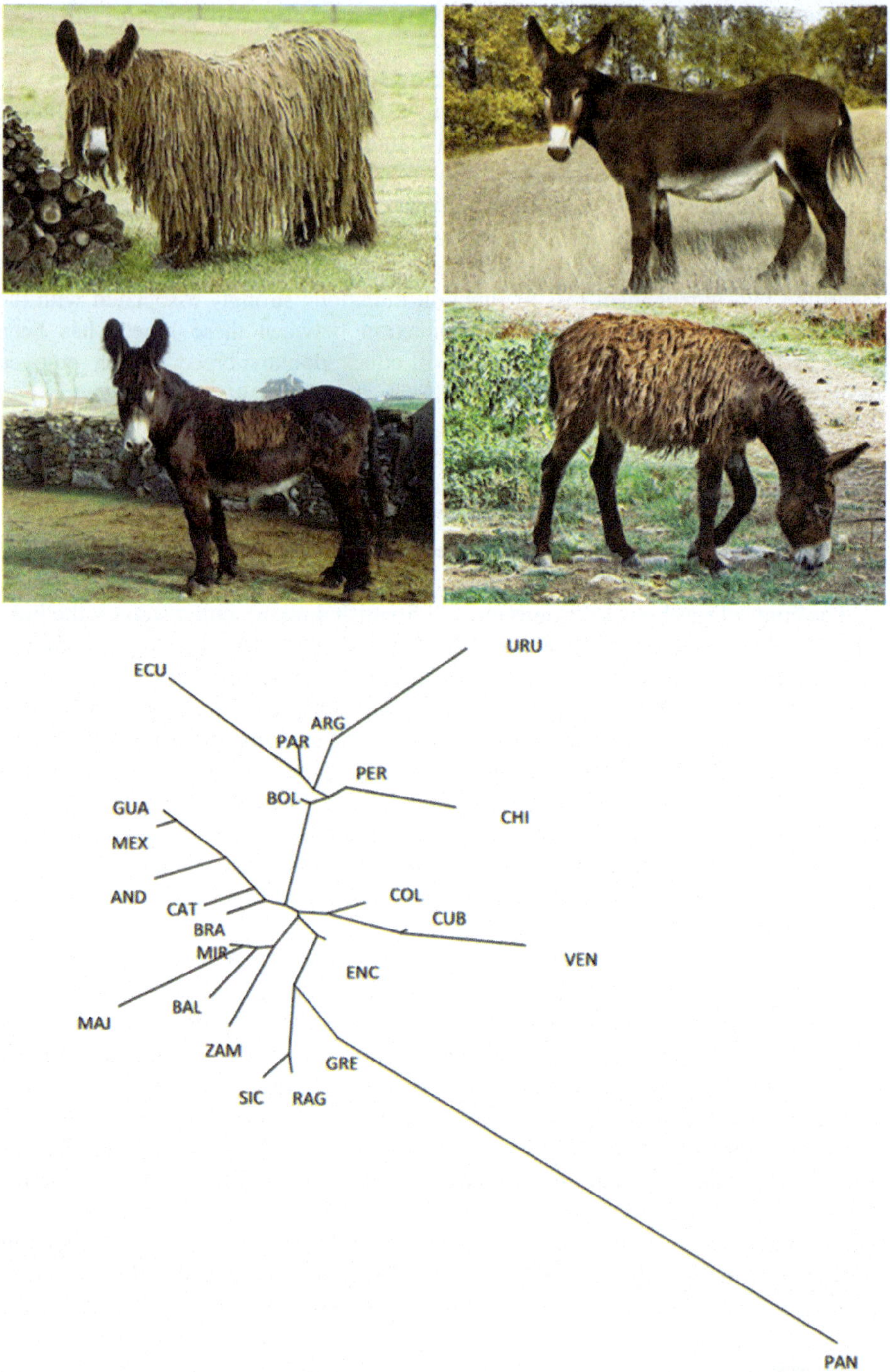

Fig. 2 Poitou donkey (upper left), Mirandese (upper right), Zamorano-Leonés (middle left) and Ragusano donkey (middle right) breeds and Unrooted tree of Reynolds genetic distances for

population and these Iberian breeds. Historical data show that Poitevin donkey individuals, also known as the Baudets du Poitou, originally from the Poitou region of western France, renowned historically for its role in producing mule hybrids through crossbreeding with mares were historically exported to Spain, where they played an important role in mule production and genetic influence, especially during the 18th and 19th centuries, but also that later on Spanish Zamorano jennies would have been introduced into France in the late twentieth century as part of conservation efforts, potentially bringing the mutations back into the Poitou gene pool (Legrand et al. 2014).

Earlier breeding practices likely facilitated the dissemination of these alleles. For example, studbooks for some French donkey breeds were only established in the late twentieth century, and early registrations were sometimes based on appearance rather than verified pedigree. This may have led to the inclusion of donkeys carrying *FGF5* mutations from Spanish or Poitou ancestry. Similarly, during efforts to recover the nearly extinct Poitou population, individuals from related or neighboring breeds were incorporated, potentially reinforcing the presence of the long-hair trait (Legrand et al. 2014).

The identification of the underlying mutations in *FGF5* has enabled the development of genetic tests to detect carriers and manage the trait in breeding programs. Because the mutations are recessive, heterozygous animals appear normal but can produce long-haired offspring if mated with another carrier. Breeders now use DNA testing to screen stock and plan matings that either preserve or avoid the long-hair phenotype, depending on breed standards and conservation goals. This is especially important for endangered breeds like the Poitou, Zamorano-Leonés, and Miranda, where maintaining genetic diversity while managing phenotypic traits is critical (Legrand et al. 2014).

4.1.3 Solid-Black Coat Color in Dezhou (Wutou) Chinese Donkeys and Their Cultural Significance

Coat color in mammals is largely determined by the balance of dark eumelanin and light pheomelanin pigments, which is controlled by the melanocortin 1 receptor (MC1R) and its antagonist, agouti signaling protein (ASIP). Binding of α-melanocyte-stimulating hormone (α-MSH) to MC1R promotes cAMP signalling and eumelanin (black/brown) synthesis (Almathen et al. 2018). In contrast, ASIP competes with α-MSH and suppresses MC1R activity, shifting pigment production toward pheomelanin (red/yellow) (Almathen et al. 2018). Thus, functional ASIP produces light "points" (pangare) on muzzles, bellies and limbs, whereas loss of ASIP function causes uniformly dark (recessive black) coats by unleashing MC1R-driven eumelanin (Almathen et al. 2018; Abitbol et al. 2015). This has been documented across species: recessive loss-of-function mutations in ASIP cause

Fig. 2 (continued) 24 American and European donkey pop-ulations obtained with the neighbour-joining algorithm (lower), ZAM Zamorano-Leonés; MIR Mirandese and RAG Ragusano. Poitou was not included

Fig. 3 Grand Noir du Berry (upper left), Normand (upper middle), Catalonian Pyrenean (lower left), Gasconian Pyrenean (lower middle) and Contentin (right) donkey Breeds from France

solid-black phenotypes in sheep, quail, foxes and other animals (Almathen et al. 2018). In donkeys, most colored individuals carry the normal ASIP allele and show the light-point (pangare) pattern (Abitbol et al. 2015). But donkeys with the "no light points" (NLP) phenotype lack these pale markings and are solid-black all over; this NLP pattern in donkeys is analogous to the classic recessive-black (a) phenotype seen in other mammals and is likewise caused by ASIP dysfunction (Abitbol et al. 2015; Almathen et al. 2018). Using sequencing of the three exons of donkey ASIP, Abitbol et al. (2015) discovered a single missense variant that perfectly cosegregates with the NLP trait. In a panel of 127 donkeys (Normand, Berry Black, Pyrenean, Poitou, Cotentin, and American Miniature breeds, Fig. 3), every individual with the NLP phenotype was homozygous for a c.349 T > C mutation in exon 3 of ASIP (Abitbol et al. 2015). This SNP causes a cysteine-to-arginine substitution at amino acid 117 (p.Cys117Arg) in the ASIP protein (Abitbol et al. 2015). All nine NLP animals (dark donkeys, Fig. 4) in the study carried the C/C genotype, whereas the 118 normal "light points" (LP) donkeys were either homozygous T/T or heterozygous C/T (Abitbol et al. 2015). Abitbol et al. showed that Cys117 lies in ASIP's conserved cysteine-rich C-terminal domain: mutagenesis studies indicate that almost all ten of these cysteines (including Cys117) are critical for ASIP activity (Abitbol et al. 2015). In particular, mouse ASIP Cys113 (the counterpart of donkey Cys117) is essential for binding MC1R; mutating it abolishes agouti function. Thus, the donkey Cys117Arg change is predicted to inactivate ASIP. Abitbol et al. conclude that the c.349T>C allele (called a^nlp) is the causative variant for the NLP phenotype (Abitbol et al. 2015).

The p.Cys117Arg substitution likely destroys ASIP's antagonism of MC1R, tilting pigmentation entirely toward eumelanin. Normally ASIP suppresses MC1R to allow pheomelanin synthesis (Almathen et al. 2018). Without functional ASIP, MC1R remains constitutively active under α-MSH, driving melanocytes to produce dark eumelanin. As noted above, loss-of-function ASIP alleles in other species cause similar "recessive black" coats (Almathen et al. 2018). In donkeys this means

Fig. 4 American NLP solid black Miniature donkeys

homozygous Cys117Arg/Arg animals cannot lighten their points: they lack the pale muzzle, eye-ring, belly and limb marks and are solid-black overall. In vitro and comparative analyses support this: the Arg substitution abolishes a conserved disulfide linkage in ASIP, and experimental mutagenesis in mouse shows that altering this cysteine yields a fully inactive agouti protein (Abitbol et al. 2015). Consequently, NLP donkeys have unchecked MC1R-driven eumelanin and no pheomelanin banding (Almathen et al. 2018). The NLP mutation behaves as a simple autosomal recessive allele. Abitbol et al. (2015) found complete association: only donkeys homozygous for the Cys117Arg allele show NLP, while heterozygotes appear phenotypically normal (LP) but can produce NLP offspring when mated together (Abitbol et al. 2015). UC Davis's Veterinary Genetics Laboratory confirms this: NLP is autosomal recessive, so carriers (NLP/normal) have normal light points, but two carriers have a 25% chance of an NLP foal. The lab notes that an ASIP DNA test is available, and that breeding two carriers yields 1/4 NLP foals and 1/2 carriers. In practice, genotyping breeding stock allows donkey breeders to identify carriers of the Cys117Arg mutation and manage matings either to propagate the black trait or to avoid it, depending on goals (UC Davis VGL 2024).

In China, solid-black Dezhou donkeys (Fig. 5) carry special cultural and economic value. The hides of Dezhou donkeys are used to produce Ejiao (donkey-hide gelatin), a traditional Chinese medicinal product reputed to boost blood and vitality. Ejiao is made by stewing and extracting collagen from donkey skin; black donkey hides are often prized for this purpose. As one review notes, Ejiao has been valued since antiquity for immunological, antioxidant and hematopoietic benefits, especially in treating anemia (Seyiti and Kelimu 2021). Consequently, demand for donkey hides for Ejiao has skyrocketed in recent decades. Prices for Ejiao and donkey skin have risen dramatically, creating an annual shortage of hides in China. This commodity boom has made donkey hide weight a critical economic trait. Indeed, Dezhou donkeys are one of China's five most important breeds, and breeding goals explicitly emphasize maximizing hide yield and quality. Producers

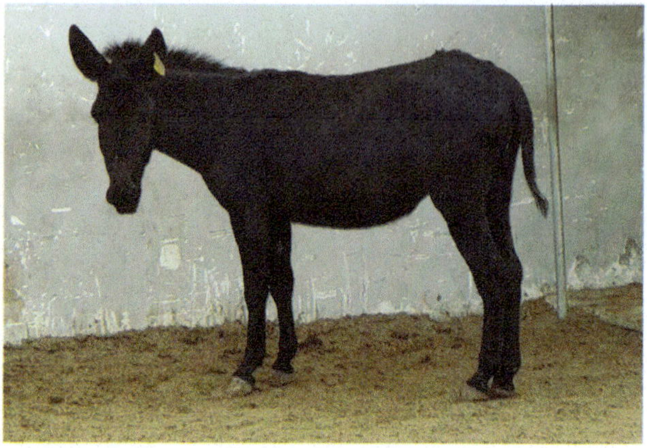

Fig. 5 Dezhou donkey breed (Liu et al. 2024)

and consumers often favor large Dezhou jacks with thick, black hides for premium Ejiao production (Huang et al. 2022; Seyiti and Kelimu 2021). In short, the NLP (solid-black) phenotype is not merely cosmetic: it is strongly linked to market demand in the Ejiao trade, elevating the status and price of black-coated Dezhou donkeys.

The discovery of the ASIP Cys117Arg variant provides a genetic marker for the NLP trait. Now that the causative SNP is known, breeders can use DNA tests to identify carriers and select matings. For example, the UC Davis VGL offers a commercial test for the donkey NLP allele (UC Davis VGL 2024). Breeders can thus choose matings to enrich or eliminate the black coat allele: pairing two carriers deliberately yields NLP foals, whereas avoiding carrier-carrier matings prevents unwanted black progeny. More broadly, modern genomic tools are being applied in Dezhou donkeys. For instance, Huang et al. (2022) identified an LCORL gene SNP associated with larger body size and heavier hide weight in Dezhou donkeys, providing a marker to breed for higher-hide-yield animals. Such marker-assisted selection (MAS) is now feasible for key traits in these donkeys.

4.1.4 Pinto or Spotted Burros; White Spotting and Dominant White Phenotypes in Donkeys

Spotted or white-patterned donkeys (Fig. 6)—commonly referred to as "pintos"—display a range of depigmentation patterns, from small white patches to nearly complete absence of pigmentation. Recent genetic studies have identified the KIT gene—already known for its role in pigmentation in horses, mice, and other mammals—as a central genetic determinant of these coat patterns in donkeys (Haase et al. 2015; Fenn et al. 2017). Two distinct mutations within the KIT gene have been associated with white spotting and dominant white phenotypes in this species.

The first is a splice site mutation (c.1978 + 2 T > A), found exclusively in donkeys exhibiting white spotting. This variant likely disrupts normal mRNA splicing, leading to impaired expression of the KIT protein, which is essential for

Fig. 6 Spotted or Pinto donkey

melanocyte development and migration. As a result, affected animals present with sharply demarcated white patches on their coats. Heterozygous donkeys (N/W^s) consistently show this phenotype, while homozygosity (W^s/W^s) appears to be embryonically lethal, a pattern analogous to certain white spotting alleles in horses (Haase et al. 2015; Tanaka et al. 2020).

The second mutation is a missense variant (c.662A > C; p.Tyr221Ser) identified in a fully white-born donkey. This mutation alters a critical extracellular domain of the KIT receptor, likely affecting ligand binding and signal transduction. The resulting phenotype includes an all-white coat with pink skin but pigmented eyes, distinguishing it from dilution phenotypes like ivory (Fenn et al. 2017). Like the splice-site mutation, this variant also appears to follow a dominant inheritance pattern, and homozygotes are presumed non-viable.

These genetic discoveries have significant practical implications for breeders. By using molecular testing to identify these KIT variants, breeders can predict coat patterns and avoid producing potentially lethal homozygous offspring (Tanaka et al. 2020; UC Davis 2025). Moreover, the identification of these alleles contributes to a deeper understanding of the genetic architecture of coat color in donkeys and highlights parallels with pigmentation genetics in other domestic species (OMIA 2025; Miller 2023).

An especially striking example of white spotting is seen in the Brazilian Pêga donkey, which exhibits the so-called "pinto" or "pega" coat. These animals typically present with large, asymmetrical patches of white and colored hair. It is likely that this pattern also involves the KIT gene, as in horses with sabino and tobiano spotting. Anecdotal evidence suggests that in some Pêga donkeys, the extent of depigmentation may increase with age, potentially due to progressive loss or dysfunction of melanocytes (Miller 2023; Haase et al. 2015).

4.1.5 Developmental Coat Lightening and Color Diversity in Donkeys: From White Hassawi Donkeys to Gray and/or Dun Andalusians and Pêgas

Donkeys display a striking range of coat colors and developmental transitions, with many breeds undergoing dramatic changes in pigmentation as they mature. This is especially evident in gray-coated breeds like the Andalusian donkey, American Mammoth or Brazilian Pêga (Fig. 7). Foals are typically born with a dark coat—ranging from deep brown to nearly black—that gradually lightens over time. As the animal matures, this juvenile pigmentation fades to reveal the soft silvery-gray or

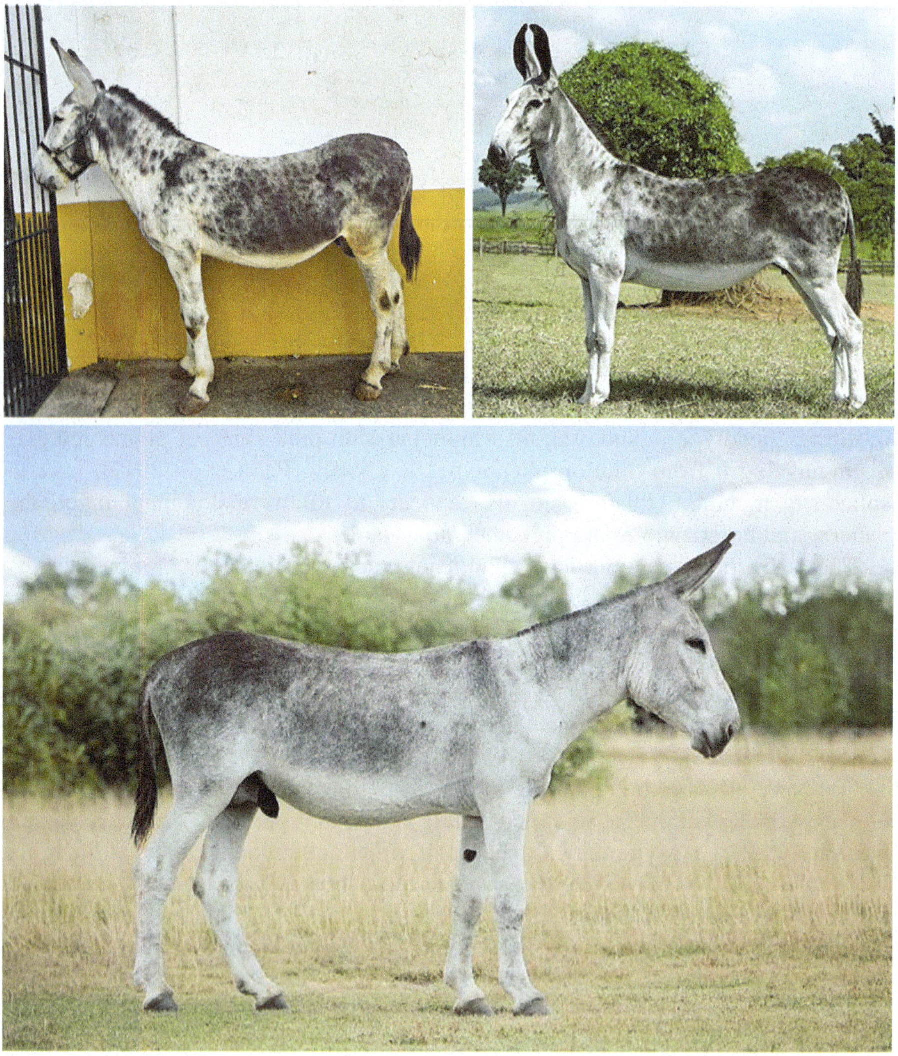

Fig. 7 Andalusian (left), Pêga (right) and American Mammoth (below) Donkey Breed

pale dove tones typical of adult Andalusians. This process is not merely cosmetic; it reflects complex developmental timing of pigment gene expression, potentially involving delayed activation of dilution pathways. Genes such as *ASIP* and *MC1R*, along with as-yet-unidentified modifiers, are likely involved, though the exact regulatory mechanisms remain incompletely understood in donkeys (Abitbol et al. 2014; Abitbol et al. 2015).

A similar lightening pattern is observed in donkeys with dun coloration, where the initial dark coat of foals conceals diluted adult hues. In these cases, the fading process reveals the hallmark features of dun dilution—pale body color often accompanied by primitive markings such as a dorsal stripe, shoulder bars, and leg striping. These features may be subtle or obscured in early life, especially in individuals born with particularly dark coats. As the animals grow, the diluted tones and striping patterns become more distinct, creating a dynamic and sometimes unexpected transformation in appearance (Wu et al. 2020).

One of the most extreme examples of progressive depigmentation is found in the Hassawi donkey of formerly Egypt and afterwards Saudi Arabia (Fig. 8). These animals are typically known for their fully white coats and pink skin in adulthood, but anecdotal evidence suggests that some individuals are born with very light cream or pale gray coats that further fade with age. This lightening process is not consistent with albinism, as eye pigmentation remains unaffected, pointing instead to a form of leucism. Genetic candidates such as *KIT* and *MITF*—both involved in melanocyte development and migration—are strong contenders for the underlying cause (Haase et al. 2015; Liu et al. 2024).

Remarkably, the Hassawi donkey may have a far older origin than commonly assumed. Historical references to similarly white donkeys can be traced back to ancient Egypt during the time of the pharaohs, suggesting that the phenotype associated with the modern Hassawi may descend from populations originally bred or valued for their pale coloration in the Nile Valley. If so, this lineage represents one of the oldest known examples of selective breeding for coat depigmentation in domestic equids (Almathen 2024; Wu et al. 2020).

Taken together, these cases of developmental coat lightening—from Andalusian gray to dun dilution and the nearly complete depigmentation of Hassawi donkeys—illustrate the complexity and diversity of pigmentation pathways in domestic donkeys. They reflect not only underlying genetic variation but also the influence of environmental pressures and millennia of human selection, creating a rich mosaic of phenotypes across geography and history (Utzeri et al. 2016; Liu et al. 2024).

4.1.6 Brown Coat with Lighter Ends in the Martina Franca, Catalonian, Balearic, de las Encartaciones and American Mammoth Donkeys

Certain donkey breeds are known for their distinctive coat color patterns, which often include a rich brown base color with lighter extremities on the muzzle, belly, and legs. This characteristic is particularly prominent in breeds such as the Martina Franca, Catalonian, Balearic, de las Encartaciones, and American Mammoth donkeys, but not only these (Figs. 3 and 9). These animals typically have deep, dark brown coats along the back and flanks, while the ends of the coat, including the

Fig. 8 Saharian (upper) Egyptian Hassawi White Riding donkey Breed (lower)

legs, belly, and face, are lightened, giving them a striking gradient appearance (Pierini et al. 2025; Shields 2025; Oklahoma State University n.d.; Ministerio de Agricultura, Pesca y Alimentación 2020; Gobierno Vasco 2020).

The coat color in these breeds results from a combination of genetic factors, primarily involving the agouti signaling protein gene (*ASIP*) and the melanocortin 1 receptor gene (*MC1R*). The *ASIP* gene regulates the distribution of eumelanin (black/brown pigment) and pheomelanin (red/yellow pigment) by locally inhibiting the activity of *MC1R*. In areas where *ASIP* expression is high—typically the ventral regions—less eumelanin is produced, leading to lighter coloration. Conversely, low *ASIP* expression dorsally allows for full eumelanin production, creating a dark brown to black back and flanks (Abitbol et al. 2015; Liu et al. 2024; Schmutz and Berryere 2007).

Fig. 9 de las Encartaciones (upper left), Ruc Catalá, Guará Catalá, Catalán or Catalonian (Middle), American Mammoth (rigth), Martina-Franca (lower left) and Ase Belear Balearic (lower middle) Donkey Breeds

Variants in the *MC1R* gene influence the type of melanin produced. Functional *MC1R* promotes eumelanin synthesis, resulting in black or dark brown pigmentation, while loss-of-function variants shift pigment production toward pheomelanin, producing chestnut or red coats. In donkeys, recessive mutations in *MC1R* and *ASIP* are associated with lighter or red phenotypes, whereas the classic brown-with-light-points pattern arises when both genes function normally and interact to create regional pigment variation (Abitbol et al. 2014; Abitbol et al. 2015; Liu et al. 2024).

The lighter extremities—commonly known as the pangaré pattern—are a widespread feature in wild and domestic equids. This dorsoventral contrast in pigmentation is thought to have evolved as a form of camouflage or thermal regulation, with the pale belly and muzzle reflecting heat and blending with ground tones in open environments. The pattern is not exclusive to donkeys; it appears across equids and other mammals and reflects deeply conserved gene expression gradients in embryonic skin development (Schmutz and Berryere 2007; Cieslak et al. 2011).

Breed standards and historical descriptions confirm the prevalence of this phenotype. The Catalan donkey has long been described as dark brown with a light muzzle and underbelly (Shields 2025), while the Martina Franca breed standard notes a bay or dark brown coat with pale markings on the abdomen and face (Pierini et al. 2025). The Balearic and Encartaciones donkeys also exhibit black or dark coats with characteristic pale points around the eyes, muzzle, and ventral areas (Ministerio de Agricultura, Pesca y Alimentación 2020; Gobierno Vasco 2020). The American Mammoth Jackstock, descended from imported Spanish breeds like the Catalonian and Majorcan donkeys, inherited this same coat pattern. Early American mule breeders, including George Washington, are known to have selected donkeys with this phenotype for use in creating sturdy, well-pigmented mules. Historical documents indicate that Washington favored Mammoth Jacks with dark coats and pale points, not only for their appearance but also for their vigor and adaptability in the American South (Cothran and Long 2011).

This brown coat with lighter extremities represents not only an aesthetic preference but also a product of complex genetic interactions and centuries of selective breeding. It reflects the interplay of gene expression, adaptation to environmental pressures, and cultural preferences that have shaped donkey breeds across continents.

4.1.7 Ancestral Markings in Donkeys: Insights from the Wild Ass Ancestors

Ancestral "primitive" markings—notably the dark dorsal stripe, shoulder cross (a dark band across the withers), and zebra-like leg bars—are characteristic of wild asses and are conserved in some domestic donkey lineages. Inndeed some sources explain that markings ae always present though they may be visible or not depending on the case, as examples of their presence can be seen on all types of coats in donkeys. These markings are rooted in equid coat-color genetics. The melanocortin-1 receptor gene (*MC1R*) and agouti signaling protein gene (*ASIP*) control base pigment and its distribution: in donkeys, loss-of-function *MC1R* alleles (e.g., *e/e*) produce a red (chestnut) coat, whereas wild-type *MC1R* alleles allow black (eumelanic) pigment (Abitbol et al. 2014). *ASIP* alleles modulate the "light points" pattern: a recessive *ASIP* missense mutation (c.349T>C) in *Equus asinus* causes a "no light points" (NLP) phenotype, eliminating the pale muzzle/eye/belly marks (Abitbol et al. 2015).

Importantly, the dilution of background coat color and emergence of dark primitive markings (dorsal stripe, leg barring, shoulder cross, etc.) are controlled by the dominant Dun locus. In donkeys this corresponds to a wild-type allele of the *TBX3* gene, which causes pigment dilution on most of the body but leaves dorsally oriented stripes and other markings intact (Imsland et al. 2016). A 1-base deletion downstream of *TBX3*, found only in domestic donkeys, reduces *TBX3* expression and abolishes the dun phenotype—producing uniformly dark or chestnut coats with few or no stripes (Wang et al. 2020). Thus, *ASIP* and *MC1R* set the basic color and light-point patterns, while *TBX3* (with its modifiers) produces the distinctive stripes and cross markings. These genes interact epistatically: for example, an *e/e* (chestnut) donkey will lack any visibly black dorsal stripe or leg barring regardless of *TBX3* status, and heterozygotes at *ASIP* may show intermediate light-point patterns. Environmental factors (seasonal coat changes or nutrition) can further modulate pigment intensity, but the striped pattern is genetically determined.

Among African wild asses, primitive markings are pronounced and vary by subspecies. All three extant (including the extinct Atlas) subspecies of *Equus africanus* exhibit a dark dorsal stripe (Librado and Orlando 2021). The Atlas wild ass (*E. a. atlanticus*, now extinct) is believed to have had both leg stripes and a prominent shoulder cross, as depicted in Roman mosaics (Librado and Orlando 2021). The Nubian wild ass (*E. a. africanus*) typically has a gray coat with a shoulder cross but no leg stripes, while the Somali wild ass (*E. a. somaliensis*) has sandy-gray fur with black horizontal leg barring but generally lacks a shoulder cross (Librado and Orlando 2021; Wang et al. 2020) (Fig. 10). These patterns match expectations from Dun genetics: all wild asses are dun (diluted) and hence show at least the dorsal stripe, while specific alleles inherited in each lineage produce the cross or leg bars observed (Imsland et al. 2016; Wang et al. 2020). Domestic

Fig. 10 Nubian Wild Ass (upper left), Atlas or Algerian Wild Ass (upper middle) and Petroglypth (Upper right), Sokotra Wild Ass, a Nubian Wild Ass Subspecies (lower left), Somali and Nubian Wild Asses (lower middle) and Algerian Wild Ass (lower right)

donkeys (*E. asinus*), descendants of these wild ancestors, can display the same markings. Surveys of wild asses show that dun coloration (with stripe) is the ancestral wild type, whereas the more pigmented non-dun forms occur only in domestic and feral stock (Wang et al. 2020; Librado and Orlando 2021).

The adaptive significance of these markings has been widely discussed. Dun patterns—dark dorsal stripes and leg bars—likely served as camouflage in open, grassy environments by disrupting the animal's outline and helping it blend into striped shadows or tall grasses (Cieslak et al. 2011), but also to prevent the attack of flies which may be prevented from gathering around this sensitive areas (Caro et al. 2019). Similar camouflage effects are noted in wild horses, and dorsal striping in equids is thought to assist in heat regulation and concealment from predators (Cieslak et al. 2011; Imsland et al. 2016). In social contexts, contrasting markings like the shoulder cross or facial masks may have aided individual or group recognition within herds, though this remains speculative. In sum, stripes and crosses likely evolved as camouflage and thermoregulatory traits in wild asses, persisting in domestic donkeys unless suppressed by selection for novel coat colors.

In modern donkey breeds, expression of ancestral markings varies due to breeding and genetic interactions. Many domestic donkeys carry non-dun *TBX3* alleles and thus lack clear primitive marks; however, dun variants still occur, especially in populations with stronger wild-ass ancestry (Wang et al. 2020). Even *non-dun1* genotypes may allow faint striping, while *non-dun2* (complete loss of *TBX3* function) abolishes them altogether (Imsland et al. 2016). Epistasis plays a role: for example, red (*e/e*) donkeys cannot visibly express black dorsal stripes, and *ASIP* interactions may influence visibility of the shoulder cross. Environmental influences (seasonal molt, diet) may modulate pigment intensity, but the pattern itself remains genetically determined.

Paleogenetic and evolutionary studies confirm the ancient origins of these traits. Dun-type coloration with striping predates equid domestication. For instance, ancient DNA from ~43,000-year-old wild horses showed both dun and non-dun alleles, confirming that *TBX3*-related striping is ancestral (Librado and Orlando 2021). Mitochondrial DNA and archaeological remains trace donkey domestication to African wild asses ~7000–6000 years ago, indicating that primitive markings were inherited from wild ancestors into early domestic populations (Wang et al. 2020; Librado and Orlando 2021). Altogether, the distinctive dorsal stripes, leg barring, and shoulder crosses visible in many donkeys today are a clear genetic legacy of their wild ass forebears, shaped by both natural selection and domestication.

4.1.8 Coat Color Inheritance Issues in Standardized Breeds: How Two Gray Donkeys Can Produce a Black Foal

At first glance, it may seem counterintuitive that two gray-coated donkeys can produce a black offspring (Fig. 11). However, when viewed through the lens of equine coat color genetics, this outcome is entirely plausible. The key lies in understanding that *gray* in donkeys, as in horses, is not a base coat color but rather the result of a dominant graying gene (*G*) that causes progressive depigmentation of the hair over time (Sponenberg 2003; Rieder et al. 2008).

A gray donkey is genetically a non-gray base color (such as black or bay) with the gray allele (*G/g*) acting on top of that base. The gray allele causes the donkey to gradually lose pigmentation, often beginning around the eyes and muzzle and spreading with age. However, genetically, the donkey retains the original base coat color beneath the graying effect (Bellone 2010).

For example:

- A gray donkey with genotype Gg Ee aa is genetically black (*aa* = black base) but appears gray due to the presence of the *G* allele.

Fig. 11 Dark coated Andalusian Foal Born to an Officially Registered Couple of Mating Parents Displaying Standardized Andalusian Coats

Fig. 12 Normand Donkey Breed. Bay Normand donkeys occasionally give birth to no light points (NLP) donkeys that are not officially recognized by the studbook (NLP donkeys, middle)

- If both parents are heterozygous for the gray gene (*Gg*), there is a 25% chance that their foal will inherit gg, meaning no gray gene at all and thus will not undergo progressive graying (Bowling 1996).

If the foal also inherits the black base coat alleles (*aa* at the Agouti locus and *E_* at the Extension locus), it will express the black coat color fully, unaffected by gray. Example of Possible Parental Genotypes:

- Parent 1: *Gg Ee aa* (gray donkey, genetically black)
- Parent 2: *Gg Ee Aa* (gray donkey, genetically bay or black carrier)
- Offspring: *gg Ee aa* → non-gray black foal

This outcome follows Mendelian inheritance patterns:

- Gray (*G*) is dominant over non-gray (*g*)
- Black (*aa*) is recessive to bay (*A_*) and must be homozygous to be expressed
- Pigment presence (*E*) is dominant and required for black or bay to appear; *ee* results in red (though this is extremely rare in donkeys) (Sponenberg 2003).

In summary, two gray donkeys—each genetically black but grayed by one copy of the *G* allele—can produce a black foal if that foal inherits *gg* (non-gray) and *aa* (black) from its parents. This event, while uncommon, is a well-understood genetic possibility, which occurs but which is not normally registered in the studbooks of breeds such the Andalusian or the Normand donkey breeds given Black Light Points individual sor NLP individuals, respectively fall out of the oficial standards of these breeds (Figs. 11 and 12).

4.1.9 Melaos, Melados or Honey-Coloured Pêgas, Mammoth and Andalusians

The term melao denotes a distinctive donkey coat pattern characterized by an intermixture of dark (black), reddish/chestnut, and white hairs. Literally derived from the Spanish melado ("molasses" or "honey-colored"), melao implies a rich amber or honey-brown hue. In practice, a melao donkey has a bay or chestnut base overlaid with white ticking and darker hairs, producing a speckled or roan-like appearance. This mixed-color "salt-and-pepper" pattern is analogous to the horse ruano (roan) phenotype: horses described as ruanos have black, red, and white hairs intermingled. In donkey breeding, melao is an informal term for such roanish coats (sometimes called "molasses-colored"), which stand out against the more typical solid or grulla coats. The melao pattern is relatively rare and has attracted interest in recent years in breeds like the Andalusian and Brazilian Pêga donkeys (Diccionario de Americanismos 2010).

Melao is a colloquial abbreviation of melado, the Spanish word for molasses or syrup. By extension, it has come to describe a "honey-colored" brown shade akin to the color of molasses or dark caramel (literally "color miel"). The Diccionario de Americanismos notes that melao is used in various Latin American dialects (Cuba, Dominican Republic, Puerto Rico, Colombia, Bolivia, etc.) as a variant of melado. In Iberia and Latin America, melao is applied informally by breeders to describe a sorrel/chestnut coat with unusual roaning. The Portuguese language lacks a direct equivalent term—melado exists, but "melao" as such is Spanish. Thus in Brazilian contexts the Spanish-derived "melao" may be understood via contact with Latin dialects, though official Pêga breed standards simply use terms like rua or tordilha for roan patterns (Cavalus 2020) (Fig. 13). In any case, melao literally evokes the idea of a sweet, syrupy brown color, which aptly describes the rich reddish-brown hairs of this coat (Fig. 14).

A melao donkey coat typically exhibits a roan-like pattern: mixed dark, reddish, and white hairs. The base coat is usually bay or black, overlain by white ticking or interspersed light hairs. The result is a mottled effect often described as chestnut or brown flecked with white. In terms of classical color terminology, this resembles the horse ruano (roan) pattern, defined as "with black, red and white hairs interspersed". Equid color authorities note that true roan arises from an even spread of white hairs over a base color. In melao donkeys one sees the same phenomenon: the dark mane, tail, muzzle, and lower legs may remain relatively dark, while the body shows a mixture of white and chestnut hairs. The overall impression is of a "molasses" or mahogany-tinted coat, sometimes described poetically as alazano-melao (chestnut-molasses). Individual melao donkeys may vary in how evenly the colors mix—some look uniformly roaned, others have patches of chestnut or darker roaning. Unlike dominant spotting patterns, melao does not create large white areas, but a fine salt-and-pepper texture.

Donkey coat colors are governed by genes analogous to those in horses. A key factor in melao is the red (chestnut/sorrel) locus: like horses, donkeys carry a recessive MC1R mutation that produces a uniformly red body and mane ("red factor") (UC Davis Veterinary Genetics Laboratory 2024). Homozygosity for this e/e allele yields an overall reddish-chestnut donkey. In melao coats, this red base is present but modified. Another major gene is Agouti, which determines distribution of black pigment (bay versus black base), but its role in melao is unclear. Importantly, melao requires an additional mechanism to introduce the white hairs. In horses, the classic roan gene (dominant Rn) has been mapped, but in donkeys a specific roan locus has not been conclusively identified. It is likely that melao

Fig. 13 Jumentos Pêga or Pêga Donkey breed coat colour variability

involves polygenic dilution or perhaps a yet-unknown allele that spreads white hairs into the red/black coat. Donkeys do have known "white spotting" genes (dominant white W alleles) that can cause white patterns, but melao is distinct from the large white areas of, say, pinto spotting. Instead, it may be analogous to the fine ticked or frosted effect of Sabino or roan.

While donkeys do not display the vivid "bright sorrel" often seen in horses, breeders note that the donkey red color tends to lean toward brownish tones. In fact, research on donkey coat genetics confirms that donkey red (MC1R e/e) typically produces a duller, mouse-brown shade rather than a brilliant red (UC Davis Veterinary Genetics Laboratory 2024). This inherent brown cast likely combines with roan-like dilution to create the melao phenotype. Thus genetically, melao can be thought of as the overlap of the red factor with an interspersed white hair pattern. Testing of Pêga and Andalusian donkeys has identified the MC1R mutation for red but has not reported a separate roan locus; the expectation is that the phenotype arises from known alleles acting together.

The pure Andalusian, Lucentinan or Cordobesian donkey is traditionally described as gray/roan (tordo). Official breed sources note a short, fine coat that is "predominantly tordo-rucio (gray-roan), ranging from light to dark" and often nearly

Fig. 14 Melaos donkeys; from upper left Andalusian, upper right; Pêga, below American Mammoth donkey breeds, displaying also different degrees of expresión of Melao coat

white (Asnopra 2019). Andalusian donkeys typically carry ample white hairs mixed with dark ones—essentially a classic roan pattern. Solid chestnut or bay coats are not standard for the breed. However, rare chestnut individuals do occur. For example, a famous Andalusian-cross donkey named "Caramelo" drew attention for its unusual alazano, melao color. In a published interview, trainer Carlos Román explicitly calls Caramelo's coat "alazano, melao" and notes that it is "very rare" (Horsepress.com 2019). This suggests that melao happens occasionally in the Andalusian gene pool but is considered a novelty. In general, Andalusian breeders prize the characteristic grey/roan look, so a chestnut-roan (melao) donkey would stand out. No official standard recognizes melao separately, but in breeding records or ads one sometimes sees terms like "alazán melao" to describe such animals, and these are registered officially at least in the Andalusian Donkey Breed.

Observed in some American Mammoth donkeys is likely inherited from their Spanish ancestors, particularly Andalusian and Majorcan donkeys. These Iberian breeds, brought to the Americas during colonial times, carried genetic variants associated with diluted coat colors. As Mammoth donkeys were developed in the United States from large European breeds—including Spanish donkeys—these dilution traits were preserved and occasionally expressed. The presence of melao coloring in modern Mammoths suggests the persistence of these ancestral alleles, possibly linked to specific dilution genes. Ongoing genetic studies aim to clarify the molecular basis of this phenotype and confirm its Iberian origin.

The Brazilian Pêga donkey, developed in Minas Gerais, has a more varied palette. Breed literature lists acceptable colors as ruça, tordilha, pelo-de-rato claro, etc., always with dark hooves and skin (Cavalus 2020; UFRRJ Equideocultura 2023). In context, ruça and tordilha are types of roan (mixed) coats, while pelo-de-rato means a mouse-colored dun. Thus Pêga donkeys commonly exhibit roaning or dun patterns with a dorsal stripe and leg barring. For example, a Pêga of ruço type has a brownish base with light hairs interspersed (similar to melao), while tordilha is a gray-roan. White markings (star, socks) are allowed, and a white belly or spotted pattern can occur but are not typical.

The Pêga breed standard does not explicitly use the term melao, but a chestnut-roan would fall under ruça ("bay-roan") in practice. In fact, Pêga registries explicitly recognize roan variants; any Pêga with a white-overlaid roan pattern (even mostly white) is included under ruã or ruça. Unlike the Andalusian, the Pêga standard places strong emphasis on gait and the presence of a dorsal stripe ("faixa crucial") and shoulder stripe. A pure solid chestnut color without roaning or stripes is unusual and not mentioned as a breed hallmark. Thus, if a Pêga donkey with melao-like coloring exists, it would likely be catalogued simply as ruço com mesclas brancas (roan with white) within the breed registry.

The melao pattern seems to be chiefly noted in Iberian and Brazilian contexts. Other donkey breeds (e.g. Italian, Miranda, Sicilian) have their own color traditions, but melao specifically is rarely mentioned. For instance, the Italian donkeys (Pantesco of Pantelleria or Grigio Viterbese, for example), Fig. 15) do not present the "pampa" coat or spotted white coat patterns, and indeed pampa coat is not even officially considered typical of Italian donkeys, contrary to what occurs in Brazilian Pêga. No major donkey registry uses melao as a formal category. It may occur sporadically in mixed-breed animals wherever Andalusian or Spanish donkey stock was introduced. In summary, aside from Andalusian and Pêga contexts, melao-type coats are anecdotal and not codified in breed standards.

Because melao coats are uncommon, local breeders often regard them as curiosities. In Andalusia, the striking appearance of a chestnut-roan donkey like Caramelo has been highlighted in media, underscoring its novelty (Horsepress.com 2019). The trainer himself noted the rarity of Caramelo's alazano-melao color. This suggests that while not a traditional trait, a melao donkey can be of special interest or aesthetic value. Breeders of pure Andalusian stock typically aim for the classical grey-roan, so a red-roan (melao) might be seen as "unusual" rather than desirable for show. In Brazil, Pêga breeders primarily value conformation and gait, though color

Fig. 15 Pantesco de Pallenteria (left) and Grigio Viterbese (right) Donkey Breeds

rules favor roans. The Pêga standard explicitly praises ruças e tordilhas, so a ruço-com-manchas-brancas (bay-roan, akin to melao) would be accepted. However, since emphasis is less on novelty, Brazilian sources do not single out melao as a special category. In both regions, there is no superstitious stigma attached to melao; it is simply a rare coat that some may find beautiful. On online forums and sales ads (e.g. Spanish classifieds) the word "melao" is used descriptively ("burro melao para cubrir") when such color appears, indicating general familiarity.

Photographic evidence of melao donkeys is limited to a few media reports and breeder archives. The Andalusian donkey "Caramelo" was widely covered in Spanish press with images clearly showing its reddish-roan coat (Horsepress.com 2019) (Fig. 16).

Similarly, photographs of Brazilian Pêga herds often include roan-patterned jacks and jennies, though melao would appear simply as a ruço. The two images above illustrate typical examples: an Andalusian donkey in performance training (with a chestnut-roan coat) and a Brazilian Pêga mare and foal (grey/bay-roan). Equid photo galleries and breed publications occasionally caption ruan-pattern donkeys as "alazán melao" or "ruço." However, there is no standard photographic compendium for melao specifically. Illustrations in older livestock manuals do not depict it distinctly. In recent years, some breeders have uploaded melao donkey pictures on social media or sales sites, using the term in Spanish, but these are unvetted. In summary, while visual examples exist, melao remains under-documented in the scientific literature—most evidence comes from the equestrian press and breed enthusiasts.

The melao phenotype in donkeys represents a rare roan-related coat pattern with cultural roots in Iberian and Latin American terminology. Defined by its molasses-

Fig. 16 Caramelo and Carlos Román

hued mix of chestnut, black, and white hairs, it highlights the genetic complexity of donkey coloration. Modern genetics confirms that donkey red (chestnut) is a recessive trait (MC1R mutation), and the white interspersion likely arises from an as-yet-unmapped roan or dilution effect. Breeders recognize melao as a curiosity: in Andalusia it has led to famous specimens like Caramelo, while in Brazil it falls under accepted roan classes. Scientifically, melao underscores how compound patterns can emerge in domestic breeds. Culturally, it connects livestock lore (the word melao itself) with equine genetics. As conservation efforts for rare donkey breeds continue, awareness of unusual phenotypes like melao may encourage preservation of genetic diversity. Further research—including genetic testing of melao individuals—could clarify the inheritance of this distinctive coat.

4.1.10 Pointed Ears in Pêga and Andalusian Donkeys

Morphology also reflects adaptation to environmental pressures. Donkeys in arid regions—including the Hassawi—tend to have long, narrow ears that curve inward at the tips. This ear structure may serve adaptive purposes such as enhanced heat dissipation or improved sound localization, while also functioning as a distinguishing feature in regional breed identification (Almathen 2024; Liu et al. 2024).

These pointed, "elf-like" ears—marked by a sharper taper and enhanced erectness—have become a distinctive feature in certain donkey breeds such as the Brazilian Pêga and Spanish Andalusian breeds. Though specific genetic studies in donkeys are limited, comparative evidence from donkey transcriptomics reveals several candidate genes (e.g. HIC1, PRKRA, KMT2A) implicated in ear morphology, suggesting a developmental genetic basis for ear shape that extends beyond simple Mendelian inheritance (Xie et al. 2015). The inheritance pattern likely follows a polygenic–quantitative model, wherein multiple genes affecting cartilage structure, growth regulation, and possibly modifier loci contribute cumulatively to ear shape variation. Analogous traits in other mammals—including pinna morphology influenced by multiple loci in sheep and humans—support this view. These genes impact not only the length and stiffness of the ear but also the taper of the apex, hence the pointed appearance. While detailed heritability analyses and genome-wide mapping remain to be performed in donkeys, breeders' observations that pointed-eared Pêga and Andalusian individuals often produce similarly shaped offspring suggest moderate to high heritability with additive genetic effects. Environmental factors such as prenatal nutrition and fetal positioning, along with age-dependent cartilage maturation, likely modulate expressivity. In sum, pointed ears in these donkey breeds are best understood as a quantitative, heritable trait governed by multiple structural genes with cumulative effect; molecular dissection of this trait in donkeys remains a promising area for future research.

4.1.11 Coat Color and Adaptive Traits in Asiatic Wild Ass Hybrids

The phenotypic diversity of donkey populations, particularly in coat color and adaptive traits, reflects complex interactions between domestication history, environmental pressures, and recent hybridization events. A notable contribution to this field is the study by Dong et al. (2022), which employed whole-genome sequencing to explore the genetic basis of coat color, immune function, and productive traits in the Asiatic wild ass (*Equus hemionus hemionus*) and its hybrids with domestic donkeys in China (Fig. 17). The findings provide a foundational reference for understanding genetic diversity, identifying targets of selection, and guiding breeding strategies in donkeys.

Mongolian kulans, a subspecies of the Asiatic wild ass, display a distinctive yellowish-brown coat and are well adapted to arid environments. In contrast, the Xinjiang donkey, one of China's native domestic breeds, is typically gray and smaller in body size. The offspring resulting from crosses between Mongolian kulans (sires) and Xinjiang donkeys (dams) exhibit a coat color closely resembling the wild-type phenotype of the kulan, with remarkable consistency across hybrid individuals. This visual uniformity prompted an investigation into the genomic regions underlying pigmentation.

Genome-wide selection scans using $\theta\pi$-ratio, XP-EHH, CLR, and nucleotide diversity metrics identified strong positive selection signatures in the *KITLG* (KIT ligand) gene region. This gene, located on chromosome 2, is a well-established regulator of pigmentation in mammals and has been associated with coat color variation in donkeys, horses, dogs, and humans. In the Kulan hybrids, the allelic pattern at *KITLG* significantly diverged from that observed in Guanzhong donkeys, a black-coated domestic breed, providing compelling evidence of its involvement in the inheritance of the yellowish-brown coat color typical of the Asiatic wild ass. These results align with previous studies that have identified *KITLG* as a major candidate gene in the formation of diluted or dun-like coat phenotypes in equids.

Beyond pigmentation, the genomic analysis also revealed signatures of selection in genes associated with immune competence. Several defensin genes (*DEFA1*,

Fig. 17 Guanzhong donkey (upper left), Mongolian Kulan (upper right), A male donkey x Asiatic wild ass hybrid *(Equus heminous khur)* (lower left) and Heitian Gray Xinjiang donkey (lower right)

DEFA5, *DEFA7*), immunoglobulin components (*IGLC1*, *IGLL5*), and regulators of lymphocyte development (*GIMAP1*, *GIMAP4*) were identified as under positive selection within the hybrid population. These genes are implicated in both innate and adaptive immune responses, suggesting that the hybrids may have inherited enhanced immunological robustness from their wild ancestors. The identification of *GZMB*, a granzyme involved in cytotoxic immune responses and natural killer cell function, further supports the hypothesis of improved disease resistance in these hybrids. Given the harsh, disease-prone environments in which the Asiatic wild ass evolved, these traits may offer significant adaptive advantages in hybrid offspring.

In addition to coat color and immunity, the study uncovered genomic regions associated with traits linked to meat quality and muscle development. Notably, genes such as *PDE1B* and *MYLK2*, associated with calcium signaling and actin cytoskeleton regulation, were detected in sweep regions. These genes have been previously correlated with carcass quality and muscular traits in other livestock species, indicating that hybrid animals may offer improved meat productivity. While phenotypic evaluations are necessary to validate these functional predictions, the genetic evidence points to the potential utility of Kulan hybrids in meat-focused donkey breeding programs.

Finally, the *LCORL* gene, a known determinant of skeletal frame and stature across multiple livestock species, was also identified in sweep analyses. Its presence reflects underlying differences in body size between hybrid animals and domestic breeds, such as the smaller Xinjiang donkey (Fig. 21) and the larger Guanzhong donkey (Fig. 21). The Kulan hybrids (Fig. 21) appear to retain the more compact stature of their dam line, despite the introgression of wild ass ancestry.

In summary, the integration of wild genetic resources through hybridization with domestic donkeys offers a novel route for enhancing phenotypic and functional diversity in equids. The findings by Dong et al. provide critical insights into the genetic mechanisms underlying pigmentation, immune resilience, and potentially productive traits in donkey populations. These results emphasize the importance of conserving wild ass lineages, not only for biodiversity preservation but also for their potential to enrich domestic breeding programs with adaptive and economically valuable traits.

4.2 Diseases and Conditions

4.2.1 Dwarfism in Miniature Donkeys

Dwarfism in miniature donkeys (Fig. 18) is a congenital skeletal dysplasia that produces disproportionate short stature with characteristic limb and craniofacial malformations (Boegheim et al. 2017). A study by the American Donkey and Mule Society (ADMS), initiated in the late 1990s and continuing into the early 2000s, aimed to establish average body proportions in Miniature Donkeys to assist in identifying potential cases of dwarfism using objective measurements. Data were collected through registration booklets and published calls in *The BRAYER* magazine. Drawing on a large sample of owner-submitted measurements and referencing earlier comparative research by W.R. Willoughby in *Empire of Equus* (1974), the ADMS focused particularly on the head-to-leg length ratio. They found that

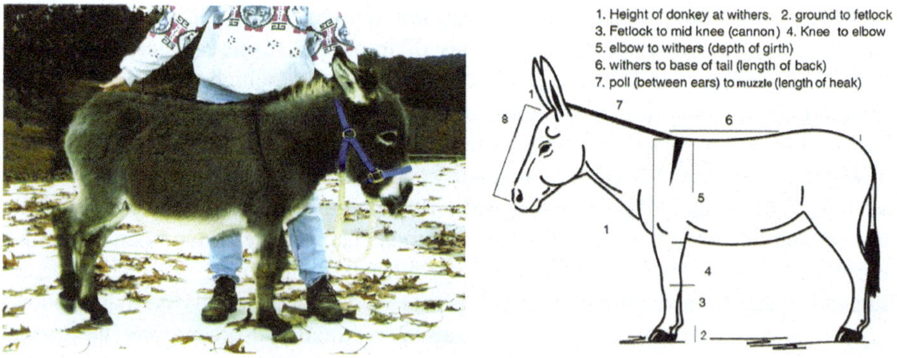

Fig. 18 Dwarfism in a Miniature Donkey and Scheme by the American Donkey and Mule Society (ADMS) to score and measure conformation and proportions

Miniature Donkeys with head-to-leg ratios between 0.75 and 0.85, averaging 0.799, fell within the normal range, while those with ratios above 0.95 often exhibited visual signs of dwarfism, such as enlarged heads and shortened limbs. When compared to Willoughby's findings in other equids—such as Arabians (0.653), Thoroughbreds (0.692), Draft horses (0.741), and Grevy's zebras (0.798)—Miniature Donkeys demonstrated unusually high ratios. These findings suggest a breed tendency toward extreme proportions and emphasize the importance of standardizing morphological assessments to better detect and manage dwarfism in Miniature Donkeys.

Affected foals typically have markedly shortened legs (micromelia) relative to body length, with bowed or angular limb deformities. The skull often appears domed or brachycephalic, with a shortened muzzle (retruded maxilla), mandibular brachygnathia (underbite), and a high-arched or cleft palate. Eyes may bulge and the nasal bridge is flattened. An oversized abdomen and hernias are common due to the severely reduced length of vertebrae and ribs. Dental malocclusion ("monkey mouth") frequently develops as the small jaw bones are misaligned, predisposing to overgrowth of incisors and cheek teeth. Gastrointestinal problems (chronic diarrhea, poor weight gain, bloating) have been observed in dwarf donkeys, likely reflecting malabsorption or abdominal overcrowding by organs. These signs mirror chondrodysplastic dwarf phenotypes documented in miniature horses and cattle (Jacinto et al. 2020). Respiratory compromise may occur if a cleft palate or narrow nasal passages are present (Metzger et al. 2017). Behaviorally, dwarfed donkeys may be lethargic or weak, presumably due to their musculoskeletal deficits, although formal studies are lacking.

Management of dwarfism in donkeys centers on supportive, multidisciplinary care. Early and frequent dental evaluation is critical, because mandibular/maxillary deformities cause malocclusion. Owners and veterinarians should monitor for hooks, ramps, and wave mouth and perform routine floating to maintain normal occlusion (Boegheim et al. 2017). Orthopedic care includes trimming or corrective shoeing to mitigate bowed or short limbs, along with weight control to minimize joint stress. Regular exercise or physical therapy (gentle exercise, physiotherapy, swimming) can help maintain joint mobility and muscle mass. Because dwarfs often tolerate cold poorly and have low body fat, shelter and warmth are also advisable. Respiratory issues deserve special attention: anesthetic risk is higher in dwarfed equids due to potential airway obstruction. In one reported miniature horse with dwarfism, severe breathing difficulty was managed during anesthesia (Metzger et al. 2017); similarly, any dwarf donkey undergoing sedation should be managed with attention to airway patency. Nutritionally, affected foals may need high-quality diets fed in small, frequent meals to aid digestion and prevent hyperlipemia. Diets rich in fiber and moderate in calories and protein help prevent obesity and liver lipidosis in miniature equids. Vitamin and mineral supplementation (e.g., vitamin E, biotin, omega-3 fatty acids, joint nutraceuticals) is often added to support muscle and joint health. Close monitoring and prompt veterinary intervention for infections or orthopedic injuries is advised.

Dwarfism in miniature donkeys is believed to be hereditary, though no specific donkey gene has been identified to date. By analogy with other species, likely candidates include genes regulating cartilage and bone growth. FGFR3 encodes fibroblast growth factor receptor 3, a negative regulator of growth plate chondrocytes. Activating mutations of FGFR3 cause human achondroplasia and severely limited bone growth (Pauli 2019). An FGFR3 gain-of-function mutation blocks chondrocyte proliferation in the growth plate and produces the classic short-limbed dwarf phenotype. In contrast, loss-of-function FGFR3 mutations cause skeletal overgrowth in animals. FGFR3 is highly conserved among mammals, and no FGFR3 sequence variation was found even among dog breeds of vastly different size, underscoring its critical role (Smith et al. 2008).

COL2A1 encodes type II collagen, the main collagen of cartilage. Mutations in COL2A1 cause severe chondrodysplasia, such as "bulldog calf" syndrome in cattle (Jacinto et al. 2020). Bulldog calves have disproportionate dwarfism with a severely shortened spine and limbs, compact body, and striking facial dysmorphism. This parallels the facial and vertebral abnormalities in dwarf equids. ACAN encodes aggrecan, the large aggregating proteoglycan in cartilage. Mutations in ACAN are well known to cause chondrodysplasia: e.g., Dexter cattle with a frameshift ACAN mutation exhibit "bulldog" dwarfism, and several mutations in ACAN (D1–D4 alleles) cause dwarfism in American Miniature horses (Boegheim et al. 2017). These ACAN dwarf mutants share many features with donkey dwarf phenotypes. Whole-genome sequencing of a dwarf Miniature Shetland pony identified a causative missense variant in ACAN (Metzger et al. 2017), further implicating this gene in equine dwarfism.

SHOX is a gene on the pseudoautosomal region of sex chromosomes that is important in bone growth. SHOX mutations in humans cause Léri–Weill dyschondrosteosis and idiopathic short stature. In equids, deletions affecting the region downstream of SHOX cause a skeletal atavism syndrome in Shetland ponies, suggesting that defects in SHOX or its regulatory elements could underlie dwarf phenotypes in miniature donkeys (Boegheim et al. 2017).

GH1 and GHR are genes in the growth hormone–IGF1 axis. Mutations in GH1 or GHR cause proportionate dwarfism. For example, Brooksville dwarfism in Miniature Brahman cattle is due to a GH1 missense mutation that markedly reduces GH activity (Rao et al. 1997). A comparable human GH1 mutation produces mild growth hormone deficiency. Mutations in the GH receptor cause GH insensitivity (Laron syndrome) in people and dwarfism in chickens (Lin et al. 2018). In GH/GHR defects, serum GH may be normal or high, but IGF-1 is low, impairing growth.

The genes implicated in dwarfism are highly conserved across species, and mutations cause remarkably similar syndromes. FGFR3 gain-of-function causes achondroplasia in humans and has been observed in mice and dogs. COL2A1 mutations that cause bulldog calves correspond to human type II collagenopathies. ACAN mutations cause chondrodysplastic dwarfism in cattle and miniature horses. SHOX haploinsufficiency causes growth disorders in humans and controls growth in pony skeletons. The GH1/GHR axis is also conserved: GH1 mutations produce proportionate dwarfs in cattle and pigs, and GHR mutations cause Laron syndrome in humans and dwarfism in chickens. Thus, each of the candidate genes is known from other species to control cartilage and bone growth.

All of these mutations converge on disrupted cartilage biology. FGFR3 normally limits chondrocyte proliferation in the growth plate; activating mutations prematurely signal the chondrocytes to differentiate and exit the cell cycle, impairing endochondral ossification (Pauli 2019). COL2A1 loss prevents normal cartilage matrix formation, yielding fragile cartilage and dysplasia. ACAN provides cartilage with resistance to compression; without it, cartilage zones collapse and bone cannot lengthen properly (Boegheim et al. 2017). SHOX regulates osteoblast/chondrocyte cell cycling; its deficiency leads to mesomelic shortening. The GH–IGF1 axis regulates systemic growth through IGF-1, which promotes chondrocyte proliferation; mutations in this pathway result in global growth failure (Lin et al. 2018).

Because many forms of inherited dwarfism are caused by single-gene mutations, DNA tests can be invaluable in breeding programs. In horses and cattle, PCR-based assays can detect carriers of known mutations. Identification of a GH1 mutation allowed testing in Brahman cattle to avoid carrier matings (Rao et al. 1997). A similar approach could be taken for dwarf miniature donkeys: once the responsible mutation(s) are discovered, genetic testing would allow carriers to be identified and bred carefully or not at all, reducing the incidence of this debilitating condition.

4.2.2 Jaw Malocclusion in Zamorano-Leonés Donkeys

Jaw malformations in equids include parrot mouth (mandibular brachygnathism, or overshot jaw), monkey or sow mouth (mandibular prognathism, or undershot jaw), and enognathism (shortened maxilla). In parrot mouth, the lower jaw is abnormally short relative to the upper jaw (Merck Veterinary Manual 2024a, b), whereas in monkey mouth the lower jaw protrudes beyond the upper (Merck Veterinary Manual 2024a, b). Enognathism is essentially the converse—a congenitally short upper jaw—leading to a relative underbite. These terms are often used interchangeably with class II and class III dental malocclusions, respectively.

In donkeys, parrot mouth is seen occasionally and is usually not clinically significant unless extreme (WikiVet 2023). By contrast, prognathism is very uncommon and usually causes no functional problem (WikiVet 2023). For comparison, parrot mouth is the most common congenital oral deformity in horses, whereas prognathism is often seen in Miniature and Arabian breeds (Merck Veterinary Manual 2024a, b). In all equids, severe jaw misalignment can impair grazing and lead to secondary dental overgrowths, but mild cases often go undetected without careful examination.

Malocclusion in equids is generally considered a multifactorial, complex trait rather than a simple Mendelian condition. Most cases are thought to have polygenic or environmental components. For example, donkey parrot/monkey mouth has been described as having a multifactorial etiology (Rodrigues et al. 2013). In horses, the inheritance of brachygnathism is unknown and can result from in utero toxins, such as griseofulvin (Merck Veterinary Manual 2024a, b). Maxillary prognathism in horses appears genetically complex; one major locus was found but is not the only risk factor (Signer-Hasler et al. 2014). In contrast, monogenic causes have been confirmed in other livestock. For instance, a recessive frameshift mutation in *WNT10B* causes brachygnathia in Brown Swiss cattle (Widmer et al. 2023), and a recessive syndrome involving brachygnathia in Merino sheep maps to ovine chromosome OAR2 (Shariflou et al. 2013). These findings indicate a polygenic/multifactorial pattern in equids, with single-gene forms only confirmed in cattle and sheep.

Some human studies of mandibular prognathism have implicated genes like *MATN1* and *SSX2IP*, suggesting conserved pathways across species (Rodrigues et al. 2013). In equids, only a few loci are known. A 500 kb region on horse chromosome ECA13 is associated with maxillary prognathism, but no specific gene within that region has been identified (Signer-Hasler et al. 2014). *MATN1* itself was not implicated in horses, despite being a marker for prognathism in humans (Rodrigues et al. 2013).

The only large genetic study of malocclusion in donkeys was conducted by Rodrigues et al. (2013). In Zamorano-Leonés donkeys, they compared 30 parrot-mouth cases, 30 monkey-mouth cases, and 30 normals. They sequenced the *MATN1* gene and identified one intronic SNP (g.503G > A) significantly associated with mandibular prognathism: the A allele was underrepresented in cases, suggesting a protective effect. No association was seen with parrot-mouth. The *MATN1* SNP appears to influence undershot jaw in this donkey population, though the effect is modest, and the trait remains multifactorial.

No other donkey genes have been reported. The *MATN1* variant remains the only known genetic association, and it does not establish a simple Mendelian model. Instead, the findings suggest a quantitative-trait model with possibly protective alleles.

In horses, both parrot mouth (overjet) and sow mouth (underjet) are recognized but rarely mapped genetically. Signer-Hasler et al. (2014) performed a genome-wide association study in Swiss Franches-Montagnes horses. Among 96 parrot-mouth cases and 763 controls, they found two SNPs on the distal end of chromosome 13 with genome-wide significance. The region, previously unknown, contains over 40 genes without obvious craniofacial function. The authors proposed that this ECA13 region harbors an unknown major gene for parrot mouth. However, only a fraction of cases carried the risk allele, and no other loci reached significance, reinforcing the complex inheritance.

Other sources treat horse malocclusion as heterogeneous and likely polygenic. Parrot mouth is described as the most common congenital deformity in horses, but many cases may result from environmental causes (Merck Veterinary Manual 2024a, b). Prognathism is noted to be common in Miniature and Arabian breeds (Merck Veterinary Manual 2024a, b). No specific horse gene, aside from the ECA13 locus, has been confirmed.

In cattle, brachygnathia inferior (overbite) has well-characterized genetic forms. In Brown Swiss cattle, a *WNT10B* frameshift mutation causes brachygnathism in homozygotes, and DNA-based testing is now used to reduce incidence (Widmer et al. 2023). Recessive cases also occur in red Angus and Simmental breeds (Merck Veterinary Manual 2024a, b). Underbite in cattle is rare and may result from viral causes or syndromes (Merck Veterinary Manual 2024a, b).

In sheep, jaw malformations are generally uncommon, but Merino sheep can suffer from a recessive syndrome—brachygnathia with cardiac and renal defects. This BCRHS syndrome maps to a 1.1 Mb region on ovine chromosome OAR2 (Shariflou et al. 2013). A causative mutation has not been published, but the inheritance is consistent with a single recessive locus.

In humans, mandibular prognathism is a complex trait with many implicated genes of small effect, such as *MATN1*, *SSX2IP*, and *PLXNA* (Rodrigues et al. 2013). In cats and dogs, jaw malformations are usually breed-typical features rather than genetic defects.

Only cattle and sheep have confirmed single-gene forms of jaw malocclusion. In equids, the traits are polygenic. The *MATN1* association in donkeys and the ECA13 locus in horses are the only mapped factors, and each explains only part of the risk (Rodrigues et al. 2013; Signer-Hasler et al. 2014).

No breed-specific predisposition has been reported for donkeys beyond the Zamorano-Leonés and Mirandês populations studied (Rodrigues et al. 2013). In horses, Miniature breeds and Arabians are noted for underbite (Merck Veterinary Manual 2024a, b). Breed registries consider these conditions faults. For example, the Miniature Mediterranean Donkey standard excludes animals with more than ¼″ misalignment (Miniature Donkey Association 2023).

No genetic tests are available for donkey or horse malocclusion, so breeding management is the primary tool. Affected animals should generally not be bred, or should be outcrossed to reduce risk. For example, breeders are advised to change mating pairs if a foal shows parrot mouth (ToothFest 2023). In contrast, cattle breeding programs now utilize DNA tests for brachygnathia (Widmer et al. 2023).

Regular dental care is also essential. In parrot-mouthed equids, unopposed incisors may overgrow and trap the opposite teeth, worsening alignment (WikiVet 2023). Preventive rasping and corrective dentistry can reduce complications. While environmental contributors (e.g. fetal position, toxins) are relevant, many cases have a heritable basis (Merck Veterinary Manual 2024a, b). Pedigree control and genetic screening are key to long-term reduction in incidence.

Parrot mouth, monkey mouth, and enognathism in donkeys and horses have confirmed but complex genetic components. The *MATN1* variant in donkeys (Rodrigues et al. 2013) and the ECA13 locus in horses (Signer-Hasler et al. 2014) support a multifactorial etiology. The Merck Veterinary Manual (2024a, b) concurs that these conditions are congenital and influenced by both genetic and environmental factors. In contrast, cattle and sheep have true Mendelian forms (Widmer et al. 2023; Shariflou et al. 2013). Jaw malocclusion in donkeys is likely polygenic, with no single major mutation identified. Breeders should avoid propagating severe cases, uphold registry standards, and ensure routine dental care. Larger genome-wide scans across diverse donkey breeds would help clarify the genetic architecture and improve selection strategies.

4.2.3 Disorders of Sex Development in Miniature Donkeys

In veterinary medicine, Disorders of Sex Development (DSDs) are congenital conditions in which chromosomal, gonadal or phenotypic sex do not correspond to each other (Rentea and Mehmood 2023). The outdated terms "hermaphrodite" and "pseudohermaphrodite" have been largely replaced by DSD terminology (Rentea and Mehmood 2023). By classical definition, a true hermaphrodite has both ovarian and testicular tissue (often in the form of bilateral ovotestes or one ovary and one testis) and ambiguous external genitalia (Tibary 2021). In contrast, a pseudohermaphrodite (sometimes called a sex-reversal syndrome) has only one type of gonad (ovary or testis) but external genitalia of the opposite sex (Tibary 2021). For example, an XY "male pseudohermaphrodite" has testes internally but typically female-like external genitalia, whereas an XX "female pseudohermaphrodite" has ovaries but virilized male-like genitalia (Tibary 2021). In horses and related species, most DSDs involve XY individuals lacking a functional SRY gene or downstream pathway, resulting in testis that fail to masculinize tissues (Tibary 2021) (Fig. 19).

An historical case is described by Mateos Nevado (1955) of an Andalusian donkey foal presenting pseudohermafrodistism (Fig. 19). This report described a rare case of male pseudohermaphroditism in an Andalusian donkey male foal (*Equus asinus*) observed in El Real de la Jara (Seville), Spain. The animal, approximately six years old with a height of 1.04 meters, presented ambiguous external genitalia that prompted veterinary attention. Though externally resembling a female—with a vulva-like perineal opening and two well-developed mammary glands in the inguinal region—it exhibited distinctly male behavior, including libido and mounting attempts toward female donkeys. Clinical examination revealed the absence of a clitoris or other true female structures within the vulvar cavity, no descended testes, and a palpable cylindrical structure beneath the perineum identified as a rudimentary internal penis. Scrotal rudiments appeared to form the labia majora-like folds surrounding the opening. Based on these findings and in reference to the works of

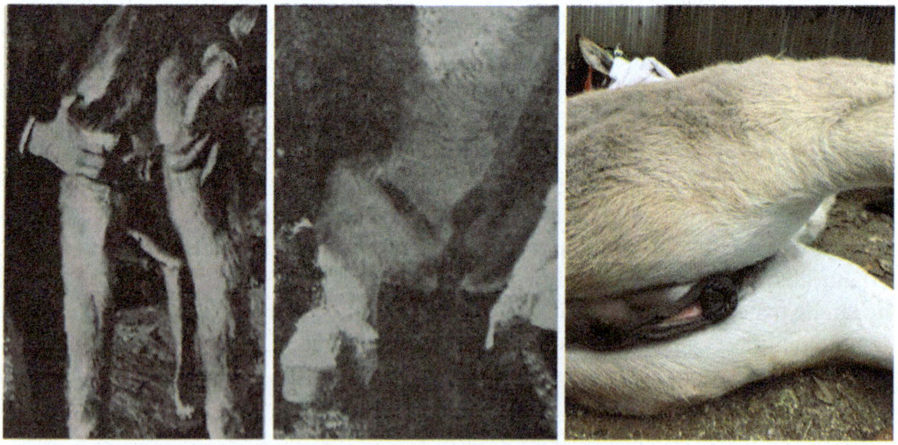

Fig. 19 Pseudihermaphroditism perineal regi6n and ginecomastia in an Andalusian donkey jack (left and central) 63,XX, SRY-negative, pseudohermaphroditism in a Somali wild ass (*Equus asinus somalicus)*hyperdeveloped clitoris (right)

Lesbre (1927), Marañón (1930), and Da Costa (1945), the case was diagnosed as male pseudohermaphroditism, characterized by rudimentary penile development, bilateral cryptorchidism, gynecomastia, incomplete uro-genital groove closure, and marked hypospadias that simulated a narrow vaginal vestibule. Although anatomohistological analysis could not be completed due to ethical limitations, the case aligns with historical descriptions of similar anomalies in cattle by Lecoq (1827) and Van de Pas et al., highlighting the developmental complexity of sexual differentiation in equids.

Diagnosis of an intersex condition relies on a combination of clinical examination, imaging, laboratory and genetic tests: Careful external inspection often reveals ambiguous genitalia (e.g. an enlarged clitoris or small penis-like structure, incomplete fusion of labial folds, absence of a normal vagina or cervix) and/or male-like behavior in a genetic female (or vice versa) (Knobbe et al. 2011).Transrectal or abdominal ultrasound and palpation can locate internal gonads. For example, ultrasound may reveal gonads that resemble testes (oval with echogenic texture) located in the abdomen or inguinal canal (Knobbe et al. 2011). Measurement of sex steroid hormones and gonadotropins helps distinguish cases. Serum testosterone (and dihydrotestosterone) is often high in intact male pseudos and low/undetectable in normal females. Anti-Müllerian hormone (AMH) is produced by Sertoli cells, so detectable AMH in an animal without testes suggests persistent male tissue. An hCG stimulation test (to assess testicular Leydig cell function) may be used to confirm testis presence (Knobbe et al. 2011). Karyotyping lymphocytes will reveal sex chromosome complements (e.g. 64,XY vs 62,XX in donkeys). This can detect cases like XXY or mosaicism. Genetic testing (PCR or FISH) can test for the presence of Y-specific markers including the SRY gene. For instance, both of the intersex racehorses in one report had 64,XY karyotypes with SRY present, confirming an XY DSD (Knobbe et al. 2011). Sequencing can look for known mutations (e.g. SRY, AR, FOXL2). Currently no specific mutation is documented for donkey DSDs, but breeders or researchers might screen SRY or other candidate genes if indicated.

Definitive diagnosis often requires examining gonadal tissue. During laparoscopy or at gonadectomy, submit gonads for histology. True hermaphrodites will show both ovarian follicles and testicular tubules in the same gonad (ovotestis), whereas pseudohermaphrodites will show normal ovarian or testicular tissue only. In a Somali wild ass case, bilateral ovaries with corpora lutea were found despite an enlarged clitoris, and karyotyping confirmed a normal XX female genotype (Gillis et al. 2020).

Combining these approaches—clinical exam, imaging, hormone assays and cytogenetics/histology—provides a clear diagnosis.

Mammalian sex determination is orchestrated by a network of master regulatory genes that drive the embryonic gonads toward testicular or ovarian development. SRY (on the Y chromosome) is the seminal testis-determining gene; when present and functional, it upregulates downstream factors to initiate the male pathway. SRY upregulates SOX9, a transcription factor critical for Sertoli cell differentiation (Huang et al. 2017). SRY sets off a feed-forward loop with SOX9 and FGF9, reinforcing testis development and actively repressing female pathways (Huang et al. 2017). In short, SRY → SOX9 → testis formation. Loss-of-function mutations in SRY or SOX9 (or failure to express them) causes XY individuals to develop ovaries (male-to-female sex reversal) (Huang et al. 2017).

In the female pathway, RSPO1 and WNT4 are critical pro-ovarian factors. RSPO1 activates WNT4/β-catenin signaling to promote ovarian development and suppress the testis pathway early on. For example, RSPO1/WNT4 signaling in XX gonads inhibits SOX9 expression so that granulosa cells form instead of Sertoli cells. FOXL2 is a key ovarian-maintenance gene expressed in granulosa cells: it actively represses the testis-determining network. In females, FOXL2 (often with estrogen receptors) inhibits SOX9 expression in the ovary (Huang et al. 2017). Studies show that loss of FOXL2 (especially combined with loss of WNT4 or RSPO1) can cause an XX ovary to transdifferentiate into a testis. In effect, FOXL2 acts like a "guardian of the ovary," continuously suppressing male pathways in ovarian cells (Huang et al. 2017).

Another player, DMRT1, is important for male development and maintenance of testis identity. DMRT1 is expressed in early Sertoli cells and germ cells. In several mammals, DMRT1 is haploinsufficient: deletion of one DMRT1 allele can cause XY individuals to partially feminize. For instance, humans with DMRT1 mutations (or deletions on chr9p affecting DMRT1) can exhibit XY male-to-female gonadal dysgenesis. In mice, loss of DMRT1 in adult testes causes Sertoli cells to revert toward granulosa-like cells, indicating that DMRT1 "locks in" the male fate. Thus, DMRT1 promotes and maintains the testis pathway, and without it even an established male gonad can "switch" to a female program (Huang et al. 2017).

Finally, the Androgen Receptor (AR) gene is essential for responding to male hormones (testosterone, DHT). AR is on the X chromosome; mutations in AR cause androgen insensitivity syndrome. In a complete AIS case, an XY individual has functional testes that produce hormones, but target tissues cannot respond, so the external phenotype is fully female. In partial AIS, individuals are XY with ambiguous genitalia. Although we found no specific reports of AIS in donkeys, it is a well-known DSD mechanism in mammals and should be considered when an XY donkey/horse has testicular tissue but appears female.

4.2.4 Osteochondrosis in Poitou Donkeys

Osteochondrosis (OC) is a developmental orthopedic disease arising from a focal failure of normal endochondral ossification in growing cartilage (Grissom et al. 2023; Merck, and Co., Inc. 2024). In affected foals, blood supply to the cartilage canals can collapse, causing necrosis of growth cartilage and disturbed cartilage maturation (Riddick et al. 2012; Van Cauter et al. 2023). This disruption leads to osteochondral lesions (loose cartilage flaps or fragments) within joints. In horses, these lesions most often occur in the fetlock (metacarpophalangeal), hock (tarsus), and stifle (femoropatellar) joints (Van Cauter et al. 2023). Lesions can change over time as the animal grows: for example, one study found that about 25% of lesions appeared or resolved between 12 and 24 months of age (Van Cauter et al. 2023). Without intervention, chronic OC can progress to arthritis and permanent lameness.

There are no published studies specifically on OC in Poitou donkeys, but analogies can be drawn from horses. Some horse breeds (e.g. Standardbreds, Warmbloods) are known to be predisposed (Merck, and Co., Inc. 2024). One large survey found OC prevalence varies by breed (e.g. ~23% in Thoroughbreds, ~49% in Lusitanos, ~40% in Warmbloods) (Van Cauter et al. 2023). The Poitou donkey is a very large draft breed (about 145 cm at the withers) (Ahmed et al. 2016). Large body size and rapid growth are risk factors for OCD in horses (Merck, and Co., Inc. 2024), so by analogy heavy breeds like Poitou may share vulnerability, although data are lacking. In all breeds, OC is multifactorial (involving genetics, growth rate, nutrition, biomechanics) (Merck, and Co., Inc. 2024; Russell et al. 2017), and incidence can be higher in populations with intense selection or inbreeding. Because the Poitou is endangered, maintaining genetic diversity and avoiding mating of affected individuals are important preventative measures.

Genetics plays a significant role in OC, although the trait is complex. Estimates of heritability for OCD range from low (~0.1) to moderate (~0.4) depending on breed and joint (Distl 2013; Naccache et al. 2018). For instance, one review reported heritabilities up to 0.29–0.46 for hock OC in Warmbloods and trotters, but much lower in Thoroughbreds (Naccache et al. 2018). Genome-wide studies have identified multiple risk loci (e.g. on equine chromosomes 3, 14, 27, 29) for OCD (Naccache et al. 2018). One notable SNP near the LCORL gene on chromosome 3 is strongly associated with tarsal (hock) OCD (Naccache et al. 2018). Because many genes of cartilage and bone development are candidate factors, researchers have examined their expression in OC lesions. Notably, early OC cartilage shows downregulated expression of cartilage matrix genes: COL2A1 (type II collagen) and ACAN (aggrecan) are significantly reduced in lesion cartilage compared to normal (Grissom et al. 2023). In contrast, COMP (cartilage oligomeric matrix protein) expression was not significantly changed in one study (Grissom et al. 2023). Enzymes involved in matrix turnover are upregulated: MMP13 (matrix metalloproteinase-13) is elevated in OC cartilage, reflecting increased collagen degradation (Riddick et al. 2012). Moreover, key growth-regulating signals are altered: Indian hedgehog (IHH) expression is increased in OC cartilage, and parathyroid hormone-related peptide (PTHrP, product of the *PTHLH* gene) is elevated at the osteochondral junction in affected foals (Riddick et al. 2012). These findings implicate disruption of chondrocyte maturation pathways (IHH–PTHrP axis) in the pathogenesis of OC.

Given the genetic component, breeding management is recommended. Breeders are advised to avoid breeding animals with a history of OC. Selection programs using estimated breeding values (and potentially genomic selection) could reduce OC incidence (Distl 2013). Distl (2013) suggests incorporating OC into breeding plans by selecting against affected individuals or alleles. In the small Poitou population, preserving genetic diversity is also crucial, since inbreeding could inadvertently increase inherited risk.

Preventing OC focuses on optimizing growth and joint health. Key measures include balanced nutrition (avoid high-energy diets and ensure proper calcium–phosphorus and trace mineral ratios) and adequate exercise (regular pasture turnout promotes healthy cartilage loading). Overly rapid growth can be mitigated by restricting concentrates and feeding fiber-based diets; Merck notes that high-glycemic diets may predispose to OC (Merck, and Co., Inc. 2024). Copper and zinc supplements are often given to support cartilage maturation (Merck & Co., Inc., 2024). Maternal factors also matter: Russell et al. (2017) found a significant dam effect on OC risk, indicating that mare nutrition and uterine environment influence foal joint health. Thus, maintaining mares at proper body condition and avoiding nutrient imbalances during gestation can be beneficial (Russell et al. 2017).

For breeding, rigorous records of any OC findings should inform mating decisions. Because OC heritability is moderate, choosing stallions and mares without OC lesions can gradually reduce prevalence (Distl 2013). Genetic testing (for example, available SNP tests in horses) is still experimental for equine OC, and no validated OC gene tests exist specifically for donkeys. In conservation breeding of Poitou donkeys, priorities include avoiding close inbreeding and screening young stock radiographically before high-intensity work or breeding. One practical guideline (based on donkey skeletal maturation studies) is that domestic donkeys reach epiphyseal closure much later than horses: most stifle and hock growth plates in donkeys only ossify around 4 years of age (Ahmed et al. 2016). Therefore, it is recommended to delay strenuous work or breeding until at least 4–5 years old (Ahmed et al. 2016) (some experts advise waiting until 5–6 years). In summary, comprehensive prevention combines good nutrition, controlled growth, appropriate exercise, and judicious breeding practices to minimize OC in Poitou donkeys and other equids.

4.2.5 Congenital Weakness in the Spine and Hips in Poitou Donkeys

Poitou donkeys, like many breeds with distinct genetic lineages, are predisposed to congenital weaknesses, particularly in the spine and hips. These congenital conditions are often linked to the breed's unique structural build, which includes a heavy, muscular body and relatively short limbs. While the Poitou donkey's build is a hallmark of the breed, it also makes them more vulnerable to certain health issues, especially in the skeletal system. The genetic factors that contribute to their size and shape can lead to an increased risk of spinal deformities, hip dysplasia, and muscle weakness in the areas surrounding these structures. These congenital issues can significantly impact their mobility, overall comfort, and quality of life (Merck Veterinary Manual 2025a).

Recent research in equids and other mammals has identified several genes that may play a pivotal role in the congenital weaknesses observed in Poitou donkeys, especially those affecting skeletal development and joint formation. Among the key genes implicated are COL1A1 and COL1A2, which encode the collagen type I protein that forms the structural backbone of bone and connective tissue. Mutations or polymorphisms in these genes can compromise bone integrity, leading to skeletal deformities, spinal abnormalities such as scoliosis or lordosis, and joint laxity (Enderli et al. 2018). GDF5 (Growth Differentiation Factor 5) is another critical gene, as it governs joint and cartilage development; variations in GDF5 have been associated with hip dysplasia in dogs and humans and are likely to affect joint stability and resilience in donkeys (Buxton and Francis-West 2001). Similarly, ACAN (Aggrecan) is vital for cartilage structure, with mutations leading to early-onset joint degeneration, reduced cartilage resilience, and disproportionate growth, making affected equids more susceptible to musculoskeletal deficiencies early in life (Abranches de Andrade et al. 2020). The transcription factors PAX1 and PAX9 play essential roles in vertebral column development, and their dysregulation or mutation can result in kyphosis, scoliosis, or a swayed-back conformation (Samarkhanova et al. 2025). Additionally, RUNX2, a key regulator of osteoblast differentiation and bone morphogenesis, is critical for the proper development of the axial skeleton, with aberrant expression leading to weakened or malformed bones (Thaweesapphithak et al. 2024). Together, these genes highlight the complex genetic basis underlying congenital skeletal and joint anomalies in Poitou donkeys and underscore the importance of targeted genomic and breeding strategies to mitigate such conditions. These genetic abnormalities are typically inherited in a polygenic manner, meaning multiple genes interact to influence the severity and type of congenital weaknesses. Environmental factors, such as nutrition and early life stressors, can further modulate gene expression through epigenetic mechanisms, including DNA methylation and histone modification.

Spinal deformities in Poitou donkeys can manifest as scoliosis, lordosis, or a swayed back (kyphosis), where the spine develops an abnormal curvature. This can cause imbalances in posture, leading to discomfort and increased strain on other parts of the body. The spine, being a central support structure, is crucial for overall stability and movement. When it is weakened or misshapen, donkeys may struggle to move efficiently, especially when walking uphill or carrying weight. In addition to spinal deformities, Poitou donkeys are also prone to hip dysplasia, a condition in which the hip joint does not develop properly, leading to instability and wear-and-tear on the joint surfaces. This can result in pain, stiffness, and difficulty walking, and over time may lead to arthritis or other degenerative joint diseases (Gardiner 2019). These conditions can severely limit the donkey's ability to perform normal activities and engage in the physical tasks typically expected of them, such as grazing, socializing, or even moving around their living space.

The early signs of spinal and hip weaknesses in Poitou donkeys are often subtle and may not be immediately noticeable to owners. However, over time, as the conditions progress, affected donkeys may exhibit difficulty standing, walking, or moving with the fluidity and grace they once had. They may show signs of discomfort, such as reluctance to stand for extended periods or resist certain movements, particularly those that require bending or twisting of the back. A swayed back is one of the more prominent signs of spinal weakness, where the donkey's back appears to dip downward in the middle, often accompanied by an exaggerated arch in the neck. Some donkeys may also exhibit stiffness in their movements or show signs of limping due to pain in the hips or spine. In severe cases, a donkey may avoid physical activities that they previously enjoyed, such as running, jumping, or playing, and might become more lethargic and withdrawn.

Veterinary examination is crucial for diagnosing spinal and hip weaknesses in Poitou donkeys. This typically involves a thorough physical assessment, where the veterinarian evaluates the donkey's posture, gait, and mobility. The vet will also examine the spine and hips for any signs of abnormal curvature, swelling, or pain. In some cases, imaging techniques such as X-rays, ultrasounds, or CT scans may be necessary to get a detailed view of the internal structure of the spine and joints, allowing the veterinarian to assess the extent of any deformities, joint damage, or muscular weaknesses (Merck Veterinary Manual 2025b). Genetic testing is not yet common in donkeys but may become increasingly important as equine genetic research progresses. Early diagnosis can be critical for managing the condition, as it allows for targeted interventions that may prevent further degeneration and improve the donkey's quality of life (American Association of Equine Practitioners 2024).

Once diagnosed, managing Poitou donkeys with spinal and hip weaknesses requires a multifaceted approach, with an emphasis on gentle care and support. A carefully planned and supervised exercise routine is essential to help maintain muscle tone, support joint function, and reduce stiffness without causing additional strain on the back and hips. Low-impact exercises, such as walking on soft ground or engaging in slow, controlled movements, are ideal for Poitou donkeys with these conditions. In some cases, owners may need to limit or avoid activities that put excessive pressure on the joints and spine, such as climbing steep inclines, carrying heavy loads, or engaging in strenuous play (Merck Veterinary Manual 2025a).

The living environment of Poitou donkeys with spinal and hip weaknesses should be carefully managed to reduce stress on their joints. Providing a soft, even surface for resting and daily activities is important to prevent additional strain. Surfaces like soft dirt, grass, or sand are ideal, as they provide cushioning and reduce the risk of joint irritation. Owners should also avoid hard, uneven surfaces that can exacerbate the donkey's discomfort, especially during walking or standing for long periods.

Weight management plays a significant role in minimizing the impact of spinal and hip weaknesses. Excess weight can put added pressure on the joints and exacerbate the symptoms of hip dysplasia and spinal deformities. Maintaining a healthy body condition is essential for reducing strain on vulnerable areas of the body. A controlled diet, in consultation with a veterinarian or equine nutritionist, can help ensure that the donkey maintains an appropriate weight without gaining excess fat that could put additional strain on their joints and spine. Special consideration should be given to feeding practices, with owners ensuring that their donkeys are receiving the proper balance of nutrients, including those that promote bone health and joint function, such as calcium, phosphorus, and glucosamine (Merck Veterinary Manual 2025a).

For Poitou donkeys with more severe spinal or hip issues, additional treatments may be necessary. Anti-inflammatory medications, such as non-steroidal anti-inflammatory drugs (NSAIDs), can help manage pain and reduce inflammation in the affected joints and spinal areas. These medications can provide temporary relief and improve mobility, but should be used under veterinary supervision to ensure proper dosing and to avoid long-term side effects. In addition to medication, physical therapies such as massage, stretching exercises, and physiotherapy can also help reduce muscle tension, improve flexibility, and strengthen the muscles around the spine and hips. These therapies are beneficial for improving mobility and providing comfort to donkeys experiencing pain from congenital spinal and hip weaknesses.

As with other hereditary conditions, responsible breeding practices are a critical component in managing the prevalence of spinal and hip weaknesses in Poitou donkeys. Selective breeding that prioritizes healthy conformation and strong skeletal structures can help reduce the incidence of these congenital issues in future generations. Breeders should be diligent in assessing the health history and physical condition of breeding stock, ensuring that they are not passing on genetic defects that could predispose offspring to these debilitating conditions (American Association of Equine Practitioners 2024). Genetic screening for markers in genes such as COL1A1, GDF5, and RUNX2, when available, could become an important tool in selecting animals less likely to transmit these conditions. By carefully selecting breeding pairs based on their genetic soundness, breeders can help reduce the likelihood of spinal and hip weaknesses in Poitou donkeys and contribute to the long-term health of the breed (Livestock Conservancy 2015).

While congenital weaknesses in the spine and hips present significant challenges for Poitou donkeys, with proper care, these donkeys can still lead happy, fulfilling lives. Owners must be vigilant in monitoring their donkeys' health, adapting their care routine to meet the needs of the affected animals, and seeking veterinary advice when necessary. By providing gentle exercise, appropriate nutrition, and a supportive living environment, Poitou donkeys with spinal and hip weaknesses can continue to enjoy companionship, gentle activity, and quality care. With responsible breeding practices and ongoing research into genetic mechanisms, the impact of these congenital conditions can be reduced in future generations, ensuring that Poitou donkeys remain a beloved and healthy breed.

4.2.6 Limb Impairment in Mammoth Jackstocks

Mammoth Jackstock donkeys (over 56 inches at the withers) are large draft animals often used for breeding and heavy work. Such donkeys are less studied than horses and are especially prone to musculoskeletal injuries and lameness (Almohamad 2024). Their size and workload predispose them to a range of limb problems, including joint instability, tendon and ligament injuries, and progressive arthritis.

Joint instability in donkeys often arises after ligament or tendon damage. Ligaments are dense bands of mostly type I collagen that firmly attach bone to bone, providing stability to joints. When a ligament ruptures or tears, the joint can become unstable.

For example, complete rupture of a cruciate ligament (as in the knee) causes catastrophic instability and severe lameness in equids. Similarly, collateral ligament tears or chronic suspensory desmitis (tendon–ligament injury) lead to joint laxity and pain.

Treatment of acute cruciate injuries typically includes prolonged rest, nonsteroidal anti-inflammatory drugs (NSAIDs), and sometimes intra-articular corticosteroid injections to reduce inflammation. Tendons—such as the superficial and deep digital flexor tendons—can also be strained or torn under overload.

In donkeys, an unusual cause of tendon injury is parasitic infection: for example, three donkeys were reported with chronic suspensory ligament desmitis due to *Onchocerca* nematodes, causing progressive lameness and ligament enlargement (Paraschou et al. 2021). In those cases, NSAIDs only provided partial relief and the lameness ultimately progressed. This underscores that tendon strains in donkeys may respond poorly to conservative therapy alone, requiring rest and sometimes surgical interventions.

Osteoarthritis (OA)—a chronic, progressive degeneration of joints—is a common cause of lameness in aging or heavily used donkeys, much as it is in horses. OA is characterized by cartilage erosion, osteophyte formation, synovitis, and pain. Early signs include stiffness after rest and difficulty rising. On a molecular level, OA involves breakdown of the cartilage extracellular matrix (ECM). Cartilage ECM is rich in type II collagen and aggrecan proteoglycan (encoded by COL2A1 and ACAN, respectively). In an induced donkey model of OA, articular cartilage degeneration was accompanied by a tenfold increase in COL2A1 and caspase-3 expression, reflecting matrix remodeling and chondrocyte apoptosis. In healthy cartilage, the balance of collagen synthesis and breakdown is tightly regulated by chondrocytes, but in OA this balance shifts toward degradation. Matrix metalloproteinase-13 (MMP-13), a collagenase produced by chondrocytes, is normally low but becomes highly upregulated (approximately 10-fold) in osteoarthritic joints, actively digesting type II collagen. Risk factors for OA in donkeys include age, joint conformation abnormalities, obesity, or chronic injuries. In practice, OA management aims to slow progression and reduce pain.

Genetic factors influence connective tissue composition and susceptibility to musculoskeletal disease. Key structural genes include COL1A1 and COL2A1 (type I and type II collagen) and ACAN (aggrecan). COL1A1 encodes collagen type I, the main collagen of bone, tendon, and ligaments, while COL2A1 encodes collagen type II, the predominant collagen in cartilage. Articular cartilage is ~80% type II collagen by content, with aggrecan the major proteoglycan. Chondrocytes are known to lay down an extracellular matrix (ECM) mixture of aggrecan and collagen type II during cartilage development (Sun et al. 2021).

Other important genes include GDF5 (growth differentiation factor 5) and IGF1. GDF5 is a growth factor essential for cartilage formation; genetic variants in GDF5 have been linked to osteoarthritis risk in multiple species. Indeed, GDF5 supplementation has shown protective effects in experimental joint disease. IGF1 (insulin-like growth factor 1) is another anabolic factor: it promotes chondrocyte proliferation and matrix production and inhibits cell death. IGF1 expression is crucial for maintaining cartilage, and reduced IGF1 signaling is associated with cartilage damage. Finally, catabolic enzymes are genetically encoded: for example, the MMP13 gene produces collagenase 3, which degrades collagen in cartilage. In OA, MMP13 expression increases dramatically, driving matrix breakdown. While specific genetic studies in donkeys are scarce, these genes are conserved across mammals. Variations or altered expression of COL2A1, ACAN, GDF5, IGF1, MMP13, and other genes would similarly affect donkey joint resilience. In summary, donkey musculoskeletal health depends not only on environment and care but also on inherited traits in structural and inflammatory gene networks.

4.2.7 Cancer Risks in Light-Coated Donkeys: Focus on Andalusian and Pega Donkeys

Light-colored donkeys with pink or white skin have reduced melanin, leaving them more vulnerable to UV-induced DNA damage. Melanin is a broad-spectrum UV absorber and antioxidant in the skin, and epidemiological studies consistently show that darker-pigmented individuals have much lower skin cancer rates than fair-skinned individuals (Brenner and Hearing 2008). In equids, squamous cell carcinoma (SCC) almost exclusively arises on unpigmented (pink) skin exposed to the sun. For example, in horses SCC is most common around the eyes and genitals of grey, palomino, and paint horses (all with white skin), and prolonged UV exposure is a major predisposing factor (Morton 2023). By analogy, lightly coated donkeys likely face an increased SCC risk in non-pigmented areas due to melanin deficiency. In contrast, animals with dark, pigmented skin (and eyes) are generally at lower risk of UV-related skin cancer (Brenner and Hearing 2008).

Donkeys most commonly develop equine sarcoids, benign fibroblastic skin tumors driven by bovine papillomavirus (BPV) infection (Davis et al. 2016). In a large study of North American and Canadian donkeys, sarcoids accounted for about 72% of all tumors and 82% of cutaneous tumors (Davis et al. 2016). Soft-tissue sarcomas were the next most frequent skin tumors. By contrast, malignant melanomas and squamous cell carcinomas are uncommon in donkeys; retrospective reviews note the near-absence of SCC in donkey tissues and no typical "grey horse" melanomas in donkeys (Davis et al. 2016). A few isolated cases have been reported (for example, a dermal melanoma in an Iranian donkey) (Mosbah and Awadin 2016), but overall donkeys have far fewer melanocytic tumors than grey horses. This difference may reflect that donkeys generally lack the gray-coat STX17 mutation that drives melanoma in horses; in horses, the STX17 intronic duplication (with associated NR4A3 overexpression) strongly predisposes to melanomagenesis (Rosengren Pielberg et al. 2008), whereas donkeys—even grey ones—have no such mutation. Thus, donkey melanomas are exceedingly rare, and TP53 or BCL2 gene alterations (which are central to human cancers) have not been implicated in equine melanomas, suggesting different genetic pathways in equids. Similarly, equine lymphosarcoma (lymphoma) is very rare in donkeys (Davis et al. 2016).

Since UV light drives equine SCC, management should emphasize sun avoidance and protection, as this could promote epigenetic mutation mechanisms. Providing ample shade and turning donkeys out during early morning or evening can reduce UV exposure. Equine-specific UV-blocking fly masks and sheets (blocking about 90% of UV) are highly effective; one study found that using a ≥90% UV-blocking mask was the only factor significantly lowering ocular SCC recurrence in horses (Larson 2019). White or pink areas of skin (muzzle, eyelids, genitals) can be coated with zinc oxide or equine sunscreen to further block sunburn (Schraer 2021). Veterinarians also recommend regular inspection of all unpigmented skin for early lesions. In sum, routine use of UV-protective gear (fly masks, protective blankets) and limited sun exposure are key preventative measures.

Donkeys appear to have a natural resistance to some skin cancers seen in horses. In horses, the "Grey" coat mutation—a 4.6 kb duplication in STX17 (with increased

NR4A3 expression)—causes near-universal melanoma in older greys (Rosengren Pielberg et al. 2008). No analogous mutation is known in donkeys, which likely contributes to their low melanoma incidence. Indeed, in comparative studies, donkey tumors were almost never melanomas or SCCs, whereas these are common in horses (Davis et al. 2016). Furthermore, classic tumor suppressor/oncogene pathways (e.g. TP53, BCL2) are well-recognized in human/murine cancer but have not been definitively linked to donkey neoplasia. Current evidence suggests that donkey cancer susceptibility reflects these species differences in coat-color genetics and viral exposure more than mutations in TP53/BCL2 (Davis et al. 2016).

Equine sarcoids (including those in donkeys) are strongly linked to bovine papillomavirus infection. BPV types 1 and 2 (and possibly BPV-13) have been detected in most equine sarcoids (Borzacchiello et al. 2008). Histologically, sarcoids show fibroblastic overgrowth with viral DNA present; studies find BPV DNA in about 80% of sarcoids and expression of the viral E5 and E7 oncoproteins in the majority of tumors (Borzacchiello et al. 2008). The E5 protein binds and activates the host platelet-derived growth factor β–receptor (PDGF-βR) on equine cells, driving proliferation (Borzacchiello et al. 2008). These viral oncoproteins and growth factor interactions explain why sarcoids are persistent and locally aggressive despite being non-metastatic. In summary, equine (and donkey) sarcoids are mediated by cross-species BPV infection, causing fibroblast transformation via viral E5/E7 dysregulation of host cell growth pathways (Borzacchiello et al. 2008).

4.2.8 Photoprotective Adaptation in Asinara Donkeys: Role of Blood Serum Retinol

Studies find that Asinara white (albino) donkeys maintain higher blood retinol (vitamin A) levels than gray Sardo donkeys. In one comparative trial, Asinara donkeys had mean serum retinol ~0.63 μg/ml vs. 0.49 μg/ml in Sardo donkeys (Cappai et al. 2017). Importantly, during the summer (positive photoperiod) Asinara retinol rose to ~0.738 μg/ml versus 0.486 μg/ml in Sardo (Cappai et al. 2017). These data confirm significantly elevated circulating vitamin A in albino Asinara donkeys relative to pigmented peers, particularly under intense sunlight.

Retinol and its carotenoid precursors have potent antioxidant functions. Vitamin A's long conjugated double-bond structure allows it to quench singlet oxygen and neutralize free radicals, stabilizing membrane lipids (Palace et al. 1999). Burrai et al. note that "Vitamin A behaves as a strong antioxidant, capable to preserve cell membrane integrity" (Cappai et al. 2017). In Asinara donkeys, high retinol (retinyl-ester) along with vitamin E (tocopherol) accumulates in blood, bolstering antioxidant defenses against UV-induced oxidative stress (Cappai et al. 2017). Thus, elevated retinol is well-suited to mitigate UV-driven ROS damage when melanin is absent.

Asinara donkeys show clear seasonal increases in vitamin A during times of long daylight. Cappai et al. report that Asinara serum retinol climbs markedly in the summer: in June (positive photoperiod) mean retinol was 0.738 μg/ml versus only 0.522 μg/ml in autumn (negative photoperiod) (Cappai et al. 2017). By contrast, gray Sardo donkeys showed no significant seasonal change (Cappai et al. 2017). This "pulsed mobilization" of stored vitamin A in summer presumably helps cope with high UV exposure. In the authors' words, retinol mobilization "reflects the need to cope with seasonal exposure to daylight" and may "increase circulating levels during the positive photoperiod" as an alternative photoprotective strategy (Cappai et al. 2017).

Normally, melanin provides powerful UV shielding and antioxidant protection. Melanin absorbs UV and limits DNA damage, acting as an endogenous antioxidant (StatPearls 2021). Without melanin, albinos face well-documented risks: StatPearls notes that "albinism leads to increased risk of sun-damage (solar lentigines, actinic keratoses, solar erythema) and UV-associated malignancies (especially squamous cell carcinomas)" (StatPearls 2021). Ocular structures also suffer: albinism disrupts foveal development and optic nerve routing, leading to visual impairment (StatPearls 2021). Thus, Asinara donkeys, lacking melanin, must rely on biochemical photoprotection (e.g., elevated retinol) to offset these UV hazards.

Vitamin A metabolic genes and stress-response pathways: Key genes in vitamin A transport and metabolism are implicated in this adaptation. RBP4 encodes retinol-binding protein 4, the serum carrier of retinol. RBP4, synthesized in liver, complexes with transthyretin (TTR) to mobilize retinol from the liver and deliver it to tissues (Steinhoff et al. 2021). LRAT (lecithin:retinol acyltransferase) esterifies retinol into stored retinyl esters in hepatic stellate cells (Steinhoff et al. 2021). CYP26B1, a cytochrome P450 enzyme, catalyzes degradation of all-trans-retinoic acid, tightly regulating RA levels in tissues (Isoherranen and Zhong 2019).

Vitamin A and its metabolites also interact with stress-response genes. All-trans-retinoic acid (ATRA) is known to activate TP53. Clinical studies show that isotretinoin (13-cis-RA) treatment upregulates TP53 expression in human skin and sebaceous glands (Abdelmeniem et al. 2022). In cell models, ATRA drives transcription of TP53 and related apoptosis factors. Moreover, retinoids induce antioxidant enzymes: ATRA robustly increases SOD2 (mitochondrial superoxide dismutase) via NF-κB signaling, making SOD2 a retinoid-responsive gene (Kiningham et al. 2008). NQO1 (NAD(P)H:quinone oxidoreductase 1) is another link: it stabilizes the p53 protein under oxidative conditions (Asher et al. 2002), coupling vitamin A signaling and detoxification. Thus, Asinara donkeys likely leverage these pathways (e.g., RA-driven TP53/SOD2 activation and NQO1-mediated p53 protection) to counter UV-induced oxidative damage.

The Asinara findings suggest a general strategy whereby albino or light-coated animals may upregulate vitamin A metabolism to compensate for melanin loss. Cappai et al. emphasize that "retinol may be an adaptive metabolic key to overcome the higher susceptibility to sun radiation of albino animals" (Cappai et al. 2024). In other words, raising serum retinol during bright seasons could furnish endogenous UV defense. This "photoprotection... achieved through higher levels of available retinol" in the bloodstream provides a plausible mechanism for albinos to mitigate UV damage in lieu of pigmentation (Cappai et al. 2024). Such metabolic adaptation may extend to other lightly pigmented species living in sunny environments, highlighting the vital role of vitamin A in protecting albinistic organisms from solar stress.

4.2.9 Cresty Neck: Physiological or Pathological?

Dorsal neck crest deformation (DBDC), or "cresty neck," is a condition increasingly recognized in donkeys as a marker of altered fat deposition, metabolic status, and potential compromise of animal welfare (Mendoza et al. 2018; Żeromski et al. 2021) (Fig. 20). Unlike its more anecdotal treatment in earlier literature, recent research has characterized DBDC as a multifactorial syndrome involving structural, physiological, and possibly genetic determinants (Mendoza et al. 2024). Particularly in breeds such as the Andalusian and Catalan donkeys, the condition is prevalent and often observed under intensive or semi-intensive management systems (Valle et al. 2017).

Histopathological analysis of DBDC in donkeys reveals progressive lipomatosis and muscular degeneration, with adipocytes infiltrating subcutaneous tissues, the nuchal ligament, and cervical muscles (Martin-Giménez et al. 2016). In more advanced stages (Scores 3–5), desmin immunoreactivity is lost, indicating irreversible myofiber damage. This pathology underpins the classification of DBDC in donkeys as not merely cosmetic but truly pathological, associated with pain, reduced mobility, and potentially compromised welfare (Mendoza et al. 2018).

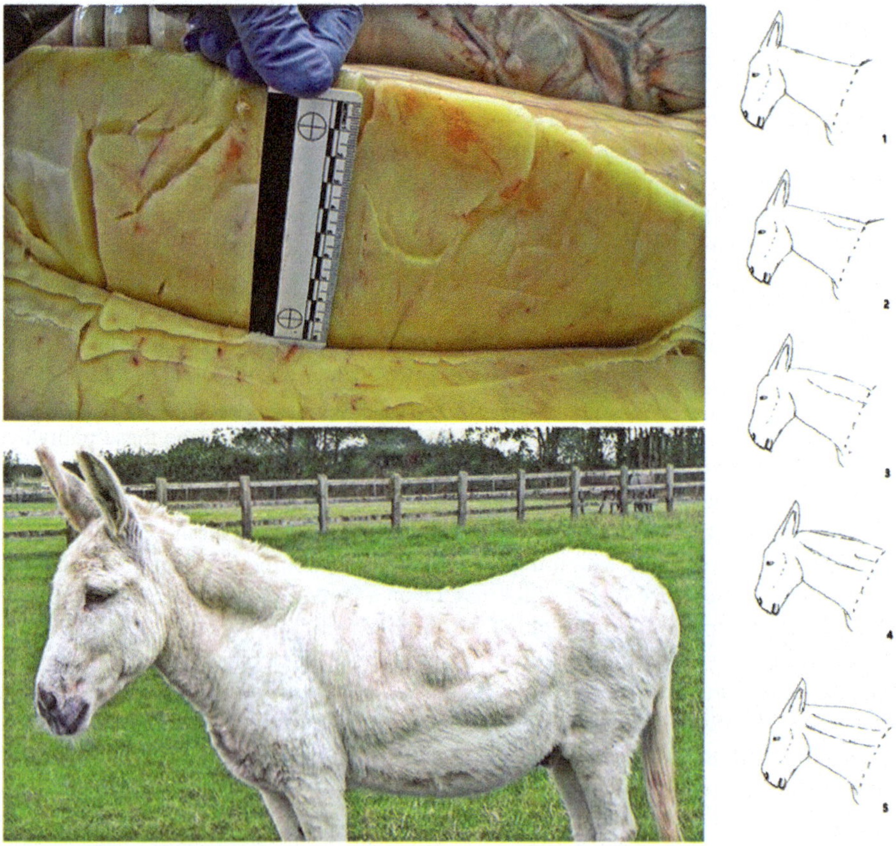

Fig. 20 Cresty Neck in donkeys; diseection, external appearance and scoring scale

While DBDC may have originated from an adaptive trait favoring energy storage in the neck region—an advantageous trait for survival in environments with scarce and unpredictable food resources—it becomes maladaptive in modern conditions where energy-rich diets and reduced exercise prevail (World Animal Data n.d.). Thus, the condition illustrates a classic case of evolutionary mismatch, where a trait that was once beneficial has become a liability under altered ecological and management contexts (Mendoza et al. 2024).

The deformity is partially reversible through dietary management and increased exercise, especially in early stages before fibrotic changes set in (Valle et al. 2017). Weight loss and nutritional correction can reduce crest size; however, in severe cases, structural changes—including replacement of muscle by adipose and fibrous tissue—limit reversibility and hamper recovery (Mendoza et al. 2018).

Diagnostic methods for DBDC include the Fatty Neck Score (FNS), a five-point scale assessing regional adiposity, which has shown a stronger correlation with dorsal profile deformation than general body condition score (BCS) (Żeromski et al. 2021). Geometric Morphometrics (GM), a landmark-based method, captures the shape and curvature of the dorsal profile. In a recent study involving 40 donkeys, GM confirmed that FNS and BCS significantly affect dorsal profile shape and size (FNS $p < 0.0001$; BCS $p < 0.0001$), with FNS exerting the strongest effect (Żeromski et al. 2021). This morphometric approach has been validated as an objective indicator of both conformation and welfare status, supporting its inclusion in comprehensive assessments of donkey health (Valle et al. 2017).

Several extrinsic and intrinsic factors predispose donkeys to DBDC. The condition is more common in males and its incidence increases with age, especially after ten years (Mendoza et al. 2024). Obesity and a positive energy balance are primary risk factors, while feeding practices (e.g., elevated feeders) may affect neck biomechanics, and reduced activity promotes localized adipose accumulation (Mendoza et al. 2018). Dental condition may indirectly influence feeding behavior and overall nutritional status, although its direct role in DBDC remains limited (Valle et al. 2017).

While no single gene has yet been conclusively linked to DBDC in donkeys, several candidate genes and pathways have been proposed based on comparative studies across species. *PPARG* (Peroxisome Proliferator Activated Receptor Gamma) plays a central role in adipogenesis and lipid metabolism, making it a strong candidate for involvement in the lipomatosis characteristic of DBDC (Udhaya Kumar et al. 2020). Similarly, genes associated with energy balance and adiposity—such as *MC4R* (Melanocortin 4 Receptor) and *FTO* (Fat Mass and Obesity Associated)—are being explored for their potential role in localized adipose tissue accumulation (Udhaya Kumar et al. 2020). Additionally, *ASIP* (Agouti Signaling Protein), although primarily implicated in coat color, may have pleiotropic effects linked to regional adiposity and metabolism (Udhaya Kumar et al. 2020).

Emerging evidence from multivariate analyses suggests that the heritability of conformational and metabolic traits in donkeys may be moderate, supporting the hypothesis of a genetic component in DBDC susceptibility (Mendoza et al. 2024).

However, no genome-wide association studies (GWAS) or SNP-based analyses have yet been published for this condition in donkeys. Given the high phenotypic variability across breeds and the increasing incidence of DBDC in endangered donkey populations, genomic mapping and candidate gene studies are an urgent priority (World Animal Data n.d.). Understanding the genetic and molecular basis of DBDC will be pivotal for guiding breeding programs aimed at reducing its incidence, especially in autochthonous donkey breeds where body conformation and animal welfare are critical conservation priorities (Valle et al. 2017).

Morphological deformation of the neck and associated changes in the dorsal profile can impair mobility, affect thermoregulation, and increase susceptibility to musculoskeletal strain (Mendoza et al. 2018). Beyond its physical effects, DBDC may serve as a chronic indicator of welfare compromise when combined with limited movement and surplus energy intake (Mendoza et al. 2024). The combined use of GM, FNS, and BCS as complementary tools provides a more holistic assessment of the health, metabolism, and management adequacy of donkeys (Żeromski et al. 2021).

5 Functionality

5.1 Ambling: The Hassawi–Andalusian–Pêga Lineage

The term ambling refers to smooth, intermediate-speed four-beat gaits (lateral or diagonal) in equids. Certain domestic donkey breeds—notably the Spanish Andalusian (Cordoban) donkey and the Brazilian Pêga donkey—owe much of their heritage to North African stock, especially the characteristically white Hassawi (Egyptian) donkeys of antiquity. Historical records and genetic studies both trace the Andalusian donkey's ancestry to Egyptian and other North African asses introduced into Iberia by migrating peoples (the ancient Camites/Iberians) around the sixth century BCE (Orlando et al. 2013; Royo et al. 2005). Spanish sources explicitly note that the Andalusian breed descends from an "Egyptian donkey (the Hassawi breed)" brought from North Africa (MAPA 2021). It is thus considered one of Europe's oldest domestic donkey breeds, with a pedigree extending roughly 3000 years (MAPA 2021; Rivas-López et al. 2023).

Similarly, modern genomic analyses place the Brazilian Pêga donkey firmly within the Iberian clade, mirroring the colonial-era import of Spanish and Portuguese donkeys into Brazil. In one comprehensive study, the Pêga donkey from Brazil "is nested within Iberia," reflecting this historical origin (Gonçalves et al. 2020). Together, these data imply that the North African (including Egyptian) "Hassawi" donkeys contributed foundational genes to both Andalusian and Pêga lines.

Genetically, gaitedness in equids is strongly influenced by the DMRT3 "gait-keeper" gene. A loss-of-function mutation in DMRT3 (a premature stop codon) disrupts normal spinal locomotor circuits, enabling alternate lateral gaits in horses (Andersson et al. 2012a, b). This variant is famously associated with ambling gaits in many horse breeds. However, in donkeys, the DMRT3 mutation is virtually absent. In a survey of Brazilian gaited donkeys and mules, 97% of Pêga donkeys were homozygous for the ancestral (CC) allele and none carried the homozygous mutant (AA) allele (Herman et al. 2023). Likewise, DMRT3 genotypes in donkeys showed no strong association with gait type (Herman et al. 2023).

This suggests that while gaited horse breeds carry the DMRT3 mutation, the donkey's contribution to ambling in these lines comes mainly via its own conformation and through outcrossing with gaited horses. Indeed, the Brazilian mule-breeding industry intentionally crosses Pêga donkeys with gaited horses (e.g., Mangalarga Marchador, Campolina) to produce comfortably ambling mules. The rare DMRT3A allele observed in these donkeys and mules likely derives from the mares, not the asses (Herman et al. 2023). In short, donkeys themselves are almost uniformly DMRT3 wild-type, implying that other genes or body-plan features underlie any smooth-gaited proclivities they exhibit. Some donkey owner also relate the ability of donkeys to switch from ambling to pace at will or depending on the situation promoting such change, for instance a threat or terrain characteristics.

Phenotypically, certain anatomical traits of riding donkeys may also affect gait smoothness. The donkey hoof is notably narrower and more U-shaped than a horse's, with very little expansion on impact. This construction—including upright hoof walls and a hard keratin tubule structure—gives donkeys a reputation for sure-footedness but also a "bouncy" locomotion (Nizou et al. 2015). As one review explains, "A donkey works at slow paces. The walk and trot are his favored gaits. They have a bouncy gait, as shock absorption is not as pronounced as in horses" (IFCE 2021). In practice, a riding donkey (especially one with a shorter back and stout conformation) will exhibit more vertical oscillation in trot than an ambling horse.

That said, breeders of gaited donkey lines likely select for any traits that dampen this bounce. For example, by favoring lateral intermediate gaits (singlefoot or stepping pace) in which one foot is always on the ground, a donkey can achieve a smoother ride than by trotting (which has a suspension phase). The hallmark of 4-beat ambling gaits is constant ground contact (no suspension), which eliminates the jarring impact felt at the end of each trot stride (Williams et al. 2020). Thus, even though pure donkeys tend toward a springy trot, the combination of their calm temperament, sturdy build, and a trained lateral gait can yield a comfortable gaited ride.

In Brazilian herds, Pêga jacks are indeed valued for producing gaited mules under saddle, indicating that the donkey's build can accommodate and complement the horse-derived gait pattern (Herman et al. 2023). In summary, White Hassawi (Egyptian) donkeys contributed a robust "riding ass" phenotype to Iberian stock, and in modern gaited breeds their legacy survives in coat color, conformation, and the base over which gaited genetics from horses can operate.

5.2 The Case of Balkan Breeds Group, Banat Donkey and Donkey Cheese Production in Zasavica Reserve

The convergence of genetic conservation and artisanal food production offers a valuable pathway for safeguarding endangered livestock breeds while fostering sustainable rural economies. Nowhere is this more evident than in the case of the Balkan donkey and, more specifically, the critically endangered Banat donkey (*Equus asinus*), whose milk is used to produce the world-renowned pule cheese at the Zasavica Special Nature Reserve in Serbia (Fig. 21).

The Balkan donkey represents a diverse group of traditional landraces distributed across the countries of the former Yugoslavia, Albania, Bulgaria, Croatia and parts of Greece. These donkeys are generally well adapted to mountainous and semi-arid environments and have played a vital historical role in transport and subsistence agriculture. However, modern agricultural mechanization, rural depopulation, and lack of structured breeding programs have placed many of these populations at risk (Ivanković et al. 2002; Radosavljević et al. 2021).

Fig. 21 Pule Cheese (upper), Banat (lower left) and Balkan (lower right) Donkey Breeds

Among them, the Banat donkey stands out due to its extreme rarity and cultural value. Native to the Pannonian Plain of northern Serbia, this large-framed donkey has been subject to genetic erosion for decades and is now considered one of the rarest domestic breeds in Europe. With fewer than 200 individuals remaining, most housed in semi-conservation settings such as Zasavica, the breed requires urgent genomic management and strategic valorization.

Genetic studies of both the Banat and broader Balkan donkey populations reveal low levels of heterozygosity and reduced effective population sizes, pointing to historical bottlenecks and recent isolation. Microsatellite and mtDNA analyses show that despite a shared ancestry, individual Balkan donkey populations exhibit unique haplotypes, reflecting localized domestication histories and gene flow patterns. For example, genetic differences between Croatian, Serbian, and Bosnian donkey groups have been documented, suggesting regional diversification shaped by geography and husbandry practices (Ivanković et al. 2002; Radosavljević et al. 2021).

The Zasavica initiative leverages this genetic heritage by producing donkey milk and cheese from the Banat donkey, linking biodiversity conservation with high-value gastronomy. Donkey milk is compositionally unique—rich in lactose, low in fat, and containing potent antimicrobial proteins such as lysozyme and lactoferrin. These traits are underpinned by genes including LYZ, LTF, and LALBA (-*α-lactalbumin*), which are expressed at high levels in donkey mammary tissue (Giorgis et al. 2018). The milk's low casein content—especially in genes such as CSN1S1, CSN2, and CSN3—poses a technological challenge for cheese making, often requiring up to 25 liters of milk to produce 1 kg of cheese (Popović et al. 2020). Nevertheless, this scarcity enhances its market value, while also serving as an economic driver for conservation.

In terms of adaptation and fitness, the genetic portfolio of Balkan donkeys includes several promising candidate genes. HSP70 is associated with heat tolerance, while MC1R may contribute to thermoregulation through coat color variation. Genes such as MTNR1A (linked to photoperiod sensitivity and seasonal breeding) and PRLR (prolactin receptor, associated with lactation) may offer insight into reproductive traits under natural and low-input systems. Moreover, TLR2 and TLR4, key members of the innate immune response pathway, are particularly relevant to early survival and colostrum-mediated immunity—traits critical in populations with minimal veterinary oversight.

Collectively, the Banat and Balkan donkeys represent a living reservoir of genetic, cultural, and agricultural heritage. Conservation programs must combine in situ management, genomic characterization, and value-added production models such as donkey cheese to ensure their long-term survival. The Zasavica model—rooted in ecological protection, culinary innovation, and cultural storytelling—provides a blueprint for preserving under-characterized breeds while positioning them within dynamic rural economies. It exemplifies the emerging paradigm in agrobiodiversity conservation, where genetics and gastronomy align to sustain endangered breeds and revitalize their roles in contemporary landscapes.

6 Conclusions

The synthesis of archaeological evidence, mitochondrial haplotyping, and whole-genome analyses supports a unified model of donkey domestication rooted in a single African origin approximately 6000–7000 years ago. This process involved a genetically diverse source population, resulting in the persistence of two major maternal haplogroups in modern donkeys. Post-domestication expansion facilitated rapid dispersal across Africa, Eurasia, and eventually the Americas, accompanied by local adaptation, introgression from wild asses, and selective breeding for traits such as coat color, size, and hybrid vigor for mule production. Molecular studies highlight substantial genetic diversity within certain populations, contrasting with sharp declines and bottlenecks in others, emphasizing the urgent need for targeted conservation strategies. The identification of genetic variants linked to phenotypic traits—such as albinism, long-hair, and pigmentation patterns—underscores the interplay between cultural preferences, functional adaptation, and evolutionary history. Modern genomic tools now offer powerful means to characterize underexplored populations, develop breed-specific conservation programs, and preserve the genetic legacy of donkeys as a critical component of global livestock biodiversity. Sustaining these efforts will require integrating genomic knowledge with socio-economic considerations to ensure the resilience and continued utility of this historically indispensable species.

Self-Evaluation Questionnaire

1. Which environmental factor is most closely linked to the domestication of donkeys in Northeast Africa?
   A. Expansion of tropical forests during the Holocene
   B. Increasing aridity of the Sahara during the mid-Holocene
   C. Seasonal flooding of the Nile Valley creating new grazing lands
   D. Proliferation of disease vectors reducing cattle populations
   Answer: B
2. Which mitochondrial haplogroups define the two major maternal lineages in domestic donkeys?
   A. Haplogroups X and Y
   B. Haplogroups A and B
   C. Haplogroups C and D
   D. Haplogroups L and M
   Answer: B
3. What evidence most strongly supports a single African origin of donkey domestication?
   A. Uniform coat color patterns across all ancient populations
   B. High diversity of Y-chromosome haplotypes worldwide
   C. Genome-wide phylogeographic clustering into African, Asian, and European groups
   D. Complete absence of gene flow between wild and domestic populations
   Answer: C

4. Which gene is associated with the non-dun phenotype in donkeys?
   A. KIT
   B. ASIP
   C. TBX3
   D. MC1R
   Answer: C
5. What is the primary reason for reduced Y-chromosome diversity in modern donkeys?
   A. Strong natural selection for disease resistance
   B. Male-biased breeding practices favoring a few elite sires
   C. Geographical isolation of female lineages
   D. Complete eradication of ancestral paternal haplotypes during domestication
   Answer: B
6. Why are feral donkey populations in regions such as Australia considered important from a genetic standpoint?
   A. They originate exclusively from Nubian wild asses
   B. They retain primitive coat color patterns linked to the TBX3 locus
   C. They harbor novel haplotypes and preserve genetic diversity lost in standardized breeds
   D. They have no history of admixture with other equid species
   Answer: C

References

Abdelmeniem IM, Talaat IM, Zaki II, Sabah RM, Melnik BC (2022) Isotretinoin treatment upregulates the expression of p53 in the skin and sebaceous glands of patients with acne vulgaris. Sci Rep 12:20446

Abitbol M, Legrand R, Tiret L (2014) A missense mutation in melanocortin 1 receptor is associated with the red coat colour in donkeys. Anim Genet 45(6):878–880

Abitbol M, Legrand R, Tiret L (2015) A missense mutation in the agouti signaling protein gene (ASIP) is associated with the no light points coat phenotype in donkeys. Genet Sel Evol 47(1): 28

Abranches de Andrade DG et al (2020) Evaluation of a new variant in the aggrecan gene potentially associated with chondrodysplastic dwarfism in Miniature horses. Sci Rep 10:15238. nature. comnature.com

Ahmed NE, Habib MA, Al-Azmy HH (2016) Radiographic specification of skeletal maturation in donkeys: defining the ossification time of donkey growth plates for preventing irreparable damage. Asian J Anim Vet Adv 11(3):204–216

Almathen F (2024) Mitochondrial DNA analysis of Hassawi donkeys reveals genetic diversity and global maternal lineages. Genet Mol Res 23(3):gmr2381

Almathen F, Qureshi AU, Al-Rasheid KAS, Almathen Z, Roman U, Khanshour AM et al (2018) Polymorphisms in MC1R and ASIP genes are associated with coat color variation in the Arabian camel (Camelus dromedarius). J Hered 109(6):700–706

Almohamad Z (2024) Ultrasonography of the normal donkey tarsus (Equus asinus). Sci Rep 14: 10425

American Association of Equine Practitioners (2024) AAEP position on genetic defects. Retrieved from https://aaep.org/resource/genetic-defects/aaep.orgaaep.org

Andersson LS et al (2012a) Mutations in DMRT3 affect locomotion in horses and spinal circuit function in mice. Nature 488(7413):642–646

Andersson LS, Larhammar M, Memic F, Wootz H, Schwochow D, Rubin C-J, Patra K, Arnason T, Wellbring L, Hjälm G, Imsland F, Petersen JL, McCue ME, Mickelson JR, Cothran G, Ahituv N, Roepstorff L, Mikko S, Vallstedt A, Lindgren G, Kullander K (2012b) Mutations in DMRT3 affect locomotion in horses and spinal circuit function in mice. Nature 488(7413): 642–646

Asher G, Lotem J, Kama R, Sachs L, Shaul Y (2002) NQO1 stabilizes p53 through a distinct pathway. Proc Natl Acad Sci 99(5):3099–3104

Asnopra (2019) Características morfológicas del burro andaluz. Asociación Nacional de Criadores de la Raza Asnal Andaluza

Bellone RR (2010) Pleiotropic effects of pigmentation genes in horses. Anim Genet 41 (Suppl 2):100–110

Beja-Pereira A, England PR, Ferrand N, Jordan S, Bakhiet AO, Abdalla MA et al (2004) African origins of the domestic donkey. Science 304(5678):1781–1783

Boegheim IJM, Leegwater PAJ, van Lith HA, Back W (2017) Current insights into the molecular genetic basis of dwarfism in livestock. Vet J 224:64–75

Borzacchiello G, Russo V, Della Salda L, Roperto S, Roperto F (2008) Expression of platelet-derived growth factor-β receptor and bovine papillomavirus E5 and E7 oncoproteins in equine sarcoid. J Comp Pathol 139(4):231–237. https://pubmed.ncbi.nlm.nih.gov/18814884/

Bowling AT (1996) Horse genetics: an introduction. Vet Clin North Am Equine Pract 12(1):1–18

Brenner M, Hearing VJ (2008) The protective role of melanin against UV damage in human skin. Photochem Photobiol 84(3):539–549. https://pmc.ncbi.nlm.nih.gov/articles/PMC2671032/

Buxton PE, Francis-West P (2001) Growth/differentiation factor-5 (GDF-5) and skeletal development. J Bone Joint Surg Am 83-A(Suppl 1 Pt 1):S23–S30. pubmed.ncbi.nlm.nih.gov

Cappai MG, Lunesu MGA, Accioni F, Liscia M, Pusceddu M, Burrai L, Nieddu M, Dimauro C, Boatto G, Pinna W (2017) Blood serum retinol levels in Asinara white donkeys reflect albinism-induced metabolic adaptation to photoperiod at Mediterranean latitudes. Ecol Evol 7(1): 390–398

Cappai MG, Nieddu M, Boatto G, Pinna W, Cappai M, Senes A, Accioni F, Liscia M, Pusceddu M, Burrai L, Melis MP (2024) Albinism and blood cell profile: the peculiar case of Asinara donkeys. Animals 14(18):2641

Caro T, Argueta Y, Briolat ES, Bruggink J, Kasprowsky M, et al. (2019) Benefits of zebra stripes: Behaviour of tabanid flies around zebras and horses. PLOS ONE 14(2):e0210831

Cavalus (2020) Características da raça Pêga. Retrieved from https://cavalus.com.br

Cieslak M, Reissmann M, Hofreiter M, Ludwig A (2011) Colours of domestication. Biol Rev 86(4): 885–899

Clutton-Brock J (1992) Horse power: a history of the horse and the donkey in human societies. Harvard University Press

Cothran EG, Long TL (2011) The role of donkeys in the early American south: mammoth Jackstock and George Washington's breeding program. In: McGowan JD (ed) American equine history revisited. University of Kentucky Press, pp 112–125

Da Costa AC (1945) Elementos de embriología (Trad. esp.). Librería Hachette S. A

Davis CR, Valentine BA, Gordon E, McDonough SP, Mauldin GN, Moore BE, Musser JM, Martin KR, Stokol T, Crisman MV (2016) Neoplasia in 125 donkeys (Equus asinus): literature review and a survey of five veterinary schools in the United States and Canada. J Vet Diagn Invest 28(6):619–629

Diccionario de Americanismos (2010) Melao. Asociación de Academias de la Lengua Española. Retrieved from https://www.asale.org/damer

Distl O (2013) The genetics of equine osteochondrosis. Vet J 197(1):13–18

Dong H, Dong Z, Wang F, Wang G, Luo X, Lei C, Chen J (2022) Whole genome sequencing provides new insights into the genetic diversity and coat color of Asiatic wild ass and its hybrids. Front Genet 13:818420

Enderli TA et al (2018) Animal models of osteogenesis imperfecta: applications in clinical research. J Bone Miner Metab 36(2):209–222. pmc.ncbi.nlm.nih.gov

Fenn DJ, Raudsepp T, Cothran EJ, Hamilton NA, Haase B (2017) Validation of a candidate causative mutation for white spotting in donkeys. Anim Genet 48(1):124–125

Gardiner J (2019, April 24) Hip Dysplasia. UC Davis School of Veterinary Medicine Animal Health Topics. Retrieved from https://healthtopics.vetmed.ucdavis.edu/health-topics/canine/hip-dysplasiahealthtopics.vetmed.ucdavis.eduhealthtopics.vetmed.ucdavis.edu

Gillis J, Bercier M, Rattner FK, Citino SB (2020) 63,XX, SRY-negative, pseudohermaphroditism in a Somali wild ass (Equus asinus somalicus). Vet Rec Case Rep 8(3):e001164. https://doi.org/10.1136/vetreccr-2020-001164researchgate.net

Giorgis V, Rolla G, Raie A, Geuna M, Boita M, Lamberti C et al (2018) A case of work-related donkey milk allergy. J Investig Allergol Clin Immunol 28(3):197–199

Gobierno Vasco (2020) Asno de las Encartaciones: raza autóctona en peligro de extinción. Departamento de Desarrollo Económico e Infraestructuras

Gonçalves RF et al (2020) Phylogeography of donkeys suggests a single domestication event in Africa. Sci Rep 10:4562

Grilz-Seger G, Utzeri VJ, Ribani A, Taurisano V, Fontanesi L, Brem G (2020) Known loci in the KIT and TYR genes do not explain the depigmented white coat colour of Austro-Hungarian Baroque donkey. Ital J Anim Sci 19(1):739–743

Grissom SK, Semevolos SA, Duesterdieck-Zellmer K (2023) Role of cartilage and bone matrix regulation in early equine osteochondrosis. Bone Reports 18:101653

Haase B, Rieder S, Leeb T (2015) Two variants in the KIT gene as candidate causative mutations for a dominant white and a white spotting phenotype in the donkey. Anim Genet 46(3):321–324

Herman M, Caceres AM, Albuquerque AL, Leite RO, Araújo CET, Delfiol DJZ, Curi RA, Borges AS, Oliveira-Filho JP (2023) DMRT3 allele frequencies in batida- and picada-gaited donkeys and mules in Brazil. Animals (Basel) 13(24):3829

Horsepress.com (2019, February) Caramelo, un burro de alta escuela. Retrieved from https://www.horsepress.com/caramelo-burro-alta-escuela

Huang J, Zhao Y, Shiraigo W, Li D, Xu W, Zhou C et al (2015) Draft genome of the domestic donkey. Nat Commun 6:10416

Huang S, Ye L, Chen H (2017) Sex determination and maintenance: the role of DMRT1 and FOXL2. Asian J Androl 19(6):619–624. https://doi.org/10.4103/1008-682X.194420

Huang B, Kou X, Liang H, Wang C, Chai W (2022) A novel A>G polymorphism in the intron 1 of LCORL gene is significantly associated with hide weight and body size in Dezhou donkey. Animals 12(23):3618

IFCE—Institut Français du Cheval et de l'Équitation (2021) Donkey locomotion and biomechanics. Equipédia. https://equipedia.ifce.fr

Imsland F, McGowan K, Rubin CJ, Henegar C, Sundström E, Berglund J, Schwochow D, Hellmén E, Cothran G, Andersson L (2016) Regulatory mutations in TBX3 disrupt asymmetric hair pigmentation that underlies dun camouflage color in horses. Nat Genet 48(2):152–158

Isoherranen N, Zhong G (2019) Biochemical and physiological importance of the CYP26 retinoic acid hydroxylases. Pharmacol Ther 204:107400

Ivanković A, Kavar T, Caput P, Mioč B, Pavić V, Dovč P (2002) Genetic diversity of three donkey populations in the Croatian region. Anim Genet 33(3):169–177

Jacinto JGP, Häfliger IM, Letko A, Drögemüller C, Agerholm JS (2020) A large deletion in the COL2A1 gene expands the spectrum of pathogenic variants causing bulldog calf syndrome in cattle. Acta Vet Scand 62:49

Jordana J, Casas M, Loarca A, Martínez López OR, Canelón JL, Stemmer A, Aguirre L, Lara MAC, Álvarez LA, Llambí S, Gómez N, Gama LT, Martínez RD, Pérez E, Guastella AM, Marletta D, Landi V, Martínez A, Delgado JV (2012) Diversidad y relaciones genéticas entre poblaciones asnales iberoamericanas. Trabajo presentado en el Congreso AICA 2012. Retrieved from https://www.researchgate.net/publication/260786786

Jordana J, Ferrando A, Miró J, Goyache F, Loarca A, Martínez López OR, Canelón JL, Stemmer A, Aguirre L, Lara MAC, Álvarez LA, Llambí S, Gómez N, Gama LT, Nóvoa MF, Martínez RD, Pérez E, Sierra A, Contreras MA, Guastella AM et al (2016) Genetic relationships among American donkey populations: insights into the process of colonization. J Anim Breed Genet 133(2):155–164

Kimura B, Marshall F, Chen S, Rosenbom S, Mohandesan E, Bendrey R et al (2011) Ancient DNA from Nubian and Somali wild asses highlights the multiple domestication origins of the African donkey. Proc R Soc B Biol Sci 278(1712):50–57

Kiningham KK, Cardozo Z-A, Cook C, Cole MP, Stewart JC, Tassone M, Coleman MC, Spitz DR (2008) All-trans-retinoic acid induces manganese superoxide dismutase in human neuroblastoma through NF-κB. Free Radic Biol Med 44(8):1610–1616

Knobbe MG, Maenhoudt C, Turner RM, McDonnell SM (2011) Physical, behavioral, endocrinologic, and cytogenetic evaluation of two Standardbred racehorses competing as mares with an intersex condition and high postrace serum testosterone concentrations. J Am Vet Med Assoc 238(6):751–754. https://doi.org/10.2460/javma.238.6.751researchgate.net

Larson E (2019, May 22) Protect horses from sun to prevent SCC recurrence. The Horse. https://thehorse.com/172341/protect-horses-from-sun-to-prevent-scc-recurrence/

Legrand R, Tiret L, Abitbol M (2014) Two recessive mutations in FGF5 are associated with the long-hair phenotype in donkeys. Genet Sel Evol 46:65

Librado P, Orlando L (2021) Genomics and the evolutionary history of equids. Annu Rev Anim Biosci 9:81–101

Lin S, Li C, Li C, Zhang X (2018) Growth hormone receptor mutations related to individual dwarfism. Int J Mol Sci 19(5):1433

Liu S, Su J, Yang Q, Sun M, Wang Z, Yu J, Jafari H, Lei C, Sun Y, Dang R (2024) Genome-wide analyses based on a novel donkey 40K liquid chip reveal the gene responsible for coat color diversity in Chinese Dezhou donkey. Anim Genet 55(1):140–146

Livestock Conservancy (2015) Baudet du Poitou (Poitou donkey) conservation success story. FAO DAD-IS report. Retrieved from https://www.fao.org/dad-is/success-story/detail/en/c/1193627/:contentReference [oaicite:48]{index=48}

Lopez CL, Alonso R, De Aluja AS (2005) Study of the genetic origin of the Mexican creole donkey (Equus asinus) by means of the analysis of the D-loop region of mitochondrial DNA. Trop Anim Health Prod 37(Suppl 1):173–188

Ma X-Y, Ning T, Adeola AC, Li J, Esmailizadeh A, Lichoti JK et al (2019) Potential dual expansion of domesticated donkeys revealed by worldwide analysis on mitochondrial sequences. Zool Res 40(1):36–45

MAPA—Ministerio de Agricultura, Pesca y Alimentación (2021) Programa de cría del asno andaluz. https://www.mapa.gob.es

Marshall F (2007) Donkeys in the African past. In: Harris DR, Hillman GC, Fuller DQ (eds) Rethinking agriculture: archaeological and ethnoarchaeological perspectives. Left Coast Press, pp 115–143

Martin-Giménez T, de Blas I, Aguilera-Tejero E, Díez de Castro E, Aguirre-Pascasio CN (2016) Endocrine, morphometric, and ultrasonographic characterization of neck adiposity in Andalusian horses. Domest Anim Endocrinol 56:57–62

Mateos Nevado (1955) Un caso de seudohermafroditismo en la especie asnal (Nota previa) [A case of pseudohermaphroditism in the species Equus asinus]. Boletín de Zootecnia 11(114):56–59

Mendoza FJ, Toribio RE, Pérez-Ecija A (2018) Donkey internal medicine—part I: metabolic, endocrine, and alimentary tract disturbances. J Equine Vet Sci 65:66–74

Mendoza FJ, Toribio RE, Pérez-Ecija A (2024) Metabolic and endocrine insights in donkeys. Animals 14(4):590

Merck & Co., Inc (2024) Osteochondrosis in horses. In: The Merck veterinary manual (11th edn)

Merck Veterinary Manual (2024a) Congenital and developmental anomalies of the mouth and dentition in horses. https://www.merckvetmanual.com/

Merck Veterinary Manual (2024b) Congenital and inherited anomalies of the mouth in animals. https://www.merckvetmanual.com/
Merck Veterinary Manual (2025a) Congenital and inherited anomalies of the musculoskeletal system in horses. Retrieved from https://www.merckvetmanual.com/musculoskeletal-system/congenital-and-inherited-anomalies-of-the-musculoskeletal-system/congenital-and-inherited-anomalies-of-the-musculoskeletal-system-in-horses merckvetmanual.commerckvetmanual.com
Merck Veterinary Manual (2025b) Developmental orthopedic disease in horses (Horse Owners section). Retrieved from https://www.merckvetmanual.com/horse-owners/bone-joint-and-muscle-disorders-in-horses/developmental-orthopedic-disease-in-horses merckvetmanual.com
Metzger J, Gast AC, Schrimpf R, Rau J, Eikelberg D, Beineke A, Hellig M, Distl O (2017) Whole-genome sequencing reveals a potential causal mutation for dwarfism in the Miniature Shetland pony. Mamm Genome 28(3–4):143–151
Miller W (2023) Genetic insights into dominant white and sabino patterns in horses and donkeys. Equine Color Genetics Review
Miniature Donkey Association (2023) Breed standards. https://www.miniaturedonkeyassociation.com
Ministerio de Agricultura, Pesca y Alimentación (2020) Catálogo Oficial de Razas de Ganado de España: Raza Asnal Mallorquina. Gobierno de España
Morton C (2023, August 16) Squamous cell carcinoma in horses. PetMD. https://www.petmd.com/horse/conditions/cancer/squamous-cell-carcinoma-horses
Mosbah E, Awadin W (2016) Diagnosis and management of a malignant dermal melanoma in a donkey (Equus asinus). Iran J Vet Surg 11(1):45–50. https://www.researchgate.net/publication/315644999
Naccache F, Metzger J, Distl O (2018) Genetic risk factors for osteochondrosis in various horse breeds. Equine Vet J 50(5):556–563
Nizou A et al (2015) Donkey hoof conformation and biomechanics. Vet J 204(1):77–84
Norris C, Redmond I (2019) The global trade in donkey skins and its impact on welfare and sustainability. People Anim 3(1):12–18
OMIA (Online Mendelian Inheritance in Animals) (2025) Coat colour, white spotting, KIT related in Equus asinus. Retrieved from https://omia.org
Orlando L et al (2013) Recalibrating Equus evolution using ancient DNA. Proc Natl Acad Sci 110(29):11736–11741
Palace VP, Khaper N, Qin Q, Singal PK (1999) Antioxidant potentials of vitamin a and carotenoids and their relevance to heart disease. Free Radic Biol Med 26(5–6):746–761
Paraschou G, Adako GM, Priestnall SL, Burden FA (2021) Suspensory ligament desmitis caused by Onchocerca sp. in three donkeys. Vet Pathol 58(2):401–404
Pauli RM (2019) Achondroplasia: a comprehensive clinical review. Orphanet J Rare Dis 14:1
Pierini C, Landi V, Ciani E, Maggiolino A, Campanile D, Carpio MG, De Palo P (2025) Conservation and genetic analysis of the endangered Martina Franca donkey using pedigree data. Front Anim Sci 6:1588467
Piras FM, Nergadze SG, Magnani E, Bertoni L, Attolini C, Khoriauli L, Raimondi E, Giulotto E (2010) Uncoupling of satellite DNA and centromeric function in the genus Equus. Genome Biol 11(4):R71
Popović V, Stevanović J, Zivković D, Gagrčin M (2020) Economic potential and marketing of donkey milk and cheese in the Zasavica Special Nature Reserve. Biotechnol Anim Husb 36(2): 195–204
Radosavljević V, Šmit B, Stevanović J (2021) Genetic monitoring and conservation of Balkan donkey populations: status, trends, and future directions. J Anim Feed Sci 30(2):138–147
Rao E, Weiss B, Fukami M, Rump A, Niesler B, Mertz A, Rappold G (1997) Pseudoautosomal deletions encompassing a novel homeobox gene cause growth failure associated with Léri–Weill dyschondrosteosis. Nat Genet 16(1):54–63
Rentea RM, Mehmood KT (2023) Ambiguous genitalia and disorders of sexual differentiation. In StatPearls [Internet]. Treasure Island: StatPearls Publishing. ncbi.nlm.nih.govncbi.nlm.nih.gov.

Riddick TL, Duesterdieck-Zellmer K, Semevolos SA (2012) Gene and protein expression of cartilage canal and osteochondral junction chondrocytes and full-thickness cartilage in early equine osteochondrosis. Vet J 194(3):319–325

Rieder S, Stricker C, Joerg H, Burger D (2008) Genetic analysis of coat colour variation in donkeys (Equus asinus). J Anim Breed Genet 125(5):304–310

Rivas-López J et al (2023) Genetic relationships among Spanish donkey breeds using microsatellite markers: implications for conservation. Animals 13(2):317

Rodrigues JB, Araújo S, Guedes-Pinto H, San Roman F, Viegas C, Bastos E (2013) Analysis of new Matrilin-1 gene variants in a case-control study related to dental malocclusions in Equus asinus. Gene 522(1):70–74

Rosenbom S, Barlow A, Worsøe Havmøller R, Bendrey R, Collins M, Buckley M et al (2023) The genomic history of the domesticated donkey. Science 381(6653):eabo5159

Rosengren Pielberg G, Golovko A, Sundström E, Curik I, Lennartsson J, Seltenhammer MH, Druml T, Binns M, Fitzsimmons C, Lindgren G, Sandberg K, Baumung R, Vetterlein M, Strömberg S, Grabherr M, Wade C, Lindblad-Toh K, Pontén F, Heldin C-H et al (2008) A cis-acting regulatory mutation causes premature hair graying and susceptibility to melanoma in the horse. Nat Genet 40(8):1004–1009

Rossel S, Marshall F, Peters J, Pilgram T, Adams MD, O'Connor D (2008) Domestication of the donkey: timing, processes, and indicators. Proc Natl Acad Sci 105(10):3715–3720

Royo LJ et al (2005) Analysis of genetic diversity and conservation priorities in Spanish donkey breeds using microsatellite markers. Genet Sel Evol 37(3):339–356

Russell J, Matika O, Russell T, Reardon RJM (2017) Heritability and prevalence of selected osteochondrosis lesions in yearling thoroughbred horses. Equine Vet J 49(3):282–287

Samarkhanova D, Zhabagin M, Nadirov N (2025) Reviewing the genetic and molecular foundations of congenital spinal deformities: implications for classification and diagnosis. J Clin Med 14(4):1113. mdpi.commdpi.com

Schmutz SM, Berryere TG (2007) Genes affecting coat colour and pattern in domestic dogs: a review. Anim Genet 38(6):539–549

Schraer K (2021, September 13) Photosensitization in horses. PetMD. https://www.petmd.com/horse/conditions/skin/photosensitization-horses

Seyiti S, Kelimu A (2021) Donkey industry in China: current aspects, suggestions and future challenges. J Equine Vet Sci 102:103642

Shariflou MR, Wade CM, Kijas JW, McCulloch R, Windsor PA, Tammen I, Nicholas FW (2013) Brachygnathia, cardiomegaly and renal hypoplasia syndrome in Merino sheep maps to a 1.1-megabase region on ovine chromosome OAR2. Anim Genet 44(2):231–233

Shields C (2025, January 20) The Catalan donkey: A national symbol, and endangered breed. Catalan News. https://www.catalannews.com/life-style/item/the-catalan-donkey-a-national-symbol-and-endangered-breed

Signer-Hasler H, Neuditschko M, Koch C, Froidevaux S, Flury C, Burger D, Leeb T, Rieder S (2014) A chromosomal region on ECA13 is associated with maxillary prognathism in horses. PLoS One 9(1):e86607

Smith LB, Bannasch DL, Young AE, Grossman DI, Belanger JM, Oberbauer AM (2008) Canine fibroblast growth factor receptor 3 sequence is conserved across dogs of divergent skeletal size. BMC Genet 9:67

Spinage CA (2012) African ecology: benchmarks and perspectives. Springer

Sponenberg DP (2003) Equine color genetics, 2nd edn. Iowa State Press

Starkey P, Starkey M (2004) Regional and world trends in donkey populations. In: Fielding D, Pearson P (eds) Donkeys, people and development. Animal Traction Network for Eastern and Southern Africa (ATNESA), pp 1–15

StatPearls (2021) Albinism—StatPearls. Retrieved from https://www.ncbi.nlm.nih.gov/books/NBK519018/

Steinhoff JS, Lass A, Schupp M (2021) Biological functions of RBP4 and its relevance for human diseases. Front Physiol 12:659977

Sun K, Guo J, Yao X, Guo Z, Guo F (2021) Growth differentiation factor 5 in cartilage and osteoarthritis: a possible therapeutic candidate. Cell Prolif 54(3):e12998

Tanaka J, Grahn R, Bellone RR (2020) Evidence supports white spotting in donkeys as a homozygous lethal condition. Anim Genet 51(5):840–842

Thaweesapphithak S et al (2024) The impact of RUNX2 gene variants on cleidocranial dysplasia phenotype: a systematic review. J Transl Med 22:1099. translational-medicine.biomedcentral.comtranslational-medicine.biomedcentral.com

Tibary A (2021) Intersex conditions of animals. In: Merck Veterinary Manual (Veterinary professional version). Merck & Co. merckvetmanual.commerckvetmanual.com

Todd ET, Tonasso-Calvière L, Chauvey L, Schiavinato S, Fages A, Seguin-Orlando A, Orlando L (2022) The genomic history and global expansion of domestic donkeys. Science 377(6611): 1172–1180

ToothFest (2023) Foal Dental Problems and Breeding Advice. https://www.toothfest.com

UC Davis Veterinary Genetics Laboratory (2024) Donkey coat color and hair length panel. Retrieved from https://vgl.ucdavis.edu

UC Davis Veterinary Genetics Laboratory (2025) Dominant White and White Spotting Testing in Donkeys. Retrieved from https://vgl.ucdavis.edu

Udhaya Kumar S, Rajan B, Kumar TD, Preethi AV, Abunada T, Younes S et al (2020) Involvement of essential signaling cascades and analysis of gene networks in diabesity. Genes 11(11):1256

UFRRJ Equideocultura (2023) Jumento Pêga. https://institucional.ufrrj.br/equideocultura/plantel/jumento-pega/? utm_source=chatgpt.com

Utzeri VJ, Bertolini F, Ribani A, Schiavo G, Dall'Olio S, Fontanesi L (2016) The albinism of the feral Asinara white donkeys (Equus asinus) is determined by a missense mutation in a highly conserved position of the tyrosinase (TYR) gene deduced protein. Anim Genet 47(1):120–124

Valle E, Cavallarin L, Pace S, Minero M (2017) A functional approach to the body condition assessment of lactating donkeys as a tool for welfare evaluation. PeerJ 5:e3093

Van Cauter R, Serteyn D, Lejeune J-P, Rousset A, Caudron I (2023) Evaluation of the appearance of osteochondrosis lesions by two radiographic examinations in sport horses aged from 12 to 36 months. PLoS One 18(5):e0286213

Wang C, Li H, Guo Y, Huang J, Sun Y, Min J, Wang J, Fang X, Zhao Z, Wang S, Zhang Y, Liu Q, Jiang Q, Wang X, Guo Y, Yang C, Wang Y, Tian F, Zhuang G, Fan Y, Gao Q, Li Y, Ju Z, Li J, Li R, Hou M, Yang G, Liu G, Liu W, Guo J, Pan S, Fan G, Zhang W, Zhang R, Yu J, Zhang X, Yin Q, Ji C, Jin Y, Yue G, Liu M, Xu J, Liu S, Jordana J, Noce A, Amills M, Wu DW, Li S, Zhou X (2020) Donkey genomes provide new insights into domestication and selection for coat color. Nat Commun 11:6014

Widmer S, Seefried FR, Häfliger IM, Signer-Hasler H, Flury C, Drögemüller C (2023) WNT10B: a locus increasing risk of brachygnathia inferior in Brown Swiss cattle. J Dairy Sci 106(12): 8969–8978

WikiVet (2023) Incisor Teeth Disorders Donkey. https://en.wikivet.net/Incisor_Teeth_Disorders_-_Donkey

Williams SB et al (2020) Locomotor biomechanics of gaited horses and donkeys. Vet Clin North Am Equine Pract 36(2):281–297

Wu DD, Li S, Zhou X, Wu F, Wang X, Zhang Y, Zheng Q et al (2020) Donkey genomes provide new insights into domestication and selection for coat color. Nat Commun 11:3807

Xie F-Y, Feng Y-L, Wang H-H, Ma J-Y et al (2015) de novo assembly of the donkey white blood cell transcriptome and a comparative analysis of phenotype-associated genes between donkeys and horses. BMC Genomics 16:1–13

Zeder MA (2012) The domestication of animals. J Anthropol Res 68(2):161–190

Żeromski M, Jedrzejewski T, Zamroziewicz R, Wezyk M (2021) Characteristics of the donkey's dorsal profile in relation to its functional body condition assessment. Animals 11(11):3095

Zhang YS, Yang XY, Wang XB, Zhang CM, Qin F, Zhou ZH, Lan XY, Chen H, Lei CZ (2010) Cytochrome b genetic diversity and maternal origin of Chinese domestic donkey. Biochem Genet 48(7–8):636–646

Donkey, Mule and Hinny Quantitative and Molecular Genetics: Multi-Omics Insights into Breeding for Functional Traits

Francisco Javier Navas González and Carlos Iglesias Pastrana

Abstract

This chapter provides an integrative review of quantitative and molecular genetic insights into donkey, mule, and hinny breeding, emphasizing the role of multi-omics in improving functional traits. It explores genomic, transcriptomic, proteomic, and epigenetic findings related to meat and milk production, skin and coat characteristics, growth performance, locomotion, fertility, and disease resistance in donkeys. Key candidate genes (e.g., *LCORL, NCAPG, HOXC8, WNT7A, SCD, LEPR*) and pathways underlying skeletal growth, muscle composition, fat deposition, and milk protein variation are identified, alongside breed-specific genomic signatures. Heritability estimates for functional traits such as gait, fertility, temperament, cognition, and disease resistance are discussed, providing a framework for genetic evaluations and selection indices. The chapter also examines hybrid systems (mules and hinnies), highlighting mitochondrial inheritance, sex-linked effects, and the potential of genomic tools for assessing hybrid performance despite sterility constraints. Collectively, these insights advance the application of molecular breeding strategies, marker-assisted selection, and functional genomics in equid breeding programs aimed at enhancing productivity, welfare, and adaptation.

F. J. Navas González (✉)
Department of Genetics, University of Cordoba, Cordoba, Spain

Worldwide Donkey Breeds Project, Cordoba, Spain
e-mail: fjnavas@uco.es

C. Iglesias Pastrana
Andalusian Institute for Agricultural, Fishing, Food and Ecological Production Research and Training (IFAPA), Córdoba, Spain

A. K. McLean et al. (eds.), *Donkey and Mule Medicine: Foundations, Breeds, and Management*, https://doi.org/10.1007/978-3-032-12726-6_5

Keywords

Genomic selection · Quantitative genetics · Meat quality · Intramuscular fat (IMF) · Milk production · Milk protein polymorphisms · LCORL · NCAPG · HOXC8 · WNT7A · SCD · LEPR · IGF1 · NR6A1 · BMP7 · MSTN · Gait genetics · DMRT3 · RELN · STAU2 · Coat color genetics · MC1R · ASIP · TYR · KIT · TBX3 · Collagen synthesis · LOXL2 · Heritability · Functional traits · Temperament · Cognition · Reproductive traits · Twinning · CD44 · Disease resistance · Immune tolerance · Cutaneous habronematidosis · Hybrid vigor · Mitochondrial DNA · Reciprocal crosses · Epigenetics · Marker-assisted selection · Genomic tools · Equid breeding programs

1 Introduction

The domestication and breeding of donkeys (*Equus asinus*), mules, and hinnies have long played a crucial role in human societies, particularly in agriculture, transportation, and animal-assisted therapies (Blench 2000; Brosnahan 2019). Donkeys, with their remarkable stamina, adaptability, and hardiness, have historically been essential in arid and mountainous regions worldwide. They are valued for their resilience in harsh conditions and their ability to perform heavy work where other animals struggle to survive (Burden and Thiemann 2015).

Mules and hinnies—hybrid offspring from the crossbreeding of donkeys and horses (*Equus caballus*)—inherit desirable traits from both parents, often displaying hybrid vigor that makes them stronger, more enduring, and more resilient than either parent in certain contexts (Franco et al. 2016). However, due to their intermediate chromosome number ($2n = 63$), they are generally sterile, which complicates the use of conventional genetic breeding strategies (Benirschke et al. 1962). Despite this limitation, their utility in industries such as draft work, transportation, and therapy underscores the need for ongoing improvements in breeding practices to optimize performance and welfare (McLean et al. 2014).

Advances in genomic and molecular genetics have revolutionized our understanding of the genetic architecture of these equids. The application of multi-omics approaches—encompassing genomics, transcriptomics, proteomics, and epigenetics—has significantly expanded our ability to investigate the genetic basis of functional traits in donkeys, mules, and hinnies (Ram et al. 2025; Zhu et al. 2025). These traits are central not only to survival and productivity but also to adaptability in diverse environments and roles in human society.

For example, elucidating the molecular mechanisms underlying meat and milk production, locomotion and gait, growth performance, disease resistance, and fertility can have profound implications for breeding and conservation, as well as for optimizing their use in varied applications (Zhang et al. 2024; Wang et al. 2024a, b, c, d).

This chapter provides an in-depth exploration of the genetic and molecular basis of functional traits in donkeys (*Equus asinus*), mules, and hinnies, highlighting recent research advances and the tools available for genetic evaluation. From meat quality and milk protein polymorphisms in donkeys to the molecular pathways underlying muscle composition, skeletal growth, and fat deposition, genetic studies are increasingly identifying specific genes and markers that can be targeted for selective breeding (Zhang et al. 2024; Zhu et al. 2025). These insights have the potential to enhance traits such as intramuscular fat (IMF) content, tenderness, and overall meat quality, while also improving milk yield and composition in donkey milk—a product valued for its hypoallergenic properties and nutritional profile (Guo et al. 2019).

One of the most significant advances in equine genetics is the identification of genes influencing growth performance and skeletal structure. For example, LCORL, NCAPG, and HOXC8 have been implicated in regulating body size and skeletal conformation, directly shaping desirable physical traits (Metzger et al. 2013; Sevane et al. 2017). In parallel, studies of muscle metabolism and fiber composition in donkeys have revealed that genes such as SCD, LEPR, and TPM3 play key roles in IMF deposition, a determinant of meat tenderness and juiciness (D'Alessandro and Zolla 2013). Proteomic analyses have further advanced understanding by mapping protein networks central to these processes, offering breeders molecular tools to improve traits linked to performance in both agricultural and working contexts (Zhu et al. 2025).

Beyond meat and milk production, donkeys serve as valuable models for studying disease resistance and fertility. They often display increased tolerance to diseases that severely affect horses, including Equine Infectious Anemia (EIA), African Horse Sickness (AHS), and various viral infections (Lazić et al. 2023). At the molecular level, donkeys exhibit unique immune responses that are less prone to severe disease outcomes, providing important clues for developing improved animal health strategies (The Donkey Sanctuary 2024). Their resistance to conditions such as cutaneous habronematidosis and other parasitic diseases is of growing interest, as identifying the genetic basis of such tolerance could inform breeding programs aimed at enhancing disease resistance in both donkeys and hybrids.

In terms of fertility, donkeys show considerable variation in reproductive traits affecting herd productivity, including gestation length, age at first reproduction, and incidence of multiple births—traits influenced by both genetic and environmental factors (Canisso et al. 2019). Heritability estimates for fertility traits in donkeys are moderate, suggesting that selective breeding could improve reproductive performance. While mules and hinnies are typically sterile due to their intermediate chromosome number, recent research has examined how hybrid vigor (heterosis) and the interaction between horse and donkey genomes influence hybrid performance, offering insights into optimizing the productivity and utility of these animals (Hanot et al. 2019).

One of the most complex and fascinating areas of research in equid genetics lies in the study of locomotion and gait. Donkeys are recognized for their distinct gait patterns, which can vary significantly across breeds and even among individuals. In some cases, donkeys exhibit unique gaits such as the lateral amble, valued for its smoothness and comfort in riding therapy. The molecular basis of these gait patterns has been linked to specific genetic variants, most notably mutations in the DMRT3 gene, which plays a critical role in regulating locomotion in horses and other equids (Herman et al. 2023; Andersson et al. 2012a, b). Understanding the genetic underpinnings of gait variation in donkeys opens new avenues for breeding animals with targeted locomotor traits, enhancing their suitability for work, transportation, and therapeutic applications.

The study of coat color in donkeys presents another intriguing area of genetic exploration. Genes such as MC1R, ASIP, and TYR govern coat pigmentation, and allelic variation in these loci contributes to the wide diversity of coat colors and patterns observed in donkeys today (Liu et al. 2024; Abitbol et al. 2015; Rieder et al. 2001). This research not only deepens our understanding of the genetic architecture of coat color but also supports the development of marker-assisted selection tools for breeding animals with desirable physical traits, including coat quality and resilience.

Hybrid breeding of mules (*E. asinus* ♂ × *E. caballus* ♀) and hinnies (*E. caballus* ♂ × *E. asinus* ♀) introduces additional complexity, particularly regarding mitochondrial inheritance and the effects of reciprocal crosses. Because mules inherit mitochondrial DNA (mtDNA) from their horse mothers, while hinnies inherit it from their donkey mothers, the mitochondrial genome plays a key role in influencing performance traits such as endurance and metabolic efficiency (Franco et al. 2016; McLean et al. 2014). Genomic studies are beginning to unravel how mtDNA variation shapes the adaptability and work capacity of these hybrids, despite the challenges posed by their sterility.

The growing role of genomic technologies—such as single nucleotide polymorphism (SNP) genotyping, whole-genome sequencing, and CRISPR-based genome editing—is transforming donkey, mule, and hinny breeding programs. These tools, when integrated with traditional breeding methods, enable more precise selection for productivity, health, and welfare traits (Li et al. 2025; Gambini et al. 2025; Meuwissen et al. 2016).

In conclusion, the integration of multi-omics and genomic tools in the study of functional traits in donkeys, mules, and hinnies represents a transformative shift in equine breeding. Advances in molecular genetics are revolutionizing breeding practices and offering novel insights into the genetic mechanisms that govern performance, health, and behavior. By leveraging these insights, breeders can develop more effective programs that optimize the utility of these animals and ensure their long-term sustainability in diverse roles across society.

2 Donkey Multi-Omics and Functional Genetics

2.1 Meat Production and Quality

Recent multi-omics studies have begun to elucidate the genetic underpinnings of donkey meat quality. Whole-genome and transcriptome surveys have identified numerous candidate genes (e.g. ACTN3, BMP7, HOXC8, LCORL, NR6A1, WNT7A, TPM2, TPM3) linked to growth, skeletal conformation and meat traits (Zhu et al. 2025; Wang et al. 2024a, b, c, d). In particular, genes influencing skeletal development and body size—such as NCAPG, WNT7A, BMP7, HOXC8, and LCORL—appear to determine carcass yield and vertebral count (Ma et al. 2023). Transcriptomic (RNA-seq) comparisons between high- and low-IMF Guangling donkeys identified hundreds of differentially expressed genes (DEGs), including lipid metabolism regulators such as SCD, LEPR, CIDEA, DGAT2, DLK1, and WNT10B, which are enriched in adipogenesis and fat deposition pathways (Ma et al. 2023; Zhu et al. 2025).

Proteomic profiling using data-independent acquisition (DIA) uncovered dozens to hundreds of differentially expressed proteins between donkey muscles. These proteins were enriched in key pathways including fatty acid biosynthesis, TGF-β, FOXO, mTOR signaling, oxidative phosphorylation, and amino acid metabolism—directly impacting intramuscular fat deposition, protein turnover, and flavor precursor formation (Wang et al. 2024a, b, c, d). Lipidomic analyses of *longissimus* muscle confirmed that higher IMF is associated with increased monounsaturated fatty acids, especially oleic acid, and distinct volatile compound profiles such as 2-acetyl-2-thiazoline and 2-pentylfuran (Ma et al. 2023). Because IMF content positively correlates with tenderness and juiciness, genes like SCD and LEPR that regulate IMF variation play key roles in meat quality (Zhu et al. 2025).

Similarly, another transcriptome study found that genes regulating muscle fiber type and glucose metabolism (e.g. MYH1, MYH7, TNNC1, TNNI3, TPM3, ALDOA, ENO3, PGK1) were differentially expressed and associated with meat tenderness (Ma et al. 2023; Zhu et al. 2025). Differential gene expression across muscle types is also pronounced: transcriptomes of fast-twitch (LD) versus slow-twitch (*gluteus*) muscles differ in genes like ENO3, MYH1, MYH4, TNNI3, PGK1, and ALDOA, reflecting muscle fiber composition and texture variation (Zhu et al. 2025).

Regional and breed differences further highlight genetic influences. Proteomic profiles differ by cut: for instance, LT versus ST versus GM muscles show distinct pathway enrichments (MAPK/GnRH vs fat digestion/actin cytoskeleton) (Wang et al. 2024a, b, c, d). Lipidomics also revealed higher triglyceride and phospholipid species in the *longissimus* relative to other muscles (Ma et al. 2023). Cross-breed comparisons show clear effects: North African donkeys (NAD) had higher IMF and tenderness than Masri donkeys, attributable to genetic background and selective pressures (Zhu et al. 2025).

Moreover, CRISPR-Cas-based assays have recently been developed to enable rapid authentication of donkey meat and may soon allow functional validation of candidate genes (Zhu et al. 2025). Together, these integrated multi-omics findings highlight specific genetic markers and molecular pathways (e.g. ACTN3, SCD, WNT7A, HOXC8, LEPR) that could be targeted to improve donkey meat traits such as IMF, tenderness, and carcass yield (Zhu et al. 2025; Wang et al. 2024a, b, c, d; Ma et al. 2023; Polidori et al. 2020, 2021).

2.2 Milk Production and Quality

Donkey milk possesses a unique protein profile, with genetic and proteomic studies revealing numerous variants. Its casein fraction (mainly αs1-, αs2-, β-, and κ-caseins) is closely related to human milk, with a low casein-to-whey ratio and small micelle size (Auzino et al. 2022). Indeed, donkey β-lactoglobulin (a major whey protein) is monomeric (unlike the dimer in ruminants), and total protein content is low, contributing to its hypoallergenic properties (Auzino et al. 2022). Although allergies to donkey milk in humans have been reported, these are anecdotal

Fig. 1 Amiata or Amiatina donkey breed

(Giorgis et al. 2018). Proteomic surveys combining mass spectrometry and cDNA sequencing have uncovered extensive microheterogeneity in donkey caseins. For example, Amiata donkey milk yielded ~30 αs2-casein proteoforms (from two CSN1S2 genes with complex splicing) and ~20 phosphorylated αs1-casein isoforms per sample (Auzino et al. 2022) (Fig. 1).

Genomic polymorphisms have been detected in casein and whey loci: novel allelic variants were found for β- and αs-caseins, lysozyme C, and β-lactoglobulin I (Auzino et al. 2022). A previously unreported β-lactoglobulin I isoform (with an N-terminal extension) was identified by tandem MS (Auzino et al. 2022), and a new β-lactoglobulin II allele of very low expression has been described (Criscione et al. 2018). Notably, PCR amplification from donkey milk somatic-cell DNA confirmed the presence of the κ-casein gene (CSN3) in donkey, even when κ-casein protein is scarce (Singh et al. 2024). Thus, molecular characterization of donkey milk has revealed extensive genetic diversity at milk protein loci (β-LG I/II, lysozyme, αs1-/α s2-/κ-casein), underlining both the novelty of its proteome and the potential for selecting variants with desirable properties.

Genome-wide association studies (GWAS) have begun to connect DNA variants with milk yield traits. A recent GWAS in Xinjiang donkeys (112 individuals, whole-genome re-sequenced) identified several candidate genes for milk production (Fig. 2). In particular, validated SNPs in *GPIHBP1*, *FLII*, *TOP1MT* (plus *TNK1*, *PLK1*, *DRG2*) showed significant association with average and total daily milk yield (Fang et al. 2025). These genes are implicated in lipid metabolism and mammary function (e.g. GPIHBP1 binds lipoprotein lipase, FLII is an actin-remodeling protein, TOP1MT is mitochondrial) and mirror findings in other dairy species (Fang et al. 2025). No GWAS for detailed milk composition (fat, protein, lactose content) in donkey have yet been reported, but candidate loci from other species (e.g. *GHR*, *DGAT*, casein cluster variants) may be relevant.

Analytical technologies have been crucial for donkey milk genomics and proteomics. Matrix-assisted laser desorption/ionization time-of-flight (MALDI-TOF) MS profiling is widely used for milk biomarker analysis and authenticity testing (Šebela

Fig. 2 Xinjiang donkey breed

2022). In equine milk studies, MALDI-TOF can quickly generate peptide/protein "fingerprints" to distinguish species or detect adulteration, and has been applied in differentiating donkey vs goat or cow milk. Genomic DNA can be efficiently extracted from milk somatic cells, offering a noninvasive source of DNA for genotyping. For example, Singh et al. (2024) reported the first isolation of DNA from Halari donkey milk somatic cells and successfully amplified the CSN3 (-κ-casein) gene (Singh et al. 2024) (Fig. 3). This enables genetic analysis without blood sampling, facilitating marker-assisted selection and breed traceability based on milk.

Comparatively, donkey milk is more similar to human or mare milk than to ruminant milk. Like horse milk, it has a high lactose (~7%) and whey proportion, and lower fat/protein content than cow's milk (Pastuszka et al. 2016). It is remarkably rich in lysozyme (≈0.3–1.1 g/L) and immunoglobulins, imparting antimicrobial and anti-inflammatory properties (ruminant milks contain only trace lysozyme) (Pastuszka et al. 2016). The major allergens of cow's milk (β-lactoglobulin and αs1-casein) are present at much lower levels or in less allergenic forms in donkey milk, explaining its hypoallergenic status (Pastuszka et al. 2016). Clinically, donkey milk has proven to be a valid substitute for cow's milk in infants with cow-milk protein allergy when appropriately supplemented with fats (Sarti et al. 2019).

These molecular insights have important implications for donkey dairy development. Polymorphisms in milk protein genes may affect milk composition and processing qualities; for instance, distinct κ-casein or β-lactoglobulin alleles could influence cheese yield and whey characteristics. GWAS-identified genes like *GPIHBP1*, *FLII* and *TOP1MT* could serve in marker-assisted selection to improve

Fig. 3 Halari donkey breed

milk yield and composition in donkey. Technologically, MALDI-TOF MS and other omics tools allow rapid authentication of donkey milk and its products. Finally, the unique proteomic profile—closer to human milk and low-allergen—underpins new product development (hypoallergenic infant formula, functional dairy foods). Ongoing sequencing and proteomics work will continue to refine our understanding of donkey milk genetics, enabling targeted breeding for enhanced dairy traits and expanding the utilization of donkey milk in human nutrition.

2.3 Phaneroptics

Donkey skin is a robust, collagen-rich organ adapted for protection and durability. The epidermis is stratified into four layers (stratum basale, spinosum, granulosum, and corneum), similar to horses (Magdy et al. 2025). Melanocytes in the basal layer produce melanin for pigmentation (Cappai et al. 2024). Basal keratinocytes secrete structural proteins (e.g. keratins like KRT14 and filaggrin) and tight-junction components to form a barrier against water loss and pathogens (Magdy et al. 2025; Cappai et al. 2024). Donkey skin comprises approximately 16% of body mass and serves as the first line of defense against the environment (Magdy et al. 2025).

Beneath the epidermis lies a thick dermis packed with collagen fibers. Histological surveys confirm that donkey skin is exceptionally collagen-dense, giving it high tensile strength (Magdy et al. 2025). Magdy et al. (2025) showed that the back (thoracic) skin of adult donkeys is the thickest and richest in collagen, while limb skin is thinner despite relatively high collagen content in the thoracic region. This regional variation likely reflects functional demands. Donkey hides are especially valued for gelatin ("Colla corii asini", or ejiao) production; the skin's high collagen content is exploited in pharmaceutical and cosmetic industries (Magdy et al. 2025).

Transcriptomic and epigenetic studies have begun to reveal the molecular underpinnings of this collagenous matrix. In a comprehensive RNA-seq analysis of Dezhou donkey skin, Wang et al. (2024a, b, c, d) identified COL1A1 and COL3A1 as among the most highly expressed extracellular-matrix genes. These genes, along with COL6A5, emerged as key genes involved in collagen synthesis in donkey dermis. Several long noncoding RNAs (lncRNAs) were found to cis-regulate COL1A1 and other collagen genes, suggesting fine-tuned transcriptional control in skin development. Likewise, the enzyme LOXL2 was highlighted for its role in collagen maturation: "our study has unveiled the role of the LOXL2 gene... shedding light on its involvement in collagen synthesis and modification" (Wang et al. 2024a, b, c, d).

Epigenetic surveys in donkeys also link collagen genes to developmental age. In an integrative study combining RNA-seq with whole-genome bisulfite sequencing, Wang et al. (2024a, b, c, d) examined Dezhou donkey skin at embryonic, juvenile, and adult stages. They found numerous differentially methylated genes impacting skin maturation and collagen deposition, including COL1A1, COL1A2, COL3A1, COL4A1, and COL4A2. Notably, COL1A1 promoter regions were hypomethylated in early (embryonic) skin correlating with high expression. By adulthood, many collagen genes and LOXL2 were downregulated and hypermethylated, consistent with a slowing of new matrix synthesis with age. These findings clarify how DNA methylation modulates gene expression during donkey skin maturation.

While collagen gives donkey skin its strength, coat color is determined by melanocyte biology and pigment genes. As in other mammals, the type of melanin is controlled by the MC1R–ASIP signaling pathway. Active MC1R stimulates tyrosinase and drives eumelanin production, whereas loss-of-function MC1R or antagonism by ASIP shifts to pheomelanin. In donkeys, Abitbol et al. (2014) showed that a recessive missense mutation c.629T>C (p.Met210Thr) in MC1R causes the chestnut (red) coat color. This variant disrupts a conserved site, yielding a nonfunctional receptor and a pheomelanic phenotype.

Abitbol et al. (2015) also identified a recessive missense mutation in the ASIP gene (c.349T>C, p.Cys117Arg) that abolishes the typical "pangare" or light-point pattern. This variant destroys a conserved disulfide bond, so ASIP cannot antagonize MC1R, resulting in a solid black/dark coat.

Beyond MC1R/ASIP, other loci affect coat color. Mutations in TYR cause oculocutaneous albinism: Utzeri et al. (2016) discovered that all Asinara (Sardinian white) donkeys share a homozygous c.604C>G (p.His202Asp) mutation in TYR. This amino acid change disrupts a copper-binding site of tyrosinase and abolishes melanin synthesis, explaining the OCA1-type phenotype. Recent work by Cappai et al. (2024) confirms Asinara albinism is TYR-linked.

Haase et al. (2015) analyzed white and spotted coat phenotypes in donkeys and found two KIT mutations. One was a missense change c.662A>C (p.Tyr221Ser), seen only in a completely white-born donkey, and the other a splice donor mutation (c.1978+2T>A) found in donkeys with piebald spotting. These variants co-segregated with their respective phenotypes and were absent in solid-colored controls.

Wang et al. (2020) performed whole-genome scans and showed that the non-dun coat is caused by a 1 bp deletion downstream of TBX3. This deletion reduces TBX3 expression in skin, lifting its inhibitory effect on pigment deposition and yielding uniformly dark coats. The TBX3 deletion was strongly selected in domestic donkeys, consistent with breeders favoring solid-colored animals.

Asinara albinos exhibit classic OCA1 albinism. TYR loss-of-function (c.604C>G) ablates melanin despite UV exposure (Utzeri et al. 2016; Cappai et al. 2024). Asinara donkeys survive well in sunlight, likely via compensatory adaptations such as enhanced antioxidant defenses.

The Austro-Hungarian White Baroque donkey breed also includes all-white animals, but with a different history. Grilz-Seger et al. (2020) genotyped the Austro-Hungarian albino donkeys and found none of the known mutations (MC1R, ASIP, TYR) were present. Neither the Asinara TYR mutation nor any of the KIT variants was found in that population. The Austro donkey's white phenotype is genetically distinct from both Asinara albinos and standard coat colors; its causal mutation remains unknown.

The Dezhou donkey of China is a melanistic breed prized for its black/brown coat. This animal exhibits strong eumelanin pigmentation. The domestic donkey genome scan identified the TBX3 deletion associated with non-dun color (Wang et al. 2020). Transcriptomic surveys in Dezhou skin emphasize ECM and skin development genes (COL1A1, LOXL2, etc.) rather than pigment pathways (Wang et al. 2024a, b, c, d), reflecting the breed's focus on hide quality. Unpublished data suggest melanogenic genes like MITF, TYRP1, PMEL, DCT are upregulated in these dark-coated donkeys.

Donkey skin anatomy and genetics reflect their dual role as work animals and sources of high-quality hide. The dermis is collagenous, with transcriptional programs (COL1A1, COL3A1, LOXL2, etc.) tightly regulated during growth. Pigmentation varies from the Asinara albinos (TYR mutation) to Austro albinos (unknown cause) to solid-colored breeds (MC1R, ASIP, TBX3 variants). These findings integrate classical coat-color genetics with modern omics analyses to explain the skin and color diversity of domestic donkeys.

2.4 Growth and Performance

Donkeys (Equus asinus) exhibit considerable variation in body size, growth rate, skeletal structure and muscle performance across breeds. Recent genomic analyses have begun to uncover the genetic architecture of these traits. For example, whole-genome scans and selection-sweep studies in Chinese donkey populations identified strong signals at loci known from other mammals. Liu et al. (2022) found that the LCORL–NCAPG locus, along with FAM184B, TBX3 and IHH, showed divergent selection between large and small Chinese donkeys, and pinpointed a stop-gain mutation in the coding sequence of LCORL (Chr3:112,664,848) in large individuals. Similarly, Wang et al. (2022) reported that the NCAPG–LCORL region on chromosome 3 is under selection in Liangzhou donkeys, consistent with a role in limiting body size. In a broader survey of six Chinese breeds, Zhou et al. (2020) mapped 11 selective sweeps and highlighted body-size genes such as LCOR (likely LCORL), NCAPG, TBX3 (body frame) and ACSL4, as well as GABPA implicated in "motion capacity". Notably, LCORL and NCAPG—genes homologous to cattle and horse height loci—repeatedly emerge in donkey studies of stature. In donkeys, as in other species, these genes likely influence skeletal growth and limb length via effects on cell proliferation and chromatin structure.

LCORL and NCAPG regulate body frame: LCORL encodes a transcriptional cofactor and NCAPG a condensin subunit. In donkeys, polymorphisms in this region associate with withers height and growth. Liu et al. (2022) found a deleterious stop-loss mutation in LCORL that segregated with larger size, suggesting truncation of the LCORL protein alters growth signaling. The same study also identified selection at NCAPG–LCORL between large and small breeds. Zhou et al. (2020) similarly flagged LCOR and NCAPG in selective-sweep analyses of six Chinese breeds. In a GWAS of Xinjiang donkeys, LCORL/NCAPG variants were again implicated in chest circumference and overall conformation. These results mirror findings in horses and cattle, where LCORL/NCAPG loci are major determinants of adult size.

Homeobox genes pattern the vertebrate body plan. In donkeys, HOXC8 has been linked to vertebral number and body length. Shi et al. (2022) sequenced the entire HOXC8 locus in Dezhou donkeys and identified 12 SNPs; two of these (g.15179224C>T and g.15179674G>A) were significantly associated with skeletal traits. The C>T SNP was negatively correlated with carcass weight and lumbar vertebra length, while G>A was positively correlated with the number of lumbar vertebrae. Thus HOXC8 variation appears to influence vertebral count and overall trunk length in donkeys, as it does in other mammals.

NR6A1 is a nuclear receptor gene and known "homeotic" locus for vertebra number in pigs. In Dezhou donkeys, Liu et al. (2022) detected seven novel intronic SNVs in NR6A1. One SNV (g.18114954C>T) was strongly associated with thoracolumbar count and carcass traits: animals with the T allele had more lumbar vertebrae and greater carcass weight than the C homozygotes. This parallels pig studies where NR6A1 variants increase vertebra number and body length. These associations suggest that NR6A1 polymorphisms in donkey may modulate spine length and thus contribute to overall size and meat yield.

Insulin-like growth factor 1 (IGF1) is a key endocrine regulator of growth. In donkeys, IGF1 is broadly expressed and its polymorphisms influence size. Lai et al. (2021) measured IGF1 expression in multiple tissues of growing and adult donkeys: muscle IGF1 was higher in young animals, whereas heart, liver, and others were higher in adults. They also found two IGF1 SNPs (IGF1-1 and IGF1-2) whose alleles (the GG genotype) were associated with larger chest circumference, especially in Dezhou jackasses. Haplotype analysis confirmed that the GG–GG combination tended to yield the largest body dimensions. This establishes IGF1 as a quantitative trait gene in donkey growth, similar to its role in cattle and horses.

Other growth regulators likely operate as well: for example, in cattle and horses, HMGA2 and NR6A1 affect height and vertebrae, suggesting these genes should be considered candidates in donkeys.

A potent negative regulator of muscle mass, MSTN is best known for "double-muscling" mutations in cattle and race distance in horses. In donkeys, however, polymorphism appears limited. A survey in Turkish donkeys found only a single synonymous SNP (g.4183919G>A) in exon 2, causing no amino-acid change. The authors speculate that this Leu→Leu silent change could subtly increase myostatin expression (perhaps by altering mRNA stability), but no loss-of-function alleles were detected. Thus, unlike Thoroughbreds (where an MSTN promoter SNP dramatically affects optimal racing distance), domestic donkeys carry no obvious MSTN variants known to produce hypertrophy. This may partly explain their relatively modest musculature.

BMP7, a growth factor in the TGF-β family, regulates bone and cartilage development. In genomic screens, BMP7 has surfaced as a candidate for body dimension. Zhu et al. (2025) included BMP7 among genes associated with donkey skeletal growth traits. In a summary of donkey "omics" studies, BMP7 was listed alongside NCAPG, Wnt7a, HOXC8 and LCORL as significantly associated with body dimensions (e.g. withers height, chest girth). These hints align with BMP7's known role in vertebrate skeletogenesis.

Donkey performance also depends on muscle fiber composition and metabolism. Advanced transcriptome/proteome analyses have identified networks of muscle-expressed genes. Sun et al. (2023) found that donkey skeletal muscle expresses a suite of myofibrillar genes: five troponin isoforms (TNNC1, TNNC2, TNNI1, TNNI2, TNNT1), seven myosin-related genes (MYBPC1, MYBPC2, MYH2, MYH7, MYL1, MYL2, MYL3), and tropomyosins (TPM1, TPM3). These proteins both constitute muscle fibers and regulate contraction, underscoring their dual structural and developmental roles. Notably, TPM3 was experimentally validated as a target of donkey muscle miRNA eca-miR-1, linking microRNA regulation to fiber formation. Other genes such as ENO3, PGK1, ALDOA and additional myosin heavy chains (MYH1, MYH4) showed differential expression among muscle types, suggesting they influence fiber composition, glycolysis, and thus endurance vs. power traits.

Candidate genes known from equine sports physiology may be relevant. For instance, ACTN3 encodes a fast-fiber Z-line protein: the R577X null variant in humans and horses distinguishes power vs. endurance athletes. While no donkey-specific ACTN3 polymorphism is published, Zhu et al. (2025) listed ACTN3 among growth and meat-quality genes in donkeys. Similarly, genes affecting fat metabolism (e.g. SCD, CIDEA, LEPR) are implicated in intramuscular fat deposition in donkey muscle. These will influence energy stores and stamina. In one review, enrichment of lipid-pathway DEGs (including SCD, THRSP) was linked to muscle fat in Guangling donkeys (Fig. 4). Finally, genes such as GABPA (identified under selection for "motion" in Zhou et al.) and exercise-adaptive pathways have yet to be fully characterized but may represent parallel adaptation of donkey muscles to sustained work.

Fig. 4 Guangling donkey in Linfen, North China's Shanxi province

Tissue-specific "omics" has begun to shed light on donkey muscle, tendon and bone biology. Wang et al. (2021) generated a transcriptome atlas of 16 donkey tissues and identified thousands of tissue-enriched genes. Although focused on milk and lysozyme genes, this resource provides a reference for expression of performance-related genes in donkey muscle and tendon. More targeted studies have examined muscle development directly. For example, Yu et al. (2023) used RNA-seq of Dezhou donkey muscle to identify long non-coding RNAs (lncRNAs) regulating key structural genes. They found lncRNAs that likely regulate DCN (decorin), ITM2A, MUSTN1 and ARRDC2—all genes involved in myogenesis and muscle differentiation. In parallel, microRNA profiling revealed eca-miR-1 targeting TPM3 in myoblast cells, confirming post-transcriptional control of muscle fiber formation. Proteomic comparisons have also been informative: Sun et al. (2023) contrasted donkey, cattle and goat muscle proteomes and emphasized the prominence of troponin and myosin isoforms in donkey meat, indicating conserved contraction machinery.

RNA-seq of developing donkey bone or tendon is still lacking, but analogous studies in horses implicate pathways like Wnt, BMP and IGF signaling in skeletal growth. Transcriptomic differences in donkey vertebrae or long bones have not yet been reported, but candidate genes like Wnt7a, NFATC2 and PROP1 have shown associations with chest and thoracic shape. As more "-omics" data accumulate, one can expect discovery of networks coordinating muscle fiber type, mitochondrial metabolism, and connective tissue traits relevant to endurance and strength.

Different donkey breeds show distinct growth profiles, reflecting both breeding goals and geographic adaptation. Heavy draft breeds (e.g. Italy's Martina Franca, Spain's Catalan, and Italy's Amiata) are tall and robust; light Asian breeds (e.g. Dezhou, Guangling) are medium-sized. For instance, the Martina Franca (MF) donkey is a mesomorphic large-framed breed with "tall stature, exceptional sturdiness" and marked dorsally-striped bay coloration. It arose from crosses with Catalan stock and was historically used for mules and heavy transport. In contrast, Chinese Dezhou donkeys are bred for meat and have a more compact conformation, though with remarkable thoracolumbar variability (most individuals are T18L5, but about 20% are T18L6 or T17L6). Genome studies reflect these differences: selection scans in Xinjiang and Dezhou breeds highlight body-size loci (e.g. LCORL/NCAPG) and even high-altitude genes (e.g. HBB, GLDC). Asian breeds also segregate variation in vertebra number (affecting body length). European breeds, by contrast, have seen little genomic study; pedigree analyses of the MF donkey reveal severe bottlenecks and inbreeding, suggesting limited diversity for size-related loci. Nonetheless, the MF (and by extension Catalan) likely carry the "tall" alleles of stature genes, in contrast with smaller native breeds. Overall, breed history and selection have created a mosaic: local Chinese and African breeds emphasize moderate size and adaptability, whereas Mediterranean breeds emphasize large frame and strength. Genome-wide analyses in diverse breeds show that body-height alleles segregate broadly but at different frequencies. This implies that marker-assisted selection could leverage breed-specific haplotypes for growth; for example, introgressing "tall" variants into smaller stock to enhance size, or vice versa to maintain compactness.

The variants identified often have predictable developmental effects. A stop-gain in LCORL likely truncates a key cofactor, possibly shifting growth towards precocity or limb elongation. HOXC8 SNPs alter its regulatory output: the G>A change at g.15179674 increased lumbar vertebra count, suggesting a boost in posterior axial segment identity. The NR6A1 intronic C>T polymorphism clearly elevates vertebra number and body size in heterozygotes. Such axial duplications not only lengthen the spine but also correlate with increased carcass weight. In contrast, the HOXC8 C>T variant reduced carcass weight and vertebral length, indicating it may shift development towards a more compact build or alter muscle distribution.

Polymorphisms in IGF1 produce subtler but important effects: the GG genotype at two IGF1 loci was associated with greater chest depth and circumference, reflecting enhanced thoracic growth. This likely arises from IGF1-mediated stimulation of chondrocyte and myocyte proliferation. Conversely, the silent MSTN mutation in donkeys may slightly increase myostatin levels and thus temper muscle growth. While its effect is presumptive, it aligns with the donkey's overall lack of myostatin-null "double-muscling."

Many other candidate pathways remain to be tied to functional outcomes. For instance, variants in BMP7 or Wnt7a would be expected to influence bone growth or limb bud patterning. Genes like PRKG2 and NFATC2 have been associated with body dimensions, hinting at their roles. Finally, genetic differences in metabolic regulators (e.g. SCD, LEPR, CIDEA) alter intramuscular fat and energy balance, affecting endurance and stamina. As more donkey-specific functional data accumulate, the mechanistic links between genotype and phenotype will become clearer.

2.5 Locomotion and Gait

Donkeys (*Equus asinus*) normally exhibit the three basic equid gaits—walk (a 4-beat lateral sequence), trot (2-beat diagonal), and canter/gallop—but some breeds show additional intermediate-speed gaits or "ambling" patterns. For example, Andalusian donkeys used in riding therapy fall into two distinct gait groups: one group moves with a smooth lateral four-beat ambling gait and rarely performs a standard walk, while another group shows a normal isochronous walk but not the amble (Navas González et al. 2018). In Brazil, the gaited Pêga donkey is bred to produce saddle mules with low-speed comfort gaits; these donkeys are classified by the *batida* (diagonalized) or *picada* (lateral) gait patterns, analogous to those in gaited horses (Herman et al. 2023). Intriguingly, within such breeds each animal tends to strongly favor one gait category (e.g., amble vs. walk), and genetic analysis confirms this is highly heritable (heritability ~0.53–0.67 for walk, amble, or trot in one study) (Navas González et al. 2018).

Evidence for sex-associated or sex-linked inheritance of gait traits in donkeys, particularly the ambling gait, is beginning to emerge from both observational and breeding data. Reports from Andalusian donkey breeding programs suggest that certain ambling sires produce a disproportionately high number of ambling offspring, regardless of the dam's gait phenotype. This pattern is especially notable because it is not reciprocally observed in gaited females; that is, while some females do amble, they often produce offspring with variable gait types, including non-amblers. This asymmetry implies a possible paternal influence that could be consistent with Y-linked inheritance, paternal imprinting, or sex-specific epistatic effects. Such male-biased transmission patterns resemble sex-linked inheritance mechanisms known in other species, though no specific Y-chromosomal loci have yet been identified in donkeys. The observed discordance between male and female transmission of gait reinforces the hypothesis that different genetic mechanisms—possibly involving male-specific regulatory elements or sex-limited gene expression—may underlie the inheritance of alternate gaits such as ambling. Further research using controlled pedigrees, genomic imprinting analysis, and sex-specific gene expression profiling will be essential to validate and clarify these patterns (Navas González et al. 2018).

Even heavy "mammoth" donkey strains have been reported anecdotally as "gaited" with a rocking-chair smooth ride (Medina 2017). Overall, then, donkey populations include standard locomotor types (walk/trot) and various alternate gaits (four-beat amble, lateral pace, etc.), often in a breed dependent manner (e.g., load-bearing, riding, or driving strains selected for smoothness).

At the molecular level, the best-characterized "gait gene" in equids is *DMRT3*, encoding a transcription factor expressed in spinal interneurons. In horses, a single-base nonsense mutation in *DMRT3* (Ser301Stop) dramatically alters locomotion: horses carrying the A allele (truncating mutation) are permissive for alternate gaits and excel at harness racing, whereas wild-type (C allele) horses are limited to walk/trot (Andersson et al. 2012a, b). Andersson et al. showed that *DMRT3* is expressed in a subset of spinal interneurons (the dI6 lineage) that coordinate left-right limb alternation, and *Dmrt3*-null mice lose normal gait coordination (Andersson et al. 2012a, b). In short, *DMRT3* sets up the core central-pattern-generation (CPG) circuitry for stride timing, and its mutant allele is the "gait-keeper" enabling four-beat ambling or pacing gaits.

In donkeys, however, the *DMRT3* gait mutation is extremely rare or absent. Genotyping of Brazilian Pêga donkeys found 97% carried only the wild-type CC genotype and none were AA (Herman et al. 2023); similarly, an Indian study found all sampled donkeys homozygous for the reference C allele (Sonali et al. 2023). Thus pure donkeys generally lack the truncating variant that produces gaitedness in horses. This agrees with the clinical observation that un-crossed donkeys rarely naturally pace or perform lateral ambling gaits. In mules (*Equus asinus* × *E. caballus*), by contrast, the gait variant can enter via the horse parent: Brazilian gaited mules showed ~53% heterozygous CA and 47% CC, whereas nearly all Pêga donkeys were CC (Herman et al. 2023). Interestingly, even in these gaited donkeys, the CA genotype was somewhat more common in *picada* (lateral) gaiters than in *batida* gaiters (Herman et al. 2023), hinting that the single C>A substitution in *DMRT3* has only a modest influence in the equine hybrid context. In sum, the known *DMRT3* "gait-keeper" allele has minimal prevalence in donkeys (Sonali et al. 2023), implying that donkey gait variation (ambler vs walker, etc.) must involve other genetic factors.

Outside *DMRT3*, equine genomics has begun to implicate additional genes in gait. A recent GWAS in Icelandic horses identified two new candidates: *RELN* and *STAU2* (on horse chromosomes ECA4 and 9) were strongly associated with pace ability (Sigurðardóttir et al. 2023). Haplotypes in *RELN* (reelin) were also correlated with tölt, trot, canter, and gallop performance, and *STAU2* variants influenced trot/gallop, indicating multiple interacting loci beyond *DMRT3* affect gait (Sigurðardóttir et al. 2023). Though not yet studied in donkeys, these loci (and others flagged in horses) are plausible candidates to modulate gait in equids generally. In donkeys, targeted sequencing has uncovered polymorphisms in *DMRT3* (stop-to-serine at codon 61) and numerous SNPs, but association with gait in donkeys has not been established (Sonali et al. 2023). No genome-wide study of gait has been reported in donkeys to date, but by analogy with horses one would examine *DMRT3* variants, any *RELN/STAU2*-like loci, and possibly fiber-type or muscle-pattern genes.

Advances in donkey genomics now provide background for gait studies. The donkey reference genome has been sequenced (Bertolini et al. 2015), and population genomics reveal substantial diversity across global breeds (Wang et al. 2020). Genome-wide scans have so far focused on production traits (body size, milk or meat yields, reproduction, etc.), identifying candidates like *IGF1*, *CYP4A11*, *TBX3*, and others (Song et al. 2023). However, transcriptomic resources exist: for example, a donkey "tissue atlas" profiled 16 organs by RNA-Seq (Wang et al. 2021). Muscle tissue shows a distinctive profile: genes enriched in donkey skeletal muscle (and heart) are heavily involved in contraction and calcium signaling (Wang et al. 2021), as expected for locomotor function. Comparative expression of fast vs. slow muscles (*biceps femoris* vs. *psoas*) identified thousands of differentially expressed genes (DEGs), reflecting fiber-type specialization (Shi et al. 2022). Notably, glycolytic (fast-twitch) muscle in donkeys upregulated key metabolic genes (*PKM*, *LDHA*, *ALDOA*) whereas oxidative (slow) muscle upregulated contractile/mitochondrial genes (*ATP2A2*, *MYH7*, *TNNC1*, *TPM3*, etc.) (Shi et al. 2022). Several "muscle structural" genes—e.g., *ACTN3*, *TNNT3*, *TPM2*, *TNNC2*—stood out as candidates regulating fiber composition in donkeys (Shi et al. 2022). These same classes of genes are known in other species to influence locomotor phenotypes (for instance, *ACTN3* genotypes affect sprint vs. endurance). Together, the donkey genomic/

transcriptomic data show that (a) breed differences exist at the DNA level, and (b) muscle gene expression reflects contractile properties, both of which are fundamental to gait biomechanics.

The translation of genetics to gait involves the musculoskeletal and neural architecture of movement. Spinal central pattern generators (CPGs) produce the basic rhythm and left-right alternation of limb cycling. In vertebrates, specific interneuron populations (including *DMRT3*-expressing dI6 cells) configure these networks (Andersson et al. 2012a, b). Loss of *DMRT3* disrupts left-right coordination in mice and eliminates galloping in horses (Andersson et al. 2012a, b), indicating its key role in the core locomotor CPG. Descending commands from brainstem (and cerebellum) set gait frequency and adapt patterns, while segmental reflexes (from proprioceptors in muscle spindles and Golgi tendons) fine-tune coordination. At the molecular level, proprioception depends on mechanotransduction channels (e.g., *PIEZO2* in sensory neurons) and synaptic sensors, whose genetics are less studied in equids but are likely similar to other mammals.

On the output side, gait emerges from muscle-tendon dynamics. Genetic control of muscle fibers (myosin heavy chain isoforms, troponin/tropomyosin subunits, metabolic enzymes) dictates contraction speed, endurance, and power. The donkey transcriptome confirms that its muscle is tailored for sustained contraction: calcium handling and adrenergic signaling genes are highly represented (Wang et al. 2021). Connective tissues (tendons and ligaments) provide elastic storage; genes for collagens and extracellular matrix proteins influence limb stiffness and stride length, although specific studies in donkeys are lacking. Overall, gait patterns depend on integration of spinal rhythms (genetically specified by factors like *DMRT3*), descending modulation, feedback from proprioceptors, and the contractile properties of muscles (shaped by fiber-type genes) and elastic tendons. For example, a trotting versus ambling gait requires different timing of ipsilateral vs. contralateral limb lifts—a feature ultimately determined by how the CPG is wired (*DMRT3*, etc.) and how limb mechanics are tuned by muscle and tendon properties.

Because direct molecular data on donkey gait are sparse, much inference comes from horse and mammalian models. Equine genetics shows that gait is a complex trait: besides *DMRT3*, recent horse GWAS suggest many loci contribute. In Icelandics, *RELN* and *STAU2* haplotypes have sizable effects on pace performance (interacting with *DMRT3* to explain ~27% of pace variance) (Sigurðardóttir et al. 2023). It is plausible that orthologs of these or similar genes influence gait in donkeys. Likewise, the set of muscle genes differentially expressed in donkey fibers (e.g., *ACTN3*, *TNNT3/TNNC2*) mirror those known to affect locomotion in other species (Shi et al. 2022). Given the evolutionary conservation of locomotor circuits, many neurodevelopmental genes (*Hox*, *Pax*, *Lbx1*, etc.) that pattern the spinal cord and limbs in mammals likely act similarly in donkeys, although specific studies are lacking. Thus, in absence of direct donkey data, the working model is that donkey gait diversity is underpinned by the same basic vertebrate neuromuscular architecture—spinal CPGs, proprioception, muscle contractile apparatus—with variation arising from allelic differences in those components. For example, a donkey lacking the *DMRT3* truncation is constrained to walk/trot by default, but other genetic variations (perhaps in muscle metabolism or neuromodulator genes) might explain why some donkeys spontaneously amble or pace when performance-selected. Future studies (e.g., donkey GWAS or QTL mapping using specialized breeds, or transcriptome analysis of spinal cord or limb muscle in gaited vs. non-gaited donkeys) will be needed to test these hypotheses.

2.6 Fertility and Prolificacy

Donkeys are monotocous (normally one foal per gestation) with a long (~11–14 month) pregnancy; jennies typically first conceive around 3–4 years of age and can reproduce well into middle age (up to ~12–15 years) under good management (Navas-González et al. 2019a, b). Field studies of endangered populations reveal surprisingly low herd fertility: for example, only about 54% of mature jennies in one Portuguese breed ever foaled, and most of those gave birth just once (Quaresma et al. 2014). Although donkeys tend to ovulate multiple follicles (reported as high as 61% in some breeds) (Quaresma et al. 2014), true twinning is rare—only ~3% of conceptions result in twins, and roughly 40% of those twins die perinatally (Quaresma et al. 2014). Average neonatal mortality in the first month is on the order of 8–9% (Quaresma et al. 2014). Inbreeding and small population size exacerbate these issues: donkeys have very narrow genetic bases, and inbred herds show declining fertility over generations (Quaresma et al. 2014). In practice, this manifests as lower conception rates, early embryonic losses and reduced lifetime foal output in highly inbred stock (Quaresma et al. 2014). Inbreeding can also impair immune competence. Genetically compromised jennies may produce foals with weaker passive immunity (e.g. reduced colostral antibody transfer), making neonates prone to sepsis, pneumonia and other infections (Turini et al. 2021). Inbred foals accordingly suffer higher mortality and morbidity, and the herd is more susceptible to disease outbreaks. These findings underscore that inbreeding depression in donkeys affects both fertility and immune health (Quaresma et al. 2014; Turini et al. 2021). Managing related matings and introducing genetic diversity are therefore critical in donkey breeding programs to preserve reproductive performance and foal survival (e.g. avoid mating close kin and replenish diversity in bottlenecked breeds) (Quaresma et al. 2014).

Donkeys have distinctive reproductive physiology. Gestation is unusually long (≈365 day, 12 months to give month during the 13th) and typically only one follicle develops per estrus, whereas multiple follicle development occurs in only 5–38% of cycles (Navas-González et al. 2019a, b). Nevertheless, some jenny populations have high twinning rates—for example, ~9.7% of births in an Andalusian donkey survey were twins (Navas-González et al. 2019a, b). This exceeds typical values in horses (e.g. ~1–5% in many breeds; ~3.5% in Thoroughbreds) (Wolc et al. 2006) and supports anecdotal reports that twin foals at term occur "higher than in horses." As in horses, twin pregnancies in donkeys are high-risk and often end in abortion or neonatal loss. Equid twin pregnancies have only ~10% success for producing two live foals (Wolc et al. 2006), so natural reduction to a single embryo is common. In fact, up to 85% of unilateral twin conceptuses self-reduce by ~day 40. Biologically, twinning in equids results from multiple ovulations triggered by a prolonged LH surge. In horses, twin rates vary with breed, age, season, and likely genetic predisposition. Although no heritability estimates exist for donkeys, Thoroughbred studies suggest moderate heritability ($h^2 \approx 0.24$–0.29) for twinning (Wolc et al. 2006), implying a genetic component. Indeed, in horses a major-locus effect was suggested by fluctuations in genetic trends (Wolc et al. 2006).

Few molecular studies have addressed twinning in donkeys. A recent genome-wide selection scan in Chinese Dezhou donkeys compared 11 jennies with twin colts to 10 with single colts, identifying several candidate genes under strong selection for the twinning trait (Xie et al. 2022). Key candidates included CD44, ENO2, PTPN11, and SOD2 (Xie et al. 2022). These genes have plausible roles in ovarian or follicular biology: for example, CD44 mediates cumulus–oocyte complex expansion via hyaluronic acid binding and gap-junction regulation of meiosis (Xie et al. 2022). ENO2 (neuronal enolase), PTPN11 (a tyrosine phosphatase), and SOD2 (mitochondrial superoxide dismutase) are involved in cellular metabolism and signaling, which could affect follicle development or oocyte maturation (Xie et al. 2022). In the Dezhou study, CD44 showed the strongest divergence (highest F_{ST}) between twin-bearing and single-bearing jennies (Xie et al. 2022). No large-effect gene (like ovine BMPR1B FecB) is yet known in equids, but these candidate loci suggest that variation in follicle dynamics underpins donkey twinning.

A related finding was that jennies with multiple ovulations had higher circulating estradiol (E_2) levels than those with single ovulations (Navas-González et al. 2019a, b). This aligns with physiology: more growing follicles produce more estrogen. Other reproductive hormones (FSH, LH, progesterone) showed no consistent differences in that study (Navas-González et al. 2019a, b). Overall, the limited data imply twinning involves enhanced folliculogenesis rather than upstream pituitary changes.

2.7 Disease—Resistance and Hypersensitivity

Donkeys encounter many of the same pathogens and stressors as horses, but genetic and immune responses often differ. Notably, donkeys tend to show greater tolerance or milder disease to certain infections. For example, Equine Infectious Anemia (EIA) in donkeys is usually subclinical, whereas horses often develop severe anemia and thrombocytopenia. The Merck Veterinary Manual reports that infections in donkeys and mules are most frequently subclinical, unidle the often-severe symptoms in horses. Similarly, donkeys tend to experience less severe courses of Equine Herpesvirus, African Horse Sickness, and West Nile virus infections (Navas González et al. 2018). Experimental infection with Equine Hepacivirus (EqHV) demonstrated identical viral kinetics in horses and donkeys, but donkeys exhibited a dampened transcriptomic immune response, characterized by reduced interferon signaling and innate immune activation (Gömer et al. 2022). These findings imply species specific immune modulation in donkeys, potentially contributing to their disease tolerance.

At the genomic level, while systematic GWAS or QTL mapping for disease resistance in donkeys remains underdeveloped, comparative data from horses suggests relevant pathways. For example, MHC class II haplotypes are associated with susceptibility to insect-bite hypersensitivity (IBH) in horses (Schurink et al. 2011). In donkeys, IBH-like allergic dermatitis has been reported, particularly in grey-coated breeds like the Andalusian donkey, indicating a likely shared immune mechanism (Pusterla et al. 2003; White 2013). Specific MHC-II alleles such as ELA-W23 and ELA-Be8 have been linked to allergic responses in horses (Citek et al. 2017), and may have analogs in donkeys, although this has yet to be confirmed by molecular studies.

A recent comparative transcriptomic study revealed that donkeys possess a complete but subtly distinct immunoglobulin heavy-chain (IGH) repertoire compared to horses, with 117 VH genes and variations in V–J usage and junctional diversity. Further, the publication of a multi-tissue donkey transcriptome atlas has laid the groundwork for future identification of immune regulatory genes and non-coding RNAs (Wang et al. 2021).

Donkeys are also known for their remarkable tolerance to gastrointestinal parasitism. Field data indicates that donkeys often harbor strongyle egg counts exceeding 1000 EPG without clinical signs, a level that would typically warrant treatment in horses (Kornaś et al. 2015). Although no donkey-specific loci for parasite resistance have been identified, horse GWAS studies have revealed associations with immune function and mucosal barrier integrity genes (Gutiérrez-Gil et al. 2010). Genetic divergence between *Parascaris* populations infecting donkeys and horses supports the theory of host-specific adaptation (Schurink et al. 2011).

Recent work by Navas González et al. (2018) has addressed the genetic basis of hypersensitivity to cutaneous habronematidosis (CH) in donkeys. This condition, caused by larval stages of spirurid nematodes deposited by flies, is especially prevalent in warm climates. Using Bayesian models and a dataset of 765 Andalusian donkeys, the heritability of CH hypersensitivity was estimated at 0.0346, indicating a low but non-zero genetic influence. The study also identified extensive management as a protective environmental factor, reducing disease incidence compared to intensive systems. These findings advocate for integrated control strategies combining genetic selection with optimized husbandry.

Donkeys also encounter a range of shared zoonotic and protozoan infections. Equine herpesviruses (EHV-1, EHV-8) and protozoa such as *Giardia*, *Cryptosporidium*, and *Toxoplasma* pose notable risks (Khan et al. 2025). These pathogens have been detected in donkeys and are recognized zoonoses. Clinical signs may be mild or subclinical, but comparative severity versus horses remains unclear. In bacterial infections, such as *Streptococcus equi* (strangles), immune responses in donkeys are presumed analogous to horses, although formal genetic studies are lacking.

Allergic and autoimmune diseases are another frontier. Beyond IBH, donkeys may suffer from chorioptic mange and photosensitivity, but the genetic underpinnings are unstudied. Histological analysis of allergic dermatitis in donkeys shows eosinophilic infiltration and intense pruritus, suggestive of a Th2-skewed cytokine profile similar to that in horses (Eriksson et al. 2008). However, cytokine gene expression in donkeys has yet to be quantified. Unlike horses, donkeys appear less prone to metabolic syndrome and autoimmune laminitis, which may reflect species-level differences in immune-metabolic regulation.

As genomic resources expand, the potential for donkey-specific GWAS and functional genomic studies increases. The availability of a reference genome, SNP chips, and transcriptomic datasets now enables disease association mapping in donkeys. Studies in horses have linked genomic variants to tumor susceptibility (equine sarcoid), IBH, and viral encephalopathies, providing a methodological blueprint (Citek et al. 2017; Schurink et al. 2011). Transcriptomic profiling, such as that used in EqHV studies, reveals interspecies differences in immune activation that could be explored further under bacterial or parasitic challenges.

3 Heritabilities for Functional Traits

Genomic evaluations rely fundamentally on prior knowledge of heritability estimates, as these inform the extent to which genetic variation contributes to phenotypic traits and determine the expected accuracy of genomic predictions. The estimation of heritabilities and variance components for key functional traits provides essential insights into the genetic architecture of the donkey. Table 1 summarizes the additive genetic variance, environmental variance, total phenotypic variance, and resulting heritability values for a range of traits relevant to reproductive performance, behavior, locomotion, cognition, and disease resistance. Notably, the Andalusian donkey was the first donkey breed worldwide for which an official set of genetic evaluations for functional traits—beyond morphological or production characteristics—was developed and implemented. These pioneering evaluations form the foundation for designing effective selection strategies in conservation and breeding programs aimed at improving both productivity and the breed's aptitude for human-interaction roles, including animal-assisted therapy and education.

3.1 Fertility and Prolificacy

Genetic analysis of Andalusian donkey fertility and prolificacy traits indicates moderate heritability. For example, Navas González et al. (Navas-González et al. 2019a, b) reported heritabilities of 0.18–0.24 for multiple-birth occurrence and related fertility measures. These estimates imply a significant additive genetic component: nearly one-fifth of the phenotypic variance is due to inheritable factors. Moderate-to-high genetic correlations (about 0.50–0.85) were also found among related reproductive traits (Navas-González et al. 2019a, b), suggesting that selection on one trait (e.g. twinning rate) would largely carry over to others. Biologically, such moderate heritabilities mean that selection can gradually improve fertility and prolificacy, although environmental and management factors will also exert strong influence. In breeding programs, these results support using estimated breeding values to select jennies with genetically higher fecundity. Over generations, even modest heritability allows cumulative gains in reproductive output, which is

Table 1 Heritabilities and variance components for functional traits in the Andalusian Donkey Breed

Trait category	Trait	Additive Genetic Variance (σ^2_a)	Environmental Variance (σ^2_e)	Phenotypic Variance (σ^2_p)	Heritability (h^2)
Reproductive	Cumulative Foal Number	0.0966	0.306	0.4026	0.24
	Maximum Foal Number per Birth	0.0287	0.1169	0.1456	0.20
	Multiple Birth Number	0.0225	0.1025	0.125	0.18
Behavioral	Response Type (Hypo-/Hyperreactivity)	0.034	0.157	0.192	0.18
	Mood/Attitude	1.698	6.491	8.189	0.21
	Degree/Intensity of Response	0.398	1.490	1.888	0.21
Locomotive	Amble	0.3819	0.2952	0.6771	0.56
	Walk	1.0789	0.9572	2.0360	0.53
	Trot	0.4139	0.2348	0.6163	0.67
Intelligence	Concentration	0.2574	0.6448	0.9022	0.29
	Curiosity	0.1218	0.6418	0.7636	0.16
	Memory	0.0487	0.6525	0.7012	0.07
	Stubbornness	0.1537	0.9919	1.1456	0.13
	Docility	0.0856	0.6247	0.7103	0.12
	Alertness	0.0617	0.2424	0.3041	0.20
Cognition	Dependence	0.1806	0.6717	0.8523	0.21
	Trainability	0.1845	0.6908	0.8753	0.21
	Cooperation	0.0815	0.7242	0.8057	0.10
	Emotional Stability	0.1304	0.5669	0.6973	0.19
	Perseverance	0.0534	0.4764	0.5298	0.10
	Get In/Out of Stables	0.1882	0.3067	0.4949	0.38
	Ease at Handling	0.0874	0.8049	0.8925	0.10
Health	Cutaneous Habronematidosis Hypersensitivity	0.1232	0.3418	0.4650	0.03

important both for conserving an endangered breed and for production objectives (e.g. milk or livestock draft). Moderate genetic correlations also highlight that a multi-trait selection index could simultaneously improve several reproductive traits.

3.2 Coping Styles or Reactivity

Behavioral reactivity in donkeys—often characterized as flight vs. freeze or "problem-focused" coping styles—also shows heritable variation. Genetic parameter estimation revealed moderate heritabilities (about 0.18–0.21) for coping-style scores (Navas-González et al. 2018a). This indicates that innate temperament traits (such as tendency to freeze or flee in response to stimuli) are partly genetically determined. Importantly, phenotypic and genetic correlations among reactivity metrics were substantial. For instance, genetic correlations between reaction intensity and other temperament descriptors were moderately high and negative (~−0.46), while correlations between mood/emotion and response type were very high (~0.95) (Navas-González et al. 2018a). In practical terms, this means that selecting for donkeys that exhibit milder (lower-intensity) reactions will also tend to produce animals that have calmer moods and consistent responses across situations. From a breeding perspective, these findings imply that selection for a neutral or "flight-to-freeze" coping style (which is considered safer and more trainable) is feasible because these traits are heritable. In animal-assisted therapy settings, calmer, more predictable donkeys are preferred, and the observed heritabilities suggest progress can be made by choosing less reactive animals as parents. However, the moderate magnitude of heritability indicates that environmental factors (training, handling, housing) will also play a major role in shaping temperament, so selection should be complemented by careful management and training.

3.3 Locomotion and Gaits

Donkey gait traits have particularly high heritability, meaning genetic selection can strongly influence locomotion. In a study of Andalusian donkeys bred for assisted therapy, heritability estimates were 0.53 for walk, 0.56 for amble, and 0.67 for trot (Navas-González et al. 2019a, b). These are among the highest trait heritabilities reported, indicating that well over half the variation in gait quality is genetic. Thus, selective breeding can effectively improve gait performance (e.g. smoother, more regular walk or specific ambling gaits) in relatively few generations. Genetic correlations among the gaits were moderate (0.28–0.42) (Navas-González et al. 2019a, b), suggesting that each gait has largely independent genetic control but with some shared genes. For breeding, this means it is possible to develop lines specialized for particular gaits. Indeed, the cited work used these genetic parameters to construct a locomotion selection index and proposed breeding separate gait-specific lines for therapy purposes (Navas-González et al. 2019a, b). Biologically, the high heritabilities reinforce the idea that gait patterns in donkeys are strongly genetically determined and relatively canalized, possibly because gait is closely linked to musculoskeletal structure. In practical terms, breeders can confidently select for superior gait genotypes to match production or therapy needs (for example, preferring an even, smooth amble for rider comfort). The moderate correlations mean that improving one gait will have some effect on others, but multi-trait selection can optimize each gait's performance as needed.

3.4 Cognition and Intelligence

Psychometric testing of donkey intelligence reveals a genetic component to cognitive abilities. Navas González et al. (2019a, b) used a nonverbal problem-solving test to derive an animal "IQ" score, finding that this latent cognitive factor explained over 62% of total cognitive variance. Crucially, they also reported heritabilities for individual cognitive processes ranging from 0.06 up to 0.38 (Navas-González et al. 2019a, b). Although some traits had low heritability (0.06), others reached moderate levels (up to 0.38), and the overall pattern was statistically significant. This indicates that genes contribute substantially to variation in learning, memory, curiosity, and related traits in donkeys. The authors note that these heritability estimates mirror patterns seen in humans and other animals, suggesting conserved genetic architecture of cognition (Navas-González et al. 2019a, b). In a breeding context, these results imply that selecting for more trainable, curious, or intelligent donkeys is possible, albeit with modest response since heritabilities are not very high. In the animal-assisted therapy role, cognitive capacity is valuable for training and adaptability, so incorporating intelligence-related scores into selection criteria could improve trainability. The evidence of a "human-analogous heritable component" to donkey intelligence (Navas-González et al. 2019a, b) suggests that welfare and management practices can be tailored to individuals' cognitive profiles, and that over time, populations can evolve enhanced cognitive traits if desired.

3.5 Disease Resistance (Cutaneous Habronematidosis)

By contrast, resistance to cutaneous habronematidosis (a parasitic skin disease) appears minimally heritable. A Bayesian analysis of hypersensitivity to habronema larvae found an estimated heritability of only 0.0346 (Navas-González et al. 2018b). This value is very low, with an extremely small additive genetic variance (about 0.1232) (Navas-González et al. 2018b). Such a low heritability means that almost all variation in susceptibility is due to non-genetic factors—for example, environmental conditions, husbandry system, and exposure risk. Consistent with this, the study found that extensive (pasture-based) management significantly reduced disease risk, whereas factors like age and sex had negligible effects (Navas-González et al. 2018b). The authors conclude that because genetic contribution is so small, genetic selection alone would not be effective in reducing incidence (Navas-González et al. 2018b). Instead, improving environmental and management conditions (e.g. fly control, pasture hygiene) is necessary to improve resistance.

3.6 Genetic Correlations and Breeding Implications

Across these domains, the generally moderate-to-high heritabilities imply substantial potential for selective breeding of Andalusian donkeys. Traits with higher heritability (gaits, some cognitive scores) will respond more rapidly to selection, while moderate traits (fertility, temperament) will improve more slowly but still measurably. Genetic correlations between traits are key for designing selection goals. For example, the near-unity genetic correlation between mood and response type in coping styles (Navas-González et al. 2018a) suggests a single underlying dimension of temperament that can be targeted. In reproduction, strong genetic correlations among fertility traits (Navas-González et al. 2019a, b) mean that selecting for increased twinning may also improve overall fecundity or vice versa. Conversely, low or moderate correlations (such as between gaits; Navas-González et al. 2019a, b) indicate that multi-trait selection is needed to improve each trait independently.

Practically, breeding programs can use selection indices that weight traits according to their economic or functional importance. For dual-purpose programs (production and therapy), one might include fertility and growth on the production side, and temperament, gait, and trainability on the therapy side. The moderate heritabilities support the use of estimated breeding values (EBVs) for each trait: animals can be ranked genetically, and best breeders chosen. For therapy-related lines, traits such as neutral reactivity and desirable gaits (high heritability) would have high weight. The presence of genetic variance in cognitive abilities also means that selection for intelligence or problem-solving (perhaps via cognitive test scores) could gradually shape a more trainable population. However, low-heritability traits like habronema resistance will require mostly non-genetic interventions; breeders should not exclusively rely on genetics to improve these health traits. In all cases, understanding the genetic architecture (heritability and correlation) informs risk: correlated responses must be monitored to avoid inadvertently affecting unselected traits. Overall, the findings from these studies suggest that Andalusian donkey traits important for both productivity and therapy use have enough heritability to justify systematic genetic improvement efforts (Navas-González et al. 2019a, b).

Understanding the heritability of functional traits in donkeys—such as fertility, milk yield, disease resistance, or docility—not only informs selection strategies but also plays a critical role in interpreting molecular signals of breed differentiation. Since heritability is a population-specific parameter, shaped by both genetic structure and environmental variance, estimates obtained within distinct donkey breeds provide valuable insights into the extent of genetic control over key traits. This contextualization is especially relevant when analyzing molecular differentiation, such as SNP or mtDNA haplotype patterns, as it helps to distinguish between neutral population divergence and functionally relevant genetic variation. In breeds with documented genetic isolation or unique adaptive histories linking heritability data to genomic divergence may help prioritize traits and populations for targeted gene discovery. Thus, heritability studies lay the groundwork for identifying candidate genes under selection and for implementing genomic evaluations tailored to the unique genetic architecture of under-characterized donkey populations.

4 Breeding Evaluations on Molecular and Quantitative Traits in Mules and Hinnies

The genetic evaluation of quantitative traits in mules and hinnies, the hybrid offspring of horses (*Equus caballus*) and donkeys (*Equus asinus*), requires distinct approaches compared to purebred populations. These hybrids are generally sterile due to their intermediate chromosome number ($2n = 63$), which disrupts meiotic pairing and gamete formation, preventing the traditional use of multi-generational data that forms the backbone of conventional breeding value estimation (Franco et al. 2016). As a result, breeding evaluations must rely heavily on understanding the genetic contributions of the parental species and the interactions between these species in producing desirable hybrid characteristics (Biology Insights 2025).

The complexity of hybrid evaluation is further compounded by reciprocal crosses, where differences in maternal and paternal genetic contributions influence the phenotype, partly due to genomic imprinting and mitochondrial DNA inheritance (Fan et al. 1992). For example, mules (male donkey × female horse) and hinnies (male horse × female donkey) differ subtly in morphology, physiology, and behavior, reflecting both nuclear and cytoplasmic genetic effects. Molecular tools such as multiplex-PCR targeting the mitochondrial D-loop region have been developed to distinguish maternal origin (Franco et al. 2016), while biochemical markers like esterase isozyme patterns have been used to study segregation in rare fertile hybrids and backcross progeny (Fan et al. 1992). This chapter explores the genetic underpinnings of quantitative traits in mules and hinnies, including the role of diallel analysis, genomics, mitochondrial DNA, and gonosome-linked traits, and highlights emerging challenges and future research directions in hybrid breeding.

4.1 Genetic Evaluation in Hybrid Systems

In hybrid breeding systems like those for mules and hinnies, traditional methods of genetic evaluation—such as those used in purebred livestock—are often not applicable. A key challenge in evaluating mules and hinnies is the sterility of the hybrids, which prevents the use of multi-generational data and hinders the calculation of genetic parameters across generations (Franco et al. 2016). In purebred populations, genetic evaluations rely on breeding values derived from performance data over several generations, but in hybrids, these evaluations must be based on the performance of the first-generation offspring alone (McLean et al. 2014).

One effective method used to evaluate hybrid genetic performance is diallel mating designs, which help assess how different combinations of parental lines perform when crossed. Diallel analysis partitions the genetic effects into general combining ability (GCA)—the average genetic contribution of each breed—and specific combining ability (SCA), which refers to the exceptional performance of certain breed combinations (Griffing 1956). For mules and hinnies, reciprocal crosses (where the sire and dam roles are swapped) can produce different results, mainly due to maternal effects, cytoplasmic inheritance, and sex-linked genetic factors (Fan et al. 1992). These crossbred analyses are essential for identifying which parent combinations yield the best hybrid traits, such as size, endurance, and work capacity (McLean et al. 2014).

However, this approach relies heavily on accurate pedigree information. In hybrid populations, where horses and donkeys may not always be purebred, the genetic contribution of each parent must be accurately quantified. This is often done by estimating breed composition, which requires the identification of the genetic makeup of both the horse and donkey parents. When breed purity is uncertain or unknown, genomic tools—such as mitochondrial DNA analysis and breed-specific SNP panels—can help confirm the precise genetic composition of the parents, thereby improving the reliability of breeding evaluations (Franco et al. 2016).

4.2 The Role of Genomics in Hybrid Breeding

Genomics is transforming the way breeders evaluate hybrids like mules and hinnies. With the availability of reference genomes for both horses (*Equus caballus*) and donkeys (*Equus asinus*), modern genomic tools—such as Single Nucleotide Polymorphism (SNP) genotyping and whole-genome sequencing—allow breeders to perform genetic evaluations that go beyond traditional pedigree analysis (Jiang et al. 2020; Wade et al. 2009).

Genomics can be used to:

- Accurately determine breed composition, even when pedigree records are incomplete or unreliable. This enables breeders to identify which specific genes from horses and donkeys contribute to desired traits in the hybrid (McCue et al. 2012).
- Estimate heterosis (hybrid vigor) at the genomic level. Heterosis is often expressed in traits such as growth, stamina, and disease resistance—key qualities for hybrids like mules and hinnies that serve as working animals in challenging environments (Ren et al. 2022).
- Identify loci associated with incompatibility or suboptimal performance. Certain gene combinations from the two species may interact poorly, leading to reduced fitness or performance; genomic analysis can detect such incompatibilities early in the breeding process (Ren et al. 2022).
- Predict phenotypic traits through genomic selection, enabling breeders to choose parents based not only on observed performance but also on genetic potential, thereby improving the efficiency of hybrid breeding (Meuwissen et al. 2001).

In hybrids, where performance traits often involve complex interactions between genes from both species, genomics offers a precise means of understanding these interactions. Integrating genomic information with traditional performance data allows breeders to predict and enhance hybrid traits—such as endurance, size, and work capacity—more accurately than relying on pedigree alone (McCue et al. 2012; Wade et al. 2009).

4.3 Mitochondrial DNA and Reciprocal Crosses

A significant difference between mules and hinnies arises from the inheritance of mitochondrial DNA (mtDNA), which is strictly maternally inherited (Griesbach 1986; Franco et al. 2016). Mules inherit the mitochondrial genome of their horse mothers, while hinnies inherit the mitochondrial genome of their donkey mothers. Mitochondria are essential for cellular energy production, and emerging evidence suggests that maternal mtDNA may influence traits such as endurance, metabolic efficiency, and thermoregulation (Allen and Short 1997).

In practical terms, mules—which inherit horse mtDNA—may have a metabolic advantage in stamina and sustained aerobic performance, traits highly desirable in working animals (McLean et al. 2014). Hinnies, conversely, may display endurance characteristics shaped by donkey mtDNA. The interaction between the nuclear genomes of the two species and their respective mitochondrial genomes likely plays a key role in determining efficiency under environmental stressors such as heat or high-altitude work (Franco et al. 2016).

Understanding the role of mtDNA in hybrid performance is an emerging research area, and genomic tools now allow detailed examination of mitochondrial–nuclear genome interactions in mules and hinnies (Ren et al. 2022).

4.4 Sex Chromosome-Linked Traits

In addition to mitochondrial inheritance, sex chromosomes contribute to differences between mules and hinnies. Male mules inherit the Y chromosome from their donkey sire and the X chromosome from their horse dam, while male hinnies inherit the Y chromosome from their horse sire and the X chromosome from their donkey dam (McCarthy 2006).

This difference in X and Y chromosome origin can influence sex-linked traits, particularly those carried on the X chromosome. For example, some coat color genes are X-linked, meaning male mules and hinnies may display different coat colors depending on whether the X chromosome came from a horse or a donkey (Allen and Short 1997). Behavioral traits influenced by sex chromosomes could also show maternal or paternal biases due to X-linked gene expression patterns.

The role of sex chromosomes in hybrid traits remains under-researched, especially regarding their influence on temperament and behavior.

4.5 Desirable Traits in Mules and Hinnies

Mules and hinnies are bred primarily for their utility in work environments. Highly valued traits include:

- Endurance and Work Capacity—Ability to work long hours in harsh conditions, including heat tolerance and metabolic efficiency.

- Strength and Size—Important for draft work such as pulling carts or carrying heavy loads.
- Surefootedness and Agility—Critical for mountainous or rocky terrains, ensuring safety for both animal and handler (Biology Insights 2025).
- Gait and Rideability—Smooth, comfortable gaits are desirable for saddle mules, especially those from gaited horse breeds.
- Temperament and Behavior—Calm, patient demeanors are preferred, though these can vary with genetics and environment.

  Trait expression is influenced by both parental species and specific breed characteristics. For example, mules from draft horses may excel in strength, while those from lighter, gaited breeds may be favored for endurance and rideability.

4.6 Epigenetic Effects in Hybrid Breeding

While genetic inheritance is central to mule and hinny traits, epigenetic factors—heritable changes in gene expression without altering DNA sequence—also play a role. Mechanisms such as DNA methylation, histone modification, and non-coding RNAs can influence gene expression patterns (Bird 2007).

In hybrids, the interaction between horse and donkey genomes may trigger epigenetic changes affecting growth rate, stress response, and metabolism (McLean et al. 2014). The maternal environment can shape expression of stress-related genes, influencing behavior and resilience. Differences in epigenetic regulation between horse and donkey genes may also contribute to hybrid vigor, particularly in endurance and disease resistance (Ren et al. 2022).

5 Future Directions

The rapid advancement of multi-omics and genomic technologies in donkey and hybrid breeding offers significant potential for transformative improvements in genetic research, breeding efficiency, and trait optimization. Future efforts will focus on several key areas that collectively aim to enhance productivity, health, and adaptability in these species.

5.1 Next-Generation Approaches to Modeling Genetic Parameters

The estimation of heritability and genetic parameters remains a cornerstone for designing effective breeding strategies in donkeys, mules, and hinnies. Although current approaches—primarily based on linear mixed models (LMM) and restricted maximum likelihood (REML)—have been instrumental in advancing our understanding of functional traits, they also present limitations when confronted with incomplete pedigrees, small population sizes, and complex trait architectures. Looking forward, several key directions are likely to shape the future of genetic evaluation and heritability estimation in equid breeding.

One critical step will be the integration of genomic information into traditional models. Pedigree-based evaluations are often constrained by gaps in ancestry data, whereas genomic relationship matrices derived from SNP genotyping or whole-genome sequencing provide a far more accurate representation of genetic connectedness. Techniques such as Genomic Best Linear Unbiased Prediction (GBLUP) and single-step GBLUP are already demonstrating improved predictive power by merging phenotypic, pedigree, and genomic data within a single framework. Furthermore, Bayesian approaches, which allow for flexible modeling of marker effects, are likely to become more prevalent as genomic datasets expand.

Beyond single-trait evaluations, multi-trait and longitudinal models will gain prominence, especially for traits that are dynamic or correlated, such as growth rate, lactation performance, and reproductive efficiency. Random regression models can track changes over time, while reaction norm models enable researchers to quantify genotype-by-environment interactions—a particularly relevant factor given the adaptability of donkeys to diverse ecological conditions. Such models will not only improve the accuracy of heritability estimates but also provide insights into the stability of genetic merit across environments.

The next frontier lies in the integration of multi-omics data into prediction frameworks. Transcriptomics, proteomics, and metabolomics studies are identifying candidate pathways underlying key functional traits. Incorporating these signals into models through weighted genomic predictions or multi-kernel approaches could enhance the resolution of genetic evaluations, enabling breeders to target biological pathways directly rather than relying solely on statistical associations.

Another dimension involves accounting for epigenetic modifications and non-additive genetic effects, which are often overlooked in conventional heritability estimates. Incorporating DNA methylation profiles and exploring dominance and epistatic interactions will help explain additional variance and shed light on the inheritance of complex traits. This is particularly important for hybrids such as mules and hinnies, where heterosis plays a significant role in performance.

As data collection accelerates, machine learning (ML) and artificial intelligence (AI) tools are emerging as complementary approaches to classical quantitative genetics. Algorithms capable of modelling non-linear relationships and performing automated feature selection can leverage high-dimensional genomic and phenotypic datasets, improving predictions for traits with low heritability or sparse phenotyping. However, these methods must be integrated carefully to maintain interpretability and biological relevance.

Finally, the development of simulation platforms and validation pipelines will be essential for benchmarking models and guiding breeding decisions. Stochastic simulations can explore different selection strategies under variable heritability assumptions, while robust cross-validation frameworks will help compare traditional REML-based models against genomic and ML-driven methods. Equally important will be the creation of open-access databases and user-friendly computational tools, ensuring that advanced methodologies are accessible to researchers and breeders worldwide.

In sum, the future of heritability estimation in equids will be defined by a transition from pedigree-based evaluations toward integrated, data-driven approaches that combine genomics, omics, epigenetics, and advanced computational tools. This evolution holds the promise of improving the accuracy of selection, enhancing genetic progress, and safeguarding the adaptability and welfare of donkeys and their hybrids in a rapidly changing world.

5.2 Functional Validation of Candidate Gene

Although numerous candidate genes have been associated with economically important traits—such as *SCD* and *LEPR* for meat quality, *LCORL* and *NCAPG* for growth, and *DMRT3* for locomotion—most findings remain at the associative stage. Experimental validation through cutting-edge tools such as CRISPR-Cas9 genome editing and gene expression modulation will be critical to establish causality. Furthermore, exploring gene–gene and gene–environment interactions will provide insights into the complex architecture of traits, paving the way for more precise selection strategies.

5.3 Development of Donkey-Specific Genomic Resources

Unlike horses and cattle, donkeys currently lack comprehensive genomic resources tailored to their genetic diversity. The development of breed-specific SNP arrays, high-resolution whole-genome databases, and reference panels will improve the accuracy of genomic predictions, particularly for traits with moderate to high heritability, such as gait, fertility, and temperament. Such resources will serve as the foundation for implementing genomic selection in donkey populations worldwide.

5.4 Integration of Epigenomics and Microbiomics

Emerging evidence highlights the critical roles of epigenetic mechanisms and gut microbiota in shaping phenotypic outcomes. Future research should investigate DNA methylation patterns, histone modifications, and microbiome composition to uncover how these factors interact with the host genome. Understanding these layers of regulation will enhance strategies for improving performance, health, and adaptability, especially under climate and nutritional stress conditions.

5.5 Precision Breeding for Dual-Purpose Roles

Modern breeding programs increasingly aim to combine productivity traits, such as meat and milk yield, with behavioral attributes, including docility and trainability.

These traits are particularly valuable for donkeys involved in animal-assisted therapy or leisure activities. Developing multi-trait genomic indices that integrate economic, welfare, and behavioral considerations will enable precision breeding for diverse market demands.

5.6 Hybrid Breeding Optimization

Mules and hinnies exhibit hybrid vigor, but the underlying genetic mechanisms—such as mitochondrial–nuclear genome compatibility, sex chromosome-linked traits, and genomic imprinting—remain poorly understood. Advances in this area will facilitate the design of optimized crossbreeding systems, enhancing hybrid performance for specific working or athletic purposes. Looking ahead, several promising areas of research could enhance the breeding and performance evaluation of mules and hinnies.

As genomic tools continue to advance, increasingly sophisticated breeding programs based on genomic prediction could be developed. These programs would integrate phenotypic and genotypic data to enable more efficient selection of parents, potentially accelerating genetic gains in hybrid populations (Meuwissen et al. 2001; McCue et al. 2012). The use of high-density SNP arrays and whole-genome sequencing in equids provides the foundation for such predictive models (Wade et al. 2009).

Further research is needed to understand the interactions between nuclear and mitochondrial genomes in mules and hinnies. These interactions likely have significant effects on traits such as endurance, metabolic efficiency, and thermoregulation (Allen and Short 1997; Ren et al. 2022). Given that mules inherit horse mtDNA and hinnies inherit donkey mtDNA, the compatibility between mitochondrial and nuclear genomes may influence hybrid vigor and adaptability to environmental stressors.

– Advances in Assisted Reproductive Technologies (ARTs), including artificial insemination, sperm cryopreservation, oocyte retrieval, embryo transfer, and in vitro embryo production, could help overcome reproductive challenges in equine hybrids and their parent species (Gambini et al. 2025). These technologies may also facilitate the preservation of genetic diversity in endangered donkey breeds and optimize parent selection for hybrid production.

Expanding research into epigenetic regulation in hybrids could reveal how DNA methylation, histone modifications, and non-coding RNAs influence gene expression and performance traits (Bird 2007). Identifying epigenetic markers linked to endurance, disease resistance, or temperament could provide breeders with new selection criteria beyond DNA sequence variation.

Combining genomic, epigenomic, and environmental data into predictive models could allow breeders to forecast hybrid performance under specific working conditions, such as high-altitude transport or hot-climate agriculture (Martins-Bessa and McLean 2024).

5.7 Genetic Strategies for Disease Tolerance

Improving disease resilience through genetic selection will remain a priority. Identifying immune-related markers and haplotypes associated with parasite resistance and viral tolerance will enable marker-assisted breeding for health robustness. Coupled with enhanced biosecurity, these strategies will significantly reduce disease burden and improve herd sustainability.

5.8 Application of Artificial Intelligence and Big Data

The integration of machine learning and big data analytics into breeding programs offers unprecedented opportunities for accelerating genetic progress. Predictive models leveraging genomic, phenotypic, and management data will support real-time decision-making, optimize mating plans, and minimize inbreeding risks, ultimately improving efficiency across the breeding pipeline.

6 Conclusions

The integration of quantitative genetics and multi-omics technologies has ushered in a new era for donkey, mule, and hinny breeding, providing unprecedented insights into the molecular foundations of functional traits. This chapter has highlighted how genomic, transcriptomic, proteomic, and epigenetic approaches have expanded our understanding of critical attributes, including meat and milk quality, growth performance, locomotion, fertility, disease resistance, and behavioral traits. By identifying key candidate genes (*LCORL, NCAPG, HOXC8, IGF1, SCD, LEPR, DMRT3*, among others) and their associated molecular pathways, researchers can now develop targeted breeding strategies that combine traditional selection with genomic tools, paving the way for precision breeding in equids. Heritability estimates for locomotion, temperament, cognitive ability, and reproductive traits demonstrate that selective breeding can effectively improve both production and behavioral traits. Conversely, traits with low heritability, such as disease resistance to cutaneous habronematidosis, require integrated strategies that combine genetic insights with optimized management practices. Importantly, the review of hybrid systems underscores the complexity and potential of mule and hinny breeding, where mitochondrial inheritance, sex-linked traits, and heterosis play pivotal roles. While sterility limits multi-generational selection, genomic evaluations and molecular analyses can still inform crossbreeding decisions and enhance hybrid vigor. Looking forward, the convergence of genomic selection, marker-assisted breeding, and advanced technologies such as CRISPR-Cas will enable the functional validation of candidate genes and accelerate genetic improvement in donkeys. These tools will also support conservation efforts for endangered breeds, maintaining genetic diversity while optimizing economically and functionally important traits. Moreover, understanding the genetic basis of temperament, trainability, and gait will expand the role of donkeys in animal-assisted therapy and other welfare-oriented applications. Ultimately, the application of multi-omics in equid breeding represents more than a technological advancement; it is a paradigm shift toward evidence-based, sustainable breeding programs. By integrating molecular data with quantitative evaluations, breeders can achieve dual objectives: improving productivity and resilience while preserving the unique adaptive traits that make donkeys, mules, and hinnies indispensable in diverse human contexts. Continued research, international collaboration, and practical implementation of these findings will ensure the long-term viability and welfare of these species in an evolving global landscape.

Self-Evaluation Questionnaire

Q1. Which genes are most strongly associated with skeletal growth and body size in donkeys?

A. MC1R, ASIP, TYR
B. LCORL, NCAPG, HOXC8
C. DMRT3, RELN, STAU2
D. IGF1, MSTN, ACTN3

Answer: B

(LCORL and NCAPG control stature; HOXC8 influences vertebral number and body length.)

Q2. Why is donkey milk considered hypoallergenic compared to cow's milk?

A. It lacks casein genes
B. It contains high lysozyme and immunoglobulin content
C. It has low levels or less allergenic forms of β-lactoglobulin and αs1-casein
D. It has higher fat and protein than ruminant milk

Answer: C

(Major allergens are present in low or less allergenic forms.)

Q3. Which genetic variant is responsible for the non-dun coat color pattern in donkeys?

A. c.629T>C in MC1R
B. c.604C>G in TYR
C. A 1 bp deletion downstream of TBX3
D. c.1978+2T>A in KIT

Answer: C

(Deletion near TBX3 decreases its expression, eliminating primitive markings.)

Q4. Why are mules usually sterile despite showing hybrid vigor?

A. They lack Y chromosomes
B. They have an intermediate chromosome number ($2n = 63$), disrupting meiosis
C. They possess incompatible mitochondrial DNA
D. They overexpress myostatin

Answer: B

(Odd chromosome number prevents proper gamete formation.)

Q5. Which factor primarily determines endurance and metabolic performance in mules and hinnies?

A. Number of lumbar vertebrae from the donkey sire
B. DMRT3 mutation inherited from the horse dam
C. Mitochondrial DNA from the maternal species
D. High intramuscular fat from both parents

Answer: C

(Maternal mtDNA controls oxidative capacity and energy efficiency.)

Q6. Which trait category in donkeys shows the highest heritability, indicating strong genetic control?

A. Fertility and prolificacy
B. Coat color
C. Locomotion and gait
D. Cognitive abilities

Answer: C

(Gait traits like trot ($h^2 \approx 0.67$) and amble ($h^2 \approx 0.56$) have very high heritability.)

References

Abitbol M, Legrand R, Tiret L (2014) A missense mutation in melanocortin 1 receptor is associated with the red coat colour in donkeys. Anim Genet 45(6):878–880

Abitbol M, Legrand R, Tiret L (2015) A missense mutation in the agouti signaling protein gene (ASIP) is associated with the no light points coat phenotype in donkeys. Genet Sel Evol 47(1): 28

Allen WR, Short RV (1997) Interspecific and extraspecific pregnancies in equids: anything goes. J Hered 88(5):384–392

Andersson LS, Larhammar M, Memic F, Wootz H, Schwochow D, Rubin C-J et al (2012a) Mutations in DMRT3 affect locomotion in horses and spinal circuit function in mice. Nature 488(7413):642–646

Andersson LS, Larhammar M, Memic F, Wootz H, Schwochow D, Rubin CJ et al (2012b) Mutations in DMRT3 affect locomotion in horses and spinal circuit function in mice. Nature 488(7413):642–646

Auzino B, Miranda G, Henry C, Krupova Z, Martini M, Salari F, Cosenza G, Ciampolini R, Martin P (2022) Top-down proteomics based on LC-MS combined with cDNA sequencing to characterize multiple proteoforms of Amiata donkey milk proteins. Food Res Int 160:111611

Benirschke K, Brownhill LE, Beath MM (1962) Somatic chromosomes of the horse, the donkey, and their hybrids, the mule and the hinny. J Reprod Fertil 4((3)):319–326. https://doi.org/10.1530/jrf.0.0040319

Bertolini F, Scimone C, Geraci C, Schiavo G, Utzeri VJ, Chiofalo V, Fontanesi L (2015) Next generation semiconductor based sequencing of the donkey (Equus asinus) genome provided comparative sequence data against the horse genome and a few millions of single nucleotide polymorphisms. PLoS One 10(7):e0131925

Biology Insights (2025) Horse donkey hybrid: what makes mules and hinnies different. https://biologyinsights.com/horse-donkey-hybrid-what-makes-mules-and-hinnies-different/

Bird A (2007) Perceptions of epigenetics. Nature 447(7143):396–398

Blench R (2000) The history and spread of donkeys in Africa. Cambridge University Press

Brosnahan MM (2019) Genetics, evolution, and physiology of donkeys and mules. Vet Clin North Am Equine Pract 35(3):457–467. https://doi.org/10.1016/j.cveq.2019.08.003

Burden F, Thiemann A (2015) Donkeys are different. Equine Vet Educ 27(10):519–523. https://doi.org/10.1111/eve.12353

Canisso IF et al (2019) Reproductive physiology and breeding management of jennies. Theriogenology 132:180–189. https://doi.org/10.1016/j.theriogenology.2019.03.027

Cappai MG, Nieddu M, Boatto G, Pinna W, Cappai M, Senes A, Accioni F, Liscia M, Pusceddu M, Burrai L, Melis MP (2024) Albinism and blood cell profile: the peculiar case of Asinara donkeys. Animals 14(18):2641

Citek J, Vostry L, Vostra-Vydrova H, Brzakova M, Prantlova V (2017) Genetic parameters of insect bite hypersensitivity in the Old Grey Kladruber horse. J Anim Sci 95:53–58

Criscione A, Cunsolo V, Tumino S, Di Francesco A, Bordonaro S, Muccilli V, Saletti R, Marletta D (2018) Polymorphism at donkey β-lactoglobulin II locus: identification and characterization of a new genetic variant with a very low expression. Amino Acids 50(6):735–746

D'Alessandro A, Zolla L (2013) Meat science: from proteomics to integrated omics towards system biology. J Proteomics 78:558–577. https://doi.org/10.1016/j.jprot.2012.10.023

Eriksson S, Grandinson K, Fikse W, Lindberg L, Mikko S, Broström H et al (2008) Genetic analysis of insect bite hypersensitivity (summer eczema) in Icelandic horses. Animal 2(3):360–365

Fan G, Zong E, Zhao Z (1992) A genetic study of esterase activity in mules, hinnies and their backcross progeny. Heredity 68(1):89–92

Fang C, Farnir F, Liu L, Xiao H (2025) Unveiling genetic markers for milk yield in Xinjiang donkeys: a genome-wide association study and kompetitive allele-specific PCR-based approach. Int J Mol Sci 26(7):2961

Franco MM, Silva EC, Gouveia JJS, Coelho EGA, Sereno JRB (2016) Quick method for identifying horse (*Equus caballus*) and donkey (*Equus asinus*) hybrids. Genet Mol Res 15(3): gmr15038895

Gambini A, Smith JM, Gurkin RJ, Palacios PD (2025) Current and emerging advanced techniques for breeding donkeys and mules. Animals 15(7):990. https://doi.org/10.3390/ani15070990

Giorgis V, Rolla G, Raie A, Geuna M, Boita M, Lamberti C et al (2018) A case of work-related donkey milk allergy. J Investig Allergol Clin Immunol 28(3):197–199

Gömer A, Puff C, Reinecke B, Bracht S, Conze M, Baumgärtner W, Steinmann J, Feige K, Cavalleri JMV, Steinmann E, Todt D (2022) Experimental cross-species infection of donkeys with equine hepacivirus and analysis of host immune signatures. One Health Outlook 4(1):9. https://doi.org/10.1186/s42522-022-00065-y. PMID: 35527255; PMCID: PMC9082851

Griesbach RJ (1986) That reciprocal cross—is it a mule or hinny? Awards Q 17(3):149–150

Grilz-Seger G, Utzeri VJ, Ribani A, Taurisano V, Fontanesi L, Brem G (2020) Known loci in the KIT and TYR genes do not explain the depigmented white coat colour of Austro-Hungarian Baroque donkey. Ital J Anim Sci 19(1):739–743

Griffing B (1956) Concept of general and specific combining ability in relation to diallel crossing systems. Aust J Biol Sci 9(4):463–493

Guo HY et al (2019) Donkey milk and its potential functional components. J Dairy Sci 102 ((5)):3833–3848. https://doi.org/10.3168/jds.2018-15606

Gutiérrez-Gil B, Pérez J, de la Fuente LF, Meana A, Martínez-Valladares M, San Primitivo F, Rojo-Vázquez FA, Arranz JJ (2010) Genetic parameters for resistance to trichostrongylid infection in dairy sheep. Anim 4(4):505–512

Haase B, Rieder S, Leeb T (2015) Two variants in the KIT gene as candidate causative mutations for a dominant white and a white spotting phenotype in the donkey. Anim Genet 46(3):321–324

Hanot P, Herrel A, Guintard C, Cornette R (2019) Unravelling the hybrid vigor in domestic equids: the effect of hybridization on bone shape variation and covariation. BMC Evol Biol 19:188. https://doi.org/10.1186/s12862-019-1520-2

Herman M et al (2023) DMRT3 allele frequencies in batida- and picada-gaited donkeys and mules in Brazil. Animals 13(24):3829. https://doi.org/10.3390/ani13243829

Jiang Y, Xie M, Chen W, Talbot R, Maddox JF, Faraut T, Wu C, Muzny DM, Li Y, Zhang W, Stanton JA, Brauning R, Barris W, Hourlier T, Aken BL, Searle SMJ, Adelson DL, Bian C, Cam GR et al (2020) The sheep genome illuminates biology of the rumen and lipid metabolism. Science 344(6188):1168–1173

Khan MZ, Ji Y, Fan X, Liu Y, Liu W, Wang C (2025) Equine herpesvirus infections: treatment progress and challenges in horses and donkeys. VetY Sci 12(11):1082. https://doi.org/10.3390/vetsci12111082

Kornaś S, Sallé G, Skalska M, David I, Ricard A, Cabaret J (2015) Estimation of genetic parameters for resistance to gastrointestinal nematodes in pure blood Arabian horses. Int J Parasitol 45(5): 237–242

Lai Z, Wu F, Li M, Bai F, Gao Y, Yu J, Li H, Lei C, Dang R (2021) Tissue expression profile, polymorphism of IGF1 gene and its effect on body size traits of Dezhou donkey. Gene 766: 145118

Lazić S et al (2023) Serological examinations of significant viral infections in domestic donkeys. Animals 13(13):2056. https://doi.org/10.3390/ani13132056

Li D, Hu X, Peng Y (2025) The classification method of donkey breeds based on SNPs data and machine learning. Front Genet 16:1496246. https://doi.org/10.3389/fgene.2025.1496246

Liu Y, Li H, Wang M, Yang G, Li S, Ji C, Liu G, Wang Y, Chen N, Lei C, Dang R (2022) Genetic architectures and selection signatures of body height in Chinese indigenous donkeys revealed by next-generation sequencing. Anim Genet 53(4):487–497

Liu X et al (2024) Coloration in equine: overview of candidate genes associated with coat color phenotypes. Animals 14(12):1802. https://doi.org/10.3390/ani14121802

Ma Q, Kou X, Yang Y, Yue Y, Xing W, Feng X, Liu G, Wang C, Li Y (2023) Comparison of lipids and volatile compounds in Dezhou donkey meat with high and low intramuscular fat content. Foods 12(17):3269

Magdy Y, Abo-Ahmed A, Abumandour M, El Shafey A, El-Kammar R, Al-Mosaibih MA, Fayad E, Ahmed O (2025) Elucidation of collagen content in different anatomical regions of the dermis of donkeys (Equus asinus): histomorphometric and ultrastructural study. BMC Vet Res 21:310

Martins-Bessa A, McLean AK (2024) Advances in donkey and mule research. Animals 14(15): 2238

McCarthy EM (2006) Mammalian hybrids. Oxford University Press

McCue ME, Bannasch DL, Petersen JL, Gurr J, Bailey E, Binns MM, Distl O, Guérin G, Hasegawa T, Hill EW, Leeb T, Lindgren G, Penedo MCT, Røed KH, Ryder OA, Swinburne JE, Tozaki T, Valberg SJ, Vaudin M et al (2012) A high density SNP array for the domestic horse and extant Perissodactyla: utility for association mapping, genetic diversity, and phylogeny studies. PLoS Genet 8(1):e1002451

McLean AK, Brandão J, Navas FJ (2014) Comparing the physiological and biochemical parameters of mules and hinnies to horses and donkeys. In: Proceedings of the 7th International Colloquium on Working Equids. University of London, Royal Holloway

Medina CE (2017) Gaited mammoth donkeys are unique breed. AP News (An article in the Associated Press)

Metzger J et al (2013) Expression levels of LCORL are associated with body size in horses. PLoS One 8(2):e56497. https://doi.org/10.1371/journal.pone.0056497

Meuwissen THE, Hayes BJ, Goddard ME (2001) Prediction of total genetic value using genome-wide dense marker maps. Genetics 157(4):1819–1829

Meuwissen T, Hayes B, Goddard M (2016) Genomic selection: a paradigm shift in animal breeding. Anim Front 6(1):6–14

Navas González FJ, Jordan Vidal J, Camacho Vallejo ME, León Jurado JM, de la Haba Giraldo MR, Barba Capote C, Delgado Bermejo JV (2018) Risk factor meta-analysis and Bayesian estimation of genetic parameters and breeding values for hypersensibility to cutaneous habronematidosis in donkeys. Vet Parasitol 252:9–16

Navas-González FJ, Jordana-Vidal J, León-Jurado JM, McLean AK, Pizarro-Inostroza G, Delgado-Bermejo JV (2018a) Genetic parameter estimation and implementation of the genetic evaluation for gaits in a breeding program for assisted-therapy in donkeys. Vet Res Commun 42(2): 101–110

Navas-González FJ, Jordana-Vidal J, León-Jurado JM, Arando-Árbulo A, McLean AK, Delgado-Bermejo JV (2018b) Genetic parameter and breeding value estimation of donkeys' problem-focused coping styles. Behav Process 153:66–76

Navas-González FJ, Jordana J, McLean AK, León JM, Barba CJ, Arando A, Delgado JV (2019a) Modelling for the inheritance of multiple births and fertility in endangered equids: determining risk factors and genetic parameters in donkeys (Equus asinus). Res Vet Sci 126:213–226

Navas-González FJ, Jordana-Vidal J, León-Jurado JM, McLean AK, Delgado-Bermejo JV (2019b) Dumb or smart asses? Donkey's (Equus asinus) cognitive capabilities share the heritability and variation patterns of human's (Homo sapiens) cognitive capabilities. J Vet Behav 33:63–74

Pastuszka R, Barłowska J, Litwińczuk Z (2016) Allergenicity of milk of different animal species in relation to human milk. Postępy Hig Med Dośw 70:1451–1459

Polidori P, Vincenzetti S, Pucciarelli S, Polzonetti V (2020) Comparison of carcass and meat quality obtained from mule and donkey. Animals 10(9):1620

Polidori P, Cammertoni N, Santini G, Klimanova Y, Zhang J-J, Vincenzetti S (2021) Effects of donkeys rearing system on performance indices, carcass, and meat quality. Foods 10(12):3119

Pusterla N, Watson JL, Wilson WD, Affolter VK, Spier SJ (2003) Cutaneous and ocular habronematidosis in horses: 63 cases (1988–2002). J Am Vet Med Assoc 222(7):978–982

Quaresma MNP, Martins AMF, Rodrigues JB, Colaço J, Payan-Carreira R (2014) Viability analyses of an endangered donkey breed: the case of the Asinina de Miranda (Equus asinus). Anim Prod Sci 55(9):1184–1191

Ram P et al (2025) Unraveling the genetic and physiological potential of donkeys: Insights from genomics, proteomics, and metabolomics. Mamm Genome 36:10–24. https://doi.org/10.1007/s00335-024-10083-y

Ren X, Liu Y, Zhao Y, Li B, Bai D, Bou G, Zhang X, Du M, Wang X, Bou T, Shen Y, Dugarjaviin M (2022) Analysis of the whole-genome sequences from an *Equus* parent-offspring trio provides insight into the genomic incompatibilities in the hybrid mule. Genes 13(12):2188

Rieder S et al (2001) Mutations in the agouti (ASIP), the extension (MC1R), and the brown (TYRP1) loci and their association to coat color phenotypes in horses. Mamm Genome 12(6): 450–455. https://doi.org/10.1007/s003350020017

Sarti L, Martini M, Brajon G, Barni S, Salari F, Altomonte I, Ragona G, Mori F, Pucci N, Muscas G, Belli F, Corrias F, Novembre E (2019) Donkey's milk in the management of children with cow's milk protein allergy: nutritional and hygienic aspects. Ital J Pediatr 45(1):102

Šebela M (2022) Biomolecular profiling by MALDI-TOF mass spectrometry in food and beverage analyses. Int J Mol Sci, 23(21):13631

Schurink A, Ducro BJ, Heuven HC, van Arendonk JAM (2011) Genetic parameters of insect bite hypersensitivity in Dutch Friesian broodmares. J Anim Sci 89(4):1286–1293

Sevane N, Dunner S, Boado A, Cañon J (2017) Polymorphisms in ten candidate genes are associated with conformational and locomotive traits in Spanish Purebred horses. J Appl Genet 58(3):355–361

Shi X, Li Y, Wang T, Ren W, Huang B, Wang X, Liu Z, Liang H, Wang C (2022) Association of HOXC8 genetic polymorphisms with multi-vertebral number and carcass weight in Dezhou donkey. Animals 12(11):1349

Sigurðardóttir H, Boije H, Albertsdóttir E, Kristjansson T, Rhodin M, Lindgren G, Eriksson S (2023) The genetics of gaits in Icelandic horses goes beyond DMRT3, with RELN and STAU2 identified as two new candidate genes. Genetics Selection Evolution 55(89)

Singh P, Chhetri KBP, Thangaraj SL, Ram D, Mahnot G (2024) Milk somatic cell DNA: a surrogate DNA source for the successful genomic breeding scheme in Halari donkeys. Heliyon 10(2):e24991

Sonali B, Unnati S, Nayan V, Legha RA, Bhattacharya TK, Pal Y, Giri SK (2023) Identification and characterization of single nucleotide polymorphisms in the coding sequence of donkey (Equus asinus) DMRT3 gene. Anim Biotechnol 34(9):4910–4920

Song S, Wang S, Li N, Chang S, Dai S, Guo Y, Wu X, Cheng Y, Zeng S (2023) Genome-wide association study to identify SNPs and candidate genes associated with body size traits in donkeys. Front Genet 14:1112377

Sun Y, Wang Y, Li Y, Li M, Zhu M, Wang C, Ren W (2023) Comparative transcriptome and proteome analyses of the longissimus dorsi muscle for explaining the difference between donkey meat and other meats. Anim Biotechnol 34(7):3085–3098

The Donkey Sanctuary (2024) African horse sickness. https://www.thedonkeysanctuary.org.uk/for-owners/owners-resources/african-horse-sickness

Turini L, Bonelli F, Nocera I, Meucci V, Conte G, Sgorbini M (2021) Evaluation of different methods to estimate the transfer of immunity in donkey foals fed with colostrum of good IgG quality: a preliminary study. Animals 11(2):507

Utzeri VJ, Bertolini F, Ribani A, Schiavo G, Dall'Olio S, Fontanesi L (2016) The albinism of the feral Asinara white donkeys (Equus asinus) is determined by a missense mutation in a highly conserved position of the tyrosinase (TYR) gene deduced protein. Anim Genet 47(1):120–124

Wade CM, Giulotto E, Sigurdsson S, Zoli M, Gnerre S, Imsland F, Lear TL, Adelson DL, Bailey E, Bellone RR, Blöcker H, Distl O, Edgar RC, Garber M, Leeb T, Mauceli E, MacLeod JN, Penedo MCT, Raison JM et al (2009) Genome sequence, comparative analysis, and population genetics of the domestic horse. Science 326(5954):865–867

Wang C, Li H, Guo Y, Huang J, Sun Y, Min J, Wang J, Fang X, Zhao Z, Wang S, Zhang Y, Liu Q, Jiang Q, Wang X, Guo Y, Yang C, Wang Y, Tian F, Zhuang G, Fan Y, Gao Q, Li Y, Ju Z, Li J, Li R, Hou M, Yang G, Liu G, Liu W, Guo J, Pan S, Fan G, Zhang W, Zhang R, Yu J, Zhang X, Yin Q, Ji C, Jin Y, Yue G, Liu M, Xu J, Liu S, Jordana J, Noce A, Amills M, Wu DW, Li S, Zhou X (2020) Donkey genomes provide new insights into domestication and selection for coat color. Nat Commun 11:6014

Wang Y, Miao X, Zhao Z, Wang Y, Li S, Wang C (2021) Transcriptome atlas of 16 donkey tissues. Front Genet 12:682734

Wang G, Wang F, Pei H, Li M, Bai F, Lei C, Dang R (2022) Genome-wide analysis reveals selection signatures for body size and drought adaptation in Liangzhou donkey. Genomics 114(6):110476

Wang L, Qu H, Wang X, Wang T, Ma Q, Khan MZ, Zhu M, Wang C, Liu W, Chai W (2024a) Data-independent acquisition method for in-depth proteomic screening of donkey meat. Agriculture-Basel 14(12):2102

Wang Z et al (2024b) Reproductive disorders in donkeys: current evidence and update. Animals 14(17):2494. https://doi.org/10.3390/ani14172494

Wang X-R, Ren W, Peng Y-D, Khan MZ, Liang H-L, Zhang Y-G, Liu X-T, Chen Y-H, Kou X-Y, Wang L-Y, Wang C-F, Zhan Y-D (2024c) Comprehensive transcriptomic analysis unveils the interplay of mRNA and lncRNA expression in shaping collagen organization and skin development in Dezhou donkeys. Front Genet 15:1335591

Wang X-R, Ren W, Peng Y-D, Khan MZ, Liang H-L, Zhang Y-G, Liu X-T, Chen Y-H, Kou X-Y, Wang L-Y, Wang C-F, Zhan Y-D (2024d) Elucidating the role of transcriptomic networks and DNA methylation in collagen deposition of Dezhou donkey skin. Animals (Basel) 14(8):1222

White SD (2013) Donkey dermatology. Vet Clin N Am Equine Pract 29(3):703–708

Wolc A, Bresińska A, Szwaczkowski T (2006) Genetic and permanent environmental variability of twinning in thoroughbred horses estimated via three threshold models. J Anim Breed Genet 123(3):186–190

Xie T, Zhang S, Shen W, Zhang G, Guo R, Zhang W, Cao Y, Pan Q, Liu F, Sun Y, Liu S (2022) Identification of candidate genes for twinning births in Dezhou donkeys by detecting signatures of selection in genomic data. Genes 13(10):1902

Yu J, Yang G, Li S, Li M, Ji C, Liu G, Wang Y, Chen N, Lei C, Dang R (2023) Identification of Dezhou donkey muscle development-related genes and long non-coding RNA based on differential expression analysis. Anim Biotechnol 34(7):2313–2323

Zhang W et al (2024) Factors affecting the quality and nutritional value of donkey meat: a comprehensive review. Front Vet Sci 11:1460859. https://doi.org/10.3389/fvets.2024.1460859

Zhou Y, Liang H, Li Y, Shi X, Liu Z, Ren W, Wang X, Huang B, Wu F, Li M, Liu G, Ji C, Wang Y, Chen N, Lei C, Wang C (2020) Whole-genome sequencing reveals selection signatures associated with important traits in six donkey breeds. Genes 11(6):652

Zhu Q et al (2025) Application of omics in donkey meat research: a review. Animals 15(7):991. https://doi.org/10.3390/ani15070991

Wild and Feral Asses and Mules

Sarah R. B. King

Abstract

Donkeys (*Equus asinus*) were domesticated from African wild asses (*E. africanus*) about 5000 years ago and were subsequently dispersed around the world by trade and colonization. After industrialization the need for donkeys declined and feral populations established from released or escaped individuals. Due to their ability to adapt to arid environments, high fecundity, and low mortality, feral donkeys became abundant in desert regions of Australia, North America, and South America. While feral donkey populations currently account for millions of animals, populations of their wild ass relatives (African wild asses, Asiatic wild asses (*E. hemionus*), and Tibetan wild asses (*E. kiang*)) are mostly in decline and in need of conservation.

The aim of this chapter is to provide an overview of feral donkey ecology and management. It describes how donkeys were domesticated and how their African wild ass ancestry has been retained in terms of their arid-adaptability and social structure. Donkeys are compared and contrasted to wild ass species in a framework of global range, population abundance, habitat use, and population management and conservation.

Keywords

Wild · Feral · Domestication · Management · Wild ass · Habitat use

S. R. B. King (✉)
Natural Resource Ecology Laboratory, Colorado State University, Fort Collins, CO, USA
e-mail: sarah.king@colostate.edu

A. K. McLean et al. (eds.), *Donkey and Mule Medicine: Foundations, Breeds, and Management*, https://doi.org/10.1007/978-3-032-12726-6_6

1 Introduction

There are seven extant species of wild equid: three species of zebra (plains zebra *Equus quagga*, Grevy's zebra *E. grevyi*, and mountain zebra *E. zebra*), three species of ass (African wild ass *E. africanus*, Asiatic wild ass *E. hemionus*, and the kiang or Tibetan wild ass *E. kiang*), and Przewalski's horse (*E. przewalskii*). Donkeys and mules are derived from the African wild ass and a sister species of Przewalski's horse. Donkeys were domesticated from the African wild ass in North Africa around 5000 years BCE, and subsequently were dispersed around the globe, particularly to relatively arid countries (Todd et al. 2022). The domestication of the horse is less understood, with several domestication events likely occurring in Eurasia around 5000 years BCE. The direct ancestor of the domestic horse is not the Przewalski's horse, but rather they both had a common ancestor. Mules have been bred for at least 2200 years and have been detected from archeological remains as early as the Iron Age, and in Roman and Byzantine Turkey sites (Fages et al. 2019).

While domestic and feral equids increased their global range with human expansion, the range of wild asses has decreased since the 1500s. Once found throughout Eurasia, Asiatic wild asses are now restricted to just a few populations in central Asia, with only 28,000 adults globally. The largest population is found in southern Mongolia (Kaczensky 2023). African wild asses were historically found throughout northern Africa, but currently only about 500 remain in Ethiopia and Eritrea (Moehlman et al. 2015). Similarly Przewalski's horses became extinct in the wild in the 1970s and were only kept extant as a species due to captive breeding in zoos. Reintroduction efforts since the 1990s has led to a growing wild population of around 1000 horses in Mongolia and China (Šimek 2024).

Humans have taken horses and donkeys to almost every country in the world, where they were used for meat, milk, transport, and to assist agriculture. Post-industrialization, the utility of these animals waned leading many to be released into a wild state to join others that had previously been intentionally or unintentionally released. A domesticated species with no owner, living in a wild state where it becomes subject to natural selection is considered feral. Wild equids are now threatened with extinction across their range, but feral equids are numerous. Populations of feral donkeys are particularly large in Australia (up to five million; Australian Government (2011)), Brazil (around 800,000; Queiroz et al. (2021)), and the United States (over 15,000; BLM (2024)), with smaller populations found in island nations in the Caribbean and Pacific and Indian oceans (Clancy et al. 2021).

As they are non-reproducing there are no populations consisting purely of feral mules, although they have been reported in US Herd Management Areas where both horses and donkeys are present. Despite differences in both genetics and social organization among species, all equids are capable of hybridization. Grevy's and plains zebra produce hybrids in Kenya (Cordingley et al. 2009), and plains zebra, mountain zebra and donkeys all hybridize in Namibia (Dalton et al. 2017; Uiseb 2024).

Feral populations exist without direct human intervention but can be subject to management to control their populations. This may take the form of gathers and removals or application of fertility control, or can be through lethal means.

Population control is considered necessary due to the potential impact of donkeys and mules on vegetation in areas where they are not native. African wild asses evolved in Africa, and were not present on the American or Australian continent (or islands) before being introduced by colonizers. Donkeys have been shown to have both beneficial and adverse impacts on habitats where they are found, and like all species will have greater impacts when they are most numerous. Wild asses are only found within their native range, and are reliant on conservation activities to ensure that they are protected from poaching and that they maintain access to pasture and water sources.

In this chapter I will discuss the biology and ecology of feral donkeys within the context of their wild ass relatives. I will consider the historical and geographical context of these species, and provide information on their ecology and how they are managed.

2 Key Concepts

Glossary:

Feral—A species that has been previously domesticated and subject to artificial selection, living in a wild state where it is subject to natural selection.

Wild—A species that has never been domesticated.

Domestic—A species that has been domesticated (traits have been selected through controlled breeding by humans) and that lives with and is directly maintained and managed by humans.

Wild ass—A hemione species of equid, phylogenetically separated from horses and zebras. Extant species are African wild ass (*Equus africanus*), Asiatic wild ass (*E. hemionus*) and the kiang or Tibetan wild ass (*E. kiang*).

BLM—United States Department of Interior Bureau of Land Management

WHBPA—Wild Horse and Burro Protection Act, 1971

IUCN—International Union for the Conservation of Nature

Additional resources:

Books:

Ransom, J. I., and Kaczensky, P. (Eds). 2016. Wild Equids: Ecology, Management, and Conservation. Johns Hopkins University Press.

Scientific articles:

Burden, F., Thiemann, A., 2015. Donkeys Are Different. Journal of Equine Veterinary Science 35, 376–382.

Choquenot, D., 1991. Density-dependent growth, body condition, and demography in feral donkeys testing the food hypothesis. Ecology 72, 805–813.

Clancy, C.L., Kubasiewicz, L.M., Raw, Z., Cooke, F., 2021. Science and Knowledge of Free-Roaming Donkeys—A Critical Review. The Journal of Wildlife Management 85, 1200–1213.

Esmaeili, S., King, S.R.B., Schoenecker, K.A., 2023. Browsers or Grazers? New Insights into Feral Burro Diet Using a Non-Invasive Sampling and Plant DNA Metabarcoding Approach. Animals (Basel) 13.

Esmaeili, S., Schoenecker, K.A., King, S.R.B., 2024. Resource availability and heterogeneity affect space use and resource selection of a feral ungulate. Ecosphere 15.

Kaczensky, P., Buuveibaatar, B., Payne, J.C., Strindberg, S., Walzer, C., Batsaikhan, N., Bolortsetseg, S., Victurine, R., Olson, K.A., 2020. A conservation strategy for khulan in Mongolia: background and key considerations. NINA Report 1889, p. 72.

King, S.R.B., Schoenecker, K.A., 2025. Effects of phylogeny and habitat on social structure and behavior of two equid species. Behavioral Ecology and Sociobiology 79, 80.

Moehlman, P., 1998. Behavioural patterns and communication in feral asses (*Equus africanus*). Applied Animal Behaviour Science 60, 125–169.

Rubin, E.S., Conrad, D., Harding, L.E., Russo, B.M., 2024. Associations between a feral equid and the Sonoran Desert ecosystem. Wildlife Monographs 215.

Schoenecker, K.A., King, S.R.B., Messmer, T.A., 2021. The Wildlife Profession's Duty in Achieving Science-Based Sustainable Management of Free-Roaming Equids. The Journal of Wildlife Management 85, 1057–1061.

Todd, E.T., Tonasso-Calvière, L., Chauvey, L., Schiavinato, S., Fages, A., Seguin-Orlando, A., Clavel, P., al., e., Orlando, L., 2022. The genomic history and global expansion of domestic donkeys. Science 377, 1172–1180.

Websites:

Bureau of Land Management Wild Horse and Burro Program website: https://www.blm.gov/programs/wild-horse-and-burro/about-the-program/about-wild-horses-and-burros. IUCN Red List website: https://www.iucnredlist.org

3 Historical Framework

Wild asses once occurred throughout the plains of eastern Europe, Arabia, and central Asia to India and the Himalayas (Asiatic wild asses *E. hemionus* and kiang, *E. kiang*), in western Europe and south-west Asia (European ass *E. hydruntinus*; extinct since the late Holocene), and across northern Africa (African wild ass *E. africanus*). Asiatic wild asses and kiang have never been domesticated. Due to hunting and conversion of land to agriculture their range has been reduced from historically being contiguous throughout central Asia to current isolated populations, with the largest population in southern Mongolia and on the Tibetan Plateau of India and China. African wild ass range has been reduced to only being found in the Horn of Africa. Historically there were three African wild ass subspecies. *Equus africanus atlanticus* was found in the Atlas Mountains, but has been extinct since Roman times (Kimura et al. 2013). The Nubian wild ass subspecies

(*E. a. africanus*) was found in Sudan and Eritrea until the 1970s, and is now considered extinct (Kimura et al. 2011). The Somali wild ass subspecies (*E. a. somaliensis*) is still present in low numbers (around 500 individuals) in Ethiopia and Eritrea, but it is not known if any still remain in Somalia (Moehlman et al. 2015).

As donkeys are rarely eaten and are not herded or penned when not in work, there are few archeological remains of this species to show exactly when domestication occurred (Kimura et al. 2013; Shackelford et al. 2013). However, both the few remains and genetic evidence point to donkeys being domesticated from African wild ass in northern Africa around 5000 BCE (Beja-Pereira et al. 2004; Todd et al. 2022). Donkeys were likely domesticated by pastoralists so they could become more mobile in response to desertification of the Sahara happening at that time (Marshall and Weissbrod 2011; Shackelford et al. 2013; Rosenbom et al. 2015). Donkeys were subsequently spread into Eurasia around 2500 BCE, with subsequent return of genetic material to Africa (Todd et al. 2022). This has resulted in two genetic clades in modern donkeys that cluster as African and mostly non-African (i.e., European and Asian) (Todd et al. 2022), which corresponds to mitochondrial DNA evidence of Nubian wild ass ancestry of African donkeys and an unknown ancestor distinct from the Somali wild ass (Beja-Pereira et al. 2004; Kimura et al. 2011).

While they may be considered similar-looking at a glance, donkeys and African wild asses have key morphological differences: African wild ass are larger than native donkeys, with longer legs which have distinctive striping patterns. Donkeys have been considered essential animals for transport for millennia, with their use little changed over time in many parts of the developing world. Through domestication, donkeys were selected for load carrying and traction, while horses were selected for speed in battle. As well as beasts of burden, donkeys were used for ploughing, threshing, and pulling carts (Marshall 2007). As pack animals, donkeys are typically loaded with most weight placed over their forelimbs (Shackelford et al. 2013). This resulted in skeletal changes in donkeys compared to African wild ass. Working donkeys make slower and more linear movements than African wild asses, resulting in lower axial and bending strains in their limb bones and more elliptical forelimb diaphysis, especially in the humerus (Shackelford et al. 2013). These changes are more likely caused by restricted gait in donkeys used for transport, rather than effects of load-bearing (Shackelford et al. 2013). Donkeys quickly became a vital part of the infrastructure of the ancient world, with remains dated 4800 to 5000 BCE from tombs in Egypt, Syria, Iran and Iraq (Rossel et al. 2008). Their domestication allowed for redistribution of food and a greater reach of trade than had been previously available (Rossel et al. 2008), as they were used to carry heavy loads of precious goods from ivory to grain on long-distance desert routes, where little water was available (Shackelford et al. 2013).

All equids can hybridize, with natural hybrids of feral donkeys and feral horses observed in the United States (P. Farris, pers. comm.), hybrids of Grevy's and plains zebras observed in Kenya (Cordingley et al. 2009), and hybrids among plains zebras, mountain zebras, and donkeys in northern Namibia and Angola (Dalton et al. 2017; Uiseb 2024; P. van Pinto, pers. comm.). Pastoralists often desire interbreeding between domestic and wild populations (Marshall et al. 2014), and admixture modeling suggests ongoing hybridization between African wild ass and donkeys, and even between donkeys and kiang in China (Todd et al. 2022). Importantly, resulting hybrids are thought to remain in the domestic population. The earliest reported donkey hybrids are from ~4500 years ago, where kungas (donkey x Asiatic wild ass hybrids) were bred as an animal of high status and prestige in Mesopotamian society, pre-dating arrival of the horse to the region (Bennett et al. 2022). Mules were vital to the Roman empire as a beast of burden and to pull military chariots (Lepetz et al. 2021). From no evidence of mules in France in the Iron Age, they made up to 40% of equine bone assemblages in Roman times (Lepetz et al. 2021), contrasting with few donkey remains. Archeological remains of a Roman donkey breeding farm in France from around 1500 BCE found giant donkeys, likely used to produce mules of a large and uniform size (Todd et al. 2022). By medieval times donkeys had become more popular than mules, which were more common in the south and mountainous regions compared to more donkeys in northern France (Lepetz et al. 2021).

Donkeys continue to play a key role in the lives of pastoralists in Africa, carrying water and moving household goods and family members. Donkeys provide a vital means for mobility, allowing herders to track good pasture rather than being restricted by proximity to permanent water sources (Marshall and Weissbrod 2011). With donkeys, pastoralists can make frequent moves over distances from 10 to >100 km (Marshall 2007). Donkeys are considered women's animals as they are used for daily tasks and do not have the symbolic status of cattle in pastoralist communities (Marshall and Weissbrod 2011). They are used for collecting firewood and water, with Maasai donkeys carrying loads of 50 kg of water over 9–14 km (Marshall and Weissbrod 2011). In modern times donkeys are still rarely eaten, but can provide an emergency source of meat and milk (Marshall 2007).

Donkeys make useful pack animals in arid environments as unlike cows they do not require time for rumination and are more drought tolerant (Maloiy 1970; Maloiy and Boarer 1971), and they have an energetically efficient walking gait (Marshall 2007). In East African pastoralist societies donkeys are managed extensively. Although they are sometimes penned within the settlement thorn fence at night, donkeys' ability to dig for water and protect themselves from predators means that they are also often left loose (Marshall 2007; Marshall and Weissbrod 2011). Donkeys are valued for strength, calmness, and disease and drought resistance, but are not strategically bred for these traits. Instead they are allowed to follow natural mating patterns, with animals from numerous households mixing during the wet season (Marshall and Weissbrod 2011). Notably all donkeys in these pastoralist societies do have an owner, so are not feral. Female donkeys are valued above males, with males controlled through culling and castration. Donkey herds tend to be small (1–6 animals), and grow slowly due to foals being born on average every 2 years and loss from predation, disease, and drought (Marshall and Weissbrod 2011).

Donkeys and mules were spread throughout the New World as it was colonized, maintaining their utility for transport especially in arid and mountainous regions (Bough 2006). In modern times donkeys were vital for farming, mining, and infrastructure development, as well as transportation. With industrialization they became less used, and donkeys that escaped or were intentionally released established populations in Australia, the Americas, and the Caribbean (Clancy et al. 2021; Schoenecker et al. 2021). The traits of adaptability and resilience that made donkeys such good pack animals in Africa allowed them to persist in feral populations (Celermajer and Wallach 2019).

4 Worldwide Context

Wild ass species are found in their historic range in Africa and Eurasia. African wild ass are currently known to be present in only a few locations in Ethiopia and Eritrea, around the Danakil depression (Moehlman et al. 2015; Fig. 1). Asiatic wild asses are found in small and isolated populations in Iran, Turkmenistan, Kazakhstan, Uzbekistan, and China, with a larger, more contiguous population in the Gobi desert of southern Mongolia (Kaczensky 2023; Fig. 2). A reintroduced population of

Fig. 1 African wild ass on the Messir Plateau, Eritrea. © Redae Tesfai

Fig. 2 Asiatic wild ass in the Gobi Desert, Mongolia. © Petra Kaczensky

Asiatic wild ass is established in Israel, and reintroductions are currently ongoing in Jordan and Saudi Arabia. Kiang are considered numerous on the Tibetan Plateau in both India and China.

Wild equid species have never been domesticated or subject to artificial selection by humans. A feral animal is defined as a domestic species living in a wild state and now subject to natural selection. In addition to domestic donkeys managed extensively, feral donkeys are reported from across the Old World: reported living in a wild state in Egypt, South Africa, Sri Lanka, and Cyprus (Santiapillai et al. 1999; Hamrick et al. 2005; Clancy et al. 2021; Soultan et al. 2023), and used for rewilding or otherwise managing vegetation in mainland Europe (Couvreur et al. 2005; Couto et al. 2016).

The largest population of feral donkeys is in Australia, with five million estimated in arid central Australia, the Kimberley in Western Australia, and the Top End in the Northern Territory (Australian Government Department of Sustainability 2011), however this number is not based on any cited scientific population estimation. In South America the largest population of feral donkeys is in the Caatinga ecoregion of northeastern Brazil ("Nordestina" breed), with smaller populations in Argentina, Chile, and Mexico (Clancy et al. 2021). Feral donkeys are numerous in the United States (Fig. 3), with 14,568 estimated on Bureau of Land Management Herd Management Areas in Arizona, California, Nevada, Oregon, and Utah (as of March, 2024; BLM 2024), where they are referred to as wild burros. Additional feral donkeys in the United States are found on military lands and National Parks in the desert southwest. Feral donkeys were eradicated from the Galapagos islands (Carrion et al. 2006), but small populations exist on Caribbean islands including St Johns, Bonaire, Inagua, Turks and Caicos, Antigua and Barbuda, and Saint Kitts and Nevis (The Donkey Sanctuary, unpublished data).

The donkey's ability to adapt to arid landscapes and survive on a variety of plant types means that in Australia and North and South America feral donkeys are largely found in more desert areas that are far from human population centers and used only for low-density cattle grazing (Dias et al. 2019; Clancy et al. 2021). In the Caribbean where donkeys have a more restricted range and there is a higher human density,

Fig. 3 Feral donkeys in Utah, USA. © Sarah King

impacts of donkeys are related to their impact on the native habitat, conflict over encroachment on gardens and crops, and vehicle collisions. Several of the islands have set up sanctuaries to protect donkeys, such as the Donkey Sanctuary of Bonaire and the Antigua and Barbuda Humane Society. In others their presence remains contentious as both a tourist draw and a problem for local people. Organizations such as The Donkey Sanctuary, Peaceful Valley Donkey Rescue, and The Brooke Action for Working Horses and Donkeys work to maintain welfare of all donkeys around the world, but little research or action is devoted to most feral populations due to their remote locations.

5 Roles in Various Environments

In their native ranges African and Asiatic wild asses are found across vast arid landscapes, naturally limited in their movements only by the distribution of water sources. African and Asiatic wild asses have been observed to dig for water in dry water courses, a trait that was inherited by feral donkeys (Lundgren et al. 2021). Asiatic wild asses in Mongolia have nomadic movements as they track the most green forage available, moving within a radius of 13.5–21 km from a water source (Kaczensky et al. 2008; Nandintsetseg et al. 2016). Asiatic wild asses drink every 1.5 to 2 days in the hot and dry season but only every 2–3 days in cooler months (Kaczensky et al. 2010; Payne et al. 2020) when they are freed from the need to find surface water by the presence of snow. In Israel where permanent water sources were available, Asiatic wild asses visited water more than once a day in summer, and less than once a day in winter, normally at dusk or overnight (Giotto et al. 2015). In the dry season African wild ass females with foals travel 9 km every day to drink at water, with non-lactating females and males only drinking every 5–10 days (Tesfai et al. 2021a). In the rainy season females with foals are able to drink twice a day, but they only drink at night and disperse up to 7 km away to avoid livestock (Tesfai et al. 2021a). Similar to their wild relatives, feral donkeys in California visited water around every 22 hours during the hot-dry season, but only every 40–46 hours during the cool-wet seasons (Karish et al. 2023).

Asiatic wild ass in Israel used areas of 23.56 ± 5.29 km^2 in summer and 31.27 ± 7.06 km^2 in winter, moving between areas of high vegetation cover that were mostly near water sources, and use of high elevation areas (Giotto et al. 2015). Home ranges of Asiatic wild asses in Mongolia were much larger: 4449–6835km^2 (Kaczensky et al. 2008), reflecting the difference in both vegetation resources and water availability. Similarly feral donkey home ranges reflect the availability and distribution of resources in their habitat. In the Mojave desert, California, ranges were around 254 km^2, with no difference across seasons (Karish et al. 2023). In this habitat donkeys selected areas closer to water, especially in the dry season, and selected for green forage and low elevations (Karish et al. 2023). In the Colorado Plateau, Utah, where there is large seasonal and spatial variation in resources particularly in terms of access to water, feral donkeys had annual ranges around 99 km^2 again with no difference across seasons (Esmaeili et al. 2024). In the Sonoran

Fig. 4 Feral donkeys in Lake Pleasant Herd Management Area, Arizona, USA. © Sarah King

Desert, Arizona, annual ranges were around 35 km^2. Sonoran Desert donkeys had larger ranges in the moderate season between the hot-dry and hot-wet seasons, although they had similar resource selection patterns across the year (Esmaeili et al. 2024). The relatively small ranges in the Sonoran Desert population is likely because they had access to a permanent water source in the form of a freshwater lake (Fig. 4). In both the Mojave and Sonoran Deserts feral donkeys selected for areas near humans, either urban areas on a military base or recreational areas near the lake, due to more resources being available in these locations in the form of green lawns or human trash (Karish et al. 2023; Esmaeili et al. 2024). Within their ranges, asses use topography where they can shelter from weather extremes. Asiatic wild asses and feral donkeys use high elevations and northern exposures to enhance exposure to the wind and to shelter from solar radiation (Giotto et al. 2015; Haddy et al. 2020; Esmaeili et al. 2024), using valleys and southern aspects to shelter from colder weather (Nowzari et al. 2013; Esmaeili et al. 2024).

Although negative effects of feral donkeys on vegetation and sympatric wildlife are often assumed rather than proven (Clancy et al. 2021), several studies in desert ecosystems have shown both direct and indirect effects of feral donkey presence. In the Grand Canyon prior to feral donkey extirpation there was a significant difference in vegetation between donkey grazed areas and those where donkeys were not present, with an increase of unpalatable shrubs and forbs adapted to intensive grazing in donkey areas (Carothers 1977). Additionally feral donkeys contributed to soil compaction, which accentuated erosion of soil in areas of high use (Carothers 1977). In Arizona there was 30% less ground cover, lower plant density and smaller plant size, and less foliage density in areas where feral donkeys were present, although there was a higher plant species richness closer to water in donkey areas (Rubin et al. 2024). The saguaro cactus is a key figure in this landscape (Fig. 5), but 63% lower cactus recruitment was found in areas with donkeys, and lower foliage densities of yellow palo verde and desert ironwood. Feral donkeys and cattle also damaged columnar cacti in Argentina (Malo et al. 2011), which affected the reproductive output of these plants (Peco et al. 2011). Rubin et al. (2024) found no effect

Fig. 5 Feral donkey with saguaro cacti in the background. © Sarah King

of donkey presence on densities of white bursage (also known as burrobush),but found lower densities far from water in areas used by burros and Hanley and Brady (1977) found lower canopy cover of bursage in areas used by donkeys. Bursage has a high nutrient content in the spring compared to other browse species available at that time (Hanley and Brady 1977) and was highly selected as a winter food source by donkeys (Woodward and Omhart 1976). Bursage is also a primary food source for Agassiz's desert tortoise, which had a lower occurrence in areas more used by feral donkeys (Berry et al. 2020), as well as common side-blotched lizards being 26% less abundant in areas where donkeys were present (Rubin et al. 2024). Feral donkeys indirectly affect small mammals and birds by affecting the plants that they need for food and shelter (Carothers 1977; Rubin et al. 2024).

Benefits of asses can be seed dispersal in arid environments (Treitler et al. 2017; Ghasemi et al. 2024). While donkey feces did not promote germination of *Prosopis* sp., it did lead to increased plant height and dry weight (Sánchez de la Vega and Godínez-Alvarez 2010). In Mexico where *Prosopis* is native this is considered a benefit to the ecosystem, however in India where it is a non-native invasive species consumption and spread of *Prosopis* by Asiatic wild ass is leading to a reduction in availability of grasses, their main food source (Vazeed Pasha et al. 2015).

Competition between feral donkeys and native ungulates can be explored through similar use of vegetation and/or habitat. No effect of feral donkey abundance was found on recruitment by desert bighorn sheep in California, but due to feral donkeys and bighorn sheep having around 50% diet overlap (Seegmiller and Ohmart 1981), in times of drought feral donkey abundance was negatively related to mountain sheep abundance (Marshal et al. 2008). There was limited overlap in diet and habitat use between mule deer and feral donkeys in California; although the diets of both species was over 70% browse and had high species overlap in summer, donkeys ate more grasses and forbs than deer and used the habitat differently, limiting competition (Marshal et al. 2012). There was spatial and dietary overlap between Argentinian feral donkeys and guanacos and vicuñas, although donkey diet differed from vicuñas as donkeys ate a higher proportion of wetland vegetation rather than steppe (Borgnia et al. 2008; Ovejero et al. 2011). Competition was avoided among these species through habitat partitioning, with donkeys being found closer to springs and roads and avoiding more rocky areas, and wild camelids avoiding areas used by livestock (Borgnia et al. 2008; Ovejero et al. 2011; Acebes et al. 2012). In Argentina donkeys are at low densities, so did not interfere with water hole use by native guanacos (Acebes et al. 2016). In the Atacama Desert, Chile, feral donkeys prefer different habitat to the native camelids, selecting slopes with tall shrubs or herbaceous communities near water, whereas vicuñas and guanacos select grassy areas on valley bottoms or upper slopes respectively, thus avoiding competition (Malo et al. 2016).

Wild asses tend to avoid humans and their livestock, often being displaced from access to water sources (Kaczensky et al. 2010; Kebede et al. 2014). Feral donkeys in Argentina were intermediate between vicuñas and livestock in terms of proximity to humans (Borgnia et al. 2008), and in the Caatinga of Brazil were found further from ranches than feral cattle (Dias et al. 2019). In Arizona and California donkeys selected areas where there was greater human use, probably reflecting both species using areas where there were available resources (Karish et al. 2023; Esmaeili et al. 2024). Wild asses are often perceived as competing with livestock for forage. In Ladakh, India, pastoralists consider there are too many kiang, but livestock numbers have increased and kiang are at low density compared to livestock (kiang density is 0.24 km^2 compared to livestock at 11.03 km^2 including 0.04 km^2 donkeys; Bhatnagar et al. 2006). In Iran onagers may come into conflict with both livestock herders and agricultural farmers either due to perceived conflict for vegetation or actual crop raiding (Esmaeili et al. 2019). Where feral donkeys are close to humans similar conflict can occur, with 25% of donkey groups detected in agricultural lands during surveys in Cyprus (Hamrick et al. 2005), and feral donkeys considered direct competitors with livestock for forage in South Africa (Muller and Bourne 2018).

6 Physiology and Adaptations

Donkeys are adapted to water scarcity thanks to their African wild ass ancestry. They can survive dessication of up to 30% of their body weight and lose less water in their feces when exposed to high temperatures and/or dehydration (compared to horses;

Maloiy 1970). As well as sweat, donkeys can employ respiratory evaporative heat loss to maintain homeothermy (Ayo et al. 2014). Donkeys can compensate for fluid deficits the same as horses, but unlike horses following thirst they can rehydrate by quickly drinking (within 3 to 60 minutes) sufficient water to return blood constituents to normal levels, without overcompensation (Maloiy 1970; Maloiy and Boarer 1971; Jones et al. 1989).

While donkeys are considered to rely on browse (consuming trees and shrubs), feral donkeys consume mostly graminoids and forbs through most of the year, selecting plants depending on their most nutritious growth stage and eating more browse only during the winter when green vegetation is less available (Woodward and Omhart 1976; Seegmiller and Ohmart 1981; Abella 2008; Marshal et al. 2012; Reus et al. 2014). Donkeys are more efficient at digesting low nutritional quality fiber than horses (Burden and Bell 2019) and adapt to what is available locally, consuming 175 plant species in the Mojave desert, California (Abella 2008). In the Sonoran desert of Arizona they had a less diverse diet consisting mostly of shrubs compared to a more diverse diet with more forbs and grasses in the pinyon juniper habitat of the Colorado Plateau in central Utah (Esmaeili et al. 2023). In Eritrea grass is an important diet component for both African wild ass and local domestic donkeys. In the rainy season African wild ass diets had lower nutrient levels than domestic donkeys, containing less fecal nitrogen and phosphorus, as herders displace the native wild ass with their livestock (Tesfai et al. 2021b).

Like domestic donkeys, feral donkeys have a gestation period of 365–370 days (Grinder et al. 2006), although African wild ass in captivity had a longer gestation period of 378–406 days (Kozlowski et al. 2018). In captivity, African wild ass estrous cycle length was 27.2 ± 1.2 days, and they cycled throughout the year; they did not express a foal heat, cycling 32.6 ± 5 days after delivery with no lactational suppression of estrous (Kozlowski et al. 2018). In feral donkeys, birth seasonality varies among populations: in Sri Lanka and on Ossabaw Island, Georgia, estrus behavior and foal births were observed in every month of the year (McCort 1980, Santiapillai et al. 1999). In Australia and the United States most foals are born during the summer or when forage is most nutritious, but births are observed throughout the year and not necessarily related to rainfall or forage parameters (Moehlman 1974; Woodward and Omhart 1976; Norment and Douglas 1977; McCool et al. 1981; Seegmiller and Ohmart 1981; Ruffner and Carothers 1982; Choquenot 1991; Rudman 1998).

Reproduction in feral donkeys can start as early as 18 months old (Woodward 1979, Johnson et al. 1987), with all females over two years old being reproductive (McCort 1980, McCool et al. 1981). Fecundity rates in feral donkey populations are high, with 60% to 89% of females pregnant in populations in the US and Australia based on observations, necropsies, and blood tests (McCool et al. 1981, Choquenot 1991, Ruffner and Carothers 1982, Johnson et al. 1987, Perryman and Muchlinski 1987, Wolfe et al. 1989). Specifically, mean fecundity of feral donkeys in southern California was 0.17 foals/female for 2–3 year old females and 0.58 foals/adult for 4 year old and older females (Gedir et al. 2021). More than half of adult female feral donkeys give birth every year (McCool et al. 1981; Moehlman 1974), and 60% of the pregnant jennies necropsied by Johnson et al. (1987) were also lactating. Natality was estimated at 20–23% in Arizona (Seegmiller and Ohmart 1981), with 47–63% of females observed with foals in Death Valley (Moehlman 1974; Norment and Douglas 1977), and 17% seen with foals in Cyprus where there was 1 foal for every 2.95 females (Hamrick et al. 2005). In Australia fecundity was high independent of density, but populations were regulated by density dependence in the form of food-related juvenile mortality (Choquenot 1991). At high density juvenile mortality was three times greater than when at low density (Choquenot 1991) due to females being in poorer condition as the only forage remaining was mineral-poor and fiber-rich (Freeland and Choquenot 1990). Survival rates of adults in feral donkey populations are high (0.96 ± 0.003 for adults 2–9 years old; (Gedir et al. 2021), with no sex-specific differences in mortality (Perryman and Muchlinski 1987). Adult deaths are rarely observed (McCort 1980, Moehlman 1974, Seegmiller and Ohmart 1981, Norment and Douglas 1977). In Australia the highest mortality was seen in the first six months of life where populations were regulated by density dependence (Choquenot 1991). In the arid deserts of southern California, foal survival was 0.90 ± 0.016 (mean ± variance) and 0.98 ± 0.001 for yearlings (Gedir et al. 2021), plus some foals are lost due to abortions or dying post-partum (Moehlman 1974, Norment and Douglas 1977). The high fecundity rates and low adult mortality rates result in high population growth rates of feral donkeys. In America feral donkey populations were estimated to grow by 20–25% annually (Woodward and Ohmart 1976, Norment and Douglas 1977, Perryman and Muchlinski 1987), and had a growth rate of around 23% in Australia (Choquenot 1990).

All ass species have a 'fission-fusion' social structure, where the only permanent social bonds are between a female and her offspring up to about a year old, which represent the smallest unit of social organization (Hoffmann 1983; Klingel 1998; Moehlman 1998b; Feh et al. 2001; King and Schoenecker 2025; Kaczensky et al. 2008). Besides females and their offspring, feral donkeys can be observed in a variety of social groupings: groups with at least one adult of each sex plus offspring (similar to a horse harem group), all female groups with their offspring, all male groups of all ages (bachelor groups), and solitary post-reproductive females (Ruffner and Carothers 1982, Moehlman 1998b, Moehlman et al. 1998), but these groupings differ from horses in that none are permanent. Asiatic wild asses can be found in large groups of more than 80 individuals (Kaczensky et al. 2019), with social group size and intergroup distance determined by vegetation productivity and density (Shah and Qureshi 2007; Nowzari et al. 2013). In most ass populations some males hold territories (i.e., seen repeatedly in specific areas where they defend females against other males) in areas frequented by females, such as areas of good forage or near water (Woodward 1979; Klingel 1998; Moehlman 1998a; Shah and Qureshi 2007; Kannan and Parsons 2017). This social system has evolved as an adaptation to arid environments where food and water are widely distributed (Rubenstein 2010). Lactating females and non-lactating females and males have different nutrition and water needs: lactating females require more frequent access to water whereas non-lactating females are free to explore food patches further away, resulting in different habitat use by different age/sex cohorts (Nowzari et al. 2013).

In arid environments males are likely to increase their fitness if they remain in areas where females are likely to pass through, including females in estrus, rather than trying to associate with females of differing resource needs, so males defend territories near limited resources such as water holes. Territories also benefit females by sheltering them from male harassment (Rubenstein 1994). Within a territory only the dominant male who holds the territory mates (Woodward 1979; Moehlman 1998a), whereas outside a territory an estrous female may attract up to 18 males, who will compete for her and mate often (Woodward 1979; Moehlman 1998b).

7 Management Strategies

Management strategies for feral donkeys vary depending on the country and land ownership where they are found, as well as management goals. In Australia feral donkeys were first described as pests in the 1930s, and eradication programs began in the Kimberly in 1949 using shooting, trapping and poisoning (Bough 2006). The high population growth rate of this species means that they can quickly recover from control efforts: despite 40,000 donkeys being killed in the Kimberly between 1964 and 1965 the population persisted, and although 83,000 were killed in the Northern Territory between 1981 and 1984 the population recovered when eradication efforts stopped (Bough 2006). Due to the remoteness and difficulty of the terrain, in Australia the most effective and practical method to control donkeys is to shoot them from helicopters. A 'Judas' technique has been used since 1994 to make this more efficient when donkeys are at low densities (Woolnough et al. 2012). The Judas technique uses the tendency of feral donkeys to find each other and congregate. Single donkeys are immobilized remotely from a helicopter and fitted with a telemetry collar while recumbent. The collared donkeys are then released to lead shooters to a group of donkeys which are shot, leaving the Judas donkey to find a new group. This method may have reduced the Kimberly population by 500,000 donkeys since the 1970s (Bough 2006).

In the United States feral donkey management on public land currently consists of gathers (round ups) and removals for private adoption, with some use of fertility control. Historically, shooting was used to control feral donkeys on military land and in National Parks, resulting in the eradication of feral donkeys from the Grand Canyon. Aerial surveys to estimate the population size of donkeys on Bureau of Land Management (BLM) Herd Management Areas (HMAs) are carried out regularly (Griffin 2015). The BLM aims to manage feral horses and donkeys within HMAs at a set Appropriate Management Level (AML) which allows for multiple uses of the public land, conducting gathers or applying fertility control to reduce population sizes when they exceed AML. Gathers consist of capturing animals using bait traps or rounding them up with helicopters. Bait traps are small corrals baited with water, salt, or forage. As bait traps rely on the donkeys finding and entering the traps, bait trapping can take some time. Helicopter gathers are more time efficient as a helicopter can quickly find and herd animals to a central corral. Although there are risks to donkeys in both types of gathers, these are less than in other wildlife capture techniques (Scasta 2020). Once captured, donkeys are either administered fertility control and released, or removed to off-range holding facilities where they are made available for public adoption. Adoption rates of donkeys are high, meaning that few remain in long-term holding facilities (BLM 2024).

Fertility control options for on-range management of feral horses currently consist of the immunocontraceptive vaccines GonaCon and different formulations of porcine zona pellucida (PZP), and intra-uterine devices (Bechert et al. 2023). Only results of PZP have been reported for feral donkeys in the literature. Porcine zona pellucida vaccines stimulate antibodies to the zona pellucida, thus preventing sperm binding to the oocyte and subsequent fertilization. Adjuvants are used to increase immune response to the antigen. In horses, three different PZP-based vaccines have been used (ZonaStat, PZP-22, and SpayVac); ZonaStat H was approved by the EPA for use in feral donkeys in 2012 (Bechert and Fraker 2018). All types of PZP require an initial dose, followed by a booster dose 2–4 weeks later to induce one year of infertility. Subsequent annual boosters are generally required to maintain antibody concentrations sufficiently to prevent sperm binding and fertilization. In vast desert landscapes the challenge can be to find donkeys to re-administer the vaccine: even when fitted with telemetry collars, only 67% of females that received an initial PZP dose were located for administration of the annual booster (Gedir et al. 2021). Among 13 female feral donkeys treated with PZP on St. John, US Virgin Islands, none became pregnant in the following year compared to 54% of controls; 46% of the treated females returned to fertility in the next year (Turner et al. 1996). A study was conducted on St Kitts to examine effects of both recombinant (reZP; n = 9) and native porcine (PZP; n = 8) vaccines formulated with Freund's modified complete (primary) and Freund's incomplete (boosters) adjuvants (French et al. 2020). Both groups showed a good antibody response, with only 2 and 1 of the reZP and pZP vaccinated females becoming pregnant respectively, and a delay of conceptions of 100 days observed compared to controls. Side effects of this treatment were injection site abscesses, and some females had prolonged ovarian shutdown after administration (French et al. 2020). A caution of PZP use with feral donkeys is that although pregnancy is prevented the female continues to exhibit estrus cycles and reproductive behaviors (Madosky et al. 2010), which may mean increased female harassment by males due to the social system of this species.

Wild ass species are at such low population sizes that conservation is needed rather than management. In most countries they are protected under local laws, and international trade is controlled under CITES. Poaching was a severe threat to Asiatic wild ass in Mongolia in the 1990s, and caused the extinction of the last remaining native population in Turkmenistan (Kaczensky et al. 2006). The current main threat to wild asses is pasture degradation from increasing numbers of livestock in central Asia, and competition with livestock for pasture and access to water (Kebede et al. 2014; Kaczensky et al. 2020; Tesfai et al. 2023). Human-wildlife conflict can be mitigated with education of local communities: in areas where there was greater knowledge about wild asses there was more support for their presence and less conflict (Esmaeili et al. 2019; Mohammadi et al. 2021). Asiatic wild asses are also threatened with increasing linear infrastructure development in Mongolia and China. This nomadic species needs vast areas in order to find sufficient forage, but development of roads and railways, which are often fenced, form barriers to their movements (Kaczensky et al. 2011; Ito et al. 2013).

8 Common Findings

Feral donkeys are found in large numbers in Australia, South America, and the United States. Their populations can grow quickly, and recover from population control events, as they have low adult mortality and can produce a foal almost every year. The home range size of feral donkeys depends on the resources available in the habitat: ranges are larger when there are fewer or more dispersed resources and smaller when resources are richer or more contained, with donkeys requiring access to water and shelter from the sun or weather. Feral donkeys will consume a variety of vegetation, mostly grasses and forbs but also shrubs and trees, particularly in the winter. The large populations and varied diet mean that donkeys have the potential to impact their habitat in fragile desert ecosystems. If populations get too large they may experience density dependent effects such as higher juvenile mortality. The social behavior of feral donkeys is flexible. Although a fission-fusion system is seen in all populations, some populations are more stable than others, likely also related to habitat. Knowledge of their social system should inform use of fertility control as a management option, and welfare concerns should inform both lethal control and use of gathers.

9 Welfare Concerns

Like feral horses, feral donkeys often fall into a grey area where they are not considered pets or livestock, but also are not considered or managed as wildlife (Hennig et al. 2023). This affects the ethical implications of their management. Wildlife populations are controlled through hunting or through natural cycles of predator-prey interactions, and livestock are controlled through direct management of their movements and reproduction. In Australia there are no predators of donkeys, and there is no evidence that predators are controlling feral donkey populations in north or south America.

Management of feral donkey populations is necessary to both protect the habitat from a species that has a capacity for fast population growth, and to maintain welfare of the feral populations in terms of good body condition. In Australia a feral donkey population was at carrying capacity and experiencing density dependent effects. This manifested as high juvenile mortality, and lower body condition of adult female donkeys as only poor quality vegetation remained (Freeland and Choquenot 1990). Females in a high density population had lower levels of calcium, phosphorus, and potassium in their feces compared to a lower density population, and at high density they consumed clay to help digest their poor quality high fiber diet (Freeland and Choquenot 1990). Feral donkey populations are mostly found in arid habitats where water is not widely available. This both leads to increased impacts of donkeys on the habitat near water, but also means that donkeys are at risk of dehydration during droughts.

While the donkey skin trade has affected domestic donkeys in Africa and Brazil (Gameiro et al. 2021), no information is available on whether feral donkey populations have yet been exploited. Exporting feral donkeys for their skins has been suggested as a potential use for feral donkeys in Australia (Celermajer and Wallach 2019), but there are concerns about the welfare of donkeys prior to slaughter (Goodrum et al. 2022).

10 Rules and Regulations

In Australia feral donkeys are considered a pest due to their fast population growth rates and impacts on the native fauna (Choquenot 1990; Freeland and Choquenot 1990). There is no Australian federal agency to manage the donkeys, but states can use lethal methods to control populations. In the United States feral donkeys are protected under the Wild Free Roaming Horses and Burros Act of 1971 on Herd Management Areas managed by the U.S. Department of Interior's Bureau of Land Management. While humane euthanasia is allowed under the Act, management actions currently consist of rounding up donkeys ("gathers") and removing them from the range for public adoption, or using fertility control. On military lands and in National Parks feral donkeys are not afforded protection under the Act.

11 Conclusion

Donkeys were domesticated from the African wild ass around 5000 years ago. They rapidly became essential to many cultures in the ancient world, and were then transported around the world as a vital part of human infrastructure. After industrialization the need for donkeys and mules as working animals declined, although they maintain their importance in much of the developing world. As donkeys were used less they were released or escaped and survived in the wild, becoming feral (i.e., a domesticated animal that is not owned and is subject to natural selection). Currently large populations of feral donkeys are present in Australia, South America, and the United States, and feral mules have been reported in a few places where both feral donkeys and feral horses occur. Feral donkey populations can have high growth rates, due to high fecundity and low mortality. There is therefore concern about their impact on the habitat, especially near water sources. There is little evidence that feral donkeys compete with native ungulates, as they have different temporal and spatial utilization of the habitat, however as feral donkeys have a wide diet, consuming mostly grass and forbs but also browse species, they can have ecosystem-level habitat effects. The size of feral donkey ranges depend on availability of forage and dispersion and availability of water sources. Management of feral donkeys ranges from aerial culling in Australia to gathers and removals in the United States. Fertility control by remote vaccination is a tool which is available but not currently widely used, and needs to account for feral donkey social structure. As new fertility control modalities emerge future research should focus on both their efficacy in feral donkeys, and any effect on donkey reproductive behavior. Despite their abundance, feral donkeys remain under-studied. There was a suite of research in the 1970s and a few studies in the 2020's, but there is scope for a lot more research into feral donkey behavior and ecology, and interactions with their habitat.

Autoevaluation

Recommended further reading is provided in Sect. 2.

Self assessment:

1. How many species of wild equid currently exist and where are they found?
   There are 7 species of wild equid: Przewalski's horse (Equus przewalskii) found in Mongolia and China, Plains zebra (E. quagga) found through much of Sub-Saharan Africa, Grevy's zebra (E. grevyi) found in Kenya and Ethiopia, Mountain zebra (E. zebra) found in South Africa and Namibia, African wild ass (E. africanus), Asiatic wild ass (E. hemionus) and the Tibetan wild ass (E. kiang). It is important to note that these are phylogenetically different species, with more phenotypic and genotypic differences between them than breeds of domestic animals.

   The domestic horse (E. caballus) and domestic donkey (E. asinus) are not wild equids, but are found in feral populations.
2. What are similarities and differences between the donkey and African wild ass?
   African wild ass stand taller than East African donkeys. African wild ass always have a grey/brown body and distinctive striping patterns on their legs that can be used for individual identification, whereas domestic donkeys can have a variety of colors and markings depending on the breed. There are skeletal differences with African wild asses having better bone robusticity and more circular diaphysis of the humeri. Both species are adapted to hot and dry conditions.
3. What are some of the reasons why feral donkeys are abundant but most wild asses are threatened with extinction?
   Feral donkeys have high fecundity, producing a foal almost every year, and low mortality. In many areas where they are abundant donkeys have few or no predators and little competition for resources. Wild ass species have lower fecundity, starting reproduction at a later age and producing fewer foals. Wild asses live with predators across most of their range, and are often in competition with livestock for forage and water. While donkeys are often found in areas used by humans and their livestock, wild asses will avoid them. Most wild ass species are protected in their range countries, but poaching still occurs.
4. Do mules occur naturally?
   Yes. They are found in areas where both feral horses and feral donkeys coexist.

References

Abella SR (2008) A systematic review of wild burro grazing effects on Mojave Desert vegetation, USA. Environ Manag 41:809–819

Acebes P, Traba J, Malo JE (2012) Co-occurrence and potential for competition between wild and domestic large herbivores in a south American desert. J Arid Environ 77:39–44

Acebes P, Malo JE, Ovejero R, Traba J (2016) Sympatric guanacos and livestock share water resources in drylands of Argentina. Mammalia 80

Australian Government Department of Sustainability, Environment, Water, Population and Communities, 2011. Feral horse (*Equus caballus*) and feral donkey (*Equus asinus*), p. 4

Ayo JO, Dzenda T, Olaifa F, Ake SA, Sani I (2014) Diurnal and seasonal fluctuations in rectal temperature, respiration and heart rate of pack donkeys in a tropical savannah zone. J Equine Sci 25:1–6

Bechert U, Fraker M (2018) Twenty years of SpayVac® research: potential implications for regulating feral horse and burro populations in the United States. Human-Wildiife Interactions 12:117–130

Bechert U, Turner JW Jr, Baker D, Eckery D, Bruemmer J, Lyman CC, Prado TM, King SRB, Fraker M (2023) Fertility control options for management of free-roaming horse populations. Human-Wildiife Interactions 16:179–216

Beja-Pereira A, England PR, Ferrand N, Jordan S, Bakhiet AO, Abdalla MA, Mashkour M, Jordana J, Taberlet P, Luikart G (2004) African origins of the domestic donkey. Science 304: 1781

Bennett EA, Weber J, Bendhafer W, Champlot S, Peters J, Schwartz GM, Grange T, Geigl EM (2022) The genetic identity of the earliest human-made hybrid animals, the kungas of Syro-Mesopotamia. Sci Adv 8:eabm0218

Berry KH, Yee JL, Lyren LM (2020) Feral burros and other influences on desert tortoise present in the Western Sonoran Desert. Herpetologica 76:403–413

Bhatnagar YV, Wangchuk R, Prins HH, Van Wieren SE, Mishra C (2006) Perceived conflicts between pastoralism and conservation of the kiang Equus kiang in the Ladakh trans-Himalaya, India. Environ Manage 38:934–941

Borgnia M, Vilá BL, Cassini MH (2008) Interaction between wild camelids and livestock in an Andean semi-desert. J Arid Environ 72:2150–2158

Bough J (2006) From value to vermin: a history of the donkey in Australia. Aust Zool 33:388–397

Burden FA, Bell N (2019) Donkey nutrition and malnutrition. Vet Clin North Am Equine Pract 35: 469–479

Bureau of Land Management. (2024). Program Data. https://www.blm.gov/programs/wild-horse-and-burro/about-the-program/program-data. Accessed 3 Mar 2025/

Carothers SW (1977) Biology and ecology of feral burros (*Equus asinus*) at grand canyon National Park. Final Research Report for the National Park Service, Arizona, p 171

Carrion V, Donlan CJ, Campbell K, Lavoie C, Cruz F (2006) Feral donkey (Equus asinus) eradications in the Galápagos. Biodivers Conserv 16:437–445

Celermajer D, Wallach A (2019) The fate of the illegible animal: the case of the Australian wild donkey. Anim Stud J 8:229–258

Choquenot D (1990) Rate of increase for populations of feral donkeys in Northern Australia. J Mammal 71:151–155

Choquenot D (1991) Density-dependent growth, body condition, and demography in feral donkeys testing the food hypothesis. Ecology 72:805–813

Clancy CL, Kubasiewicz LM, Raw Z, Cooke F (2021) Science and knowledge of free-roaming donkeys—a critical review. J Wildl Manag 85:1200–1213

Cordingley JE, Sundaresan SR, Fischhoff IR, Shapiro B, Ruskey J, Rubenstein DI (2009) Is the endangered Grevy's zebra threatened by hybridization? Anim Conserv 12:505–513

Couto M, Santos AS, Laborda J, Nóvoa M, Ferreira LM, Madeira de Carvalho LM (2016) Grazing behaviour of Miranda donkeys in a natural mountain pasture and parasitic level changes. Livest Sci 186:16–21
Couvreur M, Cosyns E, Hermy M, Hoffman M (2005) Complementarity of epi- and endozoochory of plant seeds by free ranging donkeys. Ecography 28:37–48
Dalton DL, Zimmermann D, Mnisi C, Taplin M, Novellie P, Hrabar H, Kotzé A (2017) Hiding in plain sight: evidence of hybridization between Cape Mountain Zebra (Equus zebra zebra) and plains zebra (Equus quagga burchelli). Afr J Wildl Res 47
Dias DM, Massara RL, Bocchiglieri A (2019) Use of habitats by donkeys and cattle within a protected area of the Caatinga dry forest biome in northeastern Brazil. Perspect Ecol Conserv 17:64–70
Esmaeili S, Hemami MR, Goheen JR (2019) Human dimensions of wildlife conservation in Iran: assessment of human-wildlife conflict in restoring a wide-ranging endangered species. PLoS One 14:e0220702
Esmaeili S, King SRB, Schoenecker KA (2023) Browsers or Grazers? new insights into feral burro diet using a non-invasive sampling and plant DNA metabarcoding approach. Animals Basel 13
Esmaeili S, Schoenecker KA, King SRB (2024) Resource availability and heterogeneity affect space use and resource selection of a feral ungulate. Ecosphere 15
Fages A, Hanghøj K, Khan N et al (2019) Tracking five millennia of horse management with extensive ancient genome time series. Cell 177:1419–1435.e31.
Feh C, Munkhtuya B, Enkhbold S, Sukhbaatar T (2001) Ecology and social structure of the Gobi khulan *Equus hemionus* subsp. in the Gobi B National Park, Mongolia. Biol Conserv 101:51–61
Freeland WJ, Choquenot D (1990) Determinants of herbivore carrying capacity plants, nutrients, and *Equus asinus* in Northern Australia. Ecology 71:589–597
French H, Peterson E, Schulman M, Roth R, Crampton M, Conan A, Marchi S, Knobel D, Bertschinger H (2020) Efficacy and safety of native and recombinant zona pellucida immunocontraceptive vaccines in donkeys. Theriogenology 153:27–33
Gameiro MBP, Rezende VT, Zanella AJ (2021) Brazilian donkey slaughter and exports from 2002 to 2019. Braz J Vet Res Anim Sci 58:e174697
Gedir JV, Cain JW, Lubow BC, Karish T, Delaney DK, Roemer GW (2021) Estimating abundance and simulating fertility control in a feral burro population. J Wildl Manage 85:1187–1199
Ghasemi A, Hemami M-R, Karimi S, Iravani M, Senn J (2024) Potential seed dispersal by Persian Wild Ass in South Central Iran. Rangel Ecol Manag 92:73–79
Giotto N, Gerard JF, Ziv A, Bouskila A, Bar-David S (2015) Space-use patterns of the Asiatic wild ass (Equus hemionus): complementary insights from displacement, recursion movement and habitat selection analyses. PLoS One 10:e0143279
Goodrum F, Theuri S, Mutua E, Carder G (2022) The donkey skin trade: challenges and opportunities for policy change. Global Pol 13:304–309
Griffin P (2015) Estimated abundance of wild burros surveyed on Bureau of Land Management Lands in 2014: U.S. Geological Survey Open-File Report 2015–1084, p. 42
Grinder MI, Krausman PR, Hoffmann RS (2006) Equus asinus. Mamm Species 794:1–9
Haddy E, Burden F, Proops L (2020) Shelter seeking behaviour of healthy donkeys and mules in a hot climate. Appl Anim Behav Sci 222:104898
Hamrick RG, Pirgalioglu T, Gunduz S, Carroll JP (2005) Feral donkey Equus asinus populations on the Karpaz peninsula, Cyprus. Eur J Wildl Res 51:108–116
Hanley TA, Brady WW (1977) Seasonal fluctuations in nutrient content of feral burro forages, lower Colorado River Valley, Arizona. J Range Manage 30:370–373
Hennig JD, Duchardt CJ, Esmaeili S, Fuhlendorf SD, Beck JL, Francisco TI, Scasta JD (2023) A crossroads in the rearview mirror: the state of United States feral equid management in 2023. Bioscience 73:404–407
Hoffmann R (1983) Social organization patterns of several Feral horse and Feral ass populations in Central Australia. Zeitschrift für Säugetierkunde 48:124–126

Ito TY, Lhagvasuren B, Tsunekawa A, Shinoda M, Takatsuki S, Buuveibaatar B, Chimeddorj B (2013) Fragmentation of the habitat of wild ungulates by anthropogenic barriers in Mongolia. PLoS One 8:e56995

Johnson RA, Carothers SW, McGill TJ (1987) Demography of feral burros in the mohave desert. J Wildl Manage 51:916

Jones NL, Houpt KA, Houpt TR (1989) Stimuli of thirst in donkeys (Equus asinus). Physiol Behav 46:661–665

Kaczensky P (2023) Status and conservation of the Asiatic Wild Ass across its range. Technical Workshop for the Asiatic Wild Ass (Khulan). UNEP/CMS/AWATW/Doc.3

Kaczensky P, Sheehy DP, Walzer C, Johnson DE, Lhkagvasuren D, Sheehy CM (2006) Room to roam? The threat to khulan (wild ass) from human intrusion, Mongolia discussion papers, East Asia and Pacific environment and social development department. World Bank, Washington D. C

Kaczensky P, Ganbaatar O, von Wehrden H, Walzer C (2008) Resource selection by sympatric wild equids in the Mongolian Gobi. J Appl Ecol 45:1762–1769

Kaczensky P, Dresley V, Vetter D, Otgonbayar H, Walzer C (2010) Water use of Asiatic wild asses in the Mongolian Gobi. Erforsch biol Ress Mongolei (Halle/Saale) 11:291–298

Kaczensky P, Kuehn R, Lhagvasuren B, Pietsch S, Yang W, Walzer C (2011) Connectivity of the Asiatic wild ass population in the Mongolian Gobi. Biol Conserv 144:920–929

Kaczensky P, Khaliun S, Payne J, Boldgiv B, Buuveibaatar B, Walzer C (2019) Through the eye of a Gobi khulan—application of camera collars for ecological research of far-ranging species in remote and highly variable ecosystems. PLoS One 14:e0217772

Kaczensky P, Buuveibaatar B, Payne JC, Strindberg S, Walzer C, Batsaikhan N, Bolortsetseg S, Victurine R, Olson KA (2020) A conservation strategy for khulan in Mongolia: background and key considerations. NINA Report 1889:72

Kannan PM, Parsons MH (2017) Social class and group size as predictors of behavior in male Equus kiang. Animal Behavior and Cognition 4:442–454

Karish T, Roemer GW, Delaney DK, Reddell CD, Cain JW (2023) Habitat selection and water dependency of feral burros in the Mojave Desert, California, USA. J Wildl Manage 87

Kebede F, Moehlman P, Bekele A, Evangelista PH (2014) Predicting seasonal habitat suitability for the critically endangered African wild ass in the Danakil, Ethiopia. Afr J Ecol 52:533–542

King SRB, Schoenecker KA (2025) Effects of phylogeny and habitat on social structure and behavior of two equid species. Behav Ecol Sociobiol 79(7). https://doi.org/10.1007/s00265-025-03602-7

Kimura B, Marshall FB, Chen S, Rosenbom S, Moehlman PD, Tuross N, Sabin RC, Peters J, Barich B, Yohannes H, Kebede F, Teclai R, Beja-Pereira A, Mulligan CJ (2011) Ancient DNA from Nubian and Somali wild ass provides insights into donkey ancestry and domestication. Proc Biol Sci 278:50–57

Kimura B, Marshall F, Beja-Pereira A, Mulligan C (2013) Donkey domestication. Afr Archaeol Rev 30:83–95

Klingel H (1998) Observations on social organization and behaviour of African and Asiatic Wild Asses (*Equus africanus* and *Equus hemionus*). Appl Anim Behav Sci 60:103–113

Kozlowski CP, Clawitter HL, Thier T, Fischer MT, Asa CS (2018) Characterization of estrous cycles and pregnancy in Somali wild asses (Equus africanus somaliensis) through fecal hormone analyses. Zoo Biol 37:35–39

Lepetz S, Clavel B, Alioğlu D, Chauvey L, Schiavinato S, Tonasso-Calvière L, Liu X, Fages A, Khan N, Seguin-Orlando A, Der Sarkissian C, Clavel P, Estrada O, Gaunitz C, Aury J-M, Barme M, Boulbes N, Bourgois A, Decanter F, Foucras S, Frère S, Gardeisen A, Jouanin G, Méla C, Morand N, Nieto Espinet A, Perdereau A, Putelat O, Rivière J, Robin O, Salin M, Valenzuela-Lamas S, Vallet C, Yvinec J-H, Wincker P, Orlando L (2021) Historical management of equine resources in France from the Iron Age to the Modern Period. J Archaeol Sci Rep 40:103250

Lundgren EJ, Ramp D, Stromberg JC, Wu J, Nieto NC, Sluk M, Moeller KT, Wallach AD (2021) Equids engineer desert water availability. Science 372:491–495

Madosky JM, Rubenstein DI, Howard JJ, Stuska S (2010) The effects of immunocontraception on harem fidelity in a feral horse (Equus caballus) population. Appl Anim Behav Sci 128:50–56

Malo JE, Acebes P, Giannoni SM, Traba J (2011) Feral livestock threatens landscapes dominated by columnar cacti. Acta Oecol 37:249–255

Malo JE, González BA, Mata C, Vielma A, Donoso DS, Fuentes N, Estades CF (2016) Low habitat overlap at landscape scale between wild camelids and feral donkeys in the Chilean desert. Acta Oecol 70:1–9

Maloiy GMO (1970) Water economy of the Somali donkey. Am J Physiol 219:1522–1527

Maloiy GMO, Boarer CDH (1971) Response of the Somali donkey to dehydration: hematological changes. Am J Physiol 221:37–41

Marshal JP, Bleich VC, Andrew NG (2008) Evidence for interspecific competition between feral ass *Equus asinus* and mountain sheep *Ovis canadensis* in a desert environment. Wildl Biol 14: 228–236

Marshal JP, Bleich VC, Krausman PR, Reed M-L, Neibergs A (2012) Overlap in diet and habitat between the Mule Deer (Odocoileus hemionus) and Feral Ass (Equus asinus) in the Sonoran Desert. Southwest Nat 57:16–25

Marshall F (2007) African pastoral perspectives on domestication of the donkey: a first synthesis. In: Denham T, Iriarte J, Vrydaghs L (eds) Rethinking agriculture: archeological and Ethnoarcheological perspectives. Left Coast Press, Walnut Creek, pp 371–407

Marshall F, Weissbrod L (2011) Domestication processes and morphological change. Curr Anthropol 52:S397–S413

Marshall FB, Dobney K, Denham T, Capriles JM (2014) Evaluating the roles of directed breeding and gene flow in animal domestication. Proc Natl Acad Sci USA 111:6153–6158

McCool CJ, Pollitt CC, Fallon GR, Turner AF (1981) Studies of feral donkeys in the Victoria River-Kimberleys area. Obdervations on behavior, reproduction and habitat and some possible control strategies. Aust Vet J 57:444–449

McCort WD (1980) The Behaviour and Social Organization of Feral Asses (Equus asinus) on Ossabaw Island, Georgia. The Pennsylvania State University.

Moehlman P (1974) Behavior and ecology of feral asses (*Equus asinus*). University of Wisconsin, p 277

Moehlman P (1998a) Behavioural patterns and communication in feral asses (*Equus africanus*). Appl Anim Behav Sci 60:125–169

Moehlman P (1998b) Feral asses (*Equus africanus*): intraspecific variation in social organization in arid and mesic habitats. Appl Anim Behav Sci 60:171–195

Moehlman P, Fowler A, Roe JH (1998) Feral asses (*Equus africanus*) of volcano Alcedo, Galapagos: behavioural ecology, spatial distribution, and social organization. Appl Anim Behav Sci 60:197–210

Moehlman PD, Kebede F, Yohannes H (2015) Equus africanus. The IUCN red list of threatened species 2015: e.T7949A45170994. https://doi.org/10.2305/IUCN.UK.2015-2.RLTS. T7949A45170994.en. Accessed on 3 Mar 2025

Mohammadi A, Almasieh K, Wan HY, Nayeri D, Alambeigi A, Ransom JI, Cushman SA (2021) Integrating spatial analysis and questionnaire survey to better understand human-onager conflict in Southern Iran. Sci Rep 11:12423

Muller H, Bourne A (2018) Minimum population size and potential impact of feral and semi-feral donkeys and horses in an arid rangeland. Afr Zool 53:139–144

Nandintsetseg D, Kaczensky P, Ganbaatar O, Leimgruber P, Mueller T (2016) Spatiotemporal habitat dynamics of ungulates in unpredictable environments: the khulan (Equus hemionus) in the Mongolian Gobi desert as a case study. Biol Conserv 204:313–321

Norment CJ, Douglas CL (1977) Ecological studies of feral burros in Death Valley. Cooperative National Park Resources Studies Unit. University of Nevada, Las Vegas, p 141

Nowzari H, Hemami M, Karami M, Kheirkhah Zarkesh MM, Riazi B, Rubenstein DI (2013) Habitat use by the Persian onager, Equus hemionus onager (Perissodactyla: Equidae) in Qatrouyeh National Park, Fars, Iran. J Nat Hist 47:2795–2814

Ovejero RJA, Acebes P, Malo JE, Traba J, Mosca Torres ME, Borghi CE (2011) Lack of feral livestock interference with native guanaco during the dry season in a South American desert. Eur J Wildl Res 57:1007–1015

Payne JC, Buuveibaatar B, Bowler DE, Olson KA, Walzer C, Kaczensky P (2020) Hidden treasure of the Gobi: understanding how water limits range use of khulan in the Mongolian Gobi. Sci Rep 10:2989

Peco B, Borghi CE, Malo JE, Acebes P, Almirón M, Campos CM (2011) Effects of bark damage by feral herbivores on columnar cactus Echinopsis (=Trichocereus) terscheckii reproductive output. J Arid Environ 75:981–985

Perryman P, Muchlinski A (1987) Population dynamics of feral burros at the Naval Weapons Center, China Lake, California. J Mammal 68:435–438

Queiroz MR, Gameiro MBP, Zanella AJ (2021) The population of donkeys and mules in Brazil according to agricultural censuses from 1960 to 2017. Braz J Vet Res Anim Sci 58:e174365

Reus ML, Cappa FM, Andino N, Campos VE, de los Ríos C, Campos CM (2014) Trophic interactions between the native guanaco (*Lama guanicoe*) and the exotic donkey (*Equus asinus*) ni the hyper-arid Monte desert (Ischigualasto Park, Argentina). Stud Neotrop Fauna Environ 49: 159–168

Rosenbom S, Costa V, Al-Araimi N, Kefena E, Abdel-Moneim AS, Abdalla MA, Bakhiet A, Beja-Pereira A (2015) Genetic diversity of donkey populations from the putative centers of domestication. Anim Genet 46:30–36

Rossel S, Marshall F, Peters J, Pilgram T, Adams MD, O'Connor D (2008) Domestication of the donkey: timing, processes, and indicators. Proc Natl Acad Sci USA 105:3715–3720

Rubenstein D (1994) The ecology of female social behaviour in horses, zebras and asses, Horses, Zebras and Asses, pp. 13–28

Rubenstein DI (2010) Ecology, Social Behavior, and Conservation in Zebras, Behavioral ecology of tropical animals, 231–258

Rubin ES, Conrad D, Harding LE, Russo BM (2024) Associations between a feral equid and the Sonoran Desert ecosystem. Wildl Monogr 215

Rudman R (1998) The social organisation of feral donkeys (*Equus asinus*) on a small Caribbean Island (St. John, US Virgin Islands). Appl Anim Behav Sci 60:211–228

Ruffner GA, Carothers SW (1982) Age structure, condition and reproduction of two Equus asinus (Equidae) populations from Grand Canyon National Park, Arizona. Southwest Nat 27:403

Sánchez de la Vega G, Godínez-Alvarez H (2010) Effect of gut passage and dung on seed germination and seedling growth: donkeys and a multipurpose mesquite from a Mexican inter-tropical desert. J Arid Environ 74:521–524

Santiapillai C, Wijeyamohan S, Ashby KR (1999) The ecology of a free-living population of the ass (Equus africanus) at Kalpitiya, Sri Lanka. Biol Conserv 91:43–53

Scasta JD (2020) Mortality and operational attributes relative to feral horse and burro capture techniques based on publicly available data from 2010–2019. J Equine Vet Sci 86:102893

Schoenecker KA, King SRB, Messmer TA (2021) The wildlife profession's duty in achieving science-based sustainable management of free-roaming equids. J Wildl Manag 85:1057–1061

Seegmiller RF, Ohmart RD (1981) Ecological relationships of feral burros and desert bighorn sheep. Wildl Monogr 78:3–58

Shackelford L, Marshall F, Peters J (2013) Identifying donkey domestication through changes in cross-sectional geometry of long bones. J Archaeol Sci 40:4170–4179

Shah N, Qureshi Q (2007) Social organization and determinants of spatial distribution of Khur (*Equus hemionus khur*). Erforsch. biol. Ress. Mongolei (Halle/Saale) 10:189–200

Šimek, J. (2024): Przewalski's horse (Equus przewalskii, Poljakov 1881) International Studbook 2024, current until Feb 29, 2024. Prague Zoo, Czechia

Soultan A, Darwish M, Al-Johani N, Abdulkareem A, Alfaifi Y, Assaeed AM, El-Bana M, Browne S (2023) Feral donkey distribution and ecological impacts in a hyper-arid region. Animals 13

Tesfai RT, Parrini F, Owen-Smith N, Moehlman PD (2021a) African wild ass drinking behaviour on the Messir Plateau, Danakil Desert, Eritrea. J Arid Environ 185:104327

Tesfai RT, Parrini F, Owen-Smith N, Moehlman PD, Moore B (2021b) How spatial and dietary overlap with domestic livestock affect African wild ass nutrition on the Messir Plateau (Eritrea). J Mammal 102:1174–1185

Tesfai RT, Parrini F, Moehlman PD, Young NE, Evangelista PH (2023) Predicting suitable habitat for the critically endangered African wild ass Equus africanus in the Danakil Desert of Eritrea. Oryx 57:1–8

Todd ET, Tonasso-Calvière L, Chauvey L, Schiavinato S, Fages A, Seguin-Orlando A, Clavel P, Orlando L (2022) The genomic history and global expansion of domestic donkeys. Science 377: 1172–1180

Treitler JT, Drissen T, Stadtmann R, Zerbe S, Mantilla-Contreras J (2017) Complementing endozoochorous seed dispersal patterns by donkeys and goats in a semi-natural Island ecosystem. BMC Ecol 17:42

Turner JW, Liu IKM, Kirkpatrick JF (1996) Remotely delivered immunocontraceprion in free-roaming feral burros (*Equus asinus*). J Reprod Fertil 107(Suppl):31–35

Uiseb K (2024) Are mountain and plains zebra hybridising in north-west Namibia? In: Sullivan S, Dieckmann U, Lendelvo S (eds) Etosha Pan to the Skeleton Coast. Conservation histories, policies, and practices in north-West Namibia. Open Book Publishers, pp 271–287

Vazeed Pasha S, Satish KV, Sudhakar Reddy C, Jha CS (2015) Massive invasion of Mesquite (Prosopis juliflora) in Wild Ass Wildlife Sanctuary, India. Nat Acad Sci Lett 38:271–273

Wolfe ML, Ellis LC, MacMullen R (1989) Reproductive rates of feral horses and burros. J Wildl Manage 53:916

Woodward S (1979) The social system of feral asses (*Equus asinus*). Z Tierpsychol 49:304–316

Woodward SL, Omhart RD (1976) Habitat use and fecal analysis of feral burros (*Equus asinus*), Chemehuevi Mountains, California, 1974. J Range Manag 29:482–485

Woolnough AP, Hampton JO, Campbell S, Lethbridge MR, Boardman WSJ, Sharp T, Rose K (2012) Field immobilization of feral 'Judas' donkeys (*Equus asinus*) by remote injection of medetomidine and ketamine. J Wildl Dis 48:435–443

Part B

Roles and Industry

Working Donkey and Mule Medicine

Jay Merriam

Abstract

Working donkeys and mules represent an indispensable yet underappreciated labor force, integral to the functioning of many societies, particularly in the developing world. These animals perform critical tasks under conditions that would challenge both human and other animal laborers, carrying heavy loads and navigating arduous terrain. Despite their contributions, their role in sustaining human livelihoods remains largely invisible, absent from mainstream discourse, media coverage, and global development priorities. Veterinary care is limited to nonexistence, similar to health care for the people working with these animals. When care is many challenges from the efficacy of medications are available, legislation preventing certain medications from entering the country to training professionals to treat many of the same diseases caused by poor body condition scores and ill-fitting harness. This chapter will explore challenges and solutions for working abroad as a veterinarian with limited supplies, many bureaucratic, cultural practices and challenges.

Keywords

Working equid · Beast of burden · Working donkey · Working mule · Harness · Packing

J. Merriam (✉)
Equitarian Initiative and American Fonduk, Fez, Morocco

A. K. McLean et al. (eds.), *Donkey and Mule Medicine: Foundations, Breeds, and Management*, https://doi.org/10.1007/978-3-032-12726-6_7

1 Introduction

Historical evidence underscores the longstanding significance of working equids, especially donkeys and mules. Representations appear in the archaeological record of ancient Egypt, classical Greece, and Rome. They are deeply embedded in the cultural and agricultural histories of Europe and Asia. In the Americas, their introduction catalyzed agricultural expansion, particularly in cotton production, and they remain critical to subsistence economies in many rural regions (The Donkey Sanctuary 2025; Brooke 2025a, b, c, d).

From carrying supplies to isolated mountain villages in the Atlas Mountains of Morocco to transporting food and goods in post-earthquake Haiti, these equids serve as vital links in supply chains, especially during crises. Their work underpins economic stability, providing essential transport for agriculture, commerce, and aid. They are central to disaster response and recovery efforts, contributing to human survival and resilience in regions with limited infrastructure (SPANA 2023a, b, c; ICWE 2023).

However, working donkeys and mules face profound welfare challenges, including overwork, malnutrition, disease, and limited access to veterinary care. Traditional care practices, such as nostril splitting and invasive firing, persist in some regions, often causing further harm. Veterinary interventions are frequently unavailable or reliant on sporadic "parachute" care provided by external organizations. Even where local veterinary systems exist, economic constraints often preclude routine treatment for working equids (Burn et al. 2022; Pritchard et al. 2005).

Emerging approaches are addressing these challenges. Sustainable healthcare models prioritize local leadership and community engagement, integrating modern veterinary practices with traditional knowledge. Programs focused on training indigenous veterinarians and caregivers have demonstrated success in improving equid health while fostering local capacity for long-term care (Kay 2015). Institutions like the American Fondouk in Morocco and SPANA in Egypt have pioneered hospital-based research, producing significant advancements in equid health management, such as improved treatment protocols for tetanus and parasitic diseases (SPANA 2023a, b, c; Madigan 2019). The integration of telemedicine and digital tools has further extended the reach of veterinary care to remote regions (Burn et al. 2022).

These initiatives highlight the critical role of working donkeys and mules not only as laborers but as enablers of community resilience and economic stability. By advancing veterinary care, promoting sustainable practices, and addressing the welfare needs of these animals, we can ensure their continued contribution to human livelihoods while aligning with broader One Health principles. Working equids are not merely beasts of burden; they are pivotal partners in human survival and development.

2 Challenges and Advances in Veterinary Medicine for Working Donkeys and Mules

Veterinary practice involving working equids, particularly donkeys and mules, presents unique challenges that stem from a combination of medical, nutritional, and socio-economic factors. These animals often operate in regions with limited access to veterinary care, harsh working conditions, and a lack of understanding regarding their proper management. Common welfare issues include malnutrition, dehydration, parasitic infections, lameness, and integumentary trauma caused by poor harnessing and handling practices (Brooke 2025a, b, c, d; Watson et al. 2023). In many cases, traditional treatment methods and harmful practices persist, exacerbating health problems and delaying proper care (Gichure and Olayide 2024). Furthermore, the absence of trained veterinary professionals and the economic constraints faced by owners contribute to inadequate or delayed treatment, reinforcing a cycle of poor welfare and limited resilience among equid populations (Watson et al. 2023).

2.1 Epidemiological and Socioeconomic Context

Working donkeys and mules often operate in rural, economically disadvantaged regions where they are crucial for agriculture, transportation, and tourism. These animals support livelihoods by transporting goods, water, and people, often in areas with limited infrastructure and few alternatives (Brooke 2025a, b, c, d; Village Earth 2025). In many such regions, their health and welfare are severely compromised by inadequate nutrition, parasitism, and improper handling, which are exacerbated by limited access to veterinary services and owner education (El Refugio del Burrito 2017) (Fig. 1).

Fig. 1 Donkey used for transport in mobility limited persons. This man rode 4 hours with his grandson to receive care for his donkeys' back

Shifts in local economies, such as the transition from agriculture to tourism, bring both opportunities and challenges. For example, as tourism increases, there is greater demand for animals with improved physical condition and temperament (Brooke 2025a, b, c, d). However, this also introduces the risk of overworking animals unsuited for these roles, especially when welfare regulations are weak or unenforced. In places like Mijas, Spain, efforts have been made to improve welfare standards for donkeys used in tourism, including weight limits and veterinary oversight, but challenges persist in ensuring consistent and humane treatment (El Refugio del Burrito 2017).

2.2 Key Medical Challenges in Working Equids

1. **Parasitic Infestations**
   Heavy burdens of external and internal parasites are common in working equids, with severe impacts on their health and productivity. Tick infestations are especially problematic in tropical and subtropical regions, where donkeys and mules often carry hundreds of ticks, leading to significant anemia, skin irritation, and secondary infections (Knottenbelt 2005). These infestations are particularly prevalent in areas with tall grass or wooded terrain, where equids are frequently exposed to parasitic vectors. While tick-borne diseases are less frequently reported in donkeys and mules compared to other livestock, severe infestations exacerbate overall morbidity, especially when compounded by poor nutrition and limited veterinary intervention (The Donkey Sanctuary 2020a, b, c).
   - **Intervention Strategies**: The introduction of antiparasitic medications, such as ivermectin and pyrantel, along with education on tick management, has shown notable success in improving the health and productivity of working equids (Purdy 2005; Elsheikha 2019). These drugs are widely used to treat gastrointestinal parasites and lungworms, which are common in donkeys and mules, especially in tropical and subtropical regions. However, careful dosing and monitoring are essential to prevent the development of anthelmintic resistance, which is an emerging concern in equid populations (Elsheikha 2019).

   Community-based approaches to parasite control, including the use of locally available acaricides and pasture management techniques, have helped sustain these improvements. Strategies such as rotational grazing, manure removal, and targeted deworming have proven effective in reducing parasite loads while promoting long-term herd health (Purdy 2005). These interventions are most successful when combined with owner education and locally adapted veterinary support, ensuring that treatment protocols are both practical and sustainable.
2. **Nutritional Deficiencies**
   Malnutrition is a widespread problem, particularly during periods of drought or economic hardship. In regions affected by climate shocks, such as East Africa, working donkeys have been identified as critical agents of resilience, yet their nutritional needs are often overlooked in emergency planning (Proops et al.

2024). Donkeys and mules are often less affected by malnutrition compared to horses due to their efficient metabolism and ability to thrive on low-quality forage, but prolonged nutrient deficits still compromise their health and working capacity (Burden and Bell 2019; World Horse Welfare 2025). Deficiencies in essential vitamins and minerals, especially during dry seasons, can lead to weight loss, reduced stamina, and increased susceptibility to disease, ultimately undermining their ability to support human livelihoods.

- **Intervention Strategies**: The introduction of rapidly growing tropical grasses in a "chop and carry" system has proven effective in addressing malnutrition in working equids, particularly in regions where grazing is limited or seasonal. Species such as Napier grass (Pennisetum purpureum) and Brachiaria have been successfully cultivated and used to improve forage availability and nutritional intake (Pearson 2005). These systems allow owners to harvest and deliver fresh forage directly to animals, reducing reliance on poor-quality grazing and minimizing exposure to parasites in contaminated pastures.

  Feeding improvements have also reduced the prevalence of saddle sores and internal parasites by enhancing body condition and resilience. Better nutrition leads to increased muscle mass and fat reserves, which cushion pressure points and reduce friction from harnesses and saddles (World Horse Welfare 2025; Smith and Burden 2016). Additionally, well-nourished equids show improved immune responses, making them less susceptible to parasitic infections and more responsive to treatment protocols.

3. **Musculoskeletal and Harness-Related Injuries**

   Improper use of harnesses and excessive workloads often result in chronic injuries, including saddle sores, pressure wounds, and musculoskeletal issues. These injuries are frequently observed in working donkeys and mules due to inadequate tack design, poor fit, and prolonged load-bearing without rest (Cousquer 2011; Mitiku et al. 2022). Animals with a body condition score (BCS) of 2 or less are especially vulnerable, as reduced fat and muscle mass offer little protection against friction and pressure from harnesses (Smith and Burden 2016). In regions like Ethiopia and Morocco, studies have shown that poorly maintained or improvised saddles contribute significantly to lesion formation, particularly on the withers, ribs, and pelvic bones (Mitiku et al. 2022). Addressing these issues requires not only improved equipment but also education for owners on proper harnessing techniques and the importance of maintaining healthy body condition (Fig. 2).

   - **Intervention Strategies**: Community education on proper harnessing and the provision of affordable, well-fitted saddles and saddle pads have significantly reduced injuries such as saddle sores, pressure wounds, and musculoskeletal strain. Programs led by organizations like Brooke and SaddleAid have demonstrated that training owners in harnessing techniques and distributing welfare-friendly tack can dramatically improve equid comfort and performance (Brooke 2025a, b, c, d; SPANA 2025). Locally produced saddle blankets and pads—often made from accessible materials like jute, felt, or recycled textiles—have further improved outcomes by reducing friction and

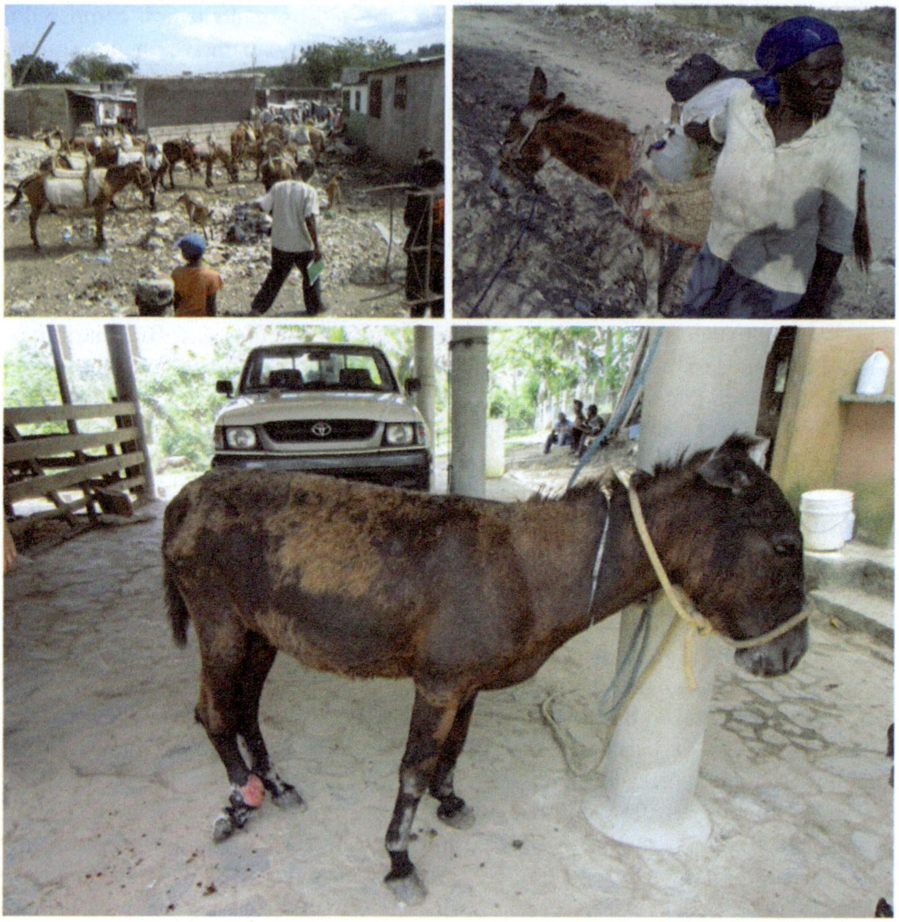

Fig. 2 Mule trails and derived injuries of overwork on difficult and harsh terrains in Haiti

distributing weight more evenly (Pearson et al. 2003). These interventions are most effective when paired with nutritional remedies, as animals with low body condition scores are more susceptible to harness-related injuries due to lack of muscle and fat padding (Smith and Burden 2016).

4. **Hernias**
   Hernias, particularly scrotal and lateral abdominal wall hernias, are a common concern in working donkeys and mules. These injuries are often caused by overloading young or underdeveloped animals, especially before their musculo-skeletal systems have fully matured (Marzok and Moustafa 2021). Acquired inguinal and scrotal herniation has been documented in working donkeys subjected to prolonged physical stress, with clinical signs including swelling, discomfort, and reduced performance. In many cases, surgical intervention—

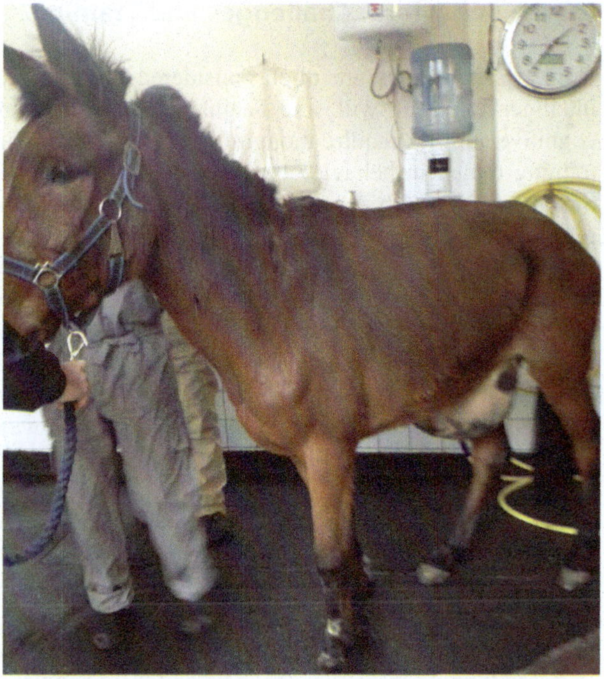

Fig. 3 Hernia case in a mule. Several attempts to repair vis application of hot irons have left the pigmented scars whilst actually increasing the volume of the hernia. Massive surgical intervention reduced it enough for him to return to work

such as herniorrhaphy with mesh reinforcement—is required to restore function and prevent recurrence (Amare and Haben 2020). These conditions highlight the importance of age-appropriate workload management and early detection through routine veterinary screening (Fig. 3).

- **Case Study**: Scrotal hernias are frequently observed in young equids worked too soon, before their musculoskeletal systems are fully developed. In one documented case involving working donkeys, surgical correction of a scrotal hernia with a lobulated testicular mass was successfully performed using herniorrhaphy and mesh reinforcement, allowing the animal to return to work post-recovery (Marzok and Moustafa 2021). These procedures typically involve unilateral orchiectomy and closure of the superficial inguinal ring using polypropylene mesh, with recovery periods averaging 2–3 weeks.

  Conversely, lateral abdominal wall hernias in mules often result from overloading or trauma, especially in rugged terrain or during prolonged transport. Surgical interventions, such as the placement of Marlex™ mesh, have had mixed success due to complications like infection, recurrence, or poor integration with surrounding tissue (Molloy et al. 1991). In some cases, long-term management with pressure bandaging and conservative care is preferred, particularly when surgical resources are limited or the hernia is non-incarcerated and reducible.

2.3 Behavioral Challenges and Training

Donkeys and mules are often considered more challenging to train due to their intelligence, strong will, and heightened sensitivity to environmental cues. Their cognitive abilities require thoughtful, consistent handling, and they respond best to training rooted in behavioral science rather than force or repetition (The Donkey Sanctuary 2020a, b, c). Traditional methods of training these animals in some regions still involve coercive or violent techniques, such as beating, excessive restraint, or punishment-based conditioning. These approaches not only compromise welfare but can result in untrustworthy or aggressive behavior, particularly among mules, who are known for their heightened reactivity and memory retention (European Commission 2022). Evidence-based training—emphasizing positive reinforcement, gradual desensitization, and clear communication—has proven far more effective in building trust and cooperation between equids and handlers.

- **Intervention Strategies**: Imprinting techniques for mule foals, introduced alongside training programs for local handlers, have demonstrated long-term positive effects on behavior, trust, and trainability. Early handling during the critical imprinting window—typically within the first few hours and days after birth—can foster stronger human-animal bonds and reduce fear responses later in life (Miller 2022; Hodges 2022a, b). These methods, when paired with community-based training initiatives, have helped local handlers raise calmer, more cooperative animals suited for work and tourism.
  Proper training methods also improve the desirability of mules for tourism-related activities, where they are now valued higher than similarly sized horse colts due to their endurance, temperament, and sure-footedness in rugged terrain (Lucky Three Ranch 2022). In regions such as southern Spain and parts of Latin America, tourism operators increasingly prefer mules for trail rides and pack excursions, citing their reliability and reduced maintenance costs compared to horses.

2.4 Surgical Interventions and Preventative Medicine

Surgical interventions, though less common in some rural areas, are critical for addressing specific conditions in working equids. Scrotal hernias and traumatic injuries often require immediate attention to restore an animal's working capacity and prevent complications such as strangulation or infection (Moustafa et al. 2022). Umbilical hernias, particularly in foals and young donkeys, can be treated using a non-surgical "banding" technique under brief general anesthesia, which typically requires minimal aftercare and has shown favorable outcomes in uncomplicated cases (University of Minnesota 2025). Lateral abdominal wall hernias, often resulting from blunt trauma or overloading, can be successfully managed with pressure bandaging and stall rest when surgical resources are unavailable or the hernia is reducible (Amare and Haben 2020).

Preventative medicine has also played a vital role in improving equid health. Vaccination campaigns targeting diseases such as rabies, tetanus, and equine influenza, along with regular deworming protocols, have significantly reduced morbidity and mortality in working donkeys and mules (The Donkey Sanctuary 2020a, b, c; Mobile Veterinary Services 2025). These measures not only protect individual animals but also enhance herd-level resilience, especially in regions prone to outbreaks and limited access to emergency care.

- **Community Engagement**: Sustained veterinary care projects have emphasized the importance of returning regularly to build trust among local communities. Long-term engagement fosters relationships between animal owners and veterinary teams, encouraging greater adoption of preventative and therapeutic practices. For example, The Donkey Sanctuary's multi-year intervention in Bahir Dar, Ethiopia, focused on controlling epizootic lymphangitis (EZL) in cart mules through recurring visits, community education, and participatory planning. Over 7 years, the project saw a dramatic reduction in disease prevalence and a measurable increase in owner compliance with welfare recommendations (Duguma et al. 2021). This model demonstrated that consistent presence and collaboration—not one-off interventions—are key to sustainable improvements in equid health and community resilience.

2.5 Lessons Learned and Future Directions

1. **Community-Centered Approaches**
   Engaging local leaders and building trust through consistent visits has proven essential for successful interventions in working equid welfare. Case studies from organizations such as World Horse Welfare and Animal Nepal demonstrate that participatory approaches—where communities help identify problems and co-develop solutions—lead to more sustainable outcomes and stronger local ownership of welfare practices (Wild 2023).
2. **Education and Training**
   Long-term improvements in working equid welfare require educating both owners and local veterinarians in proper care, management, and modern veterinary techniques. Initiatives like the Equitarian Project and Brooke's veterinary manuals have shown that training programs targeting both technical skills and communication strategies significantly improve clinical outcomes and owner compliance (Merriam et al. 2012; Brooke 2025a, b, c, d).
3. **Sustainability of Interventions**
   The transition to locally available medications and feed solutions has ensured the sustainability of veterinary projects. Responsible prescribing practices and the use of regionally sourced treatments reduce dependency on external aid and mitigate risks such as antimicrobial resistance and environmental contamination (Higham 2022; Kisslinger 2025).

4. **Tourism's Role in Animal Welfare**
   As tourism expands, demand for healthier and well-trained equids increases, creating economic incentives for better animal care. Guidance from the EU Platform on Animal Welfare highlights how tourism operators are increasingly held accountable for equid welfare, with standards for workload, hydration, and rest becoming central to ethical tourism models (Hanlon 2023a, b, c).

3 Diseases and Disabilities in Working Donkeys and Mules

The conditions and disease entities affecting working equids represent a complex array of challenges that could individually merit extensive discourse in specialized texts on diagnosis and treatment. Despite this depth, the focus on effective treatment and eventual elimination of these conditions remains inadequately addressed. The clinical manifestations of these diseases—whether viral, bacterial, parasitic, or management-induced—are well-documented, and in many cases, present minimal diagnostic ambiguity (Evans et al. 2022; Câmara et al. 2020). The widespread availability of the internet and dedicated online information platforms has significantly enhanced the dissemination of knowledge concerning these conditions, yet one must be cautious of the source and clinical applicability of such information (Brooke 2025a, b, c, d).

From a clinical perspective, there is a risk of focusing exclusively on the symptomatic presentation of diseased individuals while neglecting the broader social, political, mechanical, or environmental determinants that perpetuate poor outcomes. The interplay between acute disease management and these overarching factors requires closer scrutiny. Early intervention targeting disease and injury is imperative and should be guided by evidence-based protocols tailored to low-resource settings (Brooke 2025a, b, c, d; WOAH 2024). However, sustainable health improvements in working equid populations necessitate addressing the underlying determinants that impair treatment efficacy or drive immunosuppression, thereby predisposing these animals to opportunistic infections (Warren 2016; Thunes 2017).

For instance, socio-political barriers such as limited access to veterinary care, the economic constraints of equid owners, and the impact of natural disasters disrupt preventive and therapeutic measures (Kubasiewicz et al. 2023). Additionally, systemic issues like poor nutrition, inadequate rest, and overexertion must be recognized as critical contributors to reduced immune competence and increased disease susceptibility (Warren 2016; Kentucky Equine Research 2013). The role of mechanical factors, including improper harnessing and equipment use, in exacerbating physical and systemic health problems also warrants thorough investigation (Merridale-Punter et al. 2024; Brooke 2025a, b, c, d).

To achieve population-wide improvements in the health and welfare of working equids, a multidimensional approach is essential. This approach should integrate immediate clinical care with long-term strategies addressing root causes such as poverty, education deficits in animal management, and infrastructural challenges. Only by aligning veterinary intervention with broader socio-economic and environmental considerations can enduring solutions be realized. Future studies should aim to evaluate the effectiveness of integrated interventions in enhancing both equid welfare and owner livelihoods (Tadich 2023; ICWE 2024).

4 Resistance and Sensitivity of Working Donkeys and Mules to Common Equid Diseases and Conditions

Donkeys and mules, as members of the Equidae family, demonstrate notable physiological predispositions that can confer resistance or increased susceptibility to specific diseases compared to horses. While many diseases affecting equids are well-documented in horses, fewer studies have systematically addressed their prevalence and impact in donkeys and mules. This gap in understanding arises partly from their smaller global population and different utilization patterns, particularly in low-income and rural settings where they are primarily used for labor rather than sport or leisure (Câmara et al. 2020). Despite their importance, donkeys and mules are often treated as "small horses," with limited consideration given to their unique metabolic, immunological, and behavioral traits (Câmara et al. 2020; The Donkey Sanctuary 2023a, b). These differences influence disease presentation, response to treatment, and overall resilience, especially in the context of working animals exposed to environmental stressors, poor nutrition, and limited veterinary care. As such, targeted research and species-specific protocols are essential to improve health outcomes and welfare standards for these equids.

1. Rabies (Fig. 4)
   - Sensitivity: Working donkeys and mules are frequently exposed to rabies due to their presence in rural areas where vaccination programs are sparse or nonexistent. Their susceptibility aligns with other equids, and non-vaccination significantly increases the risk. In Morocco, 40–70 human rabies cases are reported annually, often linked to bites from unvaccinated dogs (Barkia 2021).
   - Preventive Strategy: Routine rabies vaccination should be implemented universally to mitigate this threat (Barkia 2021).
2. Tetanus
   - Sensitivity: Donkeys and mules are highly sensitive to *Clostridium tetani* infections due to environmental exposure, especially in poorly sanitized or traumatic conditions associated with harness wounds, hoof injuries, and contaminated environments. The bacterium is commonly found in soil and manure, and infection typically occurs through deep puncture wounds or abscesses (The Donkey Sanctuary 2017; PetShun 2024). Clinical signs include muscle stiffness, spasms, hypersensitivity to sound or touch, and difficulty breathing, often progressing rapidly if untreated.

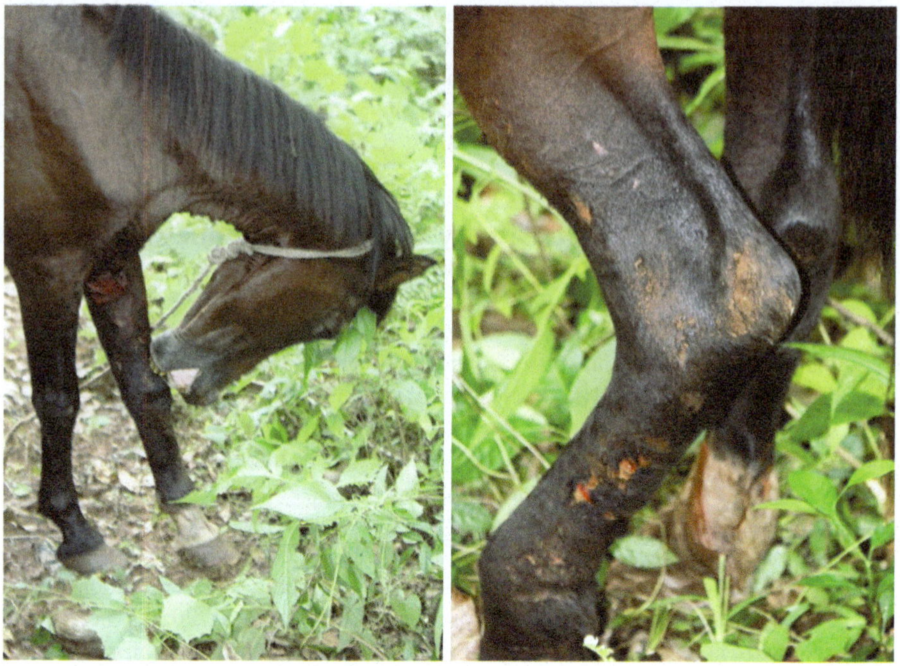

Fig. 4 Self-Injuring in a Rabbid horse

- Preventive Strategy: Regular vaccination is both effective and critical. Foals can begin vaccination at 5–6 months of age, and booster shots should be maintained throughout life. In cases of injury where vaccination status is uncertain, administration of tetanus antitoxin (TAT) is recommended. Proper wound care and hygiene in working environments are essential to reduce risk (The Donkey Sanctuary 2017).
- Treatment: When presented in the early stages, intensive care has proven effective in up to 80% of tetanus cases seen at the American Fondouk. Animals are assessed and sedated with acepromazine; their ears are stuffed with cotton and a modified bandana is placed over the eyes to reduce sensory stimulation. The patient is then stalled and supported in a sling. A nasogastric (NG) tube is passed to initiate oral hydration, and intrathecal tetanus antitoxin (TAT) is administered. Wounds are cleaned and opened if necessary, and appropriate doses of penicillin are given twice daily. The stall is darkened, and treatment is continued "to effect," with most animals able to eat and drink within 3 days. Despite these efforts, fatalities may still occur, often within 48 hours of symptom onset (G. Kay, personal communication, 2025).

3. Piroplasmosis
   - Sensitivity: This tick-borne disease, caused by Babesia caballi and Theileria equi, is endemic in tropical and subtropical regions. Donkeys and mules demonstrate variable resistance but often suffer severe outcomes due to the

absence of approved vaccines and the complexity and cost of treatment. Infected equids may become lifelong carriers, and clinical signs range from fever, anemia, and lethargy to weight loss and poor performance (AAEP 2021; WOAH 2025). The disease is particularly problematic in working animals due to their frequent exposure to tick vectors and limited access to diagnostic services.

- Preventive Strategy: Tick control and early diagnostic protocols are essential. Measures include the use of acaricides, pasture management, and avoiding reuse of blood-contaminated equipment. Research is urgently needed to develop effective vaccines and more affordable treatment options (Díaz-Sánchez et al. 2020).

4. Strangles
   - Sensitivity: Communal living conditions, urban crowding, and shared watering points significantly increase susceptibility to Streptococcus equi infections. Donkeys and mules are equally vulnerable as horses, although they may exhibit less severe clinical signs such as reduced nasal discharge or lymph node abscessation (Lascola and Boone 2024; AAEP 2020). The bacterium is highly contagious and spreads through direct contact or contaminated equipment, water sources, and handlers' clothing.
   - Preventive Strategy: Improved hygiene, isolation of infected animals, and consistent vaccine use—despite limited efficacy—can reduce outbreaks. Vaccines like Strangvac have shown promise in reducing clinical severity and bacterial shedding, especially when used alongside biosecurity measures (Royal (Dick) School of Veterinary Studies 2024).
5. African Horse Sickness (AHS)
   - Sensitivity: Donkeys exhibit a higher resistance to African Horse Sickness compared to horses, often surviving with milder symptoms such as transient fever and edema. Mules, however, are more similar to horses in their susceptibility, with clinical signs ranging from respiratory distress to cardiac failure depending on the form of the disease (Weyer 2025; WOAH 2025). Horses are the most vulnerable, with mortality rates reaching up to 90% in naïve populations, while donkeys typically experience subclinical or mild forms of the illness.
   - Preventive Strategy: Vaccination in endemic regions and vector control—particularly targeting Culicoides midges—are critical to limiting disease spread. Attenuated live vaccines are available for horses, mules, and donkeys, and are recommended in areas with seasonal outbreaks. Vector control measures include insect repellents, stabling during peak midge activity, and environmental management to reduce breeding sites (WOAH 2025; European Commission 2025).
6. Equine Encephalitis Viruses (EEVs)
   - Sensitivity: Donkeys and mules are as susceptible as horses to arboviral encephalitides such as West Nile Virus (WNV), Eastern Equine Encephalitis (EEE), and Venezuelan Equine Encephalitis (VEE). These viruses are

transmitted by mosquitoes and can cause severe neurological symptoms including fever, ataxia, muscle tremors, and in some cases, death. While horses are often the focus of surveillance and vaccination programs, equids of all types—including donkeys and mules—are vulnerable and may serve as amplifying hosts during outbreaks (APHIS 2025; WOAH 2021).

- Preventive Strategy: Vaccines for WNV, EEE, and VEE are highly effective and should be part of routine management, especially in endemic regions. Annual vaccination is recommended, with boosters in high-risk areas. Vector control measures such as eliminating standing water, stabling during peak mosquito activity, and using equine-safe repellents are essential to reduce transmission risk (EquiManagement 2024).

7. Equine Metabolic Syndrome (EMS)
    - Sensitivity: Donkeys are particularly predisposed to EMS due to their evolutionary adaptation to sparse, fibrous forage, which makes them metabolically efficient at storing fat. In captive or urban environments, overfeeding and lack of exercise exacerbate this issue, although it is less common in actively working equids (Thiemann 2019; Bertin and Boyle 2025).
    - Preventive Strategy: Dietary management, weight control, and regular exercise are essential to prevent and manage EMS. Monitoring insulin levels and avoiding high-sugar feeds are key components of care.
8. Arthritis and Hoof Disease
    - Sensitivity: Chronic overwork, poor hoof care, and heavy loads predispose working donkeys and mules to musculoskeletal disorders such as arthritis and hoof diseases including laminitis. These are leading causes of disability and reduced working capacity (Thiemann et al. 2023; Mad Barn Research Bank 2025).
    - Preventive Strategy: Regular hoof trimming, load management, and the use of properly fitted shoes are essential. Farriery tailored to donkey-specific hoof anatomy improves outcomes and reduces pain.
9. Tick-Borne Diseases (e.g., *Anaplasma phagocytophilum, Borrelia burgdorferi*)
    - Sensitivity: Donkeys and mules show variable resistance but are commonly affected in tick-endemic regions such as the Americas, Africa, and tropical zones. Lyme disease and anaplasmosis are among the most concerning tick-borne conditions (Williams and Toledo 2024; Springer et al. 2020).
    - Preventive Strategy: Routine tick control programs—including acaricides, pasture management, and owner education—are critical. In some communities, ticks are mistakenly considered harmless, underscoring the need for awareness campaigns (Author's field experience).
10. Harness and Tack Sores (Fig. 5).
    - Sensitivity: Donkeys and mules are highly prone to wounds caused by poorly fitted harnesses and tack. These injuries are often neglected, leading to secondary infections and chronic disabilities. Malnutrition exacerbates the issue by exposing bony prominences to friction (Cousquer 2011; Mitiku et al. 2022).

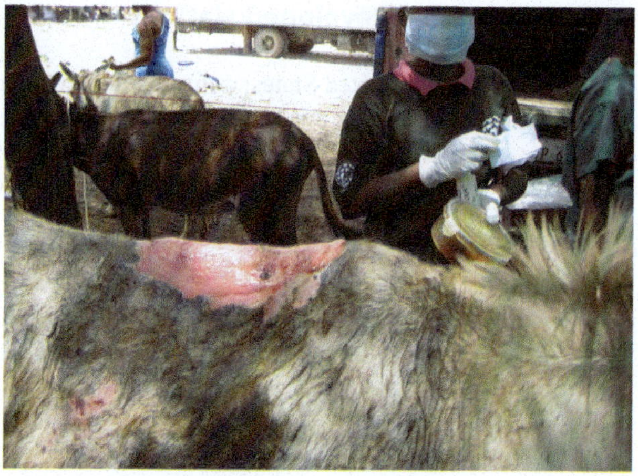

Fig. 5 Harness wounds in a donkey in Haiti. Veterinary teams meet animals at weekly markets and have supplies and trained vets to treat animals as they rest

- Preventive Strategy: Proper fitting of equipment, regular inspection, and treatment of sores are vital. Improved nutrition enhances skin resilience and healing capacity, reducing the risk of skeletal exposure.

11. Laminitis
    - Sensitivity: Overfeeding-induced laminitis is rare in working donkeys and mules, but cases linked to chronic hoof injuries, metabolic disorders, and neglect are more common. Donkeys often present subtle signs, making early detection challenging (Thiemann et al. 2023; FormaHoof 2025).
    - Preventive Strategy: Proper hoof care, balanced nutrition, and avoidance of excessive carbohydrate intake are critical. Farrier support and pain management are essential during rehabilitation.

5 Prevention Strategies in Working Equids: Challenges and Solutions in the Cases of Rabies and Tetanus

Rabies and tetanus are two of the most critical vaccine-preventable diseases impacting both humans and animals. While both diseases are nearly 100% preventable with proper vaccination, logistical, political, and economic barriers often hinder effective implementation—particularly in regions where working equids are essential to livelihoods and transportation.

5.1 Rabies Prevention Strategies

Rabies is one of the most feared zoonotic diseases due to its near-100% fatality rate, yet it remains one of the easiest to prevent through vaccination. Despite this, millions of animals and people are exposed annually, especially in developing countries (CDC 2025).

5.2 Rabies in Developed Nations

In countries like the United States, domestic animal vaccination campaigns are well-regulated, resulting in minimal exposure risks from pets. However, wildlife vectors such as raccoons (Eastern U.S.), skunks (Midwest and West), foxes, and bats continue to pose threats, underscoring the need for ongoing surveillance and public education (CDC 2025).

5.3 Rabies in Developing Nations

Rabies prevention in developing countries is more complex due to economic constraints, limited access to vaccines, and fragmented public health infrastructure. In Samana, Dominican Republic, community-based rabies vaccination clinics have operated for over 15 years, significantly reducing human exposure (Author's field experience, 2010–present). In Haiti, international collaborations—such as those led by the CDC, Humane Society International, and Christian Veterinary Mission—have supported mass dog vaccination campaigns and introduced Integrated Bite Case Management (IBCM), improving surveillance and post-exposure treatment adherence (CDC 2025) (Fig. 6).

Fig. 6 Donkey loaded with charcoal in Haiti is headed to market. His owner is working hard as well, carrying charcoal she produced to trade for food and supplies

In Morocco, despite having a national vaccine production unit, rabies control remains inconsistent. Restrictions on vaccine importation and reliance on locally produced vaccines of variable efficacy hinder progress. Additionally, national policies prohibiting euthanasia of rabies-suspect animals delay diagnosis and humane management, complicating containment efforts (Darkaoui et al. 2017; Barkia 2021; Kay, G., personal communication, 2015, American Fondouk, Morocco).

5.4 Tetanus Prevention Strategies

Tetanus, caused by *Clostridium tetani*, is a life-threatening neurological disease that disproportionately affects equids due to their high sensitivity to the tetanus neurotoxin. The bacterium is ubiquitous in soil and manure, making exposure nearly inevitable in working environments (AAEP 2025).

5.5 Sensitivity and Risk

Donkeys and mules are particularly vulnerable due to frequent exposure to wounds, hoof abscesses, and unsanitary conditions. Mortality rates in unvaccinated equids can exceed 75%, and early signs—such as stiffness, trismus, and third eyelid protrusion—often progress rapidly (Lovett et al. 2024).

5.6 Preventive Strategy

Vaccination is highly effective and should be administered as a three-dose primary series, followed by regular boosters. In cases of injury or uncertain vaccination history, tetanus antitoxin (TAT) should be administered alongside tetanus toxoid. Despite its efficacy, barriers such as lack of veterinary access, poor record-keeping, and limited owner awareness continue to impede widespread adoption (Dennis et al. 2024).

5.7 Intervention Strategies

1. **Vaccination Programs** Expanding community-wide vaccination programs for working equids and stray animals has proven effective in reducing disease transmission and improving animal welfare. In North America and parts of Latin America, coordinated campaigns targeting rabies and tetanus have significantly lowered incidence rates among equids and companion animals (Desanti-Consoli et al. 2022; USDA APHIS 2006).
2. **Education Campaigns** Educating local populations about rabies transmission and prevention remains critical. Outreach efforts that include bite prevention,

wound care, and the importance of vaccination have shown measurable improvements in both human and animal health outcomes (Project Samana 2025).

3. **Policy Reform** Advocacy for policy changes—such as permitting euthanasia of rabies-suspect animals and allowing the importation of high-quality vaccines—could save both animal and human lives. In Morocco, restrictive policies on vaccine imports and euthanasia have complicated rabies control efforts (Barkia 2021; Kay, G., personal communication, 2015, American Fondouk, Morocco).

5.8 Tetanus Prevention Strategies

Tetanus, caused by *Clostridium tetani*, is a ubiquitous but non-zoonotic disease with a nearly 100% mortality rate in unvaccinated animals. It is a significant problem in working equids due to frequent exposure to contaminated environments and untreated wounds.

5.9 Global Tetanus Burden

The disease remains prevalent in regions lacking large-scale vaccination campaigns. Globally, tetanus causes approximately 50,000 deaths annually, with the highest burden in South Asia and Sub-Saharan Africa (Our World in Data 2024; He et al. 2025). Surgical interventions in working equids are particularly high-risk due to postoperative contamination in unsanitary conditions (Khallikane et al. 2021).

5.10 Barriers to Tetanus Control

1. **Vaccine Availability** Tetanus vaccines are often unavailable in rural areas, and tetanus antitoxin (TAT) is frequently substituted despite its short duration of protection (AAEP 2025).
2. **Logistical Challenges** Maintaining a cold chain for vaccine distribution in remote regions poses significant difficulties. Innovations such as solar-powered refrigeration and controlled temperature chain (CTC) protocols are being explored to overcome these barriers (WHO 2025a, b; Adeyemo et al. 2021).

5.11 Successful Interventions

The American Fondouk Hospital in Fez, Morocco, has developed an intensive care protocol achieving up to an 80% survival rate in tetanus cases. This success is attributed to early diagnosis, antibiotic therapy, pain management, and continuous care (Kay, G., personal communication, 2015, American Fondouk, Morocco).

5.12 Recommendations

1. **Vaccination Prior to Surgery** Preoperative vaccination is more cost-effective and offers long-term protection compared to TAT.
2. **Community Education** Training local equid owners to recognize early tetanus symptoms can lead to earlier intervention and improved outcomes.

6 Logistical and Political Challenges in Vaccine Deployment

Both rabies and tetanus vaccines face logistical and political barriers that hinder their widespread use, despite being reliable and time-tested.

6.1 Logistics of Vaccine Deployment

Vaccines require a cold chain to maintain efficacy, which is challenging in remote areas. Newer heat-resistant formulations may offer future solutions (UNICEF 2021).

6.2 Political and Policy Barriers

1. **National Policies**: Some governments prioritize locally produced vaccines, which may be less effective.
2. **Charity Regulations**: Restrictions imposed by international organizations—such as requiring full population coverage before vaccinating any animals—often delay care (Blakeway 2010).

6.3 Case Study: Project Samana

In the Dominican Republic, Project Samana has successfully deployed rabies vaccines using local staffing and refrigeration infrastructure. Booster administration by trained personnel has helped circumvent logistical barriers and improve coverage (Project Samana 2025).

7 The Role of Vaccines in Shaping Policy

Rabies and tetanus vaccines serve as case studies for understanding the challenges and successes of vaccine deployment in developing regions. These vaccines are not only medically effective but also strategically important in shaping public health policy and veterinary outreach.

1. **Cost vs. Effectiveness** Vaccines are highly cost-effective compared to the resources required for treatment or outbreak management. For example, mass rabies vaccination campaigns in dogs have been shown to reduce human rabies deaths by over 90% in targeted regions, with a cost-benefit ratio favoring prevention over post-exposure treatment (WHO 2025a, b; WOAH 2023).
2. **Adaptability** Projects must adapt to local conditions and policies while balancing immediate interventions with long-term population-level strategies. Vaccine banks and regional partnerships have enabled flexible deployment models that respond to both emergency needs and sustained control efforts (WOAH 2023).

8 Ethical and Practical Considerations in Transporting Medications for Veterinary Work Abroad

Transporting undeclared or restricted medications for veterinary projects abroad presents significant ethical, legal, and logistical challenges. While the intent may be humanitarian, circumventing national regulations can undermine trust and jeopardize long-term collaboration.

8.1 Legal and Ethical Implications

Importation Laws and Regulations Each country has its own pharmaceutical import laws, often prohibiting the entry of unregistered or undeclared medications. Violating these laws—even with good intentions—can damage relationships with local authorities and compromise the credibility of the project (Verma et al. 2024).

Impact of Non-Compliance Consequences of non-compliance include:

- **Confiscation of Supplies**: Medications may be intercepted and destroyed.
- **Legal Consequences**: Personnel may face fines or legal action.
- **Erosion of Trust**: Violations can strain relationships with local partners and communities (Rebexa Group 2024).

8.2 Challenges in Medication Supply

Accessibility of Medications Essential drugs like ivermectin may be unavailable or restricted due to regulatory hurdles or limited domestic production capacity (MavenRS 2024).

Expired Medications The donation of expired medications is common but controversial. While some materials may retain efficacy beyond labeled expiration dates, their use must be guided by chemical stability data and ethical transparency (NIH OACU 2025).

8.3 Strategies for Addressing Medication Supply Issues

1. **Purchasing Locally** Local procurement ensures legal compliance and supports host economies, even if it increases costs.
2. **Partnering with International Pharmacies** Online platforms that comply with both international and national regulations offer safer alternatives for sourcing medications.
3. **Building Relationships with Local Authorities** Engaging customs and veterinary officials in open dialogue can lead to exemptions or cooperative solutions for humanitarian projects.
4. **Adhering to Expiration Guidelines** When using expired medications:
   - **Assess Stability**: Use manufacturer data or peer-reviewed studies to confirm efficacy.
   - **Label Transparently**: Clearly mark revised expiration dates and disclose usage conditions (UWM IACUC 2023).

9 Lessons Learned from Field Projects

9.1 Case Study: Ivermectin Importation

Ivermectin, a transformative antiparasitic drug, was historically restricted in many regions due to regulatory barriers and limited domestic production. Early efforts to transport undeclared ivermectin faced logistical and legal challenges, including confiscation and delayed deployment. However, increased availability through international pharmaceutical companies and improved regulatory frameworks have since alleviated these issues (European Medicines Agency 2005). Today, most projects source ivermectin through local suppliers or vetted international partnerships, avoiding the risks associated with undeclared importation.

9.2 Case Study: Customs Negotiations

In one instance, customs officials intercepted medications intended for a veterinary outreach project. After transparent discussions emphasizing the public health benefits and humanitarian nature of the intervention, a compromise was reached, and the medications were released under additional scrutiny (Project Samana 2025). This highlights the importance of proactive communication and relationship-building with local authorities.

9.3 Future Directions

1. **Strengthening In-Country Capacity** Developing local supply chains and supporting domestic pharmaceutical production can reduce reliance on imported medications and improve sustainability. Turnkey veterinary pharmaceutical projects in Africa have demonstrated success in improving access and reliability (BRAM-COR 2025).
2. **Promoting Regulatory Reform** Advocacy for streamlined importation processes for humanitarian and veterinary projects can help address urgent needs while respecting national laws. Multi-partner initiatives like GALVmed's regulatory harmonization project in sub-Saharan Africa are paving the way for more predictable and efficient access to veterinary medicines (GALVmed 2024).
3. **Standardizing Ethical Practices** Establishing international guidelines for the ethical use of donated and expired medications can enhance the credibility of veterinary projects. Institutions like the University of Michigan and UConn have published clear policies on the use of expired drugs in animal care, emphasizing transparency and safety (University of Michigan IACUC 2022; UConn IACUC 2024).

10 Exporting Knowledge and Techniques in Veterinary Practice

One of the most impactful contributions of international veterinary projects is the transfer of up-to-date medical and surgical skills to local practitioners. Unlike physical supplies, there are no regulatory restrictions on exporting knowledge, but implementation is often constrained by infrastructure, cultural practices, and economic realities.

10.1 Challenges in Adopting Modern Techniques

1. **Local Practices and Cultural Barriers**
   Traditional castration methods in some regions—such as using nylon fishing line or wire to strangulate testicles—are inhumane and associated with high morbidity. These practices result in chronic infections, non-healing wounds, and behavioral issues (The Donkey Sanctuary 2018; WikiVet 2025).
2. **Health of Working Animals**
   Working donkeys and mules often suffer from anemia, tick infestations, and underlying conditions like Equine Infectious Anemia (EIA), complicating surgical procedures. Active surveillance in Italy showed higher EIA prevalence in mules and working equids in wooded areas (Carvelli et al. 2024).

3. **Economic and Social Considerations**
   In regions where equids are vital for livelihoods, euthanasia of EIA-positive animals is rarely viable. Lack of compensation and cultural resistance create ethical dilemmas for veterinarians (Equine Veterinary Journal 2024; Lederman et al. 2021).

10.2 Innovations in Surgical Techniques

1. **Introduction of the Henderson Castration Tool**
   The Henderson Castration Tool, a stainless steel clamp attached to a cordless drill, revolutionized field castration by minimizing bleeding and improving recovery. Clinical trials have shown reduced complications and ease of use in resource-limited settings (Reilly and Cimetti 2005).
2. **Development of the Equi-Twister**
   In areas without electricity, the Equi-Twister—a manually powered alternative—was developed to replicate the Henderson tool's functionality. It has proven effective in field conditions, with simplified technique and minimal complications (Turner 2015).
3. **Adaptation for Donkeys and Mules**
   Donkeys and mules are physiologically more prone to bleeding during castration than horses, particularly from superficial skin vessels. To address this, modified clamping techniques and closed castration methods have been developed to reduce hemorrhage and improve outcomes. Despite this predisposition, **modern techniques have significantly lowered complication and mortality rates** (Abou-Khalil et al. 2020; Kilcoyne et al. 2013).

10.3 Training and Knowledge Dissemination

1. **Global Training Programs**
   Veterinarians trained in advanced castration techniques have conducted workshops worldwide, sharing methods that improve animal welfare and reduce mortality. The Donkey Sanctuary's CLNC111 course and Equitarian outreach programs have trained hundreds of professionals in safe field castration (The Donkey Sanctuary 2023a, b).
2. **Regional Adaptations**
   Cultural practices influence the adoption of castration:
   - **Mexico & Central America**: Donkeys used for packing often remain intact to maintain herd behavior and dominance.
   - **North America**: Donkeys are routinely gelded due to their use as companion animals and for safety (Canadian Veterinary Medical Association 2024).

3. **Ethical Considerations**
   The decision to castrate is often shaped by local economic and cultural factors. Intact males may be preferred for breeding mules or herding behavior, while gelding is prioritized in tourism and pet contexts (Donkey Sanctuary 2018; Gongora et al. 2021).

10.4 Lessons Learned and Recommendations

1. **Adapt Techniques to Local Conditions**
   Innovative tools like the **Henderson Castration Tool** and the **Equi-Twister** have revolutionized field castration. These devices reduce bleeding and are easy to use in remote settings (Turner et al. 2015; Racine et al. 2018).
2. **Recognize Health Limitations**
   Comprehensive health assessments are essential. Undernourished or parasitized animals require tailored surgical protocols to minimize risk (Baldwin 2024).
3. **Address Cultural and Economic Contexts**
   Training programs must incorporate cultural sensitivity and economic realities. The University of Sydney's curriculum framework emphasizes cultural competence in veterinary education (Gongora et al. 2021).
4. **Promote Collaboration**
   The **1st Equitarian Workshop in 2011** in Mexico marked a milestone in collaborative veterinary outreach. Co-sponsored by AAEP, HSUS, UNAM, WHW, The Donkey Sanctuary, AMVEE, and CEIGT, it unified six equine welfare organizations for a week of service and training (Equitarian Initiative 2011).

10.5 Cultural Perspectives and Machismo in Castration Decisions

1. **Reverence and Fear**
   In many cultures, jackstocks (intact male donkeys and mules) symbolize strength and masculinity. Their aggressive behavior is both revered and feared, often leading to harsh treatment. This cycle of aggression and mistreatment reinforces the need for castration, yet also fuels resistance to it (Donkey Sanctuary 2018).
2. **Machismo and Resistance to Castration**
   Castration is often viewed negatively in cultures where masculinity—both human and animal—is highly valued. This resistance is rooted in traditional beliefs about strength and utility, even when castration could improve safety and manageability. In tourism-dependent economies, intact stallions are kept for symbolic or economic reasons, despite increased risks of accidents and handling difficulties (Jiménez 2020).

10.6 Behavior Management Through Training

1. **The Role of Imprinting**
   Imprinting programs introduced to young mules and donkeys have shown significant promise in mitigating aggressive behaviors typically associated with intact males. Early handling and sensory exposure help shape predictable, manageable responses to stimuli (Hodges 2022a, b).
   - **Success in Specific Regions**: In parts of the Dominican Republic, imprinting has reduced the need for castration and led to calmer, more cooperative animals (Project Samana 2025).
2. **Training vs. Castration**
   Military mule regiments in Chile and Morocco demonstrate that consistent training can effectively manage behavior without surgical intervention. These animals are used for transport and rescue in remote areas and are safely handled by trained personnel (Cousquer et al. 2023; Defense Media Network 2013).
   - **Lessons for Tourism and Agriculture**: Applying similar training techniques in tourism-dependent economies could reduce reliance on castration, especially where cultural resistance is strong.

10.7 Necessity of Castration in Tourism and Agriculture

Despite the promise of training and imprinting, castration remains a practical necessity in many contexts. In tourism, aggressive behavior in intact stallions poses safety risks to handlers and tourists.

- **Benefits of Castration**:
  - Reduces aggression and improves manageability.
  - Enhances safety for handlers and tourists.
  - Aligns with economic goals by producing reliable working animals (EURCAW 2024; World Horse Welfare 2025).

10.8 Challenges in Scaling Behavioral Programs

While successful in specific regions, behavioral programs face barriers to broader adoption:

- Limited access to training resources and personnel.
- Deeply entrenched cultural norms.
- Economic constraints in equid-dependent communities (Brooke 2013; EURCAW 2024).

10.9 Recommendations for Addressing Machismo and Behavior Issues

1. **Education and Awareness Campaigns** Promote the welfare and safety benefits of castration and training through culturally sensitive outreach.
2. **Expand Imprinting Programs** Develop localized imprinting initiatives integrated into veterinary and agricultural programs (Hodges 2022a, b).
3. **Leverage Case Studies** Use examples from military mule regiments and tourism sectors to demonstrate the efficacy of training over forceful handling (Mills and Booth 2022).
4. **Collaborate with Local Leaders** Engage community leaders to address cultural resistance and promote humane practices (Cousquer et al. 2023; Brooke 2013).

11 Provision of Veterinary Services for Working Equid Populations

11.1 Phase 1: Crisis Response—The "Tip of the Spear"

1. **Recognizing the Need**
   Veterinary outreach often begins in response to acute crises. In Project Samana, international veterinarians intervened to treat malnourished and overworked animals used in tourism and agriculture (Merriam 2015a, b; Project Samana 2025).
2. **Initial Interventions**
   The first team arrived with limited supplies and minimal understanding of local needs. Yet, services like deworming, surgical care, and health evaluations quickly built trust.
   - **Example Success**: A life-saving hernia surgery catalyzed community engagement and sustained participation in veterinary care (Merriam 2015a, b).

11.2 Phase 2: Community Engagement and Capacity Building

1. **Transitioning to Preventative Care**
   As the project matured, the focus shifted from emergency interventions to preventative care, addressing chronic conditions and improving welfare standards. Services expanded to include body condition scoring, harness and saddle fitting, and routine deworming programs (Brooke 2022).
2. **Training Local Veterinarians**
   Collaboration with Dominican veterinarians and students became a cornerstone of sustainability. Initially limited in clinical experience, local practitioners were trained in advanced techniques and began assuming leadership roles.

- **Key Achievement**: Two veterinary students from early training cohorts are now instructors and clinicians, extending the project's reach (Project Samana 2025).

3. **Integration with Tourism**
   Engagement with tourism agencies elevated awareness of equid welfare. Routine health checks and fitness evaluations became essential to meet tourism industry standards and ensure safety (EURCAW 2023).

11.3 Phase 3: Sustained Growth and Expanded Reach

1. **Biannual Visits and Chronic Care**
   Twice-yearly visits enabled consistent follow-up on chronic conditions. The inclusion of a farrier addressed lameness and hoof care, which are common in working equids (SPANA 2023a, b, c).
2. **Academic Collaboration**
   Seminars and workshops at Dominican veterinary schools introduced modern techniques. Scholarships and mentorships provided hands-on experience for students, fostering long-term capacity (UNPHU and Project Samana 2012–present).

12 The Role of Technology in Expanding Services

1. **Improved Communication**
   Tools like WhatsApp and Zoom have revolutionized remote veterinary care. Real-time consultations allow international teams to support rural practitioners effectively (Teller and Moberly 2020).
2. **Enabling Community Collaboration**
   Cell tower expansion in rural areas has replaced hardwired systems, enabling reliable access to veterinary expertise and aligning with One Health principles (Innovation News Network 2025).
3. **Increasing Outreach**
   Digital platforms support ongoing education, resource coordination, and case discussions. AI-powered learning tools and virtual seminars have enhanced veterinary training (ECCVT 2020; Muca et al. 2022).

12.1 The Broader Impact: One Health and Community Development

Working equids are vital to rural livelihoods—serving as transport, school buses, ambulances, and links to healthcare.

- **Example**: Healthier equids have led to increased family income from tourism, reduced school absenteeism, and improved access to medical care in remote areas (Hanlon 2023a, b, c).

12.2 Recommendations for Future Projects

1. **Develop Community Partnerships** Early and sustained collaboration with veterinarians, leaders, and stakeholders is essential (Brooke 2022).
2. **Invest in Education** Training and resources for local practitioners and students ensure long-term sustainability (UNPHU and Project Samana 2012–present).
3. **Leverage Technology** Use telecommunication tools to extend veterinary services and maintain remote team coordination (Teller and Moberly 2020; ECCVT 2020).
4. **Promote Preventative Care** Shift focus from crisis response to long-term welfare through preventative health programs (Brooke 2022).

12.3 Collaboration and Divergent Approaches

1. **Building Early Partnerships**
   Early collaborations between volunteer-led initiatives and international NGOs were pivotal in shaping equid welfare programs. The **Equitarian Initiative** hosted annual conferences that united six NGOs with overlapping but distinct goals (Equitarian Initiative 2025).
   Differing Focus Areas:
   - **Volunteer Initiatives**: Prioritized immediate care, surgeries, and local veterinary training (Merriam 2015a, b).
   - **International NGOs**: Focused on long-term farrier training, saddle making, and humane care, often deprioritizing urgent medical needs (Haddy et al. 2023).
2. **Lessons from Collaboration**
   Despite differing priorities, collaborative workshops attracted over 100 participants annually and fostered lasting relationships. These events revealed the interplay between medical care, equipment use, and humane treatment, emphasizing the need for integrated approaches (Haddy et al. 2022).

12.4 Challenges of Sustaining Interventions

1. **Financial Constraints**
   Funding shortages led some organizations to embed local employees for long-term community assessments. While insightful, these strategies often failed to address immediate needs (Pritchard et al. 2018).

- **Self-Funded Success**: Volunteer-led initiatives like Project Samana continued delivering direct care without external dependencies (Merriam 2015a, b).

2. **Retention of Knowledge**
Sustainable impact requires training and retaining local practitioners. Visiting veterinarians alone rarely create lasting change unless local professionals are empowered (Haddy et al. 2021).

12.5 Key Elements of a Sustainable Model

1. **Local Leadership Development**
Empowering local leaders ensures continuity and builds trust. Best practices include deferring decisions to local stakeholders and respecting cultural norms (Haddy et al. 2021).
2. **Regular Engagement**
Consistent visits reinforce knowledge and allow follow-up on chronic conditions (Haddy et al. 2023).
3. **Integration of Telemedicine**
Tools like WhatsApp and Zoom enable real-time consultations, bridging gaps between visits and supporting rural veterinarians (Teller and Moberly 2020).

12.6 Case Studies: Lessons Learned

1. **Community-Driven Success**: Embedded local employees created long-term intervention plans (Haddy et al. 2023).
2. **Self-Funded Volunteer Projects**: Delivered swift, direct care and empowered local practitioners (Merriam 2015a, b).
3. **Technology Integration**: Telemedicine enabled ongoing case discussions and remote support (Teller and Moberly 2020).

13 The Path to One Health

Equid welfare is tightly linked to human well-being. Projects that improve equid health also enhance economic stability, education access, and healthcare in underserved areas (Bonsi et al. 2023).

13.1 Recommendations for Future Interventions

1. **Prioritize Local Empowerment** Train and equip local veterinarians and leaders (Haddy et al. 2021).
2. **Combine Approaches** Integrate emergency care with long-term capacity-building (Haddy et al. 2023).

3. **Leverage Technology** Use telemedicine to extend reach and continuity (Teller and Moberly 2020).
4. **Secure Sustainable Funding** Develop hybrid or self-funded models (Merriam 2015a, b).

14 The Working Equid Conundrum

1. **Traditional Source Medical Care**
   Practices like nostril splitting, joint firing, and bloodletting persist due to cultural norms and limited resources.
   - **Cultural Sensitivity**: Demonstrating effective outcomes builds trust and shifts perceptions (Haddy et al. 2022).
2. **External Medical Care**
   Sporadic care and "parachute projects" lack sustainability. Fee-for-service models exclude low-income owners (Pritchard et al. 2018).
3. **Shared Medical Care Projects**
   Local leadership and community engagement ensure continuity and relevance (Haddy et al. 2023).

Examples

- **Rabies Vaccination Campaigns**: Led by local veterinarians, improved public health outcomes (WHO 2018; AAEP 2025).
- **Tourism Engagement**: Clinics at tourism hubs raised awareness and improved equid welfare (Hanlon 2023a, b, c).

15 Education for Indigenous Professionals

Training local veterinarians and paraprofessionals ensures long-term care. Effective programs include seminars, workshops, and digital tools (Brooke 2022).

- **Scholarships**: AKC and WVA programs support veterinary students in underserved regions (AKC 2024; WVA 2024).
- **Cultural Adaptation**: Curricula must align with local traditions and values (Gongora et al. 2021).

16 External Research in Medicine, Nutrition, Farriery, and Equipment

16.1 Hospital-Based Research

Institutions like the **American Fondouk** in Morocco conduct advanced research and training (Fondouk 2024).

- **Tetanus Treatment**: Achieved 80% survival rate—an equid care breakthrough (Jarm 2023a, b).

16.2 Other Research Areas

- **Nutrition**: Improved forage and supplementation for undernourished equids.
- **Farriery**: Innovations in hoof care reduce lameness (Mohite et al. 2019).
- **Equipment**: Studies on saddle and harness design prevent sores and musculoskeletal issues (Merridale-Punter et al. 2024).

16.3 Recommendations for Sustainable Care

1. **Integrate Local Knowledge**
   Respecting and building upon traditional practices fosters trust and ensures cultural relevance. In Colombia, integrating local knowledge into rural veterinary education has enriched community engagement and improved outcomes (Guerra Monterroza and Consuegra Solano 2024).
   - Traditional remedies like *aceite y carbón* "oil and coil" or spider webs for wound care reflect resourcefulness but require sensitive replacement with evidence-based methods.
2. **Develop Holistic Approaches**
   Combining emergency care with long-term education and infrastructure development aligns with the One Health framework. European veterinary schools are increasingly adopting interdisciplinary curricula to prepare professionals for complex health challenges (Iatridou et al. 2020).
   - Holistic models like EcoHealth and Planetary Health emphasize the interconnectedness of animal, human, and environmental well-being (Lerner and Berg 2017).
3. **Leverage Technology**
   Telemedicine has revolutionized veterinary outreach in remote areas. Platforms like WhatsApp and Zoom enable real-time consultations, diagnostics, and training (Roberts 2023).
   - Wearable sensors and mobile apps allow continuous monitoring of equid health, improving early detection and intervention.
4. **Collaborate Across Sectors**
   Partnering with governments, NGOs, and community leaders ensures alignment of goals and resource sharing. The International Coalition for Working Equids—comprising Brooke, SPANA, The Donkey Sanctuary, and World Horse Welfare—has successfully influenced global policy and SDG implementation (Zafra 2023).
   - Sustainable NGO programming depends on tailoring interventions to local contexts and building capacity through community leadership (Haddy et al. 2023).

5. **Promote Research and Innovation**
   Hospital-based research centers like the American Fondouk in Morocco have led breakthroughs in equid care, including an 80% survival rate for tetanus cases (Jarm 2023a, b).
   - The Brooke's *Working Equid Veterinary Manual* offers field-tested clinical protocols and has been distributed to over 25 countries, supporting innovation in low-resource settings (Brooke 2024).

17 Conclusion

Addressing the challenges faced by working donkeys, horses, and mules requires a multifaceted approach that integrates traditional knowledge, modern veterinary practices, and sustainable education programs. By fostering local leadership, leveraging external support, and investing in research and training, veterinary initiatives can improve the welfare of working equids while supporting the livelihoods of the communities that depend on them. Efforts should focus on empowering local communities to take ownership of veterinary care projects. Local leadership, particularly by veterinarians or knowledgeable community caregivers, is critical for long-term success. Returning regularly to the same communities and prioritizing culturally sensitive education and training ensures continuity and builds trust. Digital tools like WhatsApp and Zoom can enhance communication and provide ongoing support between visits, enabling remote consultations and education. Partnerships between international organizations, local leaders, and community members play a key role in developing sustainable care models. Collaborative approaches should align immediate care with long-term capacity-building, addressing urgent needs while fostering lasting improvements. Hospitals dedicated to working equids, such as the American Fondouk in Morocco, demonstrate how focused research and training can advance medical practices and improve outcomes globally. Innovation in tools, equipment, and research is essential. From advancements like the Henderson Castration Tool to research on nutrition, farriery, and disease management, these developments provide practical solutions tailored to the unique challenges of working equids. Telemedicine, hospital care, and community-based interventions must all work together under the umbrella of One Health, recognizing the interconnected well-being of animals, humans, and their environments. Ultimately, the goal is to create a system where working equids and their owners benefit from shared knowledge, improved health outcomes, and economic stability. By integrating traditional practices with modern techniques, addressing cultural barriers, and leveraging technological advancements, veterinary initiatives can ensure sustainable progress in working equid welfare while strengthening the communities they serve.

Self-Evaluation and Learning Enhancement: Working Equids

This self-evaluation section is designed to help clinicians and practitioners assess and expand their understanding of the challenges and solutions involved in the care and management of working equids. It incorporates practical tips, interactive elements, and case-based learning to ensure a comprehensive and engaging review experience.

1. **Practical Tips for Self-Assessment**
   - **Study Focus Areas**:
     - Traditional and modern medical practices for working equids.
     - Challenges and solutions in veterinary care delivery in resource-limited settings.
     - Advances in technology, telemedicine, and community engagement.
     - Disease management, including parasitic infestations, tetanus, and rabies.
     - Innovations in farriery, nutrition, and equipment.
   - **Suggested Materials**:
     - Research articles on One Health initiatives and their impact on working equids.
     - Case studies from projects like Project Samana or hospitals like the American Fondouk.
     - Guidelines for culturally sensitive interventions in veterinary medicine.
     - Manuals on modern surgical techniques and tools, such as the Henderson Castration Tool.
   - **Self-Assessment Strategies**:
     - Conduct a SWOT analysis (Strengths, Weaknesses, Opportunities, Threats) of your current knowledge of equid care.
     - Review regional veterinary policies and practices where working equids are prominent.
     - Participate in workshops or webinars focused on equid healthcare.
2. **Interactive Elements**
   - **Quizzes**:
     1. Identify key challenges faced by working equids in resource-limited settings.
     2. Match traditional practices (e.g., nostril splitting, bloodletting) with their historical origins and modern alternatives.
     3. Diagnose and propose treatment plans for common conditions such as tetanus, parasitic infestations, and lameness.
   - **Interactive Case Studies**:
     1. *Case 1*: A mule presenting with a lateral abdominal hernia caused by overloading. Discuss diagnosis, treatment options, and preventative measures.
     2. *Case 2*: A donkey with chronic tick infestation and anemia. Propose a treatment plan considering resource limitations.

3. *Case 3*: A working horse showing signs of tetanus. Detail the steps for clinical management, including diagnostics, treatment, and preventive vaccination protocols.

Responses:

Case 1: Mule with Lateral Abdominal Hernia from Overloading
Diagnosis

- **Clinical Signs**: Visible bulge on the lateral abdomen, soft and reducible mass, discomfort during movement.
- **Palpation**: Assess for hernia ring, contents (intestine vs. fat), and reducibility.
- **Ultrasound (if available)**: Confirms herniated contents and rules out strangulation.

Treatment Options

- **Surgical Repair** (preferred if viable):
  - Sedation and local anesthesia or general anesthesia depending on resources.
  - Hernia reduction and closure using absorbable sutures.
  - Post-op antibiotics and anti-inflammatories.
- **Conservative Management** (if surgery not feasible):
  - Controlled rest and use of abdominal support bandages.
  - Monitor for signs of strangulation (pain, colic, non-reducibility).

Preventative Measures

- **Load Management**: Educate owners on weight limits (no more than 20–25% of body weight).
- **Harness Fitting**: Proper saddle and pack placement to avoid pressure points.
- **Regular Rest**: Scheduled breaks during work to reduce strain.

Case 2: Donkey with Chronic Tick Infestation and Anemia
Treatment Plan (Resource-Conscious)

1. **Tick Control**
   - **Manual Removal**: Daily grooming with tick combs.
   - **Topical Treatments**:
     - *Low-cost options*: Neem oil or diluted permethrin (0.5%) applied weekly.
     - *If available*: Ivermectin pour-on or flumethrin collars.
   - **Environmental Management**:
     - Clear brush and debris from resting areas.
     - Use ash or lime around shelters to deter ticks.

2. **Anemia Management**
   - **Deworming**: Albendazole or fenbendazole to address parasitic causes.
   - **Iron-Rich Diet**:
     - Supplement with molasses, beet pulp, or iron-rich forage (e.g., legume hay).
     - Salt blocks with trace minerals.
   - **Supportive Care**:
     - Minimize workload until hematocrit improves.
     - Monitor mucous membranes and energy levels.
3. **Education**
   - Train owners on tick life cycles and grooming routines.
   - Encourage community-wide tick control to reduce reinfestation.

Case 3: Working Horse with Suspected Tetanus
Diagnostics

- **Clinical Signs**: Stiff gait, erect ears, third eyelid prolapse, hyperresponsiveness, difficulty chewing/swallowing.
- **History**: Recent wound or surgical procedure without tetanus prophylaxis.
- **Diagnosis**: Primarily clinical—lab confirmation is rare in field settings.

Treatment Protocol

1. **Emergency Care**
   - **Sedation**: Acepromazine or diazepam to reduce muscle spasms.
   - **Antibiotics**: Penicillin or metronidazole to eliminate *Clostridium tetani.*
   - **Tetanus Antitoxin**: Administer immediately if available.
   - **Wound Management**: Debride and flush with antiseptic (e.g., iodine or hydrogen peroxide).
2. **Supportive Care**
   - Quiet, dark environment to reduce stimuli.
   - Soft feed and water via syringe or nasogastric tube if needed.
   - Muscle relaxants (e.g., methocarbamol) if accessible.

Preventive Vaccination

- **Primary Protocol**:
  - Initial dose of tetanus toxoid.
  - Booster in 4–6 weeks.
  - Annual revaccination.
- **Post-Injury Protocol**:
  - Tetanus toxoid + antitoxin if vaccination history is unknown or outdated.

3. **Extended Exam**
   - **Objective**: Assess comprehensive knowledge of diagnostic interpretation, treatment planning, and community engagement.
   - **Format**:
     - 20 multiple-choice questions covering medical and surgical care, traditional practices, and technological innovations.
     - 5 short-answer questions on ethical dilemmas, such as managing EIA-positive animals in subsistence communities.
     - 2 long-form questions requiring detailed treatment planning for cases involving working equids.
4. **Answers and Explanations**
   - Provide detailed answers for quizzes and exam questions.
   - Include evidence-based explanations for treatment decisions and diagnostic approaches.
   - Highlight the importance of integrating cultural and economic considerations into care delivery.
5. **Case Studies for Critical Thinking**
   - **Case-Based Questions**:
     1. How would you address resistance to castration in a culturally sensitive manner in a community reliant on intact stallions for economic or symbolic reasons?
     2. Propose a step-by-step intervention plan for introducing modern veterinary practices to a rural area dominated by traditional care methods.
   - **Expanded Scenarios**:
     1. *Scenario 1*: A local leader requests your assistance in establishing a community-led rabies vaccination program. Develop a plan that includes logistics, training, and community engagement.
     2. *Scenario 2*: A working equid hospital is proposed for a remote area with a high density of donkeys and mules. Assess the feasibility and propose a model for sustainable operations, including funding and local training.

Self-Evaluation Questionnaire: Working Equids

This questionnaire is designed to test your understanding of the challenges, solutions, and practices related to the care of working donkeys, horses, and mules. Each question has one correct answer. Use the key at the end to check your responses.

1. **Which historical practice is still observed today as a traditional treatment for working equids?**
   A. Vaccination
   B. Nostril splitting
   C. Use of antibiotics
   D. Laser therapy

2. **What is one of the main reasons for the limited availability of veterinary care for working equids in rural areas?**
   A. Lack of qualified veterinarians
   B. Focus of government resources on livestock for food and fiber
   C. High population of equids compared to livestock
   D. Laws preventing the import of veterinary medicines
3. **What innovation significantly reduced postoperative bleeding during castration in resource-limited settings?**
   A. Laser cauterization
   B. Henderson Castration Tool
   C. Equi-Ligature Device
   D. Cryosurgery
4. **What is the primary goal of introducing imprinting programs for mules and donkeys?**
   A. To improve their reproductive efficiency
   B. To reduce aggressive behavior and improve handling
   C. To increase their load-carrying capacity
   D. To enhance their immune response to diseases
5. **What was the primary reason for introducing rapidly growing tropical grasses in the Dominican Republic?**
   A. To prevent overgrazing in the region
   B. To provide a sustainable source of nutrition for working equids
   C. To attract more tourists with lush greenery
   D. To create an alternative cash crop for local farmers
6. **Why are donkeys and mules often not castrated in many parts of the world?**
   A. Lack of trained veterinarians
   B. Cultural resistance to the procedure
   C. Fear of reducing their strength for work
   D. All of the above
7. **What institution has been pivotal in advancing research and treatment protocols for working equids?**
   A. The Royal Veterinary College
   B. The American Fondouk in Morocco
   C. The World Health Organization
   D. The United Nations Equid Program
8. **What is a common barrier to humane euthanasia in resource-limited settings?**
   A. Religious or cultural beliefs
   B. Lack of appropriate euthanasia agents
   C. Environmental concerns related to carcass disposal
   D. All of the above
9. **How has technology improved veterinary care for working equids in remote areas?**
   A. It has reduced the need for physical equipment in the field.
   B. It allows real-time consultations and support through telemedicine tools.
   C. It replaces the need for local veterinary expertise.
   D. It eliminates traditional care practices.

10. **What key factor ensures the sustainability of veterinary care projects for working equids?**
    A. Regular importation of foreign veterinary teams
    B. Strong local leadership and community ownership
    C. Reliance on charity funding
    D. Exclusive focus on emergency care

Answer Key

1. **B** - Nostril splitting
2. **B** - Focus of government resources on livestock for food and fiber
3. **B** - Henderson Castration Tool
4. **B** - To reduce aggressive behavior and improve handling
5. **B** - To provide a sustainable source of nutrition for working equids
6. **D** - All of the above
7. **B** - The American Fondouk in Morocco
8. **D** - All of the above
9. **B** - It allows real-time consultations and support through telemedicine tools
10. **B** - Strong local leadership and community ownership

Use the key to assess your responses and identify areas where further study or review may be beneficial.

References

AAEP (2025) Rabies vaccination guidelines. https://aaep.orgresource/rabies-vaccination-guidelines

AKC (2024) Veterinary outreach scholarship recipients. https://www.akc.orgpress-releases/american-kennel-club-announces-2024-2025-veterinary-outreach-scholarship-recipients

APHIS (2025) Equine disease overview. Animal and Plant Health Inspection Service. Equine Disease Resource

Abou-Khalil NS, Ali MF, Ali MM, Ibrahim A (2020) Surgical vs. chemical castration in donkeys: stress and redox biomarkers. BMC Vet Res 16(310). https://doi.org/10.1186/s12917-020-02530-0

Adeyemo KS, Mbata AO, Balogun OD (2021) The role of cold chain logistics in vaccine distribution: addressing equity and access challenges in Sub-Saharan Africa. Int J Multidiscipl Res Growth Eval 2(1):893–901. https://doi.org/10.54660/.IJMRGE.2021.2.1.893-901

Amare E, Haben F (2020) Hernias in farm animals and its management technique: a review. Int J Clin Stud Med Case Rep 4(1):1–6. https://ijclinmedcasereports.compdf/IJCMCR-RW-00091.pdf

American Association of Equine Practitioners (AAEP) (2020) Strangles vaccine guidelines. https://aaep.orgwp-contentuploads202403Strangles_Vaccine_Guidelines_2020.pdf

American Association of Equine Practitioners (AAEP) (2021) Equine Piroplasmosis Guidelines. https://aaep.orgwp-contentuploads202402AAEPEquinePiroplasmosisGuidelines2021.pdf

American Association of Equine Practitioners (AAEP) (2025) Tetanus Vaccination Guidelines. https://aaep.org/resource/tetanus-vaccination-guidelines/

American Fondouk (2025) Our clinic. https://www.fondouk.orgour-clinic/

BRAM-COR (2025) Turnkey projects for pharmaceutical veterinary products. https://www.bram-cor.comenen-pharmaceutical-turnkey-projectsturnkey-projects-for-pharmaceutical-veterinary-products
Baldwin CM (2024) A review of prevention and management of castration complications. Equine Veterin Educ 36:97–106. https://doi.org/10.1111/eve.13880
Barkia A (2021) Rabies in Morocco: World Rabies Day presentation. Fondation Mérieux. https://www.fondation-merieux.orgwp-contentuploads202107world-rabies-day-2021-aziz-barkia.pdf
Bertin F-R, Boyle AG (2025) Equine metabolic syndrome. MSD Veterinary Manual. https://www.msdvetmanual.commetabolic-disordersequine-metabolic-syndromeequine-metabolic-syndrome
Blakeway S (2010) Charity vaccination policy and population coverage. The Donkey Sanctuary
Bonsi M, Anderson NE, Carder G (2023) Socioeconomic impact of diseases of working equids. Animals 13(24):3865. https://doi.org/10.3390/ani13243865
Brooke (2013) Welfare interpretation manual: Castration. https://www.thebrooke.orgsitesdefaultfilesAnimal%20WelfareWelfare%20Interpretation%20Manual-%20Chapter%202%2C%20Castration%20%281%29.pdf
Brooke (2024) The Working Equid Veterinary Manual. Veterinary manual overview https://www.thebrooke.orgour-work/working-equid-veterinary-manual
Brooke (2025a) Causes of welfare issues in working horses and donkeys. https://www.thebrooke.orgour-workwe-work-animalscauses-welfare-issues-working-horses-and-donkeys
Brooke (2025b) Invisible workers: the economic contributions of working donkeys, horses and mules to livelihoods. Brooke. Retrieved from Brooke Publications
Brooke (2025c) Stopping harness-related injury and suffering. https://www.thebrooke.orgour-work/stopping-harness-related-injury
Brooke (2025d) Working equids & sustainable development. https://www.thebrooke.orgour-work/working-equids-sustainable-development
Burden FA, Bell N (2019) Donkey nutrition and malnutrition. Vet Clin N Am Equine Pract 35(3):469–479. https://doi.org/10.1016/j.cveq.2019.08.004
Burn CC, Dennison TL, Whay HR (2022) Improving the welfare of working equids: a global perspective. Front Veterin Sci 9:1256501. https://doi.org/10.3389/fvets.2022.1256501
Canadian Veterinary Medical Association (2024) Surgical castration of horses, donkeys, and mules. https://www.canadianveterinarians.netpolicy-and-outreachposition-statementsstatements/surgical-castration-of-horses-donkeys-and-mules
Carvelli A, Nardini R, Carnio A et al (2024) Equine Infectious Anaemia: the active surveillance of an entire equid population reduces the occurrence of the infection. Transbound Emerg Dis: Article ID 3439871. https://doi.org/10.1155/2024/3439871
Centers for Disease Control and Prevention (CDC) (2025) Global Rabies Work. https://archive.cdc.govwww_cdc_govncezidstories-featuresglobal-storiesrabies-work.html
Cousquer G (2011) Management of saddle and harness sores in a pack mule. Wounds International. https://www.academia.edu25302064
Cousquer G, Alyakine H, Lindsay-McGee V (2023) The history and welfare of working mules in the High Atlas of Morocco. Front Veterin Sci 10:1256501. https://doi.org/10.3389/fvets.2023.1256501
Câmara RJF, Bueno BL, Resende CF, Balasuriya UBR, Sakamoto SM, dos Reis JKP (2020) Viral diseases that affect donkeys and mules. Animals 10(12):2203. https://doi.org/10.3390/ani10122203
Darkaoui S, Cliquet F, Wasniewski M, Robardet E, Aboulfidaa N, Bouslikhane M, Fassi-Fihri O (2017) A century spent combating rabies in Morocco (1911–2015): how much longer? Front Veterin Sci 4:78. https://www.frontiersin.orgarticles10.3389fvets.2017.00078/full
Dennis ST, El Hage CM, Brookes V (2024) A survey of veterinarians' practices and perceptions associated with tetanus prevention in horses. Aust Vet J 102(3):145–152. https://doi.org/10.1111/avj.13144

Desanti-Consoli H, Bouillon J, Chapuis RJJ (2022) Equids' core vaccines guidelines in North America. Vaccine 10(3):398. https://doi.org/10.3390/vaccines10030398
Donkey Sanctuary (2018) Guide to donkey castration. https://www.thedonkeysanctuary.org.ukfor-professionalsprofessional-resourcesguide-to-donkey-castration
Díaz-Sánchez AA, Roblejo-Arias L, Marrero-Perera R, Corona-González B (2020) Piroplasmosis equina. Revista de Salud Anim 42(1):1–12. http:/www.scielo.sld.cuscielo.php?script=sci_arttext&pid=S0253-570X2020000100002
Duguma BE, Tesfaye T, Kassaye A, Kassa A, Blakeway SJ (2021) Control and prevention of epizootic lymphangitis in mules: an integrated community-based intervention, Bahir Dar, Ethiopia. Front Veterin Sci 8:648267
ECCVT (2020) Impact of Digital Technologies & AI in Veterinary Education. Final report
EURCAW Ruminants & Equines (2023) Guidance on the welfare of working equids in tourism. Tourism welfare guidance
EURCAW Ruminants & Equines (2024) Review—Welfare of working equids in the European Union. https://doi.org/10.5281/zenodo.13693640
El Refugio del Burrito (2017, September 1) Donkey welfare in tourism. https://www.elrefugiodelburrito.orgenarticlesdonkey-welfare-in-tourism
Equine Veterinary Journal (2024) Equine infectious anaemia in Europe: Time to re-examine the efficacy of monitoring and control protocols. https://beva.onlinelibrary.wiley.comdoipdf10.1111/evj.12466
Equitarian Initiative (2011) 1st Equitarian Workshop logistics. https://www.slideshare.netslideshow2nd-equitarian-workshop10967186
Equitarian Initiative (2025) Annual conference overview. https://www.projectsamana.orgequitarian-initiative
European Commission (2022) Guide for good animal welfare practice for the keeping, care, training and use of donkeys and donkey hybrids. https://food.ec.europa.eusystemfiles2022-12/aw_platform_plat-conc_guide_equidae-donkey_training.pdf
European Commission (2025) African horse sickness: disease control measures. https://food.ec.europa.euanimalsanimal-diseasesdiseases-and-control-measures/african-horse-sickness_en
European Medicines Agency (2005) Ivermectin summary report. https://www.ema.europa.euendocumentsmrl-report/ivermectin-summary-report-1-committee-veterinary-medicinal-products_en.pdf
Evans L et al (2022) Infectious diseases. In: The clinical companion of the donkey, 2nd edn. The Donkey Sanctuary. https://www.ivis.orglibraryclinical-companion-of-donkey-2nd-editioninfectious-diseases
Fondouk (2024) Services and research at the American Fondouk. https://www.fondouk.orgour-services
FormaHoof (2025) Laminitis in donkeys: causes and rehabilitation. https://formahoof.comlaminitis-in-donkeys
GALVmed (2024) Improving access to quality veterinary products through better regulation in sub-Saharan Africa. https://www.galvmed.orgnewsnew-project-to-improve-access-to-quality-veterinary-products-through-better-regulation-in-sub-saharan-africa
Getachew M, Burden FA, Wernery U et al (2016) Common infectious diseases of working donkeys: their epidemiological and zoonotic role. J Equine Vet Sci 39(Suppl):107
Gichure MN, Olayide O (2024) Selected intervention strategies to improve health and welfare of working donkeys in Kenya: a narrative review. Asian J Res Anim Veterin Sci 7(1):69–82. https://doi.org/10.9734/ajravs2024/v7i1287
Gongora J, van Gelderen I, Vost M et al (2021) Embedding cultural competence in veterinary education. J Vet Med Educ 48(2):123–135. https://doi.org/10.3138/jvme-2021-0113
Guerra Monterroza T, Consuegra Solano EM (2024) Integration of local and traditional knowledge into the rural curriculum of Sahagún, Colombia. LATAM Revista Latinoamericana de Ciencias Sociales y Humanidades 5(4):2434–2443. https://doi.org/10.56712/latam.v5i4.2429

Haddy E, Brown J, Burden F et al (2022) Evaluating approaches to improving working equid welfare. PLoS One 17(9):e0273972. https://doi.org/10.1371/journal.pone.0273972

Haddy E, Brown J, Burden F et al (2023) Sustainability in NGO programming. SAGE Open 13(4). https://doi.org/10.1177/21582440231218523

Haddy E, Burden F, Fernando-Martínez JA et al (2021) Long-term welfare initiatives in Mexico. PLoS One 16(5):e0251002. https://doi.org/10.1371/journal.pone.0251002

Hanlon A (2023a) Guidance for competent authorities and tourism operators. EU tourism welfare guide

Hanlon A (2023b) Guidance for tourism operators on equid welfare. https://storage.googleapis.comworldhorsewelfare-cloud202504/bf889fe8-aw_platform_plat-conc_guide_equidae_work_tourism_en.pdf

Hanlon A (2023c, January 26) Guidance on the welfare of working equids in tourism. EURCAW Ruminants & Equines. https://www.eurcaw-ruminants-equines.euguidance-on-the-welfare-of-working-equids-in-tourism/

He L, Liu C, Nie X et al (2025) Global analysis of tetanus incidence and mortality in children under 5 years. BMJ Paediatrics Open 9(1):e003326. https://bmjpaedsopen.bmj.comcontent91/e003326

Higham L (2022, June 20) Using medicines responsibly. Vet Sustain. https://vetsustain.orgwork/using-medicines-responsibly

Hodges M (2022a) Imprinting: mule, donkey & horse training. Lucky Three Ranch. https://www.luckythreeranch.comtag/imprinting

Hodges M (2022b, August 27) Mule crossing: imprinting beyond birth. Lucky Three Ranch. Mule imprinting and training insights

ICWE (International Coalition for Working Equids) (2023) Working equids in disaster response. https://spana.orgpress-releasesicwe-the-international-coalition-for-working-equids-attends-beva-congress-2023

ICWE (2024) Achieving Agenda 2030: How the welfare of working animals delivers for development. https://www.icweworkingequids.orgwp-contentthemesicwepdfsdg-brochure.pdf

Iatridou D, Bravo A, Saunders J (2020) One health interdisciplinary collaboration in veterinary education establishments in Europe. J Vet Med Educ 48(4):456–470. https://doi.org/10.3138/jvme-2020-0019

Innovation News Network (2025) Transforming veterinary medicine with the One Health approach. One Health analysis

Ivey JL, Strickland L, Case T (2023) Equine diseases and vaccines. University of Tennessee Extension. UTIA Publication

Jarm B (2023a) Amira's recovery at the American Fondouk. Case study

Jarm B (2023b) Amira's recovery at the American Fondouk. https://www.fondouk.orginitiatives/amira

Jiménez R (2020) Cambios en los patrones culturales machistas. Revista Latinoamericana de Educación Inclusiva 14(2) https://www.scielo.clscielo.php?script=sci_arttext&pid=S0718-73782020000200017

Kay G (2015) Clinical practice in a hospital for working equids: Morocco's American Fondouk. In: AAEP annual convention proceedings. Retrieved from IVIS library. https://www.ivis.orglibraryaaepaaep-annual-convention-las-vegas-2015/clinical-practice-a-hospital-for-working-equids-morocco%E2%80%99s-american-fondouk

Kentucky Equine Research (2013) Nutritional support of the immune system in horses. https://ker.comequinewsnutritional-support-immune-system-horses

Khallikane S, Kbiri H, Haji K et al (2021) An emblematic case of tetanus in a young adult vaccinated. Asian J Res Rep Neurol 4(1):107–115. https://journalajorrin.comindex.phpAJORRINarticleview38

Kilcoyne I, Watson JL, Kass PH, Spier SJ (2013) Complications of castration in equids: 324 cases. J Am Vet Med Assoc 242(6):820–825. https://avmajournals.avma.orgviewjournalsjavma2426/javma.242.6.820.xml

Kisslinger T (2025, April 25) More than medicine: how veterinarians strengthen communities and ecosystems. Veterinarians Without Borders. https://www.vwb.orgsiteblog20250425/more-than-medicine

Knottenbelt DC (2005) Skin disorders of donkeys. In: Matthews N, Taylor T (eds) Veterinary care of donkeys. IVIS. https://www.ivis.orglibraryveterinary-care-of-donkeysskin-disorders-of-donkeys-0

Kubasiewicz LM, Watson T, Norris SL, Chamberlain N, Nye C, Burden FA (2023) One welfare: linking poverty, equid ownership and equid welfare in the brick kilns of India. Anim Welf 31(4): 517–528. https://doi.org/10.1017/S0962728600032504

Lascola KM, Boone L (2024) Strangles in horses. In: MSD veterinary manual. https://www.msdvetmanual.comrespiratory-systemrespiratory-diseases-of-horsesstrangles-in-horses

Lederman Z, Magalhães-Sant'Ana M, Voo TC (2021) Stamping out animal culling: from anthropocentrism to One Health ethics. J Agric Environ Ethics 34(27) https://link.springer.comarticle10.1007s10806-021-09868-x

Lerner H, Berg C (2017) A comparison of three holistic approaches to health: One Health, EcoHealth, and planetary health. Front Veterin Sci 4:163. https://doi.org/10.3389/fvets.2017.00163

Lovett AL, Riley CB, Chapman V, Bell B, Bishop B, Grierson A, Johnstone LJ (2024) Tetanus prophylaxis in horses: guidelines for New Zealand and Australia. N Z Vet J 72(5):241–255. https://doi.org/10.1080/00480169.2024.2365283

Lucky Three Ranch (2022) Mule training and tourism suitability. https://www.luckythreeranch.commule-crossing-imprinting-beyond-birth/

Mad Barn Research Bank (2025) Hoof disorders and farriery in the donkey. https://madbarn.comresearch/hoof-disorders-and-farriery-in-the-donkey

Madigan J (ed) (2019) Insights on donkey, mule, and horse welfare: causes, solutions, and prospects. Special issue in animals, MDPI. Retrieved from MDPI Animals Journal

Marzok MA, Moustafa AM (2021) Acquired non-incarcerated inguinalscrotal herniation in working donkeys (Equus asinus). Bulgarian J Veterin Med. https://www.researchgate.netpublication/349350745

Maven RS (2024) Global veterinary regulatory affairs: compliance and safety. https://www.mavenrs.comGlobal-Regulatory-Affairs-for-Veterinary-Products-Ensuring-Compliance--Safety

Merriam JG (2015a) Equitarian projects and sustainability. AAEP Annual Convention. https://www.ivis.orglibraryaaepaaep-annual-convention-las-vegas-2015/equitarian-projects-and-sustainability

Merriam JG (2015b) Equitarian projects and sustainability: twenty years of work in Samana. AAEP Annual Convention.. https://www.ivis.orglibraryaaepaaep-annual-convention-las-vegas-2015/equitarian-projects-and-sustainability-twenty-years-of-work-samana-and-why-our-presence-must-be

Merriam JG, Hernandez-Gil M, The Equitarian Initiative (2012) Training veterinarians for working equid care: the New Equitarians. AAEP Annual Convention. https://www.ivis.orglibraryaaepaaep-annual-convention-anaheim-2012/training-veterinarians-for-working-equid-care-new-equitarians

Merridale-Punter MS, Wodajo AL, Elias B, Bakos AM, Zewdu H, Tesfaye R, Hailegebreal G, Sori T, El-Hage CM, Wiethoelter AK, Hitchens PL (2024) Equipment-related wounds and associated risk factors in working equids of the Oromia national regional state in Ethiopia. Anim Welf 33:e42. https://doi.org/10.1017/awf.2024.52. PMID: 39600354; PMCID: PMC11589070

Miller RM (2022, November 7) Not just for horses: effectiveness of imprinting in other species confirms its value. Robert M. Miller Communications. Imprinting techniques overview

Mills WD, Booth C (2022) Marines need a few good mules. US Naval Institute Proceedings 148(4) https://www.usni.orgmagazinesproceedings2022aprilmarines-need-few-good-mules

Mitiku T, Alehegn A, Kefale D (2022) A review on assessment of cause and management of wound in donkey. Int J Adv Res Biol Sci 9(3):77–85. https://ijarbs.compdfcopy2022mar2022/ijarbs9.pdf

Mobile Veterinary Services (2025) How to keep donkeys and mules healthy. https://www.mvsequine.comhow-to-keep-donkeys-and-mules-healthy/
Mohite DS, Sheikh CS, Singh S et al (2019) Barriers to farriery interventions in India. Animals 9(5): 252. https://doi.org/10.3390/ani9050252
Molloy RG, Moran KT, Waldron RP, Brady MP, Kirwan WO (1991) Massive incisional hernia: abdominal wall replacement with Marlex™ mesh. Br J Surg 78(2):242–244. https://academic.oup.combjsarticle782242/6173668
Moustafa A, Elmetwally M, EL-Khodery S, Hamed M, Gomaa N, Rizk MA (2022) Abdominal hernia in equine: risk factors and repair using polypropylene mesh. J Equine Vet Sci 111: 103889. https://doi.org/10.1016/j.jevs.2022.103889
Muca E, Cavallini D, Odore R et al (2022) Are veterinary students using technologies and online learning resources? Educ Sci 12(8):573. https://doi.org/10.3390/educsci12080573
NIH Office of Animal Care and Use (OACU) (2025) Guidelines for the select use of expired medical products. https://oacu.oir.nih.govsystemfilesmediafile2025-01/b15-expiredmedicalproducts.pdf
Our World in Data (2024) Tetanus. https://ourworldindata.orgtetanus
Pearson RA (2005) Nutrition and feeding of donkeys. In: Matthews N, Taylor T (eds) Veterinary care of donkeys. IVIS. https://www.ivis.orglibraryveterinary-care-of-donkeysnutrition-and-feeding-of-donkeys-0
Pearson RA, Simalenga TE, Krecek RC (2003) Harnessing and hitching donkeys, mules and horses for work. Centre for Tropical Veterinary Medicine, University of Edinburgh. http://www.link.vet.ed.ac.ukctvmResearchDAPRTraining%20PublicationsHarness%20Hitching%20Donkeys/Harness%20Hitching%20donkeys%20Oct06.pdf
Pritchard JC, Lindberg AC, Main DCJ, Whay HR (2005) Assessment of the welfare of working horses, mules and donkeys, using health and behavior parameters. Prev Vet Med 69(3–4): 265–283. https://doi.org/10.1016/j.prevetmed.2005.02.002
Project Samana (2025) Veterinary and education outreach in the Dominican Republic. https://www.projectsamana.org
Proops L, Okumu TA, Haddy E (2024) "When the donkeys start dying, the people are next": working animals as agents of climate-related drought resilience in marginalised communities. The British Academy. https://www.thebritishacademy.ac.ukprojectswhen-the-donkeys-start-dying-the-people-are-next-working-animals-as-agents-of-climate-related-drought-resilience-in-marginalised-communities
Racine J, Vidondo B, Ramseyer A, Koch C (2018) Complications of closed castration using the Henderson tool in 300 equids. Vet Surg 48(1):21–28. https://doi.org/10.1111/vsu.12960
Rebexa Group (2024) Global regulatory affairs for veterinary products: ensuring compliance & safety
Roberts E (2023) Advancements in telemedicine for veterinary care: a comparative study. J Veterin Sci Technol 14(3). https://doi.org/10.37421/2157-7579.2023.14.180
Royal (Dick) School of Veterinary Studies (2024) Strangles vaccination: Strangvac update. https://vet.ed.ac.ukclinicalequinepracticenewsstrangles-vaccine
SPANA (2023a) Equid health and hospital-based research. https://spana.orgpress-releasesicwe-the-international-coalition-for-working-equids-attends-beva-congress-2023
SPANA (2023b) Preventing horse lameness through good farriery. Farriery report
SPANA (2023c, September 12) ICWE attends BEVA Congress 2023. https://spana.orgpress-releasesicwe-the-international-coalition-for-working-equids-attends-beva-congress-2023
SPANA (2025) Working animal harnessing. https://spana.orgworking-animalsworking-animal-healthworking-animal-harnessing/
Smith DG, Burden FA (2016) Practical donkey and mule nutrition. In: Veterinary care of donkeys. Veterian Key. https://veteriankey.compractical-donkey-and-mule-nutrition/
Springer A, Glass A, Topp A-K, Strube C (2020) Zoonotic tick-borne pathogens in temperate and cold regions of Europe. Front Veterin Sci 7:604910. https://doi.org/10.3389/fvets.2020.604910
Tadich TA (2023) Working equids: linking human and animal welfare. Veterinary Record, University of Chile. https://repositorio.uchile.clhandle2250181443

Teller LM, Moberly HK (2020) Veterinary telemedicine: a literature review. Veterin Evid 5(4). https://doi.org/10.18849/ve.v5i4.349
The Donkey Sanctuary (2017, December 4) Tetanus in donkeys. https://www.thedonkeysanctuary.org.ukfor-ownersowners-resourcestetanus-in-donkeys
The Donkey Sanctuary (2020a, December 29) Vaccinating your donkeys. https://www.thedonkeysanctuary.org.ukfor-ownersowners-resourcesvaccinating-your-donkeys
The Donkey Sanctuary (2020b, July 23) Skin diseases in donkeys. https://www.thedonkeysanctuary.org.ukfor-ownersowners-resourcesskin-diseases-in-donkeys
The Donkey Sanctuary (2020c, May 29) How to train donkeys and mules using the science of behaviour. https://www.thedonkeysanctuary.org.ukfor-ownersowners-resourceshow-to-train-donkeys-and-mules-using-the-science-of-behaviour
The Donkey Sanctuary (2021) From wildness to work: the global story of donkeys. https://www.thedonkeysanctuary.org.ukresearchpaper3204
The Donkey Sanctuary (2023a) Castration CLNC111 course. https://thedonkeysanctuary.eu-west.catalog.canvaslms.combrowsecastrationcourses/castration
The Donkey Sanctuary (2023b) Donkeys are not small horses: understanding species-specific needs. https://www.thedonkeysanctuary.org.ukresearchpaper3204
The Donkey Sanctuary (2025) Research findings and clinical companion series. https://www.thedonkeysanctuary.org.ukresearch/
Thiemann AK (2019) EMS and PPID in donkeys. The Donkey Sanctuary. https://www.thedonkeysanctuary.org.ukresearchpaper2688
Thiemann AK et al (2023) Laminitis in donkeys compared to horses: recognition, treatment and prevention. The Donkey Sanctuary. https://www.thedonkeysanctuary.org.ukresearchpaper3144
Thunes C (2017, February 15) Nutrition and immunity in the equine. IVC J. https://ivcjournal.comnutrition-immunity-equine/
Travel Doctor Network (2025) Rabies risk and prevention in Morocco. https://www.traveldoctor.networkcountrymoroccoriskrabies
Turner TA, Turoff DR, Haffner JC (2015) Simplified castration with the Equi-Twister. AAEP Annual Convention.. https://www.ivis.orglibraryaaepaaep-annual-convention-las-vegas-2015/simplified-castration-equitwister
UNICEF (2021) Immunization supply chain interventions for remote rural settings. https://www.unicef.orgmedia96611file/Immunization%20supply%20chain%20interventions.pdf
UNPHU & Project Samana (2012–present) Veterinary education partnership. UNPHU collaboration
USDA APHIS (2006) Vaccination practices on U.S. equine operations. https://www.aphis.usda.govsitesdefaultfiles/equine05_is_vaccination.pdf
University of Minnesota (2025) Hernias – Large Animal Surgery – Supplemental Notes. https://open.lib.umn.edulargeanimalsurgerychapterhernias/
University of Wisconsin-Milwaukee (UWM) IACUC (2023) Guideline for expired drugs and medical materials. https://uwm.eduanimal-carewp-contentuploadssites404202401/New-approved-8.25.23-Expired-drugs-and-materials.pdf
Verma S, Malik YS, Singh G, Dhar P, Singla AK (2024) International trade and regulations in animals and animal products. In: Core competencies of a veterinary graduate. Springer, pp 281–285. https://link.springer.comchapter10.1007978-981-97-0433-0_21
Village Earth (2025) Working equids (donkeys, mules and horses) driving sustainable livelihoods. https://villageearth.orgworking-equids-donkeys-mules-and-horses-driving-sustainable-livelihoods/
WOAH (2024) Welfare of working equids – Chapter 7.12. https://www.woah.orgfileadminHomeengHealth_standardstahccurrentchapitre_aw_working_equids.pdf
Warren LK (2016) Optimizing nutrition to improve immune function in horses. J Anim Sci 94 (suppl_5):392–393. https://doi.org/10.2527/jam2016-0817

Watson T, Kubasiewicz LM, Nye C, Thapa S, Chamberlain N, Burden FA (2023) Welfare and access to veterinary health services of mules working the mountain trails in the Gorkha region, Nepal. Aust J Veterin Sci 55(1):9–22. https://www.ajvs.clindex.phpajvsarticleview297
Weyer C (2025) African horse sickness. In: MSD veterinary manual. https://www.msdvetmanual.comgeneralized-conditionsafrican-horse-sicknessafrican-horse-sickness
Wild I (2023) Vet partnerships and community engagement case studies. WOAH – Europe. https://rr-europe.woah.orgappuploads20231008_icwe_case_study.pdf
Williams C, Toledo A (2024) Ticks, Lyme disease, and other tick-borne diseases in horses. Rutgers Equine Science Center. https://esc.rutgers.edufact_sheet/ticks-lyme-disease-and-other-tick-borne-diseases-in-horses
World Health Organization (WHO) (2025a) Essential Programme on immunization: supply chain and logistics. https://www.who.intteamsimmunization-vaccines-and-biologicalsessential-programme-on-immunizationsupply-chain
World Health Organization (WHO) (2025b) Veterinary vaccines for rabies control. https://www.who.intteamscontrol-of-neglected-tropical-diseasesrabiesvaccinations-and-immunizationveterinary-vaccines
World Horse Welfare (2025) Donkey and mule nutrition essentials. https://www.worldhorsewelfare.orgadvice/donkey-and-mule-nutrition-essentials
World Organisation for Animal Health (WOAH) (2023) Vaccine banks and global deployment strategies. https://www.woah.orgenwhat-we-offerimproving-veterinary-servicesvaccine-banks
World Organisation for Animal Health (WOAH) (2025) African horse sickness. https://www.woah.orgendiseaseafrican-horse-sickness/
Zafra S (2023) The contribution of working equids towards sustainable development: presentation for World Horse Welfare. SDG presentation

Dairy Donkey Farming: Industry Development and Insight into Milk Composition and Quality

Tommaso Danese, Federica Raspa, Emanuela Valle, Laura Cavallarin, and Marzia Giribaldi

Abstract

This chapter investigates the history, industry and importance of lactating donkeys and examines how their needs influence the quality of the milk. Key topics include:

1. **The dairy donkey industry: evolutions, breeds and geographical utilization.** This subchapter examines how the global dairy donkey industry is expanding through breed specialization and regional adaptation, driven by rising demand for hypoallergenic, functional products despite low milk yields and market development.
2. **Milk Composition and Quality**: this subchapter examines the constituents of donkey milk, such as proteins, fats, lactose, vitamins, and minerals. It discusses how diet influences these components and consequently affects milk quality and the unique properties of donkey milk; its similarity to human milk and its hypoallergenic nature are highlighted.

Keywords

Donkey milk · Dairy donkey industry · Allergy · Milk quality and composition

---

T. Danese
Department of Veterinary Sciences, University of Parma, Parma, Italy

F. Raspa · E. Valle
Department of Veterinary Sciences, University of Turin, Turin, Italy

L. Cavallarin · M. Giribaldi (✉)
Institute of Sciences of Food Production, National Research Council, Grugliasco, Italy
e-mail: marzia.giribaldi@cnr.it

A. K. McLean et al. (eds.), *Donkey and Mule Medicine: Foundations, Breeds, and Management*, https://doi.org/10.1007/978-3-032-12726-6_8

1 The Dairy Donkey Industry: Evolutions, Breeds and Geographical Utilization

1.1 Industry and Its Evolution

The dairy donkey sector is rapidly evolving within the global livestock industry. Donkeys (Equus asinus), first domesticated around 6000 to 7000 years ago in Northeast Africa, are historically significant animals. They were primarily used for transportation and agricultural work, due to their resilience and ability to thrive in tough environments (Aspri et al. 2017; Seyiti and Kelimu 2021). For many years, donkey milk was a specialty item, particularly appreciated in Mediterranean and Middle Eastern cultures for its nutritional, cosmetic, and therapeutic benefits (Aspri et al. 2017). In the nineteenth century, in Italy, France, and Switzerland, donkey milk was increasingly used as a substitute for human breast milk when unavailable, providing nourishment for orphans, and the elderly (Aspri et al. 2017). However, with the industrialization and mechanization of farming, the donkey population saw a significant decline across Europe, with reductions reported between 80% and 97% in nations such as Italy and Greece (Camillo et al. 2018). In the last decades, research has revealed the hypoallergenic qualities of donkey milk, its nutrient profile being overall similar to that of human milk, and its bioactive properties, including antimicrobial and antioxidant effects (Aspri et al. 2017). Currently, Europe and China are the primary centers for the advancement of the dairy donkey industry. In Europe, especially in Italy, Spain, and France, farms typically employ semi-extensive systems that prioritize animal welfare, guided by new best-practice recommendations (Dai et al. 2019). China has witnessed a swift industrial expansion in the donkey sector, particularly in the provinces of Shandong, Xinjiang, and Hebei, where donkey milk, meat, and cosmetic products are attracting investment (Seyiti and Kelimu 2021). Despite this growth, the average milk yield remains low (less than 1 liter daily), presenting an economic hurdle (Aspri et al. 2017). At present, the global donkey milk market is on an upward trajectory, as consumers increasingly seek functional alternatives to conventional dairy products, which are seen as nutritious, therapeutic, and hypoallergenic. The global market was valued at around USD 28.1 million in 2019 and is anticipated to reach USD 69.1 million by 2027 (Keitshweditse et al. 2024). Nonetheless, market penetration is uneven: in countries like Italy and France, consumers are willing to pay significantly higher prices for donkey milk (up to 10 to 30 times the cost of cow's milk), while in developing areas such as South Africa, the willingness to pay is considerably lower (only a 10% premium over cow's milk). It's essential to recognize that the positioning of donkey milk in the market differs by region: while the European and Chinese markets highlight its health advantages and hypoallergenic nature, in South Africa, it is seen as a luxury item. Globally, the factors driving demand for these products vary: some small niche investors are beginning to emerge as potential buyers, but broader awareness initiatives will be necessary to fully realize the sector's growth potential (Keitshweditse et al. 2024).

1.2 Breeds and Geographical Utilization

Breeds of donkeys have developed over time in a variety of environmental settings, primarily selected for their durability and strength rather than their ability to produce milk. Strategies for breed usage and preservation are currently changing to meet the demands of the dairy sector (Camillo et al. 2018). In Europe, the Italian breeds Ragusano and Martina Franca and Amiata, Baudet du Poitou from France, and Zamorano-Leonés from Spain have been maintained and are being more actively used for milk production. Conservation efforts are crucial as many of these breeds are currently considered to be under some form of threat, and there is ongoing work to highlight these breeds with selective breeding and welfare such as Romalongo breed (Camillo et al. 2018) (refer to Fig. 1).

China has a showcase of the region's agro-ecological diversity through the donkey dairy industry. Large breeds such as Dezhou and Guanzhong are dominant in the intensive dairy systems of the eastern provinces while smaller drought tolerant Xinjiang donkeys are used in the west (Seyiti and Kelimu 2021). Chinese donkeys have relatively low milk production, with lactation yields averaging between 100–150 kg over 300 days (Meena et al. 2024), far lower than conventional dairy species but improving through better management practices.

Fig. 1 Romagnolo's donkey mare's and foal. (From the author's repertoire)

Fig. 2 The morning milking of a Ragusano's donkey. (From the author's repertoire)

Recent studies on production have highlighted that both environmental and physiological elements greatly impact donkey milk yield. For example, seasonality, milking frequency, and breed selection all have a big impact on daily milk supply and composition. Under controlled circumstances, Martina Franca donkeys have demonstrated optimal yields when milked up to three times daily, while Ragusano jennies have produced up to 489 kg of milk each lactation when milked twice daily (Meena et al. 2024). However, increased milking frequency beyond this threshold negatively impacts udder health, as demonstrated by elevated somatic cell counts (refer to Fig. 2).

Geographically, dairy donkey production today is mainly clustered in the following areas:

- Europe: Northern and central Italy, southern France, and parts of Spain feature semi-extensive farming models emphasizing product quality and animal welfare (Dai et al. 2019).
- China: Shandong Province heads large-scale donkey milk production, with Xinjiang and Hebei following, typically functioning through intensive, vertically integrated systems (Seyiti and Kelimu 2021).
- South Africa: The donkey dairy sector is developing, aiming at affluent urban consumers. However, commercialization remains in early stages, primarily through small farms supplying niche markets (Keitshweditse et al. 2024).

- Other regions: In countries like Greece, Portugal and Brasil, donkey milk production remains largely traditional, supplying local markets with health and cosmetic products (Camillo et al. 2018; Messias et al. 2021).

Donkey milk is still produced in very modest quantities worldwide; it is thought to make for only 0.1% of all milk produced worldwide (Meena et al. 2024). However, demand is rising quickly due to claims about the nutritional value, hypoallergenic properties, and functional health benefits, which is pushing donkey breeds to become more specialized and optimized for dairy production. This geographic and breed-specific specialization illustrates the responsive adaptation of the donkey dairy industry to local cultural values, economic frameworks, environmental challenges, and changing consumer preferences. It also emphasizes the pressing need for customized strategies that incorporate genetic conservation, welfare standards, and innovative marketing to ensure the sustainable advancement of this distinctive dairy sector.

2 Milk Composition and Quality

Donkey milk has garnered significant attention due to its unique composition and potential health benefits. This section explores the various components of donkey milk and discusses factors that influence its quality (refer to Fig. 2).

2.1 Nutrient Composition

Donkey milk, as with ruminant's milk, is characterized by a rich nutrient profile, which includes proteins, fats, lactose, vitamins, and minerals. Donkey milk has been regarded with interest for its similarity to human milk, in regard to its nutrient composition and in the sequence and profile of proteins (Aspri et al. 2017; Altomonte et al. 2019; Cimmino et al. 2023). The compositional profile and a summary of the studies reporting its variation relative to lactation stage and farming system are reported in Table 1.

- **Energy**: Due to the content of proteins and fats, donkey milk has a low caloric value (about 310–460 kcal/kg). The lower energy content, as compared to both human and ruminants' milk, might pose some concern regarding its use in pre-weaned infant nutrition (Altomonte et al. 2019), but might represent an advantage when a reduced energy intake, coupled with good palatability, is required, as in the case of low-calorie diets (Cimmino et al. 2023).
- **Carbohydrates**: Lactose is the primary carbohydrate in donkey milk, present at concentrations of about 6–7%. This high lactose content contributes to its sweet taste and is important for calcium absorption and gut health. Moreover, it represents the most abundant nutrient in donkey milk, being responsible for most of its energy content. Other minor carbohydrates of interest are represented

Table 1 Nutrient composition of donkey milk

Class	Component	Range	References	Lactation effect	Farming system effect
Energy		1310–1939 (kJ/kg)	Altomonte et al. (2019); Meena et al. (2024)	Martemucci and D'Alessandro (2012)	
pH		7.0–7.5	Altomonte et al. (2019); Živkov Baloš et al. (2023)	Malacarne et al. (2019); Salari et al. (2019)	
Dry matter		7.5–12.5 (%)	Cavallarin et al. (2015); Martini et al. (2018); Živkov Baloš et al. (2023)	Martini et al. (2014); Malacarne et al. (2019); Salari et al. (2019)	
Ash		0.3–0.5 (%)	Živkov Baloš et al. (2023)	Martini et al. (2014); Malacarne et al. (2019); Salari et al. (2019)	
Carbohydrates	Lactose	5.9–7.2 (%)	Živkov Baloš et al. (2023)	Martini et al. (2014); Malacarne et al. (2019); Salari et al. (2019)	Valle et al. (2018)
	Sialylated oligosaccharides	25–60 (mg/L)	Licitra et al. (2019)	Licitra et al. (2019)	
Proteins	Total	1.12–2.14 (%)	Cavallarin et al. (2015); Živkov Baloš et al. (2023)	Martini et al. (2014); Malacarne et al. (2019); Salari et al. (2019)	Valle et al. (2018)
	Caseins and micelle size	0.66–1.03 (%) 30–600 (nm)	Živkov Baloš et al. (2023)	Martini et al. (2014); Malacarne et al. (2019); Salari et al. (2019)	
	Whey	0.49–0.80 (%)	Aspri et al. (2016)	Malacarne et al. (2019)	
	Lysozyme and activity	0.10–0.15 (%) 2700–5800 (U/L)	Conte et al. (2012); Martini et al. (2018); Živkov Baloš et al. (2023)		

	Non-protein nitrogen	0.18–0.41 (%)	Altomonte et al. (2019)	Malacarne et al. (2019)	
	Aminoacids	15.72 (mg/kg)	Altomonte et al. (2019)	Yang et al. (2024)	
	Urea	26–40 (mg/ml)	Salari et al. (2019)	Salari et al. (2019)	
Fats	Total	0.16–1.7 (%)	Cavallarin et al. (2015); Martini et al. (2018); Živkov Baloš et al. (2023)	Martemucci and D'Alessandro (2012); Martini et al. (2014); Malacarne et al. (2019); Salari et al. (2019)	Valle et al. (2018)
	Fat globules	1.92–2.12 um	Martini et al. (2018); Živkov Baloš et al. (2023)		
	Triglycerides	80–85 (% fat)	Altomonte et al. (2019)		
	SFA	43.9–67.7 (% fat)	Aspri et al. (2016); Živkov Baloš et al. (2023)	Martemucci and D'Alessandro (2012)	Valle et al. (2018)
	MUFA	15.3–35.0 (% fat)	Aspri et al. (2016)	Martemucci and D'Alessandro (2012)	Valle et al. (2018)
	PUFA	15.2–30.5 (% fat)	Aspri et al. (2016)	Martemucci and D'Alessandro (2012)	Valle et al. (2018)
	Omega 6/ omega 3	1.2–1.7	Živkov Baloš et al. (2023)	Martemucci and D'Alessandro (2012)	Valle et al. (2018)
	Phospholipids	42.8–97.1 (mg/L)	Altomonte et al. (2019)		
	Cholesterol	9–30 (mg/L)	Altomonte et al. (2019)		

(continued)

Table 1 (continued)

Class	Component	Range	References	Lactation effect	Farming system effect
Vitamins	B1	12–60 (ug/100 ml)	Živkov Baloš et al. (2023)		
	B2	30–97 (ug/100 ml)	Živkov Baloš et al. (2023)		
	B3	57–90 (ug/100 ml)	Živkov Baloš et al. (2023)		
	B6	5.38 (uM)	Živkov Baloš et al. (2023)		
	B9	0.83 (uM)	Živkov Baloš et al. (2023)		
	B12	0.11 (ug/100 ml)	Živkov Baloš et al. (2023)		
	C	200-5700 (ug/100 ml)	Martini et al. (2018); Živkov Baloš et al. (2023)		
	A	1.7–58 (ug/100 ml)	Živkov Baloš et al. (2023)		Valle et al. (2018)
	D	2.23–2.28 (ug/100 ml)	Živkov Baloš et al. (2023)		
	E	2.14–38.4 (ug/100 ml)	Živkov Baloš et al. (2023)		Valle et al. (2018)
Micronutrients	Calcium	330–1140 (mg/L)	Aspri et al. (2016)	Martini et al. (2014); Malacarne et al. (2019)	
	Phosphorus	320–650 (mg/L)	Aspri et al. (2016)	Martini et al. (2014); Malacarne et al. (2019)	
	Potassium	405–2010 (mg/L)	Živkov Baloš et al. (2023)	Martini et al. (2014); Malacarne et al. (2019)	
	Magnesium	54–249 (mg/L)	Živkov Baloš et al. (2023)	Martini et al. (2014); Malacarne et al. (2019)	

	Sodium	157–910 (mg/L)	Živkov Baloš et al. (2023)	Martini et al. (2014); Malacarne et al. (2019)	
	Iron	0.196–3.16 (mg/L)	Živkov Baloš et al. (2023)	Malacarne et al. (2019)	
	Zinc	1.23–3.19 (mg/L)	Aspri et al. (2016)	Martini et al. (2014); Malacarne et al. (2019)	
	Copper	0.08–0.30 (mg/L)	Aspri et al. (2016)	Malacarne et al. (2019)	
	Selenium	0.151 (mg/kg)	Živkov Baloš et al. (2023)		
	Iodine	0.096 (mg/kg)	Živkov Baloš et al. (2023)		

Units are reported in Range column. References to full papers are reported, as well as studied on the effect of lactation and farming system effects, where available by oligosaccharides, with a beneficial role in gut health. Few studies are available on donkey milk oligosaccharides, and their total concentration has not been reported to date (Hobbs et al. 2021). Mostly, donkey milk oligosaccharides occur in the sialylated form, while the neutral forms are predominant in human milk (Licitra et al. 2019).

- **Proteins**: Donkey milk contains about 1.1–2.1% proteins, primarily composed of casein and whey proteins (Živkov Baloš et al. 2023). One of the key characteristics for resemblance to human milk, besides the protein concentration, is the more similar casein to whey protein ratio in contrast to ruminants' milk (80: 20 for bovine milk, 60:40 for donkey milk, and 40:60 for mature human milk) (Altomonte et al. 2019; Luo et al. 2019; Živkov Baloš et al. 2023). Since caseins are mostly involved in the formation of gastric curd, their low total amount, coupled with the bigger size of casein micelles with respect to bovine milk (mean 300 vs 120 nm) are thought to be responsible for a soft flocculation in the stomach, that results in easier attachment by digestive enzymes. Also, differences with respect to ruminants' caseins are seen in the abundance of isoforms, with β-casein representing the major fraction in donkey milk, and a limited amount of α-caseins, as in human milk. Similarly, linear epitopes on bovine milk caseins have a low resemblance to that of equine milk, thus contributing to the low level of cross reactivity of these two products for allergic patients. The whey protein fraction is also involved in the high digestibility and bioactive properties of donkey milk: these proteins play a crucial role in immune function and have anti-inflammatory and antimicrobial effects. The main donkey whey proteins are α-lactalbumin, β-lactoglobulin, and lysozyme. β-lactoglobulin, the most recognized allergen of bovine whey, shares low sequence identity ($< 56\%$ of identity) with respect to its equine counterparts, whereas it is absent in human milk. Moreover, it has a monomeric structure, while in bovine milk the dimeric form is highly represented (Cunsolo et al. 2017). Despite a high level of identity ($> 70\%$), the linear epitopes of bovine α-lactalbumin show high difference with the corresponding domains present in the donkey's homologues, thus resulting in reduced cross-reactivity. Donkey milk protein profile is also characterized by a high amount of lysozyme, a natural antimicrobial agent, which is characteristic for equine milk, and supposed to have a strong influence on its peculiar bioactivity.
- **Fats**: The fat content in donkey milk is relatively low, ranging from 0.15–1.7%. The low-fat content, coupled with a modest cholesterol content, makes donkey milk suitable for individuals with fat metabolism disorders (Cimmino et al. 2023). The lipid composition of equine milk shows strong similarity to that of human milk. Donkey milk fat contains high proportions of mono- and poly-unsaturated fatty acids, and a higher ratio of ω-6 to ω-3 fatty acids, with respect to bovine milk (Gastaldi et al. 2010; Wu et al. 2023; Meena et al. 2024). Equine milk lipids are characterized by short chains and low saturation. Donkey milk has the closest proportion of palmitic acid located at the sn-2 position to that of human milk (Roy et al. 2020). The low size of fat globules results in a more convenient ratio between digestive lipases and milk fat globule membrane in donkey milk, thus contributing to its digestibility.

- **Vitamins and Minerals**: Donkey milk is a good source of essential vitamins such as vitamin C, and B-vitamins, in particular folic acid (B9), and niacin (B3) (Martini et al. 2018; Živkov Baloš et al. 2023). Conversely, it is quite low in liposoluble vitamins, with the exception of vitamin D. It also contains vital minerals like calcium, phosphorus, and magnesium, which are important for bone health and metabolic functions. Donkey milk has a significant amount of zinc, an important mineral with increased requirements in infancy and adolescence. Also copper and selenium are notably abundant.

2.2 Donkey Milk Bioactive Compounds and Application in Infant Clinical Nutrition

In addition to its basic nutritional components, donkey milk contains several bioactive compounds that contribute to its health benefits. Among the proteins, of specific relevance is the amount and activity of lysozyme, which has antibacterial, antiviral, and immune-boosting properties. During digestion, several bioactive peptides with antioxidants, Angiotensin Converting Enzyme (ACE)-inhibitory and antimicrobial bioactivity are released from donkey milk proteins (Tidona et al. 2011; Piovesana et al. 2015). The presence of these compounds enhances the milk's therapeutic potential, particularly in managing allergies and inflammatory conditions.

Bioactive compounds in donkey milk appear to be involved in the elimination of hydroxyl radicals, superoxide anions (Li et al. 2017) and inhibition of lipid peroxidation (Zenezini Chiozzi et al. 2016). In animal models, donkey milk improved the antioxidant defense mechanisms and detoxifying enzyme activity (Trinchese et al. 2015; Li et al. 2017; Lionetti et al. 2020), and improved colon damage and colon mucosal inflammation by enhancing the immune barrier function and promoting probiotic growth (Yvon et al. 2018; Jiang et al. 2018). These beneficial effects on the pro-oxidant status observed can also be attributed to the low n-6/n-3 polyunsaturated fatty acid (PUFA) ratio detected in this milk (Lionetti et al. 2012; Trinchese et al. 2015).

Donkey milk is recognized as a valuable alternative for childhood nutrition, given its biological similarity to human milk. One study found that the fortification of human milk with donkey milk improved feeding tolerance and reduced the gastroesophageal reflux in premature infants with low body weight, when compared to the fortification with bovine milk (Bertino et al. 2019; Cresi et al. 2020). Additionally, (Martini et al. 2021) reported various clinical studies that confirmed the positive effects of donkey milk on human health, highlighting its potential therapeutic applications. The most frequently studied population was children with allergies to cow's milk proteins, with results showing 82.6–98.5% of them tolerating donkey milk.

2.3 Factors Influencing Milk Quality

Several factors affect the composition and quality of donkey milk:

- **Stage of lactation**: Donkey milk composition has been reported to vary throughout lactation. Significant differences were found by different authors in fat, protein, mineral and vitamin content at different lactation stages (Lazarević et al. 2017; Aspri et al. 2017; Massouras et al. 2017; Martini et al. 2021; Barbosa dos Santos et al. 2023). As the lactation stage progresses, there is a decrease in total solids, solids-non-fat, protein, casein, and nitrogen fractions (Lazarević et al. 2017; Aspri et al. 2019). As in all milk types, there is also a variation in the number of specific protein fractions throughout lactation, compatible with the changes in the foal requirements (Gubić et al. 2016; Massouras et al. 2017; Aspri et al. 2019). Ongoing lactation is also responsible for changes in the relative amount of different lipid classes, including balance between saturated and unsaturated fatty acids, and a more positive Omega 6/Omega 3 ratio at end lactation (Martemucci and D'Alessandro 2012). Also, the concentration of calcium, phosphorus, and potassium, but not sodium, showed a progressive during lactation (Massouras et al. 2017; Malacarne et al. 2019). Recently, targeted metabolomic approaches confirmed most of these changes among the metabolite profiles of colostrum, transitional, and mature milk: colostrum exhibited a higher abundance of lipids, including fatty acids, glycerophospholipids, and sphingolipids; transitional milk was characterized by a significant increase in the levels of amino acids and their metabolites and organic acids and their derivatives (Yang et al. 2024).
- **Diet**: Few studies have investigated specifically the effect of the nutrition of lactating donkeys on milk quality, with respect to data available for ruminants. Increased dietary crude protein intakes during lactation was found to result in higher milk yield, protein, lactose, and solid-not-fat yields, as well as improved digestibility of various nutrients (Yue et al. 2022). A diet rich in high-quality forage and balanced with appropriate supplements ensures optimal levels of proteins, fats, and other nutrients in the milk (Liang et al. 2022). The use of corn straw in the diet of the lactating donkeys resulted in greater variety and content of triacyclglycerol species in the milk, than using wheat hulls and straw, while volatile organic compounds, including nitrogen compounds, esters, and alcohols, were significantly increased when wheat hulls or straw were supplied (Ren et al. 2023). Due to the specific high requirements of lactating donkeys in the first months of lactation, and to their lower ingestion levels, increased intakes of concentrates in the diet resulted in increased blood sugar levels and decreased de novo milk fatty acids synthesis (Salari et al. 2021). The use of specific supplements to influence the amount of specific milk components was investigated. Dietary supplementation with trace elements (Zn, Fe, Cu, Mn, I, Se and Co) was found to affect the thyroid hormone level of donkey milk, but not their serum counterpart (Todini et al. 2012), nor the content of the supplied trace elements in milk (Fantuz et al. 2013). A daily supplementation with verbascoside, a phenolic compounds-rich extract from leaves of Verbenaceae, was found to

Fig. 3 Romagnolo's donkey in their free stall, during late-morning feeding. (From the author's repertoire)

decrease saturated fatty acids and increase monounsaturated fatty acids, vitamins A and E in donkey milk, thereby affecting the nutritional quality of the milk and benefiting the oxidative status and serum lipid profile of the animals (D'Alessandro et al. 2017). Extruded linseed was recently shown not to affect milk yield and composition, while it has been shown to improve the milk fatty acid profile, increasing polyunsaturated fatty acids and decreasing saturated fatty acids (Correddu et al. 2022).

- **Health and management**: The health status of the donkey and management practices, including housing, hygiene, and stress levels, significantly influence milk quality. Healthy, well-managed donkeys produce milk with higher levels of beneficial nutrients and bioactive compounds. Extensive farming of dairy donkeys has been shown to increase the fat content and fat-soluble vitamin concentration of milk, as well as alter the fatty acid composition to a more favorable profile for human nutrition (Valle et al. 2018) (refer to Fig. 3).

3 Conclusion

A specialized dairy industry focused on this historically underutilized species has emerged as a result of the resurgence of interest in donkey milk. Donkey milk's distinct compositional profile, which includes bioactive components, hypoallergenic qualities, high digestibility, and a close resemblance to human milk, contributes to its increasing popularity in both nutritional and medicinal contexts. The industry is growing despite the low milk yield thanks to changing production methods, regional breed specialization, and health-conscious consumers.

However, maintaining the dairy donkey industry requires a careful balancing act between genetic conservation and economic viability. Donkey milk has the potential to evolve from a niche product to an important part of specialized nutrition and individualized healthcare as more data becomes available.

4 Questions

1. The average daily milk yield of a dairy donkey is typically less than 1 liter.
   True.
2. Donkey milk has a casein-to-whey protein ratio more similar to bovine milk than human milk.
   False.
3. Donkey milk is rich in vitamin C and has a significantly higher content compared to cow's milk.
   True.
4. Donkey milk contains high levels of saturated fats and cholesterol, making it less suitable for individuals with fat metabolism disorders.
   False.
5. The nutritional composition of donkey milk remains stable throughout the entire lactation period.
   False.

References

Altomonte I, Salari F, Licitra R, Martini M (2019) Donkey and human milk: insights into their compositional similarities. Int Dairy J 89:111–118. https://doi.org/10.1016/J.IDAIRYJ.2018.09.005

Aspri M, Economou N, Papademas P (2017) Donkey milk: An overview on functionality, technology, and future prospects. Food Rev Int 33(3):316–333. https://doi.org/10.1080/87559129.2016.1175014

Aspri M, Economou N, Papademas P (2017) Donkey milk: an overview on functionality, technology, and future prospects. Food Rev Intl 33(3):316–333. https://doi.org/10.1080/87559129.2016.1175014

Aspri M, Souroullas K, Ioannou C, Papademas P (2019) Physico-chemical composition and antimicrobial protein content of early lactation donkey milk. Int J Food Stud 8(1):68–75. https://doi.org/10.7455/IJFS/8.1.2019.A6

Barbosa dos Santos IC, Rangel AH d N, Ribeiro CVDM, Oliveira CA d A (2023) Donkey milk composition is altered by lactation stage and jennies age. J Food Compos Anal 115. https://doi.org/10.1016/J.JFCA.2022.104971

Bertino E, Cavallarin L, Cresi F, Tonetto P, Peila C, Ansaldi G, Raia M, Varalda A, Giribaldi M, Conti A et al (2019) A novel donkey milk-derived human milk fortifier in feeding preterm infants: A randomized controlled trial. J Pediatr Gastroenterol Nutr 68(1):116–123. https://doi.org/10.1097/MPG.0000000000002168

Camillo F, Rota A, Biagini L, Tesi M, Fanelli D, Panzani D (2018) The current situation and trend of donkey industry in Europe. J Equine Vet Sci 65:44–49. https://doi.org/10.1016/j.jevs.2017.11.008

Cavallarin L, Giribaldi M, Soto-Del Rio, M de los D, Valle E, Barbarino G, Gennero MS, Civera T (2015) A survey on the milk chemical and microbiological quality in dairy donkey farms located in NorthWestern Italy. Food Control 50:230–235. https://doi.org/10.1016/J.FOODCONT.2014.08.019

Cimmino F, Catapano A, Villano I, Di Maio G, Petrella L, Traina G, Pizzella A, Tudisco R, Cavaliere G (2023) Invited review: human, cow, and donkey milk comparison: focus on metabolic effects. J Dairy Sci 106(5):3072–3085. https://doi.org/10.3168/JDS.2022-22465

Correddu F, Carta S, Mazza A, Nudda A, Rassu SPG (2022) Effect of extruded linseed on sarda donkey milk quality. Ital J Anim Sci 21(1):1200–1210. https://doi.org/10.1080/1828051X.2022.2104176

Cresi F, Maggiora E, Pirra A, Tonetto P, Rubino C, Cavallarin L, Giribaldi M, Moro GE, Peila C, Coscia A (2020) Effects on gastroesophageal reflux of donkey milk-derived human milk fortifier versus standard fortifier in preterm newborns: additional data from the FortiLat study. Nutrients 12(7):2142. https://doi.org/10.3390/NU12072142

Cunsolo V, Saletti R, Muccilli V, Gallina S, Di Francesco A, Foti S (2017) Proteins and bioactive peptides from donkey milk: the molecular basis for its reduced allergenic properties. Food Res Int 99(Pt 1):41–57. https://doi.org/10.1016/J.FOODRES.2017.07.002

D'Alessandro AG, Vizzarri F, Palazzo M, Martemucci G (2017) Dietary verbascoside supplementation in donkeys: effects on milk fatty acid profile during lactation, and serum biochemical parameters and oxidative markers. Animal 11(9):1505–1512. https://doi.org/10.1017/S1751731117000441

Dai F, Dalla Costa E, Burden F, Judge A, Minero M (2019) The development of guidelines to improve dairy donkey management and welfare. Ital J Anim Sci 18(1):189–193. https://doi.org/10.1080/1828051X.2018.1503571

Fantuz F, Ferraro S, Todini L, Mariani P, Piloni R, Salimei E (2013) Essential trace elements in milk and blood serum of lactating donkeys as affected by lactation stage and dietary supplementation with trace elements. Animal 7(11):1893–1899. https://doi.org/10.1017/S175173111300133X

Gastaldi D, Bertino E, Monti G, Baro C, Fabris C, Lezo A, Medana C, Baiocchi C, Mussap M, Galvano F, Conti A (2010) Donkey's milk detailed lipid composition. Front Biosci (Elite Ed) 2(2):537–546. https://doi.org/10.2741/E112

Gubić J, Torbica A, Iličić M, Tasić T, Šarić L, Popović S (2016) Characterization of several milk proteins in domestic Balkan donkey breed during lactation, using lab-on--a-chip capillary electrophoresis. Chem Ind Chem Eng Q 22(1):9–15. https://doi.org/10.2298/CICEQ150105013G

Hobbs M, Jahan M, Ghorashi SA, Wang B (2021) Current perspective of Sialylated milk oligosaccharides in mammalian milk: implications for brain and gut health of newborns. Foods 10(2):473. https://doi.org/10.3390/FOODS10020473

Jiang L, Lv J, Liu J, Hao X, Ren F, Guo H (2018) Donkey milk lysozyme ameliorates dextran sulfate sodium-induced colitis by improving intestinal barrier function and gut microbiota composition. J Funct Foods 48:144–152. https://doi.org/10.1016/j.jff.2018.07.005

Keitshweditse B, Tsvakirai CZ, Mabuza ML, Tshehla M (2024) Towards the expansion of the functional dairy market: determining donkey milk value propositions and identifying possible consumers. Future Foods 10. https://doi.org/10.1016/j.fufo.2024.100467

Lazarević JM, Tasić TA, Popović SJ, Banjac VV, Duragić OM, Kokić BM, Čabarkapa IS (2017) Changes in milk composition of domestic Balkan donkeys' breed during lactation periods. Acta Period Tech 48:187–195. https://doi.org/10.2298/APT1748187L

Li L, Liu X, Guo H (2017) The nutritional ingredients and antioxidant activity of donkey milk and donkey milk powder. Food Sci Biotechnol 27(2):393. https://doi.org/10.1007/S10068-017-0264-2

Liang XS, Yue YX, Zhao YL, Guo YM, Guo XY, Shi BL, Yan SM (2022) Effects of dietary concentrate to forage ratio on milk performance, milk amino acid composition and milk protein synthesis of lactating donkeys. Anim Feed Sci Technol 292:115444. https://doi.org/10.1016/J.ANIFEEDSCI.2022.115444

Licitra R, Li J, Liang X, Altomonte I, Salari F, Yan J, Martini M (2019) Profile and content of sialylated oligosaccharides in donkey milk at early lactation. LWT. 115. https://doi.org/10.1016/J.LWT.2019.108437

Lionetti L, Cavaliere G, Bergamo P, Trinchese G, De Filippo C, Gifuni G, Gaita M, Pignalosa A, Donizzetti I, Putti R, Di Palo R, Barletta A, Mollica MP (2012) Diet supplementation with donkey milk upregulates liver mitochondrial uncoupling, reduces energy efficiency and improves antioxidant and antiinflammatory defences in rats. Mol Nutr Food Res 56:1596–1600. https://doi.org/10.1002/MNFR.201200160

Lionetti L, Cavaliere G, Bergamo P, Trinchese G, De Filippo C, Gifuni G, Gaita M, Pignalosa A, Donizzetti I, Putti R et al (2020) Diet supplementation with donkey milk upregulates liver mitochondrial uncoupling, reduces energy efficiency and improves antioxidant and antiinflammatory defences in rats. Mol Nutr Food Res 56(10):1596–1600. https://doi.org/10.1002/MNFR.201200160

Luo J, Jian S, Wang P, Ren F, Wang F, Chen S, Guo H (2019) Thermal instability and characteristics of donkey casein micelles. Food Res Int [Internet]. [accessed 2025 Apr 24] 119:436–443. https://doi.org/10.1016/J.FOODRES.2019.02.023

Malacarne M, Criscione A, Franceschi P, Bordonaro S, Formaggioni P, Marletta D, Summer A (2019) New insights into chemical and mineral composition of donkey milk throughout nine months of lactation. Animals 9:1161. https://doi.org/10.3390/ANI9121161

Martemucci G, D'Alessandro AG (2012) Fat content, energy value and fatty acid profile of donkey milk during lactation and implications for human nutrition. Lipids Health Dis 11:113. https://doi.org/10.1186/1476-511X-11-113

Martini M, Altomonte I, Salari F, Caroli AM (2014) Short communication: Monitoring nutritional quality of Amiata donkey milk: effects of lactation and productive season J Dairy Sci 97 (11):6819–22. https://doi.org/10.3168/jds.2014-8544

Martini M, Altomonte I, Licitra R, Salari F (2018) Nutritional and nutraceutical quality of donkey milk. J Equine Vet Sci 65:33–37. https://doi.org/10.1016/J.JEVS.2017.10.020

Martini M, Altomonte I, Tricò D, Lapenta R, Salari F (2021) Current knowledge on functionality and potential therapeutic uses of donkey milk. Animals (Basel) 11(5). https://doi.org/10.3390/ANI11051382

Massouras T, Triantaphyllopoulos KA, Theodossiou I (2017) Chemical composition, protein fraction and fatty acid profile of donkey milk during lactation. Int Dairy J 75:83–90. https://doi.org/10.1016/J.IDAIRYJ.2017.06.007

Meena S, Meena GS, Gautam PB, Rai DC, Kumari S (2024) A comprehensive review on donkey milk and its products: composition, functionality and processing aspects. Food Chem Adv 4. https://doi.org/10.1016/j.focha.2024.100647

Messias TBON, Araújo E d OM, Sant'ana AM d S, Lucena JEC, Pacheco MTB, Queiroga R d CR d E (2021) Challenges and perspectives for exploiting donkey milk in the Brazilian Northeast. Cienc Rural 52(3):e20210058. https://doi.org/10.1590/0103-8478CR20210058

Piovesana S, Capriotti AL, Cavaliere C, La Barbera G, Samperi R, Zenezini Chiozzi R, Laganà A (2015) Peptidome characterization and bioactivity analysis of donkey milk. J Proteome 119:21–29. https://doi.org/10.1016/J.JPROT.2015.01.020

Ren W, Sun M, Shi X, Wang T, Wang Y, Wang X, Huang B, Kou X, Liang H, Chen Y et al (2023) Effects of roughage on the lipid and volatile-organic-compound profiles of donkey milk. Foods 12(11):2231. https://doi.org/10.3390/FOODS12112231/S1

Roy D, Ye A, Moughan PJ, Singh H (2020) Composition, structure, and digestive dynamics of Milk from different species—a review. Front Nutr 7:577759. https://doi.org/10.3389/FNUT.2020.577759/XML/NLM

Salari F, Ciampolini R, Mariti C, Millanta F, Altomonte I, Licitra R, Auzino B., Ascenzi CD, Bibbiani C, Giuliotti L, Amerigo Papini R, Martini M (2019) A multi-approach study of the performance of dairy donkey during lactation: preliminary results. Ital J Anim Sci 18:1135–1141. https://doi.org/10.1080/1828051X.2019.1623094

Salari F, Roncoroni C, Altomonte I, Boselli C, Brajon G, Martini M (2021) Effects of the physiological status and diet on blood metabolic parameters in Amiata dairy donkeys. Animals 11:3292. https://doi.org/10.3390/ANI11113292

Seyiti S, Kelimu A (2021) Donkey industry in China: current aspects, suggestions and future challenges. J Equine Vet Sci 102. https://doi.org/10.1016/j.jevs.2021.103642

Tidona F, Sekse C, Criscione A, Jacobsen M, Bordonaro S, Marletta D, Vegarud GE (2011) Antimicrobial effect of donkeys' milk digested in vitro with human gastrointestinal enzymes. Int Dairy J 21(3):158–165. https://doi.org/10.1016/J.IDAIRYJ.2010.10.008

Todini L, Salimei E, Malfatti A, Ferraro S, Fantuz F (2012) Thyroid hormones in milk and blood of lactating donkeys as affected by stage of lactation and dietary supplementation with trace elements. J Dairy Res 79(2):232–237. https://doi.org/10.1017/S0022029912000118

Trinchese G, Cavaliere G, Canani RB, Matamoros S, Bergamo P, De Filippo C, Aceto S, Gaita M, Cerino P, Negri R et al (2015) Human, donkey and cow milk differently affects energy efficiency and inflammatory state by modulating mitochondrial function and gut microbiota. J Nutr Biochem 26(11):1136–1146. https://doi.org/10.1016/J.JNUTBIO.2015.05.003

Valle E, Pozzo L, Giribaldi M, Bergero D, Gennero MS, Dezzutto D, McLean A, Borreani G, Coppa M, Cavallarin L (2018) Effect of farming system on donkey milk composition. J Sci Food Agric 98(7):2801–2808. https://doi.org/10.1002/JSFA.8777

Wu Y, Sun Y, Chen R, Qiao Y, Zhang Q, Li Q, Wang X, Pan Y, Li S, Wang Z (2023) Analysis for lipid nutrient differences in the milk of 13 species from a quantitative non-targeted lipidomics perspective. Food Chem X 20:101024. https://doi.org/10.1016/J.FOCHX.2023.101024

Yang Y, Lv K, Zhang Y, Wang X, Deng L (2024) Widely targeted metabolomics reveals dynamic alterations in colostrum, transitional, and mature milk of jennies. LWT 200:116179. https://doi.org/10.1016/J.LWT.2024.116179

Yue Y, Li L, Tong M, Li S, Zhao Y, Guo X, Guo Y, Shi B, Yan S (2022) Effect of varying dietary crude protein level on Milk production, nutrient digestibility, and serum metabolites by lactating donkeys. Animals 12(16):2066. https://doi.org/10.3390/ANI12162066

Yvon S, Olier M, Leveque M, Jard G, Tormo H, Haimoud-Lekhal DA, Peter M, Eutamène H (2018) Donkey milk consumption exerts anti-inflammatory properties by normalizing antimicrobial peptides levels in Paneth's cells in a model of ileitis in mice. Eur J Nutr 57(1):155–166. https://doi.org/10.1007/S00394-016-1304-Z

Zenezini Chiozzi R, Capriotti AL, Cavaliere C, La Barbera G, Piovesana S, Samperi R, Laganà A (2016) Purification and identification of endogenous antioxidant and ACE-inhibitory peptides from donkey milk by multidimensional liquid chromatography and nanoHPLC-high resolution mass spectrometry. Anal Bioanal Chem 408(20):5657–5666. https://doi.org/10.1007/S00216-016-9672-Z

Živkov Baloš M, Ljubojević Pelić D, Jakšić S, Lazić S (2023) Donkey milk: an overview of its chemical composition and main nutritional properties or human health benefit properties. J Equine Vet Sci 121:104225. https://doi.org/10.1016/J.JEVS.2023.104225

Donkeys Meat Production

Boris Cagnin

Abstract

The donkey has always been a precious resource, accompanying the progress of human civilization. With its ups and downs of success, the donkey has also been a source of food, although most were considered a staple food for the poorest populations. The properties of donkey's milk, however, have been known since ancient times and were the exclusive preserve of the wealthier classes. Over time, the use of this meat has become established in some areas of northern Italy and, more recently, in southern France, although some typical dishes also appear here and there in other parts of northern Europe. The development of new breeding techniques, the attention paid to the quality and variety of nutrition, and the scrupulous care for the animal's well-being and health, from birth to slaughter, rigorously regulated by strict standards, have allowed the production of a meat that, in addition to possessing a flavor and delicacy that make it an exceptional ingredient, also has truly unique nutritional properties. Thus, despite a cultural resistance that still persists, there are more and more admirers of this meat which is also attracting the attention of the market and operators.

Keywords

Donkey in history · Donkey welfare · Donkey feeding · Donkey meat specialties · Nutritional properties

B. Cagnin (✉)
Bureau Plattner - Of Councel, Treviso, Italy
e-mail: boris.cagnin@bureauplattner.com

A. K. McLean et al. (eds.), *Donkey and Mule Medicine: Foundations, Breeds, and Management*, https://doi.org/10.1007/978-3-032-12726-6_9

Chapter Outline
Meat Production:

- History of donkey meat consumption
- Selective breeding of donkeys for meat production, focusing on breeds known for their meat quality.
- Adequate feeding and healthcare to ensure optimal growth and meat quality.
- Process of harvesting donkeys
- Products from donkey meat (e.g. salami, etc) and info on nutritional properties of donkey meat;

1 The Consumption of Donkey Meat from Ancient Times to Modern

The use of donkeys in human nutrition in ancient times is rather uncertain and there is no univocal historical evidence.

Certainly, however, in classical Athens the presence of markets dedicated specifically to donkey meat is documented. Evidence of the use of this meat can also be found in Greek literature: Aristophanes recounts the goodness of its bacon, while Herodotus describes the presence of oven-baked donkeys at the banquets of the more affluent (Fontaneto 2018; Bigi 2023).

The Romans' aversion to the consumption of horse meat in general and, consequently, also of donkey meat, is well known (Bigi 2023; Mitchell 2018).

But in rural areas, however, where the donkey was an extraordinary help for farmers, who used it both to pull carts and to plow fields, when the animal aged and was no longer usable, it constituted a resource too precious to be wasted and, therefore, was slaughtered and destined for human consumption. At that point, however, the meat was very tough and stringy, because it came from an animal that had been used for extremely heavy work throughout its life. The popular belief that donkey meat was considered, wrongly and of poor quality probably dates to that time. The ancient physician Galen criticized the use of meat from donkeys that were "too old" in Latin cuisine. But he also testified to its presence on the tables of Alexandria, in Egypt.

The use of donkey milk also dates to the classical world, and the ancients had already learned to know its extraordinary nutritional properties.

In fact, it was a widespread opinion in Rome that the components of donkey milk, rich in vitamins and minerals, had beneficial and rejuvenating effects, making it an ideal choice for those seeking to preserve their beauty.

Several historical sources testify that Poppea, Nero's wife, and Cleopatra, Queen of Egypt, bathed in donkey's milk to keep their skin soft and young. According to legend, Poppea was followed everywhere by a large group of donkeys, about 500, so that fresh milk was always available for her baths. Pliny the Elder, in his work "Historia Naturalis" dedicated to the medicinal uses of animal products, wrote: "It is thought that the skin of the face becomes smooth and soft from donkey's milk, and it is known that some women treat their cheeks seven times a day, carefully following this number. Poppea, wife of Prince Nero, introduced this practice, bathing in the tubs and for this reason she used to be accompanied in her travels by her herds of donkeys. Irritations of mucus on the face are eliminated by applying butter, with greater effectiveness if mixed with ceruse (a white powder used as a cosmetic), and the defects that spread are treated with barley flour, applied pure on the wounds of the face with a membrane soaked in cow's milk."

There is not much information regarding the consumption of donkey meat during the Middle Ages. However, we can assume that even in that period the donkey, especially when raised for slaughter and not intended for work, was a very precious and highly appreciated food, given that the Merovingian king Dagobert II (652–679) "as a penitential act" deprived himself of it.

About a thousand years later, in 1516, the use of donkey meat by the German troops besieged within the city of Verona is documented.

As proof that in Italy donkey meat has always played an important role, there is the testimony of a German doctor who, around 1700, wrote, with some surprise, that it was even considered "a delicacy" here (Fontaneto 2018).

And in fact, outside the borders of the Bel Paese, only the inhabitants of Silesia were known as "donkey eaters", while the Eselwurst, the traditional salami that spread in the nineteenth century in Germany and Hungary, is much more recent; in that period the consumption of donkey meat also increased in France, and especially in Provence where—as reported in the "History of Food" by Maguelonne Toussaint-Samat—it remained widespread until the outbreak of the Second World War: in Arles, even today, the traditional sausage is made with donkey and bull meat.

Today, Eastern countries such as Poland and Russia are also heavy consumers of donkey meat (Fontaneto 2018).

To remain in the Mediterranean basin, it must be said that for Jewish culture, donkey meat is among the forbidden meats and therefore there are no recipes that allow its use; however, the consideration of this animal also in the Judeo-Christian tradition is important, if it is true that it even finds space in some versions of the tenth biblical Commandment ("You shall not covet your neighbor's house. You shall not covet your neighbor's wife, nor his manservant, nor his maidservant, nor his ox, nor his donkey, nor anything that is your neighbor's", Exodus 20:17) and that later it will be included in the iconography of the Christian nativity scene. Furthermore, it should be remembered that Jesus enters Jerusalem on the back of a donkey on Palm Sunday.

2 Donkey Breeding and Slaughter and EU Regulations

As we have seen, the consumption of donkey meat has had mixed fortunes over the centuries, although it has remained rooted in the Italian culinary tradition, especially in Piedmont, lower Veneto and some areas of Emilia Romagna, where dishes based on donkey meat, once consumed almost exclusively by the poorest sections of the population, are now considered true culinary specialties to which numerous gastronomic festivals and events are dedicated, which have favored their diffusion even beyond regional borders.

So much so that the production of donkey meat, which since the Second World War was in constant decline, has experienced, if not a real recovery, at least a stabilization in recent years.

In this chapter, therefore, we will address the most important issues related to the breeding of donkeys for human consumption, their slaughter and we will also mention the main processes of carcass processing, also referring to the strict European Union legislation that must be respected also with reference to this type of meat.

In fact, to maximize its potentials (both qualitative and quantitative) requires special care and attention during breeding, during the slaughtering phase and during the final transformation process.

Looking only to Europe, in Italy are recorded the largest number of donkey breeds (11, in second place is France with 9 breeds). Among these, those that are normally used to be destined for slaughter are Amiatini, medium-sized donkeys originating from Tuscany, and Ragusani, a breed that was officially recognized only in the 50 s, widespread in Sicily that have a very robust constitution.

To have higher quality products, which is what the market requires, it is essential to have the right raw material. In other words, the production process cannot overcome shortcomings or failures in the quality of the raw material used.

So, it is essential that raw meat comes not only from healthy animals, but also from animals bred in comfort and welfare. This is because the direct correlation between the degree of stress experienced by the animal and the deterioration in the quality of the meat resulting from its slaughter has been demonstrated and verified.

In Europe, where the issue of animal welfare has a major impact on public opinion, it is worth mentioning the study project called WELFARE INDICATORS (Development, integration and dissemination of animal-based welfare indicators, including pain, in commercially important husbandry species, with special emphasis on small ruminants, equidae & turkeys, EU Program PQ7, FB7-KBBE-2010-4), funded by the EU, which highlighted the need for a science-based approach to the issue of animal welfare and developed and tested pain indicators for sheep, goats, horses, donkeys and turkeys, and generally sought to improve animal welfare science in Europe. As part of this study, the University of Milan in Italy in collaboration with the University of São Paulo in Brazil developed the 2015 AWIN assessment protocol for donkeys. This work is based on four good principles (good nutrition, good housing, good behavior and good health) and twelve criteria (absence of prolonged hunger; absence of prolonged thirst; Comfort around resting; Thermal comfort; Ease of movement; Expression of social behavior; Expression of other behaviors; Good human-animal relationship; Positive emotional state; Absence of injuries; Absence of disease; Absence of pain and pain induced by management procedures).

The use of this protocol provides the operator with valuable information about the state of health and welfare of the animals during their life, also indicating possible remedies in case of critical issues during the evaluation phase.

Fig. 1 Donge donkeys in China raised for meat and skin eating a silage diet. They are grown for approximately 2 years and then harvested if they are males. Most females go back into production

The benefits and the improvement of the quality of animals reared in compliance with this protocol (like others do) are evident and encourage the increasing of their application and even their implementation.

Donkeys intended for slaughter in Europe are generally between 3 and 4 years old and weigh around 300 kg.

To achieve this goal, the ideal diet must be based on foods rich in fiber and low in calories, to promote muscle growth and maintain good health without excessive accumulation of fat, especially if considered that the donkey tends to naturally accumulate fat in the belly area. Since donkeys are unable to synthesize them on their own, it is important to ensure that their diet contains an adequate supply of vitamins, which are essential for the animal's metabolic functions and propter growth (Bittante et al. 1990).

The goal is to obtain quality meat, avoiding weighing down the animal with an excessively rich diet (Fig. 1).

Donkeys are certainly very fond of fresh forage, whether it comes from naturally or cultivated pastures. However, the use of this feed must be carefully controlled because excessive intake can lead to laminitis, a foodborne disease that causes the proliferation of histamine, which can lead to inflammation of certain tissues in the hoof, thus compromising the animal's well-being (Bigi 2023). The need to guarantee the animal an adequate intake of fiber certainly makes it important to feed straw, especially from barley or wheat cuttings, and hay, taking care to select grasses that are not too rich in sugars.

Cereals have a high starch content and provide the animal with a high energy intake that, if not adequately compensated by physical activity, can lead to the accumulation of fat. Proteins, on the other hand, have a low biological value. The amount of food, therefore, must be very accurate, keeping in mind that the cereals that have the best ratio between proteins and energy values are oats and wheat (Baroncin 2014).

The intake of proteins of good biological value can be integrated using legume seeds such as broad beans, peas, lupins and, above all, oil seeds such as soy.

Obviously, the use of compound and integrated feeds, as needed, is recommended in small quantities.

The daily rationing of a donkey intended for slaughter, therefore, must include a mix of foods that support adequate development, considering that donkeys, during growth, have an energy requirement like that of an adult donkey but in the formulation of the daily diet, their reduced capacity for food intake must be considered (Perason 2005).

Community legislation (Regulation (EU) 2021/963) dictates strict rules that guarantee, in all states of the European Union, the traceability of each animal bred, its history, the changes of ownership and certifies all the health treatments to which it has been subjected.

More specifically, Regulation (EU) 2016/429 has established general rules on the responsibility of Member States to establish a system of identification and registration of terrestrial animals held, including equines. Member States must establish and maintain a computerized database of terrestrial animals held. Furthermore, the computerized database must record certain minimum information relating to equines, namely a unique code for the equine, the method of identification of the equine and the establishment in which the equine is habitually held.

Obligations are foreseen for operators who keep equines, who are required to ensure that such animals are individually identified by means of: the unique code, a correctly completed single identification document for life and a physical means (subcutaneous microchip) of identification or other method that unequivocally links the equine with a correctly completed single identification document for life.

Regulation (EU) n. 2019/6 on veterinary medicinal products—which includes specific rules on the administration of veterinary medicinal products to equids intended for food production—also prescribes obligations regarding the retention of records for equines and regarding the information that must appear in the single identification document for life. It should be considered that for an animal to be administered a medicine, the veterinarian must fill out an electronic prescription, which is recorded in the animal's identification documents and in a portal that is constantly updated, so that it is possible, at any time, to have the clinical history of each animal.

Depending on the type of medicine administered, a suspension period is provided by law that can range from 0 to a maximum of 40 days during which the animal cannot be slaughtered. Only when this period has expired, the system issues the mandatory document that must accompany the animal that is taken to the slaughterhouse and without which it is not possible to proceed with its slaughter.

Once the donkey has come to the end of the breeding process, it must be transported to the slaughterhouses. The time available to meat does not allow to examine all the variables that can affect the detection of right age and the right weight so that the donkey can express its maximum yield in quality and quantity of meat used for transformation.

It is also known that transportation and slaughtering may be extraordinary sources of stress and pain for animals, with tips that may affect the quality of the meat even for those donkeys that have been bred according to the best protocols illustrated above.

With the aim of protecting the animal welfare referred above, it is essential for the producer to ensure that the best practice rules are respected also during the last stage of animal life.

This constitutes, in addition to ethical commitment, a precise duty imposed by law. In fact, the European Union regulation imposes on transporters and slaughterers strict protocols for the transport and killing of animals. These regulations are all focused to avoid and minimize the anxiety and suffering of animals during the transport phase from farms to slaughterhouses and during the slaughter process.

As far as transport is concerned, Regulation (EU) n. 1/2005 requires only person authorized by the competent authorities, after verification of their specific competence, can carry live vertebrate animals through Europe. The means of transport used must also have specific characteristics to provide the animal with a minimum level of comfort in terms of space available, air, fresh water, nutrition and, in general, protection against movements and shivers caused by means of transport. There are tables, fixed directly by the legislator, which, depending on the breed and size of the animal to be transported, indicate the minimum living space that must be secured to each head. In the case of donkeys, for example, there should be at least 1.75 sq. meters per head for journeys of less than 48 hours and 2.4 sq. meters for longer journeys.

The journeys, whether short or long-range, depending on whether they last for 8 hours, must be interrupted by stops in special areas and must be planned to allow the shortest route.

According to the provision of the Regulation (EU) n. 1/2005 (Annex 1), to be declared fit for transport the animals must be healthy, have no injuries and / or disease, must have an age sufficient to withstand the fatigue of the journey and the females must not exceed 90% of the pregnancy.

The competent health authorities can examine livestock loads to verify compliance of these rules and have the power to stop the carriage and impose severe sanctions on offenders.

Once at the slaughterhouse, the Regulation (EU) n. 1099/2009 provides for the animals to be slaughtered in the shortest possible time. Where this is not possible, it is mandatory to house them in areas of adequate size and separate from the slaughterhouse. The proximity to the area where animals are slaughtered is, in fact, one of the most significant sources of stress for them.

Animal slaughtering can only be carried out by suitably trained personnel with appropriate authorizations in Europe but other regions standards may vary. The regulation authorizes the immobilization of the animal, but it must be performed immediately before stunning.

In the case of the donkey, the most used abatement technique is the one with penetrating bullet in the head, followed immediately by the cut of the jugular vein that determines, where the animal has not died due to the severe brain lesions, the death due to the fatal hemorrhagic shock.

The slaughter stage begins at that point.

The animal is preliminarily skinned and then eviscerated. It is good to remember that the whole donkey carcass is used, with no cuts and / or anatomical parts that are not suitable for consumption or intended for use as pharmaceuticals. For health and security reasons, the liver and kidneys cannot be used in the EU for human consumption and must therefore be disposed of. For the rest you can use everything.

One of the key aspects at this stage is to cool the animal's carcass as quickly as possible to stop the process of decaying and prevent the development of dangerous microbial charges. The time of this cooling process obviously depends on the technology available to the plant and the weight of the animal.

The carcass, gutted and skinned, is normally divided into two halves. It is important to remember that in the European Union it is imposed that, now of its birth, the animal is registered in a special register that certifies its origin and records the main events during breeding. Traceability of the animal is also ensured during the slaughter and subsequent cutting. As such, labels are applied to each half that allow at any time until the final transformation into the finished product to trace back to the origin of every single cut of meat.

As the slaughter phase ends, as mentioned, the halves are normally transported to the actual processing plants where experienced operators begin the dissection phase. Even during this phase, it is essential that all operations are carried out in environments that ensure the maintenance of an adequate cool temperature.

The halves are first dissected into the three main parts: front, belly and back. Subsequently, the cuts for different processing are obtained. At this stage they also proceed to the boning and to the grooming of the cuts. It must be said that donkey meat is very fat (especially in the belly area) that needs to be carefully removed to prevent that the quality of the raw material being compromised.

The yield of a donkey is much lower than other animals destined for slaughter. Although the remarkable progress of breeding and feeding techniques has allowed in recent years to considerably increase these values.

Assuming the approximately 300 kg donkey type as mentioned above, the final amount of meat for human consumption will be around 90 kg.

The donkey meat is very delicate. It must be worked and consumed fresh, and the time of hanging is very short in refrigerators at about 0 degrees.

Cuts normally intended for short baking (such as barbeque) are those obtained from the rear and the belly of the animal, and, after the grooming stage, they are immediately stored under vacuum and are ready for distribution.

The remaining cuts are divided according to their final use.

3 Products with Donkey Meat. Italian Specialities

Most of the front part is used to prepare donkey sausages. In this case the term "sausages" is not use just for the specific product that everybody knows, but it means the technique of preparation of the meat staffed into the guts (Fig. 2).

Fig. 2 Donkey salmi and sausage

This is a very characteristic preparations, often carried out with craftsmanship methods, which aim to flavor the meat, which is preliminarily minced, sometimes also with the addition of cuts of other animals, with the use of spices and aromas.

The size of the mince grain depends on the finished product you want to obtain coarser in the case of salami, finer for sausages. The processed meat is then stuffed. The casing used is the donkey gut, which protect the meat and must be removed just before eating. There is also artificial casing on the market, but of course the yield and the quality of the product are inferior from the natural one.

In the preparation of the sausage's products, of course, the maturing stage is crucial, where the natural molds, that contribute to the preservation and the transformation of the meat, play an essential role.

At this stage technology must necessarily give way to experience and tradition. In fact, sausages are stored in specially controlled temperature and humidity rooms.

The operator's experience in constantly modifying and adapting these variables promotes the onset of natural molds, controlling their growth to reach the expected maturation point of the meat.

At that right time the temperature is further lowered to block the maturing process.

The internal part of the rear legs, which is considered one of the best anatomical cuts, are salted and seasoned without being minced beforehand.

The maturation time varies depending on the type of product: for example, we will have an average time of about 60 days for salami and about 120 days for the bresaola.

The donkey meat, thanks to its texture, is well suited for preparation that requires long cooking that enhances its organoleptic characteristics without compromising its quality. In the Italian culinary tradition, one of the most renowned donkey-based dishes is called "stracotto", which literally means subject to a very long cooking (more than 5 hours).

The recipes, in this case, vary according to market demand, consumer tastes and tradition and may be suitably adapted and modified. We are talking about sauces, braised, stewed and in this case the limit is given solely by the imagination and the chef's ability that cooperate with the company for the preparation of the products destined for the gastronomy.

For some preparations, the cooking process is followed by a low temperature smoking process (28/30 degree maximum). The low temperature prevents the formation of toxic residues in the meat and preserves its digestibility. The aromas used in this case are beech and oak chips coming from biological hatcheries that certify the absence of harmful substances.

What kind of products can be done with the donkey meat with the processes that have been just illustrated? The answer is quite all the products we are used to eat coming from other animals like salami, sopressa (a biggest kind of salami), ham, sausages, wurstel (sausage in German style), mortadella (kind of ham) and stews.

The cradle of the culinary art based on donkey meat is certainly Piedmont, a region in north-western Italy, where the preparation of stews, casseroles and braised donkey meat is widespread, as well as a particular pork butchery (the characteristic dark-coloured salamis, but also the cotechini named "bale d'asino" invented in the monastery of Vasco, in the Cuneo area, and now widespread in the Monregalese area) and the revisitation of some Piedmontese classics such as the traditional "battuta" (tartare), served raw.

We cannot forget the agnolotti, which in Calliano d'Asti are prepared, precisely, with a filling of donkey meat.

Outside Piedmont, it is worth mentioning the mortadella produced in Ragusa, this one based on donkey meat, while the tradition of donkey salami is widespread in the Padua and Vicenza areas. In Verona you can also find pastissàda with donkey meat, first macerated in Amarone wine, while among the pasta dishes, the bigoli with donkey ragù are worth mentioning (Fontaneto 2018; Peretti 2020). Donkey meat in some parts of the world can be found in restaurants serving only donkey meat and its the source of main dishes (Figs. 3 and 4).

Fig. 3 Restaurant specializing in donkey in Shandong, China

Fig. 4 Donkey steak dish in China

The growing popularity of donkey meat is due, in addition to the rediscovery of ancient recipes that enhance its flavor, to studies demonstrating its remarkable nutritional value, being rich in protein (22.8 g/100 g) but low in fat (2.02 g/100 g), especially when compared to other red meats. Therefore, donkey meat has a lower caloric intake (116 kcal/100 g) (Keeton and Eddy 2004).

Examination of donkey muscle tissue shows that saturated fatty acids (SFA) are 40.93 g/100 g and monounsaturated fatty acids (MUFA) are 34 g/100 g. Regarding protein, some cuts of donkey meat contain essential amino acids exceeding 50% of total amino acids, with the highest concentrations being lysine and leucine (Polidori et al. 2009). Glycogen (0.45 g/100 g) is present in quantities very similar to that of horse meat (Polidori et al. 1995).

Donkey meat also contains minerals such as potassium (343 mg/100 g), phosphorus (212 mg/100 g), sodium (52 mg/100 g), magnesium (24 mg/100 g), and calcium (8.6 mg/100 g) (Polidori et al. 2008). The iron (3.80 mg/100 g) and zinc (3.67 mg/100 g) content is comparable to that of beef (Mulvihill 2004).

In conclusion, donkey meat can be considered a product with high nutritional value and low total fat and cholesterol content, but with a good percentage of unsaturated fatty acids, particularly PUFAs. These good qualities are accompanied by a good level of essential amino acids, a very important factor in determining the quality of the food (Polidori et al. 2009).

1. In which ancient civilization was donkey meat considered a delicacy?
2. What donkey breeds are most slaughtered? At what age and at what weight are donkeys generally slaughtered?
3. Can donkeys be allowed to freely eat fresh fodder? If no, explain the reasons.
4. According to European legislation, can animals intend for slaughter be freely administered medications? If no, briefly describe the procedure.
5. What are the main nutritional properties of donkey meat?

References

AWIN (2015) AWIN Welfare assessment protocol for donkeys. https://doi.org/10.13130/awin_donkeys_2015
Baroncin R (2014) L'asino, il mulo e il Bardotto, 2nd edn. Edagricola, Bologna
Bigi D (2023) L'asino. Allevamento, cura e nuovi ruoli aziendali e sociali. Edagricole, Bologna
Bittante G, Andrighetto I, Ramanzin M (1990) Fondamenti di zootecnica, miglioramento genetico, nutrizione ed alimentazione. Liviana Editore, Padova
EU Program PQ7, FP7-KBBE-2010-4. Development, integration and dissemination of animal based welfare indicators, including pain, in commercially important husbandry species, with special emphasis on small ruminants, equidae & turkey. cordis.europa.eu/project/id/266213
Fontaneto J (2018) Il ritorno della carne d'asino (che fa storcere il naso agli animalisti). La Repubblica 2.2.2018
Keeton JT, Eddy S (2004) Chemical and physical characteristics of meat. In: Jensen WK, Devine C, Dikerman M (eds) Encyclopedia of meat science. Elsevier, Oxford, pp 210–218
Mitchell P (2018) The donkey in human history, an archeological perspective. Oxford University Press
Mulvihill B (2004) Micronutrients in meat. In: Jensen WK, Devine C, Dikerman M (eds) Encyclopedia of meat science. Elsevier, Oxford, pp 616–623
Perason RA (2005) Nutrition and feeding of donkeys. In: Mathews N, Taylor T (eds) Veterinary care of donkeys. International Veterinary Information Service
Peretti C (2020) Storia, diffusione e funzionalità della popolazione degli asini romagnoli in Valle Camonica. University of Milan
Polidori P, Moretti VM, Valfrè F, Crispino A, Beretta G (1995) Qualità della carne fresca equina e di prodotto trasformato. Ind Aliment 7:723–727
Polidori P, Vincenzetti S, Cavallucci C, Beghelli D (2008) Quality of donkey meat and carcass characteristics. Meat Sci 80:1222–1224. https://doi.org/10.1016/j.meatsci.2008.05.27
Polidori P, Cavallucci C, Beghelli D, Vincenzetti S (2009) Physical and chemical characteristics of donkey meat from Martina Franca breed. Meat Sci 80:469–471. https://doi.org/10.1016/J/meatsci.2009.03.001

Gelatinized Donkey Skin, *Ejiao* Overview & Marketing

Mary Ting

Abstract

Ejiao, a Chinese traditional medicine product made from gelatinized donkey skin has diversified in recent decades with impacts on global donkey populations and their human communities. Once a medicinal tonic for the royal Chinese elite, the *ejiao* industry has expanded into a mainstream health and beauty product. A brief history of *ejiao*, including recent social media marketing provides a glimpse into the Chinese consumer perspective and possible solutions. As a Chinese state supported cultural product, the role of aggressive marketing of cultural traditions, nationalism, donkey meat as a byproduct of *ejiao* and the impact of demand on rural communities is discussed.

Keywords

Ejiao · Donkey skin · Donkey meat · Meat · TCM · Traditional Chinese medicine

1 Introduction to Ejiao

The domesticated donkey, *Equus asinus* has endured humankind's excruciating workloads and maltreatment since 4000 BCE. In the twenty-first century the humble donkey continues to be essential for global rural communities. Donkeys assist with farming, the transport of water, goods, and people. Donkeys with their long gestation period of approximately twelve months are not conducive to industrial farming (Canisso et al. 2019). Due to product diversification and intensive marketing, current

M. Ting (✉)
Art Department; Environmental Justice Program, John Jay College, New York, NY, USA

A. K. McLean et al. (eds.), *Donkey and Mule Medicine: Foundations, Breeds, and Management*, https://doi.org/10.1007/978-3-032-12726-6_10

demand for donkey hide glue, a traditional Chinese medicine product is unsustainable. The future of the world's most vulnerable communities and their donkeys is now in question due to a once obscure product called *ejiao.*

Ejiao 阿胶, (*colla corii asini*) is an ancient traditional Chinese medicine made from gelatinized donkey skin. The name *ejiao* or *E-glue* has its origins from Dong'e county in Shandong, China. The county claims that the highest quality ejiao is made from local large black donkeys whose hides are boiled in water from the legendary *E' Well* (Humane Society International 2018). The use of donkey hide glue is recorded in the earliest Chinese pharmaceutical work from Eastern Han 200C.E. (神农本草经) *Shennong bencao jing*, *Herbal Classic of Materia Medica* and the later Ming Dynasty, 1596, 本草纲目*Ben Cao Gangmu* Compendium of the *Materia Medica* (HSI 2018). *Ejiao* was traditionally used for gynecological problems, to stop bleeding, fight fatigue and as a medicinal tonic of the imperial court. Once the domain of royalty, *ejiao* is now a mainstream product aggressively promoted for healthy modern living with a continual stream of new convenient formats and ingredient combinations. *Ejiao* diversification and increased demand has devastated not only China's donkey population but led to theft and slaughter of donkeys in global rural communities, such as Kenya and its neighboring nations (Global Initiative Against Transnational Organized Crime, GI-TOC 2020). Likewise, the donkey populations and their human communities of Mexico, Columbia, Brazil, and Pakistan are also bearing the brunt of the donkey skin trade (The Donkey Sanctuary 2017).

2 Donkey Meat

天上龙肉, 地上驴肉

"In heaven there is dragon meat, on earth there is donkey meat" is a Chinese Ming dynasty proverb which reveals the cultural significance and appreciation of eating donkey meat. In the North, Northeastern and Northwestern provinces such as Hebei, Shandong provinces where donkey farms are prevalent, donkey meat is part of the regional cuisine. Historically donkeys were used for transport and coarse grain farming by small scale farmers. With the industrialization of transportation and agriculture, rise in standard of living and dietary changes, donkey usage has shifted to production for traditional medicine, health and wellness products as well as meat and dairy products. (Zhang et al. 2023). Today donkey meat is a byproduct of the *ejiao* industry. While pork, beef, lamb, and chicken remain the dominant meat consumption, interest in donkey meat exists from a cultural and higher perceived nutritional value.

Li (2018a, b, c). Hebei province is known for its two local styles of donkey burgers: Baoding with its round bun and meat served hot and the Hejian warm rectangular flaky bun with the meat served cold. Hejian hosts a Donkey Burger festival. In June 2011 the art of cooking donkey burgers was listed as Hebei province's intangible cultural heritage.

Westcott and Fang (2019) Zhang Haitao, founder of the Kong Fu Donkey restaurant chain, also the government sanctioned donkey burger representative thinks donkey burgers has the potential to go global. *Fat Wang's Donkey Burger* chain has twenty branches in Beijing serving their signature donkey burger and donkey hotpot. Meanwhile a Sydney, Australia Donkey Huoshao eatery opened with much fanfare in 2018 but closed after a year.

3 Chinese History & Ejiao Overview

An overview of Chinese history is provided to give context to the current emphasis on production and commodities of medicinal products for the service of the state and the Chinese market economy (Chee 2021) as exemplified by the *ejiao* expansion and global donkey slaughter phenomenon.

3.1 Century of Humiliation 1839–1949

The mandate to *Never Forget National Humiliation*, 勿忘国耻 has been embedded into the Chinese national psyche via history textbooks and patriotic education (Wang 2012; Chen 2020). The China of 1839–1949 was not seeking expansion, she was in fact trying to keep her doors closed from the bullying and invasions of others. A sampling of the "Century of Humiliation" includes: the first Opium Wars which forced China to import opium in exchange for British tea; the ceding of Hong Kong and Mao, the destruction of the Summer Palace by the British and French forces; Eight Nation Alliance; the British Invasion of Tibet; Nanjing Massacre of 1937; the Second Sino Japanese War (1937–1945) and ending with the Communist Party victory over the Nationalists (Wang 2012, pp. 39–69). This institutionalized historical memory of victimization serves to strengthen nationalistic fervor, consolidate, and validate the Communist Party (Wang 2012). While this period included achievements such as the founding of the Republic and the creation of a constitution—the official focus is on the narrative of shame. This emphasis on the bullying by Western and imperialist forces is key to understanding China's current nationalistic millennials, foreign policy, economic development, and continued defense of rare fauna usage in traditional Chinese medicine as intrinsic to its cultural heritage.

3.2 Traditional Chinese Medicine (TCM) as a Communist Project—Reclaiming National Pride

In 1949 Chairman Mao proclaimed the establishment of the People's Republic of China and her dream to reclaim national pride and status. But China was alone, isolated by an international embargo. One of the earliest nation building projects was the institutionalization of *Traditional Chinese Medicine* (TCM), a term formalized under Mao. In 1952 the Shandong Dong'e Ejiao Factory was established as the first state operated donkey hide glue manufacturer (The Donkey Sanctuary 2017), now the largest *ejiao* company, known as Dong'e Ejiao (DEEJ). The standardization and institutionalization of TCM within college education and medical practice was implemented from 1956–64 (Taylor 2005). While Mao himself did not believe in TCM, the lack of Western trained doctors made TCM a practical answer to getting healthcare to the rural millions (Taylor and Matthews 2011). Mao promoted TCM as China's *new* medicine—as a unification of Chinese medicine within modern natural sciences (Hsu 2008). By 1956, TCM colleges were set up in the cities of Beijing, Shanghai, Chengdu, and Guangdong and within four years, there were twenty such institutions, in every provincial capital (Taylor and Matthews 2011). The first TCM textbook was published in 1958, followed by a full set of 18 textbooks in 1962 (Hsu 2008). This short time frame of concentrated nation building demonstrates TCM as a "state invented tradition" (Hsu 2008; Taylor 2005).

3.3 The Great Leap Forward into Famine

Mao Zedong decreed that the masses were needed to labor and sacrifice, to catch up with the West, move mountains, bend the will of the rivers—"People must conquer nature", 人定胜天 (Shapiro 2001). An onslaught of radical revolutionary movements was unleashed from private land confiscations; the execution of one million landlords; state-operated communes, to larger agricultural cooperatives—each one more traumatic than the other (Schell and Delury 2013).

From 1958–1961, Mao threw the nation into *The Great Leap Forward*, a modernization program to catch up to the West through the relentless production of steel and grain. Farmers were compelled to abandon their fields and tear down the forests for fuel (Shapiro 2001, p. 198); and give up their own cooking ware and farm tools to be melted into steel (Shapiro 2001, pp. 67–93). Mao's quest to conquer nature raged on. The South China tiger, bears, wolves, and leopards were declared enemies of the people for eating livestock. Extermination was supported with bounties paid for by the government (Pomfret 2008). In 1958 the Eurasian tree sparrows (*Passer montanus*), were targeted for eating the "people's grain". The masses from young to old were mobilized in a mandatory synchronized national effort to eradicate the sparrows using everything from sling shots, rifles, poison, trapping, netting, banging on pots, to hand strangulation (Time 1958). The sparrows in fact, mainly ate insects and with their removal—locust and other insect infestations devastated the crops. Local officials had inflated grain production reports. Grain inspection teams searched for hidden grain and produce, beating thousands to death during the frenzy (Yu and Barr 2012, 116–118). *The Eradicate the Sparrow* campaign is considered a major contributing factor to *The Great Famine* (1959–1961) in which an estimated 36 million people starved to death (Yang and Mosher 2013). It was during this time of limited resources that the state promoted research for more readily available and affordable traditional medicine materials (Chee 2021). Alternatives to the more expensive donkeys were being sought after and the *ejiao* recipe re-examined (Yi 2023).

3.4 The Great Proletarian Cultural Revolution, the Lost Years: 1966–76

Nature was not the only nemesis. Mao's ideological movement, The Great Proletarian Cultural Revolution sought to eliminate the bourgeois element and *The Four Olds*—old things, ideas, customs, and habits. The Red Guards, Mao's fanatical students were given full rein to destroy, persecute and kill *anyone* they deemed as bourgeois with Western associations or feudal traditions. Schools were closed and the Red Guards ransacked temples, libraries, private homes, and historical sites. Millions died in the mass killings, public beatings, and suicides. 142,000 teachers under the Ministry of Education were "falsely charged and persecuted" (MacFarquhar 2011). Dissent was unthinkable—millions had no choice but to obey Mao's every wish (Shapiro 2001, p. 198).

The economy was on the verge of collapse. The Cultural Revolution ended with Mao's death in 1976.

3.5 Deng Xiaoping: Get Rich, Go Shopping and Forget

Deng Xiaoping emerged in 1978 to lead China into the Four Modernizations (agriculture, industry, science and technology and defense) and the Economic Reforms with an Open-Door Policy of free trade, special economic zones, and attracting foreign investment and private enterprise (Schell and Delury 2013). Along with this came a national forgetting and given the extreme violence, complicity, deprivation, and control of the Mao era—people were happy to just make money (Osnos 2011), their own money. Along with the reforms came the largest human migration with 130 million rural citizens rushing into the cities in search of jobs and leaving behind some 60 million children (Osnos 2011). Such upheaval brought income inequality, corruption, environmental disaster, and mental illness—with China ranking as one of the most mentally ill countries in the world (Osnos 2011). A society long deprived of consumer goods, fashion, and individuality, was now going shopping. Conspicuous display of wealth, banquets and the Chinese tradition of gift giving to curry favor returned. Prostitution became common place in the hypersexualized post Mao society (Zheng 2009). In the 80 s gendered job segregation increased with campaigns for urban women to give up their jobs and "return home", 回家 for the nation and its economic transformation (Liu 2007). By the mid 90s layoffs of women increased. The beauty economy (salons, beauty pageants, cosmetics, plastic surgery, and other related health and media industries) was part of the re-employment of women (Yang 2021). Fake products and companies proliferated in this regulation lax climate. China became the world's leading source for counterfeit products (Volodzko 2015). It is thus that most Chinese consumers prefer verifiable products, expensive and established name brands. The traditional Chinese medicine industry returned to an emphasis on historical authenticity and donkey alternatives to *ejiao* were abandoned with the shift into market economy (Chee 2021).

3.6 Poverty Eradication, Wildlife Farming, TCM

Wildlife farming was given a major role under the government's poverty eradiation campaign to reduce the divide between urbanites and rural poor. The breeding of wild animals was supported and promoted to farmers as a more lucrative, alternative to traditional crop farming (Arranz and Huang 2020). Following North Korea's techniques, bear bile farming appeared in China in the early 1980s (Meng 2011). By 1985 the fifth edition of TCM textbooks was published, which is the basis of TCM today (Hsu 2008). Tiger farming for bones and other parts followed in 1986 (EIA 2013). In China's first Wildlife Protection Law (WPL) in 1988 gave ambiguous protections while encouraging the breeding and farming of protected wild animals by providing licenses to individuals (Cao 2015). With China's entry into the global economies in the 90s, TCM medicine further commercialized and expanded into the global markets (Hsu 2008). Wildlife farming had a hiatus during the Covid pandemic but continues to date (Act Asia 2020).

4 Demand: Ejiao Expansion

In 1993 State owned Shandong Dong'e Ejiao was reorganized into a joint-stock enterprise and by 1996 was listed on the Shenzhen Stock Exchange as Dong-e Ejiao Limited by Share Ltd. (Crunchbase n.d.). Dong'e Ejiao (DEEJ) is part of China Resources (CR) Pharmaceutical Group Limited—a state-owned pharmaceutical platform along with other top brands such as 999, Double Crane, Zizhu, Pien Tze Huang Pharmaceutical Ltd., among others. The Pharmaceutical R&D Center includes a National Research Center for Gelatin Traditional Chinese Medicine Engineering and Technology (China Resources n.d.).

The original *ejiao* format was a solid bar of gelatinized donkey hide, which required cooking. In the 90s *ejiao* was modernized with an array of instant formats with added flavorful ingredients for the healthy food and beauty market. In 2008 Chairman, Qin Yufeng and his company DEEJ was awarded national intangible cultural heritage status for continuing the ancient ninety-nine-step *ejiao* production techniques (Hao 2015). Intense marketing and demand for hides resulted in the China's own donkey population, once the largest in the world, plummeting from 12.7 million in 1954, to 9.44 million in 1996 (China 2016) to 6.03 million donkeys in 2013 (Zhu 2017). DEEJ has increased the price of *ejiao* sixteen times in the last decade and requires an unsustainable demand of 4.8 million skins per year (The Donkey Sanctuary 2019). Demand clearly exceeds supply. Decline in donkeys has led to fake *ejiao* and the use of everything from cows, pigs to shoes (China Daily 2016).

As a member of Shandong's provincial legislature, Qin personally lobbied for more government support for donkey farmers in 2015 (Zhao 2015). That same year, Shandong Dong'e County received its own National Ejiao Association (Shang 2015). Qin worked with the State Administration of Traditional Chinese Medicine and the National Tourism Administration to create "national health tourism"—

bringing *ejiao* to the world (China Direct Selling News 2016). *Ejiao* tourism includes an annual *Winter Solstice Ejiao Festival* along with an *Ejiao World* tourist complex with a *Dong'e E Jiao Experience Factory, Chinese E Jiao Museum, Dong'e E Jiao City,* Dong'e Yaowang Mountain, and Donkey Kingdom (Donge ejiao 2023).

4.1 Ejiao Ads in Historical TV Dramas

Ejiao advertising within Chinese televised historical dramas has become common place. (culture.ifeng.com, 2012) Chinese netizens joked that the popular, "Zhen Huan Legend" had become "*Ejiao* Legend". Not only did the protagonist lines mention ejiao more than twenty times; the court ladies and the concubines all talking about *ejiao*; the main characters had eaten *ejiao*, but the emperor's reward was a box with the Dong'e Ejiao logo on it. A ten-million-yuan fee to place the *ejiao* ads was rumored, however the company would not disclose the contract details. Period dramas, *Empresses in the Palace* and the *Imperial Doctress* also had embedded scenes of *ejiao* gifts to courtesans also paid for by DEEJ (Nuwer 2018; Ko 2017). A new modern revolutionary TV drama series, '*Ejiao Under Heaven*" with 48 episodes of an ejiao shop owner's plight during the Japanese invasion is forthcoming (365 TV 2025).

4.2 Ejiao & Social Media

China boasts the world's highest numbers of Internet users at 54% of the population (The World Bank/Data 2020). She is also ranked as the world's worst abuser of internet freedom for six consecutive years per Freedom House (2021). The Reporter Without Borders (RSF) 2020 World Press Freedom Index ranked China at 177th out of 180, followed only by Eritrea, Turkmenistan, and North Korea. Social media platforms Facebook, WhatsApp, Instagram, Twitter, YouTube, along with several Google services and most international news outlets are blocked in China (FH 2021).

Sina Weibo, WeChat, Xiao Hongshu (Little Redbook), and Douyin are China's dominant social network platforms known for latest trends from key opinion leader (KOL). Key opinion leaders offer product reviews, recommendations and commentary that can influence the purchasing decision of a consumer. Female consumers have a strong connection to female KOLS due to a sense of trust and expertise. An estimated 77% of Chinese female consumers have purchased products because of a KOL (Jing et al. 2015). China is a decade ahead of other nations in e-commerce where 11.6% of retail purchases are made directly on social media (Achim 2021). DEEJ pursued digital expansion by partnering with retail giants, Tmall and JD and the creation of an e-commerce company (Hao 2015). At present there are more than 100 ejiao companies in China, DEEJ as the largest followed by the Fupai Group (HSI 2018).

Ejiao consumption since ancient times has been utilized to counter blood deficiency and gynecological problems (Zhang et al. 2021; China 2012). It is therefore no surprise that women appear in advertising as influencers, entrepreneurial sellers, and consumers. While understood that the true gender of the person who posts or seeks to influence on social media cannot be determined—this does not refute the significance of appearing to be of the female gender. All italicized quoted text in this and the following section indicates the author's English translation of Chinese language ejiao posts collected in 2017–2020 from Sina Weibo. Ejiao consumption in social media posts is portrayed as a "*modern lifestyle for all*" suitable "*for anyone over age 4*". Ad copy reads as if your auntie, neighbor, sister, mother, grandma is entreating you to: '*be good to yourself*", "*buy less buy wise.*" *Ejiao* is recommended as a gift for anyone, with specific mentions as appropriate for "*your mom, mother-in law, parents, elders, friends, for co-workers and as office snacks.*" Suitable for "*Chinese New Year, the Dragon Boat Festival, Nurses Day, and as Mother's Day gifts.*" Chinese women figure prominently as the sellers and promoters of *ejiao* in social media ads, on television, on gift boxes as historical courtesans, and with their shopping carts as modern health-conscious consumers.

4.3 Ejiao Diversification & Efficacy Claims in Social Media Posts

While the original raw *ejiao* bar that requires cooking is still available, the individually wrapped instant cracker snack form dominates in the ads collection. The cracker snack form has added ingredients such as walnuts, nut assortments, honey, ginger, and rose petals. Other *ejiao* infused novel formats include dates, powder drink, syrup tonic, pellet ginger tea, pill, ginger tea cube, milk tea, herbal paste, nut paste, cube instant drink and the compound *ejiao* paste patent medicine. *Ejiao* products are sold in different quantities, in individual plastic bag packages, gold zipper pouches inside tin gift boxes, glass vials, jars, canisters, and paper boxes.

The term "donkey" is not frequently mentioned in social media posts except to emphasize "*fresh, national donkey*", "*selecting good donkey skin*", "*authentic donkey*" in reference to the Shandong large black, Dong'e donkeys versus the "*fakes made of cow or mule hide*". Social media ads position *ejiao* alongside the established TCM rarefied products of desire such as bird's nest, and fish maw. Ginger in combination with ejiao has seen an increase and there is a collaborative agreement of Dong'e Ejiao and the Korean Ginseng Corp (KGC) to jointly develop new medicinal and health supplements (China Daily 2018). The actual amount of ejiao in these instant products is unknown. The DEEJ website product index now includes oral liquid seahorse, turtle shell glue, and deer antler glue sold in gift boxes of individually wrapped pieces.

Claimed efficacy were wide ranging and numerous including women's issues such as menstruation regulating, increased fertility, blood replenishment, pregnancy recovery as well as fatigue, reverse aging, beautify skin, natural diet (eat while fasting), hair loss, eyesight, insomnia, nourish liver, heart, lung, spleen, stomach kidney, cold hands and feet, boost immunity, among others. TCM products typically list efficacy for a wide list of ailments. One ad listed twenty illnesses that were treatable with ejiao. With some ads including medical reports, quotes from ancient medical treatises, lengthy efficacy listings, and healthy living advice. Young women with perfect white skin advise *to "eat one ejiao cake a day and forever appear 20."* In a society where job opportunities preference men and females under 30 (Zhang 2021), this is an effective talking point.

During COVID of 2020 social media posts deemed the *ejiao* snack cracker as "*a must have daily supplement" as "you need a very strong body to enter combat when the virus comes"*. *Ejiao* donations to essential worker are described as "*on the way to resist the epidemic, fear not wind, rain or difficult roads* and as being "*the invincible atomic bomb to fight the virus…bravely going to the front line…*" *Ejiao* is viewed as patriotic heroisms continuing the "*legacy of fighting against SARS*" and a "*new TCM cyclone around the world*". Since COVID and the rise of pandemics, interest in taking immunity boosting functional instant food products is growing and no longer a seasonal winter interest.

A sampling of social media posts provides a visual glimpse of how *ejiao* is marketed to Chinese consumers.

Source: Weibo.com

For Mom

Nourish Qi, Nourish blood, Prevent aging

Love is the most beautiful forms of energy. When we were young, we received our mother's love, so when we grow up, we learned to give and pass on love.

Let love flow out on this warm-hearted month of May, give pure handmade ejiao cake to your beloved mother.

May 13 is Mother's Day, is your gift ready?

Original Health Lifestyle Store

Happy Mother's Day

Is your Mother's Day gift ready? Only the best.

Source: weibo.com

Stock Market Red Weekly Magazine

The gem of Chinese medicine goes overseas, Compound Ejiao Paste helps Indonesia fight the pandemic." Among the confirmed cases of new crown pneumonia nationwide, 74, 187 people have used Chinese medicine, accounting for 91.5%. Clinical efficacy observations show the total effective rate of Chinese medicine has reached more than 90%. Traditional Chinese medicine has played an important role in China's fight against the pandemic, is receiving attention from the international community and its overseas market continues to increase…

5 Supply—China's Going out to Africa Policy

Under President Jiang Zemin, foreign investment was "*Coming in*" and China was also "*Going out*", most notably to Africa. In the 90s China was providing African nations with concessional subsidized loans, undertaking large-scale infrastructure projects, and encouraging partnerships between Chinese and African enterprises (Ministry 2018). Resource rich cash poor African nations with historically corrupt regimes and lax enforcement were now home to a growing number of Chinese nationals eager for opportunities. This included the first generation of the *Shuidong Syndicates*, which for decades trafficked ivory with impunity (EIA 2017). By 1999 trade between China and Africa totaled 6.5 billion U.S. dollars (Ministry 2018). The headquarters of the African Union was completed in 2011 at a cost of 200 million as a gift from China (BBC 2012; Wei 2012).

5.1 Century of Xi and the Belt and Road Initiative (BRI)

In 2013 President Xi Jinping launched his ambitious *Belt and Road Initiative* (BRI) to connect Asia by land and by sea to Europe and Africa, through massive investments and infrastructure projects with mutual benefits. While there is much debate about the motives and benefits of such investments, the environmental impacts, and the associated wildlife crime along BRI routes among conservationists are evident (Wong 2020). By 2016 China was the Sub-Saharan Africa's largest trading partner, comprising of 25% of all trade (Donkey 2017, p.10). As per the Green Belt and Road Initiative Center, Beijing website there are 150 countries who have joined the BRI and signed onto a Memorandum of Understanding (MoU) with China. This includes 53 African countries (out of 54), 29 European countries, 22 Latin American and Caribbean countries, 12 Pacific countries, 10 Southeast Asian countries, among others (green-bri.org).

The Chinese State Council on December 6, 2016, published its first white paper on policy for the development of TCM: calling for its "equal status" to Western medicine, greater investment in TCM research and education; and for the assertive introduction of TCM to the rest of the world" through its bilateral relations and presence in multilateral organizations (State Council People's Republic of China 2016). The 2019 inclusion of a TCM chapter in the World Health Organization influential 11th International Classification of Diseases (World Health Organization 2018), provided a significant boost to TCM global expansion. BRI Southeast Asian nations such as Indonesia, Vietnam, Singapore, and Malaysia are prime *ejiao* markets.

China Direct Selling News reported that in 2015 DEEJ Chairman Qin made trips to Australia, Mexico, West Africa, East Africa, North Africa, Central Asia, and South American seeking to build overseas donkey bases (China Direct 2016).

GI-TOCS (2020, p.7–11) Kenya with the opening of four donkey abattoirs (Goldex Kenya Ltd., Baringo County; Star Brilliant Donkey Abattoir, Nakuru County; Silzha Ltd., Turkana County, and Fuhai Machakos Trading Company

Ltd., Kithyoko, Machakos County) in 2016 became the African epicenter for the donkey trade to China. The constant demand for skins has led to local donkey thefts and cross border smuggling from neighboring Tanzania, Uganda, and Ethiopia. Within two years, the four plants had slaughtered 15% of the nation's donkey population.

5.2 "Without the Donkey, You Are the Donkey"

Kenya Agricultural and Livestock Research Organization (2019) Donkeys are essential in Kenya's rural communities for household chores, farming, income generation that is used for paying rent, school fees, clothes, and other household needs. The loss of a donkey impacts livelihoods, social benefits, status, food security, health, and the welfare of the entire family. For women particularly, the loss of a donkey, means you are the donkey. With the arrival of the slaughterhouses and the *ejiao* market—the price of a donkey has more than doubled—adding more financial pressure on owners when replacing and protecting their donkey.

Margaret (Left) loading her donkey with a water can at the river with help from local villagers in Nkareta Location, Olaimutiai village, Narok County, Kenya—an area that was largely affected by the donkey theft and slaughter. (Photo credit: Brooke East Africa)

Environmental pollution from the discarded carcasses in both bush and abattoir slaughter is another factor. In August 2021, Tanzania revoked the provisional permit of the Fang Hua Investment Company Ltd., Shinyanga region donkey slaughterhouse for its lack of "animal welfare, humane slaughter and abattoir hygiene" (Karashani 2021). In Botswana, a raid on the Bo Chang Group abattoir in June 2017 found 452 emaciated and starving donkeys, along with hundreds of rotting carcasses leaching into the Tati River (Nkala 2017).

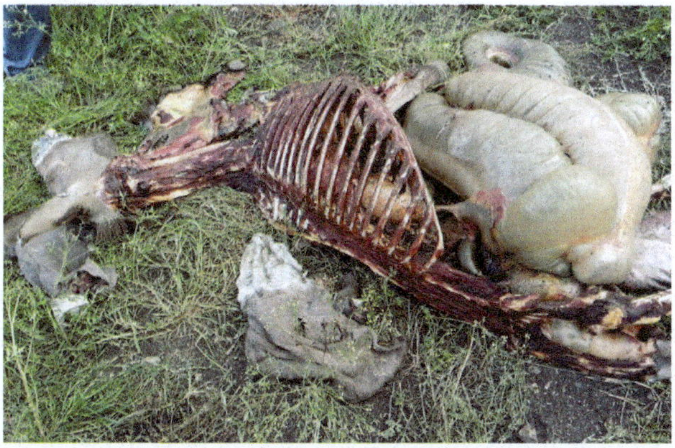

Local communities found carcasses of their donkeys skinned, at the peak of donkey theft, Baringo County, Kenya in 2018. Baringo County houses is one of the four main donkey abattoirs in the country. (Photo credit: Brooke East Africa)

Given the emphasis on the skin as the only value, there is little interest in the welfare of the donkeys sent to slaughter. The transporting of both live, sick, injured, starving cramped donkeys and skins cross borders has been associated with the increase of equine influenza, disease spread and the risk of contaminated skins (DS 2023). This is in stark contrast to the marketing of *ejiao* as a "green natural healthy product." Donkey skins have also been following the ivory trade routes and found with other illegal wildlife products (DS 2023; Nkala 2020).

Women donkey owners from Kitui and Machakos County, Kenya holding peaceful protests calling for their local county government to shut down the Machakos Fuhai Company donkey abattoir in Kithyoko area, Masinga sub-county, Machakos county. (Photo credit: Brooke East Africa)

The Brooke (2021) Women led Kenya wide community protests resulted in banning donkey skin export and the closing of the abattoirs in February 2020. However, the Chinese owned abattoirs brought legal challenges, and the High

Court overturned the ban in May 2021. On August 26, 2021, a petition signed by 30,000 donkey owners was sent to the Kenyan government to reinstate the ban on donkey slaughter for skin and meat. The 2022 Pan African Donkey conference opened the doors for the 2024 the African Union 55-member nation fifteen year ban on the donkey skin trade.

Mshelia (2025) However the continent-wide ban has proved to be ineffective—no enforcement, insufficient deterrents, low fines, weak borders enable the continuance of illegal bush slaughter. Cross border smuggling and theft in West African sub-regions combined with mistreatment during transport and inhumane bush slaughter increases the risks of disease transmission(Equine influenza, Equine piroplasmosis Theileria haneyi). Nigeria is now the epic-center of the illegal trade with the donkey meat sold in local markets as "beef". Public health risks loom with the consumption of uninspected meats slaughtered in unhygienic conditions, inappropriate disposal, contamination of water sources and transfer of infection to other species. There is also the genetic loss of indigenous donkey strains, "We are not just killing animals, we are killing a future, one we didn't even document."

Rural communities with large donkey populations in Kyrgyzstan, Mexico, Columbia, Peru, Brazil, and Pakistan are likewise devastated by this unsustainable industrial scale slaughter for *ejiao*. Pakistan is now the number one donkey skin trader to China, following the African wide ban and Pakistan's own reversal of its earlier ban. If current donkey slaughter rates continue, the impact on global donkey populations along with the "several hundreds of millions of people losing part of all, of all their livelihoods" will be catastrophic (GI-TOC 2020; KALRO 2019). Donkey skins are also both a smuggled item and a vehicle for smuggling as traffickers take advantage of the ambiguous legality status as a cover for parallel trafficking of wildlife, drugs, contraband and even human parts (Donkey 2022).

6 Advocacy and Conclusion

Government policy under Mao promoted TCM, the destruction of nature, and the establishment of DEEJ. Post-Mao Economic Reforms has advocated the utilization and breeding of nature as products. China as the source of demand and production could change the *ejiao* story. Xi Jinping and the 14th Five-Year Plan has emphasized the goal of an ecological civilization in harmonious co-existence with nature (Xinhua 2021a, b). China could follow its own policy—halt the importation of donkey skins and pursue alternatives. Given its advanced technologies in cellular regeneration, a slaughter free *ejiao* might be possible. Some ejiao companies are already exploring this (Donkey Sanctuary 2019). However, this is still unlikely to change the situation while there is steadfast support from the state for ejiao as an intangible cultural heritage in its authentic donkey form going global furthering the Chinese dream.

Meanwhile Chinese consumers are oblivious of *ejiao*'s devastating impact on global human communities and their donkeys. They believe the product is derived from local Chinese donkeys. However in tandem to the increase of disposable income is the rise of pet animal "companionship" and animal welfare concerns. Chinese animal activists and animal welfare groups now number in the hundreds and exist in nearly every province (Jeffery 2014). Chinese citizens have organized successfully in defense of animals. Immense public opposition successfully forced the *Fujian Guizhen Tang Pharmaceutical Company* in 2011 to withdraw its stock market launch to raise capital for the expansion of bears and its annual gallbladder powder production (Meng 2011). Chinese citizens also expressed overwhelming support in 2020 (90% of the 74, 000 surveyed) for a strict ban on wildlife consumption, trade, and exhibition (Shi et al. 2020). If made aware of the *ejiao* realities, Chinese consumers could influence the industry direction.

Chinese vegan restaurants are popping up in the big cities. The Shanghai based *Pure Vegan* restaurant co-founder, Ran Lau featured her *brown sugar cake*, made of brown sugar, Xinjiang red dates, Ningxia goji berries and black sesame—touting "the same health benefits of *ejiao*, without the donkey" (Yau 2021). Ran spoke out against the cruelty of *ejiao* cake as "Donkey keepers flay the animal [before its slaughter] so its skin is infused with blood," (Yau 2021). Educated Chinese women are the potential leaders of conscious compassionate consumption (Zhang 2020; Jing 2020). **It is thereby critical to educate and empower Chinese female consumers and budding East Asian and Southeast Asian consumers of *ejiao* realities. A Chinese social media influencer emphasizing harmony with nature and non-animal based products might hold sway, especially in these shaky economic times where consumers are buying less and being more strategic in their spending. But this is sensitive due to China's tight censorship on the media (Xu and Albert 2017) and the association of TCM as a patriotic product.**

Donkeys as the beast of burden lack both the endangered species status and the wild charismatic appeal of their fellow imperiled TCM fauna product counterparts—the tiger, lion, bear, pangolin, or rhino. As such the plight of working donkeys and their owners has not received the attention it deserves. BRI nations are reluctant to raise the issue to China, making demand reduction even more critical. **Support for these vulnerable communities and their donkeys is urgently needed.**

Brooke Action for Working Horses and Donkeys has launched petitions to the Amazon company to ban ejiao products from its website (Brooke n.d.). The United States Congressman Don Beyer (D-VA-08) on September 10, 2021, introduced the *Ejiao* Act (HR5203), which aims to ban the sale or trade of *ejiao* products using donkey skin in interstate or foreign commerce (Congress.gov). **Bans on the donkey skin trade need to be supported, replicated and enforced on a global scale, and at each step of the supply chain.**

The *ejiao* story demonstrates the ramifications of a state supported TCM fauna-based diversification and expansion into Belt and Road nations. It is a tragic irony that the donkey skin trade with human and environmental impacts under the label of "beauty foods". And continues expanding in this time of climate crisis, bio-cultural diversity loss, deforestation, and zoonotic diseases. The plight of donkeys and their rural communities is a **cautionary tale** and an **urgent global social, environmental justice, and animal welfare issue**.

Acknowledgments Special thanks to Dr. Amy McClean, Dr. Kimberly Spagnol, and Dr. Helen Agu; research assistants, Yingyin

Lu and Yichun Luo; photography permissions provided with assistance from Samuel Theuri, The Brooke East Africa.

Conflict of Interest The author declares no conflict of interest.

Questions and Answers

1. What is ejiao?
   Ejiao is a traditional Chinese medicinal product made from gelatinized donkey skin.
2. How has this product changed in recent years?
   Product diversification and extensive advertising has pushed ejiao into a mainstream wellness product - and unable to supply the demand.
3. Which communities have been impacted by the development of ejiao?
   Rural communities in the global South that depend on donkeys for their transportation, living, and well-being of their families.
4. Name a few countries outside of China with donkey populations
   Kenya, Nigeria, Pakistan, Mexico.
5. Is this product sustainable, if not, why?
   Demand decimating global donkey populations. Donkey reproduction rates are not suitable for intensive farming.

References

365 Television Series. 天下阿胶剧情介, Ejiao Under Heaven press release Retrieved 31 Aug 2025. (In Chinese). http://m.yue365.com/tv/8277/

Achim AL (2021) Why China is a decade ahead in social-driven sales. Jing Daily. Retrieved from https://jingdaily.com/china-social-commerce-wechat-facebook/

Act Asia (2020) China's revised law paves the way for wildlife to become farm animals. Retrieved from https://www.actasia.org/news/chinas-revised-law-paves-the-way-for-wildlife-to-become-farm-animals/

BBC News (2012) Africa Union opens Chinese-funded HQ in Ethiopia. Retrieved from https://www.bbc.com/news/world-africa-16770932

Arranz A, Huang H (2020) China's wildlife trade. South China Morning Post. Retrieved from https://multimedia.scmp.com/infographics/news/china/article/3064927/wildlife-ban/index.html

Canisso IF, Panzani D, Miró J, Ellerbrock RE (2019) Key aspects of donkey and mule reproduction. Vet Clin North Am Equine Pract 35(3):607–642. https://doi.org/10.1016/j.cveq.2019.08.014. PMid:31672204

Cao D (2015) Animals in China: law and society. Palgrave Macmillan, Hampshire

Chee LPY (2021) Mao's bestiary: medicinal animals and modern China. Duke University Press, Durham and London

Chen Q (2020) Why Beijing constantly talks about China's 'century of humiliation'. Inkstone News. Retrieved from https://www.inkstonenews.com/society/inkstone-explains-why-beijing-constantly-talks-about-chinas-century-humiliation/article/3089559

China Daily (2016) Shortage of donkey skin breeds TCM fakes. Retrieved from https://www.chinadaily.com.cn/china/2016-01/28/content_23287841.htm

China Daily (2018) Donge Ejiao joins hands with Korean Ginseng Corp. http://shandong.chinadaily.com.cn/2018-11/27/c_294501.htm

China Direct Selling News (2016) Big craftsman Qin Yufeng: National cultural heritage. Retrieved from http://en.cndsn.com.cn/headline-news/2016/0415/2584.html

China Resources Pharmaceutical Group Ltd. (n.d.) Retrieved from https://www.pharmaceutical-technology.com/contractors/pharma-wholesalers/china-resources-pharmaceutical-group-ltd/

China Today (2012) Water, heroes and donkeys: the unlikely history of Liaocheng City. Retrieved from:http://www.Chinatoday.com.cn/english/china/2012-10/24/content_492281_3.htm

Crunchbase (n.d.) Shandong Dong'e Ejiao organization. Retrieved from https://www.crunchbase.com/organization/shandong-dong-e-e-jiao

Dong'e Ejiao (2023) website. Retrieved from https://www.dongeejiao.com/item4/2023-06-21/358358.html

EIA (2013) Hidden in plain sight: China's clandestine tiger trade. Retrieved from https://eia-international.org/report/eia-online-library-for-hidden-in-plain-sight-report/

EIA (2017). The Shuidong connection: exposing the global hub of the illegal ivory trade. Retrieved from https://eia-international.org/report/shuidong-connection-exposing-global-hub-illegal-ivory-trade/

Freedom on the Net China. Retrieved: 15 Feb 2021. https://freedomhouse.org/country/china/freedom-net/2020

Global Initiative Against Transnational Organized Crime (2020) Civil Society Observatory of Illicit Economies in Eastern and Southern Africa Risk Bulletin. Issue 10,7–11. Retrieved from https://globalinitiative.net/analysis/esaobs-risk-bulletin-10/

Green Belt and Road Center (2025) website: green-bri.org

Hao N (2015) Inheritor of ejiao secrets spreads TCM around the world. China Daily. https://www.chinadaily.com.cn/cndy/2015-12/28/content_22836093.htm

Hsu Y, Serpell JA (2007) Development and validation of a questionnaire for measuring behavior and temperament traits in pet dogs. Journal of the American Veterinary Medical Association

Hsu E (2008) The history of Chinese medicine in the People's Republic of China and its Globalization. East Asian Sci Technol Soc 2(4):465–484. https://doi.org/10.1215/s12280-009-9072-y

Humane Society International (2018) Ejiao: an investigative report. Retrieved from https://hsi.org.au/uploads/publication_documents/HSI_DHG_Investigation_WEB_AUST.pdf

Jeffrey E (2014). The rise of China's animal activists. Animals Asia. Retrieved from https://www.animalsasia.org/au/media/news/news-archive/the-rise-of-china's-animal-activists.html

Jing Daily (2020). Presented by Tong: why luxury needs China's impact-conscious consumers. Retrieved from https://jingdaily.com/china-tong-tracking-trends-2021-perfect-diary/

Jing A, Zheng R, Zeng X (2015) Research on the impact of purchasing behavior of Online opinion leader for female consumers. Bus Econ Stud 36:59–60

Karashani B. Tanzania shuts down donkey abattoir. The East African (Nairobi). 28 Aug 2021. https://www.theeastafrican.co.ke/tea/business/tanzania-shuts-down-donkey-abattoir-3529250

Kenya Agriculture and Livestock Research Organization—KALRO (2019) The status of donkey slaughter in Kenya and its implications on community livelihoods [Internet]. [cited 2020 Sep 26]. Available from: https://www.thebrooke.org/sites/default/files/KalroReport-Final.pdf

Ko D (2017) The making of a traditional medicine: Donkey-hide gelatin (ejiao) in late imperial and modern china. In Wang DD (Ed), A new literary history of modern China (pp. 913–918). Belknap Press of Harvard University Press

Li P (2018a) Animal rights activism in China. Ohio State University, Modern Chinese Literature and Culture Resource Center. Retrieved from https://u.osu.edu/mclc/2018/03/07/animal-rights-activism-in-china/

Li Y (2018b) Hejian residents get a kick out of 2nd China donkey burger festival. China Daily. Retrieved from https://www.chinadaily.com.cn/a/201805/09/WS5af25b9ba3105cdcf651cd89_1.html

Li Y (2018c) Inside China's massive crackdown on wild animal farms. Sixth Tone. Retrieved from https://www.sixthtone.com/news/1006770/inside-chinas-massive-crackdown-on-wild-animal-farms

Liu J (2007) Gender and work in urban China. Women workers of the unlucky generation. Routledge, London/New York

Lu S (2020) Pretty lady cadres: new data shows the limits of women's advancement in China's leadership. Chinafile. Retrieved from https://www.chinafile.com/reporting-opinion/features/pretty-lady-cadres-china

MacFarquhar R (2011) The politics of China: the eras of Mao and Deng. Cambridge Press, Cambridge

Meng S (2011) Fighting for China's battery bears. China dialogue. Retrieved from https://chinadialogue.net/en/nature/4162-fighting-for-china-s-battery-bears/

Ministry of Foreign Affairs of the People's Republic of China (2018) A Long-term stable China-Africa relationship of All-around Cooperation. Retrieved from https://www.fmprc.gov.cn/mfa_eng/ziliao_665539/3602_665543/3604_665547/t18036.shtml

Mshelia P (2025) Impact of donkey skin trade/export on donkey population in Africa; case study in some AU MSs. [Conference presentation] 2nd Pan African Donkey Conference, Abidjan, Cote d'Ivoire. Retrieved from https://panafricandonkeyconference.org/

Nkala O (2017) Inside Botswana illegal donkey trade. Oxpeckers Investig Environ JIsm. https://oxpeckers.org/2017/08/botswana-donkey-trade/Brooke, Help Stop the Sale of Ejiao on Amazon. https://www.brookeusa.org/amazonCongress. Ejiao Act. https://www.congress.gov/bill/119th-congress/house-bill/5544

Nkala O (2020) Donkey rustlers profit off wildlife trafficking routes. Oxpeckers Reporters. 8 Sept 2020. Retrieved from https://oxpeckers.org/2020/09/donkey-rustlers-and-wildlife-trafficking-pipelines/

Nuwer R (2018) To sate China's demand, African donkeys are stolen and skinned. The New York Times Retrieved from https://www.nytimes.com/2018/01/02/science/donkeys-africa-china-ejiao.html

Osnos E (2011) Meet Mr. Freud: does psychoanalysis have a future in an authoritarian state? The New Yorker Retrieved from https://www.newyorker.com/magazine/2011/01/10/meet-dr-freud

Pomfret J (2008) South China tigers teeter on brink of extinction. Reuters. Retrieved from https://www.reuters.com/article/us-china-tigers/south-china-tigers-teeter-on-brink-of-extinction-idUSHKG32826920080710

Schell O, Delury J (2013) Wealth and power: China's long march to the twenty-first century. Random House, New York

Shang M (2015) Shandong Dong'e gets national association for ejiao. China Daily. Retrieved from: https://www.chinadaily.com.cn/m/shandong/shandongculture/2015-12/03/content_22760978.htm

Shapiro J (2001) Mao's war against nature. Cambridge University Press, Cambridge

Shi X, Zhang X, Xiao L, Li BV, Liu J, Yang F, Zhao X, Cheng C, Zhi L (2020) Public perception of wildlife consumption and trade during the COVID-19 outbreak. Biodivers Sci 28(5):630–643. https://doi.org/10.17520/biods.2020134. Retrieved from https://biodiversity-science.Net/EN/10.17520/biods.2020134

Skippen L, Collier J, Kithuka JM (2021) The donkey skin trade: a growing global problem. Braz J Vet Res Anim Sci 58(special issue):e175262. https://doi.org/10.11606/issn.1678-4456.bjvras.2021.175262

Su L (2012)《甄嬛传》变身"阿胶传"传品牌植入费高达千万, "Empresses I n the Palace" transforms into "Ejiao Legend" with brand placement fees reportedly in the millions. Phoenix Culture. https://culture.ifeng.com/whrd/detail_2012_04/14/13885929_0.shtml

Taylor K (2005) Chinese medicine in early communist China, 1945-1963 A Medicine of Revolution. Routledge Curzon, New York

Taylor TS, Matthews NS (2011) Veterinary care of donkeys. In: Equine Veterinary Education / International Veterinary Information Service (IVIS), Ithaca, NY

The Brooke. Donkey Owners Petition Kenyan Government to Ban Slaughter for Good. 6 Sept 2021. https://www.thebrooke.org/news/donkey-owners-petition-kenyan-government-ban-donkey-slaughter-good

The Donkey Sanctuary (2017) "Under the skin: the emerging trade in donkey skins and its implications for donkey welfare and livelihoods." Retrieved from https://www.thedonkeysanctuary.org.uk/what-we-do/issues/donkey-meat-skin-and-milk/full-report

The Donkey Sanctuary (2019) Under the skin update on the global crisis for donkeys and the people who depend on them. Retrieved from https://www.thedonkeysanctuary.org.uk/sites/uk/files/2019-11/under-the-skin-report-revised-2019.pdf

The Donkey Sanctuary (2022) The Global trade in donkey skins: a ticking time bomb. Retrieved from: https://www.thedonkeysanctuary.org.uk/sites/default/files/2024-06/global-trade-in-donkey-skins-a-ticking-time-bomb-report-1-2022.pdf

The State Council Information Office of the People's Republic of China (2016) China issues first white paper on traditional Chinese medicine. Retrieved from www.scio.gov.cn2016-12-06

The World Bank/Data. Retrieved 15 Dec 2020. https://data.worldbank.org/indicator/IT.NET.USER.ZS?locations=CN

TIME (1958) RED CHINA: Death to Sparrows. Retrieved from http://content.time.com/time/subscriber/printout/0,8816,863327,00.html

Volodzko D (2015) China's addiction to counterfeiting. The Diplomat. Retrieved from https://thediplomat.com/2015/10/chinas-addiction-to-counterfeiting/

Wang Z (2012) Never forget national humiliation: historical memory in Chinese politics and foreign relations. Columbia University Press, New York

Wei L (2012) Building for the future. China Africa. Vol 4. Retrieved from: http://www.chinafrica.cn/china_report/txt/2012-01/30/content_422287.htm

Westcott B, Fang N (2019) Chinese food has conquered the world. Are we ready for the donkey burger? CNN travel. https://www.cnn.com/travel/article/beijing-donkey-meat-china-intl-hnk

Wong RWY (2020) Shadow operations in wildlife trade under China's Belt and road initiative. China Inf 35(2):201–218. https://doi.org/10.1177/0920203X20948680

World Health Organization 2018. WHO releases new international classification of diseases (ICD 11) https://www.who.int/news/item/18-06-2018-who-releases-new-international-classification-of-diseases-(icd-11)

Xinhua (2021a) Full text: Speech by Xi Jinping at a ceremony marking the centenary of the CPC. http://www.xinhuanet.com/english/special/2021-07/01/c_1310038244.htm

Xinhua (2021b) Xi focus-quotable quotes Xi Jinping on building eco-civilization. http://www.news.cn/english/2021-08/12/c_1310123552.htm

Xu B, Albert E (2017) Media censorship in China. Council on Foreign Relations. Retrieved from https://www.cfr.org/backgrounder/media-censorship-china

Yang P (2021) Squad of women fight discrimination in Chinese work culture. Inkstone News. Retrieved from https://www.inkstonenews.com/society/squad-women-fight-discrimination-chinese-work-culture/article/3122744

Yang J trans. Mosher S (2013) Tombstone: the great Chinese famine: 1958–62. New York: Farrar, Straus, Giroux

Yau E (2021) Chinese New Year delights with a vegan twist. Inkstone News. Retrieved from https://www.inkstonenews.com/health/chinese-new-year-delights-vegan-twist/article/3121155

Yi W (2023) Ejiao in flux: a modern traditional Chinese medicine and the global trade of donkey hides. Global China Pulse, Vol. 2 Issue 1. https://thepeoplesmap.net/globalchinapulse/ejiao-in-flux-a-modern-traditional-chinese-medicine-and-the-global-trade-of-donkey-hides/

Yu H Trans. Barr AH (2012) China in ten words. Anchor Books, 116–118

Zhang S (2020). How young women are powering China's consumption growth. Inkstone News. Retrieved from https://www.inkstonenews.com/business/how-young-women-are-powering-chinas-consumption-growth/article/3096746

Zhang P (2021) Squad of women fight discrimination in Chinese work culture. Inkstone News. Retrieved from https://www.inkstonenews.com/society/squad-women-fight-discrimination-chinese-work-culture/article/3122744

Zhang L, Xu Z, Jiang T, Zhang J, Huang P, Tan J, Chen G, Yuan M, Li Z, Liu H, Gao D, Xiao L, Feng H, Xu J, Xu H (2021) Efficacy and safety of ejiao (Asini Corii Colla) in women With blood deficient symptoms: A randomized, double-blind, and placebo-controlled clinical trial. Front Pharmacol 12:718154. https://doi.org/10.3389/fphar.2021.718154. PMID: 34707496; PMCID: PMC8542698

Zhang Z, Huang B, Wang Y, Zhu M, Liu G, Wang C (2023) A survey report on the donkey original breeding farms in China: current aspects and future prospective. Front Vet Sci 10:1126138. https://doi.org/10.3389/fvets.2023.1126138. PMID: 37008357; PMCID: PMC10060844

Zhao R (2015) Donkey Industry calls for financial kick from govt. China Daily. Retrieved from https://www.chinadaily.com.cn/china/2015-02/04/content_19483612.htm

Zheng T (2009) Red lights: the lives of sex workers in postsocialist China. University of Minnesota Press, Minneapolis and London

Zhu W (2017) Dongeejiao Pharmaceutical to launch donkey food brands. China Daily. Retrieved from https://www.chinadaily.com.cn/business/2015-12/17/content_22734077.htm

Working Donkeys, Mules, and Hinnies Population, Roles in Production and Skin Trade Impact

Philip W. Mshelia

Abstract

Working equids serve essential roles for human development in worldwide agricultural production and transport service together with labor-intensive production fields worldwide. This segment summarizes the worldwide numbers of working equids combined with their multiple functions and healthcare obstacles as well as trade and breeding patterns. The work animals play an essential economic role since mechanization is difficult to implement in certain regions thus advancing both local earnings and sustainable growth. The chapter explores the controversial donkey skin trade which produces significant consequences on worldwide donkey populations together with their welfare oversight and sociological stability. This analysis demonstrates the importance of donkeys and mules and hinnies through detailed evaluation so the chapter asks for increased welfare standards with sustainable management and policy changes that uphold their evolving and historical value. This discussion uses historical and present-day perspectives to create a more comprehensive understanding of rarely studied animals and to demonstrate why worldwide collaboration remains essential for their future preservation and well-being.

Keywords

Donkeys · Mules · Hinnies · Working equids · Equine healthcare · Animal welfare · Global population · Equine trade · Donkey skin trade

P. W. Mshelia (✉)
Faculty of Veterinary Medicine, Department of Veterinary, Ahmadu Bello University, Zaria, Nigeria
e-mail: pwmshelia@abu.edu.ng

A. K. McLean et al. (eds.), *Donkey and Mule Medicine: Foundations, Breeds, and Management*, https://doi.org/10.1007/978-3-032-12726-6_11

1 Overview of Working Equids

1.1 Importance of Working Equids in Different Work Settings

The global economy relies heavily on working equids that include donkeys and mules and hinnies since these animals provide essential transport functions and labor in areas that lack modern technology. The Food and Agriculture Organization documented 44 million donkeys and 11 million mules across the world in 2017 (FAO 2017). Cultural and economic communities depend on working equids because they provide affordable and sustainable transport and power options in contexts where motorised or mechanised solutions are not feasible, thereby improving market access and supporting social and economic development (Starkey 2001; Starkey and Starkey 2000).

The working relationship between people and donkeys and mules together with hinnies has sustained human development across rural African marketplaces and South American mountain regions for hundreds of years. The working equid population exceeds seven million in Ethiopia while donkeys represent most of this number according to Behnke and Metaferia (2011). Their essential work remains out of view when economists compile economic development records. This chapter addresses an oversight by analyzing how these equines maintain basic human survival in areas where machine-driven agricultural methods remain inaccessible.

1.2 Purpose and Scope of the Chapter

This chapter undertakes a thorough investigation of the worldwide demographic statistics and functional aspects, medical provisions, manufacturing along with trading of donkeys, mules and hinnies. The discussion begins with their historical use and continues through present-day working scenarios before analyzing health difficulties along with the social and economic repercussions in detail, especially about the contentious donkey skin market trade. The chapter demonstrates these elements to support policy changes, welfare standards enhancement and sustainable practices which recognize the essential functions of these animals.

2 Key Concepts

2.1 Glossary of Terms

- **Donkey** (Equus asinus) represents a domesticated horse family member that people tend for agriculture, transportation and intensive manual work.
- A **mule** emerges through the connective tissue combination of a male donkey (jack) and a female horse (mare) which produces an animal characterized by its remarkable power capabilities and strong-long lasting characteristics.

- The inherited offspring of mating a horse stallion with a jenny donkey results in a **hinny** which turns out smaller and less sturdy than a mule.
- The **Working Equid** exists as equids which support various manual operations such as transport and cultivation activities.
- The worldwide donkey skin commercial market exists to create **ejiao** traditional Chinese medicine products.

It is essential to understand all possible specific meanings of these terms. Different societies interpret the donkey in two opposing ways because they see it as representing either hardship and endurance or gentleness and loyalty.

3 Historical Framework

Previous civilizations used donkeys and mules effectively for agricultural development and trading activities and transportation purposes. The domestication of donkeys began in North Africa about 5000 years ago before they became essential for economic operations in ancient Egypt and Mesopotamia (Clutton-Brock 1992). Their massive value involved carrying pyramid construction materials and delivering heavy stone blocks as well as merchandise though extensive regions.

Mules served as a vital resource for military operations, agricultural work and mining since these animals demonstrated excellent capabilities to transport heavy loads across difficult terrains.

3.1 Evolution of Veterinary Care

Veterinarians have made substantial progress in providing medical care to working equids yet this development has not been consistent across different geographical areas. Developed countries have developed their equine veterinary practices through which they provide full working equid care including specialized medical treatments and prevention plans. Contemporary horse practice utilizes extensive specialize epertise for internal medicine, nutrition, and horse horse reproduction, and animal welfare standards (Reed et al. 2004; National research Council 2007; McKinnon et al., 2011; WOAH 2021).

Rural areas of developing regions experience restricted veterinary service availability because of financial barriers alongside poor infrastructure and inadequate training among veterinary personnel. The lack of proper health resources for these animals leads to unnecessary diseases and injuries which diminish their lifetime and diminish their productivity. International organizations are working to close this division through programs that deliver aid and knowledge to help local communities (Brooke 2021).

3.2 Cultural Significance

Working equids remained unspoken observers as human beings pursued their both jubilant and difficult endeavors throughout time. Ancient Egypt together with Mesopotamia revered the donkey as a trading symbol and prosperous sign while preserving it in religious mythology and artworks. Numerous ancient documents together with historical artifacts show that the donkey occupied an important position in ancient societies.

The evolution of human civilization coincided with the changing tasks of these equids who transported Roman commercial and military resources through roads while helping European farms manage fields and logistics. Donkeys and mules play an essential role in human society because they appear throughout the folklore and literature of many different cultures.

Historical documents have always honored mules for their powerful characteristics and solid traits which helped them make essential contributions to agricultural and transport sectors and military operations. The hybrid animals performed crucial service during World War I and II because their rugged terrain capabilities allowed them to pull supplies through difficult conditions. Research published by Anthony J. Nocella II et al. in their book "Animals and War: Confronting the Military-Animal Industrial Complex" (2013) shows that mules serve crucial functions in military forces. The African-American folklore places mules into symbolic representation through Zora Neale Hurston's "Mules and Men" which was published in 1935. Throughout history mules gained recognition as dependable workers that earned the "living machines" status because of their excellent utility and long-lasting endurance.

> The history of mankind is carried on the back of a donkey. — Ancient Proverb

4 Worldwide Context

4.1 Global Population and Distribution of Donkeys, Mules, and Hinnies

4.1.1 Global Population and Distribution of Donkeys, Mules, and Hinnies

Donkeys (Equus asinus) together with mules and hinnies have played a vital role in human civilization since ancient times because they served multiple purposes including providing transportation and animal companionship alongside their dietary and medicinal value across different cultural societies. These work animals serve crucial functions for basic living activities and economic development mostly within developing nations. The worldwide numbers of donkeys and their equine offspring reveal essential information about worldwide farming customs and economic progress and various existing cultural traditions. The document includes data about donkey populations which is supported by both credible sources and numerical evidence.

Table 1 Top 5 countries by donkey population (FAO 2017)

Rank	Country	Donkey population
1	China	11,000,000
2	Ethiopia	7,000,000
3	Pakistan	5,000,000
4	Mexico	3,000,000
5	Nigeria	2,000,000

Most donkeys, mules and hinnies across the globe exist in Asia, Africa and Latin America because these regions rely on them for supporting the income of hundreds of millions of people. The Food and Agriculture Organization of the United Nations (FAO) documented the global donkey population at 44 million in 2017 and China owned 11 million donkeys while Ethiopia had seven million and Pakistan owned five million (FAO 2017). About 11 million mules existed in the world with substantial populations found in China and Mexico followed by Brazil) (refer to Table 1).

Working donkeys, horses and mules constitute a major livestock population in low- and middle-income countries and play a critical role in transport, agriculture and household economies. It is estimated that approximately 600 million people worldwide depend directly or indirectly on working equids for their livelihoods (McKenna 2007). Different industries face major obstacles when these equids do not provide their substantial support.

4.2 Donkey Population

Region/ country	Estimated number of donkeys	Notes
Africa	Largest number globally	Donkeys are essential for transporting goods, water, and people in rural communities.
Ethiopia	Over seven million	(FAO 2017; Starkey and Fielding (2004)
Asia	Significant populations in China, Pakistan, and India	China's donkey population is declining due to demand for donkey-derived products like *ejiao* (Zhang 2019). Despite having about 11 million donkeys, numbers are decreasing.
Latin America	Substantial populations in Mexico and Brazil	Used extensively in agriculture and transport in rural areas.

4.3 Mule Population

Region/ country	Estimated number of mules	Notes
Global	Around 10–11 million	(FAO 2017)
China	Over three million	Largest mule population globally; valued for work in mountainous terrains (FAO/FAOSTAT 2024.

(continued)

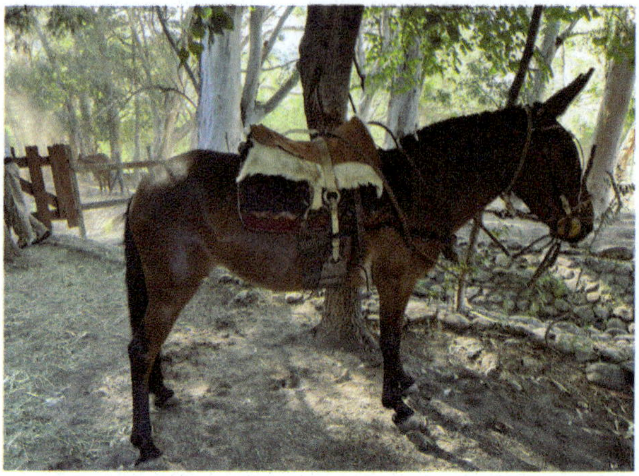

Fig. 1 The mule exhibits its characteristic features showing good structure and glossy coat along with its distinctive upright ear shape. (Photo credit: Dr. Amy McLean, San Marcos Sierra, Argentina)

Region/ country	Estimated number of mules	Notes
Latin America	Significant populations in Mexico and Brazil (Fig. 1)	Integral to agriculture and transportation, especially in regions unsuitable for vehicles.

4.4 Hinny Population

Region/ country	Estimated number of hinnies	Notes
Global	Tens of thousands	Precise numbers are difficult to ascertain (Svendsen 2008).
Distribution	Scattered globally	Not prevalent in any specific region due to rarity and less desirable working traits compared to mules.

The animals provide essential support to households by fulfilling both domestic needs and commercial purposes for everyday needs and monetary activities. Through their work and by enabling market access and service availability the working equids earn monetary benefits. They are pivotal in:

The animals play a role in agricultural operations by helping farmers with plowing fields and seed planting and crop collection. And in Transportation- moving goods and people in areas inaccessible to vehicles.

Owners gain financial advantages because working equids provide low-cost transport and traction where motorized transport and agricultural mechanization are often unaffordable or inaccessible (FAO 2014; Valette 2015).

4.5 Factors Affecting Populations

4.5.1 Economic Development

The development of economies significantly impacts donkeys and mules as well as hinnies. The improvement of countries combined with urbanization leads to the reduction of traditional working methods which were historically supported by these laboring animals. The progress of agriculture together with construction and transportation automation leads to decreased utilization of equids for activities including plowing and hauling as well as transportation of people. Scientific records demonstrate that mechanization affects developing nations' rural economic structures. Sub-Saharan African populations have decreased their draft animal numbers because of agricultural mechanization according to Starkey (2000). Rural parts with minimum mechanization maintain stable donkey along with mule and hinny communities which serve as fundamental resources in local subsistence activities according to analysis by Blench (2000). These animals sustain regional economic systems because modern industrial progress has not reached their geographic locations.

4.5.2 Demand for Donkey Products

The growing Chinese market for donkey skins has been identified as the main reason behind recent population reductions as this demand serves the production of ejiao gelatin products used in traditional medicine. Donkey slaughter rates have reached dangerous heights because of the steady increase in demand for donkey products according to the world-wide ejiao market has created a dramatic reduction in donkey populations mainly affecting African regions from which China obtains its donkeys. The expanding global donkey skin trade, driven by demand for ejiao in China, has contributed to significantpressures on donkey populations in Africa, creating threats to the sustainability of working equids and the livelihoods of communities that depend on them (The Brooke 2022; Brooke 2025; PADCo 2022) discovered that many countries face regulatory weaknesses which permit unethical trading practices to expand freely into the market. Rural areas have suffered severely because the disappearance of working animals disrupts agriculture alongside transportation operations.

4.5.3 Welfare Concerns

Donkey and mule and hinny welfare ranks as a vital element which affects these populations' numbers. Working equids must endure painful circumstances because they get denied veterinary attention while facing nutritional needs inadequacy alongside excessive work demands. According to Brooke (2022), the combination of substandard diets with strenuous labor decreases the lives of equine species. The workload pressures faced by these animals enhance their vulnerability to medical issues that subsequently limit their ability to work and shorten their productive life. Working equids maintain a vital relationship with human survival rates primarily in African and Asian rural areas because they constitute the backbone of local agricultural productivity. The Brooke (2022) along with other welfare organizations has continuously highlighted animal ethical concerns regarding working animals by supporting better management systems while advocating for enhanced veterinary care and strengthened legal protection against exploitation and neglect. (*See list of equine welfare charities and organizations assisting working equids at the end of the chapter*).

Sustained activities need to track working equids in order to check their welfare. Successfully protecting working animals from exploitation and neglect requires both collecting precise numbers and distribution data systematically and establishing protective policies for their welfare. The essential task requires international organizations alongside governments and local communities to collaborate about addressing economic and cultural practice transformations. The advancement of donkey, mule, and hinny survival requires organizations to focus on humane practices and veterinary service accessibility and awareness campaigns for these animals.

Humanford development together with cultural maintenance defines the destiny of donkeys, mules, and hinnies today. Caring for donkey populations serves more purposes than animal welfare because it supports people whose livelihood depends on them while keeping traditional cultural practices alive and contributing to sustainable development strategies. The continued development of modernization requires simultaneous protection of essential animal species. Active understanding and planned solutions to their present obstacles will make certain their essential role in global society stays intact until the future.

4.6 Socio-Economic Influences on the Utilization of Donkeys, Mules, and Hinnies

4.6.1 Cultural Significance and Perception of Working Equids

Working equids throughout the world receive their usage and treatment procedures based on the cultural beliefs of their communities. The communities of Africa and Asia view donkeys as inferior animals that people connect with hardship and poverty (Starkey 2010). Such a dismissive view leads to mistreatment and insufficient care because these animals fail to earn the abundant care typical of renowned herd members.

The population of Latin America holds mules in high esteem because of their exceptional physical capabilities. The population values donkeys over donkeys leading their owners to provide better care, treatment and management for them (Pearson et al. 2003). Society shows mules great respect because these animals remain highly important tools which enable sustainability in difficult travel areas.

Having donkeys and mules as possessions demonstrates both financial success and enhanced social acceptance in particular cultures. The local population of Ethiopia bestows special value on mules because these special animals are rare and expensive (Getachew et al. 2010). A mule's ownership provides people with improved social position which results in greater attention directed toward the animal.

4.7 Economic Conditions

The economic situation decides how widespread donkeys, mules and hinnies operate throughout the regions. People who live in poor areas use these equids as affordable substitutes for vehicles and agricultural machines. Smallholder farmers together with urban residents constantly rely on working equids to fulfill their transportation needs and maintain power requirements at low sustainability and cost levels. Working equids make it possible for communities to connect with markets while they carry goods and complete vital agricultural duties which boosts local economies (The Brooke 2025).

Such economic constraints generate conditions that harm the well-being of these animals. Clipped financial circumstances of owners prevent them from getting the required supplies, veterinary attention and appropriate facilities for caring for their equids. The combination of excessive work with inadequate nutrition and dangerous environmental factors creates negative health conditions as well as poor productivity outcomes for these animals (Burn et al. 2008).

4.7.1 Local Regulations and Policy Gaps

Working equids are directly affected by the policies which control their usage while determining their wellness status. These developing countries show insufficient laws for animal care so their animals face serious mistreatment because of abandoned regulations. Animal welfare laws remain unenforced because countries lack sufficient enforcement resources along with insufficient national interest in equid welfare protection (The Brooke 2021).

Working equids need to be part of systematic livestock policies and development programs because this integration will improve their welfare while unlocking their complete economic potential. The proper valuation of working animals should lead to better systems of animal care through improved veterinary accessibility and owner training and enforceable laws for animal kindness.

An essential path for donkey and mule and hinnie welfare improvement involves resolving economic difficulties and strengthening regulatory guidelines. Supportive policies which acknowledge donkeys and mules' essential roles will help maintain their well-being so these animals can continue supporting human existence.

5 Roles in Various Work Settings

5.1 Agricultural Settings

Donkeys together with mules perform essential roles in agricultural operations through their work in plowing and tilling activities as well as goods transportation. Donkeys and mules prove indispensable because they can operate in demanding agricultural areas although small regional farmers do not have available machines for farming tasks. The population of rural areas in sub-Saharan Africa depends heavily on animals for pulling plows and other agricultural tools and donkeys lead as the main animal type in most regions according to Starkey (2010). These animals ease

Fig. 2 Market traders use donkeys as their primary means of transporting goods because donkeys remain the best method to transport goods to and from markets across remote locations where vehicles cannot reach. (Photo credit Dr. Amy McLean, Segou, Mali)

seed planting and weed control tasks along with goods transportation operations thus providing valuable support to families by reducing their labor requirements.

Donkeys function as the main means of market supply transportation throughout Ethiopia particularly serving places where automobile transportation is impossible. Donkeys transport cargo weighing 100 kg while traversing between 15 and 20 km according to Girma and Mekuriaw (2017). The transportation abilities of farmers can significantly affect their income levels and food safety since they determine their ability to attend markets for sales and purchases of daily necessities (Fig. 2). The water and firewood collection duties foster an essential contribution that donkeys make to rural communities.

5.2 Urban Settings

Urban areas employ donkeys for waste collection as well as goods carting and transport service provision. Across African cities together with South Asian cities donkeys function as essential components in informal economic activities. Research indicates that Lahore Pakistan contains more than 70,000 donkeys which work within brick kilns and transportation services as well as construction sites (Khan et al. 2019). The urban working equids that today work in cities experience harsh living environments because they must perform long hours without proper nutrition and veterinary healthcare services (Fig. 3).

5.2.1 Specialized Work Environments

Working equids perform special tasks when they assist during search and rescue missions and serving police work and playing ceremonial roles.

5.3 Military and Pack Operations

Military personnel choose mules for their strength in mountainous environments when vehicles cannot function. The Muleskinners unit of the U.S. Army continues to

Fig. 3 Mule Transporting Goods in Markets Fe, Morocco. (Photo credit: Dr. Amy McLean)

Fig. 4 World War II saw donkeys working as supply handlers in military operations. (The original image derives from November 1943 (3)—CBI by Pacific Paratrooper, 2016. https://pacificparatrooper.wordpress.com/2016/03/11/november-1943-3-cbi/)

employ mules for carrying heavy loads through mountain areas because of their proven suitability for pack transport (Davies 2014) (Fig. 4).

5.4 Search and Rescue

Areas with demanding and difficult terrain employ donkeys and mules for search and rescue operations because mechanized vehicles prove ineffective. These animals bring reliability to search operations through their excellent sense perception and balanced walking abilities.

5.5 Ceremonial and Cultural Roles

The participation of equids occurs during ceremonial processions and religious festivals because people value their cultural importance. The Romería del Rocío pilgrimage occurring in Spain showcases donkeys and mules wearing decorative accessories that symbolize their vital importance for traditions along with community identity (Fernández and Delgado 2018). Horses playing a part in Sallah Durbar rites as well as multiple cultural ceremonies throughout Northern Nigeria receive lavish decorative treatment (Mshelia et al. 2020).

6 Impact of the Donkey Skin Trade

6.1 Global Demand and Drivers

A worldwide market for donkey dermis primarily used in the creation of ejiao has led to destructive developments for donkey health and demographics across the globe. Traditional Chinese medical practitioners use Ejiao as a gelatin product which supplies multiple health advantages including presumed anti-aging benefits. The worldwide market value of donkeys exceeds $7.8 billion each year (Liu 2020).

Donkey skin demand which stems from increasing Chinese prosperity and growing ejiao product demand exceeds local supply so the market turns to international sources. The growing international trade market now predominantly focuses on donkeys obtained from African and Asian and South American regions.

6.2 Impact on Donkey Populations

Industrial donkey slaughter has increased throughout Africa along with various other regions because of surging market demand. Numbers of donkeys in Kenya dropped more than 30% over the 2009 to 2019 timeframe because of verified as well as undocumented animal killings (Wadri et al. 2020). The same population decline patterns prevail in Burkina Faso and Nigeria as well as Niger (see Table 2).

Reduced donkey numbers puts at risk the income sources of families who depend on them for survival. When these animals are suddenly removed from use the labor force experiences a significant gap that lacks viable alternative solutions.

Table 2 Decline in donkey populations due to skin trade

Country	Population decline (%)	Years
Kenya	30	2009–2019
Burkina Faso	20	2012–2017
Nigeria	22	2010–2018

6.3 Socio-Economic Consequences

Communities facing donkey population decreases experience negative economic effects because of their dependency on these animals. The people living in rural Africa strongly rely on donkeys to transport water resources alongside firewood and agricultural items. Family members are left with two options when animals disappear: doing the work themselves or paying extra expenses which worsens their poverty condition and diminishes their access to basic services (Ayele et al. 2018) (see Fig. 5).

The majority of impacts strike female and child family members because they carry donkey duties that resulted in educational setbacks and health complications. Food insecurity becomes a problem while decreased opportunities to earn income are recorded.

6.4 Welfare Concerns

Donkeys experience tremendous suffering during slaughter because handlers lack the ability to stun them properly. Donkeys face high stress and death because they experience long journeys to slaughterhouses that lack food and water as well as necessary rest periods (Brooke 2021) (refer to Fig. 6). The absence of regulations in various facilities enables the occurrence of extreme abuse which weakens worldwide initiatives to enhance equid welfare.

The illegal trade of donkeys poses biosecurity threats to animal and public health since it enables the spread of diseases like anthrax and equine influenza according to All Africa (2017).

Donkeys face high stress and death because they experience long journeys to slaughterhouses that lack food and water as well as necessary rest periods (Brooke 2021). The absence of regulations in various facilities enables the occurrence of extreme abuse which weakens worldwide initiatives to enhance equid welfare (see Figs. 7 and 8).

Fig. 5 Nomadic communities together with rural populations employ donkeys to transport essential equipment comprising water storage containers and baskets and equipment between locations. The donkeys move with their load of merchandise between rural paths to help nomads carry everything over extensive paths

Fig. 6 Donkeys experience tremendous suffering during slaughter because handlers lack the ability to stun them properly

Fig. 7 The photograph displays extremely overcrowded donkeys where a donkey from the group has already passed away. The working donkeys face multiple risks including excessive labor along with inadequate nutrition, dehydration, extreme population density, and transport that lasts too long

The illegal trade of donkeys poses biosecurity threats to animal and public health since it enables the spread of diseases like anthrax and equine influenza according to All Africa (2017).

6.5 Global Efforts to Combat the Donkey Skin Trade

The Donkey Sanctuary together with Brooke and SPANA operate as leading organizations that run international campaigns against the donkey skin trade. These organizations deliver fundamental veterinary care and publicize the economy-destroying effects of animal trafficking as they maintain partnerships with national administrations for implementing defensive regulations. African nations like Botswana and Mali along with Senegal have established export bans for donkey skins in order to protect the donkey population and sustain their indigenous local communities (FAO 2020).

A continent-wide ban on donkey skin commercialization was established by African Union Heads of State during their General Assembly meeting in February 2024. A growing understanding of how donkey populations suffer from the trade has led governments to establish this important ban. Through the moratorium African

Fig. 8 Donkeys are being transported in cramped, inhumane conditions, bound tightly with little room to move, many lying on the ground, immobilized by their legs. Their visible distress and exhaustion highlight the cruelty of this transport method

leaders intend to suppress the consumer demand for donkey skins. The combined efforts of African leaders strengthen the preservation of essential donkey roles in African communities (African Union 2024).

The same time international organizations persist with consumer advocacy work in China because the market for ejiao stays large. The organizations advocate for ethical choices while underscoring both environmental harm and economic mistreatment that develops from the trade. These combined efforts of legislation and education along with advocacy aim to lower demand thus establishing sustainable practices for protecting donkeys across African regions (Brooke 2025; The Donkey Sanctuary 2023; SPANA 2023).

6.6 Policy and Future Directions

A solution to the problems surrounding donkey skin trade will arise from implementing advanced legislation combined with stronger animal welfare standards and economic systems which endure over time.

The government should make sure that trade restrictions are properly enforced through effective penalties against illegal practices. The implementation of sustainable community activities that generate alternative income will decrease the motivation for people to turn their donkeys over for slaughter. The required solutions require international collaboration between countries that supply donkeys and those which demand them to solve the fundamental problems. By promoting knowledge about ejiao's welfare issues and alternative herbal or synthetic products the general public will become less interested in this product.

The United Nations together with the World Organisation for Animal Health (OIE) should establish systematic procedures to tackle both ethical and ecological and economic components associated with the trade.

7 International Efforts and Organizations

Organizations around the world support the improvement of working equids' welfare through their offerings of essential services and their advocacy work and education programs.

7.1 The Donkey Sanctuary

Since its UK foundation in 1969 The Donkey Sanctuary runs across more than 40 countries where they provide complimentary veterinary treatment and coordinate animal rescue missions and establish equid management education programs (The Donkey Sanctuary 2020).

The organization focuses on two core activities to save donkeys in need and provide them with necessary rehabilitation services. Community Programs where owners are trained on animal care and welfare. The organization drives advocacy through its dual efforts to fight against the donkey skin trade while advocating for new policies.

7.2 Brooke

Since its founding in 1934 Brooke has delivered its services to working horses, donkeys and mules specifically in Asia, Africa and the Middle East together with Latin America. Brooke delivers care to 1.5 million working animals throughout each year while also training their owners (Brooke 2021). Their initiatives include providing treatments and preventive care (see Fig. 9). Training local veterinarians and para-veterinarians. The organization conducts research to enhance knowledge about equid welfare through various scientific studies.

7.3 SPANA (Society for the Protection of Animals Abroad)

SPANA delivered its founding year in 1923 to serve as the main organization dedicated to enhancing welfare conditions for working animals in developing countries. Through their activities SPANA delivers veterinary medical services in addition to teaching programs and disaster response capabilities (SPANA 2021).

Fig. 9 Brooke developed this lightweight cart for Senegal through which they established a donkey harness system that minimizes sores and lightens the burden of heavy carts. Donkeys benefit from this cart which spreads weight evenly through an agreeable harness that reduces physical stress on the donkey while promoting better equipment designs

7.4 World Horse Welfare

World Horse Welfare began in 1927 to enhance horse donkey and mule welfare through hands-on services along with education and advocacy work. World Horse Welfare maintains operations across 16 different countries because they actively develop global equine welfare standards (World Horse Welfare 2021).

7.5 The American Association of Equine Practitioners (AAEP) Stance on the Donkey Skin Trade

As an organization the American Association of Equine Practitioners (AAEP) maintains a strong official position against worldwide donkey skin commerce because it perpetuates cruel treatment of these creatures and causes major population impact. The AAEP stands against the donkey skin trade because human interest in Chinese medicine and additional product demands have led to painful conditions for affected animals thus compelling the AAEP to push for better donkey treatment standards.

7.6 Inhumane Practices and Welfare Concerns

The AAEP strongly opposes the donkey skin trade since they have documented several inhumane techniques employed throughout the capture and transport of donkeys before their deaths. Studies demonstrate donkeys face regular overcrowding and poor living spaces as well as extended transport periods during which they receive no food nor water nor veterinary services. Donkeys face major hardship while restrained within harsh conditions because traders keep them there before killing them to obtain their hides which become ejiao. The AAEP states that such practices fail to respect humane animal treatment and welfare standards while it pushes for donkeys to be classified as living creatures that deserve protective measures.

7.7 Conservation and Ethical Implications

Moreover the AAEP works to understand the preservation impact on donkey populations that derive from the rawhide industry. Dramatic population decreases of donkeys occur because of the widespread demand for their skins in major donor regions. The sharp population decrease of donkeys endangers both their survival and disrupts diverse ecosystems alongside threatening communities which depend on them for tasks and friendship. The AAEP emphasizes implementing sustainable practices which will enable protecting donkey welfare together with maintaining ecological stability in their natural habitats.

7.8 Collaborative Efforts and Global Responsibility

The AAEP pushes for worldwide team effort to fight the donkey skin trade while enhancing donkey welfare across all nations. The organization pushes for enhanced laws along with stricter implementation systems to ensure safe protection of these animals from human exploitation. The AAEP establishes collaborative partnerships with international entities including ICWE to promote better welfare standards across international regulations. This collaborative initiative works toward improving donkey and mule and hinny recognition in worldwide policy provisions which guarantees proper care for these animals.

The AAEP is dedicated to improving public knowledge about donkey trade conditions because they require veterinarians and animal welfare groups to actively fight against this trade. These organizations work with national authorities together with global organizations and neighborhood groups because they see the clear relationship between animal protection and economic development while pushing forward solutions that support society and wildlife.

8 The Future of Donkeys, Mules, and Hinnies Worldwide

8.1 Emerging Challenges and Opportunities

The survival of donkeys, mules and hinnies depends on ongoing global economic developments combined with technological progress and agricultural changes together with ethical considerations and environmental alterations. These equids face future dual prospects of challenges while developing opportunities which might transform their purpose in the world. These working animals will maintain their essential labor activities in locations where traditional farming remains dominant. Donkeys along with mules and hinnies will face reduced work duties when industries adopt mechanized systems in modernized agricultural zones. The revival of these livestock species for sustainable and environment-friendly agricultural systems among transportation networks becomes possible due to sustainability initiatives. The future integration of donkeys, mules, and hinnies into conservation programs together with animal welfare ethics will determine their contribution to human life throughout the coming years.

8.2 Agricultural and Economic Roles

8.2.1 Climate Change and Sustainable Agriculture

Donkeys and mules show tremendous value as sustainable agricultural assets because they can withstand growing climate change impacts alongside unpredictable climate conditions particularly in dry areas. These livestock thrive under extreme conditions where resources are limited thus forming fundamental collaborations with smallholder farmers who need to adapt to environmental changes (Pritchard et al. 2015).

This breed demonstrates superior ability to withstand drought conditions because donkeys require only minimal water sources and limited food supplies. Multiple research studies show Namibian and Botswana farmers are turning to donkeys for plowing and movement because their local water supplies are diminishing (FAO 2019).

Equids used for plowing create less soil compaction than heavy equipment as this practice enhances soil health while minimizing erosion. Effective preservation of fragile terrains requires this approach since it sustains soil quality.

8.2.2 Economic Empowerment and Gender Roles

The economic capabilities of underprivileged populations especially women depend heavily on their access to mules and donkeys

Women in rural areas depend on donkeys for water collection and wood acquisition and market trade because these animals enhance their ability to earn an income. The physical benefits from donkey assistance help women generate income according to Fernando and Starkey (2004). Ethiopian female beneficiaries working with donkey programs report economic gains of 30% which improves their families' access to educational and healthcare services (Solomon et al. 2013)

The low cost of donkeys enables plan owners to establish small-scale business operations like cart services that create their pathway away from poverty status. Hundreds of thousands of people throughout Pakistan and India depend on donkey carts to transport freight because they serve as essential urban transportation vehicles (Valette 2014)

8.2.3 Challenges in Mechanization

Modernizing equipment has not solved all obstacles that prevent mechanical application from reaching broad adoption worldwide. Various obstacles prevent extensive mechanization including poor infrastructure and insufficient funds and unsupportive topography suited for machines. The conditions in these areas make donkeys and mules essential for their ability to perform activities machines cannot complete

The cost to acquire and sustain tractors together with vehicles remains expensive for numerous communities specifically in regions that lack adequate financial resources. Cost-efficient equids serve as a better monetary choice than alternative options. The financial requirements for equids are lower and their operating expenses are cheaper than tractor systems therefore enabling their use among small-scale farmers and rural communities (Starkey 2010). Equids remain essential to agricultural work due to their affordable cost-performance capabilities in situations where mechanization is economically impractical

Donkeys and mules reveal their worth when operating in mountain terrains because these animals adapt to challenging environments that conventional machines cannot access. Mules stand out with their exceptional sure-footed nature since they can easily traverse tough terrains. Working equids prove indispensable because of their ability to handle difficult topography such as the Andes and Himalayas which limit heavy machinery use (Smith 2005). Equids enable both regional transportation and maintain agricultural-based livelihoods together with trade operations

8.2.4 Technological and Scientific Advances

Working equids will receive unparalleled advancements in veterinary care because of significant technological breakthroughs along with scientific innovations in veterinary medicine. The scientific and technological advancements serve as essential elements for preserving donkey mule and horse productivity in regions that rely on their work and transportation functions

Probably the most significant and promising development of modern veterinary healthcare includes mobile veterinary clinics. These healthcare facilities maintain accurate diagnostic equipment which enables them to operate in locations where standard veterinary healthcare is unavailable. Mobile units deliver vital veterinary services which feature vaccinations and deworming and dental treatment and emergency medical care according to Brooke (2021). Working equids benefit from improved health standards because veterinarians provide easily accessible healthcare services

Non-governmental organizations (NGOs) which operate mobile clinics in Kenya have demonstrated outstanding success in improving working donkey welfare throughout the country. The mobile veterinary clinics operating in Kenya have cut working donkey mortality rates by 25% which prolongs their productive lifetime according to Wanyama et al. (2020). The mobile veterinary services show their ability to make significant improvements in equid welfare and sustain local communities whose economy depends on working these animals

Telemedicine has emerged as a significant veterinary medical development through its rising incorporation of telemedicine platforms. The platforms enable veterinarians to work in real-time with both animal health workers and owners at the local level thus solving the problem of not enough equine specialists available across certain regions. Remote veterinary consulting through telemedicine enables equids to get prompt correct veterinary attention in regions that lack sufficient veterinary resources (McCracken et al. 2019). Working equids benefit significantly from this innovation because it gives them expanded access to professional healthcare services

Working animals along with their communities can maximize their welfare through the combination of traditional practices with modern innovations despite existing obstacles to mechanization

8.2.5 Innovations in Equipment and Work Practices

Modern technology has become fundamental for enhancing the life quality of equids, specifically donkeys and mules, in current times. Implementations during this modern era serve to improve animal well-being while enabling productive task execution by keeping working animals both healthy and comfortable

Ergonomic Harnesses represent an outstanding advancement because researchers have developed special harness designs to distribute weight across the animal. Special harness design distributes weight to distribute it evenly across the body of working animals thus reducing the risk of injuries and sores caused by improper equipment fit. Experts showed that locally sourced materials used to make harnesses achieve good performance at affordable prices thus helping farmers whose income depends on animals (Pearson et al. 2003). This innovative approach prioritizes animal welfare through sustainable working animal care methods which enables health protection during job performance without any harm to the animals

GPS and Tracking Devices represent a major technological advancement which allows tracking of working equids. Because of the escalating threat of donkey skin theft these security devices become a crucial defense for animal owners. Every animal owner benefits from GPS trackers because they can track their pets in real time which stops stealers and increases missing animal recoveries (Leach and Pinchbeck 2015). Added to this functionality these devices provide essential data about work-rest patterns to help owners make better therapeutic choices. Taking proper measurements of animal workloads along with ensuring enough rest becomes essential for sustaining their overall health. The provided information lets owners make better work schedule adjustments to prevent equids from excessive work which ensures their well-being and productivity

The animal care and management industry is undergoing transformation because of innovative work practice and equipment advancements that include ergonomic harnesses together with GPS tracking devices. The modern developments in animal-related technology aim to create comfortable environments and safety protocols which promote better relations between working animals and their human handlers. Our support to communities that depend on these animals becomes possible through improved equid physical comfort and essential monitoring devices. The evolution of technology creates growing opportunities to improve animal husbandry systems which supports equids to maintain their essential position in agriculture and transportation

8.3 Roles in Tourism and Recreation

8.3.1 Sustainable and Ethical Tourism

The tourism industry is increasingly embracing sustainable practices, creating new opportunities for equids.

- **Eco-Tourism Ventures:** Donkeys and mules are central to eco-friendly tourism activities such as trekking, wildlife safaris, and cultural tours. In Morocco's High

Atlas Mountains, mule trekking is a major tourist attraction. Local communities benefit economically while promoting cultural heritage (Ashley et al. 2006).
- **Animal Welfare Standards:** Tour operators are adopting guidelines that ensure the humane treatment of working animals, influenced by consumer demand for ethical experiences (Holland and Low 2010).

8.3.2 Recreational Activities and Competitions

Donkeys and mules are gaining popularity in recreational settings in developed countries.

- **Therapeutic Riding Programs:** As discussed, these programs are expanding, providing benefits to participants and promoting positive perceptions of equids (Granados and Agís 2011).
- **Competitions and Shows:** Events like mule rodeos, donkey shows, and endurance rides celebrate the abilities of these animals, fostering community engagement and awareness (Levine 2015).

Donkeys along with mules and hinnies face an uncertain future because of the moral concerns related to how their well-being and preservation will be managed. These equids maintain their essential positions in worldwide cultural practices alongside agriculture and transport so protection of their welfare and their sustained conservation demand immediate action. To protect donkeys and mules and hinnies and their welfare into future generations it is necessary to have comprehensive laws and educational efforts followed by commitments within global development objectives. Elevating regulatory power and implementation measures represents a fundamental solution to solve the working equids' welfare problems. The World Organisation for Animal Health (OIE) has created international standards regarding working equid welfare which member countries should both adopt and implement (OIE 2016). The guidelines provide essential elements for governments to frame their laws protecting equids against exploitation as well as abuse. Various nations have created legislation to battle the donkey skin trafficking because it endangers donkey groups and their local economic dependents. The government of Niger made a revolutionary move in 2017 by forbidding donkey exports alongside their products for protection of local economies and donkeys from continuing exploitation (FAO 2021).

The social awareness of donkey welfare stands as a primary factor which advocates for protection of donkeys mules and hinnies. Non-governmental organizations through their workshops and media campaigns instruct owners about appropriate care as well as welfare significance. Educational programs showcase the physical requirements along with psychological considerations of animals helping owners make improvements to their care facilities thus decreasing acts of mistreatment. The power of consumer influence on the subject should not be disregarded. The rising consumer demand for ethical products drives organizations to implement humane care policies which benefit equid welfare. Better welfare standards now exist inside industries that work with working equids as consumers push for ethical practices in business operations (Hampton et al. 2018).

The conservation of equids critically depends on their cultural value to human society. The continuation of equids depends on cultural promotion which simultaneously safeguards the animals and reinforces cultural preservation and identity retention. Spain's Romería del Rocío celebrates donkeys as well as mules while events and festivals recognize their societal value (Fernández and Delgado 2018). Traditional equid-dependent cultural practices acquire economic benefits which sustain local inhabitants while safeguarding cultural natural heritage and Asiatic Wild Ass traditions. The Li people in China's Hainan Island organize cultural ceremonies to honor their donkeys because their cultural importance plays a central role in maintaining local community pride while reinforcing community stewardship (Li et al. 2015).

Environmental conservation programs that involve nearby populations guarantee their continued success. Decisions concerning conservation must engage owners of equids and members of the community through participatory approaches which create a sense of ownership so they commit to conservation objectives (Daniel and Persai 2010). Participation of communities makes them more ready to sustain conservation projects. The probability of attaining conservation success increases when conservation measures are combined with economic oversight that connects equid preservation to sustainable practices such as eco-tourism and sustainable agriculture. Community programs offering practical benefits drive sustained commitment to conservation goals because they help members support these objectives (Roe et al. 2009).

Equids serve as vital components within the United Nations' Sustainable Development Goals framework because they create comprehensive value to global development. Multiple SDGs are linked through the operations of these animals which underscores their significance for resolving important global problems. The practice of utilizing donkeys and hinnies with mules aids rural as well as urban communities to improve their economic status per SDG 1 requirements. The economic power provided by these animals contributes to both agricultural development and transportation needs which supports the achievement of SDG 2 by sustaining food systems. SDG 5 (Gender Equality) receives advancement when women gain ownership of and use equids because it creates economic independence and social opportunities for them. Equids present themselves as environmentally friendly options by producing less carbon than mechanical transport systems which follows the principles of SDG 13 (Climate Action). Nationwide development plans should integrate equids because this recognition and support enables them to keep fulfilling essential roles in sustainable development as stated by (Valette 2014).

The future outlook for donkeys along with mules and hinnies holds prospective developments that require unified action to handle encountered obstacles and seize forthcoming prospects. The usage of technological progress combined with enhanced welfare protections alongside recognition of multiple societal benefits will allow these exceptional animals to persevere in their service of human survival and cultural incentives alongside sustainable development. The ability of donkeys and mules to adapt provides essential support during current and forthcoming social and climate changes.

8.4 Study Questions

1. According to the FAO (2017), which regions and specific countries host the largest populations of donkeys, mules, and hinnies? What are the key underlying factors contributing to the high population density of working equids in these areas?
2. How do donkeys, mules, and hinnies support worldwide rural agricultural development and transport operations, and in what ways does their work help sustain smallholder farm revenue?
3. How has the expanding market for donkey skins, particularly for ejiao production, affected global donkey populations, and what specific challenges has this posed for the communities that depend on these animals for their daily livelihoods?
4. What are the main veterinary welfare problems faced by working equids in developing countries, and what barriers prevent effective interventions from being implemented?
5. How have the economic, cultural, and military roles of donkeys and mules evolved from historical times to the present, and how do modern cultural influences continue to shape their use and perception?
6. What policy steps should governments take to strengthen existing animal welfare regulations, and how can authorities effectively prevent the mistreatment and overexploitation of working equids?
7. What proposed strategies and innovations are being examined to protect donkeys and related animals amid present-day agricultural transformations and climate disruptions, despite the ongoing debates over donkey skin trade?

9 Answers

9.1 Global Populations & Density of Working Equids

FAOSTAT livestock-stock data, estimate that working equids totaled ~123 million globally in 2019 (approximately 55 million donkeys, 60 million horses and 8 million mules) and report major national concentrations in Mexico and Ethiopia, with the highest per-capita density reported for Mongolia. The clustering of working equids in these settings is widely attributed to low levels of mechanization, limited rural infrastructure and service access, poverty constraints, and sustained reliance on equids for agricultural power, transport and livelihood activities.

9.2 Role in Rural Agriculture & Transport

Donkeys, mules, and hinnies are vital for smallholder farmers, providing affordable mechanical power. They plow fields, transport water, firewood, crops, and goods to market—enabling increased productivity and sustainable income. Their contributions help families access healthcare, education, and improve food security—all without the cost and maintenance challenges associated with motorized machines

9.3 Impact of Donkey-Skin Trade (Ejiao)

The booming global demand for ejiao has resulted in ~5.9 million donkey skins being harvested annually to produce collagen-based remedies en.wikipedia.org +15worldvet.org+15spana.org+15. China's donkey population has fallen by over 80% (from ~11 M to ~2 M), and export hubs in Africa are also depleting their stocks iwmf.org+2reuters.com+2reddit.com+2. The consequences are dire: donkey theft, skyrocketing prices (3–4×), loss of essential transport, diminished income, school absenteeism, especially among women and children spana.org

9.4 Veterinary Welfare Issues & Barriers

Working equids in developing countries commonly experience

- Neglect, injury, ulcers, lameness, parasitism, malnutrition
- Laminitis, colic, hyperlipemia, dental disease

Barriers to intervention include limited veterinary access, low owner awareness, lack of monitoring, and scarce funding or infrastructure en.wikipedia.org+11pmc.ncbi.nlm.nih.gov+11researchgate.net+11.

9.5 Historical and Cultural Roles

Historically, these equids served in warfare, transport (e.g., Borax Mule Teams in California), and agriculture en.wikipedia.org. Cultural reverence of horses often overshadowed donkeys, though recent shifts in animal welfare awareness and therapeutic roles are transforming public perception and policy

9.6 Policy Recommendations to Improve Welfare

Governments should

- Legally recognize equids within livestock protection frameworks (e.g., FAO inclusions).
- Enforce bans on donkey slaughter and skin exports (e.g., African Union moratorium in 2024) onlinelibrary.wiley.com+10thedonkeysanctuary.org.uk +10thedonkeysanctuary.org.uk+10.
- Provide veterinary training to rural communities, bolster welfare services, and expand population monitoring to detect declines early.

9.7 Strategies & Innovations Amid Market & Climate Pressure

- Alternative collagen sources: lab-grown ejiao from cultured donkey or horse cells iwmf.org+15thedonkeysanctuary.org.uk+15science.org+15.
- Improved breeding programs: though long gestation (~12 months) limits scalability science.org.
- Community-based programs: working with women's cooperatives, encouraging donkey protection and local stewardship.
- Surveillance & enforcement: national customs vigilance, community education, and leveraging international aid to combat illegal trade.

Together, these initiatives across policy, innovation, and local engagement can help safeguard donkey, mule, and hinny populations, maintaining their pivotal role in global rural livelihoods while mitigating the threats posed by commercial exploitation.

Additional Resources: Organizations Helping Donkeys, Mules, and Hinnies

Organization name	Region/country	Key efforts/activities
The Donkey Sanctuary	International (Based in UK)	Rescue and rehabilitation of donkeys worldwide, providing veterinary care, education programs for owners, advocacy for donkey welfare, and supporting communities that rely on donkeys for their livelihoods.
Brooke: Action for Working Horses and Donkeys	International (Based in UK)	Offers free veterinary services, trains owners in animal care, improves animal welfare practices, and advocates for working equines (horses, donkeys, and mules) in Africa, Asia, Latin America, and the Middle East.
SPANA (Society for the Protection of Animals Abroad)	International (Based in UK)	Provides free veterinary care to working animals, including donkeys and mules; runs education programs for animal owners and children; offers emergency relief during crises in Africa and the Middle East.
American Donkey and Mule Society (ADMS)	United States	Promotes the welfare of donkeys and mules through education, research, and support for owners and enthusiasts; provides resources on care, training, and health; advocates for the recognition and appreciation of these animals.
World Horse Welfare	International (Based in UK)	Works with communities to improve the welfare of working equines, including donkeys and mules, through training, education, and advocacy; operates in countries across Africa, Asia, Latin America, and Europe.
Animal Nepal	Nepal	Improves the welfare of working equines in brick factories and other industries through veterinary care, owner education, and advocacy against animal cruelty; conducts

(continued)

Organization name	Region/country	Key efforts/activities
		rescue operations and promotes humane treatment.
Donkey Rescue UK	United Kingdom	Rescues, rehabilitates, and rehomes abused, neglected, or abandoned donkeys; provides sanctuary care; raises awareness about donkey welfare issues through community engagement and education.
Refuge de l'Arche	France	Offers sanctuary and care for various animals, including donkeys; focuses on rescue and rehabilitation; educates the public about animal welfare and environmental conservation.
Mule Rescue and Donkey Sanctuary (MRDS)	Australia	Rescues and cares for neglected or abused donkeys and mules; promotes welfare through education, community engagement, and advocacy; facilitates adoption programs.
Karuna Society for Animals and Nature	India	Rescues and rehabilitates injured or abandoned donkeys and mules; provides veterinary care; runs spay/neuter programs; advocates for animal rights and environmental protection.
Safe Haven for Donkeys	Israel and Palestinian Territories	Provides sanctuary for working and abandoned donkeys; offers free veterinary clinics; educates owners on proper care; works to improve the welfare of donkeys in the region through community programs and outreach.
Highveld Horse Care Unit	South Africa	Improves the welfare of equines, including donkeys and mules, through rescue, rehabilitation, education, and providing veterinary services in underprivileged areas; runs outreach programs and workshops for owners.
Gambia Horse and Donkey Trust	The Gambia	Enhances the health, welfare, and productivity of working horses and donkeys through veterinary care, owner education, and community development projects; focuses on sustainable solutions and capacity building.
Egyptian Society of Animal Friends	Egypt	Provides veterinary care and rescue services; promotes the humane treatment of working animals, including donkeys and mules; conducts education programs for owners; advocates for stronger animal welfare legislation.
Donkey Protection Trust	Kenya	Focuses on the protection and welfare of donkeys through rescue operations, veterinary care, and owner education; works to prevent donkey abuse and neglect; engages in community sensitization and policy advocacy.

(continued)

Organization name	Region/country	Key efforts/activities
The Donkey Sanctuary of Canada	Canada	Rescues and provides sanctuary for donkeys, mules, and hinnies; offers education programs for the public; promotes awareness about the proper care and welfare needs of these animals; facilitates adoption and foster care when appropriate.
Brooke East Africa	Kenya, Ethiopia, Tanzania	Regional branch of Brooke organization; focuses on improving the welfare of working equines through veterinary services, community training, and advocacy; works closely with local partners to implement sustainable animal welfare practices.
Sociedade Protectora dos Animais Domésticos (SPAD)	Brazil	While primarily focused on domestic animals, SPAD also engages in rescuing and rehabilitating working donkeys and mules; provides veterinary care and promotes humane treatment; collaborates with local communities and authorities.
Fundación Ayuda a los Animales (FAADA)	Spain	Works on animal protection and welfare, including efforts to rescue and rehabilitate donkeys and mules; conducts educational campaigns and legal actions to defend animal rights; promotes responsible ownership and care.
AniMedes	Greece	Provides medical care and sanctuary for donkeys, mules, and other animals; focuses on rescue operations, rehabilitation, and rehoming; engages in public education on animal welfare issues; collaborates with international organizations for support and resources.
Help In Suffering	India	Offers veterinary care and rehabilitation for working equines, including donkeys and mules; runs mobile clinics; educates owners on proper care and management; operates in urban and rural areas to reach vulnerable animal populations.
Donkey Sanctuary Aruba	Aruba	Provides a safe haven for homeless and injured donkeys; offers medical care, shelter, and nutrition; engages in public education and tourism activities to raise awareness and funds; works to preserve the donkey population on the island responsibly.
Les Donkeys de Marrakech	Morocco	Rescues and rehabilitates donkeys working in and around Marrakech; provides veterinary services; educates owners and the public about donkey welfare; collaborates with local authorities to improve conditions for working donkeys and mules.

(continued)

Organization name	Region/country	Key efforts/activities
Brooke West Africa	Senegal, Mali, Niger	Regional extension of Brooke organization; improves the welfare of working equines through community-based programs, veterinary services, and advocacy; addresses issues like overloading, improper harnessing, and lack of access to veterinary care.
Society for Animal Welfare (SAW)	Pakistan	Works towards the welfare of working animals, including donkeys and mules; provides veterinary care, runs awareness campaigns, and educates owners; collaborates with local communities to improve animal health and productivity, thereby supporting livelihoods.
Cart Horse Protection Association	South Africa	Protects working cart horses and donkeys; offers veterinary care, rehabilitation, and support to owners; provides education on animal care and welfare; advocates for the humane treatment of working equines; operates in urban and peri-urban areas where animal labor is prevalent.
Humane Society International (HSI)	International	While focusing on all animals, HSI runs specific programs to improve the welfare of working equines, including donkeys and mules; provides disaster relief, veterinary training, and advocates for stronger animal protection laws globally; collaborates with local partners to implement programs.
Donkey Sanctuary of New Zealand	New Zealand	Rescues and cares for neglected or unwanted donkeys; provides a sanctuary environment; educates the public about donkey care and welfare; facilitates adoptions to suitable homes; promotes responsible ownership.
Liberty Horses	Germany	Works with donkeys, mules, and horses to provide therapy programs; rescues and rehabilitates abused or neglected animals; offers educational workshops on equine care and welfare; uses human-animal interaction to benefit both animals and people in need.
American Fonduk	Morocco	Provide free veterinary services to working equines of Morocco and the families who depend on their labor and teaching opportunities for local veterinarian students and abroad.
Equitarian Iniatitive	Latin America (mostly based in US)	Provide veterinary services and education to vet students and working equids.

References

African Union (2024) AU Heads of State General Assembly: moratorium on the donkey skin trade. African Union General Assembly

Ashley C, Boyd C, Goodwin H (2006) Pro-poor tourism: putting poverty at the heart of the tourism agenda. Significance 51

Blench R (2000) Working animals in Africa. The Donkey Sanctuary

BThe Brooke (2022) The Donkey Skin Trade: Policy Brief on Worldwide Impact. The Brooke Hospital for Animals

The Brooke (2025) Donkey Slaughter and Skin Trade in Africa: Post-moratorium trends. The Brooke Hospital for Animals

Brooke (2025) Retrieved from www.thebrooke.org

Cosyns E, Degezelle T, Demeulenaere E, Hoffmann M (2005) Feeding ecology of Konik horses and donkeys in Belgian coastal dunes and its implications for nature management. Belg J Zool 135:129–134

Daniel C, Persai D (2010) Community-based conservation: an emerging land use at the livestock-wildlife interface in northern Kenya. Hum Dimens Wildl 15(5):343–357

FAO (2014) The role, impact and welfare of working (traction and transport) animals. Animal Production and Health Report No. 5. Food and Agriculture Organization of the United Nations, Rome

FAO (2017) FAOSTAT statistical database. Food and Agriculture Organization of the United Nations. Retrieved from http://www.fao.org/faostat/en/#data

FAO (2019) Livestock solutions for climate change. Food and Agriculture Organization of the United Nations

FAO (2020) Global efforts to combat the trade. Food and Agriculture Organization of the United Nations

FAO (2021) Donkey skin trade threatens livelihoods in Africa. Food and Agriculture Organization of the United Nations

Fernández J, Delgado JV (2018) The donkey in Spain: from traditional use to joining the forces of innovation and research in a modern society. Animals 8(11):195

Fernando P, Starkey P (2004) Donkeys and development: socio-economic issues in the use and management of donkeys. In: Starkey P, Fielding D (eds) Donkeys, people, and development. Wageningen Academic Publishers, pp 31–44

Gandini G, Agerholm JS, Puglisi C (2014) Conservation of genetic diversity in livestock: the role of cryopreservation. In: Advances in animal genomics. Springer, pp 151–164. https://doi.org/10.1007/978-3-319-03626-1_12

Granados AC, Agís IF (2011) Why children with special needs feel better with hippotherapy sessions: a conceptual review. J Altern Complement Med 17(3):191–197

Hampton JO, Jones B, McGreevy PD (2018) Social license and animal welfare: developments from the horse industry. J Vet Behav 25:27–40

Henchion M, McCarthy M, O'Reilly S (2014) Animal welfare and ethical issues in Irish agriculture: a case study of the Irish dairy industry. Int J Agric Manage 3(1):10–20

Holland J, Low J (2010) Managing donkeys in the tourism industry: guidelines for promoting animal welfare. Tour Operators Initiative

Johnson RA, Albright DL, Marzolf JR (2018) Effects of therapeutic horseback riding on post-traumatic stress disorder in military veterans. Mil Med Res 5(1):3

Kimura B, Marshall F, Chen S et al (2013a) Ancient DNA from Nubian and Somali wild ass provides insights into donkey domestication. Proc R Soc B Biol Sci 280(1750):20130824

Kimura K, Cockett NE, Agerholm JS (2013b) Genetic resistance to trypanosomiasis in livestock. Vet J 198(3):504–510. https://doi.org/10.1016/j.tvjl.2013.09.017

Leach KA, Pinchbeck GL (2015) A survey of horse owners in Great Britain regarding their horses' transport and traveling problems, behavior, and management. Equine Vet J 47(3):278–283

Levine H (2015) Celebrating the American mule: a tradition of strength and endurance. Equestrian Times 12(4):23–29
Li X, Liu Y, Li J (2015) The culture and conservation of donkeys in the Li ethnic group of Hainan Island, China. J Ethnobiol Ethnomed 11:72
McCracken S, Lalor G, Green M (2019) The use of telemedicine in veterinary practice. In Pract 41(9):409–416
McKenna C (2007) Bearing a heavy burden: the impact of working horses, donkeys and mules on the livelihoods of the poor. The Brooke, London
Moehlman PD, Kebede F, Bruche S (2008) Equus africanus. The IUCN Red List of Threatened Species
Mshelia P (2023) Mapping the trends of slaughter for donkey skin trade across some countries in Africa. Brooke, East Africa
Murray M, Trail JC, Bawden J (2016) The role of genetics in trypanosomiasis resistance in livestock. Vet Res. 47(1):31. https://doi.org/10.1186/s13567-016-0317-5
Nimer J, Lundahl B (2007) Animal-assisted therapy: a meta-analysis. Anthrozoös 20(3):225–238
Norton E (2019a) The global trade in donkey skin: a threat to the species and a welfare concern. J Anim Ethics 9(2):199–210
Norton L (2019b) Donkey skin trade and the Ejiao crisis: implications for Africa. The Donkey Sanctuary Report
OIE (2016) Terrestrial animal health code. World Organisation for Animal Health
Pan-African Donkey Conference (PADCo) (2022) Report of the Pan African Donkey Conference 2022: 'Donkeys in Africa now and in the future'. AU-IBAR & Partners
Pearson RA, Simalenga TE, Krecek RC (2003) Harnessing and animal welfare in South Africa. Animal Traction Network for Eastern and Southern Africa (ATNESA)
Pritchard JC, Lindberg AC, Main DCJ, Whay HR (2005a) Assessment of the welfare of working horses, mules, and donkeys, using health and behavior parameters. Prev Vet Med 69(3–4): 265–283
Pritchard J, Lindberg A, O'Callaghan C (2005b) Assessment of the welfare of working horses, mules, and donkeys, using health and productivity measures. Vet Rec 158(17):607–612
Pritchard JC, Lindberg AC, Main DCJ, Whay HR (2015) Utilization of working equid welfare assessment for international development: a scoping study. Animals 5(4):978–996
Roe D, Nelson F, Sandbrook C (2009) Community management of natural resources in Africa: impacts, experiences and future directions. Natural Resource Issues No. 18, IIED
Smith D (2005) Economic and social roles of donkeys in Ethiopia: a case study of Menagesha District. J Equine Sci 16(4):147–154
Solomon TB, Snyman HA, Smit GN (2013) Rangeland dynamics in South Ethiopia: implications for pastoral development. Afr J Range Forage Sci 24(1):1–12
SPANA (2023) Protecting donkeys: the campaign against donkey skin trade. Society for the Protection of Animals Abroad, Retrieved from www.spana.org
Starkey P (2000) The introduction, intensification and decline of donkey use in Africa. Animal Traction Network for Eastern and Southern Africa (ATNESA)
Starkey P (2001) Local transport solutions: People, paradoxes and progress—lessons arising from the spread of intermediate means of transport. SSATP Working Paper No. 56. Sub-Saharan Africa Transport Policy Program (SSATP),World Bank, Washington, DC.
Starkey P (2010) The contribution of working animals to agricultural development. Agricultural Research & Extension Network Paper No. 47
Starkey P, Fielding D (2000) Donkeys, people, and development. A resource book of the Animal Traction Network for Eastern and Southern Africa (ATNESA), Wageningen, The Netherlands
Starkey P, Starkey M (2000) Regional and world trends in donkey populations. In: Starkey P, Fielding D (eds) Donkeys, people and development: A resource book of the Animal Traction Network for Eastern and Southern Africa (ATNESA). ACP–EU Technical Centre for Agricultural and Rural Cooperation (CTA), Wageningen, pp 10–21
Svendsen ED (2008) The professional handbook of the donkey, 4th edn. Whittet Books Ltd

Taberlet P, Coissac E, Pansu J, Pompanon F (2011a) Conservation genomics: diversity and demography of a highly endangered wild equid, the African wild ass. Mol Ecol 20(19): 3877–3888

Taberlet P, Camus-Kulanda R, Roudot A (2011b) Preserving genetic diversity in livestock. Anim Genet Resour 49:11–19. https://doi.org/10.1017/S2078633611000084

The Brooke (2021) Animal welfare in developing countries: improving the lives of working horses, donkeys, and mules. The Brooke Report

The Donkey Sanctuary (2020) Under the skin: update on the global crisis for donkeys and the people who depend on them. Retrieved from https://www.thedonkeysanctuary.org.uk/under-the-skin

The Donkey Sanctuary (2023) Campaigning against the donkey skin trade. Retrieved from www.thedonkeysanctuary.org.uk

Trotter KS (2012) Harnessing the power of equine-assisted counseling: adding animal-assisted therapy to your practice. Routledge

Valette D (2014) Invisible workers: the economic contributions of working donkeys, horses, and mules to livelihoods. The Brooke

Valette D (2015) Invisible Workers: The economic contributions of working donkeys, horses and mules to livelihoods. The Brooke, London

Wanyama JB, Keter K, Ndung'u J (2020) Improving the welfare of working donkeys in Kenya through community action. Pastoralism 10:13

Zhang D (2019) Impact of the donkey skin trade on the welfare of the global donkey population. Equine Vet Educ 31(8):436–438

Zhang L, Luo J (2019) The Ejiao industry: supply chain impacts on global donkey populations. Front Vet Sci 6:452

Companion Donkeys and Mules

Stephen R. Purdy and Francisco Javier Navas González

Abstract

Donkeys (*Equus asinus*) and mules (*Equus mulus*) have traditionally been valued for their strength, endurance, and service in agricultural and transport systems, but they are increasingly recognized for their potential as companion animals. Their calm temperament, cognitive abilities, resilience, and adaptability make them suitable partners for families, therapy programs, and sustainable animal ownership. This manuscript explores the diverse roles of donkeys and mules as pets, interspecies companions, therapeutic agents, protectors in livestock systems, cultural and spiritual symbols, and even participants in warfare and politics. It also addresses welfare, training, health management, and ethical considerations in domestic settings. By integrating historical, scientific, and practical perspectives, the work highlights the unique capacities of donkeys and mules to contribute to human well-being across contexts, while emphasizing the responsibilities and challenges of their care.

Keywords

Companion animals · Animal welfare · Animal-assisted therapy · Livestock guardianship · Human–animal interaction · Pet ownership · Cultural symbolism

S. R. Purdy
Belchertown, MA, USA

F. J. Navas González (✉)
Worldwide Donkey Breeds Project, Cordoba, Spain

Department of Genetics, University of Córdoba, Cordoba, Spain
e-mail: fjnavas@uco.es

A. K. McLean et al. (eds.), *Donkey and Mule Medicine: Foundations, Breeds, and Management*, https://doi.org/10.1007/978-3-032-12726-6_12

1 Introduction

For thousands of years, donkeys (*Equus asinus*) and mules (the hybrid offspring of a male donkey and a female horse) have played essential roles in the development of human civilization. Their domestication marked a turning point in the evolution of transport, agriculture, and trade, particularly in arid and mountainous regions where other animals struggled to survive. From the caravan routes of the Silk Road to the battlefields of ancient empires, these equids were not only logistical assets but also cultural symbols of endurance, humility, and service. In Mesopotamia, donkeys were depicted pulling war chariots as early as 2600 BCE, while in Roman and Islamic societies, they were revered for their reliability and adaptability (Hamblin 2006; Roth 1998).

Historically viewed as utilitarian animals, donkeys and mules have often been undervalued in comparison to horses. Yet modern ethological research has challenged these assumptions, revealing that donkeys possess a high degree of cognitive sophistication, including spatial awareness, memory retention, and social intelligence. Their so-called "stubbornness" is better understood as a manifestation of caution and self-preservation—a trait that has enabled their survival in harsh environments (McLean and Heleski 2010; Svendsen 2008). Mules, through hybrid vigor, inherit the donkey's resilience and the horse's athleticism, resulting in animals that are not only physically robust but also behaviorally stable and emotionally responsive (Bessa and McLean 2024).

Beyond their physical capabilities, both donkeys and mules demonstrate a remarkable capacity for emotional attunement and interspecies bonding. Studies in animal-assisted therapy have shown that donkeys, in particular, are effective mediators in therapeutic contexts due to their calm demeanor, sensitivity to human emotion, and ability to foster non-verbal communication (Artemiou et al. 2021; Menna et al. 2019). Their presence has been linked to reductions in anxiety, improvements in social engagement, and enhanced emotional regulation in children, veterans, and individuals with neurodevelopmental disorders.

This chapter provides a comprehensive overview of donkeys and mules as companion animals, integrating perspectives from veterinary science, psychology, anthropology, and cultural history. It examines their temperament, practical care, and welfare considerations, while also exploring their evolving roles in therapy, education, and symbolic representation. By synthesizing empirical research with historical narratives, it aims to reposition these equids not merely as working animals, but as sentient beings capable of enriching human lives in profound and multifaceted ways.

2 Why Choose a Donkey or Mule as a Pet?

Donkeys (*Equus asinus*) and mules (*Equus mulus*) have long been associated with agricultural labor and transport, yet their potential as companion animals is increasingly recognized in both academic and practical contexts. These equids offer a

Fig. 1 Donkey pets: Polly in their daily life the Ferrier family and Hope and Jane

unique blend of emotional intelligence, physical resilience, and ecological adaptability that make them compelling choices for ethical and sustainable pet ownership (Fig. 1).

2.1 Temperament and Intelligence

Both donkeys and mules are known for their calm demeanor and high cognitive abilities. Donkeys, often mischaracterized as stubborn, actually exhibit strong problem-solving skills and emotional sensitivity (McLean and Heleski 2010). Mules, as hybrids of male donkeys and female horses, inherit the intelligence and caution of donkeys with the strength and responsiveness of horses. Their sense of self-preservation makes them safer and more predictable than horses in many contexts (Svendsen 2008).

These animals are capable of forming deep bonds with humans, especially when raised in environments that prioritize trust and positive reinforcement. Their trainability and memory retention make them ideal for owners seeking interactive and enriching relationships (Bessa and McLean 2024).

2.2 Physical Strength and Endurance

Mules are particularly valued for their hybrid vigor, often outperforming both parent species in terms of strength, endurance, and disease resistance (Bessa and McLean 2024). They can carry heavy loads over long distances with minimal fatigue, making them suitable for rural or mountainous environments. Donkeys, while smaller, are also remarkably strong and resilient, capable of steady work and companionship without the high energy demands of horses (Starkey 2020).

2.3 Health, Longevity, and Low Maintenance

Both donkeys and mules are relatively low-maintenance animals. Donkeys thrive on fibrous, low-sugar diets and require minimal grooming (Svendsen 2008). Mules tend to be more disease-resistant than horses and are less prone to common equine health issues such as colic. Lifespans for both species can reach 25–30 years or more, offering long-term companionship and stability (Smith 2018).

2.4 Educational and Therapeutic Value

In educational settings, donkeys and mules serve as effective tools for teaching biology, animal behavior, and sustainable agriculture. Their historical significance and behavioral traits provide rich material for interdisciplinary learning (Starkey 2020). Additionally, their empathetic nature and calm engagement with humans make them suitable for animal-assisted therapy programs (Bessa and McLean 2024).

2.5 Environmental Adaptability and Ethical Ownership

Originating from arid regions, donkeys are well-suited to dry climates and can tolerate both hot and cold conditions. Mules share this adaptability, making both species viable pets in a variety of geographic locations, including southern Spain. Choosing to adopt a donkey or mule also reflects a commitment to ethical stewardship, especially when sourced from rescue organizations or retired working stock. Their peaceful coexistence with other animals makes them excellent additions to mixed-species households or small farms.

3 Considerations Before Adopting a Donkey or Mule as a Pet

Adopting a donkey (*Equus asinus*) or mule (*Equus mulus*) as a companion animal requires careful planning and long-term commitment. While these equids offer emotional and educational value, they also present specific welfare, logistical, and financial responsibilities that must be considered before ownership.

3.1 Lifespan and Long-Term Commitment

Donkeys and mules are long-lived animals. The average lifespan of a donkey is approximately 27–30 years, with documented cases exceeding 40 years under optimal care (Svendsen 2008). Mules, benefiting from hybrid vigor, often live comparably long lives and may exhibit greater resilience to disease and injury (Martins-Bessa and McLean 2024). Prospective owners must be prepared for decades of daily care and companionship.

3.2 Social Needs and Emotional Welfare

Equids are inherently social animals. Donkeys and mules require companionship—either from other equids or compatible livestock—to maintain psychological well-being. Isolation can lead to behavioral issues, stress, and depression (McLean 2019). Owners who cannot provide consistent interaction or a herd environment should reconsider adoption.

3.3 Daily Care and Physical Activity

Like all domestic equids, donkeys and mules require daily feeding, access to clean water, and regular exercise. These needs persist year-round, regardless of weather or personal circumstances. Responsible ownership entails twice-daily care, 365 days a year, and contingency planning for travel, illness, or emergencies (Svendsen 2008).

3.4 Nutritional Demands and Economic Considerations

Donkeys and mules are large herbivores with substantial dietary requirements. While donkeys can thrive on high-fiber, low-energy forage, mules—especially those used for work or recreation—may require more nutrient-dense feed (Martins-Bessa and McLean 2024). Owners without access to pasture must budget for hay, supplements, and veterinary care. Unexpected costs such as injury, transportation, or chronic illness should be anticipated (Fig. 2).

Fig. 2 Lola rests after a performance at the Cultural centre while the headmaster of the centre, Hawa Djabali, rebukes her neighbours because of their complaints due to the noises produced by the animal

4 Choosing the Appropriate Donkey or Mule as a Companion Animal

Selecting a donkey (*Equus asinus*) or mule (*Equus mulus*) as a companion animal requires thoughtful evaluation of the individual animal's background, temperament, and compatibility with the prospective owner's experience and environment. These considerations are essential to ensure the welfare of the animal and the success of the human–equid relationship.

4.1 Understanding the Animal's History

A comprehensive understanding of the animal's life history is critical. This includes its previous roles (e.g., working, breeding, therapy), housing conditions, feeding routines, and social interactions with other animals. Changes in routine are possible, but sudden transitions without context can lead to stress or behavioral issues (McLean 2019). Knowing the animal's background helps anticipate its reactions and adapt care accordingly.

4.2 Assessing Temperament and Owner Experience

Temperament plays a central role in successful pet ownership (Fig. 3). Donkeys and mules exhibit individual differences in sociability, assertiveness, and adaptability. French (1993) identified key temperament factors in donkeys—such as obduracy and vivacity—that influence their behavior in new environments. Owners with limited experience should prioritize animals with docile, patient, and cooperative dispositions. Matching temperament to the handler's skill level reduces the risk of miscommunication and stress.

Fig. 3 Handling miniature donkey behaviour

4.3 Seeking Expert Guidance

Prospective owners are advised to consult with individuals experienced in equid care, particularly during the selection process. Expert guidance can help avoid novice errors in assessing health, behavior, and suitability. This is especially important when evaluating animals with complex histories or behavioral quirks (Svendsen 2008).

4.4 Considering Sex and Reproductive Status

Sex and reproductive status influence behavior and manageability. Female donkeys and gelded males generally exhibit more stable and predictable temperaments, making them suitable for novice owners. Intact males (jackstock) may display heightened aggression or territoriality and should only be handled by experienced individuals (McLean 2019).

4.5 The Role of Human–Animal Interaction

Ultimately, the relationship between the owner and the animal shapes its behavior over time. Consistent, respectful treatment fosters trust and emotional stability. As McLean et al. (2019) emphasizes, equids are sensitive to human cues and respond dynamically to their environment. A well-managed bond can transform even a cautious or reactive animal into a confident companion.

4.6 Communication and Family Readiness

When adopting a donkey or mule as a family pet, communication with children is essential. It is important to ensure that the animal aligns with the child's expectations and emotional readiness. A practical approach is to allow children to spend time at farm schools or equestrian centers, where they can interact with equids and gain firsthand experience. These environments provide valuable insight into the daily responsibilities and emotional dynamics of equid care.

4.7 Fraud Prevention and Veterinary Evaluation

As with any animal transaction, the risk of fraud must be considered. Buyers should visit the animal multiple times to verify its behavior, health, and consistency. A veterinary pre-purchase exam is strongly recommended. This evaluation includes a physical and behavioral assessment, and may reveal underlying health issues that could affect long term care (Bessa and McLean 2024). In some cases, a return clause may be negotiated to protect the buyer in the event of undisclosed conditions.

5 Families and Donkeys/Mules as Pets: Aspects to Consider

The increasing popularity of donkeys (*Equus asinus*) as domestic companions reflects a shift in how families perceive animal care and child development. While their gentle nature and modest size make them attractive for households with children, adopting a donkey entails more than novelty—it demands sustained commitment and thoughtful planning (Fig. 4).

5.1 Responsibility and Developmental Readiness

Introducing a donkey into a family setting should be preceded by a realistic assessment of the household's capacity to meet the animal's needs. Donkeys require consistent feeding routines, hygienic living conditions, and regular physical activity (Svendsen 2008). These tasks often exceed the capabilities of young children, placing the onus on parents to either assume full responsibility or provide structured guidance. Rather than assigning chores arbitrarily, families should integrate care routines into daily life to foster accountability and empathy.

5.2 Behavioral Suitability and Safety Measures

When considering donkeys for recreational use—such as light riding—it is essential to evaluate the animal's temperament, training history, and comfort around children.

Fig. 4 A family walking holiday in France's Rhône Valley with a friendly donkey companion. Over 7 days, families follow child-friendly trails through vineyards, forests, and meadows, with sweeping views from the Ardèche volcanoes to the French Alps. Evenings are spent in welcoming chambres d'hôtes, enjoying regional cuisine after a rewarding day's walk

Experienced adult donkeys are generally more predictable and tolerant than juveniles, especially in dynamic environments. However, no matter how docile the animal may be, unsupervised interaction between children and equids poses risks (Martins-Bessa and McLean 2024). Establishing clear boundaries and adult oversight is non-negotiable.

5.3 Emotional Growth and Social Bonding

Beyond practical considerations, donkeys can play a meaningful role in a child's emotional landscape. Their calm demeanor and responsiveness to human interaction make them ideal for nurturing patience, compassion, and non-verbal communication skills. Research in animal-assisted education highlights the value of equids in promoting psychological resilience and social learning (Artemiou et al. 2021). The relationship formed between child and donkey often transcends utility, evolving into a source of companionship and emotional grounding.

6 Basic Care of Donkeys in Domestic Settings

Domestication of equids has led to significant behavioral and physiological adaptations, yet donkeys (*Equus asinus*) retain many traits from their desert ancestry. Their integration into domestic environments requires tailored care strategies that respect their evolutionary needs while ensuring welfare and safety.

6.1 Accommodation and Environmental Management

Donkeys thrive in environments that balance freedom of movement with protection from adverse weather. When housed in pastures, each animal requires approximately 0.5 hectares of grazing land to maintain nutritional autonomy during productive seasons (Gichure and Olayide 2024). In confined settings, stables must offer a minimum of 3.3 m^2 per donkey, with dry bedding—preferably straw or wood shavings—replaced daily to maintain hygiene and prevent respiratory issues.

Due to their limited sebaceous gland activity, donkey coats lack the waterproofing found in horses, making them vulnerable to cold, wet conditions (Thiemann 2024). Shelters must provide insulation and dry resting areas, especially during rain or snow. In winter, indoor housing during colder hours is recommended, with monitored exposure to outdoor exercise on temperate days (Fig. 5).

6.2 Grooming and Coat Maintenance

Donkeys benefit from regular grooming, which supports skin health and strengthens human–animal bonds (Fig. 6). Brushing should follow the direction of hair growth

Fig. 5 Donkey shelters

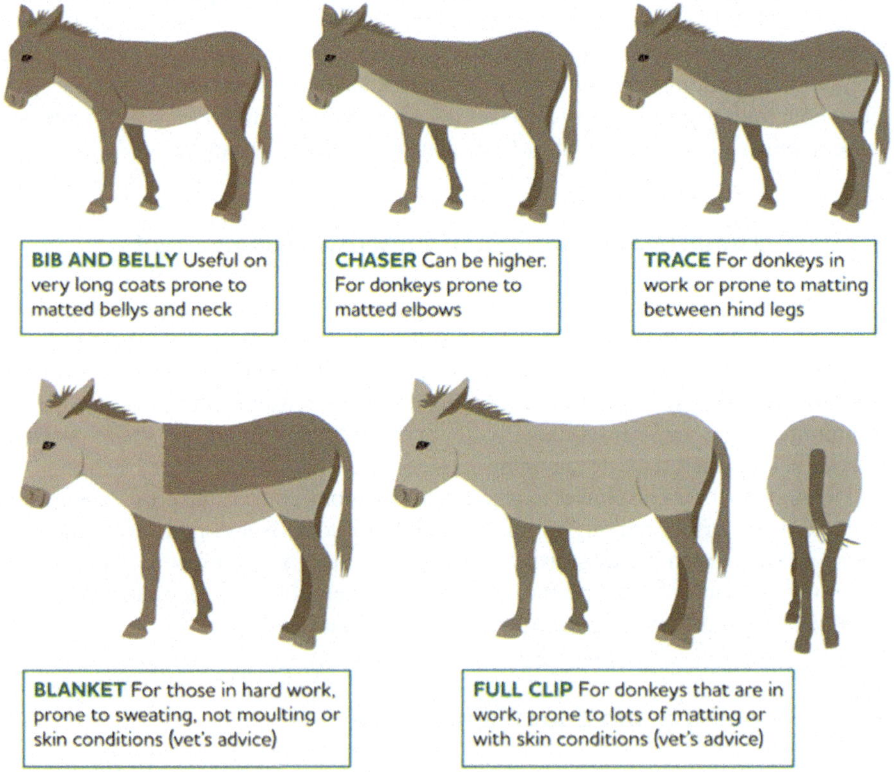

Fig. 6 Donkey grooming and clipping (The Donkey Sanctuary 2020)

and avoid sensitive areas such as the ears. During spring, shedding blades may assist in removing winter coats, though excessive grooming can disrupt insulating air pockets and increase susceptibility to cold (Sullivan 2024). Summer grooming is less critical, as donkeys naturally self-clean through dust bathing—a behavior rooted in their arid origins.

Signs of excessive rubbing, particularly near the tail dock, may indicate ectoparasite infestation and warrant veterinary evaluation. Seasonal coat care should be adapted to climate conditions, avoiding trimming unless adequate shelter is provided.

6.3 Nutritional Requirements

Donkeys possess a highly efficient digestive system adapted to fibrous, low-energy forage. Their daily intake should not exceed 1.5% of body weight in dry matter to prevent metabolic disorders such as laminitis and hyperlipemia (Raw et al. 2024). Free-grazing animals may meet their needs naturally, but those in paddocks require supplemental hay and straw, along with access to salt licks and clean water (Figs. 7 and 8).

Given their desert lineage, donkeys are predisposed to thrive on sparse vegetation. Overfeeding, particularly with rich feeds, can disrupt metabolic balance and compromise long-term health.

Fig. 7 Feeding donkeys at ground level, in their natural upright grazing position, supports healthy digestion, proper dental wear, easier breathing, and reduces strain on the neck and spine—while also encouraging natural foraging behavior. Feeding donkeys from raised nets or racks can cause uneven tooth wear, increase dust inhalation, and place strain on the neck and back, making it less suitable than ground-level feeding

Fig. 8 Salt licks

Fig. 9 Isle of Wight, Britain—14 October 2015. A donkey demonstrates the historic use of a treadmill to draw water from the well at Carisbrooke Castle, Newport

6.4 Physical Activity and Mobility

Locomotion is essential for donkey health. Regular movement supports hoof integrity, circulatory function, and gastrointestinal motility. Whether through pasture access or guided walks, physical activity reduces the risk of musculoskeletal and digestive complications (Boocock et al. 2025). Moreover, exercised donkeys exhibit improved temperament and responsiveness, facilitating handling and training (Fig. 9).

6.5 Immunization Protocols

Preventive healthcare is vital. Annual vaccinations against tetanus and equine influenza are strongly advised, with influenza immunization mandatory for donkeys

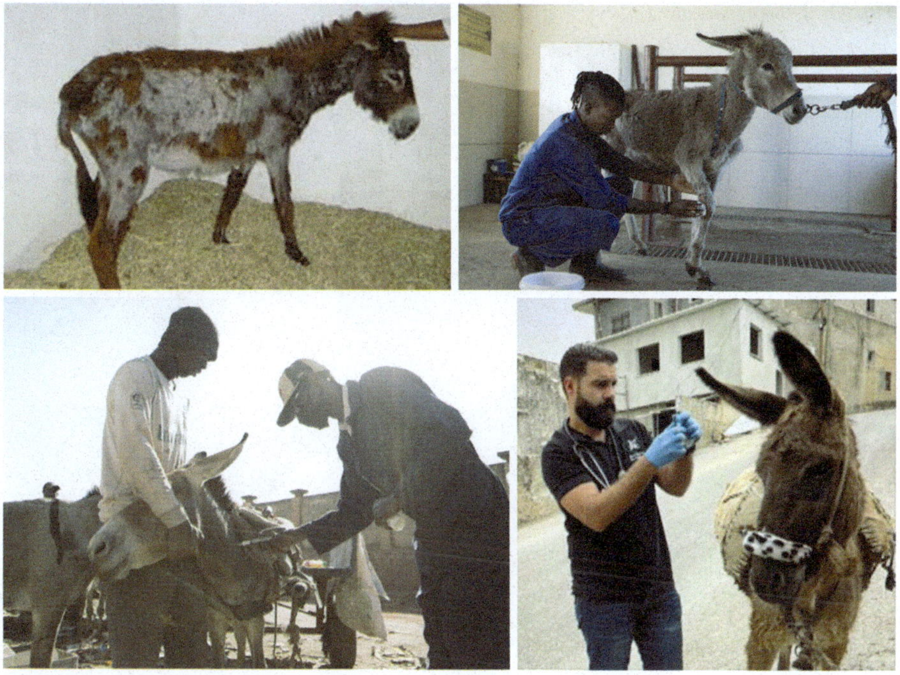

Fig. 10 Tetanus protocol implementation

participating in gatherings with other equids. Tetanus, often contracted through wounds, poses a high mortality risk without vaccination (Animal Rahat 2012). Veterinarians should tailor immunization schedules based on regional disease prevalence and individual health profiles (Fig. 10).

6.6 Parasite Control and Worming Protocols

Donkeys share many parasitic threats with horses, yet host-specific differences necessitate tailored deworming strategies. Notably, donkeys are lifelong hosts for *Parascaris equorum*, unlike horses where susceptibility is age-dependent. Additionally, donkeys often carry *Dictyocaulus arnfieldi* asymptomatically, posing a respiratory risk to cohabiting horses and ponies (Thiemann and Poore 2019).

Routine administration of broad-spectrum anthelmintics such as ivermectin or moxidectin is recommended for mixed-species environments to prevent cross-infection. Effective parasite control also hinges on synchronized treatment across all animals to avoid reinfection reservoirs. Subtherapeutic dosing should be strictly avoided due to the risk of resistance development (Gichure and Olayide 2024) (Fig. 11).

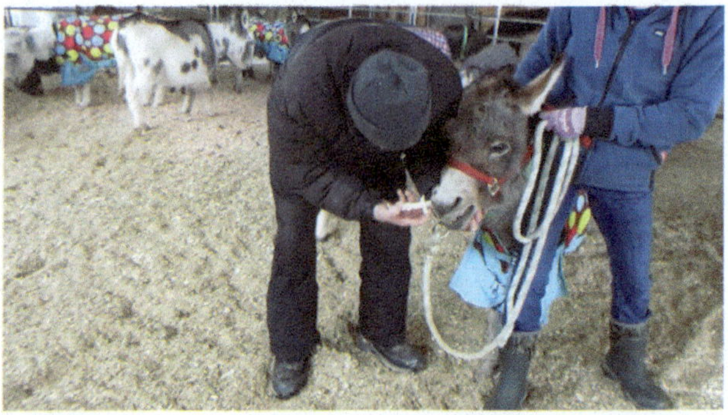

Fig. 11 Donkey worming

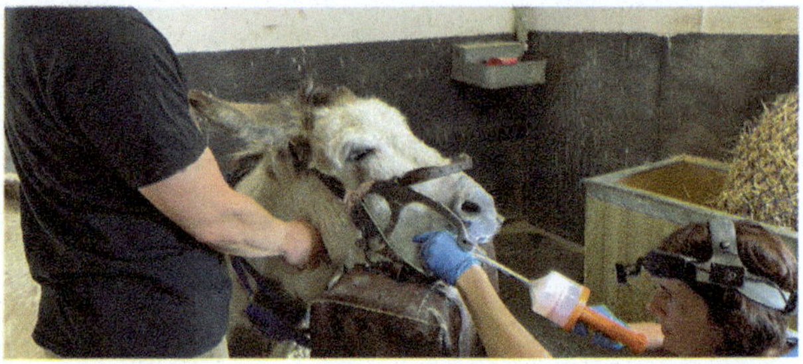

Fig. 12 Dental monitoring in a donkey

Environmental hygiene plays a pivotal role. Daily stall cleaning and weekly dung collection in pastures significantly reduce parasite loads. Rotational grazing, with strategic rest periods of 1–3 months, further disrupts parasitic life cycles. Grouping animals by age and physiological status enhances targeted management.

6.7 Dental Health and Monitoring

Donkey dentition closely resembles that of horses, though eruption timelines differ. Annual dental evaluations by qualified veterinarians are essential, with biannual checks advised for juveniles and seniors. Because donkeys typically do not wear bits, dental issues may go unnoticed until advanced stages (Fig. 12).

Key indicators of oral pathology include:

- Quidding (dropping partially chewed food)
- Halitosis due to periodontal disease
- Nasal discharge linked to dental abscesses
- Undigested fibers in feces
- Anorexia or unexplained weight loss
- Colic symptoms associated with malocclusion

Regular dental filing prevents sharp enamel points and promotes efficient mastication, contributing to overall digestive health (Raw et al. 2024).

6.8 Hoof Maintenance and Farriery

Hoof care is critical to donkey welfare. Unlike wild asses that naturally wear down their hooves through extensive movement, domestic donkeys require trimming every 6–8 weeks. Their hoof-pastern axis is more upright than that of horses, necessitating species-specific farriery techniques (Thiemann and Poore 2019) (Fig. 13).

Moist environments predispose donkeys to abscesses, thrush, and white line disease. Dry, firm ground and regular inspection mitigate these risks. Farriers must be trained in asinine anatomy to avoid misalignment and lameness.

6.9 Recognizing Illness in Donkeys

Donkeys are stoic animals, often masking discomfort until conditions become severe. Subtle behavioral changes—such as lethargy, social withdrawal, or reduced appetite—may signal underlying pathology. Common ailments include:

1. Colic
2. Hyperlipidemia
3. Laminitis and hoof abscesses
4. Respiratory infections
5. Dental and joint disorders

Fig. 13 Donkey farrier labours

Daily observation fosters familiarity with baseline behavior, enabling early detection of deviations. Emotional stressors, such as loss of a companion, can also manifest as depression-like symptoms. Prompt veterinary consultation is essential when abnormalities arise.

6.10 Basic Training and Behavioral Conditioning

Foundational training, or basic dressage, ensures safe and respectful interaction between humans and donkeys. This includes desensitization to handling, haltering, loading into transport, and discouraging aggressive behaviors like kicking.

Understanding the donkey's prey instincts is vital. Initial responses to perceived threats may include flight or defensive aggression. Building trust through consistent, calm leadership fosters cooperation and reduces fear-based reactions. The handler must establish authority while maintaining empathy, positioning themselves as a reliable decision-maker (European Commission 2022).

7 Donkeys in Livestock Management: Training, Calming, and Protective Roles

Donkeys (*Equus asinus*) have long served as versatile companions in agricultural systems, particularly in regions with extensive livestock production such as Australia and New Zealand. Their cognitive aptitude, physical resilience, and social temperament make them uniquely suited for roles that extend beyond transport and labor—into behavioral conditioning, herd protection, and interspecies companionship.

7.1 Training and Leadership in Cattle Herds

In cattle operations, trained donkeys are employed to guide and calm bulls, especially in preparation for exhibitions or sale. Their presence reduces aggression and facilitates weight gain by minimizing stress-induced behaviors. In New Zealand, geldings have been successfully used to condition breeds like Aberdeen Angus for national shows, with their value likened to that of working dogs.

Training involves tethering a young bull to a calm, experienced donkey using a wide collar and lead rope. Over several days, the bull mirrors the donkey's movements, learning to walk, graze, and rest in synchrony. This method fosters behavioral stability and reduces anxiety in unfamiliar environments such as show rings (Gichure and Olayide 2024) (Fig. 14).

Fig. 14 Cattle training

Fig. 15 Halter breaking

7.2 Calming Horses: Halter Breaking

Donkeys are increasingly used to soothe high-strung foals and colts. Their placid demeanor and non-reactive nature help regulate equine stress responses, particularly before saddling or competition. Ranchers often house donkeys with young horses for extended periods to promote emotional stability (Raw et al. 2024). Miniature donkeys are preferred in stables due to their compact size and adaptability (Fig. 15), but sizes must agree to prevent incidents derived from donkey vs foal mismatches.

Fig. 16 Donkey herd and flock guards against wolves, coyotes and big cats

7.3 Protection of Cows and Goat Herds and Sheep Flocks

Donkeys also serve as effective guardians for large and small ruminants. Their territorial instincts and strong social bonds enable them to defend flocks from predators such as foxes, wild dogs, and wolves or even big cats as reported by some experiences in Namibia. Female donkeys (jennies) and geldings are most suitable, as intact males may exhibit aggression toward lambs, kids or calves. Once bonded with the herd, donkeys integrate fully and respond to threats with vocal alerts and physical deterrence (Andelt 2004).

Unlike livestock guardian dogs, donkeys share dietary needs with sheep and goats, simplifying feed logistics. Breeds such as the Andalusian and Catalonian or similar types are recommended for large grazing areas due to their size and stamina (Fig. 16).

8 Donkey and Mule Mediated Therapy

Donkey-mediated therapy (DMT), also referred to as asinotherapy, is a specialized form of animal-assisted intervention that leverages the ethological, emotional, and physical characteristics of donkeys and mules to support individuals with a range of physical, cognitive, and psychosocial challenges. While equine-assisted therapies have long been established, DMT has gained recognition for its suitability in contexts requiring gentleness, predictability, and emotional attunement—traits that are particularly pronounced in donkeys (Borgi et al. 2016; Menna et al. 2019) (refer to Table 1).

Table 1 Modalities and therapeutic focuses of DMT (Fig. 17)

Modality	Therapeutic focus
Hippotherapy/ onotherapy	Uses the donkey's movement and body warmth to stimulate neuromotor functions
Therapeutic riding	Enhances attention, motivation, and sensory integration
Adapted riding	Recreational riding tailored to individuals with disabilities
Therapy swing	Gymnastic exercises performed on the donkey's back to improve coordination
Social equine therapy	Builds emotional bonds and supports social reintegration

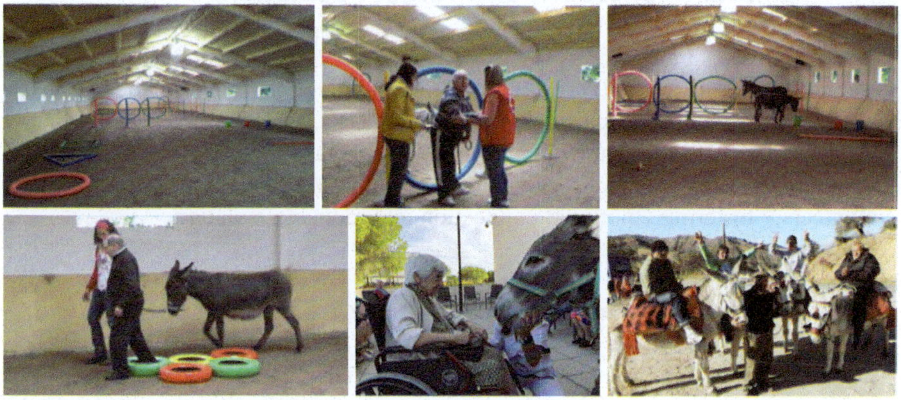

Fig. 17 Donkey mediated therapeutic interventions

8.1 Historical and Scientific Foundations

The therapeutic use of animals in clinical settings dates back to the late eighteenth century, notably with William Tuke's work at the York Retreat, where farm animals were integrated into mental health care to promote emotional stability and humane treatment (Serpell 2006). In the twentieth century, institutions such as Bethel in Germany and Green Chimneys in the United States advanced the field, while Boris Levinson's seminal work in the 1960s introduced the concept of animals as co-therapists in child psychiatry (Levinson 1969). These foundational developments laid the groundwork for structured animal-assisted therapies, including DMT.

8.2 Modalities and Therapeutic Approaches

DMT encompasses a range of modalities that address both physical and emotional rehabilitation. These include:

These approaches are grounded in principles of neuroplasticity, sensorimotor integration, and cognitive stimulation (McKinney et al. 2020).

8.3 Breed Selection and Ethological Suitability

Donkeys used in therapy are selected for their docility, intelligence, and physical characteristics. Breeds such as Miranda (Portugal), Zamorano-Leonés and Encartaciones (Spain), Andalusian, Mediterranean Miniature (USA), and Egyptian donkeys are preferred for their calm demeanor and adaptability (Borgi et al. 2016). Crossbreeds are often employed to balance size and temperament, ensuring safety and comfort for both patients and handlers.

8.4 Clinical Applications and Target Populations

DMT has demonstrated efficacy across a wide spectrum of conditions, including autism spectrum disorder (ASD), attention-deficit/hyperactivity disorder (ADHD), Down syndrome, cerebral palsy, Alzheimer's disease, and post-stroke rehabilitation (Menna et al. 2019; McKinney et al. 2020). The therapy is also used in correctional facilities, schools, and elder care environments, where donkeys act as "social volunteers," fostering emotional connection and reducing isolation.

8.5 Psychomotor and Neuropsychological Benefits

The donkey's rhythmic gait and three-dimensional movement stimulate balance, muscle tone, and coordination. Interaction with the animal activates the limbic system, enhancing emotional regulation and reducing anxiety. Structured activities improve memory, attention, and language, particularly in children and elderly patients with cognitive decline.

8.6 Communication and Emotional Development

DMT promotes both verbal and non-verbal communication. Patients often show increased self-expression, trust, and self-esteem through interaction with donkeys, whose clear body language provides intuitive feedback (Borgi et al. 2016) (Fig. 18). In Alzheimer's patients, donkeys evoke emotional responses and memory recall, while children with ASD exhibit reduced stereotypical behaviors and improved social engagement.

8.7 Donkeys and Post-Traumatic Stress Disorder (PTSD)

Donkeys have demonstrated particular therapeutic value in interventions for individuals experiencing post-traumatic stress disorder (PTSD), especially among military veterans. Their calm demeanor, nonjudgmental presence, and intuitive respect for personal boundaries create a therapeutic environment conducive to

Fig. 18 Paola De Rose (2011) Drawings of a patient affected of obsessive compulsive disorder (OCD), before the experience and at the end of the therapeutic cycle, 6 months later. Note the increase in detail and the use of color

trauma recovery. Unlike more reactive therapy animals, donkeys offer a stable and emotionally neutral presence that allows patients to engage at their own pace, fostering autonomy and psychological safety (Pendry et al. 2018).

Veterans often report a profound sense of identification with donkeys, citing parallels in loyalty, resilience, and emotional suppression. Both donkeys and veterans are frequently misunderstood and undervalued yet possess deep capacities for connection and healing. The donkey's ability to mirror emotional states without imposing expectations enables a form of silent companionship that is especially beneficial for individuals reluctant to verbalize trauma. One veteran described the experience as "reclaiming some of what PTSD took away," noting that the donkey's presence allowed him to reflect without the pressure of clinical analysis or interpersonal scrutiny (Save Your Ass Long Ear Rescue 2021).

Moreover, donkeys exhibit a form of behavioral feedback that is both subtle and instructive. When confronted with impulsive or emotionally dysregulated behavior, a donkey may disengage calmly, prompting the individual to self-regulate without experiencing shame or rejection. These dynamic fosters insight and accountability, aligning with trauma-informed therapeutic principles that emphasize safety, empowerment, and relational atonement (Pendry et al. 2018). Physiologically, the donkey's calm presence has been observed to influence autonomic regulation, with anecdotal reports suggesting synchronization of heart rhythms between donkey and patient—a phenomenon that may contribute to reductions in anxiety and improvements in emotional clarity (see Fig. 19).

8.8 Donkeys and Children with Emotional and Behavioral Challenges

Donkeys also play a transformative role in therapeutic work with children experiencing emotional, behavioral, or developmental difficulties. Their capacity for reciprocal engagement and emotional attunement makes them ideal partners in interventions aimed at building trust, resilience, and self-regulation. A widely

Fig. 19 First Sgt. John Mangels, C Company, first Battalion, 32nd Infantry Regiment, establishes a relationship with a donkey using his natural instincts Jan. 30. This technique is important since the donkey is bred to please and work hard

documented case from The Donkey Sanctuary in Birmingham, England, illustrates this dynamic: a donkey named Shocks, rescued from severe abuse, formed a therapeutic bond with Amber, a child born prematurely and not expected to walk or talk. Through sustained interaction, both donkey and child experienced significant recovery—Amber now walks and speaks, and Shocks has become a fully integrated member of his herd (The Donkey Sanctuary 2020).

Therapeutic sessions often begin with passive observation, allowing donkeys to initiate contact. This approach respects the autonomy of both child and animal, fostering mutual trust. Activities such as grooming, feeding, and pen cleaning promote empathy, responsibility, and emotional regulation. Children who were previously withdrawn or oppositional often exhibit rapid behavioral improvements, including increased verbalization, cooperation, and self-esteem (Pendry et al. 2018).

Importantly, the experience of interacting with a donkey—an animal significantly larger than the child—can recalibrate fear responses and build psychological resilience. The process of overcoming initial apprehension and forming a bond with the animal rewires neural pathways associated with fear and avoidance, equipping children with adaptive strategies for future challenges. This neurodevelopmental impact is particularly valuable in populations with histories of trauma, neglect, or social exclusion (McKinney et al. 2020). Adolescents who struggle with interpersonal conflict or emotional dysregulation often benefit from the structured, nonverbal feedback provided by donkeys, which encourages prosocial behavior and emotional accountability in a non-threatening context.

8.9 Environmental Design and Psychomotricity Circuits

Therapeutic environments often incorporate covered circuits designed to enhance spatial awareness, coordination, and memory. Donkeys are trained to respond to patients' mobility needs and environmental stimuli, ensuring a safe and enriching experience. These circuits emphasize respect for natural rhythms and multisensory engagement.

8.10 Professional Training and Protocols

Effective DMT requires specialized training in donkey ethology, behavior therapy, and natural horsemanship. Programs typically include modules on communication, therapeutic design, and safety protocols. Sessions are tailored to individual needs, ranging from 30 minutes to several hours depending on therapeutic goals.

8.11 Research and Evaluation

Empirical studies using tools such as ADAS-COG, Rey-Osterrieth Complex Figure Test, and verbal fluency assessments confirm significant improvements in motor praxis, cognitive function, and emotional regulation following DMT interventions (Menna et al. 2019; McKinney et al. 2020). These findings support the therapy's efficacy and underscore the importance of individualized, flexible protocols.

9 Donkeys and Mules as War Companions: From Sumerian Chariots to Modern Combat

Donkeys and mules have long served as indispensable allies in warfare, valued not for their speed or aggression, but for their endurance, intelligence, and adaptability. Their role as transport animals spans thousands of years, from the dawn of civilization to contemporary military operations.

9.1 Ancient Origins: The Standard of Ur and the Kuga Hybrids

The earliest known military use of donkeys dates back to ancient Sumeria, where the domestication of the donkey coincided with the invention of the wheel (Hamblin 2006). The *Standard of Ur*, a Sumerian artifact dated to 2600–2400 BCE, depicts chariots pulled by four donkeys—or more precisely, *kugas*, hybrids between domesticated donkeys and wild onagers (British Museum 2014). These animals were bred for strength and speed, ideal for pulling war-carts in skirmishes between rival city-states (see Fig. 20).

In the "War Panel" of the Standard, donkeys trample enemies while infantry wield spears and axes. Captured soldiers are paraded before a king holding a spear, underscoring the donkey's central role in early military logistics and symbolism (British Museum 2014).

Fig. 20 The British Museum (2014) "War Pannel" from the Standard of Ur, Excavated by C.L. Woolley. From Ur, southern Iraq, about 2600–2400 BC. British Museum, Room 56: Mesopotamia. 'War' shows one of the earliest representations of a Sumerian army. Chariots, each pulled by four potential *kugas* or tamed onagers, trample enemies; infantry with cloaks carry spears; enemy soldiers are killed with axes, others are paraded naked and presented to the king who holds a spear (upper row) and "Peace Pannel" from the Standard of Ur, Excavated by C.L. Woolley. From Ur, southern Iraq, about 2600–2400 BC. British Museum, Room 56: Mesopotamia. The 'Peace' panel depicts animals, fish and other goods brought in procession to a banquet. Seated figures, wearing woollen fleeces or fringed skirts, drink to the accompaniment of a musician playing a lyre. Banquet scenes such as this are common on cylinder seals of the period, such as on the seal of the 'Queen' Pu-abi, also in the British Museum (lower row)

9.2 Classical Exploitation: Bronze Age to Rome

Syrian armies used donkeys to pull war-carts and carry supplies during the Early Bronze Age. Later, Greek and Roman forces employed both donkeys and mules for pack work, riding, and chariot-pulling. Roman campaigns in desert regions posed logistical challenges, particularly in transporting water. Pompey reportedly ordered

Fig. 21 Donkeys and mules performing warfare tasks depicted in Trajan Column (Rome, Italy)

water to be carried in 10,000 skins across arid terrain, while Herod supplied Roman troops with water via donkey caravans during their march across the Sinai (Roth 1998).

Roman art and coinage reflect this reliance. Trajan's Column (completed in AD 113) shows pack animals supporting military campaigns in Dacia. Reliefs of the *carroballista*—a cart-mounted artillery weapon—also depict donkeys and mules in logistical roles (Adame Lorite 2012; VRoma Project 2012) (Fig. 21).

9.3 Modern Warfare: From Gallipoli to Afghanistan

Donkeys remained vital in modern conflicts, particularly in inaccessible terrain. During World War I, they were widely used by Australian and British forces in the Middle East. Lt. Richard Henderson famously used a donkey to carry wounded soldiers at Gallipoli, a moment immortalized in wartime photography (Gsl 2011) (Fig. 22).

In the East Africa Campaign (1916–1917), 34,000 donkeys were deployed, but only 1042 survived—many perished from tsetse fly or arsenic poisoning (Cooper 2000). In the Near East, 8000 donkeys helped build roads from Jaffa to Jericho under General Allenby. Their resilience in desert conditions proved invaluable to Allied forces.

Even today, donkeys are used in Afghanistan by both Allied forces and insurgents. Elite Special Operations units rely on them in mountainous terrain where vehicles cannot operate. Their instinct to freeze when frightened—rather than flee—is seen as an asset in stealth operations. Handlers often develop deep bonds with their animals, trusting their judgment in dangerous environments (Special Forces Use of Pack Animals 2004).

Fig. 22 Donkeys used in warfare actions during WWI and WWII

Fig. 23 Animals in War memorial, Hyde Park, London, UK. Sculptor David Backhouse

9.4 Memorial and Recognition

The enduring legacy of these animals is now recognized in monuments like the *Animals in War Memorial* in Hyde Park, London. Bronze statues of a donkey and mule, burdened with World War I equipment, stand as tributes to the hundreds of thousands of animals who served and suffered in human conflicts (Bough 2011a) (Fig. 23).

10 Donkeys and Mules as Spiritual Companions in Religion and Myth

Donkeys and mules have not only served humanity in physical labor and transport—they have also carried profound symbolic and spiritual significance across millennia. Revered, reviled, and ritualized, these animals have walked alongside gods, prophets, and pilgrims, embodying both divine humility and cosmic mystery.

10.1 Sacred Origins and Funerary Rituals

The spiritual role of donkeys dates back over 6000 years. In Archaic-period Egypt, donkeys were buried alongside elite humans, interred as “funerary furnishings” to accompany the deceased into the afterlife (Way 2011). The discovery of ten complete donkey skeletons at Abydos underscores their sacred status—possibly as offerings to gods or as companions for eternity.

Originally associated with the desert god Seth, the donkey symbolized virility and power, with its erect ears and tail representing sexual potency (Bulliet 2005). Yet as Seth’s image darkened into one of chaos and evil, so too did the donkey’s reputation. By the Middle Kingdom, donkeys were deemed “unclean,” and ritual scapegoating emerged: villagers loaded their sins onto a donkey and cast it to its death—a practice that persisted into early Coptic Christianity (Fig. 24).

10.2 Donkeys in Jewish Tradition

In contrast, ancient Semitic tribes revered the donkey for its resilience in desert life. Tacitus noted that the Israelites worshipped the donkey, recognizing its vital role in survival, especially in water transport. The Hebrew Bible reflects this reverence: donkeys are symbols of peace, prophecy, and divine presence.

The story of Balaam’s ass (Numbers 22:21–35) is particularly striking. This donkey, the only animal in the Bible to speak, perceives an angel invisible to Balaam and intervenes to protect him. Jewish tradition holds that this donkey was preserved for the Messiah’s arrival, highlighting its prophetic and spiritual stature.

The Petter Chamor (Hebrew: פטר חמור) or Redemption of the firstborn, is a mitzvah in Judaism in which a male firstborn (bechor) donkey is redeemed by the owner of the donkey, who gives a lamb or kid to a Kohen. The mitzvah of petter chamor requires redeeming a firstborn male donkey by giving a sheep, goat, or item of equal value to a cohen; otherwise, the donkey must be killed through *arifah*. Unlike firstborn kosher animals, which gain permanent *kedusha* and cannot be redeemed, the donkey’s sanctity disappears entirely once redeemed. The mitzvah parallels *pidyon haben*, sharing rules about ownership by kohanim or levi’im but applying only to donkeys.Unredeemed donkeys remain prohibited for use throughout their lives and must be buried if they die. Chazal explain this unique sanctity as a reward for donkeys carrying the wealth of Israel out of Egypt, teaching the value of hakaras hatov (gratitude) (refer to Fig. 25).

Fig. 24 Abydos donkey burials, A vignette from the Book of the Dead of Chensumose, priest of Amun-Ra. Three demons bear knives, while Chensumose confronts a trussed donkey. Papyrus. 21st Dynasty (1080–960 BCE), Third Intermediate Period, Egypt. During the New Kingdom, the "Donkey of Set" was ritually killed with a lance at the festival of Osiris [Papyrus]. and Leyden Papyrus "Figure with human body and Seth-animal head. His arms are stretched to the left and right and keeping big man standards. On his chest is 'Seth' in Coptic letters, he wears a short skirt." Greco-Roman Period, Roman, fourth century (CE 300/350) The Dutch National Museum of Antiquities (Rijksmuseum van Oudheden). In Hans Dieter Betz, you can find a translation to the spell PGM XII. 449–52, which is a "spell of separation", in which the goal is to "Separate NN, born of NN, from NN, born of NN!"

Fig. 25 Petter Chamor or Redemption of the firstborn and Balaam's speaking donkey

10.3 Donkeys in Islamic Lore

Islamic tradition is rich with tales of talking donkeys. The Prophet Muhammad's personal donkey, Ya'fūr (also rendered as Yofir or `Ufayr), often ridden without harness, was said to be the last in a prophetic lineage stretching back to Noah. According to legend, Ya'fūr spoke to Muhammad, recounting its mistreatment by previous owners and declaring its destiny to serve only the final prophet. After Muhammad's death, Ya'fūr reportedly committed suicide by leaping into a well—an act symbolizing the end of a sacred lineage.

Donkeys were also used in the spread of Islam, carrying provisions for armies and pilgrims alike. Shi'ite Imams rode donkeys to emulate prophetic humility, and figures like Ibn Tumart, founder of the Almohad movement, insisted on riding donkeys as spiritual symbolism (refer to Fig. 26) (Sindawi 2006).

10.4 Christian Symbolism and Messianic Imagery

Christianity inherited and transformed donkey symbolism. The donkey became a symbol of humility and divine grace, most famously in Jesus' triumphal entry into Jerusalem. This act fulfilled messianic prophecy and contrasted with the warhorse of conquerors—Jesus rode a humble donkey, signaling peace and spiritual kingship.

Legends surrounding Jesus' birth also feature donkeys: one bearing Mary to Bethlehem, another accompanying the flight to Egypt. The cross-shaped marking on a donkey's back is said to be the shadow of the crucifixion—a living emblem of Christian faith (Fig. 27).

Fig. 26 Depictions of donkeys across Islamic literary and artistic representations

Fig. 27 The role of donkeys in Christianity

10.5 Donkeys in Sanskrit and Buddhist Thought: Symbols of Simplicity, Reversal, and Enlightenment

While donkeys are often overlooked in the grand tapestry of Eastern religious iconography, their presence in Taoist, Buddhist, and Sanskrit traditions reveals a subtle but profound spiritual symbolism. Far from being mere beasts of burden, donkeys have served as metaphors for humility, reversal of ego, and the journey toward inner truth.

One of the most iconic figures in Chinese spiritual folklore is Zhang Guo Lao, a Taoist immortal from the Tang Dynasty. Known for riding his white donkey backwards, Zhang Guo Lao embodied the Taoist principle of paradox: that true progress often requires reversal, and that wisdom lies in simplicity and detachment (Werner 1986; Maspero 1981).

Zhang's reversed posture on the donkey was not mere eccentricity—it was a living parable. In Taoist philosophy, the path to enlightenment often involves unlearning societal norms and returning to one's original nature. Riding backwards symbolized the rejection of worldly ambition and the embrace of spiritual introspection.

When Zhang Guo Lao's donkey drank wine and collapsed, he revealed it was made of paper—animated by magic. This tale, told to the Tang emperor, illustrates the Taoist teaching that illusion and reality are intertwined, and that only truth endures.

Zhang's donkey, which he could fold and store in a box, then revive with a splash of water, also reflects the Taoist belief in impermanence and transformation. The donkey becomes a vessel for mystical travel and philosophical reflection.

In Sanskrit literature and Buddhist allegory, donkeys appear less frequently than elephants or horses, yet their symbolic presence is no less meaningful.

The Sanskrit word for donkey is *khara* (खर), which appears in classical texts such as the *Ramayana*, where the demon Khara is named after the animal—suggesting a dual nature of stubbornness and brute force. However, in later poetic and didactic literature, the donkey is used to represent ignorance that can be transformed through wisdom.

In Buddhist parables, the donkey is sometimes portrayed as a creature of burden that, despite its lowly status, carries the weight of karma and samsara. Its slow, deliberate gait is likened to the meditative path—steady, grounded, and persistent.

Zhang Guo Lao's fish drum—a bamboo cylinder with iron rods—was his constant companion, used to sing songs about Taoist philosophy. These lyrics warned against the pursuit of fame, desire, and material gain, urging listeners to "return to the truly original self." The donkey, in this context, was not just a mount but a spiritual partner in spreading wisdom (Fig. 28).

Fig. 28 A painting depicting Balaram, Krishna's elder brother, battling wild donkeys. On the left, Zhang Guo Lao riding his white donkey backwards. On the right, a painting depicting Balaram, Krishna's elder brother, battling wild donkeys

10.6 From Sacred to Stigmatized

Despite these spiritual associations, Greco-Roman traditions cast donkeys in a negative light. Linked to Dionysus and Priapus, they became symbols of lust, foolishness, and servitude. This duality—sacred in some traditions, profane in others—shaped Western attitudes and contributed to the neglect and abuse of donkeys in later centuries.

Yet literature often redeems them. In Cervantes' *Don Quixote*, Sancho Panza's donkey, Dapple, is a loyal companion and moral mirror. In Jiménez's *Platero y yo*, the donkey becomes a confidant and philosophical partner, embodying serenity and wisdom.

11 The Donkey as a Political Companion: The Birth of the Democratic Emblem

The donkey (*Equus asinus*), long associated with humility and labor, occupies a symbolic role in American political iconography. Its transformation into the emblem of the U.S. Democratic Party reflects a blend of satire, populism, and visual rhetoric that began in the early nineteenth century and was later cemented by the work of political cartoonists.

11.1 Origins in Jacksonian Politics

The association between the donkey and the Democratic Party began during the 1828 presidential campaign of Andrew Jackson. Opponents mockingly referred to Jackson as a "jackass," intending to highlight his populist defiance and perceived stubbornness. Rather than reject the label, Jackson embraced it, incorporating the donkey into his campaign imagery to symbolize his commitment to the common citizen and resistance to elite institutions such as the National Bank (Kampion 2025).

11.2 The Role of Thomas Nast

Although Jackson popularized the donkey as a campaign symbol, it was the cartoonist Thomas Nast who solidified its role as the visual emblem of the Democratic Party. Nast first used the donkey in the January 15, 1870 issue of *Harper's Weekly* in a cartoon titled *A Live Jackass Kicking a Dead Lion*, which depicted the donkey as a stand-in for the anti-war "Copperhead" Democrats (Degg 2020). Nast's repeated use of the donkey in subsequent cartoons—often to critique Democratic positions he viewed as passive or indecisive—embedded the image in the public consciousness. By the 1880s, the donkey had become a widely recognized symbol of the Democratic Party, though never officially adopted by the party itself.

Fig. 29 The donkey as the democrat emblem (Nast 1980; Paraskevas 2011)

11.3 Symbolic Interpretations

The donkey's symbolism has evolved over time. To Republican critics, it represents stubbornness and ineffectiveness. Yet for Democrats, it has come to embody humility, perseverance, and loyalty—qualities that resonate with the party's historical emphasis on working-class values and grassroots activism (Kampion 2025). The contrast with the Republican elephant, also popularized by Nast in 1874, highlights the symbolic divide between the two major U.S. political parties (Fig. 29).

12 Companions to the Afterlife

Donkeys have long occupied a liminal space in human culture—revered as indispensable companions and burden-bearers, yet also entangled in ritual sacrifice and symbolic ambivalence. Their presence in funerary contexts across ancient civilizations reveals a rich and complex narrative of utility, reverence, and transformation.

12.1 Sacred Burials in Ancient Egypt

In a landmark excavation at Abydos, Egypt, archaeologist Fiona Marshall and her team uncovered ten donkey skeletons interred in a royal mortuary complex dating to around 3000 BCE, during Egypt's First Dynasty (Rossel et al. 2008). These animals were buried in mud-brick tombs with wooden coverings, and their bones showed clear signs of vertebral stress and joint wear—evidence of their use as beasts of burden. Despite their utilitarian role, the donkeys retained many wild traits, suggesting that domestication was still in its early stages and proceeded more gradually than previously assumed (Marshall 2008).

Comparative analysis with African wild asses and modern donkeys confirmed their Nubian lineage and revealed that domestication preceded major skeletal or genetic changes. Their burial in a royal context implies high social value, likely linked to their role in enabling long-distance trade and food distribution in early state formation (Rossel et al. 2008; Marshall 2008).

12.2 Symbolism and Sacrifice in Egyptian Rituals

Over time, the donkey's image shifted in Egyptian religious thought. In the Sokar-Wesir festivals—rituals celebrating Osiris's resurrection—donkeys were symbolically destroyed alongside serpents, both seen as manifestations of Apep, the embodiment of chaos (Vandenbeusch 2017). This ritual, performed by priests behind closed doors, reflected the donkey's association with Seth, a deity linked to disorder and the desert. The donkey's symbolic inversion from sacred companion to threat underscores the fluidity of its cultural role (EgyptMythology 2024).

12.3 Hyksos Integration and Cultural Hybridity

During the Hyksos period, Near Eastern migrants introduced burial customs that included donkey sacrifices at tomb entrances, particularly in the city of Avaris (Tell el-Dab'a). Excavations led by Manfred Bietak revealed Canaanite-style tombs with donkey remains, indicating a fusion of Egyptian and Asiatic funerary traditions (Bietak 2010). These practices likely signified status and spiritual passage, echoing Mesopotamian customs and reinforcing the donkey's role as a symbolic and practical companion in death.

12.4 Roman Funerary Feasts: From Sacred to Sustenance

By the Roman era, the donkey's role had shifted again. While still depicted in funerary iconography, such as sarcophagi reliefs, it was no longer solely a spiritual escort. Roman funerals (funus) included ritual feasts where sacrificial animals—including goats, calves, and occasionally donkeys—were consumed by mourners (Fife 2012; Mitchell 2018). This transformation from sacred companion to ritual sustenance reflects the donkey's mutable status across time and cultures (Fig. 30).

13 Conclusions

Donkeys and mules embody a unique intersection of practicality, resilience, and symbolic meaning in human culture. Once regarded primarily as working animals, they are now increasingly recognized for their companionship, therapeutic value, and ability to integrate into family life. Their intelligence, adaptability, and emotional sensitivity allow them to serve not only as pets but also as guardians in livestock systems, partners in education, and mediators in therapy for vulnerable populations. Historically, these equids have shaped human societies through their roles in transport, agriculture, warfare, and religion, leaving an enduring cultural and political legacy. From the Sumerian battlefield to modern therapeutic centers, and from sacred rituals to political symbolism, donkeys and mules have continually adapted to human needs while maintaining their distinct identities. To fully realize

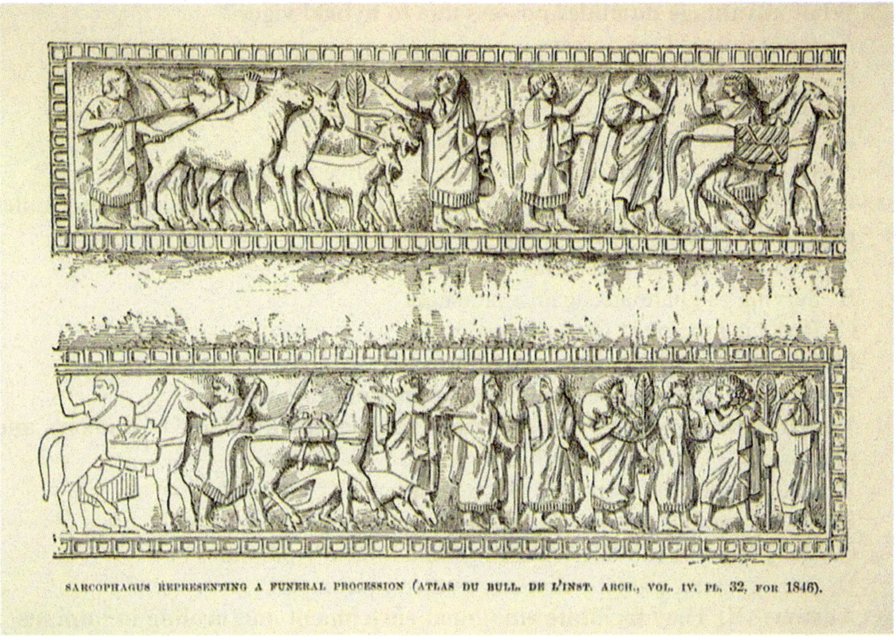

Fig. 30 Tracing of the relief of a Roman sarcophagus representing a funeral procession among which we can observe the presence of donkeys

their potential as companions, owners must embrace ethical responsibility, ensuring welfare through proper housing, nutrition, healthcare, and social interaction. Recognizing both their ancient heritage and modern relevance reframes donkeys and mules as more than beasts of burden: they are intelligent, empathetic, and sustainable companions whose partnership with humans continues to evolve across time, space, and culture.

Self-Evaluation Questionnaire

1. **Which characteristic of donkeys is often misinterpreted as stubbornness but actually reflects intelligence and self-preservation?**
   A. Aggressiveness
   B. Calm temperament
   C. Strong self-preservation instincts
   D. Lack of social bonds
   Answer: (C) Strong self-preservation instincts

2. **What advantage do mules possess due to hybrid vigor?**
   A. Reduced need for training
   B. Combination of donkey resilience and horse strength
   C. Greater reproductive efficiency
   D. Natural ability to resist all diseases
   Answer: (B) Combination of donkey resilience and horse strength
3. **In addition to companionship, what functional role can donkeys and mules play in livestock systems?**
   A. Acting as feed supplements
   B. Serving as guardians against predators
   C. Reducing the need for fencing
   D. Increasing reproductive rates of herds
   Answer: (B) Serving as guardians against predators
4. **Which of the following best describes the therapeutic value of donkeys and mules?**
   A. They reduce the workload of veterinarians
   B. They facilitate emotional enrichment and healing in humans
   C. They serve as substitutes for dogs in all therapy programs
   D. They prevent zoonotic diseases in therapy contexts
   Answer: (B) They facilitate emotional enrichment and healing in humans
5. **What is one essential responsibility of ownership emphasized in the chapter?**
   A. Minimizing social interaction to prevent dependency
   B. Providing adequate nutrition, healthcare, and housing
   C. Training them only for agricultural tasks
   D. Ensuring rapid reproduction for sustainability
   Answer: (B) Providing adequate nutrition, healthcare, and housing
6. **Beyond practical uses, what broader significance do donkeys and mules hold in human culture?**
   A. They symbolize economic decline
   B. They are primarily sources of luxury goods
   C. They serve as cultural, spiritual, and political symbols
   D. They are considered irrelevant in modern society
   Answer: (C) They serve as cultural, spiritual, and political symbols

References

Adame Lorite JF (2012) Trajan's Column
Andelt WF (2004) Use of livestock guarding animals to reduce predation on livestock
Animal Rahat (2012) Vaccination campaign against tetanus in Udgaon, Maharashtra
Apuleius (1994) The golden ass. Translated by Robert Graves. Penguin Classics
Artemiou E, Hutchison P, Machado M, Ellis D, Bradtke J, Pereira MM, Carter J, Bergfelt D (2021) Impact of human–animal interactions on psychological and physiological factors associated with veterinary school students and donkeys. Front Vet Sci 8:701302
Bessa AM, McLean AK (2024) Advances in donkey and mule research. Animals 14(15):2238

Bietak M (2010) Avaris: capital of the Hyksos
Boocock H, Flyps J, Escobar A, Redondo JI, Taylor PM, Gozalo-Marcilla M, Johnston GM, Sullivan R (2025) Donkey and hybrid anaesthetic mortality in a multicentre cohort study. Animals 15(13):1880
Borgi M, Loliva D, Cerino S, Chiarotti F, Venerosi A, Bramini M et al (2016) Effectiveness of a standardized equine-assisted therapy program for children with autism spectrum disorder. J Autism Dev Disord 46(1):1–9
Bough J (2011a) Animals in war. In: Book 2, Chapter 10
Bough J (2011b) Donkeys and religion. In: Book 2, Chapter 10
British Museum (2014) War panel from the standard of Ur
Bulliet RW (2005) Hunters, herders, and hamburgers: the past and future of human-animal relationships. Columbia University Press
Cooper J (2000) Donkeys in the near east campaign
Degg DD (2020) 150 years ago: Thomas Nast & Democrat donkey. The Daily Cartoonist
EgyptMythology (2024) The sacred donkey: the importance of set and his connection to the desert
European Commission (2022) Factsheet: hoof care and training practices for donkeys and hybrids
Fife S (2012) The Roman funeral
French JM (1993) Assessment of donkey temperament and the influence of home environment. Appl Anim Behav Sci 36(2–3):249–257. https://doi.org/10.1016/0168-1591(93)90014-G
Gichure MN, Olayide O (2024) Selected intervention strategies to improve health and welfare of working donkeys in Kenya: a narrative review. Asian J Res Anim Vet Sci 7(1):69–82
Gsl (2011) Lt. Richard Henderson and the Gallipoli Donkey
Hamblin WJ (2006) Warfare in the ancient world
Kampion H (2025) The donkey and the elephant. In: Our white house: looking in, looking out. Candlewick Press
Kampion H (2025) The donkey and the elephant. Our White House | Looking In, Looking Out. https://ourwhitehouse.org/the-donkey-and-the-elephant/
Levinson BM (1969) Pet-oriented child psychotherapy. Charles C. Thomas
Marshall F (2008) Egypt's revered donkeys
Maspero H (1981) Taoism and Chinese religion (p. 259). Amherst: University of Massachusetts Press
McKinney C, Mueller MK, Frank N (2020) Equine-assisted activities and therapies: a systematic review of empirical literature. J Altern Complement Med 26(6):499–512
McLean AK (2019) Donkey and mule behavior. Vet Clin North Am Equine Pract 35(3):493–510
McLean AK, Heleski CR (2010) Using science to improve equine welfare: the importance of ethology. Vet Clin North Am Equine Pract 26(3):481–491
McLean AK, Navas González FJ, Canisso IF (2019) Donkey and mule behavior. Vet Clin North Am Equine Pract 35(3):493–510
Menna F, Santaniello A, Todisco M, Amato A, Borrelli L, Scandurra C, Fioretti A (2019) The human–animal relationship as therapeutic intervention for children with autism spectrum disorder: a pilot study. J Vet Behav 33:17–24
Mitchell P (2018) The triumph of the mule. In: The donkey in human history: an archaeological perspective. Oxford University Press
Nast T (1870) A live jackass kicking a dead lion. Harper's Weekly
Paraskevas M (2011) Name calling: Old labels don't work anymore, at least for Democrats [Cartoon]. Dan's Papers.
Pendry P, Smith AN, Roeter SM (2018) Randomized trial examines effects of equine facilitated learning on adolescents' basal cortisol levels. Hum-Anim Interact Bull 6(1):1–19
Raw Z, Collins JA, Burden FA (2024) What is a working equid? Analysis of terminology and welfare implications. Animals 14(14):2026
Rossel S, Marshall F, Peters J, Pilgram T, Adams M, O'Connor D (2008) Domestication of the donkey: timing, processes, and indicators. Proc Natl Acad Sci 105(10):3715–3720
Roth J (1998) Roman military logistics in desert campaigns

Save Your Ass Long Ear Rescue (2021) Veteran testimonials and donkey therapy
Serpell J (2006) Animal-assisted interventions in historical perspective. In: Fine A (ed) Handbook on animal-assisted therapy. Academic, pp 3–20
Sindawi K (2006) Shi'ite symbolism and the donkey in Islamic tradition
Smith C (2018) The modern donkey: a guide to care and companionship. Oxford University Press
Special Forces Use of Pack Animals (2004) Military field manual
Starkey P (2020) Animal traction: constraints and impact among African households. In: Rural households in emerging societies. Routledge, pp 77–90
Sullivan R (2024) Hepatic disease and coat management in donkeys. Equine Vet Educ 37(2)
Svendsen ED (2008) The professional handbook of the donkey, 4th edn. Whittet Books Ltd
The Donkey Sanctuary (2020) Shocks and Amber: a healing bond
The Donkey Sanctuary (2020) Clipping your donkey: Essential care advice. The Donkey Sanctuary. https://www.thedonkeysanctuary.org.uk/for-owners/owners-resources/clipping-your-donkey
Thiemann AK (2024) Gastrointestinal disorders in donkeys. UK-Vet Equine 8:66–74
Thiemann AK, Poore LA (2019) Hoof disorders and farriery in the donkey. Vet Clin N Am Equine Pract 35(3):643–658
Vandenbeusch M (2017) Following Seth's path? The Ambivalent Nature of Ancient Egyptian Donkey-Headed Deities
VRoma Project (2012) Roman coinage and carpentum imagery
Way KC (2011) Donkeys in the biblical world: ceremony and symbol. Eisenbrauns
Werner K (1986) Perspectives on Indian religion: Papers in honour of Karel Werner (P. Connolly, Ed.; Bibliotheca Indo-Buddhica, No. 30, 253 pp.). Sri Satguru Publications. Original from the University of Michigan. Digitized June 20, 2009. ISBN 8170300282; 9788170300281.

Part C

Husbandry & Functional Aspects

Saddle Fit in Mules and Donkeys

David Alonso

Abstract

Proper saddle fit in mules and donkeys requires a nuanced understanding of their unique anatomical structures and functional needs. Often mischaracterized as difficult to ride or equip, these animals possess distinct morphological traits—such as a straighter back, narrower thoracic cavity, and more upright scapula—that set them apart from horses and necessitate specialized tack. This text examines those differences in depth, offering practical guidelines for selecting and fitting saddles, pads, girths, bridles, and other equipment in ways that minimize discomfort and injury. Emphasis is placed on choosing appropriate materials, respecting behavioral sensitivities, and implementing thoughtful training and handling protocols. A detailed saddling sequence is presented alongside strategies to address common behavioral challenges, all grounded in the principles of animal welfare and functionality. Through a multidisciplinary perspective combining biomechanics, traditional knowledge, and contemporary equestrian practice, this work reframes mules and donkeys as capable, high-value partners in recreational and working contexts, deserving of equipment and treatment tailored to their specific needs.

Keywords

Hybrid conformation · Equine equipment adaptation · Animal welfare · Morphological differences · Training protocols · Ethological considerations · Rein control · Desensitization · Equitation safety · Behavioral management · Saddle placement · Riding stability · Forage conversion · Traditional tack · Back structure · Functional biomechanics · Pressure points · Tack materials · Equestrian disciplines · Load-bearing capacity

D. Alonso (✉)
Doma Natural, Brio Natural Equestrian Center, Arteaga, Coahuila, Mexico
e-mail: ventas@brionatural.co

A. K. McLean et al. (eds.), *Donkey and Mule Medicine: Foundations, Breeds, and Management*, https://doi.org/10.1007/978-3-032-12726-6_13

1 Introduction

Properly fitting a saddle on mules and donkeys presents unique challenges and opportunities that differ markedly from those encountered with horses. Often overshadowed in mainstream equestrian discourse, these hybrid and donkey breeds exhibit a combination of traits—such as a straighter back, narrower thoracic cavity, shorter lumbar region, and more upright scapular alignment—that demand equipment specifically shaped to their conformation. As their roles expand beyond traditional draft and pack work into recreational riding, competitive sports, and therapeutic programs, understanding these anatomical distinctions becomes critical not only for performance but also for long-term soundness and welfare.

A foundational review of morphological and biomechanical principles establishes how skeletal and muscular geometry influences saddle placement, pressure distribution, and rider balance. By comparing vertebral counts, rib cage dimensions, and limb angulations between horses, mules, and donkeys, we gain insight into why standard equine saddles often create hotspots or impede natural motion in these hybrids. This section also introduces basic principles of load transfer and dynamic stability, highlighting why even small misalignments can lead to discomfort, muscle atrophy, or unsoundness over time.

Next, common pitfalls in equipment selection are examined in detail. This includes an evaluation of tree materials and shapes—rigid versus semi-flexible, wool flocked versus foam panels—and pad configurations that provide optimal shock absorption without excessive bulk. Special attention is given to how girth design, crupper placement, and breastplate attachments can either exacerbate or relieve pressure under different riding and driving scenarios. Recommendations for breathable, moisture-wicking fabrics and corrosion-resistant hardware underscore the importance of durability alongside comfort.

Since donkeys and mules often display heightened sensitivity to tactile stimuli and novel tack, the introduction also addresses ethological considerations. Desensitization techniques—ranging from systematic "tactile flooding" to positive-reinforcement clicker training—are presented to minimize stress responses during saddle fitting. Handling protocols for safe, low-stress leading and mounting positions ensure that both animal and handler develop confidence and clear communication throughout the process.

A detailed saddling sequence then guides the reader step by step: from symmetrical pad placement and incremental cinch tensioning to final bridle adjustment, cheekpiece fitting, and reins alignment.

Each stage includes practical checklists—such as ensuring two to three fingers' clearance at the withers and a smooth spine channel—to verify correct fit before mounting (Fig. 2). Interwoven throughout are discipline-specific tips, whether for western trail riding, English flat work, driving harnesses, or traditional pack saddles.

Finally, maintenance and troubleshooting strategies help practitioners adapt as animals gain condition, undergo conformation changes with age, or assume new workloads. Guidance on routine tack inspection, panel reflocking, and minor on-the-spot adjustments ensures that fit remains optimal between professional saddle-fitter visits. By blending anatomical analysis, material science, behavioral insight, and hands-on methodology, this expanded introduction sets the stage for a comprehensive exploration of saddle fit in mules and donkeys—affirming that equipment designed around the animal's body and mind is essential for longevity, comfort, and peak performance.

2 Selecting a High-Quality Riding Mule

2.1 The Modern Reappraisal of Mules

In recent years, equestrian culture has undergone a notable reassessment of the mule. Once dismissed as a mere beast of burden, the mule is now recognized for its unique blend of resilience, intelligence, and athletic potential. This paradigm shift is fostering the development of new equestrian niches, stimulating research in animal behavior, and promoting more ethical, science-based approaches to training and breeding. As these trends gain momentum, the mule's historically marginalized status is gradually being overturned.

2.2 Historical Breeding Priorities and Practices

For centuries, mule breeding was driven by utilitarian goals: breeders prioritized size and strength to meet the demands of agriculture, transport, and military logistics. Phenotypic traits such as balanced conformation, efficient gait mechanics, and amenable temperament were often overlooked. Training methods were typically based on force, fatigue, or dominance. However, well-trained mules—particularly those capable of performing riding, pack, and draft tasks—were historically prized, sometimes commanding the value of three horses. Despite these rare exceptions, the onset of the Industrial Revolution drastically reduced the mule's utility, limiting demand to niche agricultural and remote operations.

2.3 The Renaissance of the Riding Mule

Over the last 15 years, a revival of interest in mules has sparked a transformation. Increasing participation in recreational riding and competitive disciplines has prompted both the refinement of breeding practices and the development of more humane training systems. Specialized donkey breeds are now being cultivated to enhance traits that are particularly suited for saddle work. Selection criteria have evolved to include three core attributes:

- **Physical appearance** (conformation, size, and breed type)
- **Locomotor quality** (fluid gaits, joint articulation, and movement efficiency)
- **Work disposition** (trainability, temperament, and endurance)

In countries like Brazil and Colombia, mules with refined conformation and quality gaits are now fetching high prices on the market. Similarly, in Mexico and the United States, mules are gaining recognition in charro events and western-style competitions. Nevertheless, knowledge gaps remain—particularly in terms of optimal sire-dam pairings—sometimes resulting in substandard offspring and, in some cases, the re-emergence of abusive training practices to compensate for poor conformation.

2.4 Genetic Foundations and Breed Influences

The foundation of mule breeding in the Americas traces back to the early colonial period. Spanish donkeys, typically under 1.20 m tall and referred to as "criollos," were introduced for their hardiness, efficient gaits, and robust constitution. A major turning point came in the late eighteenth century when George Washington received a Zamorano–Leonés jack—standing approximately 1.52 m tall, with a black, long-haired coat and robust conformation. This stallion helped revolutionize breeding programs in the New World. Wealthy landowners began to import large European breeds, often circumventing Spanish royal bans on export, thus giving rise to the so-called "mamut" donkeys. While this size increase favored labor-intensive tasks like mining and plowing, the singular emphasis on size came at the expense of functional conformation. Consequently, the more agile and better-balanced criollo types were devalued—a bias that continues to affect many contemporary breeding programs.

2.5 Conformational Considerations for Riding Mules

Large size alone does not guarantee riding quality. In fact, many oversized donkeys exhibit conformational faults such as excessively straight backs, thick and inflexible limbs, and poor shoulder angles—traits that negatively impact gait quality and comfort. Increasingly, breeders and riders are recognizing that **well-proportioned mules with balanced, athletic builds** offer superior performance in the saddle.

Selecting an ideal riding mule requires attention to morphological details that align with the demands of the intended discipline. These include:

- **Head and neck carriage** for balance and responsiveness
- **Back length and topline** for comfort and strength under saddle
- **Limb alignment and joint structure** to support soundness and gait quality
- **Overall body balance** for agility and endurance

The same morphological standards used in evaluating horses for riding purposes should apply equally to donkeys and, by extension, to mules.

2.6 The Importance of Thoughtful Breeding Strategies

Optimal mule production hinges on the **strategic pairing of sire and dam**, guided by the end use of the animal. A focus on discipline-specific traits—whether for trail, endurance, dressage, ranch work, or showmanship—is essential. This includes combining:

- The **hardiness and gait efficiency** of traditional donkey lines (e.g., criollo, Andalusian, or North African)
- The **refinement and athleticism** of selected horse breeds
- Traits like **trainability, docility, and stamina**, which affect long-term usability

Future sections will delve into **recommended crossbreeding schemes**, offering a roadmap for producing versatile, high-performing mules tailored to diverse equestrian needs.

3 Cross-Breeding Strategies for Discipline-Specific Mule and Donkey Hybrids

3.1 The Importance of Discipline-Specific Conformation

Effective mule and donkey hybrid breeding begins with a deep understanding of **conformation traits specific to each discipline**. Traits such as **head and neck carriage**, **limb alignment**, and **body balance** not only influence athletic performance but also affect long-term health, trainability, and rider comfort. Selection must go beyond superficial traits like size and focus on functional anatomy, **ensuring structural harmony between dam and sire**. Applying equine conformation standards to both horses and donkeys is essential for producing versatile, high-performing hybrids.

3.2 Breeding for Gaited and Smooth-Riding Hybrids

Gaited hybrids offer smooth, energy-efficient locomotion that reduces concussive impact and rider fatigue—ideal for long-distance riding, leisure, and show. For these goals, breeders prioritize equine breeds with **natural lateral or diagonal four-beat gaits**.

Equine Breeds:

- **Mexican Criollo:** A tough, agile horse developed in rugged terrains. Criollos display efficient ground-covering strides and stamina, contributing both endurance and frugality to mule offspring.

- **Colombian Paso Fino:** Renowned for their natural four-beat gait ("paso fino"), which is fast, rhythmic, and smooth. These horses transmit gait quality and elegant carriage.
- **Peruvian Paso:** Known for the unique "termino" (lateral shoulder movement) and a gentle, gliding gait. They pass on comfort, soft rein contact, and showy presence.
- **Mangalarga Marchador / Campolina:** Brazilian gaited breeds offering two marching gaits—**marcha picada** and **marcha batida**—with natural rhythm, coordination, and impulsion.
- **Tennessee Walking Horse:** Celebrated for its running walk, this breed contributes ground-covering stride and docility, ideal for hybrid riding mules.
- **Missouri Fox Trotter:** Combines diagonal footfall with smooth motion. These horses transmit a calm, reliable temperament and surefootedness.

Donkey Breeds:
- **Brazilian Mangalarga Pêga:** Bred specifically for gaited hybrids. It contributes smooth movement, an upright posture, and a cooperative attitude.
- **Colombian Criollo Donkey:** Hardy and agile, this regional jack adds endurance and balance, supporting rhythmical and compact motion.
- **Selected Andalusians:** Certain Andalusian lines maintain lightness and flexibility. Their use in gaited crosses improves elegance and front-end lift.
- **Mexican Criollo and Mixteco Donkeys:** Small to medium-sized types adapted to difficult environments, transmitting vigor, resilience, and efficient gaits.

These combinations produce mules with **reduced vertical motion**, minimal jarring for the rider, and enough energy conservation for extended travel or competition.

3.3 Breeding for Speed, Agility, and Cow-Sense

Western disciplines demand animals that are **quick on their feet, mentally alert, and brave around livestock**. The goal is a mule that can anticipate movement and react with precision and minimal cueing.

Horse Breeds:
- **Quarter Horse:** The benchmark for western performance. Its compact build, powerful hindquarters, and quick reflexes make it ideal for roping and reining.
- **Pinto and Appaloosa:** Color breeds with working-horse ancestry. Known for adaptability and often used in ranch work.
- **Argentine Criollo:** The product of centuries of selection for endurance and cattle work. Highly intelligent and capable of long days in harsh terrain.
- **Chilean Corralero:** A specialized cattle horse developed for rodeo-style events. Agile, strong, and deeply responsive to rider commands.
- **Morgan Horse:** A compact and muscular breed known for its versatility, stamina, and animated motion. Morgans bring both cow-sense and elegance to hybrids.

Fig. 1 Mammoth Donkey

Donkey Breeds:

- **Mexican Criollo Donkey:** Alert, surefooted, and fast-moving. These jacks contribute balance, responsiveness, and strong hooves.
- **Mixteco Donkey:** Native to southern Mexico, valued for its strength, endurance, and cooperative temperament.
- **Selected Mammoth Donkeys:** Larger but selectively chosen for agility and limb conformation. They provide the power necessary for western maneuvers (Fig. 1).

These crosses produce hybrids with **rapid acceleration, precise stops and turns, and mental sharpness**, suitable for **competitive or working western applications** (see Fig. 2).

3.4 Breeding for Elegance, Suspension, and Show Presence

Disciplines that prioritize **suspension, elevation, and refined motion**—such as classical dressage or ceremonial presentation—require a different phenotype emphasizing expressive movement and balanced action.

Horse Breeds:

- **Lusitano:** A Portuguese baroque breed prized for its collected gaits, high knee action, and calm under pressure. Ideal for classical disciplines and aesthetically imposing.
- **Spanish Purebred (PRE):** Similar to the Lusitano, these horses offer arched necks, strong hindquarters, and a natural predisposition for dressage.
- **Morgan:** Brings animation, presence, and willingness. Their upright carriage and suspension complement Iberian lines.
- **Costa Rican Criollo:** Known for lightness and proud movement; selectively used in exhibition crosses.
- **Peruvian Paso and Friesian Sport Horse:** Add flair, lift, and dramatism under saddle. The Friesian cross increases presence, while the Paso line smooths the gait.

Fig. 2 Mule Mixteca a cross between local breed of criollo horse and Mexican Criollo donkeys or Mixteco donkey

Fig. 3 Pega donkey a gaited breed of donkey from Brazil

Donkey Breeds:

- **Brazilian Pêga Donkey:** Specifically bred to complement gaited and baroque horses. Offers lift, regularity, and mental calmness (see Fig. 3).
- **Colombian Criollo and Mixteco:** Lighter jack types that enhance front-end action and work ethic.
- **Selected Andalusians:** These may contribute frame and elegance if selected for movement traits.

Fig. 4 Lustanio mare from Portgual (figure 4)

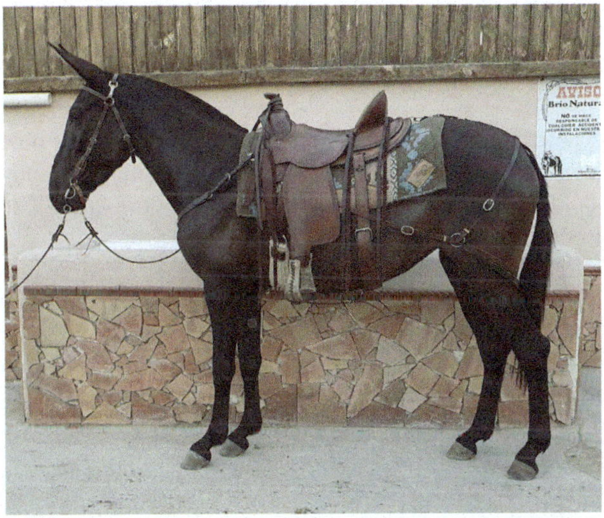

Fig. 5 Lustiano mule out of a Lustiano mare and jack

These hybrids are suitable for riders seeking **highly presentable, stylish mounts with proud carriage and elevated action**, particularly for **parades, exhibitions, and classical flatwork (see** Fig. 5).

3.5 Breeding for English Disciplines and Jumping

Jumping disciplines demand **scope, coordination, and courage**. While mules will naturally exhibit a tighter arc over fences, proper pairing can enhance their ability to excel in **hunter, jumper, or low-level eventing** contexts.

Equine Breeds:
- **Holsteiner:** Offers power and technique over fences. These mares transmit lift, roundness, and rideability.
- **Hanoverian:** Balanced and tractable, ideal for amateur riders needing reliable movement and boldness.
- **KWPN (Dutch Warmblood):** A versatile sport breed known for elegance and impulsion, highly successful in eventing and show jumping.
- **Oldenburg, Selle Français, and Westphalian:** These contribute size, strong toplines, and consistent jumping form, creating athletic and trainable offspring.

Donkey Breeds:
- **"Mammoth" Donkeys (selected lines):** While generally heavy, carefully selected individuals contribute compact strength and straight limbs.
- **Andalusian Donkeys:** Moderately sized and well-proportioned, these donkeys offer smooth gaits and trainability.
- **Mixteco Donkeys:** Often used in sport-mule lines in Mexico; they lend agility and good mental focus.

These crosses yield mules capable of **accurate foot placement, steady contact, and confident navigation of fences**, though jump style may differ from that of purebred horses.

3.6 Breeding for Driving and Harness Work

Harness animals must be **visually imposing, powerful, and even-tempered**. For carriages, processions, or agricultural work, aesthetics and strength are equally important.

Equine Breeds:
- **Percheron:** French draft horses known for endurance and intelligence. They bring calm strength and deep muscling.
- **Belgian:** Large-framed and easygoing, with a steady temperament and great pulling power.
- **Shire and Clydesdale:** English breeds famed for their feathered legs, tall stature, and majestic motion—ideal for ceremonial or public display.
- **Friesian:** Adds refinement, upright movement, and beauty. Popular in both harness and saddle work, they improve visual elegance.

Donkey Breeds:

- **"Mammoth" Donkeys:** Bred for size and strength, often over 1.40 m at the withers. Suitable for agricultural hybrids.
- **Catalán Donkey:** Indigenous to northeastern Spain, valued for its powerful frame and intelligent demeanor.
- **Zamorano–Leonés Donkey:** A long-haired, black-coated breed from northern Spain. Historically used to improve size and work capacity.
- **Andalusian and Egyptian Donkeys:** Moderate in size and active, suitable for lighter carriages or elegant harness teams.

These pairings result in hybrids that are **strong, calm, and capable of prolonged pulling**, yet sufficiently refined for **presentation in formal or touristic settings**.

3.7 Evaluating Pairings: Structural and Behavioral Soundness

Across all breeding strategies, it is critical to:

- **Assess conformation in both parents**, avoiding faults that may impair performance or comfort.
- **Evaluate gait mechanics and stride length**, ensuring suitability to the desired discipline.
- **Consider temperament and trainability**, selecting animals that are mentally stable, responsive, and willing to work.

By implementing these strategies and **tailoring crosses to the functional demands of each activity**, breeders contribute not only to mule quality, but also to the overall **advancement of modern hybrid equestrianism**.

4 Morphological Differences Among Horses, Mules, and Donkeys

Understanding the **morpho-functional differences** among horses, mules, and donkeys is crucial for breeding decisions, tack design, training strategies, and long-term health. Though genetically related, these species diverge significantly in skeletal conformation, musculature, and soft-tissue dynamics, resulting in species-specific requirements for riding and working comfort.

4.1 Cranial Structure and Bit Sensitivity

Horses typically display a proportionate skull with a straight (mesocephalic) or slightly convex profile. The **mandible** is slender, with a wide diastema that accommodates various bit designs with minimal interference.

In contrast, **donkeys** present a pronounced **roman nose** (convex nasal bone), a shorter, broader skull, and a **massive, square jaw** adapted to prolonged grinding of fibrous material. The **diastema** is reduced, and the interdental space is narrower and more reactive to bit pressure.

Mules, being hybrids, exhibit intermediate features—often with a slightly convex profile, a **thicker jawbone**, and **larger nasal passages** that contribute to efficient thermoregulation but complicate bridle fit. Their **reduced diastema and heightened oral sensitivity** demand anatomically fitted, padded or bitless solutions to avoid behavioral resistance or injury.

4.2 Cervical Spine and Poll Mechanics

The **atlas (C1)** and **axis (C2)** govern poll flexibility and vertical/lateral neck movement.

- In **horses**, these vertebrae are elongated and mobile, allowing a generous range of motion essential for bending, collection, and head carriage.
- **Mules** possess **thicker atlas wings** and **shorter axis spines**, which enhance stability but reduce lateral flexion. This gives them a stiffer, more resistant poll—an important consideration when using leverage bits or tight cavessons.
- **Donkeys** exaggerate this rigidity. Their poll region is notably immobile and densely muscled. **Headstalls, halters, or fly masks must avoid pressure on the atlas-axis junction**, and training should focus on gradual desensitization.

Poll tension and stiffness also impact **bridle placement and rein aids**, making it imperative to adapt training and tack to each animal's spinal capacity.

4.3 Withers and Thoracolumbar Region

Withers serve as a saddle anchor and an indicator of spinal slope.

- **Horses** usually have prominent, well-defined withers and a supple thoracolumbar line with good shock absorption. Their backs can accommodate **standard saddle trees** without excessive modification.
- **Mules** exhibit **low, flat, or gently rounded withers** with a straighter, shorter back. The thoracic spine is **more rigid**, which reduces flexion under saddle and disperses weight unevenly if improperly fitted.
- **Donkeys** have nearly **flat backs**, minimal thoracic mobility, and a limited "sway" capacity. Their vertebral processes are **less prominent**, making **saddle bridging and slippage common** without custom panels or cruppers.

Saddle design must account for a broader load distribution and reduced spine flexion in hybrids and donkeys to prevent pressure sores and behavioral resistance.

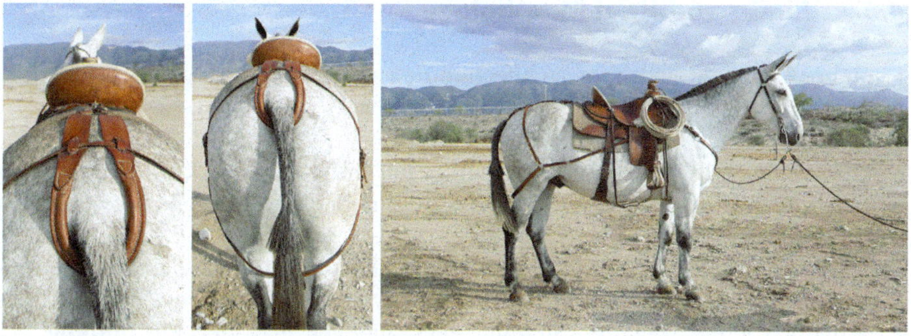

Fig. 6 Mule with American-style crupper

4.4 Ribcage, Barrel and Girth Area

Girth mechanics are profoundly influenced by **thoracic shape**:

- **Horses** possess **rounded, deep barrels** that allow conventional girths to distribute tension evenly along the sternum.
- **Mules** display a **narrower, more cylindrical thorax** with a longer rib cage and a slightly higher girth line. This means standard girths often slide forward or pinch.
- **Donkeys** have the **narrowest ribcage** of all, yet retain depth behind the elbow. Their **sternal keel is less pronounced**, increasing the risk of girths riding up into the axillae.

Breastplates, cruppers, or contoured girths are often required in hybrids and donkeys to stabilize the saddle and avoid shoulder interference (see Fig. 6).

4.5 Pelvic Structure and Croup Geometry

Pelvic shape determines **stride mechanics, jumping capacity, and hind-end engagement**.

- **Horses** typically have a **long, inclined pelvis** and a **smooth croup transition**, facilitating **extended gaits** and **arched jump trajectories**.
- **Mules** exhibit a **shorter, steeper pelvis**, producing powerful engagement but reducing the length of stride and limiting scope over fences.
- **Donkeys** intensify this pelvic structure: **a short, broad pelvis** with a sharply sloped croup improves **traction and stability**, especially on uneven terrain, but **restricts extension and vertical thrust**.

Saddles for hybrids and donkeys must allow **cantle clearance** and **flexible panel contact** to adapt to the steeper slope and avoid pressure points along the lumbar-sacral junction.

4.6 Limbs and Hoof Morphology

Limb conformation and hoof architecture determine **soundness, gait mechanics, and load absorption**.

- **Horses** generally have **four splint bones per limb** and **broad, rounded hooves** suited for diverse terrains. Their hoof wall is moderately thick and can be shod or barefoot.
- **Mules** and **donkeys** often show **only two or three splint bones**, and their hooves are **smaller, more oval or "drip-shaped"**, with **very dense horn tissue**.
- Without routine wear or trimming, these hooves may develop **"tubing"**—a vertical growth pattern that alters breakover and interferes with stride.

Donkeys also **bear weight more caudally**, increasing pressure on the **heel and quarters**. As a result, **farriery protocols must emphasize heel balance, toe angles, and sole concavity**, and may differ substantially from horse practices.

4.7 Functional Implications for Tack and Training

These morphological traits have direct implications:

- **Bits and bridles** must be selected with jaw shape and poll rigidity in mind.
- **Saddles require species-specific trees or adjustable panels** that accommodate flatter backs, narrower girths, and steeper croups.
- **Hoof care** should follow growth patterns and mechanical loading differences in hybrids and donkeys.
- **Training strategies** should recognize differences in spinal mobility and emotional reactivity, especially in donkeys and first-generation mules.

Understanding and respecting these **species-specific anatomical constraints** is essential to promoting **comfort, performance, and welfare**, and to ensuring that **mules and donkeys are not merely treated as small horses**, but as **distinct working partners** with their own functional design.

5 Ideal Equipment for Riding Mules

Mules are anatomically and behaviorally distinct from horses, requiring **specialized tack** that respects their conformational nuances. Improper equipment not only reduces performance but can provoke resistance, discomfort, and long-term injury. Each element of tack—from halter to saddle to cinching systems—must be **precision-engineered** to accommodate the mule's **flat withers**, **narrow thorax**, **sensitive poll and mouth**, and **semi-rigid back**.

5.1 Groundwork Halter (Jáquima de Nudos)

A foundational piece for training and daily handling, the groundwork halter must be light, responsive, and fitted for clarity and comfort.

- **Construction**: Made from a **¼″-diameter nylon cord** braided with **cotton yarn**, it combines strength with softness.
- **Functional Design**: The noseband sits **five inches above the nostrils**, avoiding cartilage compression, while **cheek straps contour the zygomatic arch** to ensure consistent signal transmission.
- **Purpose**: This halter enables **pressure-and-release cues** that communicate intent without force, reducing stress and improving learning outcomes. It prevents the mule's typical "pull away" behavior triggered by sudden pressure.

5.2 Bit Progression and Oral Comfort

The mule's **reduced diastema** and **reactive oral mucosa** necessitate a **graduated bitting strategy**.

- **Stage 1—Snaffle Bit**:
  - **Material**: Mild iron mouthpiece, 3/8″ in diameter.
  - **Cheeks**: Stainless-steel, 4″ long, adapted to the mule's cheek width to avoid pinching.
  - **Effect**: Promotes acceptance and educates the animal in **direct-rein pressure** without encouraging resistance.
- **Stage 2—Curb Bit**:
  - **Mouthpiece**: ½″-diameter iron with a **moderate port** or mullen mouth.
  - **Shanks**: 7–9″ with mild purchase to provide leverage when needed.
  - **Use**: Only for **experienced hands** with soft contact to avoid damaging the bars or tongue. Ideal for refined control once the mule understands basic aids.

A **mouth check** before bitting ensures no ulcers, wolf teeth, or incisor overgrowth interferes with bit comfort.

5.3 Bridle Design and Facial Fit

Bridles must respect the **mule's cranial structure**, including wide-set eyes and large ears.

- **Material**: **Supple bovine leather** (or vegetable-tanned alternatives) with **stainless-steel hardware** to resist corrosion and adjust easily.
- **Ear Accommodation**: Poll strap and browband must **curve around the ear base**, avoiding chafing at the auricular cartilage.

- **Bit Stability**: Cheekpieces should keep the bit **just above the canine groove**, forming **one soft wrinkle** at the lip—neither slack nor tight.

Mule bridles often benefit from **extra adjustment holes** and **split-ear designs** that allow for fine-tuning fit across individual head shapes.

5.4 Padding Systems: Sudadero and Carona

Padding protects the skin, distributes pressure, and regulates temperature.

- **Sudadero (Base Pad,** Fig. 7):
  - **Size**: 34″ × 34″.
  - **Material**: Densely woven **sheep or wool blend**.
  - **Function**: Absorbs sweat and prevents dust abrasion.
- **Carona (Upper Pad)** (Fig. 7):
  - **Size**: 32″ × 32″ or 32″ × 33″.
  - **Material**: Virgin wool, minimum 1″ thick.
  - **Benefits**:
    - **Thermal regulation**
    - **Moisture-wicking up to 30% of weight**
    - **Hydrophobic and flame-resistant**
    - **Naturally antibacterial**
    - **UV-resistant and biodegradable**

Pads must be cleaned regularly to maintain **loft, cleanliness, and resilience**, especially in hot or humid climates.

5.5 Mule-Specific Saddle Design

The saddle is the most critical contact point between rider and mule.

Fig. 7 Wool saddle pad a good option for a natural fiber for the back of a mule or donkey

Fig. 8 Saddle tree unsuitable for the mule based on how the tree does not sit properly on the back of the mule and lacks balance

- **Tree Construction**:
  - **Material**: Laminated wood or polymer for flexibility and strength.
  - **Design Features**:
    - Narrower **gullet width**
    - **Shorter bars** to fit shorter backs
    - **Conical skirts** to contour flanks
- **Panel Shape**:
  - Must mirror the **mule's flat thoracic surface**.
  - Avoid bridging (pressure at front and back only) or rocking (pivoting center).
  - **Shoulder clearance** is essential to avoid impinging on the scapula during movement.
- **Leather Skirts and Flaps**: Custom-fit to flank angle, particularly for **steep croup slopes** and **shorter torsos**.

The use of a **custom-fitted or adjustable tree** is recommended for professional use or high-performance mules (refer to Figs. 8 and 9).

5.6 Cinching System and Saddle Stability

Securing the saddle requires adapting to the mule's **narrow, deep rib cage** and sometimes steep terrain.

Fig. 9 Saddle tree suited to the mule's back as it sits level and evenly makes contact with the mule's back

- **Cincha (Girth)**:
  - **Material**: Sheep or alpaca wool, fleece-lined for airflow.
  - **Fit**: Should lie **posterior to the elbow**, contouring the thoracic barrel without pinching.
- **Barrigüero (Overgirth/Back Cinch)**:
  - **Function**: Prevents cantle lift during descent or collection.
  - **Placement**: Behind the true belly, anchored to the cincha, not floating independently.
- **Breastplate (Pecho-Pretal)**:
  - Prevents backward saddle migration.
  - Should align with the scapula angle and **pass under the neck** through the girth—not over it—to avoid inhibiting forelimb movement.

All components must allow **freedom of motion without rotational slippage**, particularly in trail, endurance, or working conditions (see Fig. 10).

5.7 Downhill Stability: Crupper and Breeching

Descending steep terrain poses challenges without rear anchoring systems.

- **Crupper (Baticola)**:
  - Padded under-dock strap (≥5/8″ thick).
  - Prevents saddle from sliding forward, especially in mountainous work.

Fig. 10 A mule equipped with a breast plate in the front to prevent saddle from sliding back, breeching to assist with saddle sliding or moving forward and back cinch to help stabilize the saddle on the back

Fig. 11 Saddle equipped with breast plate, breeching, crupper, and back girth to maintain stability over various terrain

- **Breeching Strap (Retranca)**:
  - Covers the **lower thigh and gaskin**, securing the saddle to the **pelvic anchor**.
  - Must be adjusted snugly without chafing.
  - Preferred in pack mules or for long downhill sections where cruppers may rotate.

Leather components should be **conditioned monthly** with natural oils or wax-based balms to avoid drying, cracking, or stiffness (refer to Fig. 11).

5.8 Hoof Protection and Maintenance

Mules generally thrive **barefoot** due to dense hoof horn and low metabolic hoof turnover.

- **Hoof Shape**: Oval or oblong, with thick outer walls and **narrower frogs**.
- **Growth Tendency**: Susceptible to **"tubing"** (vertical overgrowth) in soft pastures or dry pens.
- **Trimming**: Every 6–8 weeks, focusing on **heel angle, sole concavity, and medial-lateral balance**.
- **Shoeing**: Rarely needed except in **working disciplines** like reining, ranch work, or charrería—where **protective boots or specialized shoeing** may reduce concussion and lateral torque.

Routine care, ideally performed by a **farrier familiar with hybrid morphology**, prevents navicular stress, frog bruising, or joint compression.

6 Recommended Saddling Protocol for Mules

A systematic, step-by-step saddling protocol is essential to ensure both rider safety and the mule's comfort. The following detailed guidelines are based on anatomical principles and best welfare practices to minimize injury, improve equipment stability, and promote a cooperative relationship between mule and handler.

6.1 Initial Preparation: Grooming and Behavioral Conditioning

Before applying any tack:

- **Grooming:** Thoroughly brush the mule to remove dust, sweat, and debris that could cause abrasions under the saddle. Ensure hooves are clean and dry.
- **Behavioral conditioning:** The mule should be accustomed to standing quietly while tied, reducing the risk of sudden movements that could interfere with saddling or injure the handler.

6.2 Placement of Padding Layers: Saddle Pad and Blanket

Proper layering beneath the saddle protects the skin and distributes pressure evenly:

- **Saddle pad (mantilla):** Made of wool-woven material, it is placed symmetrically over the mule's back to absorb sweat and minimize skin friction.
- **Blanket (carona):** Positioned one inch behind the saddle pad, the virgin-wool blanket provides extra cushioning, helps regulate temperature, and aids in moisture wicking.

6.3 Saddle Placement and Positioning

Accurate saddle placement is key to avoiding discomfort and ensuring stability:

- **Set the saddle** one inch behind the trailing edge of the wool pad.
- **Gullet clearance:** Ensure the saddle's gullet fully clears the withers.
- **Skirt contact:** The saddle skirts should lie flush along the rib cage without bridging or pinching.

6.4 Girth Adjustment: Progressive Tightening and Positioning

Improper girthing can lead to pressure sores or saddle instability:

- **Incremental tightening:** Secure the girth in at least three gradual stages to distribute tension evenly and avoid discomfort.
- **Spacing:** Maintain approximately eight inches between the saddle's D-rings and girth buckles.
- **Positioning:** Mules generally require the girth to sit further back than on horses, due to their straighter dorsal line.
- **Skin sensitivity:** Avoid abrupt tightening, as the mule's flank is particularly delicate.

6.5 Rear Cinch (Barrigüero) Installation

Essential for preventing forward saddle slippage on slopes or under sudden load shifts:

- **Material:** Wool or leather is preferred for flexibility and breathability.
- **Width and connection:** The rear cinch should be broader than the main girth and attached to it to prevent shifting.
- **Positioning:** Place it directly behind the belly, about one inch behind the navel, to stabilize the cantle without irritating the skin.

6.6 Breastplate Application

The breastplate ensures the saddle remains in place during uphill climbs:

- **Anchoring points:** Attach it to the front of the saddle and secure it to the girth between the forelegs.
- **Alignment:** It should follow the scapular angle to avoid impinging on shoulder movement.

- **Fit:** Prevent pressure on pectoral muscles while maintaining forward saddle stability.

6.7 Crupper or Breeching Strap (Baticola/Nagüilla) Setup

Vital for downhill stability and protection of the lumbar region:

- **Crupper pad:** Should be at least ⅝″ thick to cushion the dock and surrounding vertebrae.
- **Nagüilla placement:** Sits just below the gaskin, covering the lower thigh without obstructing motion. If placed too high, it may be dislodged during jumping or steep movement.
- **Combined systems:** Colombian-style tack combines crupper and breeching functionality, allowing streamlined security and freedom of movement.

6.8 Bridle and Bit Fitting

Final step before riding, with emphasis on oral sensitivity:

- **Bit type:** Snaffle or curb bits are both acceptable depending on training level.
- **Positioning:** The bit should rest at the lip commissure, forming a single soft wrinkle—neither hanging loosely nor pulling tightly.
- **Oral anatomy:** Given the mule's short diastema and sensitive oral tissues, correct bit placement is critical to avoid pain or resistance (e.g., head tossing).

6.9 Conclusion: Anatomically Guided, Welfare-Centered Practice

While cultural and regional variations in tack exist, adherence to this structured protocol:

- Prevents saddle-related injuries such as chafing or pressure sores.
- Improves saddle security in diverse terrains.
- Enhances communication, trust, and performance by respecting the mule's anatomy and behavioral traits.

A properly saddled mule is not only safer and more responsive but also demonstrates better long-term physical and mental wellbeing.

7 Management Considerations for Maintaining a Mule's Readiness for Riding

Mules earn their reputation as steady, sensible mounts—but that reputation can unravel quickly if basic management practices are neglected. From diet to socialization, each facet of daily care influences a mule's physical comfort and mental outlook. The sections below explore common pitfalls and offer proactive strategies to keep your mule reliable, responsive, and ready for the saddle.

7.1 Nutrition and Body-Condition Balance

Mules are masters at extracting nutrients from coarse forage, yet modern feeding regimens often overemphasize flashy looks over functional fitness. Loading up on grain- and fat-based concentrates to produce a plump, glossy coat can backfire: excess energy with too little work leads to a "charged" mule prone to sudden bolting, bracing, or head-tossing when asked to settle.

To avoid this, make high-fiber roughage the foundation of the diet—think good-quality grass or alfalfa hay—then add concentrates sparingly, matched precisely to your mule's workload. Monitor body condition scores monthly and adjust portions so that the mule carries enough energy for its job without becoming overexcited or lethargic under saddle. For more information about mule nutrition, see nutrition chapter.

7.2 Ear and Skin Sensitivity Management

Many mules bristle at having their ears touched, a reaction that may mask simple causes: itchiness from skin dryness, minor fungal infections, or even biting insects. Left untreated, ear sensitivity can escalate into head-butting or bolting away when bridled.

A gentle desensitization routine works wonders here. Begin by lightly stroking the base of the ears, rewarding calm acceptance with praise or a small treat. Over days and weeks, increase the duration and lightness of contact, always pairing the sensation with positive reinforcement or the immediate cessation of pressure. Simultaneously, address any underlying skin issues—cleanse, moisturize, and treat fungal or parasitic irritations promptly—and your mule will learn to tolerate, even welcome, ear handling.

7.3 Mounting and Weight-Bearing Confidence

Initial balkiness around the stirrup or during the mounting process generally stems from uncertainty, not stubbornness. Overcoming that hesitation requires building trust through gradual exposure.

Start on the ground: teach your mule to accept light pressure along its barrel where the stirrup will rest. Reward any step toward yielding. Next, introduce a low mounting block and ask the mule to bear increasing weight in the stirrup, always relieving pressure the instant it shifts its stance appropriately. Finally, practice standing quietly while you swing into the saddle, again using rewards for calm behavior. Over time, mounting becomes a familiar sequence rather than a fearful obstacle.

7.4 Social Bonds and Separation Stress

Mules form tight attachments to paddock companions, and abrupt separation—whether for riding or stabling—can trigger panic reactions like frantic spinning or bolting.

Prevent over-attachment by rotating turnout groups every few days and pairing your mule with several different partners. Before a ride, choose turnout or stabling arrangements that mirror the trail group so your mule isn't caught off guard. If separation anxiety does appear under saddle—ears pinned, frantic head shake or spin—use controlled, looping patterns to refocus attention and keep the experience positive: short, calm rides that end before the mule becomes worried again.

7.5 Tack-Fit Checks on the Move

Even perfectly fitted gear can shift with movement, and a slipping girth or misaligned breeching often spooks a mule mid-trail.

Make it a habit to pause every 20–30 min and visually inspect saddle position, girth tension, and crupper fit. Listen for telltale swishes or shifts, and feel along the panels to ensure nothing has crept forward or back. Address minor adjustments immediately—add a pad here, re-tighten a strap there—so your mule can remain balanced and confident, never worrying about rogue pressure points.

7.6 Desensitization to Prevent Spooking and Kicking

When startled, a mule's first instinct may be to snap out its hind legs rather than freeze and assess. To curb this reflex, weave systematic desensitization into your regular groundwork.

Introduce innocuous stimuli—plastic bags, flapping tarps, ropes across the flank—in a low-pressure environment. Approach, allow investigation, then retreat before the mule becomes alarmed, rewarding calm curiosity at each step. Over time, move these exercises under saddle or on the trail, maintaining a relaxed pace. Your mule will learn that sudden noises or touches do not always signal danger, reducing the likelihood of dangerous kicking or bolting.

By thoughtfully managing diet, skin health, mounting routine, social dynamics, tack fit, and spook-proofing drills, you'll keep your mule physically sound and mentally tuned. A consistently comfortable, confident mule isn't just safer to ride—it's a testament to care that honors the mule's intelligence, resilience, and invaluable partnership.

8 Conclusions

The proper fitting of saddles for mules and donkeys is not a matter of convenience, but a foundational element in ensuring their health, comfort, and performance. Their anatomical uniqueness—ranging from cranial structure to hoof morphology—demands a specialized approach to tack design, saddle placement, and equipment adjustment. Recognizing these animals not as smaller horses, but as unique equine hybrids with distinct behavioral and biomechanical traits, is essential for building a more ethical and functional partnership. Through thoughtful selection of equipment, attention to body condition, consistent behavioral desensitization, and systematic saddle-fitting protocols, we can create a safer, more responsive riding experience. Furthermore, understanding the morphological differences across breeds and disciplines empowers trainers, breeders, and riders to tailor strategies that honor the capabilities and needs of each individual mule or donkey. By grounding our practices in science, tradition, and empathy, we foster not only better saddle fit but also a deeper appreciation for these often-overlooked equine partners. As equestrian disciplines evolve, so must our commitment to equipment that respects the anatomy and spirit of the animals we work with.

Self-Evaluation Questionnaire

Selecting a High-Quality Riding Mule

1. Why is a mule's conformation more important than its size when selecting for saddle work?
   A. Larger mules have longer backs, which improves comfort.
   B. Conformation determines balance, gait quality, and saddle fit.
   C. Size guarantees better load-bearing capacity.
   D. Only temperament matters, not physical structure.
   Correct Answer: B

Explanation: Balanced conformation affects movement efficiency and long-term comfort under saddle.

2. What were traditional breeding goals that limited mule riding quality?
   A. Color and speed
   B. Strength and tractability
   C. Size and pulling power
   D. Gait fluidity and rein response
   Correct Answer: C

Explanation: Breeding for agricultural strength led to conformation unsuitable for saddle performance.

Keywords Conformation, gait, size vs function, temperament, historical breeding, saddle quality

Cross-Breeding Strategies

1. What is the main reason for selecting donkey breeds like the Colombian Criollo or Brazilian Pêga?
   A. They are large and strong
   B. They produce flashy colors
   C. They improve gait rhythm and softness
   D. They are easier to train
   Correct Answer: C

Explanation: These donkeys contribute smooth movement and energy-efficient gaits to hybrids.

2. Which trait is essential when choosing parents for hybrids meant for jumping?
   A. Neck length
   B. Conformation and limb structure
   C. Hair type
   D. Ear size
   Correct Answer: B

Explanation: Structural soundness ensures safe and functional jumping performance.

3. Why is it important to match breeding pairs by discipline?
   A. To preserve rare donkey breeds
   B. To avoid behavior problems
   C. To align physical traits with intended use
   D. To reduce hoof trimming needs
   Correct Answer: C

Explanation: Function-focused crosses improve outcomes for specific riding, driving, or showing goals.

Keywords: Hybrid breeding, gaited mule, discipline-specific traits, conformation, sire-dam matching

Morphological Differences

1. Why is the standard horse saddle often problematic for donkeys?
   A. Donkeys dislike tack
   B. Their flat backs and low withers don't match equine saddle trees
   C. Donkeys prefer packs, not saddles
   D. They have long ribs and narrow shoulders
   Correct Answer: B

Explanation: Donkey anatomy requires tack that conforms to flatter, shorter backs and less defined withers.

2. What makes bit fitting more complex in mules than in horses?
   A. Bigger tongues
   B. Narrower jaws and heightened oral sensitivity
   C. Smaller teeth
   D. Less saliva production
   Correct Answer: B

Explanation: Mules' reduced diastema and thick jaws require careful bit placement and sometimes bitless options.

Keywords Spinal flexibility, withers, thoracic cavity, poll mechanics, diastema, hybrid anatomy

Ideal Equipment for Riding Mules

1. What is one advantage of using virgin wool for saddle pads?
   A. It's waterproof
   B. It reduces saddle size
   C. It regulates temperature and wicks moisture
   D. It makes the saddle heavier
   Correct Answer: C

Explanation: Wool is breathable, shock-absorbing, and keeps the mule dry and cool.

2. Why are bridles for mules different from horse bridles?
   A. Mules have larger nostrils
   B. Mules have wide-set eyes and prominent ear bases
   C. Mules chew more
   D. Horse bridles are too short
   Correct Answer: B

Explanation: Bridle design must avoid pinching around sensitive facial structures and accommodate mule head shape.

3. What is the purpose of a properly contoured breastplate on a mule?
   A. Decorative effect
   B. Restrains forward movement
   C. Prevents saddle from sliding backward
   D. Adds weight to the front end
   Correct Answer: C

Explanation: Breastplates help stabilize the saddle on narrow-chested animals during uphill or abrupt stops.

Keywords Saddle tree, wool pads, crupper, breeching, bridle fit, mule anatomy, cinch design

Saddling Protocol

1. Why is progressive tightening of the cinch necessary?
   A. To warm the back muscles
   B. To stretch the girth
   C. To avoid sudden pressure that causes discomfort
   D. To protect the cinch from dirt
   Correct Answer: C

Explanation: Mules have sensitive flanks; sudden girth tightening can lead to behavioral resistance or pain.

2. How should a crupper or breeching strap be positioned?
   A. Loose and decorative
   B. High and tight under the tail
   C. Cushioned and aligned with natural movement paths
   D. Tight enough to restrict the croup
   Correct Answer: C

Explanation: Rear tack should stabilize the saddle while respecting the animal's movement range.

Keywords Girth pressure, step-by-step saddling, saddle stability, crupper, breeching, comfort checks

Management Considerations

1. What feeding mistake can cause a mule to become overly reactive under saddle?
   A. Too much roughage
   B. Feeding close to ride time

C. Overfeeding high-energy grains without workload
D. Skipping salt blocks
Correct Answer: C

Explanation: Mules require minimal concentrates; excess energy can result in spookiness and bracing.

2. What is a practical strategy to reduce separation stress in mules?
   A. Keep the mule alone in a stall
   B. Let it ride with the same partner only
   C. Rotate companions and simulate trail groups
   D. Avoid all contact with other animals
   Correct Answer: C

Explanation: Social flexibility reduces panic when mules are ridden apart from favorite herd mates.

3. Why are regular tack checks necessary during trail rides?
   A. To correct saddle style
   B. To clean dust off pads
   C. To fix slipping gear and prevent soreness or spooking
   D. To give the mule time to graze
   Correct Answer: C

Explanation: Loose girths or sliding breeching can create fear, pain, or behavior issues mid-ride.

Keywords Behavior, diet, body condition, desensitization, social bonding, mid-ride tack check

Harnessing and Equipment Used for Working Donkeys and Mules

Mathilde Merridale-Punter, João Brandão Rodrigues, and Amadou Doumbia

Abstract

Most donkeys, mules and hinnies around the world are located in low- and middle-income countries and are used for work, supporting millions of households. Equids can be used for a variety of purposes for which they often require different types of work equipment. This may range from riding tack, such as riding saddles and bridles used for personal transport, sports or showing events, to harnesses used to pull carts and agricultural implements, or sawbucks used to carry loads without the use of a vehicle. In this chapter, we will define work equipment as the different types of harnesses and implements used for work, where the term harness refers to the equipment in direct contact with the animal. Although the specific equipment needed by the donkey or hybrid will depend on the type of work performed and on the animal's conformation and body condition score, certain key principles and design features apply to all work equipment. Respecting these principles allows the animal to work comfortably and efficiently, and helps prevent injury to both the animal and user. In working equids throughout the world, harness-related wounds are extremely common and represent a serious welfare concern. Understanding key concepts of equipment design, fit, maintenance and use can help reduce the incidence of these largely preventable wounds. In addition, adequate work practices that respect the animal's physical and psychological limits are essential to safeguard animal

M. Merridale-Punter (✉)
School of Veterinary Medicine, University of Melbourne, Parkville, VIC, Australia

Society for the Protection of Animals Abroad (SPANA), London, UK
e-mail: m.seabramerridalepunter@unimelb.edu.au

J. B. Rodrigues
The Donkey Sanctuary, Sidmouth, UK

A. Doumbia
Society for the Protection of Animals Abroad (SPANA), London, UK

A. K. McLean et al. (eds.), *Donkey and Mule Medicine: Foundations, Breeds, and Management*, https://doi.org/10.1007/978-3-032-12726-6_14

welfare, and should therefore be understood by owners, veterinarians and others working with these equids. This chapter discusses basic concepts of pulling or carrying loads, common types of harnesses, as well as common clinical presentations associated with work equipment. It aims to provide practical advice on how to recognise important features of work equipment and to equip those handling working equids with the knowledge they need to make recommendations and prevent equipment-related problems.

Keywords

Harness · Tack · Cart · Work equipment · Working equids

1 Introduction

Working donkeys and hybrids are a major source of sustainable energy worldwide, especially in low- and middle-income countries, with significant contributions to agroforestry and industrial activities (Almeida et al. 2017; Fernando and Starkey 2004; Norris et al. 2021). Good health and welfare are essential factors for donkeys, mules and hinnies to be able to contribute and effectively support millions of people who rely on them. One of the most important aspects that directly impacts on those factors is the harness. A good harness is one that fits well to the equid's body, that efficiently transmits the animal's energy while being padded and appropriately designed and is used so that the force is spread over a large area, without causing trauma (Pearson et al. 2003). Poorly designed, ill-fitted harnesses will cause unnecessary discomfort, fatigue, and the induction and exacerbation of pain, that ultimately decrease working efficiency and output (Farhat et al. 2020; Merridale-Punter et al. 2024a; Rodrigues et al. 2020). This section covers the key functions of a harness in some detail, as well as the different components needed to correctly perform those functions, always keeping in mind the unique nature of donkeys and hybrids.

2 Principles of Traction and of Equipment Design, Function and Adjustment

There are several different types of harnesses, which are used for different types of work and also based on cultural, traditional or personal preference. As such, there is no single correct harness model or design. However, regardless of the type of harness used, it should always enable the following key functions:

1. Moving the load;
2. Carrying (and balancing) the load;
3. Braking the load;
4. Communicating with the equid(s).

This section will explore in more detail what each of these functions means and what harness components and design characteristics are required to enable them.

1. **Moving the load**

When equids are used for draught purposes and are requested to move loads (with or without wheels), a collar should always be present as part of the harness, together with traces and a swingle tree. The **collar** is a key piece of equipment that makes best use of the equid's 'draught' power, whether that is a vehicle or sledge, an agricultural implement or a felled tree. Essentially, animal energy is primarily captured through the collar, providing traction and movement of the load. The two most common types of collars used are the breast collar and the full collar.

- **Breast collar**: This is a simple type of collar and the most used worldwide. It is essentially composed of a strap across the chest that the equid can push into and is combined with a series of other strap(s) that keep the breast strap (or breast collar) in place. The advantage of this type of collar is that it is usually cheaper to manufacture, can be made with more limited experience and tools and can, to some extent, be easily transferred between equids. However, these are considered less efficient when compared with full collars, due to the smaller area of contact with the body of the equid (Rodrigues et al. 2021), especially where a low draught is required (such as when pulling agricultural implements).
  A breast collar should be positioned so as not to impede motion (if placed too low) or breathing capacity (if placed too high) (Margentino 2020). This is especially important in the case of donkeys due to their narrow chest region, higher and more prominent point of shoulders and deep trachea, when compared to horses. The scapulohumeral joints and the trachea region are considered critical points and are among the most common locations for skin lesions in working equids using collars (Farhat et al. 2020; Rodrigues et al. 2020). Thus, collars should ensure a clear space in the tracheal region allowing the equid to breathe without any mechanical blockage. A good option in donkeys may be to take out a shallow arch from the straight breast strap, providing extra space in the region of the trachea, while allowing the breast strap to be correctly adjusted in the shoulders (Fig. 1a). Some models of breast collars try to solve this issue by changing the straight shape of the breast strap into a "V" shape model, with limited success (Rodrigues et al. 2021) (Fig. 1b).
- **Full collar:** This is a more complex but more efficient version of a collar (Figs. 2a and 2b). The advantage of the full collar is that it is more efficient at capturing the equid's power, due to the larger area of contact with the shoulders of the equids, allowing the animal to do more work with less effort. However, it is a skilled job to manufacture (and repair) these collars properly, and because they bed into an equid's shape they do not easily transfer between different equids unless they are of a very similar size and shape. They are also more expensive to produce. In general, full collars include a soft padded area (in direct contact with the body of the animal), and a rigid structure called hames, responsible for spreading the load

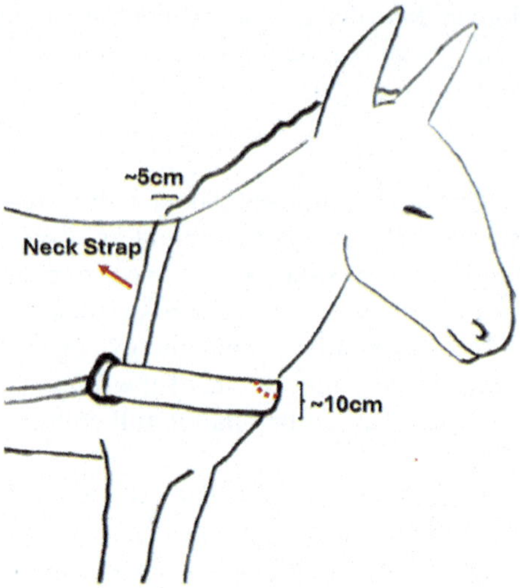

Fig. 1a Schematic illustration of a donkey using a simple straight breast collar. The dotted red line represents a potential modification to allow additional space in the region of the trachea. Note: the width of the breast strap should be of approximately 10 cm where it fits against the breast of the equid. The other part of the breast collar goes over the top of the equid's neck and is just there to hold the main strap in place. Therefore, it can be narrower such as approximately 5cm wide and should be protected where in contact with the neck. This is commonly referred to as the neck strap

Fig. 1b Schematic illustration of a donkey using a purposefully designed "V shape" breast collar, designed with the aim of giving extra space in the trachea region

evenly over the collar, which in turn means that it is evenly spread over the equid's bearing surface, reducing pressure points and preventing the collar from becoming misshaped. Hames include the hitching point or point of draught (from where the traces are attached) and the rein terrets.

Fig. 2a Mule using a full collar, a more efficient type of a collar which is also more complex, expensive and less transferable between animals. Note the presence of a padding system. Image: kindly provided by Mafalda Brito e Silva, Associação Portuguesa de Tracção Animal (APTRAN)

Fig. 2b Donkey using a home-made full collar, made from inexpensive materials. Note how this collar still respects all the key traction principles, having a layer of padding adjusted to the equid's conformation and a central rigid structure (hames) allowing for adjustment. The correct 90-degree angle of draught formed between the traces and the hames of the collar at the hitching point is illustrated by the dotted green line. Hames are the rigid structure of the full collar responsible for spreading the load evenly. The point where the traces attach to the hames is called the hitching point (red square). Note how in donkeys (b) the hitching point should correspond to approximately the lower 45% of an imaginary line between the lower trachea and the withers. However, in hybrids that have a conformation closer to that of horses (a) the hitching point should be lower, approximately at the top of the lower third of this imaginary line

Full collars may be non-adjustable or may include a series of adjustments to ensure the correct fit into the equid's body, both in terms of height or length (vertical adjustment) and width (horizontal adjustment), allowing the padding system to fit the specific anatomy of the equid (Rodrigues et al. 2021). The hitching point for the traces may also be adjustable or non-adjustable, depending on the collar. Full collars should also have a clear space over the tracheal region, even under load. Additionally, they should be adapted to the animal's individual neck conformation. Donkeys and horses have anatomical differences in fat distribution and neck morphology, with donkeys generally having an "8" transversal neck shape due to a more prominent neck crest, while horses present more of an "egg" shape. In hybrids the neck shape may vary, but such anatomical differences should always be taken into consideration when selecting full collars for different working equids (Rodrigues et al. 2021).

Regardless of the type of collar, it should always be used in combination with traces and a swingle tree, to efficiently move the load. These components connect the collar to the load, whether that is a vehicle, an agricultural implement, a log or other types of load. **Traces** are the straps, ropes or chains that connect the collar directly to the swingle tree (by being attached to the hames of the full collar at the hitching point, or by being connected to both ends of the breast collar) (Fig. 3). If the material chosen for the traces is abrasive it may cause skin lesions, so those traces should be protected by an outer non-abrasive material. The **swingle tree** is a metal or wooden bar, connected to the traces by both ends, and to the load or cart at the centre (Figs. 4 and 5). This allows the swingle tree to move freely from side to side when the animal moves, therefore also allowing the traces and collar to move freely with the movement of the animal, without causing abrasions. The swingle tree is a fundamental

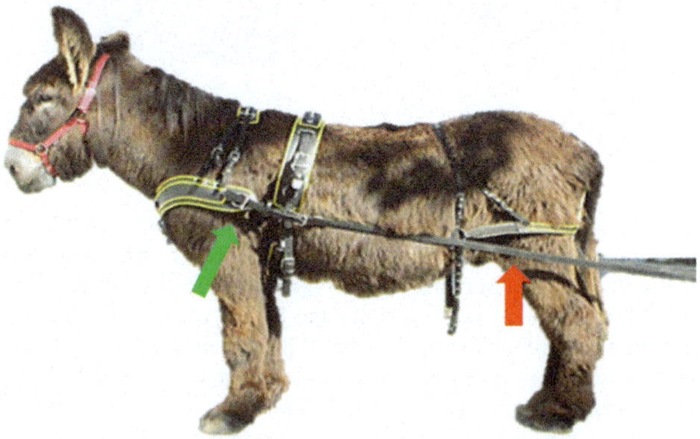

Fig. 3 Donkey wearing a breast collar harness. Traces are illustrated by the red arrow, connecting the breast collar and the swingle tree. The green arrow illustrates the point where the traces connect to the collar. Note how in this case the breast collar has a slight V shape providing additional unobstructed space over the region of the trachea. Also note the correct positioning of the collar, which should not be too high as to obstruct the trachea, nor too low as to impinge on the point of the shoulders or interfere with normal movement

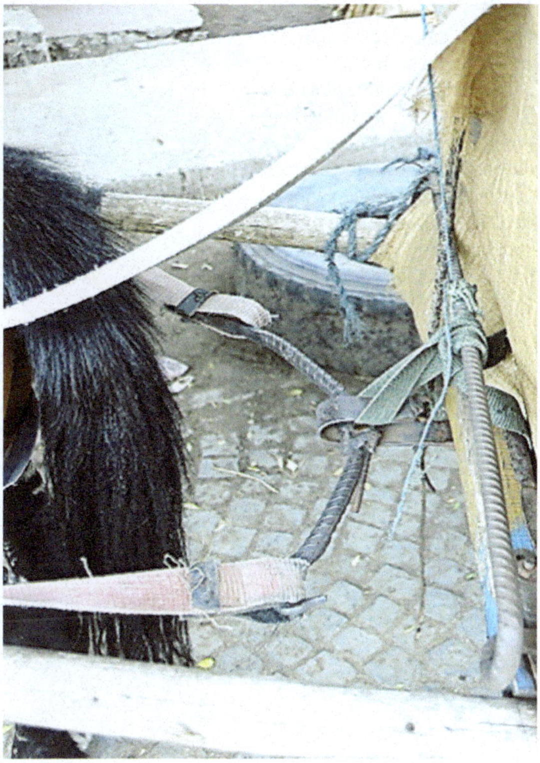

Fig. 4 Example of a metal swingle tree used to pull a cart. Note this swingle tree is made from low-cost recycled materials, which does not compromise its function or efficiency provided it is used correctly. The swingle tree can be made from different materials and have various designs, but essentially connects to the traces at either end and to the load, cart or implement at the centre, allowing for a swinging motion of the traces when the equid moves

harness component which is often missed and a major contributor to equipment-related wounds (Merridale-Punter et al. 2024a, c). The size of a swingle tree is another important aspect and should correspond to the distance between the tuber coxae of the equid, in a straight line (Fig. 5).

Important Aspects to Keep in Mind:

Collars, traces and swingle tree are the three elements needed to move the load. Correct adjustment of the harness and collar is key to make best use of the energy that equids can transmit to the harness system, and to ensure their comfort and safety while working. To achieve this there are three key aspects that should be kept in mind when using collars, traces and swingle trees:

- *Position of the hitching point/point of draught:* This point is where the traces are attached to the hames of the full collar, and its correct location is crucial to ensure that there is a full and equal contact between the padding system of the collar and the whole scapular region of the equid, when pulling a load. As a rule of thumb, the correct hitching point for a horse corresponds to the top of the lower third of a straight line that runs from the withers to the lower part of the trachea, parallel with the scapula. For donkeys and because of their anatomy, the correct hitching

Fig. 5 Two draught equids pulling a cultivator. Each animal has a swingle tree, combined with a third swingle tree attached to the implement (yellow arrows). The red dashed line shows the correct size of the individual swingle tree. The third swingle tree should be big enough to allow both animals to work side by side

point should be higher, corresponding to the point that divides the described line in a 45% (lower)/55% (upper) section (Fig. 2b). In the case of hybrids, it will depend on whether the shoulder anatomy of the animal is closer to that of a horse or a donkey (Fig. 2a). Because each equid is unique the hitching point should be adjustable, ensuring minimal but important adjustments to the anatomy of each individual equid. If the hitching point is too high, the collar can be pulled upward against the trachea, causing discomfort and affecting breathing capacity. It also increases the likelihood of causing pressure points, due to an incorrect distribution of the force in a smaller area of contact. If the hitching point is too low, then there will be extra pressure on the point of the shoulders (in the case of a full collar), and the collar may interfere with the equid's normal stride (Morrissey 2013). This is a very common situation, observed when full collars that are designed for horses are used in donkeys without correcting the hitching point. In the case of breast collars, traces are always connected on both ends of the breast strap as these collars don't have a rigid structure. However, correct positioning of the collar is still an essential consideration, where it should not be too high as to obstruct the trachea, nor too low as to impinge on the point of the shoulders or interfere with normal movement (Fig. 3). The correct position will allow optimal efficiency and can be regulated by adjusting the neck strap.

Fig. 6 Correct angle of draught in a full collar, creating a 90-degree angle between the hames and traces when in tension (illustrated in green). The correct angle of draught ensures an efficient transfer of power. Note also how the traces create a straight line of draught between the hames and the swingle tree

- *The angle of draught:* This corresponds to the angle between the hames or collar and the traces, when in tension. In the case of full collars, the most efficient transfer of power from equid to load happens when there is an angle of 90 degrees (Morrissey 2013) (Fig. 6). Keeping the correct hitching point at the hames will make it easier to obtain the correct angle of draught, avoiding pressure points, discomfort and potential skin lesions. In the case of breast collars, the correct angle should be of 180 degrees, meaning a straight line between the breast strap and the traces.
- *The position of the line of draught*: This line relates to how each trace runs from the hame to the swingle tree. This line should always be straight and not 'broken' (Morrissey 2013) (Fig. 6). A common problem seen in many parts of the world, is when traces are wrapped around the shafts of the cart (which may be to increase tension on the traces or to prevent them being caught on the animal's legs or other hazards). However, this means that the traces are no longer in a straight line between the collar and the swingle tree, as there is no free movement of the traces and swingle tree, entirely defeating the purpose of the swingle tree. This practice is easily corrected with adequate information and advice. Another way of preventing some of the issues that may lead to this practice is the use of a **trace carrier**. This is a simple strap that passes over the loin region and connects both traces, allowing the traces to be suspended when the animal is stopped or turning tightly. A similar function may also be achieved by placing loose loops around the shafts through which the traces pass, without interfering with their free movement or breaking the line between the collar and the swingle tree. While the traces are in tension, trace carriers should not interfere with the correct functioning of the traces.

2. **Carrying (and balancing) the load**

Within a harnessing system, a **saddle** is the component responsible for carrying and balancing the load. Donkeys, hybrids and horses have complex back shapes, so in ideal conditions saddles should be designed and manufactured to adapt and mirror that shape, avoiding pressure points and other welfare concerns. The same applies to all types of saddles, whether a cart harness, pack or riding saddle.

Important Aspects to Keep in Mind:

Pack, riding and cart saddles differ in their specific use but certain key design principles to ensure welfare of the equids apply equally to all:

- *The saddle must provide a gullet space or channel that keeps weight off the spine.* The withers and the spinal region are considered critical points (Merridale-Punter et al. 2024c; Rodrigues et al. 2020). All saddles should have a wide space between the saddle panels (i.e. area of contact between the saddle with the lateral back muscles of the equid), creating a channel or gullet which avoids direct contact with the bony prominences of the spinal process (Figs. 7a and 7b). The gullet space should be visible even when loaded and should be proportional to the size of the animal, in most cases approximately 8–10 cm wide, running the full length of the gullet space. Importantly, it should be ensured that any padding or blankets used under the saddle do not obstruct the gullet space and are equally adjusted in such a way that prevents direct pressure over the spine.
- *The saddle should spread the weight of any cargo evenly across the bearing surfaces of the animal,* i.e. the muscular area on either side of the spine, and should be correctly adjusted to the back of the equid (Figs. 7a and 7b). Just as described for full collars, an uneven distribution of the forces will cause pressure points, leading to pain and lesions. It is often believed that padding, layers of blankets or similar can be used to avoid pressure points, but padding only provides a little more comfort and absorbs sweat, so padding should not be viewed as a replacement of wide contact surfaces and correct saddle shape. One

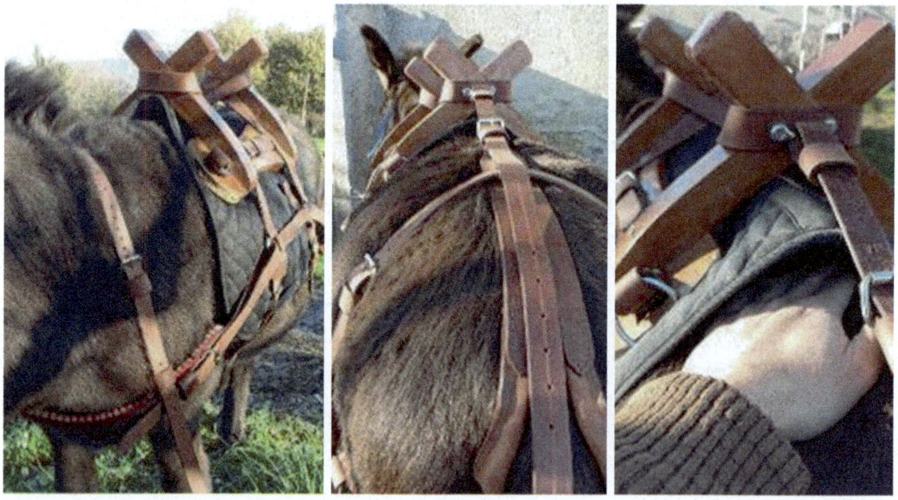

Fig. 7a Front and rear views of a donkey pack saddle demonstrating the presence of a clear gullet space. The gullet space or channel prevents direct pressure and weight from being applied directly over the bony prominences of the spine. Note how the gullet space should also be created for any layers of padding or blankets used with the saddle, and it should be present regardless of the type of saddle used (7b). Images: kindly provided by Bernardo Carreiro

Fig. 7b Schematic illustration showing other types of saddles presenting the correct gullet space (red arrow). Regardless of the type of saddle used, a gullet space should always be ensured, such as exemplified in this cart saddle (left) and riding saddle (right)

of the biggest causes of 'saddle' sores is in fact the padding, particularly if large amounts of padding are used (Merridale-Punter et al. 2024c; Rodrigues et al. 2020).

- In the case of pack and riding saddles, the animal should be protected from the cargo to prevent skin lesions due to contact with abrasive, sharp or irritating materials.

Unlike collars, saddles should be secured tightly to the body of the equid, avoiding rubbing and unwanted movement, and so should be kept in place through either one or a series of connecting straps. In the case of **riding and pack saddles**, these may be kept in place through (Figs. 8a and 8b):

- **Girth strap**: designed to keep the saddle in place, it is attached to the saddle at both ends, with at least one end being adjustable. At least one girth strap should always be present with a saddle. In mountain areas a second girth may be used, passing behind the most ventral part of the belly (Fig. 12).
- **Breast strap**: similar to a breast collar but with the function of keeping the saddle in place and preventing it from moving backwards rather than providing any type of traction.
- **Breech strap**: keeps the saddle in place avoiding forward movement of the saddle. Recommended for pack saddles.

The **cart saddle** is part of the cart harness, and it is used on the equid's back to take any weight that might be transmitted to that region during movement. In an ideal world, there would be little or no weight on the cart saddle but there often is, especially in the case of the two-wheel vehicles where balancing the cart can be difficult. As such, it is important to have a well-design cart saddle, with a gullet,

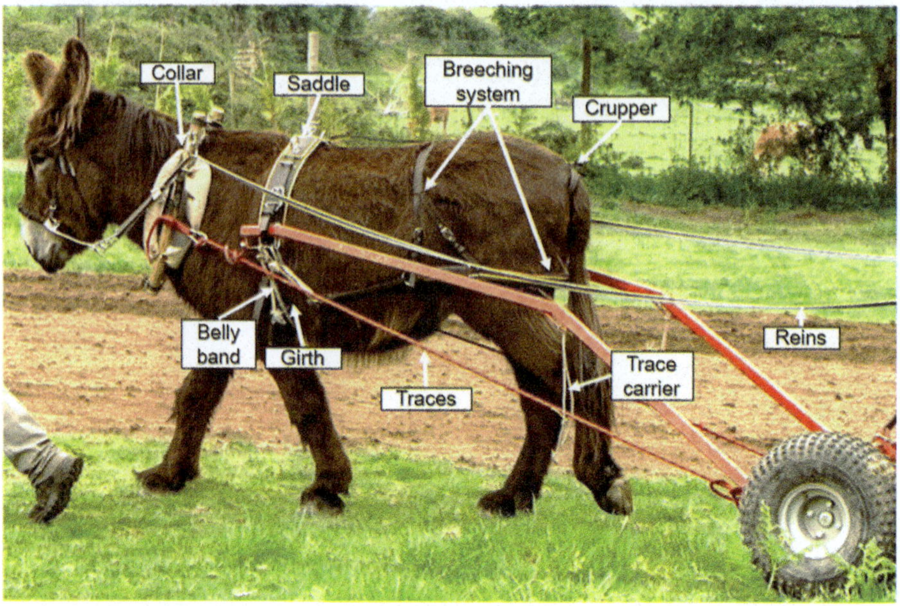

Fig. 8a Illustration of a typical cart harness and its main components. A full collar is used in this case. The cart shafts have the correct height, halfway between the ventral and the dorsal aspects of the chest of the equid at the level where the shaft straps attach. The angle of draught between the hames and traces is of approximately 90 degrees. A full collar is used including padding which adapts to the conformation of the donkey. Breeching is used, as well as a crupper. In this case, trace-carriers are also used, keeping the traces suspended and avoiding interference with the donkey's legs. Note that in this case the belly band is a rope that connects both shafts loosely under the donkey's belly practically without touching the animal. However, in situations where the belly band is likely to contact the animal, it is recommended to use a wider non-abrasive strap or covering the rope with non-abrasive material such as leather. Image: kindly provided by Mafalda Brito e Silva, Associação Portuguesa de Tracção Animal (APTRAN)

some padding and a wide area of contact (Figs. 7a and 7b). In the case of four-wheel vehicles, the four wheels allow for a better distribution of the load in the cart platform, with less weight on the equid's back, so the saddle may be smaller. The **shafts of the cart** are attached to the cart saddle through the shaft straps (shaft tugs or connecting backband). Ideally, in those cases where the shafts are fixed and do not move independently from each other, the shaft strap slides to absorb the movement of the vehicle so as not to put additional pressure on the equid's back. Where the shafts move independently (as in the case of many modern four wheeled vehicles) the shaft strap can be fixed.

Important Aspects to Keep in Mind:

- *Position of the shafts*: As mentioned, shafts connect the vehicle to the animal, attaching to the cart saddle through the shaft straps. When in a correct position, the front part of the shafts should run along the centre of the equid's body, in a

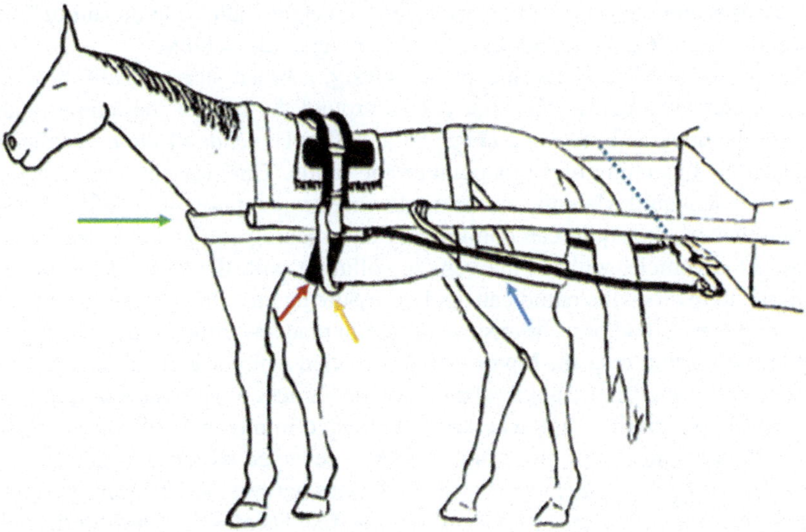

Fig. 8b Schematic illustration of a typical cart harness and its main components using a breast collar. The cart shafts have the correct height (green arrow) and the angle of draught creates approximately 180 degrees between either end of the collar and the traces. Note the presence of the **belly band** (yellow arrow), a strap that passes under the equid's belly and is attached to the shafts or shaft straps. The belly band should be loose and stops the shafts from lifting in the air on a two-wheels vehicle, if the load is unbalanced or placed too backwards. The belly band is different to the **girth or girth band** (red arrow), which attaches tightly to both sides of the saddle under the equid's belly and has the function of keeping the cart saddle in place avoiding movement and rubbing. **Breeching** (blue arrow) is used, showing the correct position in the equid and relative to the cart. In cases where there is no breeching, a **false breeching** may also be used, consisting of a band attached to the cart shafts behind the animal (dotted line)

horizontal line, halfway between the ventral and the dorsal aspects of the thoracic area and with both shafts placed at the same height (Figs. 8a and 8b).

- *Distance between the back of the equid and the cart*: The length of the shafts is another important aspect to consider. There should be enough room between the back of the equid and the front of the cart, when shafts are in place and attached to the cart harness. It is crucial that the animal can move without any contact with the cart. On the other hand, too much space could make it difficult to manoeuvre the cart, so this aspect must be assessed and advice provided accordingly.
- *Distance between the shafts and the body of the equid:* There should be a gap of about 10 cm between the shafts and the equid on both sides. This is roughly equivalent to the width of a hand between the shafts and each side of the equid's abdomen.

3. **Braking the load**

This function directly relates to the use of vehicles, as the presence of wheels requires a braking system. This is provided by a set of straps adjusted to the body of the animal and connected to the shafts of the vehicle, allowing the animal to slow or stop the load. It also allows the animal to reverse the vehicle.

Braking is enabled by the use of **breeching**, a broad strap (ideally 8 to 10 cm wide) similar to a breast collar but placed around the rear of the animal, halfway between the hock and the rump (Figs. 8a and 8b). The ends of the breech strap are connected by breeching tugs to a point on the shafts. This prevents the vehicle from overtaking the equid when the animal is slowing or going downhill. When the signal is given to stop, the equid can comfortably and securely sit back into the breeching and stop the vehicle, without the vehicle colliding with the backside of the equid. Similarly, to reverse, the equid can apply pressure against the breeching and achieve this action. The broad breeching strap is held in place by either 1, 2, or a single split strap that sits either over the hip or the loins (called a hip or loin strap) (Figs. 8a and 8b). The breeching should always be part of the harness of animals working in shafts or a central pole, but it is not necessary on dragged implements or loads as they will stop as soon as the animal stops and cannot be reversed (Rodrigues 2022).

If a breeching is not included as part of the cart harness, another option is to use a **false breeching.** This consists of a simple strap fixed across the shafts of the vehicle, closely behind the hind legs of the equid. If kept at the right height and allowing enough space it may work well, although it is less effective than a true breeching (Fig. 8b, dotted line).

The **crupper** is a strap that sits under the equids' tail and is attached to the saddle (Fig. 9). In draught animals its *sole* purpose is to stop the saddle from creeping forward over the withers and neck, and to help keep the breeching in place. It is crucial to check that it is *not* being used as a braking system, and when braking there should be no change in the tension of the back strap connecting the crupper to the saddle. In equids used for packing or riding purposes cruppers perform the same function, especially in hilly areas (Fig. 9). The pressure put on them can be substantially more in these cases, depending on the weight of the cargo or rider and the steepness of the slopes. If needed, a good option for pack equids used in mountainous areas is to add a second girth, as this will give stability to the saddle and will release tension from the crupper. The crupper sits in a very sensitive area for the equid. It must therefore be well made, smooth and fitted, and there should never be excessive pressure on it. It is one of the major causes of wounds, compounded by the fact that the wound often remains unseen even when the crupper is removed. The best crupper designs are usually quite hard, but smooth and shaped to fit comfortably.

Important Aspects to Keep in Mind:

- *Width of the breeching:* The width of the breeching should be of approximately 10 cm where it fits against the body of the equid. When adjusted correctly, it should leave a gap of 10 cm between the strap and the body of the animal when not in tension / not braking the vehicle.
- *Distance between the back of the equid and the cart:* The adjustment of the breeching combined with the length of the shafts is another important aspect to consider. It should ensure enough room between the back of the equid and the

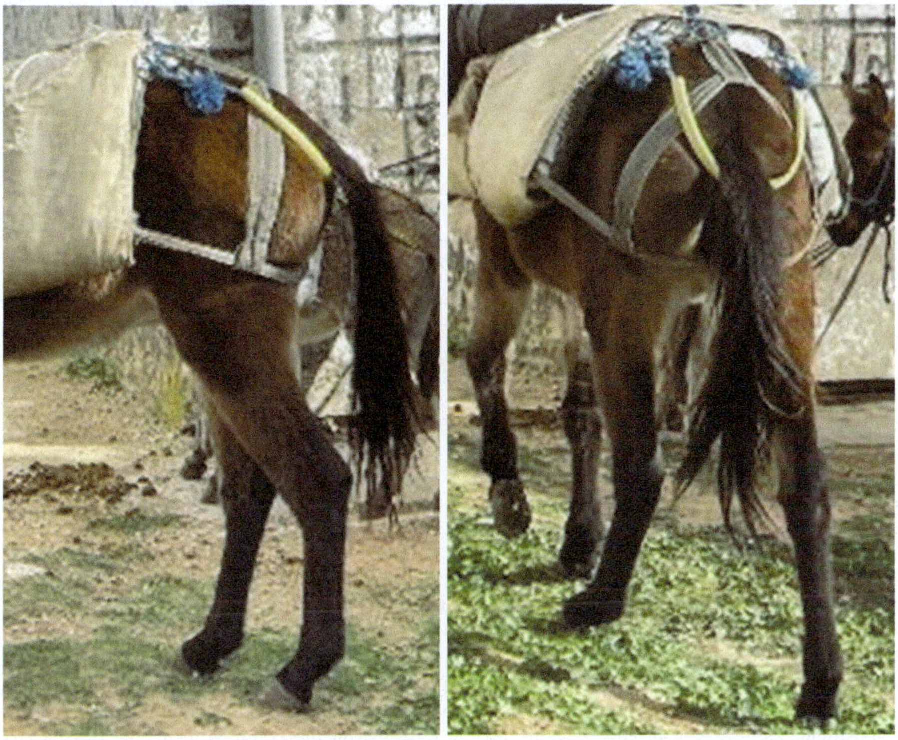

Fig. 9 Crupper (yellow band), fitted under the equid's tail. Note how the crupper is fitted loosely and is made from smooth, non-abrasive materials

front of the cart, when shafts and breeching are in place. It is crucial that the animal can move without any contact with the cart, but also without leaving excessive space as this could make it difficult to manoeuvre the cart.

4. **Communicating with the animal(s)**

The fourth key function of a harness relates to the equipment that usually goes on the animal's head and is used to communicate and secure the equid. This can consist of a variety of systems, most commonly involving a **head collar** or a **bridle**, combined with a **bit** and **reins** (Fig. 10). In many areas of the world none of this is used, and it is common to see equids guided by sticks without any equipment on their heads. In those cases, the risk of stress, fear or head injury to the animal is usually higher. Head collars and bridles can be very efficient for communication purposes but, like any other piece of equipment, can be used inappropriately and cause damage to equids. Bits and bridles that are not properly adjusted or are made of sharp materials to increase responses will also create lesions and behavioural problems (Fig. 11). One

Fig. 10 Schematic illustration showing the main components of a bridle in the correct position. The blinkers (dotted line) may or may not be present. In the opinion of the authors, blinkers should not be used. However, if used, they should not contact the eyelids of the animal, should have smooth edges and should not represent a hazard in any way. Also note the correct placement of a bit, with the space of approximately one finger fitting between the bit and the lip. With the right bit adjustment, one to two wrinkles should also be observed at the commissure of the lips. Image: Modified version from "The goof harness guide", The Donkey Sanctuary

Fig. 11 Examples of inappropriate bits. Note how these bits are designed to be excessively hard on the mouth and tongue, are kept in poor condition and are secured with wire representing a further injury hazard

problem commonly seen is the incorrect adjustment of the bit in the equid's mouth. Other areas to consider are what type of curb chain or chinstrap is being used, as it is not uncommon to see very abrasive material used under the sensitive chin of the equid. It is also important to consider the adequacy of bitless bridles such as hackamores, that can exert a lot of pressure on the bridge of the nose. This is more commonly seen in donkeys used for riding or packing purposes. Again, this piece of equipment should be properly adjusted, not too tight and made of nonabrasive material. Noseband and curb chain (both in bits and hackamores) should be smooth and not cause any lesion to the equid.

The correct positioning of the bridle should allow for sufficient free space between the different components of the bridle ensuring they are all within safe limits that don't impinge on the structures of the head. There should be enough space to fit two fingers between the different parts of the bridle and the head of the equid, and four fingers in the case of the throat latch. If the bridle is used with a hackamore, two fingers should also fit between noseband, the curb chain and head of the equid.

A **bit** consists of a bar, usually made from metal or sometimes rubber, placed in the equids mouth for the purposes of communication and control. The most commonly used bit is the snaffle. The snaffle itself comes in many forms but essentially acts on the mouth, the bars of the mouth, the tongue and sometimes the roof of the mouth or hard palate. Most donkeys will work well with a snaffle, but availability, regional or cultural preferences may mean they use other types of bit, some of which can be very hard on the mouth. Bits should be fitted snuggly up into the corners of the mouth, just tight enough to wrinkle the skin 1 to 2 times (Fig. 10). If the bit is too tight it will cause pain, and if too loose it will eventually collide into the lower incisors or canines, when present. In terms of size of the bit, one should be able to fit a finger on each side of the equid's head between the bit and the lip (Fig. 10). Bits should always be smooth, any repairs should be done carefully and weld splatter, wire, or other abrasive materials should not be allowed to cause injury (Fig. 11).

Many working equids also wear **blinkers** (Fig. 10), which are designed to reduce the peripheral vision of the animal, usually with the aim of increasing focus and reducing distractions. However, it is important to consider that blinkers are an entirely optional part of the harness and in fact many animals do not react well to blinkers which means they are not always indicated. Blinkers do not allow the animals to have a full vision of the surrounding environment. This may affect their ability to express their natural behaviour and predispose the equid to accidental injury due to a reduced ability to detect hazards in the environment. For these reasons, in the opinion of the authors, blinkers should always be avoided. However, if blinkers are used, they should be designed in such a way that they do not rub against the eyes (a concave or dished shape is recommended) and should not have any sharp edges or protrusions.

3 Different Types of Harness Used for Different Work Purposes

Now that the main functions of a harness have been discussed, as well as the different components responsible for those functions, it is important to understand the most common differences in harnesses used for different work.

- **Pack** and **riding saddles:** These saddles are designed to carry loads (Fig. 12) or people.
- **Agroforestry harness or similar**: This type of harness refers to those used to drag a load on or through the ground (Fig. 13). There is no load on the back and no need for a braking system. There is a communication system in place, and other straps may be present to ensure the adjustment of the harness to other tasks such as the use of a trace carrier to keep the traces suspended avoiding interference with the legs of the equids. If more than one animal is used to perform any agroforestry activity, there should be no changes in the harness, and only different

Fig. 12 Packsaddle with all connecting straps in place: girth strap (red); second girth (blue); breast strap (yellow); breech strap (green). Note the size of the lateral panels, designed to protect the body of the equid from the cargo, and the gullet space preventing the saddle and the padding from putting direct pressure over the spine

Fig. 13 Donkey using an agroforestry harness with a full collar to perform agriculture activities

Fig. 14 Pair of donkeys hitched in parallel, with central shaft/pole attached to the full collars via a similar swingle tree-like system (eveners)

combinations of swingle trees are used to ensure all animals work evenly (such as the example on Fig. 5).

- **Single cart harness:** designed to enable one animal to move a vehicle. This type of harness is the most complete, as it includes components to move the load, carry the load, brake the load, and usually for communication, depending on local custom (Figs. 8a and 8b).
- **Pairs or team harness:** These enable two animals to move a vehicle or implement. The most common situation is the use of two parallel equids (abreast), and in these cases a central pole or shaft is present, combined with an evener or set of eveners (similar to swingle trees), attached to the lower part of the collars (Fig. 14). When used for moving a vehicle, although there is usually less weight on the central shaft, animals carry the weight (if not balanced) using the collars and not the saddle, so cart saddles in the pairs harness tend to be smaller. A braking system is needed (when used for moving a vehicle), as well as a communication system. Team harnesses can also be used in agriculture, hitched directly to the implement. Equids can also be hitched "in tandem", with one animal in front of the other rather than side by side. In this case only the animal in between the shafts will carry the load and brake the load. Therefore, the rearmost equid will have a complete cart harness while the equid at the front only helps moving the load, combined with a communication system.

4 Considerations Regarding the Vehicle

There are many types of vehicles that can be pulled by equids, but in general we consider two types of vehicles: the two-wheeler and the four-wheeler. Sometimes, other types and combinations of vehicles such as forecarts with three wheels may also be used, although this is less common.

A **two-wheel cart** has one axle with a wheel on either end. These carts are usually cheaper to make, lighter and smaller, and are the most common type of cart seen worldwide (Fig. 15). The one great disadvantage of the two-wheeler is that it has no balance on its own. This means that the driver must distribute and balance the load so that there is minimum weight on the shafts and therefore on the equids' back. The balance of the cart is crucial for the welfare of the equid. When well-balanced, there

Fig. 15 Example of a common two-wheel cart. Note that in this case the wheels of the cart are bulky and heavy. The shafts are positioned too high and should fit approximately halfway between the ventral and the dorsal aspects of the chest of the equid. There is not sufficient space between the shafts and the sides of the animal, risking rubbing and wounds

is practically no weight on the shafts, meaning no weight on the back of the animal. In those cases where the cart is unbalanced, due to the construction design itself (wrong position of the platform in relation to the axle) or due to an incorrect distribution of the load (either too forward or too backwards), this will cause pressure on the back or in the girth area, depending on where the extra load is placed in the platform.

A **four-wheel cart** has two wheels on a fixed axle at the rear, and two wheels on a front axle. Ideally, there is a swivelling axle at the front to which the shafts are attached. This type of vehicle needs no help from the equid to stay off the ground. The great advantage of the four-wheeler is that it is stable and has no balance issues. It does not matter where or how the vehicle is loaded or unloaded, as it makes minimal difference to the weight put on the equid's back. The downside is that it generally costs a lot more to make and maintain, and it weighs significantly more than the two-wheeler. In ideal conditions more than one equid should be used to pull this type of vehicle.

Another important consideration is that of outrigger or satellite animals (Fig. 16). These are additional equids that may be attached to the cart, usually by a rope around the neck, but that are not between the shafts. This often takes place to allow for periodically replacing the main draught equid or to prevent leaving equids unattended, but it is also often done due to beliefs that it helps train young equids to the cart and that outrigger animals contribute to traction. However, it is important to know that these animals do not contribute to draught efficiency and may sometimes contribute to lesions and discomfort in the main cart-pulling equid that is working between the shafts (Shaw et al. 2007).

There are certain cart design considerations that should also be kept in mind. Donkey carts are often designed with funnelling shafts that narrow at the point of hitching (ie the shafts are not parallel and are narrower at the animal end than at the cart end). This often creates rubbing wounds at the point of the hips or other parts of the body. The wheels of the cart are usually made from pneumatic tires, although wooden or metal wheels are sometimes also used in some parts of the world. It is

Fig. 16 Example of two outrigger animals connected to a two-wheel cart either side of the main draught equid. Only the equid working between the shafts significantly contributes to moving the load

important to have good axle alignment as well as good tire inflation to reduce resistance to movement, improve draught efficiency and prevent unnecessary effort to the animal. Additionally, good wheel bearing should be ensured, to prevent friction or oscillatory movement of the wheels that may also affect the equid's performance.

5 Practical Aspects of Work Equipment Use and Maintenance

Once we understand the design and function of a harness, it is also important to remember certain practical considerations for the everyday use and maintenance of harnessing equipment. It does not matter how good or well designed the equipment is if it is not used correctly, and it is incorrect harnessing practices that lead to several of the welfare concerns we see relating to work equipment. Key questions to consider when assessing a harness are:

- **Is the harness suitable for the task performed by the equid?**
  This is a key aspect to consider. Everything discussed in the section dedicated to the functions of the harness should be kept in mind when checking how the equipment is being used by the equid owner or user. You may have a well-adjusted packsaddle that has been used to move loads, so despite the correct adjustment the harness is not suitable for the task being performed.
- **Is the harness operating in a correct, efficient and animal friendly manner?**
  Once again consider all the information provided in terms of the harness functions, including the presence of all parts needed to operate it in a correct way, such as the presence of breeching in a cart harness, or the connecting straps to keep a packsaddle in place. Also consider key aspects related with the

adjustments of the equipment. The harness should be well-adjusted to the body of the animal and should not have pressure points that may cause lesions to the animal. Any contact or pressure on critical points should be avoided (trachea, point of shoulders, spine, etc.).

- **Load distribution:**
  Another essential consideration is that of an even load distribution in the case of pack saddles and two-wheeled carts particularly. In the case of pack saddles, it is crucial to keep the load evenly distributed on both sides of the animal and close to the centre of gravity of the equid. When not balanced, the load will not only be more difficult to keep in place, but it will also cause pressure points on the side where it is heavier. When the load is kept away from the animal's centre of gravity (even when balanced) it will increase the oscillatory movement and make the load unstable, also increasing the movement of the saddle and increasing the likelihood of wounds. In the case of two-wheeled carts, even load distribution must be considered not only in relation to both sides of the animal but also in relation to the front and the back of the cart. This should include the sitting position of the cart driver and will affect how much weight is placed on the saddle and back of the equid (if the majority of the load is at the front of the cart) or on the girth, belly band and ventral belly of the equid (if the load is placed predominantly at the back of the cart). In a well-balanced cart, there should be minimal weight on the shafts and saddle. In many parts of the world, it is common practice for cart-drivers to sit on one of the shafts, which allows them to jump on and off the vehicle easily. However, it is easy to see how this can add to and unbalance the load on the equid's back, increasing the likelihood of skin and musculoskeletal lesions in working equids.
- **Materials:**
  Selection of the materials is also a key aspect to consider, when deciding if the harness operates in an animal- friendly way. Materials used for harnessing should not be abrasive or traumatic in any way. Many different types of materials can be used, but the important thing to remember is that working equids may sweat profusely during work, so it is important to use natural materials that breathe. Synthetic materials such as nylon, rubber and plastic are often cheap, strong and easy to work with, but they do not breathe, so if synthetic material is present against the equid's skin, the owner, user or handler should be advised to pad the harness with natural materials like cotton or wool. This can be made from old clothes, blankets, sheepskins or whatever else can be found locally. Materials used should also be resistant to prevent tearing or damage, and easy to clean to ensure hygiene can be maintained. Cart materials are also an important consideration. Again, it should be ensured that these are not traumatic or can cause damage to either the equid or the user. Metal structures can easily overheat in hot temperatures and should not be in contact with the animal or user.
- **Daily maintenance and cleanliness:**
  When dealing with working equids, regular care and hygiene of both animals and harness are key aspects to ensure not only the health and welfare of animals but also the durability and functionality of the materials. It is crucial to pay special

attention to those areas of direct contact with the harness, where dry sweat and dust may accumulate. Different types of materials may require different cleaning or maintenance methods. For leather or synthetic leather materials, the best way to clean it is by applying water with soap using a hand sprayer. Let the dust moisten and remove it with a brush or a towel. Apply grease to the leather material on a regular basis, keeping the material hydrated and soft. For canvas or similar materials if we decide to wash it with water and soap, we need to always keep in mind that the filling material will be wet, and it will need time to dry, in a well-ventilated area. For daily maintenance of such materials, the best option is to allow the harness to dry after use, and once dry brush it to remove all dust. Finally, daily maintenance should also include basic visual inspection, checking for any new damage, tearing or broken or sharp bolts or protrusions that may harm the animal.

6 Common Equipment Weaknesses and Socioeconomic Challenges

Working equids are used in different contexts around the world, each with their unique challenges regarding work practices and equipment. These may relate to social, economic, environmental, cultural or other factors, and can affect not only the type and quality of equipment used but also the animal's welfare condition.

Common weaknesses in equipment design include a low traction functionality, where the components required to move the load are often lacking or not correctly used. For example, it is common for donkey harnesses to lack traces or swingle trees, with a breast collar attaching directly onto the saddle or shafts. Saddles often lack a gullet space to protect the spine from direct pressure and bits are frequently in poor condition or aren't the correct size to fit the animal (Merridale-Punter et al. 2024a). Other common shortages include contact materials that are made from synthetic fibres or that are unclean or poorly maintained in such a way that causes abrasion and friction. However, several factors help explain these weaknesses. In most countries where working equids are used, the high reliance on animal traction power is often combined with limited resources. This not only affects the living and working conditions of donkeys and hybrids, but also that of their owners and users. Equipment cost and lack of access to adequate materials or manufacturers are some of the key barriers seen in low- and middle-income contexts where working equids are used (Merridale-Punter et al. 2024a). This favours the use of less expensive and often recycled equipment materials for manufacturing harnesses. Professional harness-makers are often not available, and equipment is frequently home-made by equid users. Although home-made equipment using inexpensive or recycled materials can be entirely adequate and perfectly fit-for-purpose (Merridale-Punter et al. 2024a; Rodrigues et al. 2021), owners and users may sometimes lack the necessary technical knowledge to produce functional equipment designs. Similarly, harness and cart-makers are often not professionally trained in equipment design and function principles. Other equipment-related challenges include the poor availability of maintenance and repair services, and lack of technical harness training and awareness of the importance of harness design by working equid owners and users (Merridale-Punter et al. 2024a). Several equipment and work practices have deep traditional roots, and it is important to understand and respect the priorities and motivations of working equid users. Finally, there is often no regulation regarding work practices, loading or equipment standards, and improvements to these challenges are therefore reliant on supporting positive human behaviour change.

7 Common Equipment-Related Injuries and Their Clinical Approach

The use of inadequate harnesses or carts has been linked to several welfare impacts such as lameness or wounds (Ali et al. 2016; Bazezew et al. 2014). Harness-related wounds are extremely common among working equids and one of the most important clinical concerns in equids of low- and middle-income countries (Merridale-Punter et al. 2024b; Rodrigues et al. 2020). Although the causes, type and severity of harness-related wounds will vary, the overall approach to these cases should be the same: Firstly, clinicians should aim to identify which factors in the equipment are leading to wounds. Secondly, a wound treatment plan should be decided and implemented, remembering the importance of rest for recovery. Finally, clinicians should discuss potential solutions with the owner, addressing the concerns in the equipment and discussing possibilities for improvement. Here, it is important for clinicians to work together with owners and consider local human and material resources available to arrive at solutions that are appropriate for the socioeconomic context as well as the needs and limitations of the owner.

7.1 Predisposition

When work equipment does not respect the key principles of design, fit and use discussed above, it can easily lead to injury. As such, work equipment and practices should be part of the clinical investigation when treating working equids, and likewise animals with detectable equipment flaws should be examined for signs of injury. This should be in the clinician's mind while taking a history, performing a physical examination and formulating a treatment plan. In addition to the equipment itself, other husbandry and management factors may predispose working equids to a higher risk of injury. For example, factors such as poor quality nutrition or energy imbalances frequently translate into low body condition, which can in turn be associated with a higher likelihood of having wounds or lameness (Pritchard et al. 2005), so the equid's broader management and environment should also be considered.

7.2 Clinical Signs and Critical Findings (Clinical Case Examples)

A variety of injuries can result from unsuitable work equipment or work practices. Early signs to look out for before having obvious clinical problems include behavioural issues and equids showing a reluctance to work or subtle changes to the way they normally work. However, one of the most direct manifestations of inadequate work equipment are harness-related wounds.

As discussed throughout this chapter, harness-related wounds can result from a multitude of factors relating to either the design, materials, fit or use of the equipment. Often, these wounds are not associated with a single aspect of the equipment but rather to the overall adequacy of the harnessing system and how the different components work together. However, some of the most common sites and causes of equipment-related wounds are described below (Fig. 17):

Bit wounds These are typically caused by bits that are poorly manufactured and designed, in poor condition or that are not the correct fit to the animal's size. Wounds can often be seen in the commissure of the lips, but inadequate bits can also cause injuries to the tongue leading to laceration or, in the worse cases, amputation. The

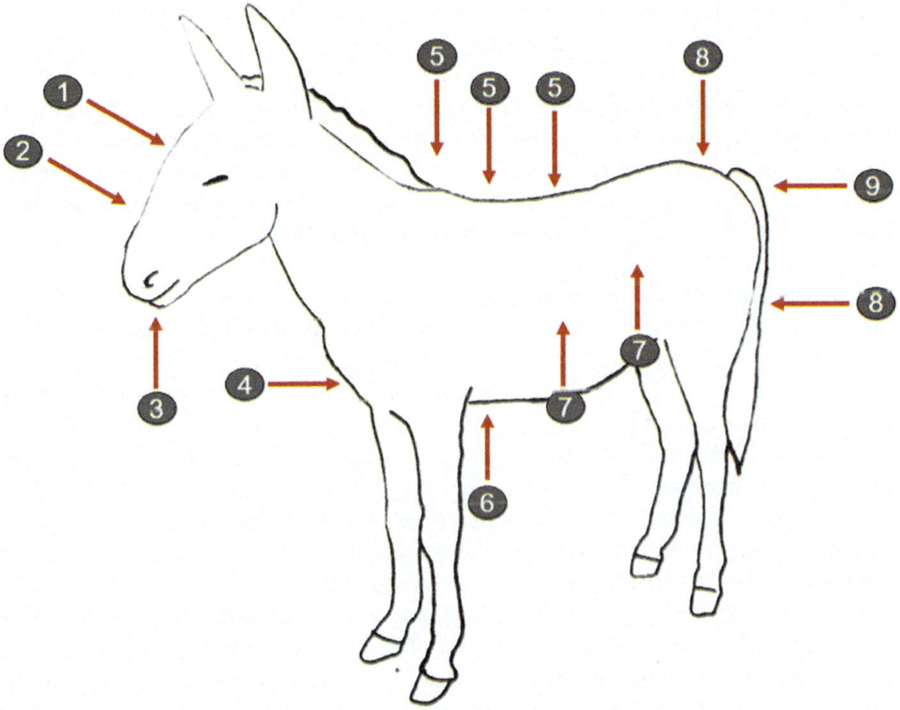

Fig. 17 Common sites for harness-related wounds: 1 blinkers; 2 noseband; 3 bit; 4 breast collar; 5 saddle; 6 girth; 7 shafts; 8 breeching; 9 tail/crupper or cart

equid may present with signs such as discomfort around the mouth, difficulty eating or reluctance to wear a bit. Wounds caused by bits are often unnoticed as they can be "hidden" inside the mouth. The bars, tongue and roof of the mouth should be regularly checked, as well as the teeth, lips and the surrounding areas of the mouth. Bit wounds are particularly common in donkeys, and often caused by the use of bits that are in poor condition or oversized for the animal (such as bits intended for horses) (Merridale-Punter et al. 2024c) (Fig. 18c, d).

Wounds to the bridge of the nose These lesions are common in animals wearing a halter or head collar that does not fit adequately or is made from inadequate materials. They are usually caused by the friction of the noseband particularly if made of metal, rope or other abrasive materials. They can also be the result of driver practices, if used by equid drivers as an element of restraint and punishment, causing trauma by vigorous or repeated pulling of the reins or lead rope (Fig. 18b).

Wounds in the chest region Chest wounds associated with the harness are usually due to the friction of the breast collar against the chest. This typically occurs due to a poorly functional traction, where either traces or a swingle tree may be absent or inefficiently used, in such a way that the normal movement of the animal causes friction against the collar.

Collar wounds may also be caused by poorly adjusted or manufactured collars or by those made with traumatic materials (Fig. 18a).

Wounds in the girth region These are often caused by the friction of the girth or the belly band, particularly when the saddle is loose and the load unbalanced (such as when the majority of weight is placed on the back of a two-wheeled cart). Alternatively, these may be caused by girths or belly bands that are too tight and often made from rope or other stiff and traumatic materials (Fig. 18f).

Lesions of the back and withers These are usually lesions caused by the friction of the saddle or saddle pad, when it is poorly designed, fitted or made with synthetic or abrasive material. They may be present at different sites along the back, depending on the saddle shape, loading and positioning on the animal. When the saddle is worn over the withers and particularly if there is no gullet space, this may lead to fistulous withers. This may manifest in a variety of ways, including bursitis, abscesses, phlegmon or fistulous wounds of the withers. These lesions are characterized by swelling of the withers with inflammation, presenting hot and painful in most cases. In cases of bursitis and abscesses of the withers the swelling is soft with points of fluctuation. On the other hand, in cases of phlegmon the swelling is rather hard. In cases of fistulous wounds which can result from the spontaneous eruption of abscesses, several fistulas can usually be observed with either a serous, mucous, mucopurulent or purulent discharge. These lesions have a high morbidity and are difficult to treat despite adequate care. They may require several months of rest from work in order to heal (Fig. 18h,i).

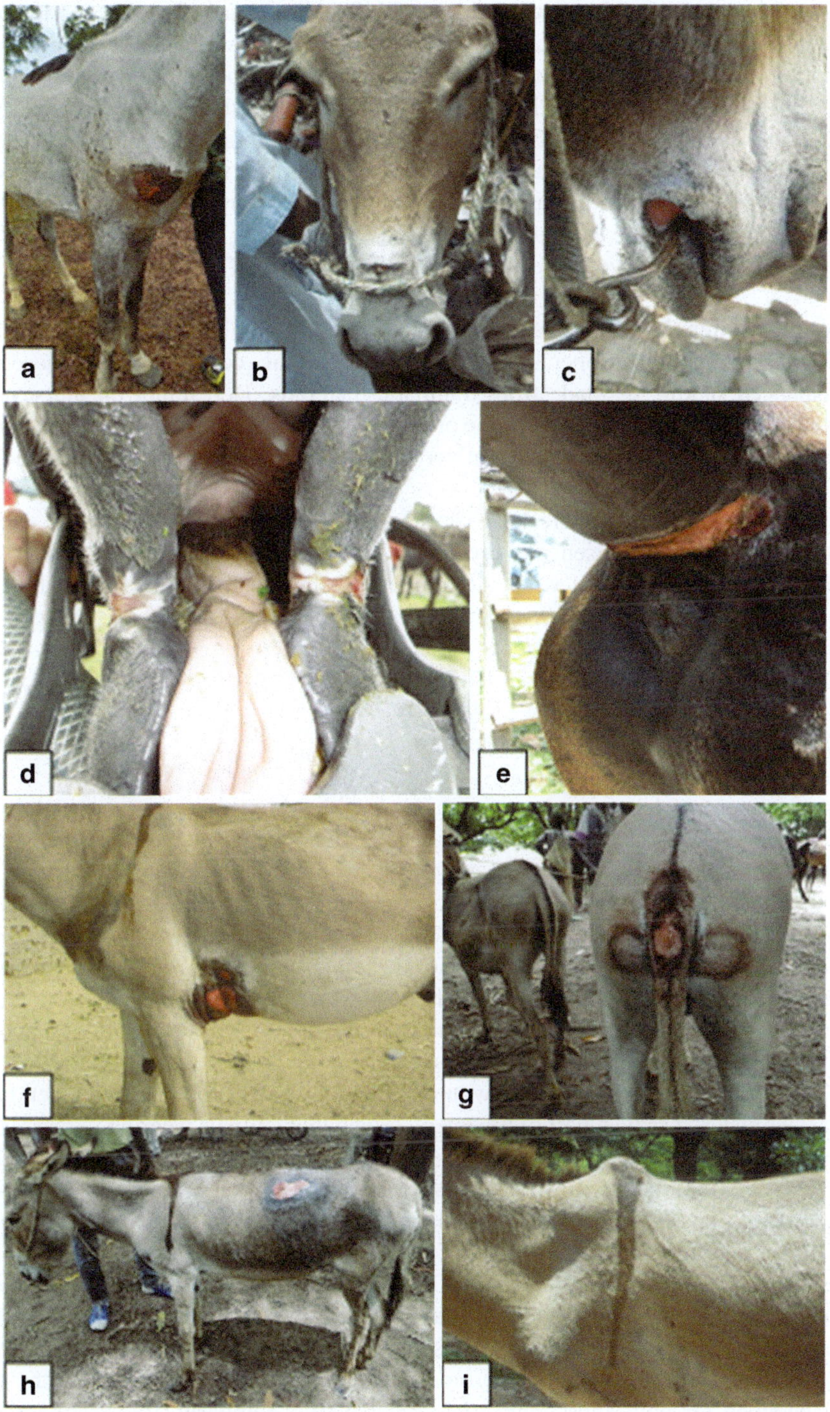

Fig. 18 Common skin lesions sites caused by harnessing equipment seen inworking equids

Wounds on the caudal aspect of the hindlimbs These are often due to friction of poorly designed breeching or a false crupper (mix between a breeching and a crupper, usually as an "oversized crupper" that resembles the breeching in its position but is attached to the saddle rather than the shafts therefore not fully achieving either function). The use of defective or abrasive equipment causing trauma to the proximal and caudal aspect of the hindlimbs, which most often appears in the form of abrasions or superficial wounds but can vary in severity. Wounds in this region may also be caused by the absence of breeching in a cart harness, and the repeated collision between the vehicle and the backside of the animal (Fig. 18g).

Wounds under the tail These wounds are most commonly caused by the continuous rubbing of the crupper, which may be too tight, connected to a loose and unbalanced saddle, or made from traumatic materials. Depending on the severity of the rubbing and the duration of the process, these wounds can affect a good part of the tail muscles, reaching the caudal vertebrae in the most severe cases (Fig. 18e).

Wounds on the base of the tail These are typically seen in cart-pulling equids, usually caused by the absence of breeching, resulting in the frequent collision of the vehicle against the animal when stopping, reversing or going downhill. They can also be caused when the shafts are too short and the animal is hitched too closely to the cart, rubbing it frequently (Fig. 18g).

7.3 Treatment and Prevention

The treatment of harness-related wounds should follow the general principles of wound treatment. This involves a thorough cleaning and debridement of the wound, with further medical or surgical therapy as appropriate. In some cases, dressing the wound may also be required. In cases of back injuries or painful fistulous withers, a more careful exploration of the wound may be necessary, often involving diagnostic imaging to investigate the extent of bone and soft tissue damage or surgery to expose the pockets of suppurations but also often to extract bone sequestrae as a potential consequence of the micro fractures caused by the pressure of the saddle. The use of drainage is often necessary in these cases.

In all cases of harness-related wounds, like any other wounds, it is essential to consider rest as a critical part of treatment. Given that animals wearing harnesses are normally used for work and may often be the main source of income for their owners, this may be a delicate conversation to have with the owner. As such, it is important to involve the owner in the planning, taking the time to explain the aims of treatment and finding a suitable plan of action that best protects both the equid's and the owner's interests. Continued work and repetitive trauma from the harness will compromise welfare, contribute to aggravating the injuries and may lead to complications and chronic disabilities. This should form part of the advice to equid owners and users, to promote understanding and cooperation with perspectives for both short- and long-term benefits. Similarly, the treatment plan should involve discussing the likely causes and equipment origins of the injury with the owner. Owner and clinician should work together to identify areas of concern in the equipment or work practices and develop constructive solutions to prevent repetition of the problem. Community engagement and communication is essential for effective and sustainable improvement to harness quality, favouring locally available and inexpensive materials that are suitable for the individual animal and owner needs.

Prevention of harness wounds and other equipment-related problems should be focused on ensuring the equipment is designed, maintained and used appropriately. In other words, ensuring not only the correct contact materials, but also an adequate design, assembly and fit for the individual animal and the conditions of work. Most importantly, the equipment must be *used* correctly. Veterinary professionals and paraprofessionals play a key role in promoting good practices such as appropriate load balancing and distribution, tire inflation, routine maintenance and others as part of injury prevention. Additionally, it must be remembered that wounds are not the only possible consequence of inappropriate work equipment, and other welfare concerns can be prevented through the correct use of equipment. For example, adequate work practices can help prevent animal stress and behavioural alterations (Bukhari and Parkes 2023), and not overloading and providing at least one rest day per week can reduce the risk of lameness (Ali et al. 2016; Merridale-Punter et al. 2022). As such, when trying to prevent equipment-related health and welfare issues, taking the time to critically observe the equipment used and asking owners and users about routine work practices is critical. Owners and animal health professionals should discuss this as part of the consultation and jointly design a suitable prevention plan.

8 Summary

Harnesses are a key part of a working equid's life and consideration should be given to work equipment as part of welfare strategies and clinical evaluations. Although there are various types and models of harnesses, they generally enable the same key functions: moving the load, carrying the load, braking the load and communicating with the animal. Understanding these functions as well as the key equipment components responsible for each of them and their main design considerations is essential to understand how to detect weaknesses in the equipment and prevent equipment-related issues. Wounds caused by harnesses are one of the main clinical presentations in working equids and a major welfare concern. Animal health professionals should therefore understand how to recognise equipment-related wounds and risk factors, so that adequate prevention and treatment advice can be given to owners. Owners and animal health professionals should work together to ensure the equipment fits the animal correctly and is fit for purpose, appropriately designed, maintained and used, to protect the welfare of the equid and the needs of its owner.

Reflection Time: Scenarios and Auto-Evaluation

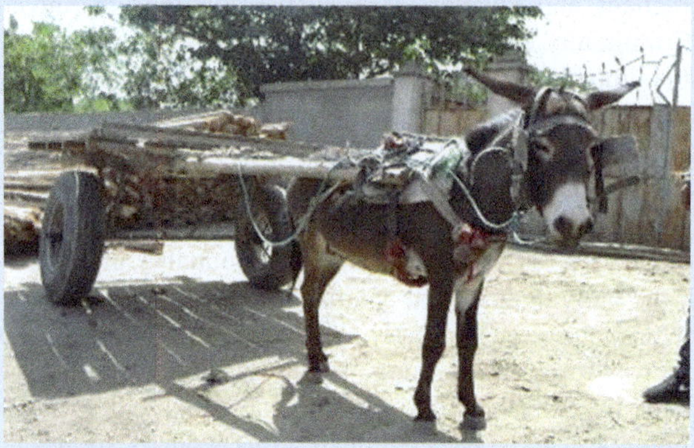

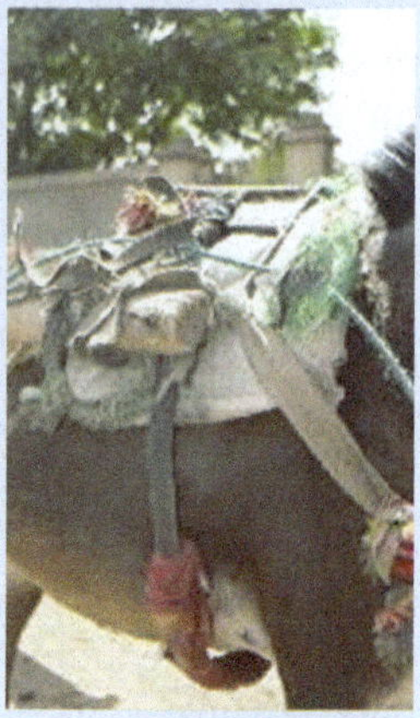

Think about...

- What concerns can you identify with this donkey's harness?
- What concerns can you identify with the cart?
- What welfare problems is this work equipment likely to cause?
- What would you change in the harness and cart to avoid these problems?

Think about...

- What concerns can you identify in this scenario?
- What can you observe regarding the materials used in these pictures?
- Where is this work equipment likely to cause lesions?
- What are the three main changes you would recommend in this case?

(continued)

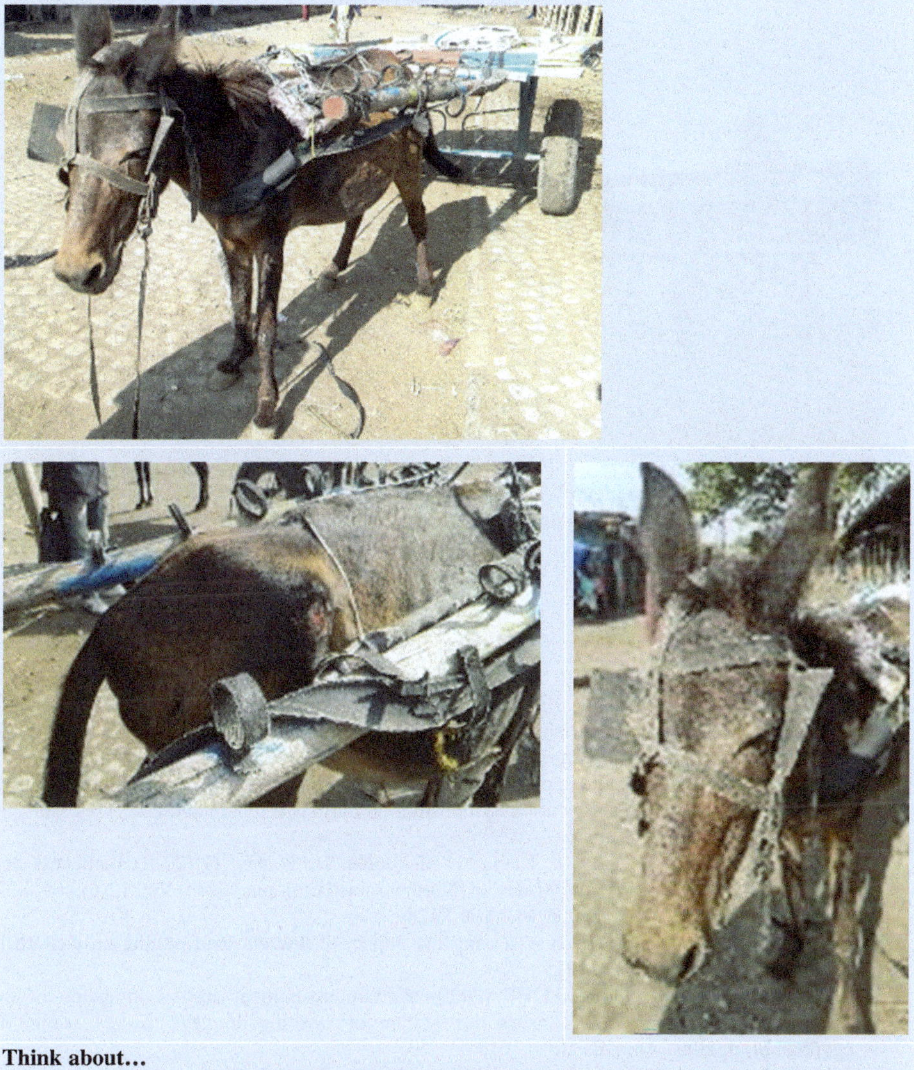

Think about…

- Identify two positive and two negative aspects of this harness.
- How could you improve efficiency of load carrying?
- What recommendations would you give to prevent equipment-related injuries?

(continued)

Think about...

- What concerns can you identify in this scenario?
- What key components do you think might be missing from this harness?
- What are the main concerns you can identify with the cart?

References

Ali A, Orion S, Tesfaye T, Zambriski JA (2016) The prevalence of lameness and associated risk factors in cart mules in Bahir Dar, Ethiopia. Trop Anim Health Prod 48:1483–1489

Almeida A, Rodrigues JB, Figueiredo T d (2017) Animal traction: new opportunities and new challenges. IX International Scientific Symposium on Farm Machinery and Processes Management in Sustainable Agriculture

Bazezew M, Chanie M, Tesfaye T, Kassa A, Mekonnen B, Wagaw N (2014) Lameness and associated risk factors in Cart Mules in Northwestern Ethiopia. Glob Vet 12(6):869–877. https://doi.org/10.5829/idosi.gv.2014.12.06.84289

Bukhari SSUH, Parkes RSV (2023) Assessing the impact of draught load pulling on welfare in equids. Front Vet Sci 10:1214015

Farhat SF, McLean AK, Mahmoud HF (2020) Welfare assessment and identification of the associated risk factors compromising the welfare of working donkeys (Equus asinus) in egyptian brick kilns. Animals 10(9):1611

Fernando P, Starkey P (2004) Donkeys and development: socio-economic aspects of donkey use in Africa. Donkeys, People and Development. A Resource Book in the Animal Traction Network for Eastern and Southern Africa (ATNESA), 15. https://citeseerx.ist.psu.edu/document?repid=rep1&type=pdf&doi=3d55b422b0454c1ebb981299f6907a05a50d0744

Margentino MR (2020) Proper harness condition and fit. Retrieved 15 November 2024 from https://nasdonline.org/904/d000744/proper-harnesscondition-

Merridale-Punter MS, Wiethoelter AK, El-Hage CM, Hitchens PL (2022) Prevalence and factors associated with working equid lameness in low-and middle-income countries: a systematic review and meta-analysis. Animals 12(22):3100. https://doi.org/10.3390/ani12223100

Merridale-Punter MS, Elias B, Wodajo AL, El-Hage CM, Zewdu H, Tesfaye R, Hailegebreal G, Sori T, Wiethoelter AK, Hitchens PL (2024a) Putting the cart before the horse: mixed-methods participatory investigation of working equid harnessing practices in three selected towns of the Oromia national regional state in Ethiopia. BMC Vet Res 20(1):113. https://doi.org/10.1186/s12917-024-03967-3

Merridale-Punter MS, Wiethoelter AK, El-Hage CM, Patrick C, Hitchens PL (2024b) Common clinical findings identified in working equids in low-and middle-income countries from 2005 to 2021. PLoS One 19(6):e0304755. https://doi.org/10.1371/journal.pone.0304755

Merridale-Punter MS, Wodajo AL, Elias B, Bakos A-M, Zewdu H, Tesfaye R, Hailegebreal G, Sori T, El-Hage CM, Wiethoelter AK (2024c) Equipment-related wounds and associated risk factors in working equids of the Oromia national regional state in Ethiopia. Anim Welf 33: e42. https://doi.org/10.1017/awf.2024.52

Morrissey J (2013) Harness variety and function. Rural heritage. Retrieved 15 November 2024 from https://www.ruralheritage.com/new_rh_website/resources/harness_tack/harness_variety_function.shtml

Norris SL, Little HA, Ryding J, Raw Z (2021) Global donkey and mule populations: figures and trends. PLoS One 16(2):e0247830

Pearson RA, Simalenga TE, Krecek RC (2003) Harnessing and hitching donkeys, mules and horses for work. Centre for Tropical Veterinary Medicine, University of Edinburgh

Pritchard J, Lindberg A, Main D, Whay H (2005) Assessment of the welfare of working horses, mules and donkeys, using health and behaviour parameters. Prev Vet Med 69(3–4):265–283. https://doi.org/10.1016/j.prevetmed.2005.02.002

Rodrigues JB, Sullivan RJ, Judge A, Norris SL, Burden FA (2020) Quantifying poor working equid welfare in Nepalese brick kilns using a welfare assessment tool. Vet Rec 187(11):445–445

Rodrigues JB, Garrett C, Norris SL, Albuquerque F, Queijo L, Cooke F, Judge A (2021) Collar pressure mapping: an evaluation of seven collar types used on working donkeys in Europe. Vet Rec 189(8). https://doi.org/10.1002/vetr.772

Rodrigues JB et al. (2022) The Donkey Academy—Harness and Working Equids (HWAT 107) https://www.thedonkeysanctuary.org.uk/form/donkey-academy. https://www.thedonkeysanctuary.org.uk/form/donkey-academy

Shaw A, Kamau A, Lewa A, Murithi E, Ochieng F (2007) Determination of risk factors contributing to priority welfare issues in working equines: an illustration using breast and shoulder lesions in Kenyan draught donkeys. Fifth International Colloquium on Working Equines. The future for working equines. Addis Ababa, Ethiopia, 30 October–2 November, 2006

Anatomic Differences of Donkeys and Mules and the Implications for Veterinary Procedures

Suzanne Burnham

Abstract

Equine veterinarians and farriers have all the necessary education for working with donkeys and mules. The procedures for examination and treatment are the same, but some of the anatomy is different. This chapter describes the clinically relevant anatomical differences. The donkey has a layer of muscle that covers the jugular vein that makes venipuncture challenging. Donkeys and some mules have a pharyngeal recess that can trap a nasogastric tube. Because of the greater vasculature through the spermatic cord, donkeys castrated without ligation bleed excessively. Aging donkeys and mules by their incisor teeth using the horse charts as a guideline is unreliable. A donkey will usually be older than expected. Donkeys and mules are more efficient at converting nutrients to fat and can have extensive disfiguring fat deposits. The location of the nasolacrimal duct orifice in donkeys and mules is highly variable. Understanding the differences will help the treatments go smoothly and the outcomes be favorable. A last section reviews additional differences with less clinical relevance that are merely interesting to know and will surely be known to the client. Donkey and mule ergots are very prominent. Donkeys have chestnuts on the forelimbs only. The composition, shape and size of donkey teeth differ from horses. Donkey mane, forelock and tail differ from horses. Mules have tail hair like their horse parent. Donkeys have multiple anatomical differences in the pharynx and larynx that allow them to create loud sounds during both inspiration and expiration of air. Donkeys and mules have very long ears.

S. Burnham (✉)
Department of Veterinary Pathobiology, College of Veterinary Medicine and Biomedical Sciences, Texas A&M University, College Station, TX, USA
e-mail: sburnham@cvm.tamu.edu

A. K. McLean et al. (eds.), *Donkey and Mule Medicine: Foundations, Breeds, and Management*, https://doi.org/10.1007/978-3-032-12726-6_15

Keywords

Donkeys · Mules · Donkey anatomy · Mule anatomy

1 Introduction

The obvious differences between horses and donkeys are ear size, tail hair, hoof shape, and voice volume. Some of the less obvious differences are behavioral, metabolic, cerebral and anatomical.

This chapter provides a practical approach to finding the anatomical differences of the donkey and mule or hinny for times when referencing horse anatomy is unreliable for performing veterinary procedures. Procedural information will not be included except for recommendations for adapting already known procedures to donkey and mule anatomy.

Horses and donkeys evolved from a common ancestor but diverged and adapted to different environments. The horse adapted to eat grasses and live in moderate to cool climates of northern Asia and throughout Europe (Evans et al. 1977); donkeys evolved browsing rougher vegetation in hot, arid, mountainous terrain of northern Africa. Most of the anatomical features identified in this chapter are adaptations to those dry conditions (Tegetmeier and Sutherland 1895).

Mules and hinnies are the hybrid offspring of any of the many breeds of donkeys with any of the many breeds of horses. The resulting variability means that there is not a "standard" population of mules or hinnies, and therefore there are few scientific papers regarding mule or hinny anatomy. Anecdotal experiences report that mules and hinnies may have anatomical features of the horse or donkey or some combination of both (Burnham 2002).

As with horses, some degree of variation may be expected in location and features of donkey and mule anatomy. For example, the nasolacrimal duct orifice of a mule might be found near the expected horse orifice, near the expected donkey orifice or more often, somewhere in between.

The information presented in this chapter has been obtained from donkey dissection in the gross anatomy lab at Texas A&M University, College of Veterinary Medicine & Biomedical Sciences by the author, from practical clinical experience treating mules and donkeys, as well as more recently from sampling and data collection at the Peaceful Valley Donkey Rescue. The information presented herein also includes a thorough literature review.

This chapter is organized by veterinary procedures that may be affected by differences in anatomical structures between horses and donkeys and their hybrids.

1. Jugular venipuncture
2. Dental examination
3. Hoof care
4. Passing of a nasogastric tube
5. Access to the nasolacrimal duct

6. Castration
7. Epidural anesthesia
8. Abdominocentesis
9. Miscellaneous anatomical differences of interest

2 Jugular Venipuncture

2.1 Reason for Procedure

In veterinary medicine, the jugular vein is used to administer medication as well as for obtaining blood samples.

2.2 Anatomic Feature

There is one important anatomical feature of donkey anatomy that challenges the success of jugular venipuncture. The cutaneous colli muscle covers the entire jugular furrow of the donkey.

- In the horse, the procedure to raise and visualize the jugular vein is relatively simple. Horses have paired cutaneous colli muscles that overlay a small ventral area of the sternocephalicus neck muscles, the fibers of which are very thin and converge with the superficial cervical fascia (Sack and Habel 1977; Clayton et al. 2005). In the horse, this thin layer may pass over the most ventral region of the jugular furrow (Figs. 1 and 2).
- By contrast, the much thicker cutaneous colli muscles of the donkey, with their ability to tense the skin of the neck, covers the entire jugular vein, which is the reason visualization of the jugular vein is more difficult in the donkey. In the foal,

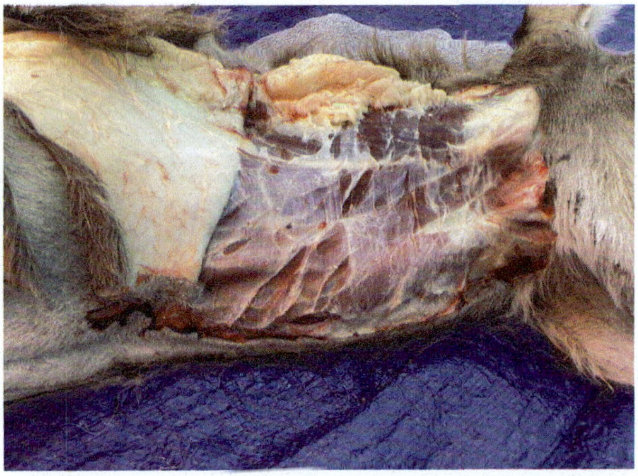

Fig. 1 Field necropsy of a mature donkey showing the cutaneous colli muscle prior to its reflection, demonstrating that it covers the external jugular vein within the jugular furrow

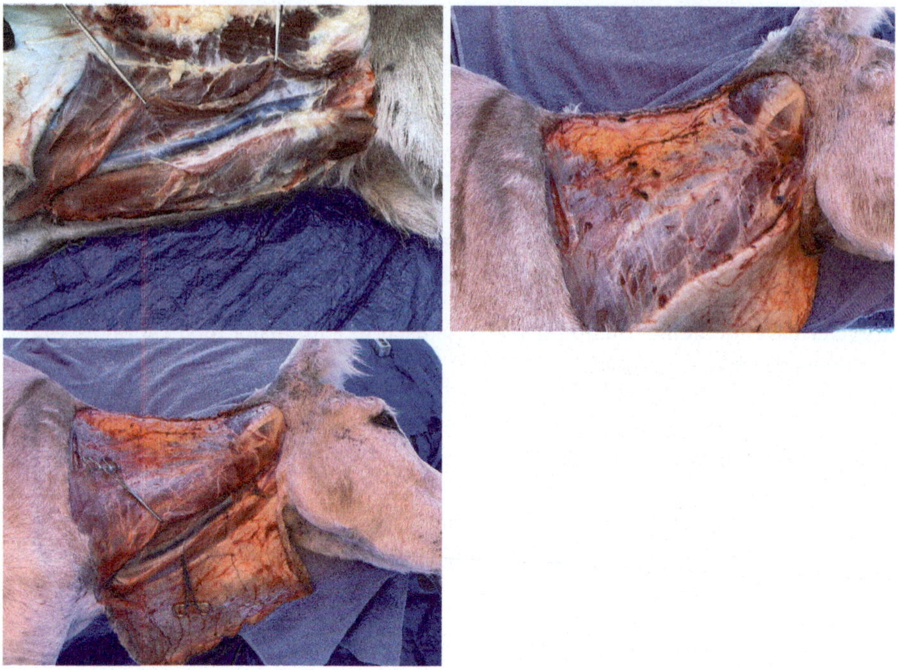

Fig. 2 The cutaneous muscle reflected away from the jugular vein

this muscle may only cover the middle one third of the jugular furrow (Burnham 2002) but in the mature donkey the cutaneous colli muscle is much better developed and will overlay the entire jugular vein and sternocephalicus muscle, converging with the brachiocephalicus muscle (Burnham field necropsy 2024, Fig. 1).

- Occasionally, the vein is more apparent in the cranial or ventral third of the jugular furrow in donkeys.
- How conformation, age, gender and conditioning may affect the development of the cutaneous colli muscle is yet to be determined.
- Mules bred from warm blooded horses are more like the horse; their veins are more easily visualized.

2.3 Recommendations

Raise and lower the vein with pressure near the thoracic inlet a couple of times to visualize the vein and aim with an increased angle than would be used for the horse. The most prominent raised area of the vein may be seen in the uppermost or lowest third of the furrow. Expect thicker, tougher skin when performing venipuncture in the donkey. As with horses, use caution, especially in the lower third, to avoid puncture of the carotid artery.

2.4 Donkey and Mule Behavior

Expect donkeys and mules to be more suspicious and more self-preserving than horses. Tip: after cleaning and prepping the area chosen for venipuncture site, take a few more seconds to habituate the area to the prick of the needle by scratching or even slightly pinching an area near the venipuncture site. While donkeys tolerate higher levels of pain, most mules and donkeys will not tolerate multiple stabs with good grace.

3 Dental Examination

3.1 Reasons for Procedure

Clinical examination of the teeth is useful for age determination and to look for signs of wear and sharp edges on the cheek teeth. Good routine dental care for donkeys determines their longevity, as poor dental condition is a main cause of declining health in donkeys. Donkeys and mules suffer the same types of dental disorders, diseases, and trauma as horses.

Donkey teeth are better suited to eating rough vegetation high in lignin and browsing on shrubs and bushes. Most of the health issues, especially dental problems suffered by domestic donkeys, come from living in non-native environments and being fed horse diets.

3.2 Anatomic Features

Donkey teeth differ from horse teeth in several clinically relevant ways making aging a donkey challenging. The composition of donkey teeth, the curve of Spee, and the donkey's longevity also differ from horses. Mules have some of the characteristics of both parents.

3.2.1 Morphology

- Donkeys have a greater degree of anisognathia (5–7%) than horses (Rodrigues and Lilly 2019; du Toit et al. 2008a, b; Buil García 2023) meaning that the distance between the cheek teeth of the maxilla is 27–30% wider than the cheek teeth of the mandible, giving the donkey jaw a greater lateral excursion when chewing browse. Anisognathia in horses is about 23% (du Toit et al. 2008a, b) (Fig. 3).
- Donkeys have a narrower mandibular space which can lead to dental overcrowding or enamel overgrowth (Fig. 4) (Thiemann and Sullivan 2019). The narrow mouth makes visualizing the molars more difficult. The narrow space between the mandibles is in part due to the higher mass of bone.

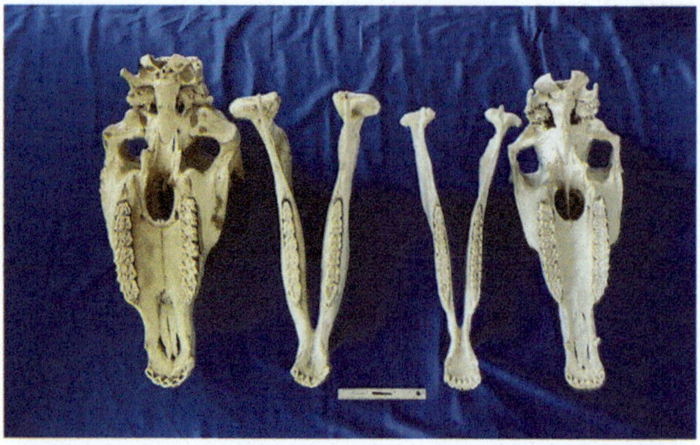

Fig. 3 Anisognathia

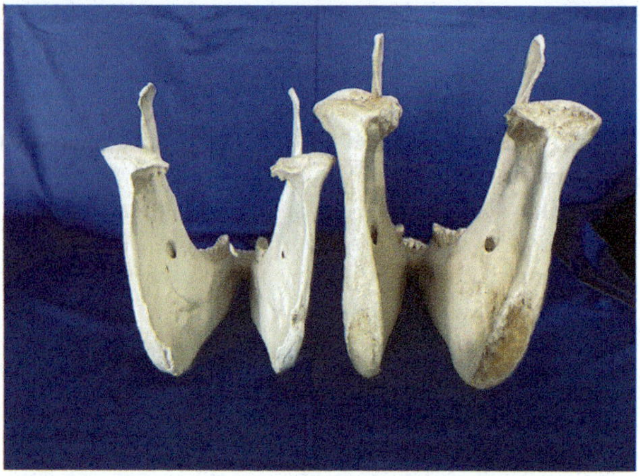

Fig. 4 Narrow mandibular space

- The curve of Spee, the natural upward curve of the caudal part of the occlusal surface of the cheek teeth, is pronounced in some breeds of donkeys and should be recognized as normal (Rodrigues and Lilly 2019; Buil García 2023) and not *corrected* with dental floats (Fig. 5).
- Important for molar extraction, in donkeys the superior cheek teeth are reported to have three roots and the inferior cheek teeth are reported to have two roots, as in the horse. There are reports that the last inferior molar (TRIADAN 311,411) in donkeys can have three roots (Misk and Seilem 1997), an extra root, when compared with the horse but this is considered to be a normal variation and may not occur in all donkeys (Buil García 2023) (Fig. 6).

3.2.2 Aging and Longevity

- Eruption of both deciduous and permanent incisors occurs three to six months later in donkeys than in horses (von Bünger and Hertsch 1981; Muylle et al. 1999) however, the appearance of dental stars on the permanent incisors are seen at a

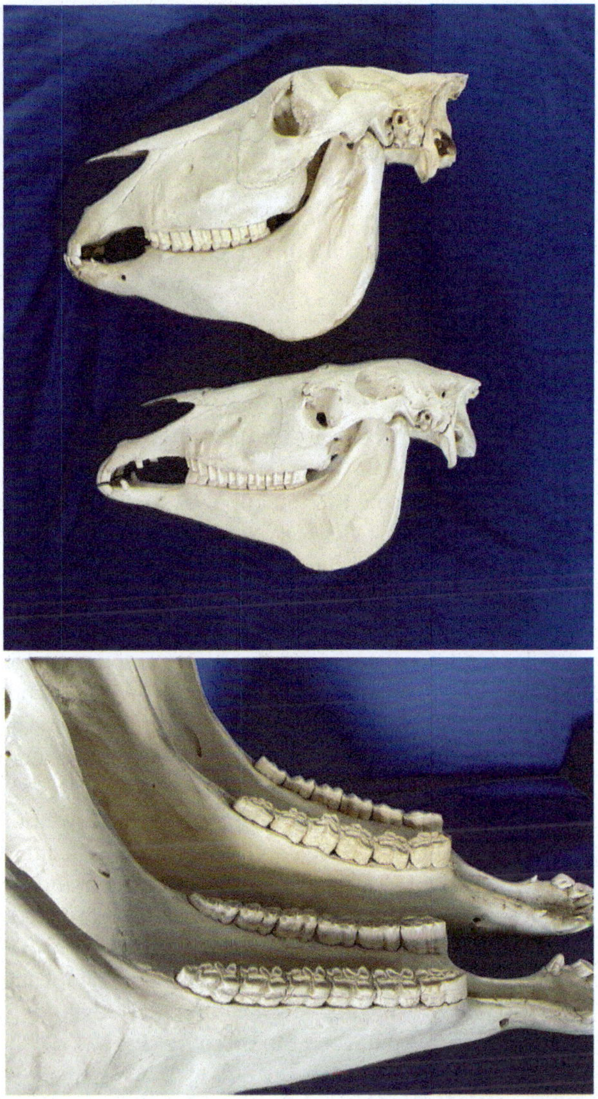

Fig. 5 Curve of Spee

younger age than in horses (Muylle et al. 1999; Toit 2008). Hopefully, the advent of microchipping, breed registries and more data will bring greater clarity to donkey aging for all breeds of donkeys and mules.

- In donkeys the disappearance of the cups, the presence of hooks on the upper corner teeth (TRIADAN 103, 203) and the presence of a Galvayne's groove are not reliable features for dental age determination (Muylle et al. 1999). Aging after 12 years is done by educated guessing.

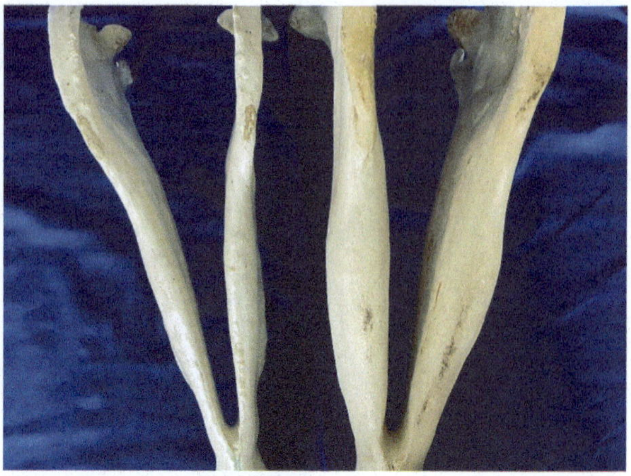

Fig. 6 Mass of bone of the mandible

- About 66% of donkeys (60 out of 91 mature donkeys) had enamel overgrowth on the corner teeth (TRIADAN 303, 403) resulting in no cup on the surface—a common site used for aging horses (Burnham unpublished 2024 investigation at a donkey rescue of 91 mature and 11 juvenile donkeys).
- In the donkey, the second permanent molar (TRIADAN 110,210,310,410) erupts 5–9 months earlier than in the horse, erupting at 15 months to 19 months of age (Mohamed 2000; Misk and Seilem 1997). The premolars and molars have all reached their maximum length by four years of age in donkeys. In horses, the cheek teeth continue to grow in length until the animal is 6–7 years of age (Misk and Seilem 1997)
- The greater longevity of donkeys and their slower rate of dental attrition was once thought to be due to the composition of their teeth (Misk and Seilem 1997; Misk and Semieka 1997), possibly having harder or more enamel thus resistant to wear, but this has been found not to be the case (Toit 2008a). Donkeys live well past 30 years (Burden and Thiemann 2015), some say 40–50 years, perhaps because of their genetics, their environment, geographic location, level of care, their ability to survive grazing for fewer hours a day than the horse on coarser diets or for some reason we have yet to discover (Fig. 7).

3.2.3 Wolf Teeth and Canine Teeth

- As with horses, jack donkeys usually have all 4 canine teeth (TRIADAN 04) visible by 4.5 years (Misk and Semieka 1997). Some reports mention that unlike mares, around 17% of jennies have 3–4 canine teeth visible compared to 5.8% in mares (Toit et al. 2008b; Buil García 2023). In the author's investigation of rescue donkeys most of the jacks (41 of 47) had all 4 canine teeth fully erupted (Table 1). However, of 55 jennies only 1 had 4 canine teeth, 6 had 2 canine teeth and 6 had only one erupted canine tooth. Most jennies examined at the donkey rescue had zero, one or 2 similar to the mare (Burnham unpublished 2024 investigation at the donkey rescue). The reason for this is unclear and further investigations are warranted (Tables 1 and 2).

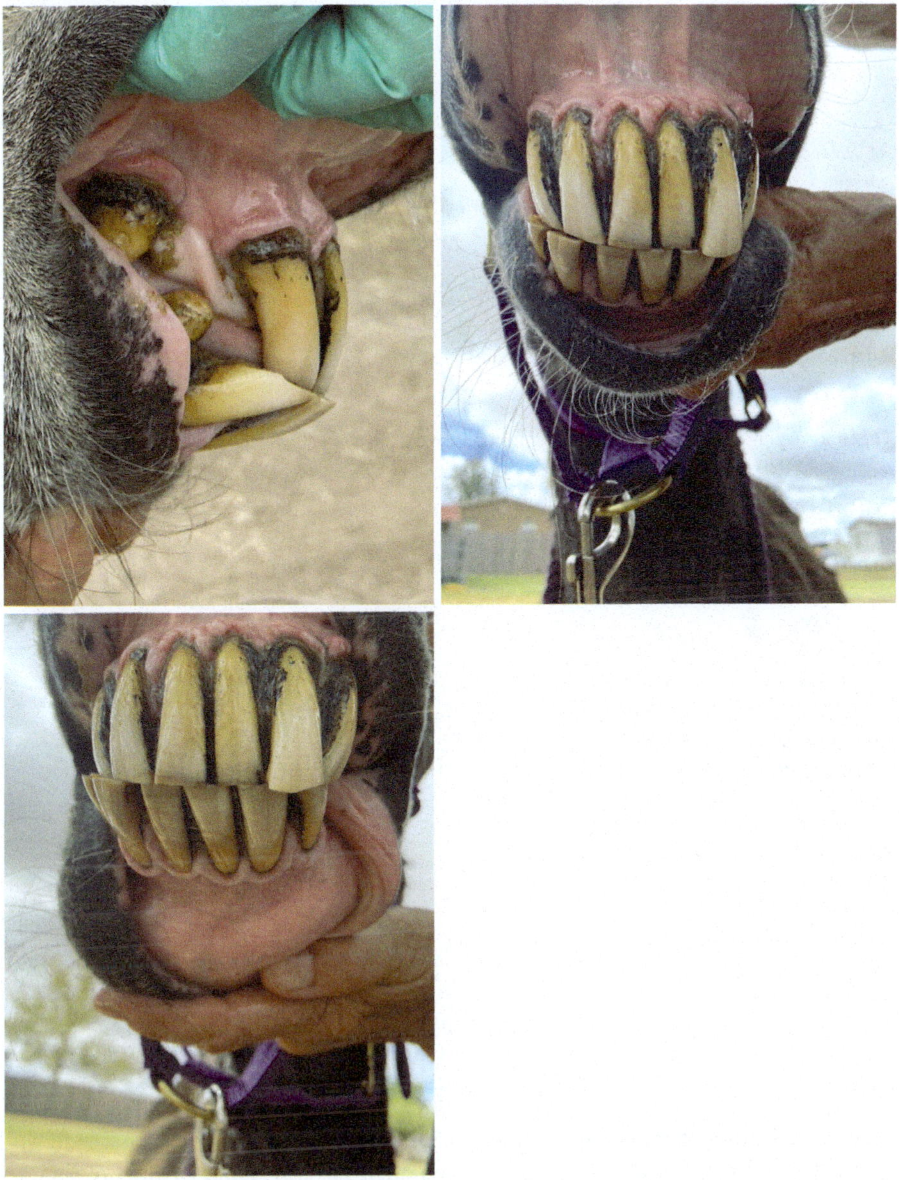

Fig. 7 Incisors of 44-year-old donkey

- The first premolar (TRIADON 05), or "wolf teeth" in donkeys have been found in both the upper and in lower arcades (Misk and Seilem 1997) and are reported to be small enough to cause no issues with bit training (Buil García 2023; Misk and Seilem 1997). Upper arcade wolf teeth erupt bilaterally as early as 6 weeks in

Table 1 Comparison of average incisor eruption schedule of the horse and donkey

	Central Incisor 01		Middle Incisor 02		Corner Incisor 03	
	HORSE	DONKEY	HORSE	DONKEY	HORSE	DONKEY
Age for eruption of deciduous incisors	0–7 days	0–14 days	4–6 weeks	8 weeks	6–9 months	12 months
Eruption of permanent incisors	2.5 years	3–3.5 years	3.5 years	4 years	4.5 years	5–5.5 years
Appearance of dental star	5 years	3.5–4 yrs	6 years	4–4.5 years	7–8 years	5.5–7 years

Adapted from Muylle et al. (1999), Evans and Jack (2007)

Table 2 Aging donkeys by dental eruption and wear

Incisor	Central	Middle	Corner	Canine
TRIADON #	01	02	03	04
Eruption of deciduous	0–14 days	1–2 months	7–12 months	
Tooth in wear, fully developed	8 months	9 months	10 months	
Shedding deciduous	2.5–3.5 years	3 years	3–4.5 years	
Eruption of permanent	3–3.5 years *Horse + 2–8 months*	4 years *Horse + 2–8 months*	5–5.5 years *Horse + 2–8 months*	4.5–6 years
Appearance of dental star on inferior arcade	3.5–4 years	4–4.5 years	5.5–7 years	
Permanent teeth in wear "full mouth"	4 years	5 years	6 years	6 years

The table was compiled from data in Muylle et al. (1999) (63 donkeys), Misk and Semieka (1997) (60 donkeys), Toit (2008), Bünger and Hertsch (1981) (3 donkeys), Buil García (2023)

90% of studied donkeys with some remaining up to nine years; lower arcade wolf teeth, if present, erupt at 6 months in less than 25% of donkeys and are shed by 2 years of age (Misk and Seilem 1997).

3.3 Recommendations

- The donkey will be older than you think based on the horse eruption schedule.
- Wolf teeth do not need to be extracted for training purposes (Buil García 2023) as they do not cause problems.
- Recommend regular dental examinations.
- Do not attempt to correct the curve of Spee when identified.
- Examine the molars for food trapped in diastemata between aging teeth.

3.4 Donkey and Mule Behavior

Donkeys and mules drink water by sipping from the top of the water source barely disturbing the surface. They do not splash nor wash their mouths the way horses can and do. Historical Note: Nineteenth century cavalry manuals recommend watering the mules first, because horses splashed and muddied the water so much that the mules would not drink. Donkeys and mules prefer clean, clear water.

4 Hoof Care

4.1 Reasons for Procedure

Donkeys and mules, like horses, require hoof care which includes regular trimming of the hoof wall and regular cleaning of the sole and sulci around the frog.

Donkeys, mules and hinnies are susceptible to the same hoof diseases that affect horses, especially if housed in damp or muddy environments or if they consume diets high in carbohydrates predisposing them to laminitis, white line disease or thrush. Donkeys suffer more often from hoof care neglect and sadly, many instances of slipper feet from overgrowth of the foundered hoof can be found in the literature and in practice.

4.2 Anatomic Features

There are many features of the donkey hoof that differ from the horse. Mules and hinnies have their own unique features as well. Donkeys evolved in desert mountains and many of the features discussed here are advantages in that type of environment.

- The donkey's hoof-pastern angle when viewed from the lateral is steeper than the horse. The fore feet are steeper than the hind feet all around the hoof. By comparison, a horse hoof is splayed gradually sloping to the ground (Fig. 8).
- Hoof-pastern axis for donkey's front feet is 61.6° (standard deviation SD, 5.24) compared to the horse front feet hoof-pastern axis of 50.5° (SD 5.03) (Burden and Thiemann 2015)

- The donkey hoof shape is oval with straight sides when viewed from the plantar surface (Thiemann and Poore) compared to the horse's rounder shape. The mule foot has a distinctly U shape, with a straight medial wall and a slightly splayed lateral wall (Figs. 9 and 10).
- The donkey hoof wall is relatively thicker than the horse hoof wall and is a constant thickness from toe to heel on the palmar and plantar surface.

Fig. 8 Donkey hoof

Fig. 9 Mule hoof

Fig. 10 Horse hoof

- The dorsal plane of the distal phalanx is convex so that the distance from the distal phalanx to the dorsal plane of the exterior hoof wall deviates from closer at the proximal and distal ends and furthest at the midpoint (Collins et al. 2011; Walker et al. 1995) as seen in radiographs.
- In the donkey there is a difference in the relationship of the coronary band to the second and distal phalanx. In the horse, the second phalanx ends at or above the coronary band. In the donkey the distal end of the second phalanx is below the level of the coronary band and thus within the hoof capsule. The extensor process of the distal phalanx is about 1 cm below the coronary band (Collins et al. 2011) when viewed radiographically.
- The frog of the donkey is a weight-bearing structure with the bulbs located behind the center of rotation behind the heels, with the cranial point of the frog more caudal than in the horse (Thiemann and Poore 2019). For this reason, heart-bar shoes do not support the coffin bone appropriately when the donkey has laminitis (Burden and Thiemann 2015) (Fig. 11).
- The sole of the donkey hoof does not slough naturally like the horse's sole. (Thiemann and Poore 2019)
- The sole of an adult donkey is thicker than the sole of an adult horse (Thiemann and Poore 2019) and responds to insult by growing thicker. Donkeys can stand and walk soundly on their sole when the hoof wall has broken away (Fig. 12).
- Donkeys evolved to have a different microstructure with a more open tubule structure than a horse which enables moisture to be drawn into the hoof. In the dry environment of the desert, donkeys can take advantage of any small amount of moisture they encounter. For this reason donkeys have many problems living in

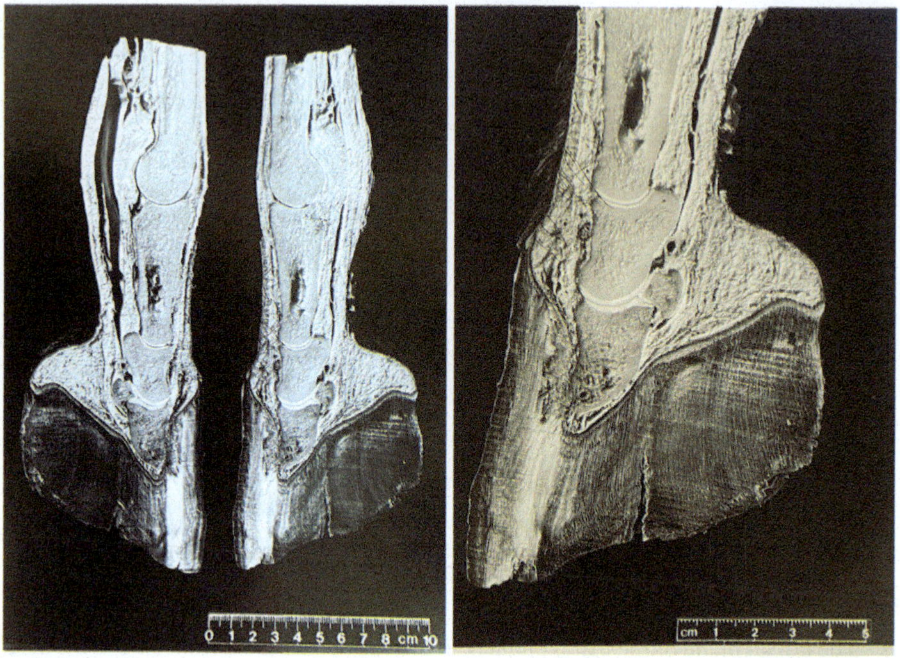

Fig. 11 Laminitic donkey hoof cross-section

wet conditions leading to white line disease, thrush and abscesses (Burden and Thiemann 2015).
- There are differences in the density and distribution of horn tubules compared with horses

4.3 Recommendations

- Ensure that the donkeys are accustomed to having their legs and hooves handled.
- When trimming excess hoof, consider the steeper hoof pastern angle and avoid removing too much heel.
- Hoof testers are not reliable on the donkey hoof because the sole may be much thicker, especially in response to insult (Taylor and Matthews 2002a, b).

4.4 Donkey and Mule Behavior

- Hoof diseases may go undetected for longer periods in donkeys due to their stoic nature and their ability to support the distal phalanx with thickening of the sole (Burden and Thiemann 2015) (Fig. 11).

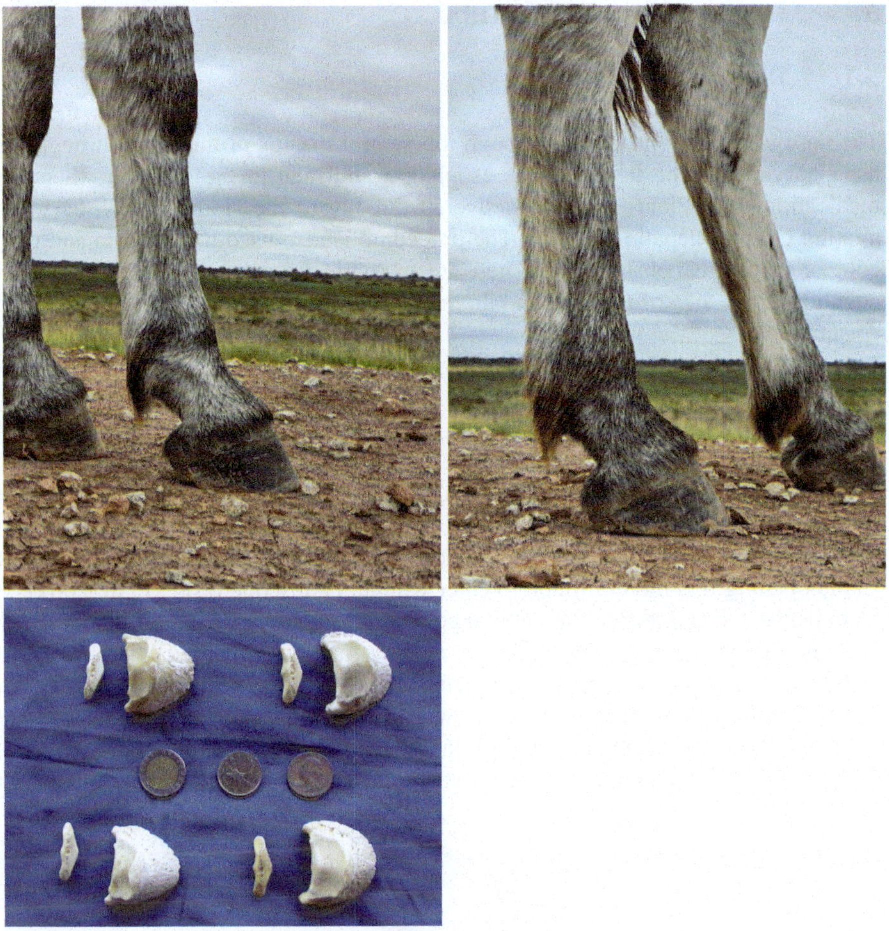

Fig. 12 The steeper angle of the hoof-pastern axis of the donkey

- Donkeys and mules have remarkable accuracy when aiming a kick when they feel threatened. Unless their owners have habituated them to frequent handling, they have low tolerance for leg and hoof treatments. Consequently, many farriers are unwilling to work on donkeys or mules. Trimming and hoof care, therefore often falls to the veterinarian or the owner.
- Donkeys and mules have long memories of ill treatment. Patience is key. Donkey and mule handlers should allow time for the animal to work out the situation and consider its options (Burden and Thiemann 2015).

5 Passing Nasogastric Tube

5.1 Reason for Procedure

The nasogastric tube is passed through the ventral meatus of the nose, through the nasopharynx and the larynx to the esophagus and then to the stomach whenever medication or nutrients must be delivered directly to the stomach such as when treating colic or hepatic lipidosis. The procedure is also used to diagnose and treat colic or esophageal obstructions. There are anatomic differences that require special attention to the size of the tube and the method of reaching the esophagus to avoid epistaxis.

5.2 Anatomic Features

- The ventral meatus of donkeys and mules is much narrower than that of a horse of similar size (Forés et al. 2001).
- Donkeys and mules have guttural pouches like the horse, but due to the angle of the opening it is more difficult to enter with an endoscope (Forés et al. 2001).
- In between and dorsal to the guttural pouches, the donkey (Hutchins 1984) and to some extent, the mule, have a recess, or deep membranous pouch 2.0–3.0 cm in diameter and 4.0–6.0 cm long, extending caudally. The pharyngeal recess or diverticulum communicates rostrally with the nasopharynx through a constricted triangular opening about 1.5 cm in diameter (Lindsay and Clayton 1986) (Fig. 13). The membranous pouch is flexible and can be drawn into the lumen

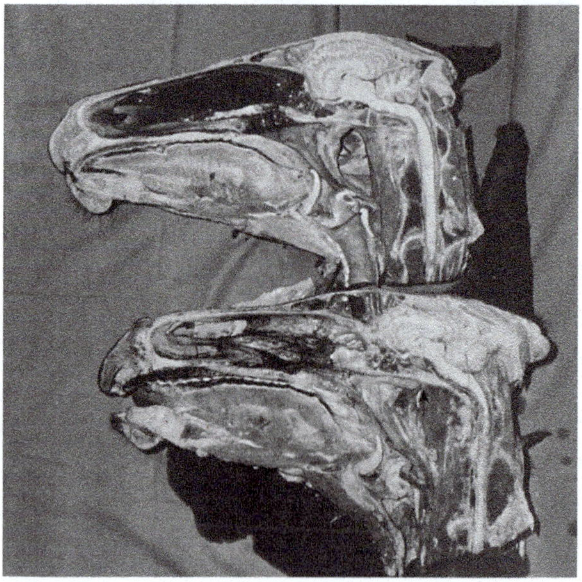

Fig. 13 Comparison of horse and donkey pharyngeal recess size

of the airspace during forced inspiration and during the initial phase of braying (Lindsay and Clayton 1986).

- The air space in the pharynx of both the horse and donkey is constricted in the middle, however in the donkey this difference in diameter is more pronounced (Lindsay and Clayton 1986) (Fig. 14).
- The angle that the nasogastric tube passes through this space in a horse is the opposite of the angle in this space of a donkey. The donkey's airway from the nasopharynx through the larynx angles caudoventrally. The degree of angulation in donkeys is greater than in horses, likely affecting the quality of vocalization (Lindsay and Clayton 1986) (Figs. 15 and 16).
- The tip of the donkey epiglottis is more pointed in shape than the horse's epiglottis.
- The donkey has additional differences in size and disposition of the laryngeal ventricles from the horse. Donkeys have a vocal saccule next to the vocal fold with heavy muscle mass around all the laryngeal components (Hutchins 1984). An excellent detailed anatomical and histological description has been provided by Lindsay and Clayton 1986, in their examination of the features of donkey vocalization.

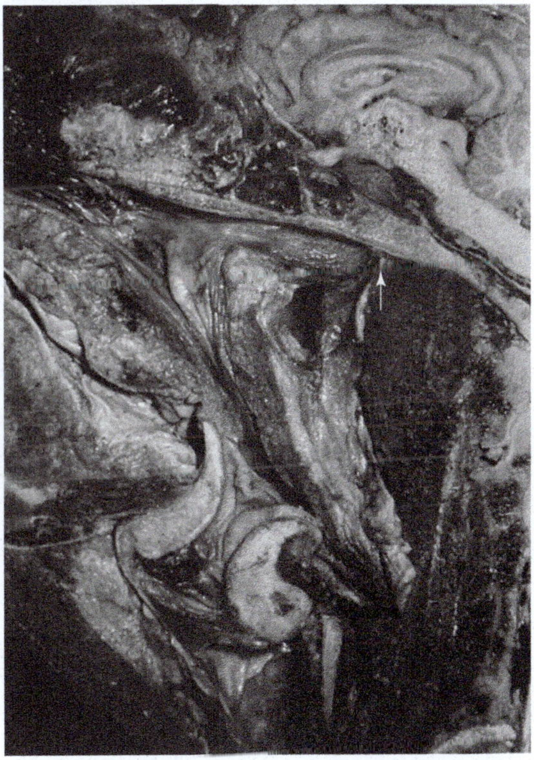

Fig. 14 Cross section of donkey head showing the pharyngeal recess

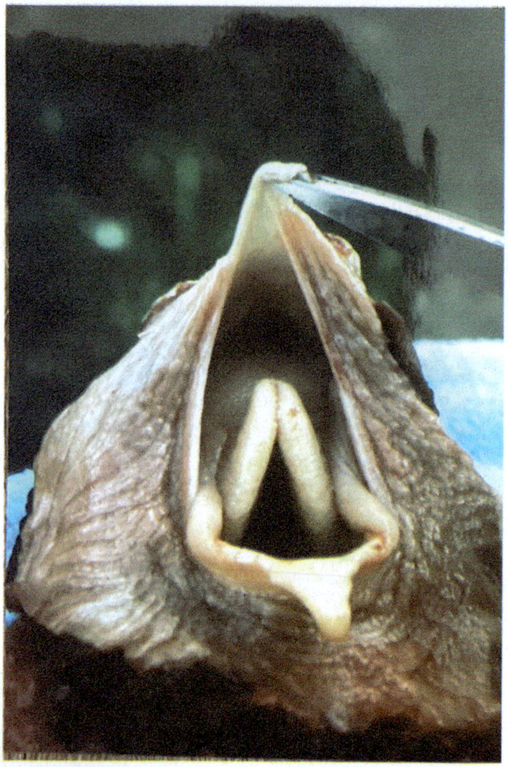

Fig. 15 Pharyngeal recess view rostral to caudal showing triangular shape

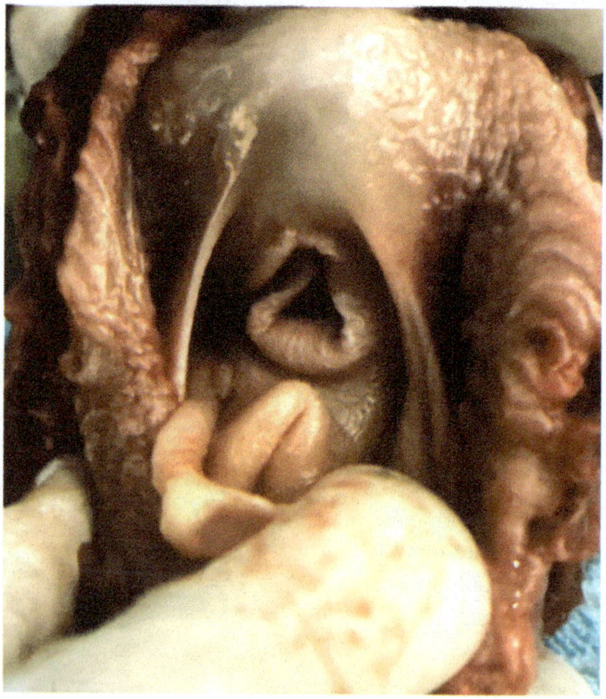

Fig. 16 Pharyngeal recess partially everted demonstrating flexible nature

5.3 Recommendation

- For most donkeys and mules, a foal size nasogastric tube works well. A medium or regular horse tube will work for larger donkeys and mules, but the fit will be tight and difficult to maneuver. The larger tube is more likely to cause epistaxis.
- Be aware of the pharyngeal recess in the caudal pharynx. If the tube is trapped by the pouch, retract the tube at least three inches and redirect the end, turning 45–90 degrees.
- If a nasogastric tube must be used for medicating mules or donkeys not accustomed to this procedure, sedation is recommended.

5.4 Donkey and Mule Behavior

Imprint training (Miller 1994) works well on donkeys and mules of all ages. Once habituated to having a finger inserted into the nostril, a mule will tolerate the passage of a nasogastric tube using no restraint or sedation (personal experience).

6 Access to Nasolacrimal Duct Orifice

6.1 Reason for Procedure

The nasolacrimal duct in donkeys or mules may become obstructed due to congenital or acquired causes (Abdelbaset-Ismail et al. 2022) such as foreign body, trauma, rhinitis, neoplasia, dental infection (Adams et al. 2013) onchocerca conjunctivitis, mycotic granuloma (Said et al. 1977), or fly strike (Burnham 2002). The nasolacrimal duct may be backflushed to restore patency at the rostral end of the duct, using a catheter inserted into the nasolacrimal orifice (NLO). The size and location of the NLO in mules and donkeys differ from the NLO size and location in horses.

6.2 Anatomic Features

- In the horse, the NLO is oval, 3–5 mm and located on the floor of the nasal cavity at the mucocutaneous junction (Sack and Habel 1977). The horse nostril is flexible and can be gently flared open to visualize the NLO. The rim of the nostril in horses is relatively thin.
- The location of the NLO in donkeys is in the cutaneous internal wall of the nostril (Burnham 2002; Adams et al. 2013) at the dorsal nasal fold (nasal diverticulum), in the region called the "false nostril".
- The location of the donkey NLO is variable; occasionally slightly dorso-medial or dorso-lateral in the dorsal fold and can vary in the distance back from the rim of

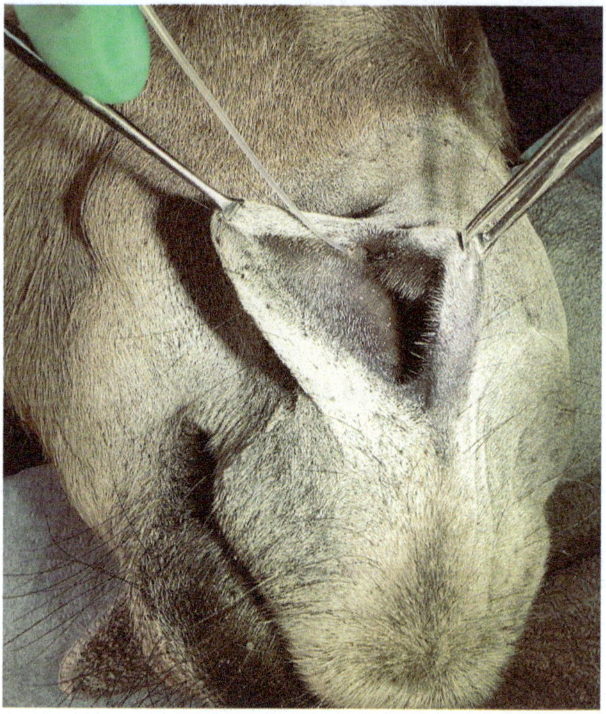

Fig. 17 Catheter placed in the nasolacrimal duct orifice of the donkey cadaver

the nostril anywhere from 0.5 to 2.5 cm (Burnham 2024; Pohlmeyer and Wissdorf 1975) (Figs. 17 and 18).

- Some sources describe the location variably as dorsal, (Said et al. 1977: 10 donkeys), closely ventro-laterally to the dorsal nasal angle and 1.0–1.5 cm back from the edge of the nostril (Pohlmeyer and Wissdorf 1975: 20 donkeys), dorso-lateral on the flare of the nostril, (Hutchins 1984: 1 donkey) or dorso-medially within the nasal vestibule (Abdelbaset 2022: 36 donkeys).
- Examination of 102 donkey NLO at the donkey rescue found 63% dorsal, (64/102), 24% dorso-lateral (24/102), 13% dorso-medial, (13/102).
- One study describes the NLO at the muco-cutaneous junction on the medial wall of the nasal vestibule approximately 12.1 mm from the dorsal commissure (Abdelbaset-Ismail et al. 2022).
- The NLO in donkeys varies in size and shape which can present as a slit, oval or elliptical shape (Said et al. 1977) ranging from 0.1 mm to 3.0 mm.
  - 2.0–3.0 mm (Said et al. 1977)
  - 3.0 +/− 0.9 mm (Abdelbaset-Ismail et al. 2022)
  - 0.1–0.2 cm (Pohlmeyer and Wissdorf 1975)

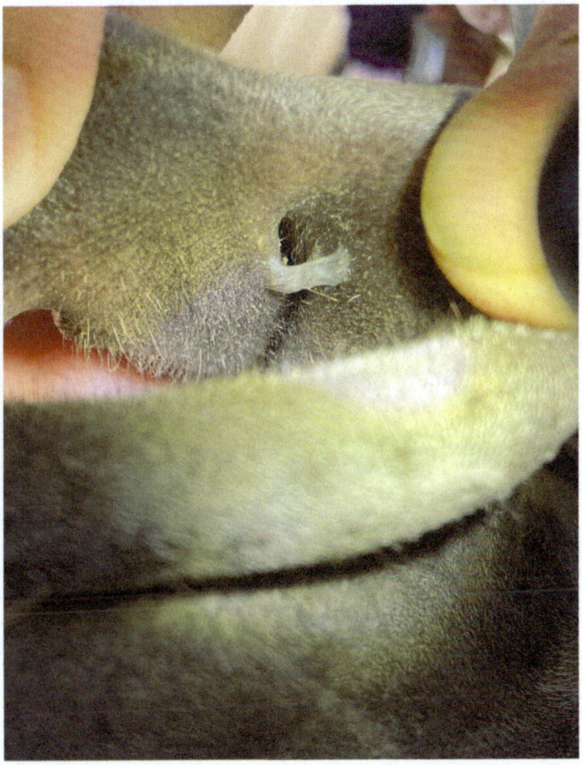

Fig. 18 The NLO opening of this donkey has a discharge that adhered to the opposite wall of the upper nasal commissure or "blind fold". The location is far from the mucosa of the nostril, high in the cutaneous, haired skin. Dusty debris from the dry environment is frequently seen around the NLO

- Said describes the NLO of donkeys histologically as being lined by stratified squamous keratinized epithelium and the lamina propria contains hair follicles, sebaceous glands, but mucous glands are absent.
- Unlike a horse, the donkey has a relatively thick rim of the nostril, especially laterally and ventrally (Fig. 18).
- In the mule the NLO can be oval or a slit and varies with the size of the mule but will be smaller than expected compared to that in a horse of similar size. In the Adams study, size of the opening in mules varied from 0.25 mm (foal mini-mule) to 4 mm, with an average of 1 mm.
- The typical location of the NLO in mules is completely different from the horse or the donkey. In one study 35 out of 40 (87.5%) mules had the NLO in the cutaneous tissue of the lateral wall of the external nares (Adams 2013) at junction where lateral wall curves to meet the floor of the nasal vestibule and about 5 cm caudal to the rim of the nares. Of the 40 mules in the Adams study, 4 (10%) had NLO on the floor of the vestibule at the muco-cutaneous junction, near the location for the horse (Fig. 19). All of these were molly mules (females). None of the mules in this study had NLO in the area common for the donkey, i.e. dorsal nares.

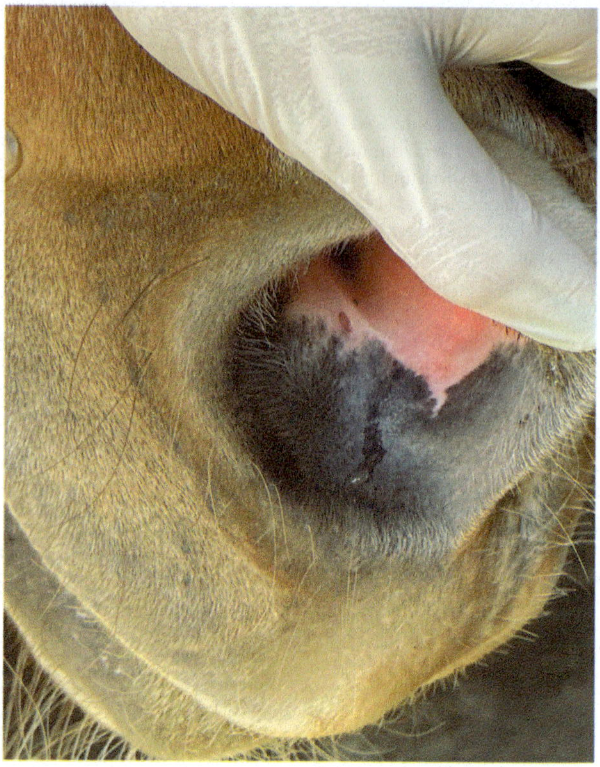

Fig. 19 NLO in horse and mule

6.2.1 Two Nasolacrimal Ducts with Two Patent Nasolacrimal Orifices Per Nostril

- Very rarely, a horse may have two openings with two separate ducts per nostril that both communicate with the lacrimal sac at the medial canthus of the eye.
- Donkeys and mules will also occasionally have two nasolacrimal duct orifices per nostril:
  - One out of 20 donkeys had 2 NLO (Pohlmeyer and Wissdorf 1975)
  - Five out of 40 mules had 2 NLO (Adams et al. 2013) and 4 of the 5 had two NLO on the lateral wall of the internal nostril 5–10 mm apart bilaterally, but not always bilaterally symmetrical in location or size. The remaining animal with dual NLO had one in the lateral wall and one on the floor of the vestibule. None of the mules in the Adams study had NLO at the location of their sire, the jack donkey (Adams et al. 2013).
  - One out of 24 horses had 2 openings (Adams et al. 2013)
  - Of 102 donkeys examined at the donkey rescue facility, one jack and one jennet had 2 NLO

6.3 Recommendation

- Locating the NLO in donkeys or mules may require a penlight, a soft cloth for cleaning and patience. Once the nostril is clean, the NLO can usually be found by locating a small mucoid plug or normal lacrimal secretion somewhere in the dorsal region of the nostril. But because some of these orifices are very close to the rim and others are deep within the nostril, caution must be used in everting the nostril to visualize the NLO as the fingers may obscure the opening (Fig. 20).
- Many donkeys living in a dry, dusty climate may have dirt accumulations near the NLO, with mucopurulent plugs some of which adhered to the opposite wall of the dorsal nares.

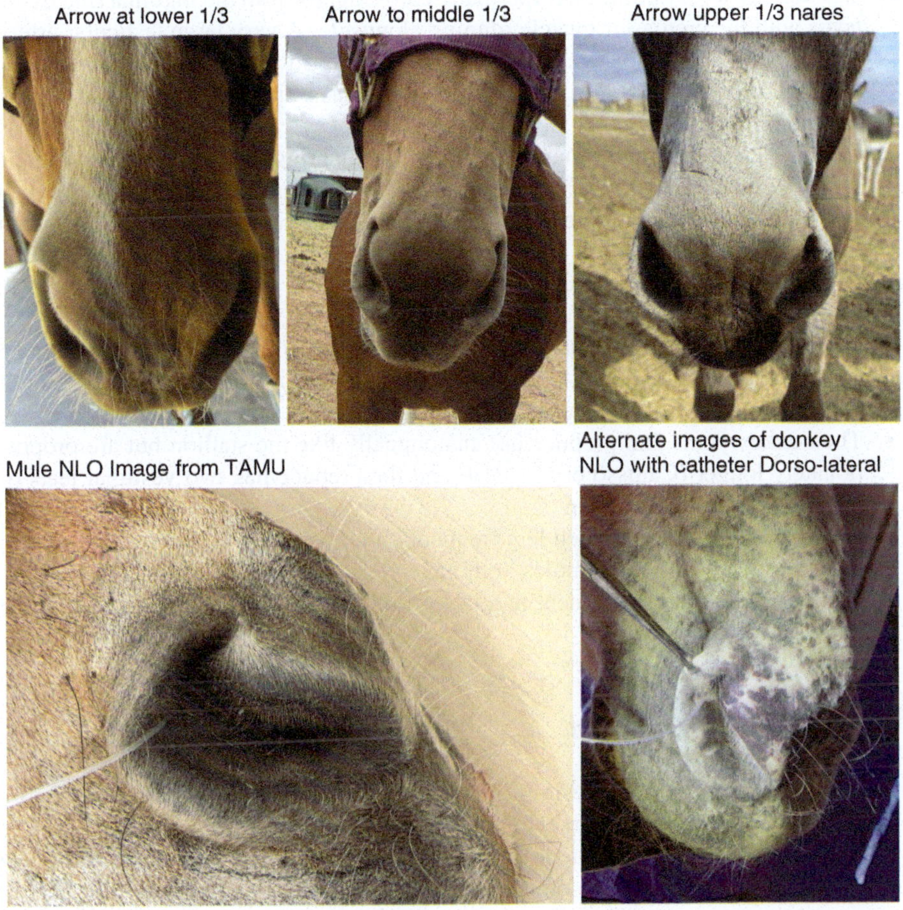

Fig. 20 Comparison of donkey, mule and horse nostrils showing the regions to find the NLO

7 Castration

7.1 Reason for Procedure

Most male mules and many jack donkeys are castrated to modulate their behavior for training purposes when they are not entering a breeding program.

7.2 Anatomic Features

- Jack donkeys have proportionately larger heavier testicles than male mules or stallions. They also have a thicker tunica vaginalis (Pathak et al. 2013).
- The donkey's very prominent spermatic cord extends from the inguinal ring to the testis with the cremaster muscle running alongside. A very large testicular artery and associated testicular veins form the pampiniform plexus. The larger donkey spermatic cord (23–28 mm) has a capsule that has a rich vasculature and a thick muscular layer (Canisso et al. 2019).
- Donkeys and some mules may have larger inguinal rings due to a larger spermatic cord and greater amount of tissue passing through them. This is a concern when castrating jack donkeys by emasculation because evisceration is always a concern with donkeys (Taylor et al. 2001).
- Donkeys have higher vascular flow through the cord than horses (Gacem et al. 2018).
- Older donkeys may have fat deposits in the scrotum which has thicker skin and a heavier median raphe with larger blood vessels. This is not true for all breeds of donkeys and is seldom observed in mules (Taylor et al. 2001).
- The donkey penis and prepuce are anatomically like the stallion but are proportionately longer (Canisso et al. 2019) and the prepuce has two vestigial teats of varying size from 1 to 3 cm.
- Cryptorchidism occurs much less frequently in the donkey (Canisso et al. 2019).
- Sheath of jacks and male mules both have vestigial teats (non-functioning)
- Epididymis is more prominent in a jack forming a button type appearance at the end of the scrotum

7.3 Recommendation

The pharmacology of anesthesia in donkeys and mules has been demonstrated by others to be different from horses. The difference in the fright response may be one factor in metabolism of anesthetic agents so care should be taken to minimize noise and threatening actions. It is common to get an incomplete response to sedation or more rapid recovery from sedation. Domesticated gentle donkeys are easier to sedate than feral or stressed donkeys and respond to horse doses. Once sedated, the castration process may proceed as with horses with the following recommendations.

- Perform open castration and ligate the cord and the cremaster muscle separately. A common complication following donkey castrations is excessive bleeding that can be life threatening. Because of the larger spermatic cord and greater vascularization, the jack's entire cord must be ligated during castration.
- A ligation high above the emasculator crush site will help prevent serious bleeding.
- The procedure should protect against evisceration.
- Deep anesthesia for untrained or unsocialized donkeys and mules does not last as long as for horses or domesticated longears. However, expect donkeys to remain in sternal recumbency much longer, until they know they can rise, usually taking up to 30 minutes before attempting to rise. Mules rest post procedure for intervals between the horse and the donkey, some will attempt to rise before they are able to stand unassisted.
- With feral or wild donkeys, it is often helpful to prevent injury to the boney prominences of the head, hip, shoulders, etc. because they cannot be safely treated after the procedure without sedation. Clean and dry the incision site as much as possible before the donkey is alert.

7.4 Behavior

Anesthesia down time is shorter, especially if the animal is frightened by unfamiliar circumstances, loud noises and if handlers do not allow time for the donkey or mule to assess the situation. However, donkeys remain in lateral or sternal recumbency much longer before attempting to rise (Taylor et al. 2001; Taylor and Matthews 2002a, b) because donkeys have a strong sense of self preservation, much more so than horses and mules. Horses and mules may attempt to rise before they have recovered from anesthesia.

Jack donkeys exhibit aggressive or sexual behavior for a long time after castration, especially if the donkey was older when castrated. Occasionally, geldings continue to attempt to breed jennets in heat until they are geriatric but most cease the behavior after a year. For this reason and others, it is often recommended that mules and donkeys be castrated at younger ages than is recommended for horses.

8 Epidural Anesthesia

8.1 Reasons for Procedure

Indications for epidural anesthesia may include symptomatic relief of pain during dystocia, surgical procedures such as repair of rectal or vaginal prolapses or treatment for trauma such as recto-vaginal tears associated with dystocia or tumors of the tail or perineal area (Fig. 21).

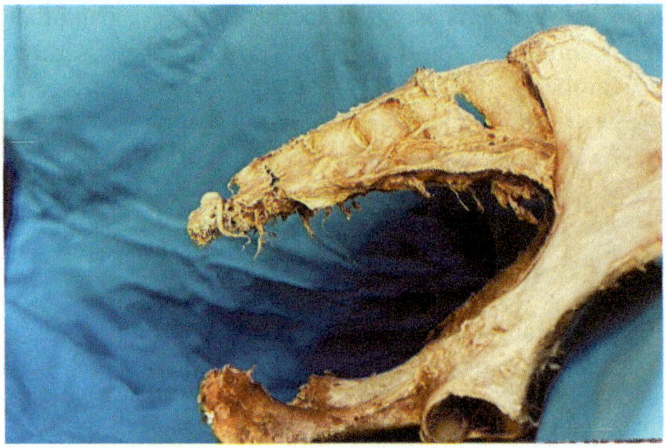

Fig. 21 Dissected specimen of donkey pelvis, sacrum and first three coccygeal vertebrae. Spines 2–5 of the sacrum are visible with the first coccygeal vertebra fused to caudal segment. A gap is evident between Ca2 and Ca3

8.2 Anatomic Features

The anatomic features of the donkey sacrum, caudal (coccygeal) vertebrae and spinal cord differ from horses.

- The horse sacrum has five segments (Sack and Habel 1977; Clayton et al. 2005)
- The preferred location for epidural anesthesia for the horse is in the wider first intercoccygeal space and the needle should be inserted at a 45-degree angle (Heath and Meyers 1972)
- The sacrum of the donkey also consists of five segments. The first caudal vertebra is often fused with the sacrum with occlusion of the sacro-coccygeal space. The sacral spines point caudally and decline in length (Shoukry et al. 1975).
- The first intercoccygeal space is not conspicuous in the donkey because of the more caudal direction of the first caudal spine and a common site of ankylosing spondylitis (Shoukry et al. 1975).
- The first intercoccygeal space is narrower than the second in the donkey. The spines of the first two caudal vertebra are easily palpated (Shoukry et al. 1975).
- The spinal cord in donkeys ends at the second sacral vertebra but the dural sheath extends as far back as the first or second caudal vertebra (Shoukry et al. 1975) (Fig. 22).
- Additionally, the donkey sacrum is curved dorso-ventrally (Doherty et al. 2022) while the horse sacrum is relatively flat (Clayton et al. 2005)
- The caudal vertebrae in donkeys are much better developed than horses and the vertebral arches of the first three caudal vertebrae are complete (Shoukry et al. 1975).

8.3 Recommendation

- Locate the space between the second and third caudal vertebrae by palpating back from the first sacral segment. The first and second caudal vertebral spines are well

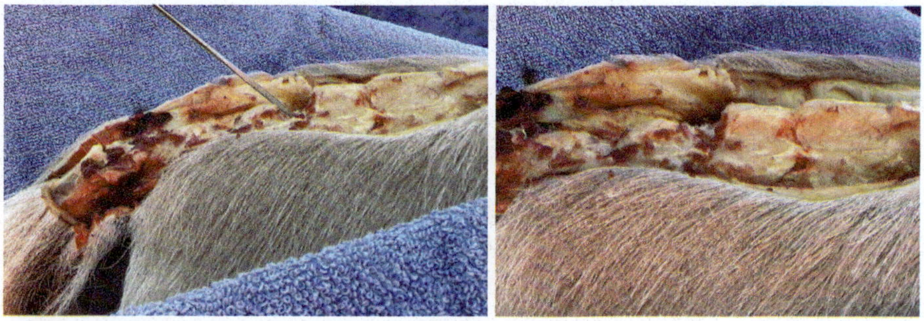

Fig. 22 Field necropsy dissection of donkey showing the angle of the approach to the second intercoccygeal space

developed and close together, while the third spine is less defined providing the landmark for injection (Shoukry et al. 1975).
- The needle will be inserted at an angle of 30 degrees (from the horizontal) ventral and cranially under the second caudal vertebral spine (Shoukry et al. 1975).

8.4 Vertebral Formulae

DONKEY: C-7, T-18, L-5, S-5, Ca-15-17 (Jamdar and Ema 1982)
HORSE: C-7, T-18, L-6, S-5, Ca-15-21 (Sack and Habel 1977)

9 Abdominocentesis

9.1 Reasons for Procedure

Abdominocentesis is performed to collect abdominal fluids for various reasons such as to determine the degree of intestinal compromise, to identify whether intestines have ruptured during a colic, or to investigate any abdominal effusion such as in cases when neoplasia is suspected.

9.2 Anatomic Features

- Domestic donkeys are often fed in excess of need and will therefore deposit fat in many locations.
- An excessively fat donkey is easy to recognize. There may be many visible fat deposits. Surprisingly, a donkey with no visible fat deposits, even with palpable ribs, may have a very thick fat pad on the ventral abdominal wall as well as in the subcutaneous space.
- Ventral abdominal fat layers can be 4–6 inches (roughly 10–15 cm) thick (Fig. 23).

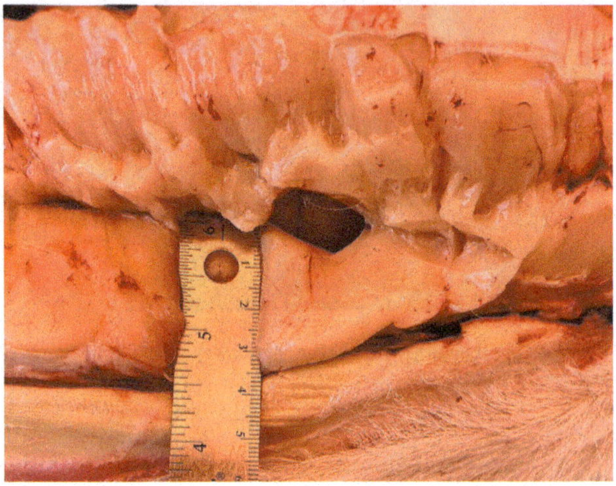

Fig. 23 Field necropsy of donkey showing layers of abdominal fat on the ventral midline

9.3 Recommendation

- Prepare for a donkey or mule abdominocentesis as you would for a horse procedure. The fat layer can be measured prior to the procedure using ultrasound. It can be helpful to have spinal needles in a variety of lengths or longer teat cannulas at your disposal to reach past the fat layers.

10.0 Miscellaneous additional anatomic features that differentiate donkeys and mules from horses

9.3.1 Ears

- Donkeys and mules have longer ears. There are sweat glands around the base of their ears, and it is possible that their blood vessels in the ears help to cool donkeys on hot days when they are at work.
- The cartilage of the bulb of the ear is attached by muscle to a separate "shoe-horn" shaped cartilage medial to the base of the ear pinna (Burnham 2024 field necropsy). The circular, concave cartilage fits over the temporal muscles and can be palpated (Fig. 24).

9.3.2 Chestnuts and Ergots

- Most horses have chestnuts above the carpus medially on the front legs and below the hocks, also medially—some ponies are the exception.
- Donkeys have chestnuts on the front limbs only, none on the hind limbs.
- Mules have chestnuts on the forelimbs but may have no chestnuts on the hind limbs or very small ones (Fig. 25).

Fig. 24 Donkey ears

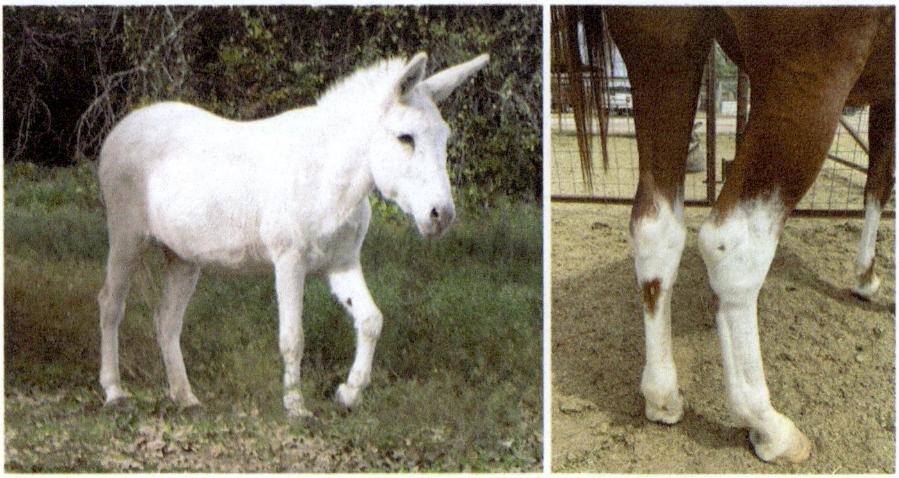

Fig. 25 Donkey and mule chestnuts

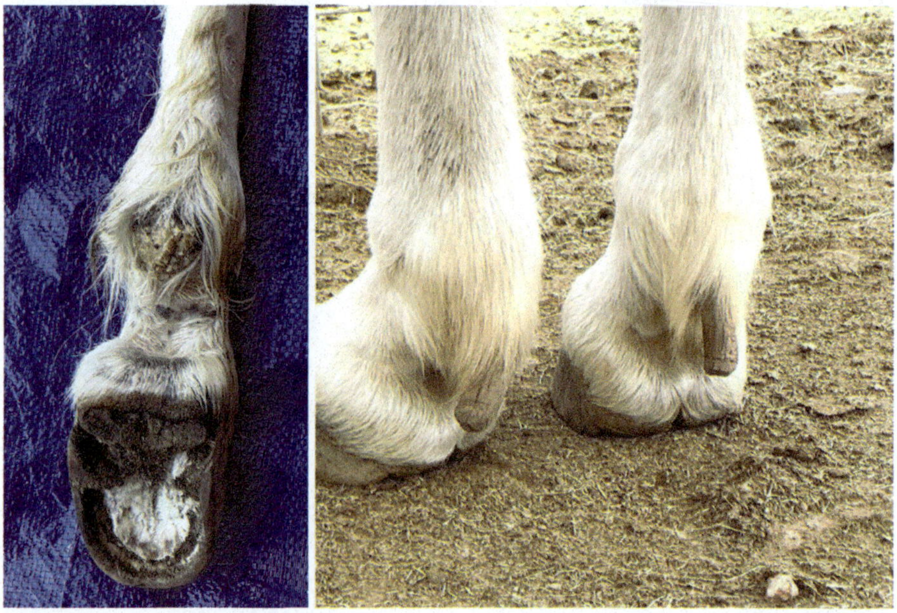

Fig. 26 Donkey and mule ergots

- Horse ergots, the rough, horny tissue found in the long hair on the caudal aspect of the pastern are very small and do not occur on all horses.
- Most donkeys have large ergots that can grow long and can be up to two inches in diameter (Fig. 26).

Fig. 27 The donkey's tail is a switch and their manes are short

9.3.3 Donkey Tails and Mule Tails

- The long hairs of the donkey tail start 5–9 inches from the top of the tail head forming a switch, similar to the cow. The rare donkey can have longer mane and tail hair (Fig. 27).
- Mules have tails like horses (Hutchins 1984) (Fig. 28).

9.3.4 Manes and Forelocks

- Horses have manes and forelocks of variable lengths.
- Donkey manes are short and begin at the back of the poll. The mane hair gets shorter as it approaches the withers. Donkeys have no forelock.

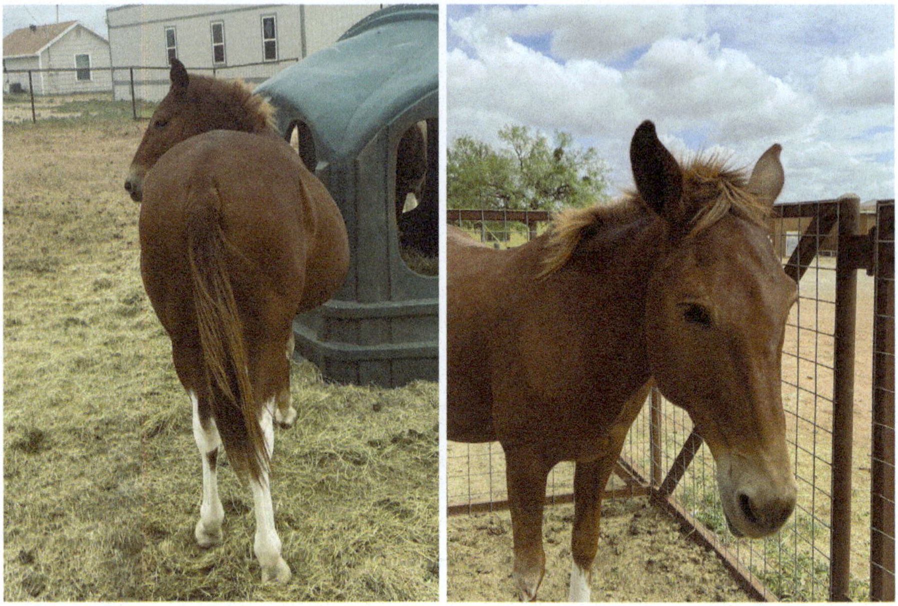

Fig. 28 The mule's tail is like a horse; their manes often stand up like a donkey's

- Mules have mane hair longer than donkeys, but often not as long as the horse. Most will have a tuft of hair that falls forward between the ears (Fig. 29).

9.3.5 Hair Pattern of the Flank

- Horses have a whorl in their hair at the flank that protects their external genitalia from rainwater.
- Donkeys do not have a whorl in their flank hair. The fold of skin and the long hair serve to shed water to protect the external genitalia (Figs. 30 and 31).

9.3.6 Heavy Neck Crest

- Donkeys can develop a fat neck crest that is so wide that it disfigures the neck (Fig. 32).
- Fat deposits on both sides of the nuchal ligament and over time fibrous connective tissue invades the fat deposit (Hutchins 1984) (Fig. 33).
- If the fat deposits have been there for an extended period, they cannot be removed with neck sweats or restricted diets (Fig. 34). Fat pads and neck crests in donkeys should be considered permanent (Taylor and Matthews 2002a, b) (Fig. 35).

9.3.7 Vocalization

- Donkeys make sounds on inspiration as well as expiration, unlike other mammals that only vocalize during expiration of air. Several anatomical features of the

Fig. 29 Mane and forelock of the mule

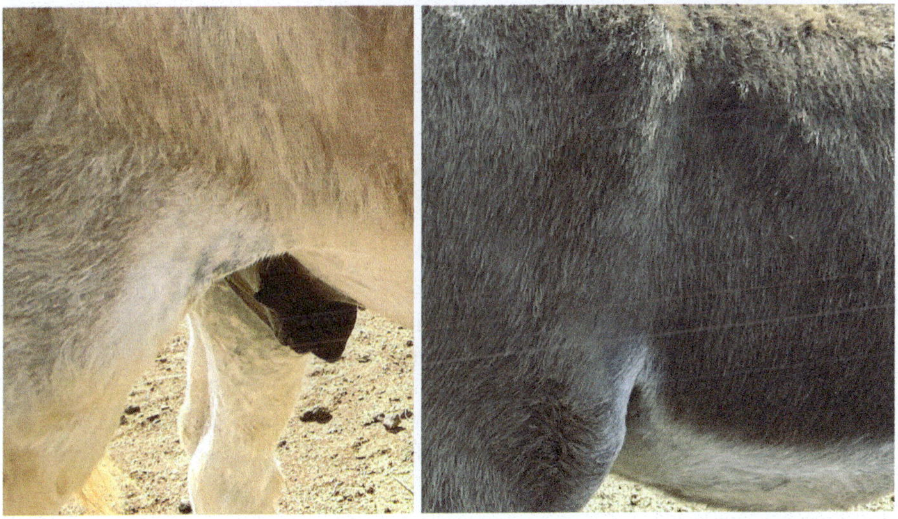

Fig. 30 Hair pattern of the flank of the donkey that also reveals the vestigial teats on the prepuce

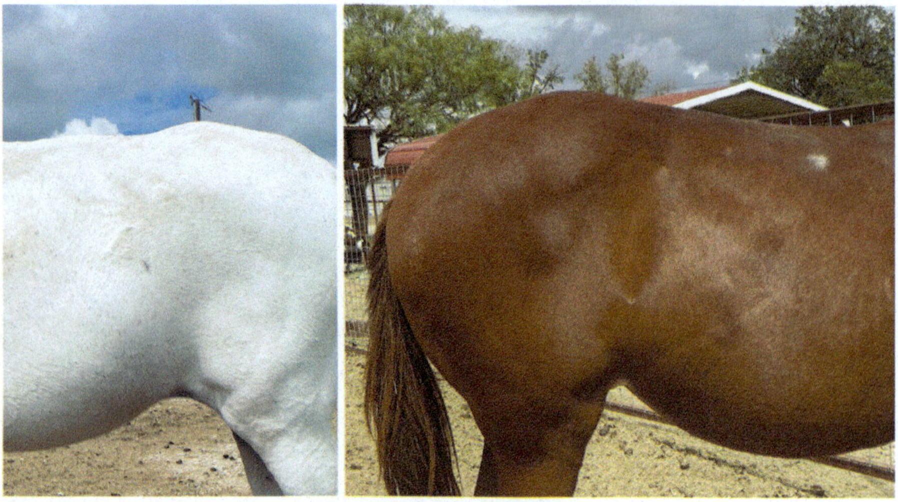

Fig. 31 Grain of hair for 2 mules showing whorl location

Fig. 32 Excessive fat deposit on neck crest of a donkey

larynx help the donkey to make very loud sounds that carry for long distances (Lindsay and Clayton 1986). Donkeys have a double vocal fold and can expand and collapse the pharyngeal diverticulum as they bray. The short nasopharyngeal airway has a pronounced constriction in the middle and is flared dorsally and ventrally that is also thought to contribute to creating big sounds (Evans et al. 2022).

9.3.8 Trachea

- The donkey's tracheal rings are oval, slightly flattened dorsoventrally and smaller in diameter than a comparably sized horse (Powell et al. 2010) This should be

Fig. 33 Fat deposits on topline

Fig. 34 Obesity is serious health problem for donkey

recognized as normal in a donkey. Procedures such as tracheostomies would be the same; endoscopies would appear different from the horse.

9.3.9 Inferior Check Ligament

- Donkeys have an extension of the accessory check ligament that extends from the deep digital flexor tendon to the superficial flexor tendon.
- Donkeys have been found to have no inferior check ligament in the hind limb (Taylor and Matthews 2002a, b; Taylor et al. 2001).

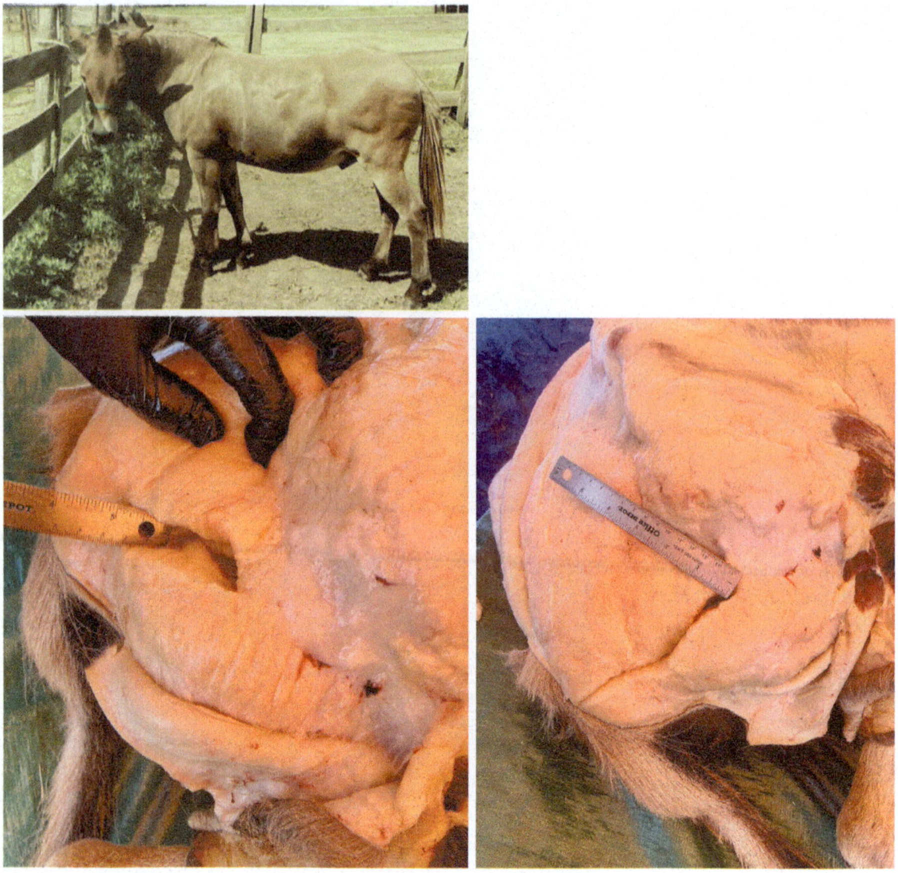

Fig. 35 Working mule with excessive fat deposits

9.3.10 Distal Colon

- The haustrations of the donkey distal colon are less defined than those of the horse. At this point in the gastrointestinal tract the fecal material is already formed and dry, as most of the water has been resorbed earlier in the tract (Doherty et al. 2022).

9.3.11 Physeal Closure

- Epiphyseal closure of the distal radius in the horse occurs between 2 and 2.5 years (Adams 1979).
- In miniature donkeys evaluated for harness training and showing, the epiphysis of the distal radius had gaps (was not completely ossified) at ages of 48–53 months but were closed at 56–60 months of age. (Burnham Data from miniature donkeys in private practice). (Burnham, S. 1999. The Asset 1999:33:18.)

- There is evidence that suggests the physis of the tarsus and stifle may close at later dates in donkeys than in the horse, not being complete until age 4 years (Abdelbaset-Ismail et al. 2016).

Anatomic features of the jenny that differ from the mare

- The jenny's reproductive tract is proportionately longer and larger than in mares or female mules of similar size. The uterus is similar in shape to the mare's uterus, and jenny ovaries are located more cranial than the ovaries in horses (Canisso et al. 2019).
- The caudal end of the reproductive tract of jennies is more tilted ventrally so that the dorsal vulva is less concealed than the ventral commissure, which makes contamination of the reproductive tract and bladder less likely (Taylor et al. 2001).
- In jennies the entire length of the vulva is below the pelvic brim and the ventral area is often tipped more cranially than the dorsal area. This may explain why jennies have less tendency to pool urine.
- Proportionally, jennies have a larger clitoris than mares. (Canisso et al. 2019)
- The jenny cervix has a narrower, tortuous lumen that makes intrauterine procedures more challenging. The vaginal caudal end of the cervix forms a point (Vendramini et al. 1998).

9.3.12 Conformation of Donkeys and Mules

Horse breed registries and show events have contributed to the breeding of horses with conformation that allows them to excel at their respective jobs. Long sloping croups, balance and symmetry of body, deep shoulders and bone structures meant for racing, riding or harness work have been selected over many generations and many years. The domesticated donkey of today has not had the same benefits, and donkeys, especially in North America, have been allowed to breed in small herds in the wild or in large pastures without selection for good hips, strong shoulders or good conformation overall. Consequently, base narrow or cow-hocks, steep or small hip conformation have been deemed "normal" traits of the donkey. However, in countries where donkey breeds are protected and registered, this is not the case. Donkeys that are bred for work and spend their lives in exercise and with appropriate nutrition schemes have a broad stance, developed pectoral muscles, stout hips and strong leg bones.

The registration of donkeys arbitrarily by size in North America has promoted donkeys of extremely small stature or very lanky tall stature. Breeding for size alone has created value for the micro donkey and mammoth-sized donkey. By comparison, the donkey in the middle is called "standard" and continues to have very little value. Yet in this "size" bracket are the remnants of the older breeds that were brought to the Americas long ago that had been bred for good conformation. Fortunately, in recent years mules are being bred for show and work. Rather than presenting a pasture jack to the "lesser" mares, mules are being bred using champion jacks to quality mares. Hopefully, the same is true in the production of hinnies for work and show. Draft horses may have a slightly cow-hocked conformation as an advantage for pulling, so too, may draft donkeys. But to say that cow-hocks are normal for all donkeys is inaccurate. Donkeys and mules are capable of hard work and can build enough muscles (gluteal, semimembranosus and semitendinosus muscles) to have "britches"!

While today's donkey and mule may have a more prominent manubrium or may have a steeper hip (taller ilia bones), we should continue to select for animals with sufficient musculature to cover the manubrium and against the donkeys with the "rafter" hip. Donkey and mule shows, pleasure driving, and mountain races give donkeys and mules the opportunity to prove that good conformation and good conditioning can be their norm.

Acknowledgements My thanks are due to my colleagues and reviewers, Dr. Tita Clark, Dr. Jeffrey Musser, and Dr. Karen Russell for sharing their expertise. I am also grateful for the support and encouragement from the Peaceful Valley Donkey Rescue Staff for making over 1,000 donkeys available for my investigations. With special thanks to Sandra Pitcox, the Chief Veterinary Technician at the donkey rescue and Rebecca Scott, the Wild Burro Project Manager for wrangling donkeys for me, for sharing their expertise and experiences without which much of the data collecting would not have been possible. My thanks to Cole Cowden, farrier and trainer at the donkey rescue who provided some of the photographs for this chapter.

Questions

1. Why is jugular venipuncture more challenging in donkeys compared to horses?
   A. Donkeys have a smaller jugular vein than horses
   B. Donkeys lack a jugular furrow entirely
   C. The cutaneous colli muscle in donkeys covers the entire jugular furrow
   D. Donkeys have thinner skin, making veins harder to locate
   Answer: C
2. Which of the following contributes to the difficulty in aging donkeys by dental examination?
   A. Donkeys have fewer teeth than horses
   B. The curve of Spee is absent in donkeys
   C. Donkeys' teeth erupt earlier than horses
   D. Donkeys show unreliable age markers such as cups and Galvayne's groove
   Answer: D
3. What anatomical feature makes epidural anesthesia in donkeys more challenging than in horses?
   A. Donkeys have fewer sacral vertebrae than horses
   B. The first intercoccygeal space in donkeys is often fused or narrowed
   C. The spinal cord ends more caudally in donkeys
   D. Donkey caudal vertebrae are poorly developed
   Answer: B

4. What is a common anatomical obstacle when performing abdominocentesis in donkeys?
   A. Lack of abdominal fat
   B. Poor development of the peritoneum
   C. Thick ventral fat deposits, even in donkeys that appear thin
   D. Difficulty identifying the midline
   Answer: C
5. Which statement accurately describes donkey vocalization?
   A. Donkeys bray only during expiration
   B. Donkeys vocalize using only one vocal fold
   C. Donkeys vocalize during both inspiration and expiration
   D. Donkeys have a larynx identical to horses
   Answer: C
6. What distinguishes the chestnuts and ergots of donkeys compared to horses?
   A. Donkeys have no ergots or chestnuts
   B. Donkeys have large ergots and no hind limb chestnuts
   C. Donkeys have chestnuts on all four limbs and no ergots
   D. Donkey ergots are smaller than horse ergots
   Answer: B
7. Why is aging miniature donkeys based on bone development more complex than in horses?
   A. Their bones develop more rapidly
   B. Their growth plates close later than in horses
   C. Their limb length prevents ossification
   D. Their bones lack normal physis lines
   Answer: B

References

Abdelbaset-Ismail A, Gugjoo MB, Ghazy A, Gomaa M, Abdelaal A, Amarpal A, Behery A, Abdel-Aal A, Samy M-T, Dhama K (2016) Radiographic specification of skeletal maturation in donkeys: defining the ossification time of the donkey growth plates for preventing irreparable damage. Asian J Anim Vet Adv 2016:1–6. https://doi.org/10.3923/ajava.2016

Abdelbaset-Ismail A, Aref M, Ezzeldein S, Eisa E, Gugjoo MB, Abdelall A, Emam H, Al Syaad K, Ahmed AE, Alshati A, El Raouf MA (2022) Ultrasound, dacryoscystorhinography and morphological examination of normal eye and lacrimal apparatus of the donkey (Equus asinus). Animals 12(132):1–12. https://doi.org/10.3390/ani12020132. https://www.mdpi.com/journal/animals

Adams OR (1979) Lameness is horses, 3rd edn, p 488

Adams MF, Castro JR, Morandi F, Reese RE, Reed RB (2013) The nasolacrimal duct of the mule: anatomy and clinical considerations. Equine Vet Educ 25(12):636–642

Buil García J (2023) Odontología en burros y sus principals diferencias con el caballo. Equinos 66(2):44–66

von Bünger I, Hertsch B (1981) Das Gewiss des Hausesels (Equus asinus asinus L.) Morphologische und röntgenologische Untersuchen (the dentition of the domestic donkey (Equus asinus asinus L.) morphologic and roentgenologic studies). Anatomia Histologia

Embryologia 10(1):61–86. https://doi.org/10.1111/j.1439-0264.1981.tb00505.x. PMID:6454364

Burden F, Thiemann A (2015) Donkeys are different. J Equine Vet Sci 35:376–382

Burnham SL (2002) Anatomical differences of the donkey and mule. In: AAEP proceedings in depth: Mule/Donkey medicine and surgery, vol 48, pp 102–109

Burnham SL (2024) Field necropsy report for Pleasant Valley donkey rescue, miles, TX unpublished Burnham, S.L. data collected 2024 from 102 donkeys at Peaceful Valley donkey rescue July 2024 unpublished

Canisso IF, Panzani D, Miró J, Ellerbrock RE (2019) Key aspects of donkey and mule reproduction. Vet Clin North Am Equine Pract 35:607–642. https://doi.org/10.1016/j.cveq.2019.08.014

Clayton HM, Flood PF, Rosenstein DS, Mandeville D (2005) Clinical anatomy of the horse. Mosby Elsevier Limited chapter 3 spinal column, pp 31–33

Collins SN, Dyson SJ, Murray RC, Burden F, Trawford A (2011) Radiological anatomy of the donkey's foot: objective characterization of the normal and laminitic donkey foot. Equine Vet J 43(4):478–486. https://doi.org/10.1111/j.2042-3306.2010.00312.x

Doherty T, Valverde A, Reed RA (2022) Chapter 36 Anesthesia of donkeys and mules: anatomic, physiologic, and behavioral differences. In: Burnham SL (ed) Manual of equine anesthesia and analgesia, 2nd edn. Wiley, pp 543–548

Evans PA, Jack N (2007) Aging horses by their teeth. Utah State University cooperative extension: equine, 6pr. Accessed online 7/18/2024. https://extension.usu.edu/equine/research/aging-horses-by-their-teeth

Evans WJ, Borton A, Hintz HF, Van Vleck LD (1977) The horse, Chapter One: The history and development of the horse, pp 3–17

Evans L, Crane M, Preston E (2022) The Respiratory System. In: The Clinical companion of the donkey, 2nd edn. https://www.ivis.org/library/clinical-companion-of-donkey-2nd-edition/respiratory-system. Accessed 8/12/2024

Forés P, López J, Rodríguez A, Harrán R (2001) Endoscopy of the upper airways and the proximal digestive tract in the donkey (Equus asinus). J Equine Vet Sci 21(1):17–20

Gacem S, Papas M, Catalan J, Bonilla S, Miró J (2018) Reproductive ultrasonography in Catalonian donkey. Reprod Domest Anim Proc 53(2):136. (abstract)

Heath EH, Meyers VS (1972) Topographic anatomy for caudal anesthesia in the horse. Vet Med/Small Anim Clin:1237–1239

Hutchins B (1984) The donkey in veterinary practice. Includes dissection notes of S.L.(Schoener) Burnham. Equine Pract 6(1):8–12

Jamdar MN, Ema AN (1982) A note on the vertebral formula of the donkey. Br Vet J 138:450–452

Lindsay FEF, Clayton HM (1986) An anatomical and endoscopic study of the nasopharynx and larynx of the donkey (Equus asinus). J Anat 144:123–132

Miller RM (1994) Imprint training of the newborn foal. Chapter 17 Mules. Gulf Publishing Company, pp 132–137

Misk NA, Seilem SM (1997) Radiographic studies on the development of cheek teeth in donkeys. Equine Pract 19(2):27–38

Misk NA, Semieka MMA (1997) Radiographic studies on the development of incisors and canine teeth in donkeys. Equine Pract 19(7):23–29

Mohamed S (2000) Anatomical studies on the upper cheek teeth and their relation with maxillary sinus in postnatal donkeys. Assiut Vet Med J 43:31–51

Muylle S, Simoens P, Lauwers H, Van Loon G (1999) Age determination mini-Shetland ponies and donkeys. J Vet Med A 46:421–429. https://doi.org/10.1046/j.1439-0442.1999.00229.x

Pathak V, Sheikh A, Rajput R, Bhardwaj RL, Sharma A (2013) Morphometric and histological observations on the testis of mule and donkey of Himalayan region. Indian J Vet Anat 25(2):99–101

Pohlmeyer K, Wissdorf H (1975) Anatomische Grundlagen zur Spülung der tränenabführenden Wege des Hausesels (Equus africanus f asinus—Mazedonischer Zwergesel). [Anatomical Basis

for the irrigation of the lacrimal ducts of donkeys (Equus africanus f. asinus—Macedonian dwarf donkey)]. Deutsche Tierätztlische Wochenschrift 82(8):314–316
Powell RJ, du Toit N, Burden FA, Dixon PM (2010) Morphological study of tracheal shape in donkeys with and without tracheal obstruction. Equine Vet J 42(2):136–141. UI: 20156249
Rodrigues JB, Lilly G (2019) Dental disorders of the donkeys. Vet Clin North Am Equine Pract 35: 529–544. https://doi.org/10.1016/j.cveq.2019.08.008
Sack WO, Habel RE (1977) Rooney's guide to the dissection of the horse. Veterinary Textbooks, New York. Part One: Neck and thorax, p vii, 2–13
Said A, Shokry M, Saleh M, Hegazi AA (1977) Contribution to the nasolacrimal duct of donkeys in Egypt. Anat Histol Embryol 6:347–350
Shoukry M, Saleh M, Fouad K (1975) Epidural anesthesia in donkeys. Vet Rec 97:450–452
Taylor TS, Matthews NS (2002a) Anesthesia of donkeys and mules: how they differ from horses. In: Proceedings of the AAEP, in depth: Mule/Donkey medicine and surgery, vol 48, pp 110–111
Taylor TS, Matthews NS (2002b) Donkey and mule scenarios: when to stop, think, read or call. In: Proceedings of the AAEP, in depth: Mule/Donkey medicine and surgery, vol 48, pp 115–116
Taylor TS, Matthews NS, Blanchard TL (2001) Introduction to donkeys in the U.S. elementary assology. N Engl J Large Anim Health:21–28
Tegetmeier WB, Sutherland CL (1895) Horses, asses, zebras and mules chapter 3 the African wild ass. Horace cox, "field" office, Windsor house, Bream's Buildings, E.C, London
Thiemann AK, Poore LA (2019) Hoof disorders and farriery in the donkey. Vet Clin North Am Equine Pract 35:643–658
Thiemann AK, Sullivan RKE (2019) Gastrointestinal disorders of donkeys and mules. Vet Clin North Am Equine Pract 35:419–432
du Toit N, Bezensek P, Dixon PM (2008a) Comparison of the microhardness of enamel, primary and regular secondary dentine of the incisors of donkeys and horses. Vet Rec 162:272–275
du Toit N, Kempson SA, Dixon PA (2008b) Donkey dental anatomy. Part 1: gross and computed axial tomography examinations. Vet J 176:338–344. https://doi.org/10.1016/j.tvjl.2008.03.003. https://pubmed.ncbi.nlm.nih.gov/18396076/
Vendramini OM, Guintard C, Moreau J et al (1998) Cervix conformation: a first anatomical approach in Baudet du Poitou asses. Br Soc Anim Sci 66:741–744
Walker M, Taylor T, Slater M, Hood D, Weir V, Elslander J (1995) Radiographic appearance of the feet of mammoth donkeys and the finding of subclinical laminitis. Vet Radiol Ultrasound 36(1): 32–37

A Practical Approach to Donkey, Mule and Hinny Behavior: Learn the ABC's of Behavior

Amy K. McLean

Abstract

Donkeys and their hybrid offspring are likely one of the most misunderstood animals due to their behavioral differences from their cousin the horse. Donkeys' natural response to fear, pain or discomfort is to freeze and a horse is to flee. Donkeys evolving from a resource poor environment, the desert may better explain why they freeze versus running away when there's limited forage and water. The mule and hinny are even more interesting in terms of behavior because they may exhibit behavioral signs of both parents, the donkey and the horse. Mules and donkeys are extremely strong and if there's little to no trust between the handler, practitioner, or owner routine procedures may prove to be extremely challenging and they will likely use their strength too. Participatory data has shown that owners who work with their mules at an early age develop a trusting and good working relationship so physical examinations or taking temperatures are not challenging. In most cases mules and donkeys respond very well to soft soothing voices, gentle touch and even food rewards. However, the best approach to mule and donkey behavior is slow and steady and proceed when your patient gives you the green light. Start at the head and work your way back when they allow you to. When observing donkeys and mules for discomfort and pain similar to working with them, pay attention to subtle changes in their body language from orbital tightening, ear position, body stance and tail swishing. Both will tell you what they are thinking, how they will react and if they are experiencing discomfort but you have to be willing to pay attention to their body language and recognize small changes as strong indicators of their current behavior.

A. K. McLean (✉)
Department of Animal Science, University of California Davis, Davis, CA, USA

World Donkey Breeds Project, Cordoba, Spain
e-mail: acmclean@ucdavis.edu

A. K. McLean et al. (eds.), *Donkey and Mule Medicine: Foundations, Breeds, and Management*, https://doi.org/10.1007/978-3-032-12726-6_16

Keywords

Behavior · Pain behavior · Learning theory · Restraint

1 Introduction

Donkey behavior is likely the reason why donkeys were domesticated prior to horses (McLean et al. 2019a). The stoic but calm nature of a donkey made it an easier target to catch versus a horse that fled or ran away at top speed. The saying we are products of our environment can easily relate to why donkeys choose to conserve their energy and fight versus flee considering they evolved and still thrive in a desert environment. Deserts are harsh environments where limited food and water are available, yet the donkey is quite adapted to live here. The adaptive traits of a donkey conserving energy and water affects every aspect of donkey physiology from taking bites of food to metabolizing and storing food and clearing medicine as well as their reproductive cycle. Donkeys browse the desert for the most nutrient available source of food and then due to an adaptive trait to conserve energy the body responds by taking up the energy and storing it in a form of fat or adipose tissue. The donkey is known for taking larger bite sizes and bigger gulps of water to maximize feeding behavior. Along the lines of energy conservation if the donkey is fearful its natural response is to freeze then possibly fight. These similar behavior characteristics are found in their hybrid offspring, mules and hinnies (donkey × horse cross). Mules and hinnies share behaviors common to donkeys and to horses (McLean et al. 2019b). They too are stoic animals and show less acute signs of pain compared to horses yet often are more expressive than donkeys. Similar to donkeys they can be cautious when working with them for routine procedures and extremely sensitive to pain such as the prick of a needle for an injection yet resilient to chronic pain from laminitis ((See Fig. 1) Tilbrook et al. 2002; Joseph and Whirledge 2007; Valsamakis et al. 2019).

The interesting and challenging aspect of working with donkeys, mules and hinnies from a behavioral point of view is one being able to handle and treat them from a veterinary perspective to being able to identify or notice behavioral signs of pain. Pain is associated with the sympathetic nervous system and how pain is expressed can vary amongst animals. It is a behavioral response at the central nervous system broadly consists of rapid activation of the sympathetic adrenomedullary system (SAM) and relatively slower engagement of the hypothalamic-pituitary-adrenal axis (HPA), which interact to build both an immediate and a longer-lasting response to perceived stressors (Rasiah et al. 2022; Godoy et al. 2018). Mule owners and handlers tend to agree through participatory surveys that they can interact easier with their animals compared to unfamiliar people (McLean et al. 2019b). This complicating factor spills over into a clinical setting. Therefore, when working with a mule or hinny for the first time, get to know them. Reach out with your hand and allow the mule or hinny to approach you (McLean et al. 2019a) and then proceed with your physical exam. Donkeys are more forgiving

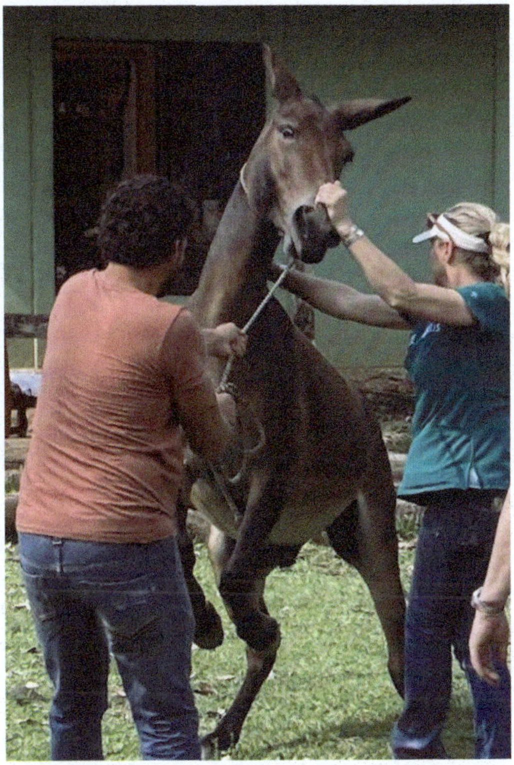

Fig. 1 Example of fearful mule when trying to take a blood sample at an Equitarian Initiative Workshop in Costa Rica. Notice the restrain of the hand on the end of the frontal bone. This may have some effect but the mule is extremely sensitive and fearful of blood draw

and freeze when they are frightened so proceeding with a physical exam, injection for sedation or any form of diagnostic testing will easily be accomplished. In donkeys, it is shown that using slow movements, rewarding good behavior with treats, tactile contact such as pets or scratches, and speaking with a quiet, calm voice all hone a positive interaction (McLean et al. 2019a) (see Fig. 2).

On the other hand, abrupt movements, the use of prods and whips, yelling, pushing, and kicking are considered negative and are likely to create a negative relationship between humans and animals.

On a global basis there's almost as many donkeys (43 Million) as horses (50 Million) and mules and hinnies account for another 13–15 million (FAO 2023; Norris et al. 2021) A recent survey in the U.S. showed the fifth largest equine breed/ population is mules and donkeys (American Horse Council 2023). Due to the increase in the number of companion and performance/recreation animals there's a growing interest for knowledge about donkeys and mules. Researchers and those who work regularly with mules and donkeys find that if. You spend time in a "*get to know you phase*" you will enhance overall experience working with donkeys and especially mules. Practitioners must keep in mind you likely won't see a patient until the disease has reached a more advanced state. So, consider how to safely restrain for exam- using both chemical and effective restraint equipment (e.g. nose twitches for

Fig. 2 Establishing a working relationship with positive human-mule interactions is the best approach to establishing trust. (Hewlin Brick Kilns, Egypt)

mules)l. Also, a bad experience will be remembered by most mules, so if you go to take a blood sample or perform an intravenous injection make sure you have your mule properly restrained because they are extremely sensitive to needles if they have not been conditioned to the pinch/poke of the needle. You may only get one chance to take your sample or administer your medication without proper restraint.

In this chapter we will cover the ABC's of donkey and mule behavior from the approach to training. The learning objectives will include increasing understanding about donkey and mule behavior compared to horses, improving understanding on how to work safely and effectively with donkeys and mules and increasing understanding about behavior possibly related to pain or discomfort.

- A-Approach
- B-body Language
- C-can you recognize and treat signs of pain
- F-Facial Grimace
- D-digestive
- P-Pain
- R-restraint
- R-reproduction
- T-Training

By following these suggestions, it will allow you as a veterinarian, student or owner to increase your understanding of how and why long ears may respond the way they do and how to improve your working relationship with these animals. Increase understanding about donkey and mule behavior compared to horses, improve understanding on how to work safely and effectively with donkeys and mules and increase understanding about behavior possibly related to pain or discomfort.

2 Approach

One of the most challenging tasks when working more with a mule or hinny versus donkeys is approaching them and getting close. With horses you may approach from the side or in some cases headfirst, but we have found with mules and donkeys they tend to do best if you approach face first and extend a hand. A person that is familiar to the mule will have more success approaching them and working their way around their body compared to a person that is unfamiliar. A participatory study showed that owners agreed that it was easier for them to work with their mules compared to strangers ($p = 0.001$) (McLean et al. 2019a). The study showed that familiar people compared to unfamiliar people could touch their forehead, neck and touch their ear compared to unfamiliar people. Unfamiliar people had very little success touching the neck and none touching their ears (McLean et al. 2019a) (see Fig. 3).

A way to gain their trust and acceptance is to extend the hand and allow them to smell and approach you. A treat or soft voice may help gain more access. Make friends and proceed when they allow you to do so (see Fig. 4).

Fig. 3 Notice the mule's body language and posture. The mule is leaning away, ears are back, eyes are large. The mule is not accepting the touch of the veterinarian student but she's doing a great job scratching the mule and maintaining a consistent stance that's not threatening to the mule and working with its fearfulness

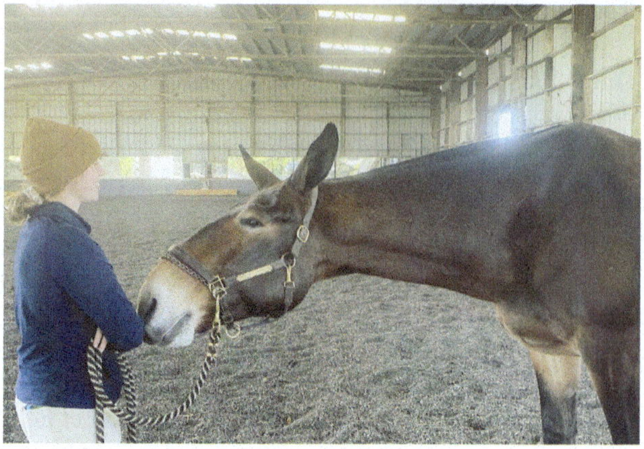

Fig. 4 The mule is reaching out to the vet technician who has cookies in her pocket but notice he is making the first move then she can proceed. The mule is undergoing an acupuncture appointment (note needles placed in the neck). By reading the facial expressions we can tell by the bright and alert eye he is curious and accepting, the same with the expression of the nostrils, lips and cheek muscles

3 Body Language

Donkeys and mules like all animals communicate to each other and to us through body communication and language versus verbal. We as their caretakers must understand how to speak donkey or mule just like we learn to speak horse. When working with donkeys and mules and thinking about where donkeys evolved from and thinking a mule or hinny has a donkey and half horse, so reactions and some behaviors may be quicker or more acute you will likely succeed in working with both animals. When working with donkeys or mules consider ear position, head and neck, body, weight on hind legs and tail movement or swishing. Their bodies from literally head to tail will rely on how they are feeling and what they are thinking (Figs. 5 and 6).

A donkey or mule's body language can also tell us if they are healthy or not healthy. Again by looking at key facial action units such as ear position, orbital tightness in the eyes, tension in the cheek muscle, tension in the nostrils and lips or in chronic pain cases, extremely droopy lower lips may suggest discomfort or pain (Fig. 7). Other indicators include excessive lying down, pawing, abnormal body stance with the head and neck lowered and back raised, legs stretched forward and or reluctance to move. A healthy donkey or mule will be curious and alert (Fig. 8). They will want to inspect you with eyes bright and ears forward. Notice the differences in healthy versus unhealthy donkeys and mules such as the figure illustrating pawing in a mule which could be related to unhealthy behavior when observing the entire behavior (Fig. 9). Further in the chapter will discuss pain behavior and pain is discussed in anesthesia as well.

Fig. 5 Notice the mule's ears are laid back and there's tension around the eyes, nostrils are tight and flared. The head is in an upright dominant position. If you were wanting to pet a mule, this should not be your choice

4 Digestive Behavior

Donkeys and mules are naturally browsers meaning they will pick through and select forages or pieces of grain or grasses they prefer to eat versus being grazers like horses. This behavior is likely related to donkeys evolving from the desert and seeking out plants with higher nutrient value or more fiber or energy content. Considering donkeys think about their food selection increases their level of cognition and likely their ability to reason as shown in a study measuring cognition comparing horses, donkeys and mules. Mules were the quickest to complete each task by finding the feed according to shape in the quickest time and had the most correct responses. Donkeys were right behind them in correct selection, but the energy conserving donkey was the slowest of the three (Proops et al. 2009). When donkeys and mules are sick they will stop eating but this occurs when the disease is extremely advanced. Studies have shown that donkeys will continue to eat even when they are extremely dehydrated, again likely a survival trait from the desert (Maoliy 1970). They naturally take larger bites, drink big gulps of water and can easily survive off of poor quality forages meaning they are high in fiber and low in

Fig. 6 In this mule example we notice its head and neck is elevated, rigid body, standing squarely on all four limbs and the tail is swishing/moving. This suggests agitation and resistances. When you look closer at the photo you notice there is a nose twitch on the upper lip of the mule. Nose twitches are very effective in mules and produce a sedative effect from the pressure point and then you can begin your examination although you will likely find an elevated heart rate and respiration but at least you can try safely to proceed. You will find the nose twitch will make it easier to administer injectable medications

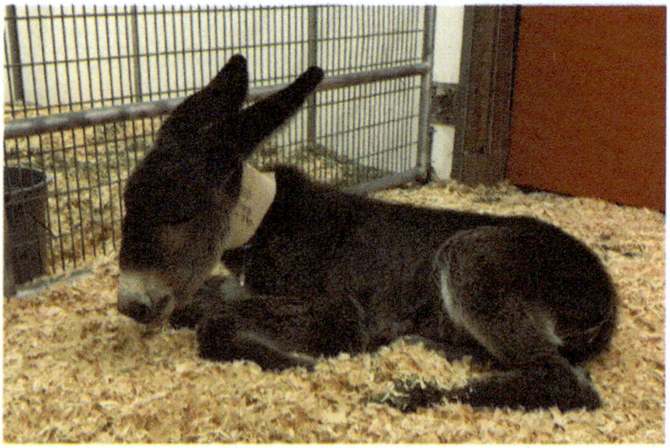

Fig. 7 The donkey foal had aspiration pneumonia due to Vit E deficiency. Notice her body language that suggests discomfort starting with her face

energy and other important nutrients. Their ability to consume and thrive off of poor-quality forages is related to likely a longer gastrointestinal transient time. Researchers have described the donkey to take over 72 in digestion and fermentation along with possible differences in the microbes found in the cecum that can assist with fermentation of high fiber diets (Kasirer-Izraely et al. 1994; Jones et al. 1989).

Fig. 8 This is a curious and healthy donkey and we can tell based on his facial body language such as the bright open eyes, ear position, and nostrils

Fig. 9 Notice the front right hoof, the mule is pawing. Some mules will naturally paw out of frustration, they have a need to be meet, such as their companion has left but when a mule is pawing in its normal environment, we must look closer at this behavior and ask why they are doing this and what has changed or what is wrong causing this behavioral response. In this case, the mule wanted to lay down but due to a right hind limb lameness he did not feel comfortable to do so. So, the behavior related to a frustration response due to discomfort

Donkeys and mules both tend to browse on bitter tasting plants which are often high in tannins and or toxins, so owners and veterinarians be aware and if the donkey or mule behavior is suggesting it is sick consider checking liver enzymes and asking or considering what the donkey and or mule has been eating. Remember when donkeys stop eating, they are typically a dead donkey! Mules tend to show signs of gastrointestinal or abdominal discomfort better than donkeys meaning they are quicker to look at their sides, flanks, roll, paw, have dull eyes, yawn and excessive blinking. Donkeys that are colic are usually more subtle in their behavioral response so you must watch carefully.

Other digestive behavior considerations include how both donkeys and mules respond to being hospitalized. Many donkeys do not adapt well to being alone and in a new setting. Consuming water may be a challenge. So, consider the color of the bucket, the depth, holding it for the donkey and even adding a sweetener to encourage them to drink (see Fig. 10). Ideally, bring a friend with a donkey patient to decrease separation stress. Mules that are hospitalized may go off feed and water like donkeys but generally respond well to food rewards or higher food valued offerings like alfalfa hay or sweet feed. Like the donkey, if they can see a horse or another equine patient this will ease their stress and help them continue to eat and drink. You want to avoid extra stress of being in the hospital so neither patient

Fig. 10 Sometimes holding up a bucket of water is the easiest way to get them to drink. You can also try different colors, adding sweeteners, and if you only have access to an automatic waterer, I highly encourage offering a second source in a bucket or pan

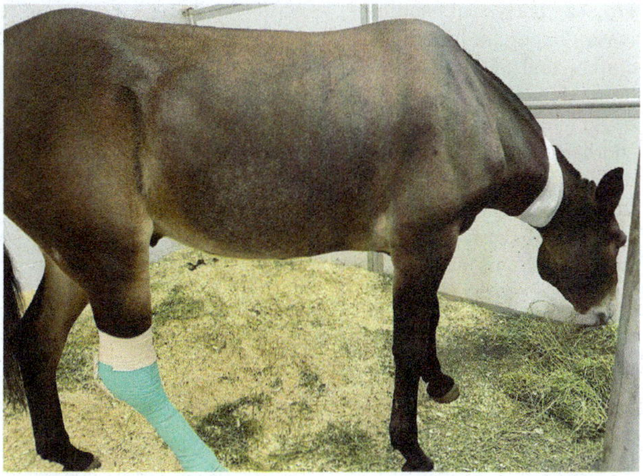

Fig. 11 Mules and donkeys can be selective eaters, so if they are not eating well in the clinic considered a higher value food option in this case the mule is eating alfalfa hay, yes, mules and donkeys can eat alfalfa, and this may encourage them to eat in the clinic. Both can develop hyperlipidemia if they stop eating and healing time will take even longer

Table 1 Specific Pain indicators to colic/abdominal discomfort

Behavior	Application	Reliable in Donkeys and Mules
Rolling	Excessive rolling	Unreliable
Kicking at abdomen	Yes	Yes, especially in mules
Flank watching	yes	Yes in both
Stretching and pawing	yes	Yes in both
Yawning and chewing	Yes	Reported in both
Dullness & depression	Only observable behavior in the donkey Much disagreement in the horse as indication of abdominal pain	YES

develops hyperlipemia. Also, keep in mind the browsing behavior and both mule and donkey patients may spread out their hay or pick through it and not completely clean it up like a horse patient (see Fig. 11). Refer to Table 1 for examples of behaviors associated with colic and abdominal discomfort.

5 Foal Handling

Encouraging breeders to handle both their donkey and mule foals from day one can help improve their welfare in the long run. Most mule breeders will agree that mule foals should be handled gently from day one such as picking up their hooves, touching their ears, mouth, anus, and picking up their hooves. From early on, positive human interactions with people will build their trust and help with routine procedures from trimming hooves to taking a temperature which is often one of the hardest tasks with a mule to become very easy to perform. Some researchers suggest

Fig. 12 Early foal handling is very important for mule and donkey foals. Once they are born it is advised that breeders begin interacting with them and building trust by touching their entire bodies, picking up their hooves, taking a temperature, and rubbing ears and mouth. This will set them up for success later and improve their welfare. Here Dr. Camillo Morals Sanz is interacting with a mule foal that was handled from the time she was born

imprint training professed by equine veterinarian Dr. Robert Miller but what's most important is early foal handling in these key areas. Handlers and owners have continued to say mules handled at an early age are easier to train and work with. Mule foals that were handled as foals generally allowed more interaction from unknown handlers (Bott et al. 2014). We have found as well through participatory interviews and research most owners would prefer to raise their own mules versus buy them because they know the animal's history and feel they are often more gentle and easier to handle (McLean et al. 2019a) (Fig. 12). Building a positive relationship from day one in mules tends to last a lifetime, there are many field examples that reflect this concept of early foal handling that increases the mule's level of trust. Mules and hinnies with such training tend to allow unfamiliar people to more easily approach them and provide treatment

6 Discomfort and Pain

Recognizing signs of pain and discomfort can be one of the more challenging aspects to treating donkeys and mules. Once you get past the mule's opposition to being examined you then have to figure out a route of diagnostics and then treatment. Some tools commonly used in horses such as hoof testers are not as successful in donkeys and mules. Their response to pain may be more subtle or in some mules that are extremely reactive and they associate the human with the discomfort making examining the patient increasingly challenging. The other concern is once pain is identified is the route of pain medication and the amount to provide. This information is covered in the pharmacology and anesthesia chapters but presents another layer of treating donkey and mule patients. However, before you start to treat, consider these various behavioral signs that may relate to nonspecific and specific types of pain. Keep in mind both the donkey and their hybrid offspring are going to

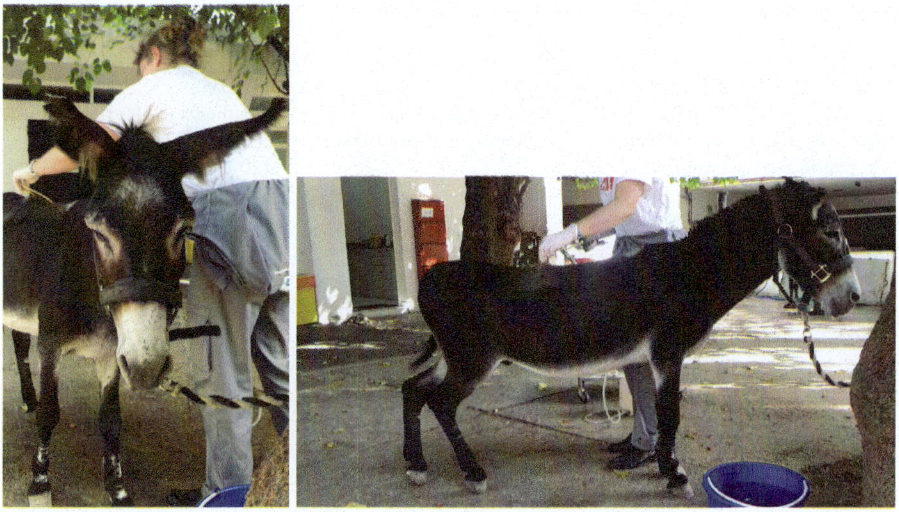

Figs. 13 and 14 Donkeys being more subtle and saving their energy may show little changes with their body language until you look closer. Consider eye blinking, ears pointing down and increased respiration. Also, look at the whole donkey (Fig. 11). When the student flushes the open lesion the donkey drops his whole back and we see abnormal stance and tail swishing (patient at the American Fonduk, Fes, Morocco)

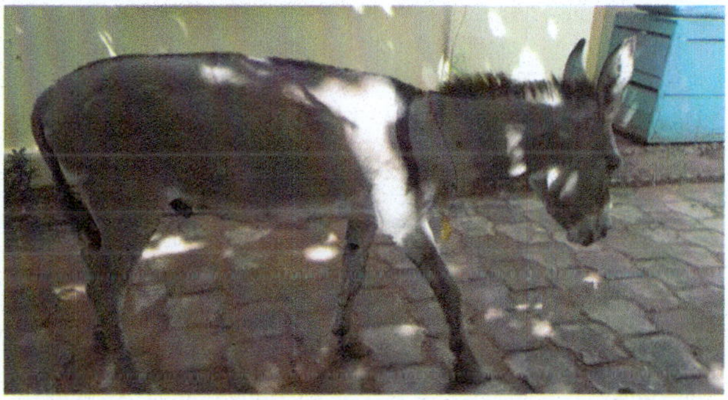

Fig. 15 Consider the behaviors you are seeing: Agitation, tail movement, fixed stare, lowered head and neck, clinched facial and jaw muscles. (Photo Dr. Amadou Dombia, SPANA Clinic, Bamako, Mali)

try and mask or not exhibit high levels of discomfort because in a wild or natural environment, a donkey that's limping or painful is an easier target for a predator (e.g. mountain lion) and one can conserve energy by not exhibiting a painful response or the response may be subtle (Figs. 13, 14, and 15, refer to Table 2).

Starting with the face such as dental pain or disease within the sinuses or guttural pouch or ocular area. Consider how the donkey or mule is holding its head, is it food packing, excessive tearing, head shaking, and excessive yawning. These can all be indicators of pain in the head or dental cavity (refer to Table 3, Fig. 16)

Diagnosing and identifying lower limb and hoof pain in both donkeys and mules may be challenging like other diseases. Consider body posture, gait abnormalities and shifting weight when standing still. Again, look at their face and consider

Table 2 Nonspecific Pain Indicators in Equids

Behavior	Application	Reliable in Donkeys & Mules	Source
1. Restlessness, Agitation, Anxiety	Commonly seen in acute pain cases Confined horses	Possibly- may notice tail switching, and eye blinking	Matthews et al. (1998), Sandford (1989), Fleming (2002), Price et al. (2003), Orth et al. (2020)
2. Rigid stance, reluctance to move	General description for many disorders Horses face away Protective Behavior	Possibly- consider condition and possibilities such as (tetanus cases and or laminitis cases)	Matthews et al. (1998), Sandford (1989), Taylor and Matthews (2002), Orth et al. (2020)
3. Lowered head carriage	Chronic pain, unrelenting	Difficult to differentiate in but consider ear and body position	Orth et al. (2020)
4. Fixed stare, dilated nostrils, clenched jaw	Facial pain nondescript lacks info on pain source	Yes	Taylor and Matthews (2002), Orth et al. (2020)
5. Aggression towards own foal or people	Pain during parturition Behavior emerges with pain	NO more subtle in donkeys	Short (1998), Sandford (1989)

Table 3 Behavioral Indicators of head or dental pain

Behavior	Application	Reliable in Donkeys
Headshaking	Possible dental or ear pain	No reported cases
Quidding, food pocketing, anorexia	Chewing on one side, dropping food, choke	YES
Yawning	Possible oral cavity pain	Yes

grimacing. In laminitis cases it's not unusual to see donkeys lifting and pointing and or hanging their limbs, resistances to move and rocking back on their hind limbs. For other lameness again pay attention to subtle signs, edema, sensitive to touch but a small reaction may be because the mule or donkey is conserving their energy. Consider sensor technology like the Lameness Locator and other sensor tools (e.g. thermography) to help with diagnostics. Like many other conditions the lameness may have to advance before you are able to diagnose or before the owner recognizes their mule or donkey is showing discomfort. Refer to the pharmacology and anesthesia chapters for specific information on doses for nonsteroidal anti-inflammatories because in many cases the medication will need to be

Fig. 16 An example of a donkey displaying head tilting. Notice ear position, position of the head and the upper lip is to the side. (Animal Nepal. Kathmandu, Nepal)

Table 4 Behavioral indicators of limb/foot pain

Behavior	Application	Reliable in Donkeys & Mules
Weight shifting-between limbs	Clear reliable pain indicator	Donkeys Observed (personal observations- action video)
Hanging or pointing limb	Pointing, lifting limb,	Not reported in literature Observed (personal observations) and see videos, Abscess
Abnormal movement	Jerky, ataxic	Yes (e.g. stringhalt in donkeys)
Reluctance to move	Constant lying down	Yes

administered at a higher dose and or more frequently or in some cases (e.g. fibrocoxin is less effective in mules and donkeys). Also, keep in mind that despite fallacies mules and donkeys can both easily get laminitis and in all four hooves, ring bone, side bone, navicular disease and may need joint injections to replace synovial fluid depending on their jobs and use. Soft tissue injuries may occur as well and refer to the chapters on orthopedics, biomechanics and podiatry to learn more about treatments (Table 4).

Fig. 17 Hinny patient at the American Fondouk Hospital a laminitis case notice ear position, orbital tightening and tight cheek muscles. All indicators of pain

Facial Grimacing is likely one of the most reliable indicators of pain in both mules and donkeys. Consider the eyes, orbital tightness, ear position, check muscles, upper and lower lips. In the following photos you will notice the lack of expression in the hinny and the donkey face (Figs. 17 and 18). The donkey has a fixed glare on his face as well and tightness in the nostrils (Fig. 18). In chronic pain cases such as the laminitic donkey and blind mule with ringbone you will see the lips may become loose but still consider the other facial areas and how they are presenting (Figs. 18 and 19). Look for excessive yawning, chewing, and blinking (Fig. 20).

7 Pain Related to Quality of Life

As a veterinarian you have an oath to caring for animals, but you cannot make all decisions for owners and handlers. Providing donkey and mule owners with educational information on how to make informed decisions about quality of life and end of the life choices is an important role. Since both mules and donkeys tend to hide or disguise chronic pain and clinical signs of disease until they are in a more advanced state, we must be realistic in the outcome. Considering how little we know about pain management in donkeys and almost nothing in mules and hinnies we must

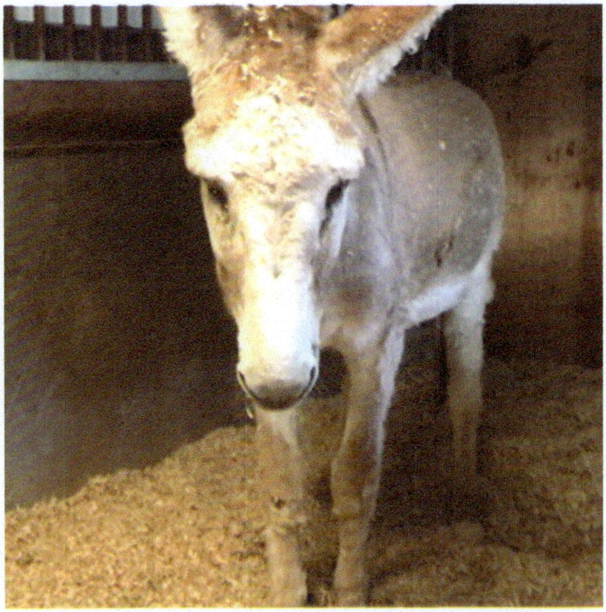

Fig. 18 Donkey that had foundered in all four hooves, chronic pain which can be observed in his facial grimacing from ear positions pointing downwards, tightness over the eyes which is sometimes hard to observe since the ocular bone is thicker in donkeys but tension in the muzzle and lips is also seen

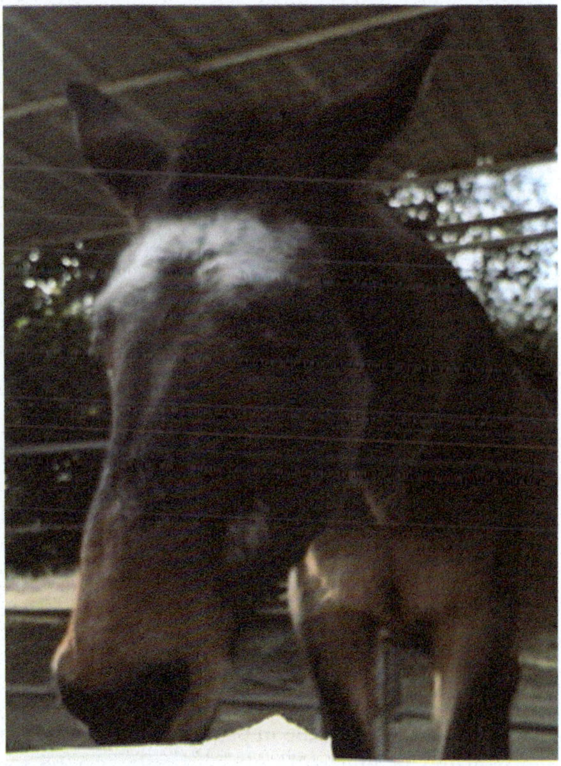

Fig. 19 The mule was blind in the right eye and had high ring bone in both front hooves like the donkey and hinny noticed the tightness in the orbital area, tension in the muzzle, nostrils and cheek bone. Also, the head and neck is lowered and the ears are to the side

Fig. 20 An acute painful condition, the mule was experiencing abdominal upset and had loose manure. He was expressing his discomfort by chewing to the right, tension in the eyes, orbital tightening, cheek muscle and ears pointing downwards. He was being administered an antibiotic after a surgery. Once the antibiotic was changed the facial expressions and grimacing was gone and his manure returned to a consistent form

Fig. 21 A donkey at Animal Nepal that couldn't stand without assistance and had started to develop open lesions from lying down so much. In some cultures, euthanasia is not allowed so consider how you will address this and other options to helping donkeys and mules have quality of life and little to no pain

consider how we will suggest or prescribe a treatment and maintenance program for long ears that allows them to live comfortably and not experiencing pain. In some cultures, euthanasia will or may not be an option so consider alternative methods for helping a donkey or mule in pain such as purchasing the animal if that's an option so the decision to end suffering and pain can be made (see Fig. 21). Consider the donkey and mule's ability to interact with other equids, its ability to stand and walk, its ability to eat, to get up and down without assistance, and overall, its ability to live without little human intervention a normal life.

8 Reproductive Behavior

In terms of reproductive behavior donkeys again are not like horses and respond and behave differently which can make jacks easy or hard to breed. Most jacks that breed mares by live cover have been raised with horses instead of donkeys. The theory behind this and there's no scientific study to prove its correct or not other than anecdotal experience, the jack thinks he is a horse and then will breed mares. Many breeders will only breed mares to their jacks versus jennies. Some fear the jack will stop breeding mares if you breed the jack back to a jenny. Again, there's no literature to support this theory. Many breeders have started to collect their jacks for artificial insemination, and this has become popular amongst those using donkeys for production either the milk industry or meat and skin. We are seeing an increase in donkeys being collected and taken to local clinics to be collected and shipped for making performance mules. When collecting jacks for artificial insemination be aware there are longer teasing times and again some jacks may need to be collected off a mare or a jenny. Others that have been trained to be collected off a phantom may still need the visual stimulation of a mare or jenny. Most jacks are slower to breed compared to horses (30–60 mins), younger jacks are slower than older jacks and keep in mind they may need multiple mounts before they have an erection (Canisso et al. 2009, 2019). Other behavioral signs to consider when collecting a jack, watch for clear fluid dripping form the nasal cavity, a slight jerk and then lowering of the head, possibly a kick out with one hind leg prior to the penis starting to drop and then become erect (see Fig. 22).

Some jacks may be very docile and gentle when breeding or mounting a mare or jenny in other cases they can be extremely aggressive. It is common to see jacks biting the neck, and lower limbs of both a jenny or mare to increase their libido and

Fig. 22 Jack behavior before breeding may include a kick out and this may also happen after ejaculation. It is part of their natural behavior and should not be discouraged because it may then cause the jack to not mount or be collected. (Photo from video of Dr. Jordi Miro, University of Barcelona, Barcelona, Spain)

Figs. 23 and 24 A jenny displaying mouth clapping, opening and shutting is a sign of estrus along with the ears laid back also seen with the mare mule in estrus and a stretched out stance of the body.

stimulate the mare or jenny to display signs of estrus (see Fig. 23). Jacks are also very vocal and often bray when entering the breeding shed and or before compilation. With shy breeders some clinicians may choose to use chemical stimulation to decrease the time to erection and then ejaculation by using Prostaglandin F2 alpha (PGF2 alpha)- Lutalyse and Estrumate (0.5 mL). You will likely notice sweating and even their hind limbs trembling after administering the prostaglandin and such techniques will be discussed later in the book in the theriogenology section (Panzani et al. 2020). Consider the environment and the staff that is assisting with the collection and try to keep the environment comfortable and without few distractions such as additional people watching or other animals all of these things may delay time to erection.

Jennies are interested in how they present when they are in estrus which can occur throughout the year. The behavior of jennies who are in estrus will open and close their mouths, lay their ears directly back and tail straight out. It's an unusual stance that we also see in mare mules and hinnies. Not all mare mules and hinnies will display signs of estrus, but they may, and they present similar to jennies in heat see Fig. 24). Another difference is jennies and mare mules will mount one another when they are in estrous Jennies will show and cycle in the winter months.

9 Restraint and Training

The most effective method to working with donkeys and mules is through training. Early foal handling helps decrease stress for routine procedures later in life from simply taking a temperature or administering oral medication. More in mules than donkeys when they were not exposed to early foal handling practicing such procedures at an adult age can become very difficult to perform for a practitioner. Your job is not to train the person's donkey or mule. Keep in mind donkeys are generally easier to interact with and perform routine procedures because they revert to their natural behavior of freezing when they are fearful or experiencing discomfort where the mule or hinny will want to flee or strongly react to the touch of a needle or thermometer. Methods of restraint that work extremely well in mules and hinnies are blind folds and nose twitches (refer to Figs. 25 and 26).

Any clinic that works on a regular basis with mules such as the American Fonduk in Fes, Morocco rely on nose twitches to interact with some of the most challenging mules to work with and provide medical treatment. Mules in many working conditions their relationship with people is work and or likely discomfort from

Figs. 25 and 26 Blind folds or covering the eye with your hand tend to be very effective on mules and may allow you to perform your exam or apply a nose twitch so chemical sedation can be utilized

Fig. 27 Example of a blind fold and nose twitch being used to trim the mule's hooves. Chemical sedation was not available

working many hours with ill-fitting harnesses so people are associated with this discomfort and it's not a positive relationship. So, the nose twitch made of rope or chain applied to the upper lip and with a longer handle can be essential for administering chemical restraint. The nose twitches like the skin or ear (which we do not recommend because mules especially become ear and head shy) act by releasing beta endorphins which have a very calming effect on mules (see Fig. 27). They enter a very tranquil state that usually allows the practitioners to then supply sedation.

Refer to the pharmacology and anesthesia sections for the proper amount and type of sedation to effectively use in mules. Donkeys tend to be more agreeable for basic physical exams and injections but if needed soft ropes and immobilizing a leg along with a blind fold and or nose twitch may work. Also, chemical or pharmacological sedation is highly recommended when it's available like the mule check doses and types of sedation and anesthesia because both equine. When a mule or donkey is sedated look for relaxed facial units, ears down or to the side, eyes closing, little tension in the muzzle or nostril area (see Fig. 28).

Also, monitor heart rate, respiration during any procedure and signs from their body or body language that the sedation or anesthesia is wearing off and you may need to administer more. The more fearful the animal is going into a procedure then likely more sedation they will need. Also, consider oral sedation but do be aware some donkeys and mules may have an allergic response to oral detomidine and even xylazine administered intravenously. We are currently trying to research why we are

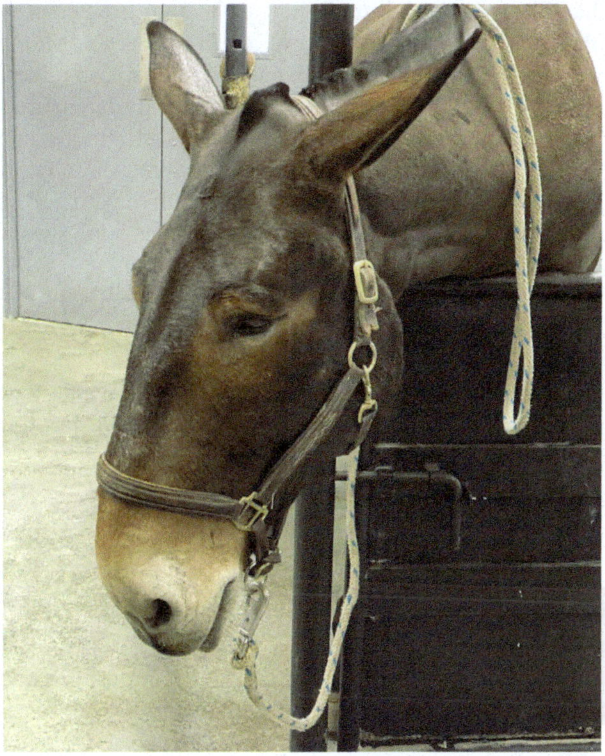

Fig. 28 When a mule or donkey is sedated consider observing the facial action units which should appear relaxed as seen in this photo of the mule. His ears are to the side, relaxed and pointing downwards, eyes are partly closed, loose muzzle and nostrils. Head and neck is lowered

seeing excessive tearing, red mucous membranes, and itchiness when alpha two receptor sedation is administered to some donkeys and mules. Again, try to read body language, make the donkey or mule your friend before proceeding, proceed slowly, confident, and accurately. Do not stress out the mule or donkey and apply the appropriate restraint mechanism as soon as possible to make this a good experience. Remember, mules especially will remember a bad event or experience.

10 Conclusion

When working with donkeys and mules, remember to be patient and gain their trust. Donkeys will come around quicker than the mules. Consider the ABC's of donkey and mule behavior and start with reading their body language and getting to know them before you proceed. Make all interactions if you can be positive. Effectively and wisely choose your restraints and medication to help with restraining your long eared patient. When working with jacks for breeding purposes consider a quiet environment with few distractions, be patient and smart about your plan of action and consider the individual personality if he's shy or aggressive. Encourage mule owners and breeders to work with their foals from day one and do early foal handling; don't wait until they are mature to interact with them. You will hear this throughout this book donkeys and mules are not horses with big ears. Horses are generally easier to work with and train but so can mules be easy to work with. It's all how they get started in life. Consider body language and facial action units when trying to identify pain and discomfort in both mules and donkeys. Take the slightest hint that they are experiencing pain as a reason to continue your exam and perform diagnostics. Make your interactions positive with a soothing voice, soft voice, a treat or cookie, if possible, a pet or scratch goes a long way.

Questions

1. You are working with a new client that wants to start breeding for mules and donkeys. They ask when is the best time to start training mules?
   A. As soon as they are born
   B. When they are weaned
   C. 2 year olds
   D. Any time
   E. All of the above
   F. None of the above
   Answers: A
2. When trying to observe pain and discomfort in mules and donkeys where would you look or how would you go about this?
   A. Their face
   B. Ear position
   C. Muzzle
   D. Body stance
   E. Tail movement
   F. All of the above
   G. None of the above
   Answers: F
3. You are hiring a new breeding manager for your clinic and the person comes with years of experience breeding Thoroughbreds, but you have a large donkey practice, will his experience be useful in dealing with jacks? (short answer)
   Possibly, since he's worked with a breed of horse that requires live cover, he will have a lot of experience handling different stallions and hopefully willing to learn how to be patient with the jack when breeding and will need to learn some of the tricks of the trade in terms of individual donkey preference for collection methods such as appropriate AV types, mare vs jenny collection and possibly when to administer medication to encourage erection and or ejaculate.
4. What are some key behavioral signs you should be aware of when working with mules and how you proceed?
   A. Tail swishing
   B. Weight shifting on hind limbs

C. Ears laid back
D. Tension in the muzzle
E. All of the above
F. None of the above
Answers: E

5. Which methods of restraint would you use for a very fearful donkey that has come into the clinic? Would it be the same as a mule? (short answer)
No, mules respond well to nose twitches, donkeys soft ropes and or blind folds, could use a blind fold on both and best restraint would be chemical restraint

References

American Horse Council (2023) National economic Impact of the U.S. Horse Industry. American Horse Council, Washington, DC

Bott R, McLean A, Heleski C (2014) Community-based participatory research interfaced with equine welfare assessment to learn about working equids and their owners in Vera Cruz, Mexico. In: Owers R (ed) Proceedings of the 7th international colloquium on working equid. World Horse Welfare, London, pp 1–15

Canisso IF, Carvalho GR, Davies Morel MCG, Guimarães JD, McDonnell SM (2009) Sexual behavior and ejaculate characteristics in Pêga donkeys (Equus asinus) mounting estrous horse mares (Equus caballus). Theriogenology 73(1):56–63. https://doi.org/10.1016/j.theriogenology.2009.07.026

Canisso IF, Panzani D, Miró J, Ellerbrock RE (2019) Key aspects of donkey and mule reproduction. Vet Clin North Am Equine Pract 35(3):607–642. https://doi.org/10.1016/j.cveq.2019.08.014

FAO (2023) FAOSTAT—Food and Agriculture Organization of the United Nations. Global livestock statistics. Retrieved from FAOSTAT database

Godoy P, Whirledge S, Tilbrook AJ (2018) The hypothalamo-pituitary-adrenal axis and the regulation of food intake. In: McKinley MJ (ed) The physiology of the gastrointestinal tract, 4th edn. Elsevier, pp 2131–2165

Jones NL, Houpt KA, Houpt TR (1989) Stimuli of thirst in donkeys (Equus asinus). Physiol Behav 46(4):661–665. https://doi.org/10.1016/0031-9384(89)90348-X

Joseph D, Whirledge S (2007) The role of glucocorticoids in the regulation of reproductive function. J Endocrinol 194(3):1–10

Kasirer-Izraely H, Choshniak I, Shkolnik A (1994) Dehydration and rehydration in donkeys: the role of the hind gut as a water reservoir. J Basic Clin Physiol Pharmacol 5:89–100

McLean AK, Navas Gonzalez FJ, Canisso IF (2019a) Donkey and mule behavior. Vet Clinics of North America 35:575–588. https://doi.org/10.1016/j.cveq.2019.09.010

McLean AK, Varnum A, Ali A, Heleski C, Gonzalez-Navas FJ (2019b) Comparing and contrasting knowledge on mules and hinnies as a tool to comprehend their behavior and improve their welfare. Animals 9(8):488

Norris SL, Little HA, Ryding J, Raw Z (2021) Global donkey and mule populations: figures and trends. PLoS One 16(2):e0247830. https://doi.org/10.1371/journal.pone.0247830

Panzani D, Quaresma M, Fanelli D, Camillo F, Moroni R, Rota A, Martins-Bessa A, Nóvoa M, Catalán J, Canisso IF, Conte G, Mirò J (2020) Hastening time to ejaculation in donkey jacks treated with the $PGF_2\alpha$ analog, Cloprostenol sodium. Animals 10(12):2231. https://doi.org/10.3390/ani10122231

Proops L, Burden F, Osthaus B (2009) Mule cognition: a case of hybrid vigour? Anim Cogn 12(1): 75–84. https://doi.org/10.1007/s10071-008-0172-1

Rasiah R, Godoy P, Whirledge S (2022) The role of glucocorticoids in the regulation of reproductive function. J Endocrinol 194(3):1–10. https://doi.org/10.1677/joe.0.1940191

Tilbrook AJ, Turner AI, Clarke IJ (2002) The hypothalamo-pituitary-adrenal axis and the regulation of food intake. In: McKinley MJ (ed) The physiology of the gastrointestinal tract, 4th edn. Elsevier, pp 2131–2165

Valsamakis G, McLean A, McGreevy P (2019) The welfare of working donkeys in developing countries: a review of the literature. Animals 9(9):1–15. link.springer.com

Maloiy GMO (1970) Water economy of the somali donkey. Am J Physiol 219:1522–1527

Matthews NS, Taylor TS, Potter GD (1998) Physiologic responses during an exhaustive driving test in donkeys: effect of conditioning. Appl Anim Behav Sci 59(1–3):31–38

Fleming P (2002) Nontraditional approaches to pain management. Vet Clin North Am Equine Pract 18:83–98

Price J, Seago Catriona, Welsh EM, Waran NK (2003) Preliminary evaluation of a behaviour-based system for assessment of post-operative pain in horses following arthroscopic surgery. Vet Anaesth Analg 30(3):124–137. https://doi.org/10.1046/j.1467-2995.2003.00139.x

Sandford S (1989) Donkeys. Macmillan Education Ltd., London

Orth E, Navas Gonzalez FJ, Pastrana CI, et al (2020) Development of a donkey grimace scale to recognize pain in donkeys (Equus asnius) Post castration. Animals, https://doi.org/10.3390/ani10081411

Taylor TS, Matthews NS (2002) Donkey and Mule Scenarios: When to stop, think, read or call. Proc Am Assoc Equine Pract 48:115–116

Short CE (1998) Pain management in large animals. Veterinary Clinics of North America: Food Anim Pract 14:83–100

Physical Examination of the Donkey and Their Hybrid Crosses

Cynthia Xue

Abstract

Physical examination of the donkey and their hybrid crosses (mules and hinnies) is performed in the same manner as in horses. Note that both mules and hinnies may be less trusting than donkeys and it may be more challenging to obtain their heart rate and temperature. Particular attention should be paid to behavioral changes as pain is not manifested as overtly in donkeys and their hybrids compared to horses. Further, their increased metabolic efficiency for nutrient absorption increases their risk for endocrinopathies, and subclinical laminitis can be overlooked even in the absence of overt lameness due to their natural stoicism. The physical examination should always be performed in a systematic manner, utilizing both distance and hands-on examination, to avoid missing subtle clues that may alter the clinical course of disease workup.

Keywords

Physical exam · Workup · Vitals

C. Xue (✉)
Department of Clinical Sciences, Ross University School of Veterinary Medicine, Basseterre, Saint Kitts and Nevis, West Indies
e-mail: cxue@rossvet.edu.kn

A. K. McLean et al. (eds.), *Donkey and Mule Medicine: Foundations, Breeds, and Management*, https://doi.org/10.1007/978-3-032-12726-6_17

1 General Approach to Physical Examination of the Donkey, Mule, and Hinny

The approach to the physical examination (PE) of donkeys and donkey hybrids (mules and hinnies) is much like that of horses. A thorough examination—beginning with the distance exam, followed by the hands-on exam—performed in the same order every time will help clinicians avoid missing key details that may alter the course of disease workup. Because donkeys and their hybrids, mules and hinnies, may not always be amenable to being approached for the hands-on portion of the PE, being systematic is of even greater importance such that "muscle memory" helps with collecting information should the clinical examination require truncation in a fractious animal for safety.

The chapter will be broken down into:

Distance examination
- Individual behavior and/or interactions with the herd
- Expressions of pain
- Body condition scoring
- Respirations
- Lameness evaluation

Hands on examination
- Vital parameters
- Specific considerations by body system

2 Before the PE, Collect Signalment and History

Examination of every patient should begin with collection of the signalment (age, breed, sex) and history (if known). This information can increase the index of suspicion for specific organ system pathology that heightens a clinician's impetus to be more meticulous in their evaluation of those specific areas on the physical examination. For example, a patient with a history of weight loss should have their teeth examined. A patient with a history of lethargy and coughing should have a rebreathing bag examination performed combined with tracheal and lower airway auscultation. Reports of stertor or "honking" should prompt careful palpation of the trachea. An older donkey with changes in hair coat and/or appetite should have particular attention paid to visualizing their feet and palpating for digital pulses (as is safely allowed by the patient).

When collecting history, also be aware of any reported changes in behavior. Sham-eating, for example, should increase a clinicians' concern for colic (Burden and Thiemann 2015). Consequently, clinicians should be more thorough on their gastrointestinal examination even if the patient does not appear actively colicky during the PE.

These recommendations are a reminder that a comprehensive history, paired with an excellent physical exam, can improve the accuracy of choosing appropriate diagnostic tests. Ultimately, all diagnostics to evaluate a patient's systemic health and/or suspected organ system of pathology should always be determined based on PE findings as owner recollection of history may be incomplete, their interpretation of clinical signs may be inaccurate, or a patient may simply have an unknown history.

3 Distance Examination

3.1 Individual Behavior and/or Interaction with the Herd

Meticulous observation of donkeys', mules', and hinnies' behavior from a distance before beginning the hands-on PE is strongly recommended. Animals may be less guarded in the absence of human intervention, and abnormal behaviors are more likely to be observed than masked. It is important to consider information about the animal's behavior in the context of any available history. Concern should be raised if a normally gregarious animal is seen acutely separating itself from the herd, less engaged with a bonded herd mate, or interacting less with the caretaker. The opposite can also be concerning—a normally flighty animal that is easier to catch, or unreactive in the presence of a veterinarian (an unfamiliar stranger) may be dull secondary to systemic disease.

Abnormal behavior is a sensitive but non-specific indicator of pathology. Rarely can behavioral changes immediately pinpoint specific organ system pathology. (The exception would be seizures that have characteristic tonic-clonic movements.) Thus, behavior changes should prompt further evaluation, but not be used as the sole assessment, of an animal's health status.

3.2 Expressions of Pain

Because overt behavioral changes may not often be evident even in severe disease, the manner in which a donkey holds tension in the face can be used as part of a comprehensive evaluation of the presence of pain (e.g. ocular widening or narrowing; nostril wrinkles; muzzle tension; ear position). However, the caveat in interpreting facial expressions as normal or abnormal is always to factor in the context of the entire clinical picture of the animal—specifically, any ongoing changes in behavior and/or abnormalities in physiologic parameters. For example:

Both of these donkeys do not have forward-facing ears, which can be indicative of discomfort (Regan et al. 2014; Regan (nee Ashley) et al. 2015; Regan et al. 2016; van Dierendonck et al. 2020; Orth et al. 2020). However, the donkey on the left immediately changed its ear position towards the handler when whistled to (not shown in the picture), and its lower lip is loose (i.e. hanging lower than its upper lip). The donkey to the right did not change its ear position and maintained tension in its muzzle such that the rostral-most margin of the lower lip was consistently pulled up higher than that of the upper lip throughout its physical exam. More significant than these indices of behavior at a single point in time, the donkey on the right was noted to have consistent muzzle and nostril tension, horizontal ear positioning, and abnormal behavior (i.e. self-isolating from the herd) for 3 days leading up to its evaluation. In contrast, the donkey on the left was interactive with the herd, frequently changed its ear position in response to auditory stimuli, and had a normal, voracious appetite in the days preceding this picture. This emphasizes that the *chronicity* in which abnormalities are seen, rather than the mere presence of abnormalities, is important to consider when using facial expressions and behavior for pain interpretation. This also underscores the importance of involving a donkey's caretaker in the initial collection of patient anamnesis as they are more likely to be aware of the subtle nuances of that animal's behavior leading up to the clinical evaluation.

Other subtle expressions of pain include sham eating, wherein a donkey may act like it is eating (even to the point of nuzzling and/or putting its head into the feed) but consume no volume of food (Burden and Thiemann 2015). It should be taken seriously any time a donkey is demonstrating sham eating, even in the absence of changes in vital parameters, given the donkey's penchant to remain stoic even in instances of severe systemic illness. Further, physiologic parameters alone without corroborating ethogram data have been shown to be an inaccurate measure of pain in donkeys (Xue et al. 2024). Donkeys are also prone to dyslipidemias; thus, any donkey that has changes in appetite should be evaluated thoroughly for any primary organ system pathology that may be inducing secondary inappetence.

3.3 Body Condition Scores (BCS) and Fatty Neck Scores (CNS)

There are various BCS scores available for donkeys, with 1 indicating emaciation and the higher end of the range (5, 7, or 9) indicating obesity (The Donkey Sanctuary 2020; Valle et al. 2017; de Oliveira et al. 2019; Pearson and Ouassat 1996). The BCS scale of 1–9 developed for donkeys (Pearson and Ouassat 1996) has also been used for mules (Kay et al. 2004). It is strongly recommended that a clinician always include the denominator of the score when referencing a patient's BCS to avoid misinterpretation (e.g. 2/5 instead of only 2; 6/9 instead of only 6). This is especially important if an animal is being managed by multiple clinicians. Fatty neck scores (analogous to horses' cresty neck scores) are also important to perform in donkeys as an animal may not grossly appear on visual inspection to have significant peripheral fat deposits but still have palpably abnormal adipose tissue in their dorsal neck (Maśko et al. 2021). In general, a higher FNS is more likely in animals with a higher BCS.

Physiologically, donkeys have evolved to be very nutritionally efficient and are more prone to build up significant adipose tissue stores even with forage that is considered low quality feed for horses. Their hybrids (mules and hinnies) also seemed to have inherited this tendency. Clinically, their more efficient metabolism is advantageous in facilitating survival in resource-limited environments. However, in the context of animals that are no longer in work and are instead kept as pets, this metabolic efficiency increases their risk for asinine metabolic syndrome with endocrinopathic laminitis as a potentially fatal sequela. Thus, appropriate body condition scoring and fatty neck scoring is a very useful aide in tracking changes over time, particularly in patients that need to lose weight but are resistant to stepping on a digital floor scale for objective weight monitoring. Weight tapes can be used as a proxy of digital floor scales but are ultimately estimates of weight compared to the weights obtained from a scale. Care must be taken not to use horse weight tapes for donkeys as donkey weight tapes have been specifically developed for their unique body morphology. There is currently no weight tape developed for donkey hybrids (mules and hinnies). While use of horse weight tapes for mules seems to produce weight estimates that match most closely with weights obtained from a digital scale (Noikhura et al. 2015), there is still concern that the weights obtained are overestimates of the actual animal's weight (Kay et al. 2004). In situations where an accurate weight is absolutely required for a mule or hinny, the author recommends relying on a digital weight scale.

3.4 Respiratory Rate and Effort

The author always recommends collecting the respiratory rate and evaluating respiratory effort from a distance as not all animals are amenable to body handling; some may not be approachable at all. Note that assessing the respiratory rate and effort from a distance does not preclude thoracic auscultation in the hands-on portion of the PE. The latter should always be attempted with the knowledge that, if not possible, at least respirations were already counted and respiratory effort noted prior to the animal becoming agitated which, in and of itself, can both elevate the respiratory rate and give off the appearance of greater respiratory effort.

3.5 Lameness Examination

Lameness examinations should begin with observations from a distance as donkeys may be recalcitrant to walk on a lead when painful, may mask their lameness while being handled, or may not be approachable at all. Look for classic signs of forelimb lameness with a head bob, and greater hip excursion with pelvic limb lameness. Being able to deduce which leg is lame from a distance can facilitate the hands-on lameness exam if a donkey masks its pain upon being handled. Because donkeys may not permit repeated attempts to elicit lameness on flexion tests, using visual information can help enhance the efficiency of the ultimate musculoskeletal exam that may be needed.

4 Hands-On Examination

4.1 Animal Handling During the Clinical Examination

Donkeys and their hybrids are highly intelligent, prey animals that can be very guarded in the presence of an unfamiliar stranger. This heightened defense mechanism can result in reactive, aggressive animals, particularly if a practitioner approaches or performs the PE too quickly. In the author's experience, a slow measured approach characterized by frequent breaks and positive reinforcement of the desired behavior yields greater opportunity for collecting accurate clinical data. Using behavioral cues to guide forward progression of the PE at an acceptable pace for the donkey, mule, or hinny is more time-efficient overall than moving too quickly through the PE, activating the patient's fight-or-flight instincts, and inadvertently creating a fractious animal that then becomes a safety concern for both clinician and handler. Where possible, this can be mitigated by having the animal handled by a familiar caretaker during the PE. The use of food rewards for positive reinforcement has also been very helpful in the author's experience and can be considered if the owner is not opposed to treats. Smaller volume treats (e.g. a single Cheerio, a few pieces of ration balancer pellets) doled out with high frequency (e.g. every 15–30 seconds while the patient is standing still or maintaining the desired behavior) is a good balance of rewarding the patient and training them to tolerate the body handling required in a PE without overfeeding species that are known to be easy keepers. Other options for positive reinforcement include gently rubbing the edge of the pinnae with an open palm or even rubbing the inside of the pinnae at the base of the ear. Both have been surprisingly effective when used by the author in establishing a good working relationship with a nervous patient. At minimum, this gentle handling allows for collection of clinical findings helpful in determining the patient's systemic status before utilizing sedation to do more extensive evaluation of a potentially painful site. It also allows serial examinations on hospitalized animals to be less stressful for both clinician and patient.

4.2 Vital Parameters—Temperature, Pulse, and Respiration (TPR)

4.2.1 Donkeys

Normal vital parameters for donkeys (The Donkey Sanctuary, 2018) are:

- Temperature: 97.7–100 °F (36.5–37.8 °C)
- Pulse: 36–52 beats/min
- Respiration: 12–28 breaths/min

4.2.2 Mules

Less has been published in the veterinary literature about the normal vital parameters for mules (horse mare × donkey jack cross). The following vital parameters are compiled from *averages* taken from two studies (Ali and Anjum 1998; McLean 2014; McLean et al. 2016)

	Ali and Anjum (1998)	(McLean 2014; McLean et al. 2016)
Temperature:	99.9 °F (37.7 °C)	99.5 °F (37.5 °C)
Pulse:	33 beats/min	44 beats/min
Respiration:	12 breaths/min	35 breaths/min

In the Ali and Anjum study, vital parameters were studied at three time points (morning, noon, evening). Although values were consistently highest at noon, each parameter had a high positive correlation ($r = >0.9$) with itself, indicating that these values can be reasonably used as a normal reference for mule TPR values regardless of time taken during the day. In the McLean study (2014), comparison between hinny, mule, donkey, and horse vital parameters revealed that mule's rectal temperature and pulse are lower than that of horses and more similar to horses. The temperature and pulse from the Ali and Anjum study results seems to align with the pattern found in McLean's. Respirations had the greatest variation with mule and hinny values not showing any closer association with either donkey or horses.

4.2.3 Hinnies

Even less peer reviewed literature has been specifically published about the vital parameters in hinnies (donkey jenny × horse stallion cross). Average values found from the McLean study are below:

- Temperature: 98.6 °F (37.0 °C)
- Pulse: 43 beats/min
- Respiration: 30 breaths/min

4.3 Cardiovascular Evaluation and Hydration Status

Evaluation of the cardiovascular and hydration status of donkeys should always include, at minimum, an assessment of oral mucous membranes, jugular refill time, skin tent and auscultation of the heart. Normal mucous membranes are pink and moist with a capillary refill time of <2 seconds. Jugular refill can be assessed by distending the jugular vein at the jugular furrow on both sides and witnessing the vein refill briskly. Note that because donkey's have a thicker cutaneous colli muscle (Burnham 2002), the jugular vein may not be as easily visible even when distended if the patient is holding tension in their neck, or their neck is twisted to one side. Therefore, the author also recommends utilizing a finger from the free hand to stroke the jugular vein parallel to the direction in which it runs (cranial to the site of digital distention) in order to feel blood within the vein ballot against the finger of the contralateral hand that is distending the jugular vein. Skin tent can be reliably evaluated using the superior eyelid. When pinched, the skin will return to its original position quickly in a euhydrated animal but remain tented in a dehydrated animal. While the skin tent can be performed on the neck, the author prefers to use the eyelid as it is possible for skin turger to be affected by dermatologic conditions. Auscultation of the heart can be performed with the stethoscope placed in the third, fourth, and fifth intercostal spaces on the left thorax (to listen to the pulmonic, aortic, and mitral valves, respectively), and around the fourth intercostal space on the right thorax (to listen to the tricuspid valve). In a moving donkey, these areas can be approximated underneath the triceps muscle between the palpable point of the elbow and the lateral thoracic vein. In those situations, the author recommends placing the stethoscope in this region and moving with the animal until a heartbeat can be heard, then shifting the stethoscope bell minutely along the thoracic wall to follow the sound of the pulse. It should become louder the closer to the valve the bell gets, and quieter the further away the bell is. Simultaneous cardiac auscultation and palpation of an arterial pulse at the facial or transverse facial artery is essential as this can help determine the presence of arrhythmias or characterize the type of murmur. Normally, a pulse is felt with every heartbeat heard, and the heart beats in a regularly regular manner. If there is asynchrony between the palpable pulse and auscultation of the heartbeat (e.g. a regularly regular pulse with a heartbeat that is irregular), this suggests an arrhythmia. If there is a murmur heard in time with the pulse, this suggests a systolic murmur (as the pulse is generated with ventricular systole). Similarly, a murmur heard opposite to the timing of a pulse suggests a diastolic murmur.

Note that correlating physiology and anatomy knowledge with that of the physical exam findings is essential in helping one rule out inappropriate differentials and utilizing diagnostic tests in a manner that would be incongruous with the patient's clinical picture. Using cardiac auscultation as an example, prioritizing pulmonic valve insufficiency would be inaccurate for a murmur that is heard on the left thorax and occurs in time with the arterial pulse. This is because these findings suggest a systolic murmur, which indicates pathology in a valve that should be closed during systole. Because the pulmonic valve is normally closed during diastole, a murmur secondary to pulmonic valvular insufficiency would be heard during diastole. Thus, utilizing echocardiography to search for a regurgitant jet over the pulmonic valve (at the expense of ignoring more reasonable findings like a regurgitant jet over the mitral valve) would be an inefficient use of imaging diagnostics. This highlights the need to always correlate the differential diagnoses list with the findings from PE.

4.4 Respiratory System

Evaluation of the respiratory system should involve visual evaluation for the presence of nasal discharge, assessment of airflow from both nares, percussion of the sinuses, palpation of the trachea, and auscultation of the trachea and lungs. Airflow from the nares against the examiner's palm should be equal from both nostrils. There should be no nasal discharge or evidence of prior nasal discharge that has dried as crusts, although clear serous nasal discharge is possible without true pulmonary disease in colder climates. Percussion of the frontal sinuses should not result in dullness, which could suggest fluid or masses within the sinuses. Tracheal palpation may normally elicit a paroxysmal cough; however, sustained coughing is abnormal and should prompt further evaluation. In smaller breed donkeys, the trachea should be gently palpated along its entire length to evaluate for evidence of tracheal collapse, which typically manifests as stertor or a honking noise subsequent to depression of the tracheal causing a narrowed lumen. In extreme cases, collapse of the trachea is complete and there is bilateral loss of airway patency. Normally, auscultation of the trachea reveals laminar flow that is louder than the bronchovesicular sounds heard on thoracic auscultation. The bronchovesicular sounds themselves should be loudest cranioventrally and dissipate caudodorsally. Although donkeys are not as athletic as their horse counterparts, and thus less evolved to have the same disproportionally large tidal volume compared to their body size, there remains the possibility of missing adventitious lung sounds in mild lower airway disease. The author recommends using a rebreathing bag in any cases where there is ambiguity on auscultation findings of the lower airways, as the rebreathing bag examination is a simple, non-invasive, and inexpensive diagnostic that can yield answers before the need to resort to more expensive diagnostics.

In an animal that has clinical signs consistent with infectious lower airway diseases (e.g. adventitious lung sounds, fever, purulent nasal discharge, tachypnea), pleurodynia may be a feature of diseases like pleuropneumonia. Percussing the thoracic wall and evaluating for signs of pain would be important to perform in those animals. Such patients may also demonstrate reluctance to walk and stand braced with elbows abducted. This highlights the importance of corroborating observations from the distance examination and PE and adjusting the extent of the physical examination that was initially performed to evaluate for further clinical signs that would be expected for a particular disease process.

4.5 Gastrointestinal System

As part of a comprehensive evaluation of the gastrointestinal system, physical exam findings should also be paired with information about the patient's appetite. Donkeys that are acutely inappetent or sham eating should be treated as emergencies, with workup of the gastrointestinal system prioritized. If no pathology is found attributed to the gastrointestinal system, other organ pathology must be considered.

Auscultation of the abdomen is performed much like in horses with attention paid to listening for borborygmi in the upper and lower left and right quadrants. Cecal contractions can be heard in the right dorsal quadrant, with a normal cecal motility rate of 1–2 contractions/minute. Stool should be formed and moist. Unlike horses, diarrhea is an uncommon presentation of colitis in donkeys (Du Toit et al. 2010). Because of this, any evidence of active diarrhea or diarrhea staining around the perineum in donkeys should prompt further evaluation for colitis, particularly if there are other referable clinical signs (lethargy, fever, evidence of systemic inflammation, weight loss, alterations in mucous membranes, dehydration) or history (recent travel, changes in management/husbandry, new feed, antibiotic or anthelmintic administration) that increases an animal's risk for development of colitis.

Comprehensive dental evaluation would necessitate a sedated speculum exam in order to visualize pathology in the cheek teeth. Although this is not necessary in all cases, if obvious malocclusion is found in the 01–03 second on unsedated physical examination, a sedated speculum examination with dental radiographs may be indicated as follow up to detect.

4.6 Other Facial Structures (Eyes, Ears, Lymphatics)

Normal eyes have clear corneas, horizontal eyelashes, and are free of ocular discharge. The pupils are horizontal and parallel to the plane of the medial and lateral canthi. The nictitans should not be exposed at the medial canthus. If it is, tetanus needs to be prioritized as a differential, particularly in animals with a history of trauma or surgery.

Within the nares, the nasolacrimal duct is dorsal to dorsolateral in donkeys, and lateral in mules (Adams et al. 2013). In the absence of available literature describing the typical location of the nasolacrimal duct in hinnies, it is presumed to be variable in hinnies.

The ears are normally quite motile and should be forward facing in a systemically, bright, alert, and responsive animal. The palpable lymphatic structures around the head and neck are anatomically located in similar positions in donkeys as in horses (Ali et al. 1987).

4.7 Dental Evaluation

A comprehensive dental evaluation will require a sedated speculum examination. However, a non-sedated examination of, at minimum, the incisors, wolf teeth, and canines (if the latter two are present) can be easily performed on the initial PE. The animal can also be aged based on this evaluation. It would be common sense to say that if obvious pathology is detected on a non-sedate exam, a sedated exam should be followed up next. Regardless a sedated dental exam is indicated in any scenario with an animal presenting for weight loss or demonstrating quidding.

4.8 Dermatologic System

A normal donkey with healthy skin should have a full coat without ectoparasites, lesions, or areas of alopecia. Common dermatologic conditions in donkeys and mules include sarcoids, dermatophilosis, dermatophytosis, ectoparasites, and pressure sores (Barrio et al. 2019). The latter is of special importance to check for as the donkey's longer coat in comparison to the horse can obscure lesions if the hair is not clipped. Note that breed differences, such as the long-haired phenotype of the Poitou donkey (Legrand et al. 2014), should not be misinterpreted as an unhealthy coat. The coat of these donkeys will grow life-long and can normally appear matted and tangled in adults.

4.9 Musculoskeletal System

Examination of the musculoskeletal system should start off with observation from afar for any obvious areas of muscle atrophy with the animal standing square, and for gait abnormalities with the animal in motion. This distance examination should then be followed up with palpation of the muscle bellies and/or further evaluation of the lame leg. Hoof testers, nerve blocks, and joint blocks may be necessary to isolate a lameness prior to additional imaging diagnostics like radiographs and ultrasonography. If the animal is not amenable to nerve blocks or joint blocks, the clinician may have to make a reasonable conjecture about the area of lameness and focus imaging diagnostics appropriately there.

4.10 Neurologic Evaluation

Mentation and the patient's general response to stimuli should always be evaluated on the PE. Further neurologic examination would be indicated based on the presence of concomitant abnormalities concerning for neurologic pathology (e.g. ataxia, facial asymmetry, vestibular signs, nystagmus, etc.)

4.11 Urogenital System

Abnormalities corresponding to the urinary system disease are not often overt on the PE. Many times, definitive diagnosis of a renal disorder is only found on necropsy (Slavik et al. 2025) given that the ante-mortem clinical signs can be vague. Non-specific clinical signs of kidney disease on PE include weight loss, inappetence, altered mentation, peripheral edema (Olsen and van Galen 2022), and may not be seen unless disease is chronic. Thus, even when clinical signs are present, they cannot preclude evaluation for other system disorders. Other clinical signs referable to the urinary system (e.g. stranguria, polyuria/polydipsia, pollkiuria) can be difficult to detect in an animal that has not been meticulously observed by the owner over a period of time. Serum biochemistry analysis should be performed next in any patient with PE findings concerning for renal disease with adjuvant imaging testing and urinalysis (preferably sterilely collected) performed based on the bloodwork results.

The external genitalia (e.g. vulva, perineal conformation, penis) should be visually assessed for obvious abnormalities that would indicate the need for more advanced evaluation of the reproductive tract. In jennies, this includes visual inspection of the udder with palpation as the patient allows. In jacks, this includes visual inspection of the scrotum for distended testes and palpation of the inguinal canal as the patient allows.

5 Use of Sedation for the Clinical Examination

Ideally, the initial PE should be performed in an unsedated patient to obtain for greatest accuracy given that pulse, respirations, and borborygmi are depressed with sedation. However, it is possible for painful donkeys and their hybrids to react in a manner refractory to completion of the PE when they become concerned about getting examined at the site of pain. This tends to be true for disease that does not cause systemic instability as systemically unstable animals generally demonstrate increased amenability to examination. Further, donkeys that are needle shy may become even more agitated in the process of a clinician attempting to administer sedation, thus increasing their overall anxiety and expression of fractious behavior that makes further progression of the PE unsafe. Ultimately, a clinician must use their clinical judgement based on patient factors, suspected disease process, and the necessity of evaluating for specific clinical findings to determine whether or not sedation should be used. If sedation is required to complete the necessary aspects of the PE, it is recommended that pulse and respiratory rate and effort be collected prior to sedation.

Should sedation be needed, common sedatives and their dosages are below and found in the pharmacology and anesthesia chapters:

- Xylazine 0.5–1.2 mg/kg IV (Lizarraga and Beths 2012; Parentoni et al. 2020)
- Detomidine 5–60 μg/kg (Mostafa et al. 1995; Parentoni et al. 2020)
- Romifidine 40–100 μg/kg IV (El-Kammar and Gad 2014; Lizarraga et al. 2017)

The high and low range of dosing was collated from the high and low range from these sources. In general, dosing ranges are similar to horses, though their duration of efficacy may be shorter.

Donkeys will commonly rub their muzzles subsequent to alpha-2 agonist sedation. The mechanism behind this behavior is unknown. Although this transient and abates once the animal has woken up from sedation, some donkeys can develop abrasions to the oral mucosa and muzzle from the severity of the rubbing. Anecdotally, some clinicians report that giving hydroxyzine (1 mg/kg PO) approximately an hour prior to planned sedation with an alpha-2 agonist significantly decreased the incidences of muzzle rubbing. There are currently no studies that support or contraindicate the use of hydroxyzine in donkeys.

Yohimbine (0.2 mg/kg IV) is anecdotally used to reverse alpha-2 sedation and is clinically efficacious in the author's experience as well. However, pharmacokinetic and pharmacodynamic studies on this drug in donkeys and their hybrids are lacking.

6 Conclusions

The principles of performing a PE in horses are the same for donkeys and their hybrids. While a definitive diagnosis is seldom immediately reached on the first PE, a meticulous, comprehensive PE should allow a clinician to narrow down the list of most likely diseased organ systems, which then greatly facilitates selection of diagnostic tests that can lead to an appropriate final diagnosis. In donkeys and their hybrids, it is particularly important to pay attention to the animal's history (if known) and to involve the client (where possible) as any past medical history and subtle changes in behavior (that may be missed by a clinician unfamiliar with the patient) can be utilized to hone in on certain aspects of the PE that would otherwise be overlooked in a species known to be stoic. Ultimately a combination of careful observation, knowledge of donkey behavior, and consistently practicing thorough, systematic evaluation of all body systems will help a clinician maximize the information that is obtained from the PE.

Questions

Question 1—Multiple Choice

Which of the following best explains why donkeys and their hybrids may mask signs of illness compared to horses?

A. They have more sensitive cardiovascular systems.
B. Their physiology limits pain receptor signaling.
C. They are naturally stoic and exhibit fewer overt signs of pain.
D. Their metabolic rate is higher, reducing disease manifestation.

Correct Answer: C

Question 2—Short Answer

List three reasons why a systematic and consistent approach to the physical exam is especially important in donkeys, mules, and hinnies.

Sample Answer:

1. They may become fractious, making it difficult to complete the exam.
2. Subtle signs of disease or pain are easily missed due to stoicism.
3. Performing the exam in a consistent order helps develop muscle memory and prevents omission of critical steps.

Question 3—Clinical Scenario (Short Answer)

You are evaluating a 12-year-old donkey that is reportedly sham eating. The owner notes normal appetite behavior, but no feed seems to be consumed. Vitals are within normal limits.

What two major organ systems should be prioritized for further evaluation and why?

Sample Answer:

1. Gastrointestinal system—Sham eating may indicate colic or other GI pathology, even without abnormal vital signs.
2. Endocrine/metabolic system—Donkeys are predisposed to dyslipidemia and metabolic disorders, particularly if appetite is altered.

Question 4—Multiple Choice

Which of the following is NOT a reliable site for hydration assessment in donkeys?

A. Superior eyelid skin tent
B. Mucous membranes

C. Jugular refill
D. Neck skin tent (in all cases)
Correct Answer: D
(Neck skin tent may be affected by dermatologic conditions and is less reliable than the superior eyelid.)

Question 5—Fill in the Blank

When scoring body condition in donkeys, the ________ score should always be reported with a denominator (e.g., 6/9) to prevent misinterpretation between different systems.
Correct Answer: Body Condition

References

Adams MF, Castro JR, Morandi F, Reese RE, Reed RB (2013) The nasolacrimal duct of the mule: anatomy and clinical considerations. Equine Vet Educ 25:636–642

Ali R, Anjum AD (1998) Rectal temperature, pulse rate, and breath rate in mules. Pak J Biol Sci 1: 271–273

Ali AMA, Saber AS, Mansour AA (1987) Lymphocenters of the head and neck of the donkey (Equus asinus). Assiut Vet Med J 18(2):22–29

Barrio E, Rickards KJ, Thiemann AK (2019) Clinical evaluation and preventative care in donkeys. Vet Clin N Am Equine Pract 35:545–560

Burden F, Thiemann A (2015) Donkeys are different. J Equine Vet Sci 35:376–382

Burnham SL (2002) Anatomic differences of the donkey and mule. AAEP Proc 48:102–109

de Oliveira EM, Canisso IF, McDonnell SM (2019) Donkey reproductive physiology and management. Journal of Equine Veterinary Science 79:57–64

van Dierendonck MC, Burden FA, Rickards K, van Loon JPAM (2020) Monitoring acute pain in donkeys with the Equine Utrecht University scale for donkeys composite pain assessment (EQUUS-DONKEY-COMPASS) and the Equine Utrecht University scale for donkey facial assessment of pain (EQUUS-DONKEY-FAP). Animals 10:354

Du Toit N, Burden FA, Getachew M, Trawford AF (2010) Idiopathic typhlocolitis in 40 aged donkeys. Equine Vet Educ 22:53–57

El-Kammar MH, Gad SB (2014) Evaluation of the sedative, analgesic, clinicophysiological and haematological effects of intravenous detomidine, detomidine-butorphanol, romifidine and romifidine-butorphanol in standing donkeys. Equine Vet Educ 26:202–207

Kay R, Hall C, Mendl M (2004) A survey of the management and welfare of donkeys in the UK. Applied Animal Behaviour Science 85(3–4):259–271

Legrand R, Tiret L, Abitbol M (2014) Two recessive mutations in FGF5 are associated with the long-hair phenotype in donkeys. Genet Sel Evol 46:65

Lizarraga I, Beths T (2012) A comparative study of xylazine-induced mechanical hypoalgesia in donkeys and horses. Vet Anaesth Analg 39:533–538

Lizarraga I, Castillo-Alcala F, Robinson LS (2017) Sedative and mechanical antinociceptive effects of four dosages of romifidine administered intravenously to donkeys. Res Vet Sci 112:46–51

Maśko M, Romaszewski M, Jasiński T (2021) Comparison of surface thermal patterns of horses and donkeys using infrared thermography. Animals (Basel) 11(2):512. https://doi.org/10.3390/ani11020512

McLean AK (2014) Comparing the physiological and biochemical parameters of mules and hinnies to horses and donkeys. In: Proceedings of the 5th international hydra mule and donkey conference, hydra, Greece. October 10–12

McLean AK, Wang W, Navas-González FJ, Rodrigues JB (2016) Reference intervals for hematological and blood biochemistry reference values in healthy mules and hinnies. Comp Clin Pathol 25:871–878. https://doi.org/10.1007/s00580-016-2276-3

Mostafa MB, Farag KA, Zomor E, Bashandy MM (1995) The sedative and analgesic effects of Detomidine (Domosedan) in donkeys. J Veterinary Med Ser A 42:351–356

Noikhura N, Samanta AK, Pal BC, Mandal GP (2015) Health and welfare status of working donkeys under rural management systems. Indian Journal of Animal Research 49(6):835–840

Olsen E, van Galen G (2022) Chronic renal failure-causes, clinical findings, treatments and prognosis. Vet Clin N Am Equine Pract 38:25–46

Orth EK, Navas González FJ, Iglesias Pastrana C, Berger JM, le Jeune SS, Davis EW, McLean AK (2020) Development of a donkey grimace scale to recognize pain in donkeys (Equus asinus) post castration. Animals 10:1411

Parentoni RN, Henrique FV, Brasil AW d L, Dias RA, de Araújo AL, Mendes R d S, Martins EF, da Costa JM, de Souza AP, da Nóbrega PI (2020) Detomidine and xylazine, at different doses, in donkeys (Equus asinus). Braz J Vet Med 42

Pearson RA, Ouassat M (1996) Estimation of the liveweight and body condition of working donkeys. Tropical Animal Health and Production 28(2):67–73

Regan (nee Ashley) FH, Hockenhull J, Pritchard JC, Waterman-Pearson AE, Whay HR (2014) Behavioural repertoire of working donkeys and consistency of behaviour over time, as a preliminary step towards identifying pain-related behaviours. PLoS One 9:e101877

Regan FH, Hockenhull J, Pritchard JC, Waterman-Pearson AE, Whay HR (2015) Clinical abnormalities in working donkeys and their associations with behaviour. Vet Rec Open 2

Regan FH, Hockenhull J, Pritchard JC, Waterman-Pearson AE, Whay HR (2016) Identifying behavioural differences in working donkeys in response to analgesic administration. Equine Vet J 48:33–38

Slavik K, Bender S, Sharkey L, Nolen-Walston R (2025) Renal lipidosis in horses and donkeys: 25 cases (2008–2022). J Vet Intern Med 39:e70072

The Donkey Sanctuary (2018) Sedation, anaesthesia and analgesia. In: Evans L, Crane M (eds) Clinical companion of the donkey, 1st edn. Troubador Publishing Ltd, Leicestershire, pp 211–227

The Donkey Sanctuary (2020) The donkey welfare assessment: A practical guide to improving donkey welfare. Sidmouth (UK): The Donkey Sanctuary. Available from: https://www.thedonkeysanctuary.org.uk

Valle E, Raspa F, Giribaldi M, Barbero R, Bergagna S, Cavallarin L (2017) A functional approach to body condition assessment of lactating donkeys as a tool for welfare evaluation. PeerJ. 2017;5:e3001. https://doi.org/10.7717/peerj.3001

Xue C, Segabinazzi L, Hall A, Dzikiti TB, French H, Gilbert R (2024) A retrospective comparison of postoperative outcomes in ovariectomised jennies (*Equus asinus*) treated with phenylbutazone or flunixin meglumine. Equine Vet J. https://doi.org/10.1111/evj.14082

Dentistry in Donkeys and Mules: Anatomical Differences, Clinical Challenges, and Preventive Care

Jose María Galafate Pica

To my parents, Jose maria y Melu

Abstract

Dentistry is a fundamental discipline within equine veterinary medicine, due to the direct impact that oral health has on the general welfare, nutrition and performance of animals. Although horses have been the main focus of study and clinical attention, there is a growing need to deepen the dental knowledge of other equine species, particularly donkeys and mules. These species present some significant anatomical and physiological differences with respect to horses, which condition the clinical presentation, handling, diagnosis and treatment of dental diseases. It is noteworthy that in recent years, dentistry dedicated to horses has developed as a specialty of its own in many parts of the world, while at the same time it has allowed the implementation of both new knowledge at the diagnostic level and treatments, which in turn has developed a technology within the reach of professionals that was previously obsolete.

Keywords

Oral health · Dental anatomy · Dental exam · Dental diseases

J. M. G. Pica (✉)
Private Practice Veterinarian, Specialist in Veterinary Dentistry and Maxillofacial Surgery (Complutense University of Madrid) Jerez de la Frontera, Cádiz, Spain
e-mail: atencion@dentequinus.com; jmgalafate@dentequinus.es; https://www.dentequinus.com

A. K. McLean et al. (eds.), *Donkey and Mule Medicine: Foundations, Breeds, and Management*, https://doi.org/10.1007/978-3-032-12726-6_18

1 Equine Differences

Although they share common ancestors, donkeys, mules, and horses possess unique dental characteristics that should be considered in clinical practice. Variations in the size and shape of the skull, the number and pattern of eruption of the teeth, and the conformation of the oral cavity, among other factors, mean that the dental approach to each species requires an adapted and specific approach (refer to Anatomy Chapter for images) (Dixon & Dacre, 2005; Easley et al., 2011; AVDC, n.d.; BEVA, n.d.).

Equids belong to the order *Perissodactyla* and are grouped within the family **Equidae**, whose only living genus is **Equus**. This genus includes a variety of domesticated and wild species, including **the horse** (*Equus caballus*), the **domestic donkey** (*Equus asinus*), and African zebras. From a phylogenetic point of view, these species share a relatively recent common ancestor, which allows the possibility of interspecific hybridization, although with reproductive restrictions.

Among these hybrids, the **mule (mula in Spanish) stands out**, the product of the cross between a **mare** (*Equus caballus*) and a **male donkey** (*Equus asinus*). The term *mulus*, of Latin origin, has been used since ancient times to refer to this animal of remarkable rusticity, resistance and functional value in agricultural and loading activities. Although it shares a genus and family with its parents, the mule has **63 chromosomes**, the result of crossing between two species with different chromosomal numbers (64 in the horse and 62 in the donkey), which generally determines its **sterility** due to alteration of the meiotic process ((Easley et al., 2011; Dixon & Dacre, 2005).

Conversely, there is also the **burdegano (Spanish for hinny) or hinny**, obtained from the cross between a **male horse** and a **female donkey**. Although genetically similar to the mule, it may present **morphological and temperamental differences** attributable to the maternal effect and mitochondrial influence of the female lineage. Both hybrid forms offer a valuable comparative model for the study of craniodental anatomy, by integrating structural characteristics of both parental species (Refer to Tables 1 and 2).

Table 1 Craniofacial Morphological Differences Among Horses, Donkeys, and Mules (Easley et al. (2011); Dixon & Dacre (2005)

Feature	Horse (Equus caballus)	Donkey (Equus africanus asinus)	Mule (♀ E. caballus × ♂ E. asinus)
Skull	Elongated, narrow; straight or slightly convex	Robust, wide; straight profile	Intermediate: longer than donkey, wider than horse
Ears	Short and mobile	Very long, less mobile	Long, shorter than donkey
Jaw	Fine, proportionate	Wide and strong	Strong, well-developed
Eye sockets	Spacious, mobile	Smaller, less mobile	Intermediate in size and mobility
Eyes	Large, expressive, panoramic vision	Slightly smaller	Intermediate size and expression

Table 2 Dental Differences Among Horses, Donkeys, and Mules (Du Toit et al. (2014); Rodrigues et al. (2013a, 2013b); Easley et al. (2011)

Feature	Horse (Equus caballus)	Donkey (Equus africanus asinus)	Mule (♀ E. caballus × ♂ E. asinus)
Type of dentition	Hypsodont (high-crowned, continuous eruption)	Hypsodont	Hypsodont
Number of teeth	36–44 (depending on canines and/or wolf teeth)	Similar to horse	Similar to horse
Tooth eruption	Premolars: 2.5–3.5 years; molars: 1–3 years	Molars erupt earlier (~5–9 months)	Variable; blends parental patterns
Jaw	Thin, symmetrical; good occlusion	Wider; prone to malocclusions	Strong, intermediate; possible asymmetry
Dental wear	Predictable; useful for aging	Less predictable; irregular wear	Variable; unreliable for age estimation
Canines	Common in males; rare in females	More frequent in females than in mares	Variable; less frequent than horses
Wolf teeth	Common, especially upper PM1	Less common than horses	May be present or absent

These anatomical differences reflect the evolutionary and functional adaptations of each species or hybrid, influenced by their genetics and the environment in which they have developed, while there are certain differences in dental anatomy (Refer to Table 2).

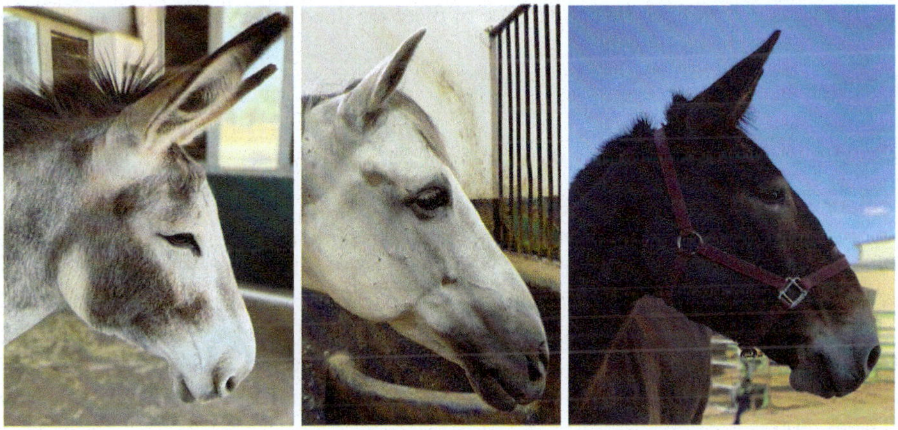

1. Donkey (BONICA. Cross breed Donkey. The donkey sanctuary. https://www.thedonkeysanctuary.org.uk), 2. horse (GERMANICO. Pure spanish breed horse. https://www.realescuela.org), 3 mule

1.1 General Clinical Considerations

Importance of dental care

- Horses, donkeys, and mules require regular dental care to prevent problems such as malocclusions, sharp points, cavities, and periodontal disease. Donkeys and mules are more predisposed to the development of very sharp enamel tips as well as abnormal occlusion patterns.
- In mules, eruption and wear patterns should be assessed individually, as they do not follow consistent rules.
- It is very important to highlight that horses tend to have better and more frequent care of their teeth than donkeys and mules, in which often the lack of this maintenance results in an increase in the pathology of dentaria .

Differences in tooth wear

- Wear patterns can vary significantly between these species, affecting how dental health is assessed and ages are estimated.
- Diets generally based on concentrates and forage in horses produce a different wear and tear than donkeys and mules, whose diets are mostly based on forage.

Need for specific tests

- Since mules have a combination of dental features, it is essential to tailor examinations and treatments to their particular needs.
- In general, age is a determining factor in both, although the longevity of many donkeys determines the appearance of diseases due to excessive wear and tear on their chewing apparatus.

2 Importance

Understanding the dental differences between these species not only allows for a more accurate diagnosis and more effective treatments, but also favors a more appropriate preventive approach. In regions where donkeys and mules are essential working animals, often with limited access to specialized veterinary services, specific clinical knowledge becomes even more relevant. This chapter seeks to provide an integrated and comparative vision that facilitates clinical decision-making in the dental management of horses, donkeys and mules. (Du Toit et al., 2014; Dixon & Dacre, 2005)

3 Key Concepts

3.1 Glossary of Terms

- **Dental arch**: A set of teeth in the maxilla or jaw.
- **Cementosis**: Pathological thickening of the dental cementum.

- **Diastema**: Abnormal space between adjacent teeth. Very common in elderly donkeys
- **Wolftooth**: Maxillary premolar, often vestigial, which may be absent and be both upper and lower (less frequent)
- **Tooth eruption**: The process by which teeth emerge through the gum.
- **Exodontia**: Extraction of a tooth.
- **Hypodontia**: Congenital absence of one or more teeth.
- **Malocclusion**: Abnormal misalignment of the teeth when closing the mouth.
- **Puntas de esmalte**: Punta o elevación del esmalte en la superficie de un diente.
- **Secondary odontogenic sinusitis**: Infection of the sinuses caused by a dental problem.
- **Apical abscess:** Infection at the root of the tooth
- **Step mouth:** Stepped arch of premolars and molars
- **Shear or scissor mouth:** Excessive angle in molars
- **Wavy mouth: Wavy** arch of premolars and molars.
- **Periodontal pockets:** Spaces that usually accumulate food and form in the gums adjacent to the teeth as a result of an infection.
- **Caninos/Canine teeth:** Fangs, more common in males
- **Clinical crown:** Visible part of the tooth
- **Accentuated transverse ridging:** Transverse ridges of enamel on the occlusal surfaces of very pronounced premolars and molars.
- **Tooth wear:** Natural wear of teeth
- **Cheek teeth**: Premolars and molars for food grinding
- EOTRH: (Equine odontoclastic tooh resorptiond and hypercementosis) Resorptive odontoclastic disease.
- **Equidae: A broader and more correct term from a zoological point of view**. It includes **horses, donkeys, mules, zebras, and other members of the Equidae family.**
- **Tooth eruption:** Sequential exit of the teeth
- **Hooks and ramps:** Shape that some PM and M take because of poor occlusion
- **Incisors/Incisors:** Six upper and six lower for cutting food
- **Infundibulum:** Invaginations of incisors and premolars and molars useful for age calculation.
- **Malocclusions** Abnormal gag contact
- **Occlusion:** Functional tooth contact
- **Odontoplasty or dental filing:** A procedure by which the dental arches are balanced.
- **Triadan System:** Standardized numbering of teeth.(Easley et al., 2011; Horse Illustrated, n.d.; Kentucky Equine Research, 2021; AVDC, n.d.)

4 Clinical Evaluation

4.1 Predisposition

In the study of dental diseases in equids, it is essential to distinguish between the different types of factors involved in the appearance, evolution and resolution of pathological processes. These factors are commonly classified as predisposing, conditioning and causal, and each one plays a different role within the etiopathogenesis.

4.1.1 Predisposing Factors

Predisposing factors are those that **increase the probability of the appearance of an alteration** without being, in themselves, its direct cause. They act by preparing the ground for a **disease or dysfunction to develop more easily** in the face of certain stimuli or conditions.

In the case of **donkeys,** they have a **dental morphology that**, compared to horses, favors the formation of **sharp enamel tips**, **diastemas** and **food retention** between the teeth, especially at advanced ages. These characteristics predispose to the development of **periodontal diseases** and other functional alterations.

Mules, due to their hybrid nature (cross between mare and donkey), can inherit dental anatomical characteristics from both parents. This makes them particularly **susceptible to complex malocclusions, asymmetrical tooth eruptions,** and other **structural abnormalities** that can compromise normal masticatory function.

In addition, **the narrow mandibular conformation observed in many donkeys**, as well as certain genetic traits present in specific breeds, such as the Zamorano-Leonesa or the Mirandesa, should also be considered relevant predisposing factors, since aumentan el riesgo de desarrollo de anomalías oclusales y desgaste dental irregular.(Rodrigues et al., 2013a; Rodrigues et al., 2013d)

4.1.2 Conditioning Factors

Conditioning factors are elements that, **without directly causing a dental pathology, influence its evolution, severity or clinical presentation**. Its role is to **modify the development of the problem once it is already present**, or to **favor its appearance in certain external circumstances**.

In donkeys and mules, **environmental and management factors** are of outstanding importance. Intensive **working conditions**, **unbalanced diet**, and the **absence of regular dental check-ups** contribute significantly to the accumulation and progression of oral pathologies. Unlike horses—which tend to receive more preventive veterinary care—these animals are usually more exposed to **abnormal wear** and tear and **poor oral hygiene**, which negatively affects their dental health.

In addition, **advanced age** represents a factor that determines weight. The greater longevity of horses as well as donkeys has now made **more frequent pathologies associated with aging**, such as **EOTRH** (*Equine Odontoclastic Tooth Resorption and Hypercementosis*), which mainly affect the incisors in geriatric animals.

Other clinically relevant conditioning factors include:

- **Presence of systemic diseases**, such as endocrinopathies, that interfere with healing or promote oral infections.
- **Improper prior dental management**, which perpetuates malocclusions or alters the normal biomechanics of chewing.
- **Compromised body or immune status**, which reduces the ability to respond to infections or postoperative complications.

Together, these elements **do not in themselves cause dental pathology**, but have a direct effect on its presentation, clinical course, and response to treatment.(Dixon & Dacre, 2005; Easley et al., 2011)

4.1.3 Causal Factors

They are directly responsible for an alteration. Its presence inevitably causes, or with high probability, the appearance of a clinical picture. Correctly identifying the cause is essential to apply an effective treatment.

Examples

- Impacted wolf's tooth causing mouth ulcers.
- Dental fracture due to direct trauma.
- Untreated apical infection due to deep caries.

Genetic Predisposition

Although there is no documented genetic load as deep as in other species, certain lines of mules have been observed to exhibit persistent wolf teeth and congenital malocclusions more frequently. In donkeys, there is a tendency to hypodontia, especially in the premolars.

4.2 Patient History

4.2.1 Social and Environmental History

Management history is essential to assess dental risk. Animals used in rural areas or in carrying tasks tend to have more injuries associated with the use of inappropriate mouths or harnesses, as well as greater exposure to maxillofacial trauma. In addition, many donkeys and mules are adopted or rescued animals, with incomplete or no medical history.

4.2.2 Nutritional History

In horses, a controlled diet can help avoid problems of uneven wear and tear. In donkeys and mules, the diet is usually more rudimentary, based on old hay or dry grass, which does not always favor balanced physiological wear. Cases of food entrapment in diastemas are also more frequent, especially in donkeys.

4.2.3 Chronicity of the Signs

Clinical signs of dental disease in donkeys and mules often go unnoticed for long periods of time, due to their pain-resistant temperament or poor observation of subtle changes by caregivers. This leads to many cases being diagnosed in advanced stages, with severe complications. We highlight again at this point, the lack of scheduled dental check-ups in these animals.

4.3 Clinical Signs

Acute vs. Chronic Presentation: Provide detailed descriptions of acute and chronic clinical signs, along with their respective timelines and implications for diagnosis and treatment.

Although certain clinical signs may be common to acute and chronic processes, their interpretation must consider the temporal context: acute symptoms are characterized by a sudden onset, while chronic symptoms manifest progressively and are maintained over time.

In donkeys, identifying clinical signs of dental pathology can be more challenging than in horses, due to their more rustic nature and greater ability to mask pain. This apparent tolerance favors the chronification of pathological processes that, in other species, would generate more evident signs in early stages. Dixon & Dacre 2005; Easley et al. (2011); Du Toit et al. (2014)

4.3.1 Acute Signs of Dental Problems

The following table summarizes the most common acute clinical signs associated with dental pathologies, their possible causes, and initial recommendations for clinical management:

Acute clinical sign	Possible cause	Recommended Initial Management
Difficulty chewing/ quidding	Sharp enamel tips, ulcers, foreign body	Complete oral examination and sedation for examination. Odontoplasty
Hypersalivation (sialorrhea)	Abscesses, tongue wounds	Oral examination under sedation, pain control and treatment of the injury.
Resistance to biting or mounted work	Wolf's teeth, hooks, lesions in the corners, bad occlusions.	Dental evaluation and bite adjustment if necessary. Odontoplasty
Facial or jaw swelling	Apical abscess, fracture, osteomyelitis	Palpation, X-ray, or CT scan if available
Unilateral nasal discharge	Sinusitis of dental origin (M1 most likely)	Nasal endoscopy, oral examination, dental radiology, CT. Probable extraction
Bruxism and Grief Attitude	Acute dental pain, pulpitis, fracture	Analgesia, complete dental examination
Anorexia or sudden reduction in intake	Fracture, pulpitis, painful entrapment (diastema)	Analgesia, evaluation under sedation
Oral bleeding or visible lesions	Nailed spikes, lacerations, trauma	Immediate examination and control of bleeding and treatment of the lesion.
Acute Halitosis	Pulp necrosis, food impaction (Distinguish if the odor comes from the oral or nasal cavity)	Oral and/or nasal examination by oroscopy or endoscopy.

4.3.2 Chronic Signs of Dental Problems

The following table summarizes the most common chronic clinical signs associated with dental pathologies, along with their possible causes and suggestions for initial clinical management:

Chronic clinical sign	Possible cause	Recommended Initial Management
Loss of body condition	Prolonged malocclusion, inefficient wear, 'enamel tips'	Complete dental evaluation and progressive correction plan. Odontoplasty
Hay balls fallen in the manger	Quidding for tooth pain or loss	Examination under sedation and odontoplasty.
Food partially chewed in the stool	Masticatory inefficiency due to occlusal alterations	Occlusion Review and Masticatory Efficacy Test
Persistent bad breath	Chronic periodontitis, necrotizing infections	Complete periodontal examination, cleaning and antibiotic therapy
Mobile or missing teeth	Advanced periodontal disease, EOTRH	X-rays, extraction if needed
Chronic jaw inflammation or facial asymmetry	Chronic apical abscess, bone or muscle remodeling	Radiological or ultrasound evaluation, rule out chronic abscesses
Chronic unilateral nasal discharge	Sinusitis secondary to PM4-M1 to M3 pathology.	Breast X-rays, revision of PM4-M1 to M3
Chronic recurrent avoidant behaviors to embouchure	Untreated dental pain, hypersensitivity, periostitis in mandibular seats.	Bite fit review, pain assessment, radiological assessment seats.
Persistent tilting of the head when chewing	Occlusal imbalance, asymmetrical wear	Functional Chewing Assessment and Occlusal Correction

4.4 Main Problems. Critical Findings

In the context of equine dentistry, there are clinical findings and diagnostic signs that should be considered **critical** due to their direct involvement in the **animal's welfare** and the **urgency of treatment**. Recognizing these signs allows the clinician to **prioritize interventions** and establish a quick and effective therapeutic plan.

Listed below are some of the most relevant findings that require immediate attention:

- **Acute facial or mandibular swelling**: May indicate apical abscesses, fractures, or osteoarticular infectious processes (TMJ)
- **Purulent unilateral nasal discharge**: Highly suggestive of sinusitis secondary to upper molar infection (especially M1).

- **Sudden loss of appetite or inability to chew:** Suggests acute tooth pain, fracture, or severe impaction.
- **Movable teeth, with radicular exposure or spontaneous bleeding**: Potential indicator of advanced periodontal disease or EOTRH.
- **Rejection of the bite or defensive behavior before its use**: In many cases associated with wolf's teeth, lesions in the corners of the mouth or in seats.
- **Persistent oral malodor (severe halitosis):** Associated with pulp necrosis, advanced periodontitis, foreign bodies or impacted food.

These signs should be interpreted as **clinical alerts** that warrant detailed evaluation under sedation, complementary diagnostic tests (x-rays, endoscopy and CT scans if available) and, in many cases, immediate treatment to avoid further complications.

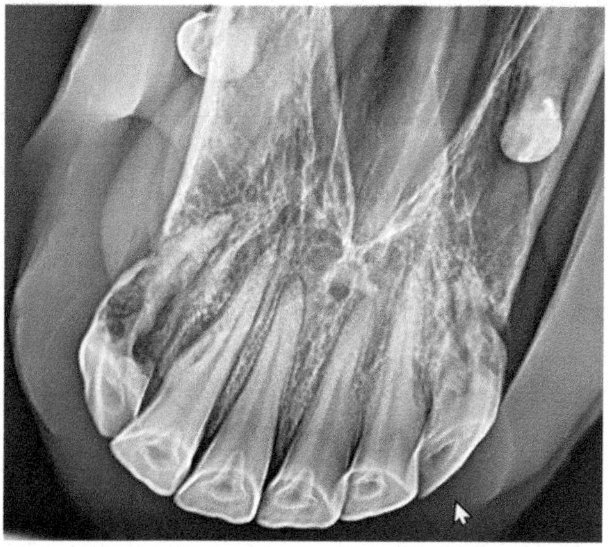

Radiograph of donkey incisors showing loss of dental material due to EOTRH

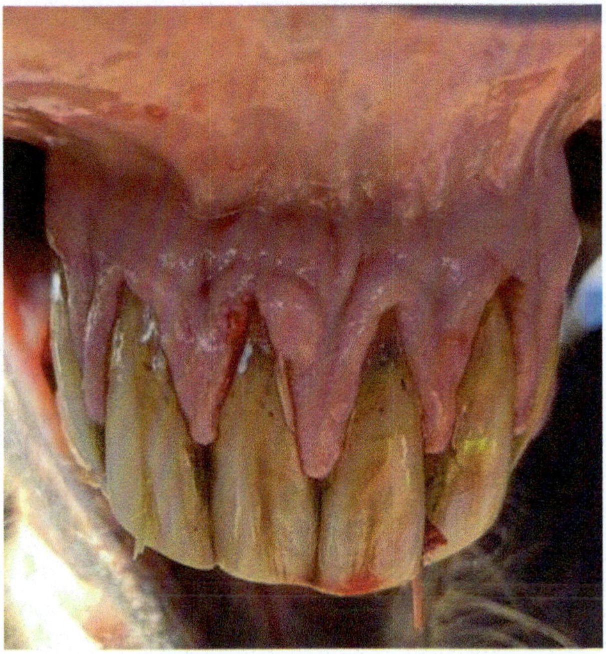

Photograph of severe gingival retraction in a donkey due to periodontal disease

4.5 Diagnostic Tests

4.5.1 Diagnostic Imaging in Equine Dentistry: Techniques Applied to Horses, Donkeys, and Mules

Diagnostic imaging has become an essential tool in equine dentistry, allowing a more accurate evaluation of dental and maxillofacial structures. Its use has spread not only in horses, but also in donkeys and mules, species that, although related, have anatomical particularities that must be taken into account when applying these techniques. The main diagnostic imaging tools used in equids are described below, with emphasis on their applications, advantages and limitations.

Oroscopy (Oral Endoscopy)

Oroscopy consists of direct visualization of the inside of the oral cavity using a rigid or flexible endoscope. It requires sedation of the animal and the use of an appropriate mouth opener. This technique allows you to inspect and evaluate in great detail:

The occlusal surface of the teeth.(Easley et al., 2011; Dixon & Dacre, 2005; AVDC, n.d.; BEVA, n.d.)

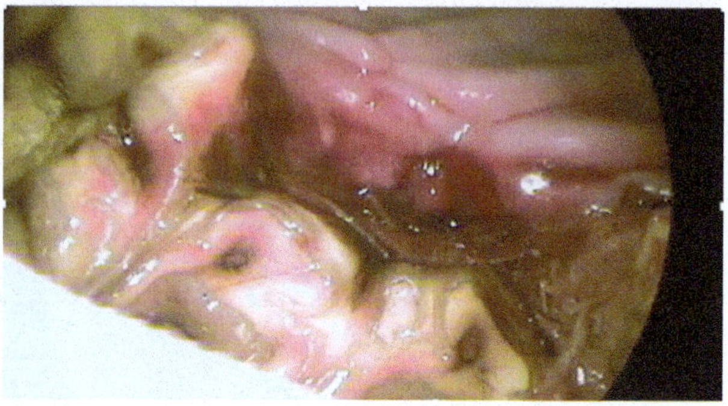

Oroscopy to evaluate a molar fracture

- Soft tissue lesions (mucosa and tongue)
- Dental fractures.
- Diastemas and food entrapments.
- Signs of periodontal disease.

Advantages

1. Minimally invasive method.
2. It allows documentation through images and videos.
3. Useful for treatment follow-up.

Limitations

1. It does not allow the assessment of internal structures of the tooth or alveolar bone.
2. It requires technical expertise for its correct interpretation.

Considerations on donkeys and mules

- Donkeys, due to their smaller oral aperture and more compact cranial morphology, may require adaptations in the type of appetizers and finer oroscopes.

Digital Radiology (DR)

Direct digital radiology (DR) is one of the most widely used tools in equine clinics, although it is a limited technique when it comes to diagnosing dental and facial pathologies. It allows two-dimensional images of teeth and cranial bone structures to be obtained. It should be noted that a good radiographic technique in terms of projection, as well as the age factor in the individual to be studied due to the different heights of the reserve crowns, is essential when assessing both dental and facial structures.

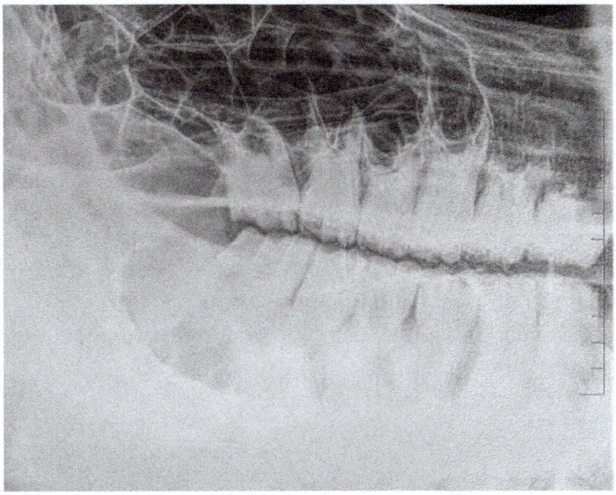

Latero-lateral radiography in a donkey to evaluate sinus structures

Evaluation of dental roots and alveolar bone

- Diagnosis of apical abscesses.
- Identification of dental or alveolar fractures.
- Study of secondary sinusitis (dental origin).
- Tooth fractures

Advantages

1. Fast, accessible and portable.
2. Allows on-site diagnostics in the field or clinic.
3. Immediate imaging.
4. Ability to share images with dental specialists for second opinions.

Limitations

1. Superposition of structures.
2. Difficult interpretation in complex areas such as the sinuses.
3. It requires good positioning technique.

Considerations on donkeys and mules

- Lower cranial mass and different angulation of the maxillary sinuses in donkeys can modify the optimal projection.
- In mules, morphological variability according to the dominant parent (horse or donkey) may require individualized adjustments.

Computed Tomography (CT) Scan

Computed tomography allows high-resolution three-dimensional images to be obtained, being the technique of choice for complex or deep pathologies. Currently, it is the technique of choice to evaluate pathologies at this level.

Applications

- Detailed diagnosis of dental and bone fractures.
- Assessment of apical infections, osteolysis and bone remodelling.
- Surgical planning of complicated tooth extractions
- Detailed study of sinus and nasal structures.

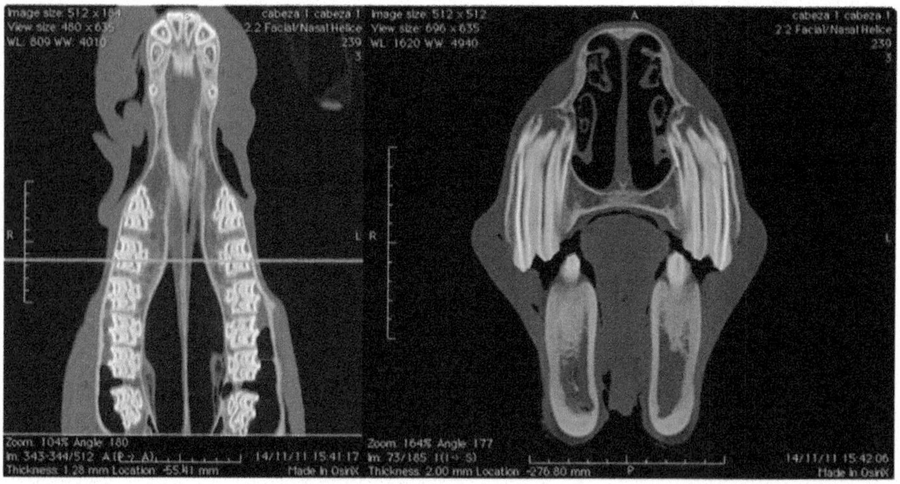

Horse Computerized Axial Tomography (CT Scan)

Advantages

1. High anatomical resolution.
2. Allows multiplanar and 3D reconstructions.
3. Useful for detecting hidden lesions on X-rays.

Limitations

1. High cost and lower availability.
2. It generally required general anesthesia, although there is now much equipment that allows the technique to be performed on the station using appropriate sedation protocols.

Considerations on donkeys and mules

- On donkeys, the more compact head facilitates complete exploration in a single acquisition.
- In mules, morphological heterogeneity may require a flexible imaging protocol.

Magnetic Resonance Imaging (MRI)

MRI offers detailed images of soft tissues and bone structures without ionizing radiation. Its use in equine dentistry is still limited, but expanding.

Applications

- Evaluation of the dental pulp.
- Early diagnosis of periodontal disease.
- Study of periodontal soft tissues, abscesses and fistulas.
- Investigation and complex cases.

Advantages

1. Excellent soft tissue resolution.
2. No radiation required.

Limitations

1. Expensive and inaccessible equipment.
2. It requires prolonged general anesthesia.
3. Lower resolution of bone structures than CT scan.

Considerations on donkeys and mules

- Due to their smaller cranial size, they may require a smaller field of vision.
- Its application is still experimental in these species.

Nasal Endoscopy

Nasal endoscopy is a very useful tool to complement the diagnosis of dental pathologies with sinus or nasal involvement. It allows direct inspection of the inside of the upper airway, from the nostrils to the nasopharynx, including the nasal meatus and sinus drainage holes. Even with the right gauge, sinus structures can be evaluated.

Applications in dentistry

- Evaluation of persistent unilateral nasal discharge
- Identification of oronasal fistulas or oroantral septal defect.
- Detection of purulent exudate from the sinuses.
- Visualization of masses, foreign bodies, or necrotic tissue.
- Eliminate causes susceptible to nasal discharge for non-dental reasons such as ethmoid hematomas or tumor masses.

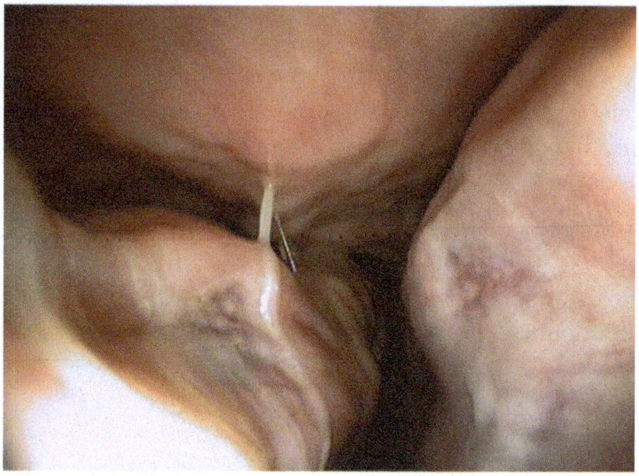

Nasal endoscopy image to evaluate unilateral nasal discharge

Advantages

1. Minimally invasive technique.
2. It allows samples to be obtained for cytology or culture.
3. It complements the information obtained by X-ray or CT scan

Limitations

- It requires sedation, in-depth anatomical knowledge, and technical prowess.
- It does not allow you to visualize dental structures directly.

Considerations on donkeys and mules

- Narrower nasal anatomy in donkeys may require smaller-caliber endoscopes.
- In mules, anatomical variability between individuals can condition the depth reached by the endoscope.

Molecular Diagnosis

Molecular diagnostics are emerging as a complementary tool in the identification of genetically-based or early-stage dental diseases. This area includes specific genetic testing and the use of molecular biomarkers to detect predispositions or subclinical alterations.

Applications in dentistry

- Identification of genetic predisposition to diseases such as EOTRH or dental dysplasias.
- Study of polymorphisms associated with craniofacial features that affect occlusion.
- Detection of biomarkers in blood or saliva that indicate chronic inflammation or periodontal infection.

Advantages

1. High specificity and sensitivity.
2. Possibility of early diagnosis before clinical onset.
3. Useful for selective breeding and prevention in predisposed genetic lines.

Limitations

1. It requires specialized laboratories and tests still in development.
2. Still limited in its general clinical application.

Considerations on donkeys and mules

- Genetic research on these species is scarce, but extrapolation from studies in horses may open up new lines of preventive diagnosis.

4.5.2 Conclusion

The right combination of diagnostic imaging techniques allows for a comprehensive assessment of oral health in equids. While oroscopy, digital radiography, and nasal endoscopy are widely available basic tools, CT and MRI represent advanced methods that offer superior diagnostic value in complex cases. In donkeys and mules, the adaptation of these techniques to their anatomical particularities is key to guaranteeing diagnostic accuracy and treatment effectiveness.

4.6 Treatment

4.6.1 Medical Management

The treatment of dental pathologies in equids includes a combination of medical, surgical and supportive approaches. In many cases, medical management is the first step to controlling inflammation, pain, and associated infections and restoring normal chewing function as soon as possible.

In donkeys and mules, the response to medical treatment may vary due to physiological differences and the frequency with which these species are treated later in the course of the disease.

1. Medication
   Nonsteroidal anti-inflammatory drugs (NSAIDs): such as flunixin meglumine, phenylbutazone, or meloxicam, which help control pain and inflammation.
   Antibiotics: indicated in active infections, especially when there are abscesses, secondary sinusitis or fistulas. Penicillins, trimethoprim-sulfonamides, or doxycycline are commonly used, depending on bacterial sensitivity.
   Topical antiseptics or mouthwashes: for irrigation of periodontal pockets or affected areas in local treatments.
   Topical therapy: chlorhexidine irrigation in affected cavities or diastemas as well as the use of chlorhexidine gels and antibiotics for the treatment of localized lesions.
2. Dietary Management
   Modification of feed texture: In horses with acute dental pain or loss of pieces, moist hay, chopped forage, or porridge may need to be administered. It is noteworthy that cut hay is widely used, especially in geriatric donkeys, due to physiological dental wear or appearance of diastemas in which food is impacted, while specific odontoplasia protocols are made
   Control of access to hard or fibrous pastures: to reduce the risk of mucosal lesions and avoid aggravation of food entrapment.
   Specific supplementation: in animals with loss of condition, supplements rich in energy, protein or vitamins can be added.

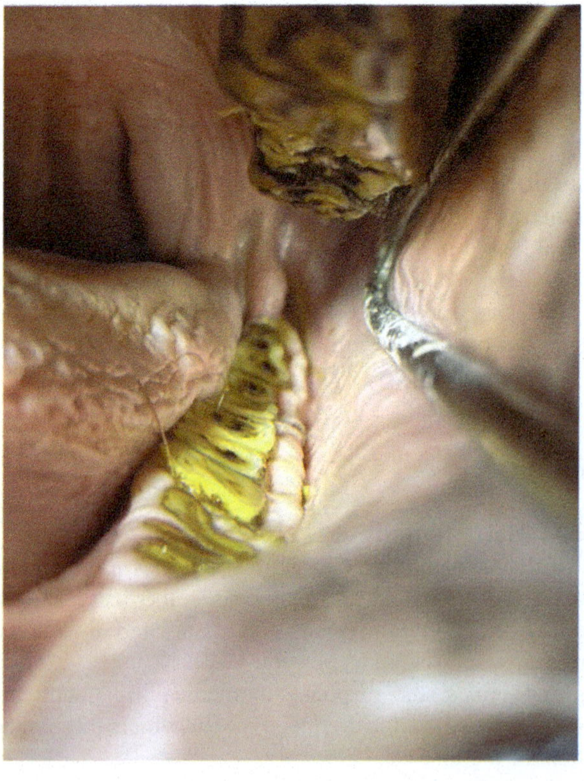

Physiological wear and tear in aged donkeys. Incompetence of the masticatory system.

3. Supportive Care
   Regular dental check-up: once the acute condition has stabilized, a check-up schedule should be established to prevent recurrences.
   Environmental management: avoiding stressful situations, improving access to water and controlling food competition in groups.
   Continuous clinical record: monitoring of weight, evolution of symptoms and response to treatment.

4.6.2 Particularities in Donkeys and Mules

In donkeys with significant tooth loss, diet adjustment should be prioritized due to their reduced chewing efficiency.

They are more stoic or may have increased pain tolerance which can delay the proper treatment. They need doses adjusted for metabolic differences (see pharmacology and pain chapters).

Mules may exhibit unpredictable temperamental reactions to pain and handling, necessitating a treatment plan that includes safe sedation and careful monitoring.

4.6.3 Surgical Interventions

Surgical interventions follow common principles in horses, but require adaptations depending on the species.

Common procedures

- **Exodontia**: extraction of pathological teeth, periodontitis, severe diastema, endodontic processes, fractures or impacted wolf teeth.
- **Treatment of odontogenic sinusitis**: may require trepanation and washing of the maxillary sinus. It is common in donkeys due to the prolonged presence of untreated infections.
- **Correction of malocclusions**: by advanced odontoplasty or, in severe cases, orthodontic surgery.

Anatomical particularities to consider

Donkeys: narrower and more compact jaw; more limited access, especially in adult animals with little oral opening.

Mules: They may have more marked craniofacial discrepancies that make surgical alignment and access difficult.

Anesthesia and sedation

In donkeys, sedation for dental procedures must be carefully adjusted, as these animals usually require **higher doses per kilogram of weight** compared to horses to obtain effective sedation. In addition, its peculiar metabolism and more reserved behavior make it advisable to use **combined protocols**, especially in prolonged procedures such as tooth extractions or oroscopic evaluations. The sedatives used in dental protocols are Alpha-2 agonists (Xylazine, detomidine, romifidine) along with analgesics such as butorphanol (refer to anesthesia chapter and pharmacology chapter).

4.7 Prevention

Diet plays a fundamental role in the prevention of dental disorders. The type of forage, the texture of the food and the feeding routine directly influence physiological dental wear.

General recommendations

Quality forage: Fresh and fine-cut hay should be prioritized, avoiding very dry hay or hay with hard ears that favor entrapment and diastema.

Avoid over-ground food: Over-crushed feed reduces chewing and therefore natural tooth wear.

Free access to grass: Stimulates regular chewing and balance between dental arches.

Specificities by species

Horses: tolerate a wider variety of diets, but benefit from a balance between forage and concentrate to maintain balanced wear.

Donkeys: they evolved to feed on poor and fibrous forages. Diets high in sugar or grain can be harmful and predispose to oral entrapments and fermentations.

Mules: inheriting traits from both species, they require a carefully balanced diet, avoiding both unsupervised excess fiber and energy oversupply.

Identification and avoidance of predisposing factors

Periodic monitoring and constant observation are key in dental prevention.

Avoidable factors

- **Prolonged use of snacks without revision**: can lead to chronic mouth lesions.
- **Lack of routine dental check-ups**: Ideally, animals should be evaluated every 6 to 12 months.
- **Poor handling of work equipment**: in mules and donkeys, poorly fitted muzzles or harnesses can cause trauma to the cheeks and jaws.

Preventive check-ups

- Examination with oral speculum and adequate light.
- Early detection of diastema, sharp points, malocclusions and tooth mobility.
- Radiological evaluation in animals over 10 years of age or with a history of weight loss or unilateral chewing.

4.8 News Insights and Future

Although most treatments in equine dentistry are still mechanical or surgical, there are ongoing lines of research with potential to improve the management of chronic dental diseases.

- **Stem cell therapy**: it is being evaluated in periodontal and dental pulp defects.
- **Regenerative biomaterials**: for cavity filling and infection prevention.
- **3D printing of personalised surgical guides**: still incipient, but with great potential in complex extractions.

4.8.1 One Health Perspective

Equine dentistry not only affects the health of the animal, but also the well-being of the people who depend on them. In regions where donkeys and mules are essential means of work, their dental health has a direct impact on the family economy, agricultural productivity and social welfare.

Promoting preventive dental care in these species is also a way of taking care of the balance between animal health, human health and the environment.

5 Conclusion

Equine dentistry, traditionally focused on the horse, needs a broader perspective that actively includes donkeys and mules as patients with **their own distinct needs**. Although all these species belong to the equidae family, their anatomical, physiological, behavioral, and management differences significantly influence the way dental pathologies manifest, diagnose, and treat.

In particular, **donkeys** have a **more compact cranial morphology**, **narrower jaws** and a **less complex tooth structure**, which makes them prone to develop **sharp points**, **diastemas**, **food entrapments** and **eruptive alterations**. However, they show **a lower incidence of complex dental fractures** than horses. Its **high resistance to pain** and **its lower clinical expressiveness** make it difficult to detect problems early, favoring in many cases the **chronification of oral pathologies**.

These differences require an **adaptation of diagnostic tools**, such as oroscopy, digital radiology, CT or nasal endoscopy, adjusting size, positioning and depth depending on the species. Clinical **management and sedation also** require adjustments, since donkeys usually need higher doses per kilogram of weight to achieve an effect equivalent to that of horses, and their more reserved nature imposes less invasive and more progressive strategies.

Dental treatment must be respectful, individualized and oriented towards well-being, minimizing stress and prioritizing regular check-ups from an early age. The implementation of **adapted preventive protocols** and the recognition of the particularities of the mule or donkey patient are essential to improve their quality of life.

Finally, it is essential to promote **awareness among professionals and owners** about the importance of dental care in donkeys and mules, species that have historically received **less clinical attention** than sport or recreational horses, despite their growing presence in therapeutic, educational or work environments.

Recognizing and addressing these differences with flexible attitude, up-to-date knowledge, and clinical sensitivity is the path to move toward **more inclusive, effective, and equitable equine dentistry**.

5.1 Multidisciplinary Approach

The joint work between clinical veterinarians, veterinary dentists, maxillofacial surgeons, ethologists, nutrition specialists and animal welfare professionals is key to guaranteeing comprehensive care. This approach makes it possible not only to solve current problems, but also to prevent pathologies, improve performance and extend the lifespan of animals.

Future prospects

- Greater inclusion of donkeys and mules in clinical studies.
- Specific prevention and treatment protocols.
- Training veterinarians in rural environments with a focus on underserved species.
- Portable technologies for diagnosis and treatment in the field.

6 Decision Flows

6.1 Clinical Diagram: Decision Flow for Dental Approach

Initial assessment

1. Does the patient have obvious oral clinical signs?
   No: Preventive control —> Check-up every 6–12 months
   Yes: Continue
2. Do you have weight loss, hay balls, halitosis, runny nose?
   No: Periodic review and owner education
   Yes: Perform detailed oral examination with an appetizer
3. Are there diastema, dental fractures or mobility?
   No: Assess odontoplasty
   Yes: Oroscopy-X-rays + treatment evaluation (Extraction-treatment of diastemas or periodontals)
4. Periapical Infection/Sinusitis/Complex Fracture?
   No: Conservative treatment or simple extraction
   Yes: Refer or perform advanced surgery

6.2 Support Material

Check-list for dental evaluation in the field

- Adequate sedation per species
- Clean and functional oral speculum
- Front light or powerful light source
- Dental mirror or endoscope
- Periodontal probe and scanner
- Water and chlorhexidine irrigation
- Portable X-ray if there is suspicion of underlying injury
- Photographic record
- Medical history and informed consent form

Questions

1. Which of the following statements about donkey dental physiology is correct?
   A. The tooth eruption pattern is identical to that of the horse.
   B. Donkeys have a higher tooth density than horses.
   C. Hypodontia is a common anomaly in donkeys.
   D. The length of the mandibular dental arch is greater than the maxilla.
   Correct answer: C
   Explanation: Hypodontia (congenital absence of teeth) is more common in donkeys than in horses. In addition, their eruption pattern differs slightly, and their tooth density is lower, which favors the formation of tips and food entrapments.
2. In an adult mule with chronic unilateral nasal discharge, what would be the first reasonable dental suspicion?
   A. Severe malocclusion
   B. Mandibular fracture
   C. Odontogenic sinusitis due to periapical infection
   D. Impacted wolf tooth
   Correct answer: C
   Explanation: Unilateral sinusitis of dental origin is common in horses and especially in donkeys and mules when there is a prolonged periapical infection. This may be related to an infected diastema or a compromised tooth root.
3. Which of the following statements is correct about wolf teeth on mules?
   A. They are always absent.
   B. They are more frequent than in horses.
   C. They present with greater variability and can cause malocclusion.
   D. They do not require clinical attention.
   Correct answer: C
   Explanation: Mules, due to their hybrid origin, have variable dentition. Wolf teeth can appear asymmetrically or with atypical rashes, which can lead to malocclusion and discomfort with biting.

4. In a geriatric donkey with progressive weight loss and hay balls in the mouth, what is the most likely diagnosis?
   A. Chronic gastritis
   B. Diastema with food entrapment
   C. Maxillary fracture
   D. Parasitic anemia
   Correct answer: B
   Explanation: Diastemas are very common in donkeys, especially at advanced ages. They cause food entrapment, difficulty chewing, weight loss, and risk of secondary infection.

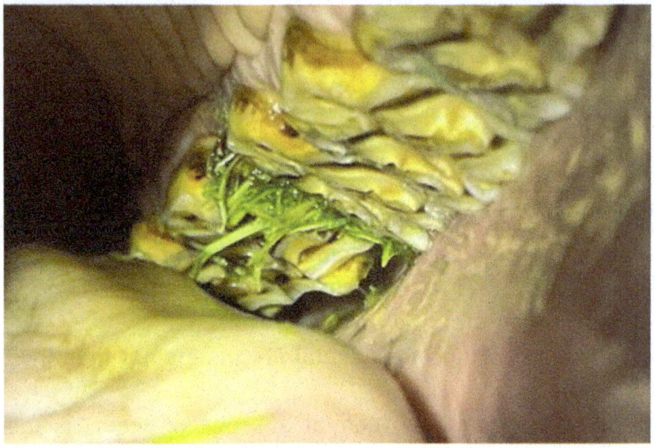

5. What morphological differences exist between the skull of a donkey and that of a horse, and how do they affect the dental approach?
   A. The donkey has a skull that is longer and wider than that of the horse, which facilitates oral exploration.
   B. The donkey has a more compact skull and narrower jaws, making it difficult to access and use standard tools.
   C. Both species have the same cranial morphology, so the dental approach is identical.
   D. The horse has a shorter and wider skull, which makes imaging diagnosis difficult.
   Correct answer: B
   Explanation: The donkey's more compact skull and narrower jaw require adapted tools and make certain diagnostic and therapeutic procedures difficult.
6. Why are dental clinical signs usually less evident in donkeys and mules than in horses?
   A. Because they have a greater capacity for bone regeneration.
   B. Because they usually have milder injuries.
   C. Because they are more resistant to pain and tend to hide clinical signs.
   D. Because their diet is softer and more varied.
   Correct answer: C

Explanation: Donkeys and mules are more rustic and have a high tolerance to pain, which can mask pathologies until they are advanced.

7. Explain at least three predisposing factors of dental pathology in mules.
   A. Soft diet, wide jaw and long roots.
   B. Hybrid origin, inherited diastemas and asymmetric eruptions.
   C. Cranial size, difficult sedation and a diet rich in fiber.
   D. Square jaw, prominent canine teeth and short tongue.

   Correct answer: B

   Explanation: Mules can inherit problems from both the donkey and the horse, making them more susceptible to diastemas and malocclusions.

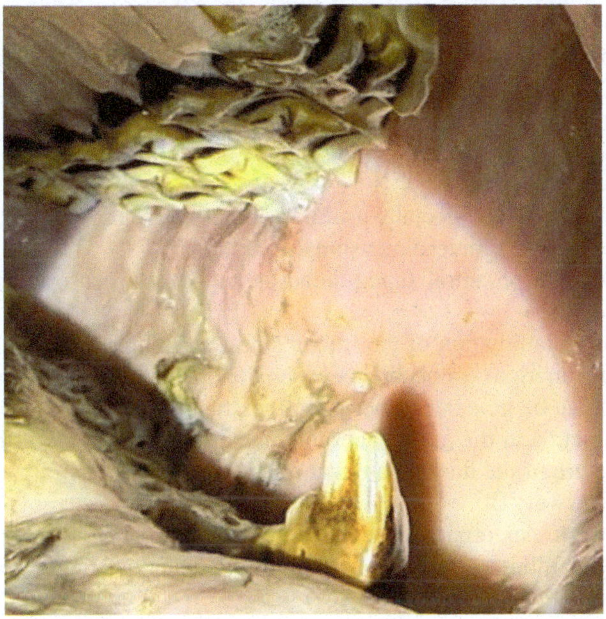

8. What special considerations should be taken into account when sedating a donkey for a dental procedure?
   A. The same doses can be used as in horses because their metabolism is the same.
   B. They require fewer doses than horses due to their lower body weight.
   C. Their metabolism is different and they usually need higher doses per kg than horses.
   D. It is not necessary to sedate them for basic dental procedures.

   Correct answer: C

   Explanation: The metabolism of the donkey is different and its response to drugs usually requires higher doses per kg to achieve effective sedation.

9. What nasal endoscopic findings could suggest a dental pathology with sinus involvement?
   A. Ciliary hypermotility and laryngeal collapse.
   B. Obstruction of the ventral meatus.
   C. Unilateral purulent secretion and abnormal contents in the middle meatus.
   D. Bilateral pharyngitis and nosebleeds.
   Correct Answer: C
   Explanation: A unilateral purulent discharge from the middle meatus suggests a possible sinus infection secondary to dental pathology.
10. In which cases is the use of CT scans justified in dental diagnosis in equids and what advantages does it offer over digital radiology?
   A. Only in cases of mild sinusitis, as it offers better resolution than oroscopy.
   B. When there is suspicion of complex fractures or deep pathologies not visible with X-rays.
   C. To check wolf's teeth in foals.
   D. It is the first choice in all clinical cases due to its low cost.
   Correct answer: B
   Explanation: CT is especially useful in complex fractures or deep infections, as it allows a detailed three-dimensional evaluation.
11. Define what an oroantral septal defect is and how it can originate in a dental procedure.
   A. It is an inflammation of the frontal sinus caused by dental infections.
   B. It is an abnormal connection between the oral cavity and the maxillary sinus, which can be generated after a tooth extraction.
   C. It is a fistula in the gum that communicates with the root of the tooth.
   D. A fracture of the hard palate that affects breathing.
   Correct answer: B
   Explanation: An oroantral septal defect occurs when a passage is formed between the mouth and the maxillary sinus, usually after the removal of an upper molar.
12. Mention at least three acute clinical signs that require immediate intervention in equine dentistry.
   A. Loss of appetite, bruxism and purulent unilateral nasal discharge.
   B. Itching on the flank, diarrhoea and laminitis.
   C. Sweating, polyuria and ptialism.
   D. Coughing, colic and muscle tremors.
   Correct answer: A
   Explanation: Purulent nasal discharge, loss of appetite, and bruxism indicate acute pain or infection requiring urgent dental evaluation.
13. What elements make up good medical management in a horse with chronic periodontal disease?
   A. Rest, fluid therapy and bed change.
   B. NSAIDs, antibiotics, dietary control and regular check-ups.
   C. Oxygen therapy, tranquillisers and liquid diet.
   D. Mandibular massages, vitamins and controlled exercise.
   Correct answer: B

Explanation: Medical management includes anti-inflammatories, treatment of infections, dietary adjustments, and periodic dental follow-up.

14. Why is it important to incorporate molecular diagnosis into the study of dental pathologies in equids?
    A. To identify respiratory viruses involved in oral inflammation.
    B. To select the ideal type of anaesthesia according to genetics.
    C. To detect genetic predispositions and biomarkers associated with dental diseases.
    D. Because it allows the sex of the animal to be identified more quickly.

    Correct answer: C

    Explanation: Molecular diagnosis allows the detection of genetic alterations and biomarkers that help to anticipate diseases or choose more appropriate therapies.

References

AgroimportHN (2023, April 13) Odontología equina: Condiciones principales. Retrieved from https://www.agroimporthn.com/2023/04/13/odontolog%C3%ADa-equina-condiciones-principales/

AVDC (American Veterinary Dental College) (n.d.) Equine dentistry resources. Retrieved from https://www.avdc.org

BEVA (British Equine Veterinary Association) (n.d.) Equine dentistry resources. Retrieved from https://www.beva.org.uk

Dixon PM, Dacre I (2005) A review of equine dental disorders. Vet J 169(2):165–187. https://doi.org/10.1016/j.tvjl.2004.03.022

Du Toit N, Burden FA, Dixon PM, Shaw DJ (2014) Dental disorders in donkeys and mules. Equine Vet Educ 26(1):48–56. https://doi.org/10.1111/eve.12047

Easley J, Dixon PM, Schumacher J (2011) Equine dentistry, 3rd edn. Saunders Elsevier

Horse Illustrated (n.d.) Glossary of equine dentistry terms. Retrieved from https://www.horseillustrated.com/images/equine-u/equine-dentistry-glossary.pdf

Horsedentistry.eu (2020) Odontología equina: Los fundamentos [PDF]. Retrieved from https://horsedentistry.eu/wp-content/uploads/2020/10/Odontologia-fonaments.pdf

Kentucky Equine Research (2021) Brush up on equine dental terminology. https://ker.com/equinews/brush-up-on-equine-dental-terminology/

Rodrigues JB, Araújo S, Sanroman-Llorens F, Bastos E, San Roman F, Viegas C (2013a) A clinical survey evaluating the prevalence of incisor disorders in Zamorano-Leonés and Mirandês donkeys (Equus asinus). J Equine Vet Sci 33(10):710–718. https://doi.org/10.1016/j.jevs.2013.01.006

Rodrigues JB, Bastos E, Sanroman-Llorens F, San Roman F, Viegas CA (2013b) Polyodontia in donkeys. Equine Vet Educ 25(7):363–367. https://doi.org/10.1111/eve.12030

Rodrigues JB, Araújo S, Guedes-Pinto H, San Roman F, Viegas C, Bastos E (2013c) Analysis of new Matrilin-1 gene variants in a case–control study related to dental malocclusions in Equus asinus. Gene 522(1):70–74. https://doi.org/10.1016/j.gene.2013.03.105

Rodrigues JB, Dixon PM, Bastos E, San Roman F, Viegas C (2013d) A clinical survey on the prevalence and types of cheek teeth disorders in 400 Zamorano-Leonés and 400 Mirandês donkeys (Equus asinus). Vet Rec. https://doi.org/10.1136/vr.101747

Rodrigues JB, Ferreira LM, Bastos E, San Roman F, Viegas C, Santos AS (2013e) Influence of dental correction on nociceptive test responses, fecal appearance, body condition score, and apparent digestibility coefficient for dry matter of Zamorano-Leonés donkeys (Equus asinus). J Anim Sci 91(10):4765–4771. https://doi.org/10.2527/jas.2013-6311

Donkey Nutrition and Feeding Strategies

Tommaso Danese, Federica Raspa, Laura Cavallarin,
Marzia Giribaldi, and Emanuela Valle

Abstract

Donkeys and mules differ from other equine in their digestive physiology and nutritional adaptations. Although they are closely related, their efficiency in digesting high-fiber and low-quality forage, their tolerance to dehydration, and their shorter gastrointestinal transit times are unique compared to horses. These characteristics render donkeys extremely well adapted to desert and marginal conditions. Donkeys can live on low-protein, low-energy roughage, attesting to their evolutionary background. Overfeeding, especially carbohydrates and proteins, however, is an issue in non-working animals and can cause obesity and metabolic disease.

This chapter will be structured as follows, to give a detailed explanation of donkey nutrition at various physiological stages:

1. General Nutrition and Feeding Strategies
   Optimal feeding strategies are examined to ensure welfare, with consideration of adjustments by body condition score, fatty neck score, and body weight.
2. Nutritional Requirements and Diet Composition
   The fundamental nutritional requirements of foals, growing donkeys, maintenance donkeys, pregnant jennies, working donkeys, and lactating donkeys are presented. Emphasis is placed on a balanced diet in terms of energy and protein, rich in fiber, minerals, and vitamins, for maintaining overall health

T. Danese
Department of Veterinary Sciences, University of Parma, Parma, Italy

F. Raspa · E. Valle
Department of Veterinary Sciences, University of Turin, Turin, Italy

L. Cavallarin · M. Giribaldi (✉)
Institute of Sciences of Food Production, National Research Council, Grugliasco, Turin, Italy
e-mail: marzia.giribaldi@cnr.it

A. K. McLean et al. (eds.), *Donkey and Mule Medicine: Foundations, Breeds, and Management*, https://doi.org/10.1007/978-3-032-12726-6_19

and physiological functions. Suitable feedstuffs, including forage (hay, haylage, and pasture), and concentrates are described, with an emphasis on the importance of good-quality forage and correct mineral-vitamin supplementation to address possible nutritional deficiencies, and avoiding, whenever possible, the use of cereal concentrates.

3. Health implications and Clinical nutrition of Donkeys
   Optimum nutrition is directly correlated with enhanced immune status, decreased incidence of disease, and improved reproductive performance. The chapter covers the clinical nutrition aspect, connecting nutritional approaches to the prevention of prevalent disorders and advocating for prime health outcomes.

Keywords

Donkey nutrition · Nutrition related diseases · Clinical nutrition · Forage-based diets

1 General Nutrition and Feeding Strategies

1.1 Nutritional Status Evaluation

The nutritional management of dairy donkeys critically impacts their nutritional status (NS), which is essential for their health, productivity, reproduction, and welfare. Evaluating NS involves multiple factors such as age, physiological status, body morphometry, and adiposity, making it a complex task. Accurate assessment methods include body condition score (BCS), fatty neck score (FNS) and body weight (BW).

1.1.1 Body Condition Score

The Body Condition Score (BCS) is a visual and palpable assessment of the amount of fat and muscle covering the bones of a donkey, which can significantly vary depending on several factors including diet, metabolism, and overall health (Raspa et al. 2019). To precisely score the BCS of a donkey, the animal needs to be well placed on four legs and observed from 2 m of distance, from the right flank, looking at key point areas for fat deposit such as neck, shoulder, dorsal spine and back, ribs and hips (Fig. 1). BCS can then be scored from 1 (severely emaciated) to 5 (severely obese), using half-point increments, or from 1 to 9 using only whole numbers, depending on the scoring chart adopted, as described in Table 1.

1.1.2 Fatty Neck Score

The Fatty Neck Score is a recently developed regional scoring system that allows for a clear definition of the adipose tissue deposited along the crest of the neck. Through the assessment of the morphometric measurement of the neck thickness and together with the BCS of donkey, it is possible to perform a correct evaluation of the nutritional status of donkeys (Valle et al. 2017; Raspa et al. 2019) (refer to Table 2).

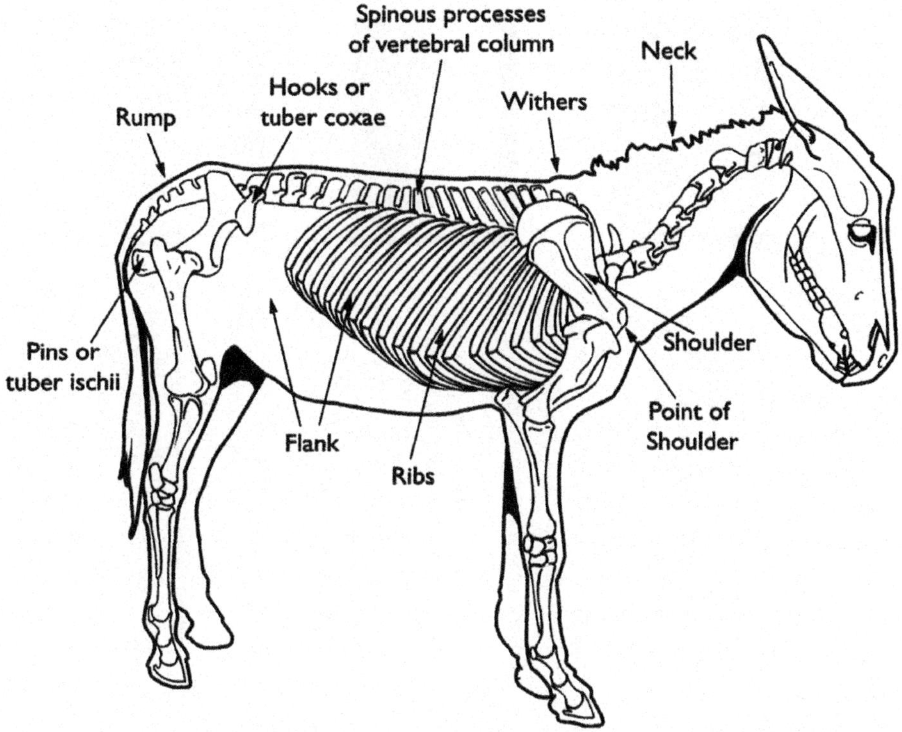

Fig. 1 Main anatomical points used to assess the BCS of a lactating dairy donkey. (Adapted from Raspa et al. (2019))

1.1.3 Body Weight

While BCS is essential, it should not replace body weight (BW) monitoring, as it reflects not only the overall condition but also factors such as gut-fill, hydration, and physiological status (Becvarova et al. 2009). Although scales provide the most accurate BW, they are often impractical in field conditions: therefore, estimating BW through body measurements is a necessary alternative. These measurements should be taken with a soft tape, ensuring adequate tension, while the donkey stands on level ground (Pearson and Ouassat 2000).

The latter author developed two equations, to apply depending on BCS:

1. For adult donkeys with a BCS of 2 to 6 on the nine-point scale and a height of 90–120 cm:

$$Live\ weight\ (kg) = \frac{\left(heart\ girth\ [cm]^{2.12} \times length\ [cm]^{0.688}\right)}{3801}$$

Table 1 Comparison of the two body condition scoring systems for donkeys

Body Condition Scoring for Donkeys				
	Scale from 1 to 9		Scale from 1 to 5	
1. Very thin (Emaciated)	Neck appears thin. The dorsal spine, ribs, backbone, and hip bone are prominently visible with minimal or no muscle mass present.	1. Poor (Very thin)	Neck is thin, with bones easily felt; shoulder bones are prominent and angular. Dorsal spine and withers are prominent and easily felt. Ribs visible from a distance, easily felt; belly is tucked up. Backbone is prominent; dorsal and transverse processes are easily felt. Hip bones visible and felt easily; little muscle cover, may have a cavity under the tail.	
2. Thin	The neck is thin with sharply defined shoulders. The dorsal spine, ribs, backbone, and hip bone are prominent and clearly defined. There is some muscle development.			
3. Less thin	There is little muscle or fat covering the bone structures, making them easily felt. The loin area and rump are concave.	2. Moderate (Underweight)	Some muscle overlying bones; slight step where neck meets shoulders. Some cover over dorsal withers; spinous processes felt but not prominent. Ribs not visible but easily felt. Dorsal and transverse processes felt with light pressure; poor muscle development. Poor muscle cover, hip bones felt with ease.	
4. Less than moderate	The dorsal and transverse processes are visible; withers, shoulders, and neck have some fat and muscle covering. The pins can be felt but are not yet visible. The hooks are rounded but still visible. The rump is flat rather than prominent.			
5. Moderate	The hooks and rump are rounded, and the pins are not visible. Some fat can be felt in the shoulder area and at the base of the neck. The ribs may be felt but are not visible.	3. Ideal	Good muscle development; bones felt under light cover of muscle/fat. Good cover of muscle/fat; withers flow smoothly into the back. Ribs just covered by a light layer of fat/muscle, felt with light pressure, belly firm with good muscle tone. Individual spinous or transverse processes felt with pressure; good muscle development. Good muscle cover, hip bones rounded in appearance, felt with light pressure.	

6. More than moderate	The spinous processes are not easily felt. The rump is convex and well-muscled. Some fat can be felt on the neck.	4. Overweight (Fat)	Neck thick, crest hard; shoulder covered in even fat layer. Withers broad; bones felt with pressure. Ribs felt with firm pressure dorsally, belly overdeveloped. Dorsal and transverse processes felt with firm pressure, may have a slight crease along the midline. Rounded, bones felt only with pressure, fat deposits evenly placed.	
7. Less fat	The spinous processes cannot be felt, and the hooks are just visible. Fat on the neck and shoulders is expanding over the ribcage. The neck is thickening.			
8. Fat	The animal appears well-covered with a rounded body, and no fat or bones are discernible. The flanks are filled, and the back is broad.	5. Obese (Very Fat)	Neck thick, crest bulging with fat; shoulder rounded and bulging with fat. Withers broad; bones felt with firm pressure. Large fat deposits cover ribs; ribs not palpable dorsally, belly pendulous. Back broad; difficult to feel individual spinous or transverse processes due to fat. Cannot feel hip bones; fat may overhang either side of the tail head, often uneven and bulging.	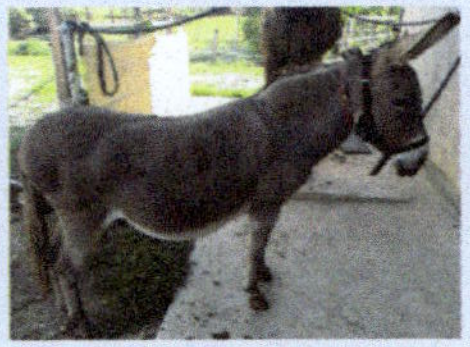
9. Very fat	The bones are buried in fat, with large accumulations on the neck, over the shoulder area, and on the ribs. The flanks are filled with fat.			

Adapted from Raspa et al. (2019), pictures from the author's archive and from "Vivi gli Animali Onlus – Donkey Refuge"

Table 2 Fatty neck score (FNS) for donkeys

Score	Illustration of the Individual Fatty Neck Score	Description	Neck Thickness Range According to FNS (in cm)
0		Neck: thin with the absence of a visible and palpable crest.	<14
1		Neck: thin with no visible crest, but a slight filling felt upon palpation.	>14–19
2		Neck: with a moderate deposition of fat. Noticeable appearance of a crest, with fat deposited fairly evenly from the poll to the withers. Crest: easily cupped in one hand and easily bent from side to side.	>19–22
3		Neck: enlarged and thickened. Crest: palpable from the poll to the withers, filling a cupped hand, and beginning to form longitudinal fat deposits on both sides of the neck.	>22–27
4		Neck: very enlarged and thickened. Crest: grossly thickened with fat deposits from the poll to the withers, forming longitudinal bands of fat on both sides of the neck. Crest cannot be bent easily from side to side.	>27–34
5		Neck: very enlarged and thickened. Crest: very thickened with hard fat deposits, rounded along both sides of the neck.	>34

Adapted from Valle et al. (2017) and Raspa et al. (2019)

2. For adult donkeys with a BCS > 6

$$Live\ weight\ (kg) = \frac{\left(heart\ girth\ [cm]^{2.575} \times length\ [cm]^{0.240}\right)}{3968}$$

3. For intractable donkeys, only using heart girth:

$$Live\ weight\ (kg) = \frac{heart\ girth\ [cm]^{2.65}}{2188}$$

The Donkey Sanctuary (2020) also proposed another equation:

$$Live\ weight\ (kg) = 0.000252 \times height\ at\ the\ wither\ [cm]^{0.24} \times hearth\ girth\ [cm]^{2.575}$$

1.2 Dental Health and Status

Dental care is a critical indicator of appropriate nutrition and overall health in donkeys. In lactating donkeys, studies have shown that those with a lower BCS of 1 or 2 out of 5 often have poorer dental conditions (Valle et al. 2018). Dental disorders in donkeys are commonly associated with weight loss and the potential onset of intestinal diseases (Rodrigues and Lilly 2019). Given the significant impact of dental health on overall welfare, it is crucial to conduct regular dental examinations, ideally at least once a year, by a qualified equine dental professional. This is especially important because dental disorders are often underrecognized and undertreated, with many donkeys showing few to no signs of pain despite significant dental issues (Burden et al. 2013). See chapter "Physical Examination of the Donkey and Their Hybrid Crosses" for more information about dentistry.

1.3 Environmental Factors

Seasonal variations can significantly impact forage availability and nutritional quality. In colder months, additional concentrates may be necessary to compensate for lower-quality forage. During warmer months, fresh pasture can provide excellent nutrition but should be managed to avoid overgrazing.

High temperatures and humidity can affect a donkey's appetite and water intake. Ensuring access to shade and cool, clean water is essential during hot weather, to avoid heat stress and to maintain hydration, and support milk production. Heat stress signs in donkeys are increased respiratory and heart rate, lethargy, excessive sweating, reduced feed intake and elevated body temperature (over 38.5 °C; 101.3 °F).

Providing access to high-quality pastures can enhance the nutritional intake, which can be very useful in the case of lactating or working donkeys, but should be used with caution in the case of maintenance donkeys due to their ability to gain weight easily. However, it is important to monitor pasture conditions and supplement with hay or concentrates as needed to ensure a balanced diet.

1.4 Regular Feeding Schedules

Maintaining a regular feeding schedule helps prevent digestive issues and supports consistent milk production. Irregular feeding can lead to fluctuations in nutrient intake, affecting both milk yield and the donkey's overall health. Regular, small meals are preferable to large, infrequent feedings. This approach mimics natural grazing patterns and helps prevent colic and other digestive disturbances; moreover, consistent feeding schedules ensure a steady supply of nutrients. Sudden changes in diet or feeding times can stress the donkey and negatively impact milk yield.

1.5 Housing

Proper housing is essential for the welfare and productivity of donkeys and mules. Housing should provide adequate shelter from extreme weather conditions, sufficient space for movement, and a clean, dry environment to reduce the risk of infections and diseases. Research has shown that adopting windproof facilities can significantly improve growth performance, enhance immune function, and maintain stable biochemical indicators in donkeys during cold weather (Gao et al. 2022). Additionally, donkeys' hair coats do not change significantly across seasons, indicating their need for extra protection in colder climates to ensure their welfare (Osthaus et al. 2017).

1.6 Bedding Material

Bedding material should also be considered when discussing housing for donkeys, specifically since edible bedding material may present a risk factor for the development of nutritional disease (Raspa et al. 2019). Avoid carboard or paper since it has been shown to be more palatable than hay or fresh forage, increasing hyperlipidemia and colic risk (Cox et al. 2007; Burden et al. 2011). Barley straw is considered to be the safest bedding material for donkeys to ingest, but it should be free of grains to avoid the risk of laminitis. Hygiene conditions must be thoroughly checked daily: whichever material is used, a clean bedding decreases the risk of respiratory and hoof disease (Bambi et al. 2018).

1.7 Social Interactions and Hierarchy in Donkeys

In feral donkey populations, social organization varies significantly between sexes. Females typically form groups with their offspring, while males often remain solitary. Environmental factors heavily influence these social structures, demonstrating donkeys' flexible social strategies that help them thrive in diverse conditions (Rudman 1998). Therefore, in donkey farming, the provision of adequate space and appropriate group sizes is vital. Donkeys tend to form pair bonds within a herd, and stable groups are typically composed of females and their offspring. Minimizing negative social interactions and maintaining stable herd structures contribute significantly to their welfare. Overcrowding can lead to increased aggression, while adequate space allows donkeys to express natural behaviors and reduces competition for resources (Raspa et al. 2019).

2 Diet Composition and Nutritional Requirements

2.1 Diet Composition

All dietary values of the chemical composition are reported on a dry matter basis (% DM).

2.1.1 Forages

Forages, such as hay, haylage, and grass, are the pillars of a donkey's diet. When compared to horses, donkeys are highly efficient at digesting poor nutritional quality fiber, specifically when digesting highly fibrous roughages, such as straw (Mueller et al. 1998). High-quality forage provides essential fiber, which is vital for maintaining healthy digestive function and promoting efficient nutrient absorption. The choice of forage—whether hay, haylage, grass, or other hay equivalents—depends on availability, nutritional quality, and the specific needs of the donkey. Understanding the differences between these options can help formulate a balanced diet that meets the energy and nutritional demands.

Hay

Hay is a staple in donkey diets due to its high fiber content and availability throughout the year. A lactating donkey typically consumes about 1.3–2.1% of body weight (BW) per day. High-quality hay should be green, leafy, and free from mold and dust. Good hay for lactating donkeys should contain around 8–10% crude protein (CP%) and 30–35% crude fiber, with a moisture content of less than 15% to ensure long-term storage stability (Salari et al. 2020), while it can be less rich in CP for maintenance donkeys (5–8% CP).

Straw

Straw, a byproduct of grain production, can be included as a fiber source in the diet of lactating donkeys, or used solely as the forage for maintenance donkeys. While it is lower in protein and energy compared to hay and haylage, it provides valuable roughage that aids in maintaining gut motility and digestive health. Typically, straw should be fed in conjunction with other forages to ensure a balanced intake of nutrients.

Haylage

Haylage, partially fermented and preserved at a higher moisture content than dry hay, can be more nutrient-dense and palatable, encouraging higher intake. Donkeys may consume slightly less haylage by weight compared to dry hay due to its higher moisture content. Haylage should have a dry matter content of 40–60% to ensure proper fermentation and preservation of nutrients (De Palo et al. 2018).

Grass

Fresh grass offers a variety of nutrients and can be an excellent component of a donkey's diet, particularly in pasture-based systems, however it should be avoided in overweight donkeys. Donkeys grazing on pasture may consume between 2 and 3 kg of fresh grass per hour, depending on pasture quality, blade height, and donkey size (Smith and Burden 2013). Monitoring grazing time and pasture condition is essential to ensure that donkeys meet their daily intake requirements without overgrazing the land. As reported by (Valle et al. 2018), the use of grass as the main forage, as in extensive farming, increases the fat content and fat-soluble vitamin concentration of milk and alters the fatty acid composition to a more favorable profile for human nutrition.

Forage Analysis

Conducting forage analysis using Near-Infrared spectroscopy (NIRs) or wet chemistry is essential to ensure the nutritional quality of the forage being fed to donkeys, particularly in semi-intensive or production systems. These methods allow for accurate determination of key nutritional components, including CP, crude fiber (CF), neutral detergent fiber (NDF), acid detergent fiber (ADF), and ash content.

- NIRS: A rapid, non-destructive method that uses infrared light to measure the chemical composition of forage. It provides quick results and is useful for routine analysis, allowing for adjustments in feeding strategies based on nutrient content.
- Wet Chemistry: A more traditional method involving chemical extraction and analysis to determine nutrient content. Although more time-consuming and labor-intensive than NIRS, wet chemistry is often used to validate NIRS results and for more precise measurements.

Forage analysis ensures that lactating donkeys receive balanced diets that meet their nutritional requirements, preventing deficiencies and optimizing milk production.

2.1.2 Concentrates: Supplementing Energy

While forage forms the bulk of a lactating donkey's diet, concentrates can be helpful, in specific situations, to meet the additional energy and nutrient demands of lactation or work, especially important when the quality of forage is insufficient to meet their nutritional needs. These concentrated feeds provide readily available energy, protein, vitamins, and minerals.

Concentrates formulated for horses are often used for donkeys due to their similar dietary needs; however, caution is needed with grain-based concentrates, as excessive use can be harmful to a donkey's health.

These concentrates can be broadly categorized into grain-based and fiber-based types:

Grain-Based Concentrates

Grain-based concentrates are energy-dense and provide readily available carbohydrates. The most common grains used in these concentrates include:

- Oats: oats are a popular choice due to their high fiber content and moderate energy level. They are easily digestible and provide a balanced source of carbohydrates.

Oats are less likely to cause digestive upset compared to other grains due to their fiber content.

- Barley: barley offers a higher energy density than oats and is often used to increase the caloric intake for donkeys requiring additional energy. It should be processed (rolled or crimped) to enhance digestibility. Barley is a good compromise between the energy density of corn and the fiber content of oats.
- Corn: corn is one of the most energy-dense grains available, providing a high level of carbohydrates. However, it should be used cautiously, as it can contribute to rapid weight gain and digestive issues if fed in excess. It is typically ground or cracked to improve digestibility.

Grain-based concentrates, if needed, should be introduced gradually and used with caution; they also need to be balanced with adequate fiber intake to prevent digestive upset and metabolic disorders, such as laminitis or colic.

Fiber-Based Concentrates

Fiber-based concentrates provide a safer alternative source of energy and nutrients, limiting the amount of starch supplied to the animals.

- Beet pulp: a highly digestible fiber source that provides energy without excessive starch. Beet pulp is a byproduct of sugar beet processing and is often soaked before feeding to increase palatability and digestibility. It provides a slow-release energy source.
- Soybean hulls: another source of digestible fiber that can supplement the energy needs of donkeys without the high starch content found in grains. Soybean hulls are a byproduct of soybean processing and offer a moderate energy source with a good balance of fiber.
- Wheat bran: a byproduct of milling wheat, which contains a good amount of digestible fiber along with moderate protein content. Wheat bran is particularly useful for balancing the fiber and energy requirements in a donkey's diet. It is also known for its palatability and its role in supporting gut health by promoting regular digestion.

Fiber-based concentrates help maintain gut health and can be used to complement grain-based concentrates or replace them entirely in diets that require lower starch levels.

2.2 Nutritional Requirements for the Maintenance Donkey

2.2.1 Energy

A healthy adult donkey at maintenance requires approximately 87 kJ of digestible energy (DE)/kg BW/day, which translates to approximately 68 kJ of net energy (NE)/kg BW/day.

Maintenance energy requirements need to be met mainly by forage, considering an approximate 1.3–2.1% of BW as dry matter intake (DMI) (Pearson et al. 2006; Burden et al. 2013), the energy requirement can be easily met with the use of a good quality forage source.

However, energy requirements can fluctuate depending on climate, housing, and individual metabolic rate.

For donkeys, energy requirements should indeed primarily be met through forage, with a balance of non-structural carbohydrates (NSC; including sugars and starch) and structural fiber (such as cellulose and lignin). The quality of the forage directly influences the proportion of NSC and fiber a donkey consumes, which is why it's essential to understand the characteristics of different forage types. Here's a breakdown of the most used forages:

- Grass hay: The primary forage available in most stables housing donkeys is meadow hay. It is by far the most commonly used roughage. However, it is essential to perform a proper nutritional analysis of the hay being used, as its nutrient composition—particularly crude protein and NSC content—can vary significantly depending on harvest conditions and botanical composition (Pornaro et al. 2019).
- Straw (oat, barley, or wheat): excellent roughage of choice for donkeys, especially for those prone to being overweight or suffering from metabolic disorders. Its lower energy content and reduced NSC levels make it suitable for managing body condition and metabolic health, while still providing adequate fiber and allowing for sufficient chewing activity to support digestive health and mental well-being.
- Fresh pasture: Although it is a natural forage, fresh pasture needs to be heavily limited for donkeys due to their metabolic predisposition, donkeys—particularly those prone to obesity—can be at risk when grazing on lush, fresh pasture. High-sugar content in young grass can exacerbate metabolic issues and lead to conditions such as laminitis: therefore, pasture intake should be strictly regulated, and in some cases, restricted entirely depending on the animal's individual risk factors.

Increased energy intake can be met by the use of concentrates, preferably fiber-based, and solely used when the metabolic conditions request it (e.g. working donkeys, lactating).

2.2.2 Fat

In maintenance donkeys, dietary fat is generally unnecessary but could be useful for donkeys who are having difficulty gaining weight or keeping on weight. However, for individuals at risk of energy deficiency or during cold weather, small amounts of vegetable oil or stabilized rice bran can be used to enhance energy density without significantly increasing starch intake.

It is important to ensure the intake of essential fatty acids, particularly linoleic acid (omega-6) and alpha-linolenic acid (omega-3), are met, since they play key roles in maintaining overall health, skin health, and metabolic function.

These essential fatty acids cannot be synthesized by the animal and must therefore be supplied through the diet. They are naturally found in fresh grass and in certain vegetable oils—flaxseed oil (linseed oil) is a valuable source of alpha-linolenic acid.

Ensuring the presence of these nutrients in the donkey's diet contributes to a balanced nutritional profile without the need for excessive fat intake.

2.2.3 Protein

Unlike horses and ponies, which mainly eliminate excess nitrogen in the form of urea through the kidneys, donkeys have the capacity to reabsorb urea, thereby allowing them to recycle nitrogen to meet their metabolic needs. Furthermore, donkeys can modulate this recycling process based on variations in dietary protein intake and their own physiological demands for nitrogen.

Thus, maintenance donkeys require lower CP content compared to horses, which average 0.4 g/kg BW of CP (Burden 2012). In practice, this will be generally satisfied, in a healthy mature donkey at maintenance, once energy requirements are met, feeding a diet around 4–8% CP.

Low-protein forages, like straw, will meet this need.

Provisional addition of protein rich feed (eg. soya bean meal, alfalfa) can be used when in need or supporting an injury recovery.

2.2.4 Minerals and Vitamins

The overall mineral and vitamin requirements for donkeys are not yet fully established, and guidelines are primarily based on extrapolation from other equids. Since the diet of a healthy maintenance donkey is typically composed of roughage, it is often beneficial to supplement it with a commercial horse mineral-vitamin balancer.

It is advised that the calcium to phosphorus ratio should be maintained at approximately 1.5/2:1, and copper to zinc at approximately 1:5–7, to ensure optimal absorption and utilization (Martin-Rosset 2018).

Plasma selenium concentrations in donkeys, ranging from 0.02 to 0.14 μg/mL, are significantly lower than those recommended for horses, suggesting that donkeys may have a naturally lower selenium requirement (Bazzano et al. 2019; Yue et al. 2022a, b). Adequate selenium is crucial for preventing oxidative stress, but excessive supplementation should be avoided to prevent toxicity. Precise toxicity levels are not presented for donkeys, but the upper limits established for horses can be considered as a reference (toxic plasmatic concentrations exceeding 1.5 μg/mL, Desta et al. 2011; toxic dietary concentrations <5 mg/kg of dry matter, NRC 2007). When evaluating dietary selenium concentrations, it is important to view these concentrations in relation to the geographical region involved, since broad differences in soil selenium concentrations, and therefore forage selenium concentrations, exist in various parts of the world (Witte and Will 1993; Spadoni et al. 2007). Confirming this, a more recent study conducted in northeast Portugal found a significant association between regional selenium levels and health status in Miranda donkey foals, including limb malformations and survival rates, thus highlighting the critical importance of regional selenium bioavailability for foal and donkey viability (Quaresma et al. 2021).

While specific data for donkeys is scarce, vitamin needs are likely similar to horses. In donkeys, vitamin A to vitamin D ratio is suggested to be maintained at approximately 1:5–10, while plasma vitamin E levels are reported between 3.29 and 12.99 μmol/L (Martin-Rosset 2018; Bazzano et al. 2019). This is particularly true for animals fed on dry forages, since green forages and pastures can increase vitamin intake (Valle et al. 2018).

Vitamin A is obtained from carotenoids in fresh forages, while vitamin D3 is synthesized through sunlight exposure, and vitamin E is found in green, leafy forages. Deficiencies in these vitamins can lead to issues such as night blindness, weakened bones, muscle weakness, and immune dysfunction. Toxic levels, particularly from excessive supplementation, can cause serious health problems: vitamin A toxicity may result in liver damage if intake exceeds 16,000 IU/kg, vitamin D3 toxicity can lead to hypercalcemia above 44 IU/kg, and excessive vitamin E (over 20 IU/kg of DM) may impair blood clotting and vitamin absorption. Crucially, blood testing can easily determine the status of some nutrients, like vitamin E and selenium, allowing for the prompt detection and remediation of deficiencies before symptoms appear, or the accurate clinical follow-up during nutritional support.

2.2.5 Water

Water intake averages 60 ml/kg BW/day (Freeman 2021), but may vary based on temperature, forage type, and physiological status. Donkeys are highly efficient at conserving water but may refuse excessively cold water, impacting feed intake.

2.3 Nutritional Requirements by Physiological Stage

2.3.1 Foals and Growing Donkeys

The intake of colostrum is critical within the first 12 hours of life, and the first feeding is best done within 2–3 hours of delivery.

A health donkey foal should be weaned gradually from about 6 months of age, when it will also start to graze and consume more straw; foals develop quickly in the initial 6 months, increasing ~0.3–0.5 kg/day. It is important to monitor the foals weight gain every 2 weeks to avoid any rapid peaks: if the growth rate is unsteady then the foal may need supplementary feeding or restricting, as appropriate (Stoneham et al. 2017).

Barley straw is the forage of choice for healthy foals with good teeth since it offers low energy content in conjunction with high fiber content, which promotes slow digestion and helps to minimize the incidence of gastric ulcers. In cooler climatic conditions or in the event of a requirement for greater energy input, supplementation of straw with hay is advisable. Too much caloric input should be avoided in young donkeys because this predisposes to musculoskeletal disease (Mendoza et al. 2018).

After weaning, ad libitum straw and a vitamin-mineral balancer that is low in calories should be supplemented until at least 2 years of age, or up to 3 years of age for larger donkey breeds; whether this balancer should remain a consistent component of the diet long-term, in order to prevent nutritional imbalances.

2.3.2 Pregnant Jennies

Energy and protein demands rise notably during the last third of gestation, which can be met by supplying a good quality forage or, if not available, alfalfa-based feeds. Digestible energy allowances should be increased by 11% above maintenance in the ninth month, 13% in the tenth month and 20% in the final month of pregnancy (Smith and Burden 2013).

Selenium and vitamin E are crucial for fetal development.

A study by (Lisuzzo et al. 2022) examined the changes in energy metabolism from late gestation to early lactation in donkeys and found that the inclusion of fats in the diet could help meet the increased energy requirements during this period; the study indicated that a balanced diet containing appropriate levels of fat helps to prevent metabolic diseases such as hyperlipemia, which can occur when energy demands exceed intake.

2.3.3 Lactating Jennies

Lactation increases nutritional needs, particularly during the first 3 months.

The estimated energy requirement to produce 5 l of milk averages 150 kJ of DE/kg BW/day, which translates to approximately 127 kJ of NE/kg BW/day (Raspa et al. 2019).

Additionally, it is important to consider the increased protein requirements during lactation, as adequate protein intake can enhance milk quality. In general, the CP requirement for lactating donkeys is estimated to be around 14% of dry matter intake (Yue et al. 2022a, b).

Special attention should be paid to the use of cereal-based concentrates. Dietary starch intake should not exceed 1 g/kg of BW per day (Hale et al. 2012), and should only be introduced if the animal's BCS is declining. In most cases, good-quality hay and controlled access to pasture are sufficient to meet the nutritional needs of lactating jennies without the need for additional concentrates.

What can truly enhance milk production and quality in lactating donkeys is access to pasture. According to the study by (Valle et al. 2018) diets that include concentrates and meadow hay actually result in lower milk quality compared to diets based solely on hay or fresh grass.

Fresh pasture, in particular, enriches donkey milk with fat-soluble vitamins—especially vitamin E—and contributes to an improved overall nutritional profile. This suggests that, whenever possible, providing access to well-managed pasture can be a valuable strategy for optimizing both the quantity and quality of donkey milk.

2.3.4 Working Donkeys

Working donkeys have increased demand for energy, electrolytes, and water depending on the intensity of their workload. DE intake ought to be adjusted depending on the work required.

In general, a forage-based diet at 2% of body weight is often insufficient to meet the demands of physical work. It is therefore important to closely monitor body condition and to supplement the diet with a balancer that provides additional protein, vitamins, and minerals.

Providing a plain salt lick is also advisable, allowing the donkey to self-regulate its intake. However, for animals engaged in more intense or prolonged physical activity, additional sodium chloride supplementation may be beneficial.

Despite these practical approaches, scientific data on the precise electrolyte requirements of working donkeys remains limited, and further research is needed to establish evidence-based recommendations.

2.3.5 Geriatric Donkeys

Older donkeys may experience dental deterioration, reduced nutrient absorption, and slower metabolism; this can often be solved by providing softer forages (e.g., chopped hay, soaked hay cubes) and highly digestible fiber and adequate protein to maintain muscle mass.

It is crucial to monitor BCS more frequently.

3 Health Implications and Clinical Nutrition of Donkeys

Nutrition is a medical therapy, and rules should be followed regarding indications, contraindications, dose adaptation for nutrients and patient monitoring. At any age and physiological state, donkeys with inadequate nutrient intake become malnourished, with a decrease or halt in milk production, altered metabolism and decreased tissue synthesis and repair (Remillard 2023). On the other hand, overweight donkeys can develop metabolic diseases or have parturition complications. Nutritional imbalances, either deficiencies or toxicities, can lead to various health issues, but these can be mitigated with appropriate dietary interventions.

3.1 Gastric Ulcers

Gastric ulceration is a prevalent yet misdiagnosed condition in donkeys, that typically develops without visible clinical signs. Clinical signs include abdominal pain, loss of appetite and colic. Although post-mortem studies have indicated a prevalence of 41% in donkeys submitted for necropsy (Burden et al. 2009), recent endoscopic examinations in live donkeys suggest even higher incidence rates. In one study, 51.3% of 39 donkeys examined endoscopically showed evidence of equine gastric ulcer syndrome (EGUS) (Sgorbini et al. 2018).

Of these, 95% of lesions were confined to the non-glandular mucosa and classified as equine squamous gastric disease (ESGD), while only 2.6% of affected animals also exhibited equine glandular gastric disease (EGGD).

Interestingly, none of the donkeys displayed clinical signs of gastric ulcers, confirming the often-subclinical nature of EGUS in this species. Additionally, no associations were found between ulcer prevalence and sex, age, or breed.

Feeding management plays a significant role in the development of ulcers. Donkeys have a continuous secretion of stomach acid, and saliva is the predominant buffering agent, which is produced in greater quantities when donkeys are fed long stem forage compared to cereal grain-based rations. Moreover, grains contain starch that, if fed in excess, ferments to form additional acids in the stomach. Therefore, large, infrequent meals of starchy concentrates (starch intake >1–1.5 g/kg BW) help increase gastric acidity while decreasing natural buffering capacity, particularly during periods of fasting when the stomach is empty and unprotected. This physiological rationale is supported by epidemiological evidence: donkeys fed cereal or high-starch rations have a much higher prevalence of gastric ulcers compared with those fed exclusively on forage (56% vs. 34%, respectively), illustrating the central role of diet composition in ulcerogenesis (Burden et al. 2009).

3.2 Small and Large Colon Intestinal Disease

Impaction colic incidence represents almost 50% of the types of colic seen in donkeys (Cox et al. 2007), and the main predisposing factors recognized are dental pathologies, limited pasture and water access, limited exercise and concentrate feeding. Colic secondary to enterolithis are also common in donkeys, specifically in certain geographical areas where there is access to hard water and alfalfa fields or hay (Mendoza et al. 2018).

3.3 Dyslipidemia

Donkeys have evolved to efficiently store and mobilize energy due to their adaptations to harsh environments with limited and poor-quality roughage. This efficiency, however, makes them prone to dyslipidemia, particularly during negative energy balance situations such as pregnancy, lactation, and stress. Dyslipidemia in donkeys, characterized by excessive lipolysis and hepatic triglyceride synthesis, can lead to severe systemic consequences like hyperlipemia, with a prevalence up to 3–5 5% in the general population (McKenzie 2011); donkeys with hyperlipemia have non-specific clinical signs such as dullness and inappetence.

One of the most insidious clinical presentations associated with the early stages of hyperlipaemia is the so-called *"dull donkey"*—a subtle behavioral change that may be the only outward indicator of an underlying pathology. Care must be taken to identify the phenomenon of *sham eating*, in which the donkey appears to engage in normal feeding behavior—approaching food, mouthing, and even chewing—but fails to actually ingest or swallow. In these cases, it is always advisable to monitor daily intake.

Dyslipidemia is prevalent in obese, pregnant, lactating, insulin resistant and older donkeys, and it is associated with high mortality rates in hospitalized animals

(Mendoza et al. 2024). Depression and anorexia are typical signs that develop in later stages of the pathology; diagnosis can be made through the observation of plasma or serum turbidity, with hyperlipidemia indicated by triglyceride (TG) concentrations higher than 500 mg/dL (Mendoza et al. 2024). Healthy donkey TG concentrations average 66.4 ± 34.2 mg/dL, with a range of 23.5–144.0 mg/dL (Burden et al. 2011) or 30–190 mg/dL in U.S. donkeys (Goodrich and Webb 2024). Also, liver and kidney enzymes can be altered.

The frequently stated threshold for hyperlipidemia (TG > 500 mg/dL) should be reexamined since donkeys naturally have higher resting triglyceride concentrations and their lipid metabolism is different from that of horses and ponies (Mendoza et al. 2024).

Proper nutritional management, including the monitoring of energy balance, is crucial to prevent excessive lipolysis and hyperlipemia. Strategies include providing high-calorie foods to stimulate insulin release (e.g. honey, molasses), gradual refeeding of underweight donkeys, and the use of therapeutic interventions such as insulin and parenteral nutrition when necessary (Mendoza et al. 2018).

3.4 Donkey and Asinine Metabolic Syndrome

Donkey or asinine metabolic syndrome (DMS) is the most common endocrine disorder in donkeys, and it is a condition that includes insulin dysregulation (ID), recurrent laminitis, and obesity. These disturbances can occur individually or together.

A link between obesity, insulin resistance, and recurrent laminitis is evident in donkeys, although a direct connection between hyperinsulinemia and laminitis, as seen in horses, has not been established. Gender differences also affect metabolic parameters: jennies show lower insulin sensitivity and gain weight more easily than males.

Diagnosing insulin dysregulation in donkeys involves measuring basal fasting glucose and insulin levels and performing dynamic tests. Factors such as fasting, stress, transport, and exercise can interfere with diagnosis. Insulin levels over 20 μIU/mL are associated with laminitis, although higher cutoffs of 50 μIU/mL are considered for diagnosis of ID (Mendoza et al. 2018).

Proper nutritional management is crucial for donkeys with DMS. Weight-loss diets should consider their efficiency in using poor-quality feed; soaking the hay can be useful in order to lose water soluble sugars (WSC). Commercial diets for horses with equine metabolic syndrome may not be suitable due to excessive calories and protein. Donkeys should have access to an open paddock to exercise, preferably with other equids to reduce stress and encourage activity. Grazing should be avoided especially during the late afternoon (even more in the spring) or after rainfall, and a muzzle could also be useful in certain conditions. Weight-loss should be gradual, aiming for a 2% bodyweight reduction per month (Mendoza et al. 2018).

Pharmacological treatments used in horses may also be applied to donkeys, although specific studies are lacking.

3.5 Donkey Pituitary Paras Intermedia Dysfunction

Pituitary Pars Intermedia Dysfunction (PPID), formerly Cushing's disease, is common in donkeys, especially older animals over 15 years of age. There are no known sex or breed predispositions. Clinical signs in donkeys are akin to those in horses, with hypertrichosis being the most prevalent symptom. However, due to the natural hair length of some breeds, this symptom may be harder to identify. Other common clinical signs include hyperhidrosis (particularly in summer), laminitis, ID, abnormal fat distribution, and muscle wasting leading to a pot-belly appearance (Mendoza et al. 2024). Diagnosis is based on resting plasma ACTH concentrations, which are different from those established for horses. Suggested upper limits for baseline plasma ACTH are 12.5 pg/ml in spring, 53.2 pg/ml in summer, 77.2 pg/ml in fall and 12.8 pg/ml in winter (Mendoza et al. 2024). Humphreys et al. (2022), who documented seasonal variation in plasma ACTH concentrations in healthy donkeys in California, support these findings. Their median values were 12.5 pg/mL in the spring, 53.2 pg/mL in the summer, 77.1 pg/mL in the fall, and 12.6 pg/mL in the winter. September had the highest monthly median ACTH (122.0 pg/mL).

Medical and nutritional therapies must be carefully synchronized for donkeys with PPID.

Nutritional management should follow the guidelines outlined for DMS if overweight, while supporting the animal's condition via low-sugar roughages, if underweight. Medical treatments should be adapted from equine PPID protocols, since specific studies are still lacking for donkeys (Gehlen et al. 2020).

3.6 Liver Disease

Liver disease, or fatty liver disease, is often more common in donkeys than expected due to its insidious nature and the liver's ability to function normally even with significant damage. The disease can range from mild liver compromise to severe hepatic encephalopathy, causing altered brain functions and unusual behavioral and neurological signs (Burden 2012). Secondary photosensitization is common in animals with chronic liver problems, with dermatological lesions often being the first indicators. In donkeys with terminal liver damage, signs include dullness, anorexia, and ataxia, differing from the excitation and head-pressing seen in other equids due to the donkey's typical nature. Treatment for chronic liver disease focuses on supportive therapy and appropriate dietary changes to support liver function without causing additional stress.

A diet rich in fiber and high in branched-chain amino acids is suggested; on the other hand, it is generally advisable to include low amounts of fat in the diet, exceptions made for severely underweight donkeys. Short-chopped or pelleted hay should be used when necessary to accommodate the donkey's dietary needs. It is also important to feed donkeys "little and often," breaking meals into at least four daily feedings. For donkeys suffering from photodermatitis, consider allowing them to graze at night. Protein intake should not be reduced; instead, provide small quantities of highly digestible, good quality protein. However, it is important to avoid feeding large quantities of high-protein alfalfa chops (The Donkey Sanctuary 2020)

3.7 Dental Issues

Dental diseases in donkeys often go unnoticed, as they are not typically used for riding. It is recommended to conduct an annual oral examination to detect and address common dental issues such as diastemata, wave mouth, and displaced teeth. Although dental abnormalities do not strictly require a nutritional solution, they can be alleviated and managed by providing soaked feed in small, frequent meals ("little and often"). Untreated dental problems can lead to complications such as weight loss, colic, and hyperlipemia, which often necessitate clinical nutritional interventions to ensure the donkey's health and well-being (Burden et al. 2013). See chapter "Physical Examination of the Donkey and Their Hybrid Crosses" for additional information.

3.8 Refeeding Syndrome

Refeeding syndrome is a common concern in donkeys that have experienced significant weight loss and prolonged anorexia, often due to underlying medical conditions. When feeding a donkey recovering from this condition, it is crucial to proceed very gradually, regardless of the animal's apparent appetite.

Close monitoring of electrolyte levels through targeted biochemical tests is essential, as imbalances are common and must be corrected through appropriate medical therapy. Dietary management must follow the principle of "Start Slowly and Go Slowly": begin cautiously and increase feed intake gradually. During the first 2–3 days, only about 30% of the animal's nutritional requirements should be provided. If biochemical parameters remain stable, intake can be gradually increased to approximately 50% of the requirements by day seven or eight. The diet should be based primarily on high-quality forage. Meadow hay may be suitable, and if the donkey shows strong appetite, small amounts of fresh grass may also be offered—provided that actual intake is carefully monitored. Once the critical phase is over, the diet can be progressively adjusted to meet 100% of the animal's nutritional needs. Weight recovery should occur through a forage-based diet with an adequate supply of protein, vitamins, and minerals to support safe and effective rehabilitation (Bookbinder and Schott 2021).

3.9 Laminitis

Nutritional laminitis can be induced by excessive dietary starch and sugar, often concurrent with other metabolic issues, as previously outlined. Additional information about diagnosis and management of laminitis can be found in Chapters 23, 24 and 25. While energy requirements must be met, NSC must be less than 10% of the daily ration. A straw-based diet can be implemented or consider soaking the hay to lose WSC. Oil supplementation can effectively provide energy, but protein, minerals, and vitamins must also be carefully balanced to avoid deficiencies, which can be done with a low caloric dietary balancer.

Oils high in omega-3 fatty acids are preferred for their potential anti-inflammatory and insulin-sensitizing properties (Van Weyenberg et al. 2009).

4 Conclusions

Donkeys' distinct physiological and metabolic traits call for a species-specific approach to nutritional management. Our knowledge of feeding practices has been shaped by decades of equine-based research, but extrapolating results from horses to donkeys is still inappropriate and may even be hazardous. Their *easy-keeping* nature, while advantageous in resource-poor settings, becomes a liability under modern management systems—particularly when donkeys are fed high-energy diets designed for horses or receive inappropriate supplementation. Donkeys differ greatly from horses and ponies in their endocrinological patterns, energy efficiency, mineral metabolism, and nutrient requirements. These misunderstandings can lead to poor management, which puts donkeys at risk for a number of dangerous clinical conditions.

Even though these distinctions are becoming more widely acknowledged, there are still a number of important gaps in our scientific understanding. Promoting the health, welfare, and longevity of a species that continues to be vital to both subsistence economies and the companion animal industries globally requires advancement in this field.

Questions

When evaluating the nutritional status of a donkey, which of the following diagnostic tools is appropriate and practical for assessing selenium and vitamin E levels?

A. Liver biopsy
B. Blood plasma analysis
C. Urinalysis
D. Skin fold thickness measurement

Answers: B

Which feeding strategy is most appropriate for a healthy adult donkey at maintenance under non-working conditions?

A. A high-energy concentrate feed twice daily and limited forage
B. Alfalfa hay ad libitum and a vitamin–mineral supplement
C. Low-calorie, high-fiber forage (e.g., straw) ad libitum and a low-energy balancer
D. A diet formulated for ponies, adjusted for body weight

Answers: C

What is a major risk of treating donkeys as small horses in nutritional management?

A. Overestimating their need for high-protein feeds
B. Overfeeding energy and inducing metabolic disorders
C. Under-supplying fiber and increasing protein digestibility
D. Providing insufficient vitamin C, which donkeys require in higher amounts

Answers: B

What is a key welfare concern that arises from mismanaging donkey nutrition in domestic environments?

A. Rapid muscle development
B. Increased risk of obesity and related metabolic diseases
C. Inadequate protein synthesis due to high forage intake
D. Heightened aggression due to nutrient excess

Answers: B

References

Bambi G, Rossi G, Barbari M (2018) Comparison between different types of bedding materials for horses. Agron Res 16(3):646. https://doi.org/10.15159/AR.18.124

Bazzano M, McLean A, Tesei B, Gallina E, Laus F (2019) Selenium and vitamin E concentrations in a healthy donkey population in Central Italy. J Equine Vet Sci 78:112–116. https://doi.org/10.1016/J.JEVS.2019.04.003

Becvarova I, Pleasant RS, Thatcher CD (2009) Clinical assessment of nutritional status and feeding programs in horses. Vet Clin North Am Equine Pract 25(1):1–21. https://doi.org/10.1016/J.CVEQ.2009.01.001

Bookbinder L, Schott HC (2021) Refeeding syndrome in a miniature donkey. J Vet Emerg Crit Care (San Antonio) 31(5):668–673. https://doi.org/10.1111/VEC.13074

Burden F (2012) Practical feeding and condition scoring for donkeys and mules. Equine Vet Educ 24(11):589–596. https://doi.org/10.1111/j.2042-3292.2011.00314.x

Burden F, Du Toit N, Thiemann A (2013) Nutrition and dental care of donkeys. In Pract 35(7): 405–410. https://doi.org/10.1136/inp.f4367

Burden FA, Gallagher J, Thiemann AK, Trawford AF (2009) Necropsy survey of gastric ulcers in a population of aged donkeys: prevalence, lesion description and risk factors. Animal 3(2): 287–293. https://doi.org/10.1017/S1751731108003480

Burden FA, Du Toit N, Hazell-Smith E, Trawford AF (2011) Hyperlipemia in a population of aged donkeys: description, prevalence, and potential risk factors. J Vet Intern Med 25(6):1420–1425. https://doi.org/10.1111/J.1939-1676.2011.00798.X

Cox R, Proudman CJ, Trawford AF, Burden F, Pinchbeck GL (2007) Epidemiology of impaction colic in donkeys in the UK. BMC Vet Res 3(1):1–9. https://doi.org/10.1186/1746-6148-3-1/TABLES/7

Desta B, Maldonado G, Reid H, Puschner B, Maxwell J, Agasan A, Humphreys L, Holt T (2011) Acute selenium toxicosis in polo ponies. J Vet Diagn Invest 23(3):623–628. https://doi.org/10.1177/1040638711404142/ASSET/B7FBEF77-8345-422A-A02A-BC341D1B197C/ASSETS/IMAGES/LARGE/10.1177_1040638711404142-FIG3.JPG

Freeman DE (2021) Effect of feed intake on water consumption in horses: relevance to maintenance fluid therapy. Front Vet Sci 8:626081. https://doi.org/10.3389/FVETS.2021.626081

Gao R, Shi L, Guo W, Xu Y, Jin X, Yan S, Shi B (2022) Effects of housing and management systems on the growth, immunity, antioxidation, and related physiological and biochemical indicators of donkeys in cold weather. Animals 12(18). https://doi.org/10.3390/ani12182405

Gehlen H, Schwarz B, Bartmann C, Gernhardt J, Stöckle SD (2020) Pituitary pars intermedia dysfunction and metabolic syndrome in donkeys. Animals 10(12):1–16. https://doi.org/10.3390/ANI10122335

Goodrich EL, Webb JL (2024) Complete blood count and biochemistry reference intervals for healthy adult donkeys in the United States. Animals 14:2018. https://doi.org/10.3390/ANI14142018

Hale C, Warren H, Hemmings A (2012) The fermentation of hay and starch when incubated in vitro with faecal inoccula from either normal healthy horses or horses with a history of laminitis. EAAP Sci Series 132(1):357–362. https://doi.org/10.3920/978-90-8686-755-4_48

Humphreys S, Kass PH, Magdesian KG, Goodrich E, Berryhill E (2022) Seasonal variation of endogenous adrenocorticotropic hormone concentrations in healthy non-geriatric donkeys in Northern California. Front Vet Sci 9. https://doi.org/10.3389/FVETS.2022.981920

Lisuzzo A, Bonelli F, Sgorbini M, Nocera I, Cento G, Mazzotta E, Turini L, Martini M, Salari F, Morgante M et al (2022) Differences of the plasma total lipid fraction from pre-foaling to post-foaling period in donkeys. Animals 12(3). https://doi.org/10.3390/ani12030304

Martin-Rosset W, Martin L (2018) Chapter 1. Nutritional principles for horses. In: Equine nutrition, pp 23–96. https://doi.org/10.3920/978-90-8686-855-1_1

McKenzie HC (2011) Equine hyperlipidemias. Vet Clin North Am Equine Pract 27(1):59–72. https://doi.org/10.1016/J.CVEQ.2010.12.008

Mendoza FJ, Toribio RE, Perez-Ecija A (2018) Donkey internal medicine—part I: metabolic, endocrine, and alimentary tract disturbances. J Equine Vet Sci 65:66–74. https://doi.org/10.1016/J.JEVS.2018.02.001

Mendoza FJ, Toribio RE, Perez-Ecija A (2024) Metabolic and endocrine insights in donkeys. Animals (Basel) 14(4):590. https://doi.org/10.3390/ANI14040590

Mueller PJ, Protos P, Houpt KA, Van Soest PJ (1998) Chewing behaviour in the domestic donkey (Equus asinus) fed fibrous forage. Appl Anim Behav Sci 60(2–3):241–251. https://doi.org/10.1016/S0168-1591(98)00171-3

National Research Council (2007) Nutrient Requirements of Horses. 6th rev. edn. National Academies Press, Washington, DC. https://doi.org/10.17226/11653

Osthaus B, Proops L, Long S, Bell N, Hayday K, Burden F (2017) Talk TITLE Weathering the weather: effects of the environment on donkey, mule and horse welfare

De Palo P, Maggiolino A, Albenzio M, Casalino E, Neglia G, Centoducati G, Tateo A (2018) Survey of biochemical and oxidative profile in donkey foals suckled with one natural and one semi-artificial technique. PLoS One 13(6):e0198774. https://doi.org/10.1371/JOURNAL.PONE.0198774

Pearson RA, Archibald RF, Muirhead RH (2006) A comparison of the effect of forage type and level of feeding on the digestibility and gastrointestinal mean retention time of dry forages given to cattle, sheep, ponies and donkeys. Br J Nutr 95(1):88–98. https://doi.org/10.1079/BJN20051617

Pearson RA, Ouassat M (2000) A guide to live weight estimation and body condition scoring of donkeys. Centre for Tropical Veterinary Medicine, University of Edinburgh

Pornaro C, Basso E, Macolino S (2019) Pasture botanical composition and forage quality at farm scale: a case study. Ital J Agron 14(4):1480. https://doi.org/10.4081/IJA.2019.1480

Quaresma M, Marín C, Bacellar D, Nóvoa M, Navas FJ, McLean A (2021) Selenium and vitamin e concentrations in miranda jennies and foals (Equus asinus) in Northeast Portugal. Animals 11: 1772. https://doi.org/10.3390/ANI11061772/S1

Raspa F, Cavallarin L, McLean AK, Bergero D, Valle E (2019) A review of the appropriate nutritionwelfare criteria of dairy donkeys: nutritional requirements, farm management requirements and animal-based indicators. Animals 9(6). https://doi.org/10.3390/ani9060315

Remillard RL (2023) Equine clinical nutrition, vol 112, 2nd edn, p 550

Rodrigues JB, Lilly G (2019) Dental disorders of donkeys. Vet Clin North Am Equine Pract 35(3): 529–544. https://doi.org/10.1016/J.CVEQ.2019.08.008

Rudman R (1998) The social organisation of feral donkeys (Equus asinus) on a small Caribbean Island (St. John, US Virgin Islands). Appl Anim Behav Sci 60(2–3):211–228. https://doi.org/10.1016/S0168-1591(98)00165-8

Salari F, Licitra R, Altomonte I, Martini M (2020) Donkey feeding during maintenance, pregnancy, and lactation: effects on body weight, milk production, and foal growth. J Equine Vet Sci 91. https://doi.org/10.1016/j.jevs.2020.103131

Sgorbini M, Bonelli F, Papini R, Busechian S, Briganti A, Laus F, Faillace V, Zappulla F, Rizk A, Rueca F (2018) Equine gastric ulcer syndrome in adult donkeys: investigation on prevalence, anatomical distribution, and severity. Equine Vet Educ 30(4):206–210. https://doi.org/10.1111/EVE.12747

Smith DG, Burden FA (2013) Practical donkey and mule nutrition. In: Equine applied and clinical nutrition: health, welfare and performance, pp 304–316. https://doi.org/10.1016/B978-0-7020-3422-0.00016-X

Spadoni M, Voltaggio M, Carcea M, Coni E, Raggi A, Cubadda F (2007) Bioaccessible selenium in Italian agricultural soils: comparison of the biogeochemical approach with a regression model based on geochemical and pedoclimatic variables. Sci Total Environ 376(1–3):160–177. https://doi.org/10.1016/J.SCITOTENV.2007.01.066

Stoneham SJ, Morresey P, Ousey J (2017) Nutritional management and practical feeding of the orphan foal. Equine Vet Educ 29(3):165–173. https://doi.org/10.1111/EVE.12546

Valle E, Pozzo L, Giribaldi M, Bergero D, Gennero MS, Dezzutto D, McLean A, Borreani G, Coppa M, Cavallarin L (2018) Effect of farming system on donkey milk composition. J Sci Food Agric 98(7):2801–2808. https://doi.org/10.1002/JSFA.8777

Valle E, Raspa F, Giribaldi M, Barbero R, Bergagna S, Antoniazzi S, McLean AK, Minero M, Cavallarin L (2017) A functional approach to the body condition assessment of lactating donkeys as a tool for welfare evaluation. PeerJ 2017(3):e3001. https://doi.org/10.7717/PEERJ.3001/SUPP-1

Van Weyenberg S, Hesta M, Kalmar ID, Vandermeiren J, Janssens GPJ (2009) Nutritional management of laminitis in a horse. Vet Rec 164(22):694–695. https://doi.org/10.1136/VR.164.22.694

Witte ST, Will LA (1993) Investigation of selenium sources associated with chronic selenosis in horses of western Iowa. J Vet Diagn Invest 5(1):128–131. https://doi.org/10.1177/104063879300500133

Yue Y, Li L, Tong M, Li S, Zhao Y, Guo X, Guo Y, Shi B, Yan S (2022a) Effect of varying dietary crude protein level on milk production, nutrient digestibility, and serum metabolites by lactating donkeys. Animals 12(16):2066. https://doi.org/10.3390/ANI12162066

Yue Y, Yan S, Li L, Tong M, Shi B, Zhao Y, Guo X, Guo Y (2022b) Comparative analysis of haematological, biochemical and nutritional parameters of Dezhou donkey with healthy and weak foals (Equus asinus). Ital J Anim Sci 21(1):1140–1150. https://doi.org/10.1080/1828051X.2022.2097454

The Donkey Sanctuary (2020) Feeding donkeys with liver disease. https://www.thedonkeysanctuary.org.uk/forowners/owners-resources/feeding-donkeys-with-liver-disease

Mule Nutrition and Considerations

Elaine Norton

Abstract

Although published guidelines exist for horse and donkey nutrition, specific guidelines for mules are lacking, and their nutritional needs remain poorly studied. The limited published information on mule nutrition is based on author experience and recommends feeding them like donkeys, with a diet high in fiber and low in energy (Smith DG, Burden FA, Chapter 16: Practical donkey and mule nutrition. In: Geor RJ, Harris PA, Coenen M (eds) Equine applied and clinical nutrition. W.B. Saunders, pp 304–316. https://doi.org/10.1016/B978-0-7020-3422-0.00016-X, 2013; Burden FA, Bell N, Vet Clin N Am Equine Pract 35: 469–479, 2019; The Clinical Companion of the Donkey. The Donkey Sanctuary, Leicester, 2021). While this approach is appropriate for some mules, it may oversimplify others metabolic and nutritional needs by overlooking the influence of environmental and maternal factors. Further, as with horses and donkeys, the development of a nutrition program for mules should account for individual energy needs based on growth, level of work, and metabolic status. For instance, growing mules or those engaged in intense work need more energy, protein, and water than mature, companion, or lightly worked mules.

Keywords

Mule nutrition · Weight tape · Body condition score (BCS) · Metabolic syndrome · Obesity · Cresty neck score (CNS) · Obesity · Non-structural carbohydrates (NSC)

E. Norton (✉)
College of Veterinary Medicine, Mississippi State University, Starkville, MS, USA
e-mail: emn149@msstate.edu

A. K. McLean et al. (eds.), *Donkey and Mule Medicine: Foundations, Breeds, and Management*, https://doi.org/10.1007/978-3-032-12726-6_20

1 Metabolic Efficiency and Digestive Adaptations in Mules

Mules exhibit greater metabolic efficiency compared to horses, a trait which has been attributed to hybrid vigor and inherited from the donkey. Donkeys have adapted to living in semi-arid climates and feeding on sparse poor quality forage while performing high work-loads (The Clinical Companion of the Donkey 2021; Starkey and Starkey 2004). This adaptation is predominantly due to their longer gastrointestinal retention time, allowing for higher digestibility of dry matter by increasing enzymatic digestion in the small intestine and more extensive fiber fermentation and volatile fatty acid production in the hindgut (Pearson and Merritt 1991; Suhartanto et al. 1992; Faurie and Tisserand 1994; Tisserand et al. 1991). This is supported by studies which demonstrated that when fed restricted or *ad libitum* amounts of forage, donkeys retained feed longer and had higher fiber digestibility compared to horses and ponies (Pearson and Merritt 1991; Suhartanto et al. 1992; Pearson et al. 2001). The differences in apparent digestibility of dry matter were notably higher when high-fiber forages were offered *ad libitum*, which was attributed to the fact that donkeys are browsers and will selectively consume the most digestible portions of feed (Pearson et al. 2001). Another key adaptation is the unique ability for donkeys to recycle nitrogen from urea, a process which they can up or down-regulate to conserve nitrogen and maintain digestible energy intakes when fed diets low in crude protein (Pearson et al. 2001; Izraely et al. 1989).

Mules have been found to have similar gastrointestinal transit times and fiber digestibility as donkeys (Bala et al. 2017), and are presumed to have the same ability for urea recycling. Further, although mules are reported to both graze and browse (McLean et al. 2019), they tend to favor browsing behavior (Personal Communication 2025). As a result, mules can efficiently select and extract digestible energy from coarser fiber feeds and obtain more energy from the same quantity of feed compared to horses despite being a similar size. This is supported by mules which are able to maintain weight on a lower-quality forage in developing countries where feed resources are limited (Bala et al. 2017; McLean et al. 2019; Mota-Rojas et al. 2021). Further, sedentary mules or those engaged in low levels of exercise are at a higher risk of obesity when fed the same quantity of feed as horses or when consuming diets high in non-structural carbohydrates (NSC). This has led to the current dietary recommendations of feeding mules similar to donkeys, however, this approach may not work for growing mules or those with higher energy needs and the nutritional program for each mule should be approached individually (Smith and Burden 2013; Burden and Bell 2019; The Clinical Companion of the Donkey 2021; Burden 2012).

2 Metabolic Considerations and EMS Risk in Mules

Due to their enhanced metabolic efficiency, mules are considered a high-risk group for equine metabolic syndrome (EMS), which is a combination of metabolic derangements including insulin dysregulation, dyslipidemia, obesity, and regional adiposity (Fig. 1) (Durham et al. 2019).

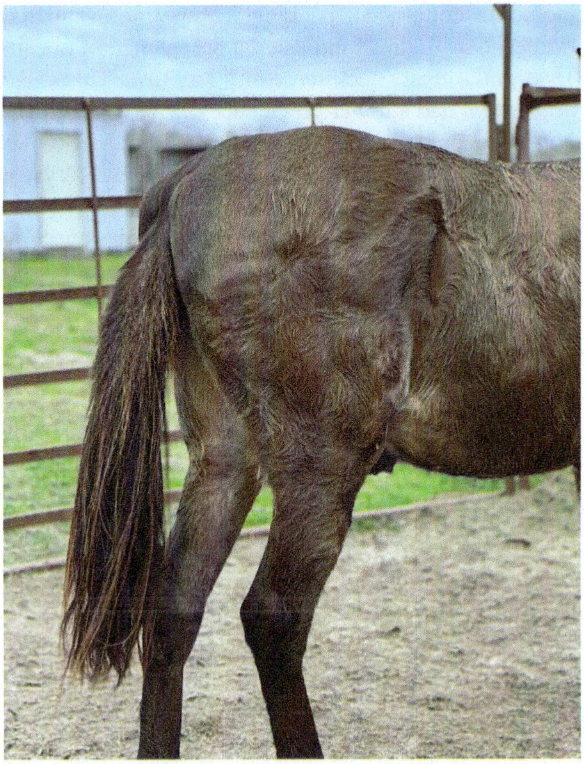

Fig. 1 Mule with metabolic syndrome with regional adiposity over the hind end. (Photo credit: Lost Meadows Mule Refuge, Snook, Texas)

Due to prolonged levels of hyperinsulinemia, EMS can lead to hyperinsulinemia-associated laminitis, the leading cause of laminitis in equids (Wylie et al. 2011; Karikoski et al. 2011). Equine metabolic syndrome is of particular concern in the modern mule, as they are increasingly transitioning from working animals to leisure and recreational use in industrial countries and are often fed higher quality feeds rich in NSC.

Although initial literature on EMS has been based on horses, it has been shown that there are inherent differences in metabolic profiles between horses and donkeys. This has led to the development of the term asinine metabolic syndrome, or donkey metabolic syndrome (further discussed in this text, in the donkey nutrition and donkey endocrinology chapters), to emphasize species differences, as well as donkey-specific reference intervals and management recommendations (Gehlen et al. 2020a; du Toit et al. 2011; Humphreys et al. 2022; Gehlen et al. 2020b). Unfortunately, this same progress has not been made in mules and it is currently unknown how common EMS is in the mule population and which parent the mule more closely resembles in terms of metabolic profiles.

To assess the prevalence of EMS risk factors in mules, my lab recently conducted an epidemiological study using a population of 159 mules and 31 control ranch horses (Johnston et al. 2025). The mules in the study represented a range of uses, including endurance, packing, competition, and recreation (e.g., trail or pleasure riding). Our findings indicated a prevalence of obesity (body condition score ≥ 8) at 15%, and a cresty neck score of ≥3 at 32%. This suggests that, although mules in this population had a lower prevalence of obesity compared to what has been reported in horse breeds considered at high risk for EMS (Giles et al. 2014; Kosolofski et al. 2017; Potter et al. 2016; Thatcher et al. 2012; Wyse et al. 2008), they have a similar prevalence of regional adiposity (Giles et al. 2015).

We also found that serum triglyceride concentrations were higher in mules compared to the control horses, suggesting innate differences in fatty acid metabolism between mules and their horse parent. Additionally, non-esterified fatty acids were found to be higher in mules used for recreation compared to those used for packing or competition indicating a significant influence of environment and management for lipid metabolism in these groups.

We also found that draft mules had higher insulin levels post oral sugar challenge using 0.15 ml/kg light Karo® syrup compared to Thoroughbred or Quarter horse mules. This finding aligns with research in horses, where certain breeds have more exaggerated responses to an oral sugar challenge and show a greater predisposition to EMS (Bamford et al. 2014; Norton et al. 2019), and suggest that maternal lineage may influence both nutritional requirements and EMS risk in mules. Given that many mules in our cohort had unknown pedigrees, further investigation in a larger population is needed to confirm these results.

Therefore, although EMS is a significant disease in mules, further research is needed to determine the population prevalence across working groups and mule breeds. Importantly, there are both lean unhealthy metabolic phenotypes and healthy obese metabolic phenotypes, indicating that a diagnosis of metabolic syndrome cannot be made solely off body condition score. Therefore, mule-specific reference ranges and diagnostic cut-offs are required for accurate diagnosis. Nevertheless, monitoring fasting insulin concentrations and postprandial insulin responses can still provide valuable insights into how a mule is responding to its current dietary management.

3 Current Nutritional Recommendations for Mules

When addressing the lack of current knowledge on mule nutrition, it can be challenging to determine the best starting point for developing a feeding program. However, general guidelines can help establish an initial feeding plan.

Unless medically contraindicated, fibrous forage should form the primary basis of a mule's diet at 1.25–2% body weight (as-fed) per day. Therefore, the total percentage of forage fed can be increased or decreased as needed to maintain an ideal body condition, but a veterinarian or nutritionist should be consulted if feeding below 1.25% body weight. When creating a nutritional plan, a forage analysis is essential as NSC and fiber content vary based on geography, cut and maturity. This is particularly important when adding straw into the diet as some straws can have NSC equivalent or higher than mature grass hay. The portion of hay to straw incorporated in the diet will depend on forage availability, nutrient composition, weather conditions, and the mule's digestible energy requirements.

Companion or Recreation Mules For mules at maintenance, performing low levels of work, or those considered easy-keepers, nutritional guidelines for donkeys are generally more appropriate. In these cases, a high-fiber, low non-structural carbohydrate (NSC $\leq$ 13%) forage, such as mature grass hay, should be offered at a starting rate of 1.5% of body weight (as-fed) per day to help prevent excessive weight gain and reduce the risk of metabolic disturbances. Grazing should also be limited or restricted due to the propensity for these mules to become obese on native or improved pastures. Strip grazing has been suggested as a way of limiting grazing while still allowing mules to have access to turnout and herd socialization.

Working Mules For mules at a higher level of work, such as packing, endurance riding, racing, and some competition mules, a more nutrient dense, higher quality hay should be provided to meet energy demands. Concentrate feeds may be incorporated into a mule's diet to support higher energy needs, but they should be introduced gradually, closely monitored, and fed in small more frequent meals to avoid an excessive postprandial insulin response. Alternatively, fat sources may be utilized to provide additional calories without added NSC but body condition and serum triglyceride concentrations should be monitored.

Growing Mules Mules tend to develop more slowly than horses, with their most significant growth phase occurring between 3 and 4 years of age. However, they may not reach full physical and musculoskeletal maturity until they are 7–9 years old. Throughout this extended developmental period, young mules should receive adequate energy to support steady, controlled growth, while also meeting the additional caloric demands of training or work. Juvenile mules should have free-choice access to high-quality forage and be supplemented with appropriate feed concentrates during periods of rapid growth or increased energy demand, such as cold weather or weaning. Inadequate energy intake can lead to stunted development, whereas excessive energy can result in accelerated growth. In foals under 1 year of age, rapid growth is particularly concerning, as it increases the risk of developmental orthopedic diseases such as osteochondrosis dissecans (OCD). These risks may be further compounded by imbalances in the calcium-to-phosphorus ratio, which is critical for proper bone development and mineralization. Notably, there are no published feeding guidelines specific to growing mules, and research is needed to establish evidence-based nutritional recommendations for each stage of their development.

Obese and Metabolic Mules For mules that are obese, diagnosed with metabolic syndrome, or have a history of laminitis, forage with a <10% NSC content should be fed. If forage exceeds this threshold, it should be soaked in water for 1 h to reduce NSC levels, then thoroughly drained before feeding. Since soaking can leach essential micronutrients, a low-NSC ration balancer should be added to the diet to ensure nutritional adequacy. The recommended quantity for the ration balancer should be divided into two to three meals to minimize post-prandial insulin spikes. Additionally, the NSC content of ration balancers should be carefully reviewed, as this value can vary significantly amongst products labeled as low-NSC.

Grazing should also be restricted as fluctuations in pasture NSC content can trigger episodes of hyperinsulinemia-associated laminitis. For mules where dry lots are not available for turnout, a grazing muzzle can be utilized although many mules become adept at removing them. If the mule is physically able, regular exercise should also be incorporated in their management regimen to help improve insulin sensitivity.

For obese mules, a combination of diet and exercise are essential to promote adequate weight loss. Weight loss programs should target a 2–4% reduction in body weight per month. A nutritional analysis should be performed as many obese mules are being fed according to nutritional guidelines for horses. In these cases, removal of grain, concentrates, and excessive calories, and switching to a lower quality hay, may be sufficient to promote weight loss. For mules that are on an appropriate forage but still obese, forage intake can be reduced up to 1.25% of their current weight or 1.5% of their ideal weight. I typically start with reductions in increments of 0.15–0.2% and monitor serum triglyceride levels to ensure the mule is not shifting into negative energy balance. For example, a mule being fed 1.5% of body weight would be initially reduced to 1.3–1.35% of body weight for the first 2–4 weeks and then reassessed. This group is also more likely to benefit from including straw in their diet to provide additional satiety and to prevent potential behavioral vices due to boredom and lack of fiber intake, starting at 25% of their total forage and increasing as needed. However, a feed analysis should be performed to ensure the targeted digestible energy and NSC are being met. For mules that are resistant to weight loss, total forage can be reduced to <1.25% of body weight; however, a nutritionist or veterinarian should be consulted prior to this reduction to ensure the mule is obtaining balanced nutrition.

Although mules typically require less protein than horses, mules diagnosed with pituitary pars intermedia dysfunction (PPID; previously named equine Cushing's disease) may require additional protein to support the restoration of atrophied muscle. This need can typically be met by incorporating a low-NSC ration balancer that provides higher levels of essential amino acid, such as leucine, which is the first limiting amino acid in equid diets. Further, since PPID is a disease of aged equids, many of these mules will have a natural age-associated insulin dysregulation and may require forage at <10% NSC, similar to mules with EMS, to maintain a healthy body condition score.

Underweight Mules For mules that need to gain weight, a clinical assessment should be performed by a veterinarian to rule out underlying medical or dental issues that may contribute to weight loss. If no primary medical concerns are identified, a nutritional analysis should be performed to guide dietary adjustments. As an initial step, I will estimate the amount of weight the mule needs to gain and assess its current workload. I then estimate the mule's digestible energy requirements starting at 70–80% of the values published for horses in the *National Research Council's Nutrient Requirements of Horses* (Council, N. R 2007). While this estimate provides a practical starting point, reflecting the mule's lower basal metabolic rate compared to horses, it may need to be adjusted based on individual response and ongoing body condition monitoring. Analysis of available feed should also be performed as supplementing or switching to a higher quality hay may be the only modification required to promote weight gain.

To meet nutritional requirements, the percentage of forage can be increased up to 2.5% of current body weight on a dry matter basis, making changes in increments of 0.15–0.2%. When fed ad libitum, mules have been observed to consume up to 1.9–2.5% of their bodyweight (Bala et al. 2017). However, some owners report that mules may not consistently consume these higher percentages of forage, especially when fed a higher fiber forage. Therefore, supplementing a higher quality or more energy dense hay can be utilized to provide the same energy at a reduced percentage of forage.

Concentrate feeds can also be added to the mule's diet to provide additional energy, but they should be introduced gradually, closely monitored, and fed in small more frequent meals to avoid an excessive postprandial insulin response. Alternatively, fat sources may be utilized to provide additional calories without added NSC but body condition and serum triglyceride levels should be monitored.

In cases of starvation or emaciation (BCS of 1/5 or <3/9), it is critical to involve a veterinarian to develop a carefully controlled refeeding plan. Rapid reintroduction of calories, particularly carbohydrates, can trigger refeeding syndrome, a potentially fatal condition characterized by electrolyte imbalances, fluid shifts, and multi-organ dysfunction. Mules, like other equids, are susceptible to this syndrome if a controlled, restricted feeding regimen is not implemented (Schneider et al. 2021).

Vitamins and Minerals While mules are generally more metabolically efficient than horses, we assume they have similar requirements for vitamins and minerals, but the author acknowledges that there are no studies available for mules on vitamin and mineral needs. For mules on a hay or straw-based diet, supplementation with a high-quality, low NSC ration balancer is likely necessary to ensure the diet meets all trace minerals. Mules should also be provided with loose table salt (2 tablespoons per day for 454 kg equid) as well as free choice access to a salt block.

Monitoring Close monitoring of body condition is key to ensuring a mule is on the appropriate level of nutrition. Notably, as with horses and donkeys, often mules will retain deposits of adiposity, particularly above the nuchal ligament, despite weight loss, and the entire body should be considered when assessing body condition. Separate body condition score systems are available for horses and donkeys; however, a mule specific score has not been established (The Clinical Companion of the Donkey 2021; Burden 2012; Henneke et al. 1983). In general, mules tend to fall in between the horse and donkey, where the rump resembles the horse but the neck and barrel resembles the donkey (Burden 2012). Therefore, these differences must be

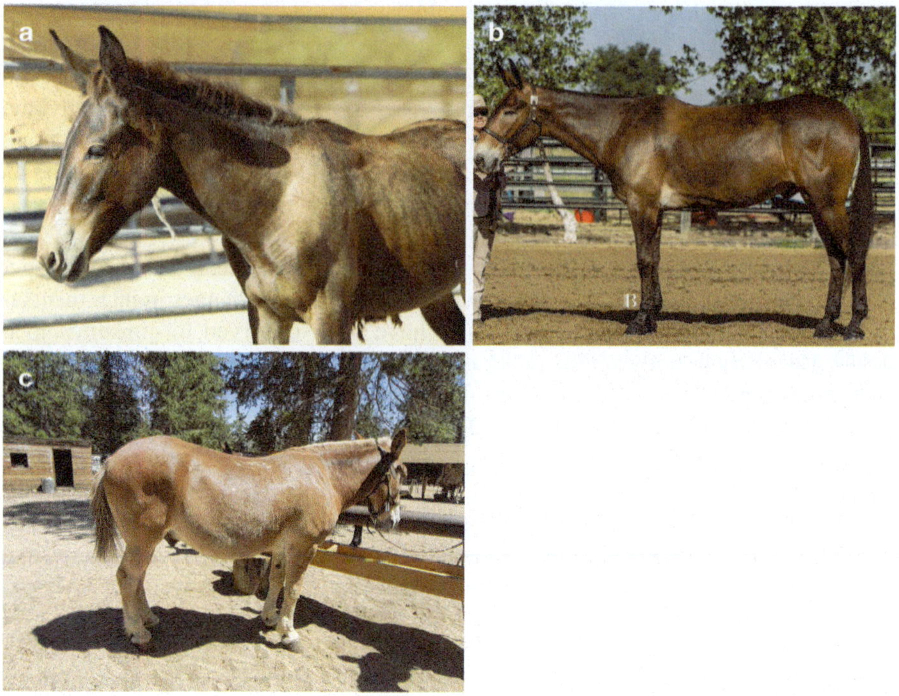

Fig. 2 Examples of body condition scores (BCS) in mules using a modified 9-point scale. (**a**) Example of a mule with a BCS of 3. The mule's ribs and the spinous process are visible and there is minimal fat coverage over the shoulder, points of the hips, and at the base of the tail head. (**b**) Examples of a mule with a BCS of 5. The mule's ribs are not visible but can be palpated along the thoracic. The base of the tail head is fleshy and there is also adequate fat coverage over the shoulder and points of the hips, but regional adiposity is not present. (**c**) Example of a mule with a BCS of 8. The mule's ribs are not visible and are not easily palpated across the thorax. There are also noticeable regions of adiposity over the shoulder, back, hind end, points of the pelvis, and in the sheath

accounted for when body condition scoring a mule and consistency is key to follow-up monitoring. I score based on a 9-point scale that combines the Henneke body condition scoring system for horses (Henneke et al. 1983) with a modified approach based on the donkey scale (Burden 2012)—scoring the rump using equine standards and assessing the neck and barrel using donkey standards (Fig. 2a–c).

More objective measurements of body mass in mules can be obtained using horse weight tapes (Fig. 3) or predictive equations based on morphometric measurements such as height, length, and neck and girth circumference.

However, most of these methods were originally developed for horses and may not accurately reflect mule body weight. As a result, they can either overestimate or underestimate a mule's true weight, and studies differ in their conclusions about which method is most accurate for use in mules (Noikhrua et al. 2016; Andersen

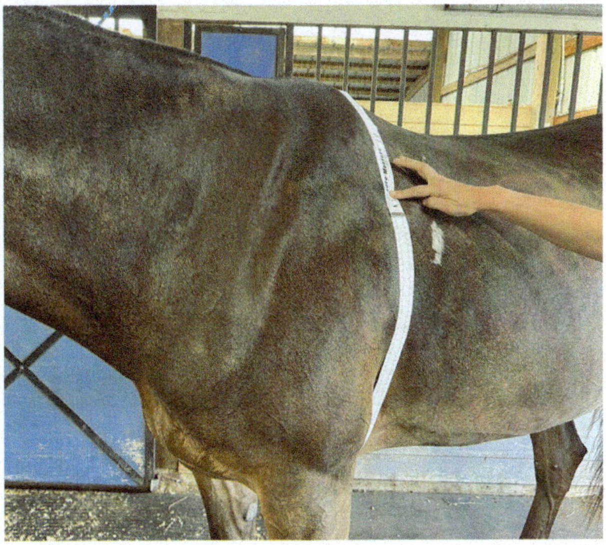

Fig. 3 Placement of an equine weight tape for estimating body weight in mules. The mule should be standing square and on a flat surface. The weight tape should be started at the zero end and then carefully draped over the withers and behind the elbows until both ends of the weight tape meet, completing a circle around the girth. Ensure the weight tape lies flat and is snug but not tight. The estimated body weight is read at the point where the zero end intersects the numbered side of the tape

et al. 1996). One predictive equation was developed specifically using mule data (Kay et al. 2004) where:

$$\text{mule live body weight (kg)} = -33 + 2 - 8 * \text{heart girth (cm)} + 1 - 36 * \text{length (cm)}$$

An example of landmarks for measurements of heart girth and length are presented in Fig. 4.

While this equation may offer improved accuracy, it may not be practical for routine use by owners. Therefore, for monitoring purposes, a consistent method that is easily implemented by owners can still be valuable for tracking trends in body weight and neck circumference over time and identifying whether a mule is gaining or losing weight. Photographs taken from the side and behind the mule can be valuable for documenting changes in body condition.

Water Provision Similar to donkeys, mules are considered thirst tolerant; storing water longer than horses to help maintain hydration. However, they still have the same water requirements as similar sized horses, drinking 10–15 gallons of water per day for a 1000-pound mule. However, water requirements increase with workload and during hot, arid weather. Periods of dehydration impair performance, disrupt electrolyte balances, and increase the risk of impaction colic or esophageal obstruction. Therefore, mules should be provided with free access to a source of clean, fresh water (Fig. 5).

For mules working over extended periods, water should be offered a minimum of every 4 hours to prevent dehydration. Additionally, some mules may be sensitive to cold water and become reluctant to drink during winter months when ice is present.

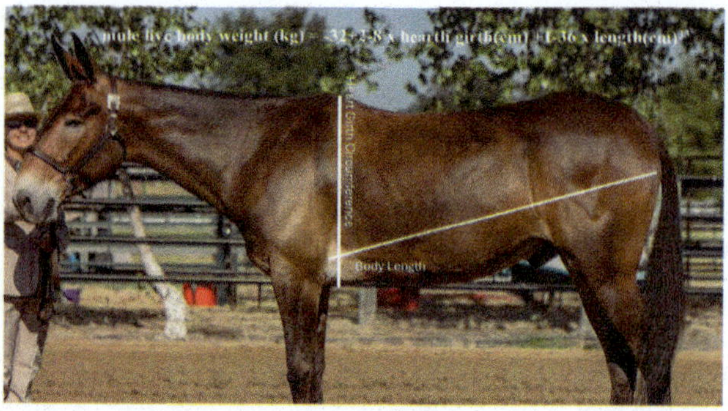

Fig. 4 Body measurements for use in the mule predictive equation created by Kay, et al. (2004). Heart girth is measured starting at the zero end of the measuring tape which is then carefully draped over the withers and behind the elbows until both ends of the measuring tape meet, completing a circle around the girth. Length is measured starting from the point of the elbow (olecranon process) to the point of the hip (tuber ischii). Measurements should be recorded to the nearest centimeter. It should be noted that different predictive equations may have their own specified landmarks for morphometric measurements and the specific landmarks should be utilized if using a different predictive equation to estimate body weight

Fig. 5 Mules provided with free choice access to a fresh, clean water source. (Photo credit Lost Meadows Mule Refuge, Snook, Texas)

Using heated water sources or adding electrolytes to encourage drinking can help reduce the risk of dehydration-related health issues.

4 Conclusion

In conclusion, as mules gain popularity in recreation and competition riding, and remain essential as pack, trail, and working animals, targeted research will be crucial to understanding the distinct nutritional requirements of these diverse groups. Understanding how factors such as workload, environment, and maternal breed influence mule metabolism will support the development of evidence-based feeding strategies. These insights will not only enhance their health and performance but also contribute to their longevity and overall well-being.

Questions

1. What should be taken into consideration when developing a nutritional program for a mule?
   A. Workload
   B. Age
   C. Current feed available
   D. All of the above
   Answer: D
2. At minimum, how often should a mule be offered water when in work for extended periods of time?
   A. 4 hours
   B. 8 hours
   C. 12 hours
   D. 24 hours
   Answer: A
3. How should a NSC of $<10\%$ be achieved in a mule diagnosed with equine metabolic syndrome (EMS)?
   A. Allowing 24-hour access to pasture
   B. Feeding a straw-based diet
   C. Feeding warm-season grass hay
   D. Soaking hay for 1-hour prior to feeding
   Answer: D
4. Why are mules at a higher risk of hyperinsulinemia-associated laminitis?
   A. Ability to recycle urea
   B. Increased feed efficiency
   C. Slower growth rate
   D. Higher digestible energy requirements
   Answer: B

References

Andersen P, Bray R, Cogger E, Wickler S, Lewis C et al (1996) Validation of the equine weight tape for mules. J Anim Sci 74:243

Bala PA, Legha RA, Dedar RK, Kumar V (2017) Nutrients intake and utilization in mules during maintenance. Vet Practitioner 18(2):297–299

Bamford NJ, Potter SJ, Harris PA, Bailey SR (2014) Breed differences in insulin sensitivity and insulinemic responses to oral glucose in horses and ponies of moderate body condition score. Domest Anim Endocrinol 47:101–107

Burden F (2012) Practical feeding and condition scoring for donkeys and mules. Equine Vet Educ 24:589–596

Burden FA, Bell N (2019) Donkey nutrition and malnutrition. Vet Clin N Am Equine Pract 35:469–479

Council, N. R (2007) Nutrient requirements of horses: sixth, Revised edn. The National Academies Press, Washington, DC. https://doi.org/10.17226/11653

du Toit N, Shaw DJ, Keen JA (2011) Adrenocorticotropic Hormone. In: Domestic Donkeys – reference values, seasonality and Association With Laminitis | IVIS. Liverpool, pp 85–86

Durham AE et al (2019) ECEIM consensus statement on equine metabolic syndrome. J Vet Intern Med 33:335–349

Faurie F, Tisserand J (1994) Comparative study of the cellulolytic activity of caecum microbes in ponies and donkeys. Ann Zootech 43:281–281

Gehlen H, Twickel S, Stöckle S, Weber C, Bartmann CP (2020a) Diagnostic orientation values for ACTH and other parameters for clinically healthy donkeys and mules (insulin, triglycerides, glucose, fructosamines, and γ-GT). J Anim Physiol Anim Nutr (Berl) 104:679–689

Gehlen H, Schwarz B, Bartmann C, Gernhardt J, Stöckle SD (2020b) Pituitary pars Intermedia dysfunction and metabolic syndrome in donkeys. Animals (Basel) 10:E2335

Giles SL, Rands SA, Nicol CJ, Harris PA (2014) Obesity prevalence and associated risk factors in outdoor living domestic horses and ponies. PeerJ 2:e299

Giles SL, Nicol CJ, Rands SA, Harris PA (2015) Assessing the seasonal prevalence and risk factors for nuchal crest adiposity in domestic horses and ponies using the Cresty Neck Score. BMC Vet Res 11:13

Henneke DR, Potter GD, Kreider JL, Yeates BF (1983) Relationship between condition score, physical measurements and body fat percentage in mares. Equine Vet J 15:371–372

Humphreys S, Kass PH, Magdesian KG, Goodrich E, Berryhill E (2022) Seasonal variation of endogenous adrenocorticotropic hormone concentrations in healthy non-geriatric donkeys in Northern California. Front Vet Sci 9:981920

Izraely H, Choshniak I, Stevens CE, Shkolnik A (1989) Energy digestion and nitrogen economy of the domesticated donkey (Equus asinus asinus) in relation to food quality. J Arid Environ 17:97–101

Johnston A, McLean A, Xue C, Jacob S, Norton E (2025) Establishing metabolic parameters in mules [abstract]. J Equine Vet Sci 148:111

Karikoski NP, Horn I, McGowan TW, McGowan CM (2011) The prevalence of endocrinopathic laminitis among horses presented for laminitis at a first-opinion/referral equine hospital. Domest Anim Endocrinol 41:111–117

Kay G, Pearson R, Ouassat M (2004) Estimation of the liveweight of working mules in Morocco from their body measurements. Vet Rec 154:85–88

Kosolofski HR, Gow SP, Robinson KA (2017) Prevalence of obesity in the equine population of Saskatoon and surrounding area. Can Vet J 58:967–970

McLean A, Varnum A, Ali A, Heleski C, Navas González FJ (2019) Comparing and contrasting knowledge on mules and hinnies as a tool to comprehend their behavior and improve their welfare. Animals (Basel) 9:488

Mota-Rojas D et al (2021) The use of draught animals in rural labour. Animals 11:2683

Noikhrua K, Tangjitjaroen W, Peansukmanee S, Lawananon P (2016) Body weight estimation in mules by a horse weight tape provided an estimated value close to a real body weight obtained from a digital scale. Vet Integr Sci 14:85–94

Norton EM et al (2019) Heritability of metabolic traits associated with equine metabolic syndrome in Welsh ponies and Morgan horses. Equine Vet J 51:475–480

Pearson RA, Merritt JB (1991) Intake, digestion and gastrointestinal transit time in resting donkeys and ponies and exercised donkeys given ad libitum hay and straw diets. Equine Vet J 23:339–343

Pearson RA, Archibald RF, Muirhead RH (2001) The effect of forage quality and level of feeding on digestibility and gastrointestinal transit time of oat straw and alfalfa given to ponies and donkeys. Br J Nutr 85:599–606

Personal Communication: McLean, Amy. Telephone conversation. 15 Mar. 2025

Potter S, Bamford N, Harris P, Bailey S (2016) Prevalence of obesity and owners' perceptions of body condition in pleasure horses and ponies in South-Eastern Australia. Aust Vet J 94:427–432

Schneider LG, Cox Self A, Hines MT, Lin-Zambito Ivey J (2021) Clinical factors associated with survival outcomes in starved Equids: a retrospective case series. J Equine Vet Sci 101:103370

Smith DG, Burden FA (2013) Chapter 16: Practical donkey and mule nutrition. In: Geor RJ, Harris PA, Coenen M (eds) Equine applied and clinical nutrition. W.B. Saunders, pp 304–316. https://doi.org/10.1016/B978-0-7020-3422-0.00016-X

Starkey P, Starkey M (2004) Regional and world trends in Donkey populations. Animal Traction Network for Eastern and Southern Africa (ATNESA). Technical Centre for Agricultural and Rural Cooperation ACP-EU (CTA), The Netherlands

Suhartanto B, Julliand V, Faurie F, Tisserand J (1992) Comparison of digestion in donkeys and ponies. In: Pferdeheikund Sondersgabe, proceedings of the 1st European conference on horses. Hannover, pp 158–161

Thatcher CD, Pleasant RS, Geor RJ, Elvinger F (2012) Prevalence of Overconditioning in mature horses in Southwest Virginia during the summer. J Vet Intern Med 26:1413–1418

The Clinical Companion of the Donkey (2021) The Donkey Sanctuary, Leicester

Tisserand J-L, Faurie F, Touré M (1991) Comparative Study of Donkey and Pony Digestive Physiology. Donkeys, Mules and Horses in Tropical Agricultural Development, edited by D. Fielding and R. A. Pearson, Centre for Tropical Veterinary Medicine, University of Edinburgh, 1991, pp. 67–72

Wylie CE, Collins SN, Verheyen KLP, Richard Newton J (2011) Frequency of equine laminitis: a systematic review with quality appraisal of published evidence. Vet J 189:248–256

Wyse CA, McNie KA, Tannahill VJ, Murray JK, Love S (2008) Prevalence of obesity in riding horses in Scotland. Vet Rec 162:590–591

Clinical Parasitology in Horses, Donkeys and Mules

Mihajlo Erdeljan, Tijana Kukurić, and Stanislav Simin

Abstract

Horses, donkeys and mules share common parasites but the clinical response, such as diarrhea and poor body condition seen in some species with endoparasitism, is less common in donkeys and mules. Donkeys exhibit a higher prevalence of intestinal nematodes than horses and a greater resilience to clinical signs of infection. Common internal parasites are strongylids, from the subfamilies Strongylinae and Cyathostominae: *Strongyles* spp., *Dictyocaulus arnfieldi*, *Oxyuris equi*, *Parascaris* spp., and *Anoplocephala* spp. External parasites such as mosquitoes, ticks, lice, and flies affect the skin, suck blood and can transfer many pathogens. The survival and transmission of parasites are influenced by environmental conditions, social structures and population density, co-grazing between species, overuse of anthelmintics and overall management. Fecal egg count (FEC), serological tests, ELISA or PCR are diagnostic testing modalities used in the field of parasitology, along with microscopic analysis of skin scrapings to diagnose ectoparasite infestations. Good management includes grooming, the use of insect repellents, strategic deworming programs, and maintaining clean and dry environments. Commonly used anthelmintic drug classes in equids are benzimidazoles, tetrahydropyrimidines, macrocyclic lactones and praziquantel for internal parasites. Treatment used against ectoparasites include organophosphates, pyrethrin, pyrethroids, insect growth regulators, acid salt and biological control. Owners usually control parasites themselves which leads to errors in therapy, the use of incorrect doses, and inadequate selection or formulation of medicines. Parasitic resistance to drugs in horses is an emerging concern, leading to reduced efficacy of treatments. Plants

M. Erdeljan (✉) · T. Kukurić · S. Simin
Department for Veterinary Medicine, Faculty of Agriculture, University of Novi Sad, Novi Sad, Serbia
e-mail: erdeljanm@polj.uns.ac.rs

A. K. McLean et al. (eds.), *Donkey and Mule Medicine: Foundations, Breeds, and Management*, https://doi.org/10.1007/978-3-032-12726-6_21

or plant extracts can be used as therapeutic alternatives in antiparasitic treatment. Parasitology is comprised of the biology of both hosts and parasites, the interactions between them, and the pharmacokinetics of the drugs used to control them. The complexity of the field and the emerging threat of anthelminthic resistance make additional education of breeders, owners and veterinarians in all aspects of parasitology in horses, donkeys and mules. Advances in the scientific literature will be necessary to expand the knowledge of the pharmacokinetics, diagnostics, resistance and alternative treatments in the field of parasitology.

Keywords

Parasites · Diagnostics · Management · Therapy · Resistance · Plant therapy

1 Introduction

Donkeys are known for their stoicism, greater pain tolerance, and consequently a different clinical response to the same infection or disease. Wrongly, as a consequence, some people think that donkeys don't need veterinary medical attention. This misconception can lead to neglect of their health needs (Smith and Burden 2013). Mules, being hybrids, inherit characteristics from both parents. They tend to have a stronger immune response than horses, which can influence their reaction to parasitic infections. However, they can still act as carriers of parasites, potentially infecting other equines. Regular health checks and parasite management tailored to their unique physiology are essential to maintain their health (Matthews 2008).

In general horses, donkeys and mules share common parasites and can act as reservoirs for each other. However, the clinical response, such as diarrhea and poor body condition associated with endoparasitism, is much less commonly seen in donkeys and mules. Notable differences in parasitic infections exist between horses and donkeys, especially regarding prevalence and response to infections. Donkeys generally exhibit a higher prevalence of intestinal nematodes (92% in donkeys vs. 63% in horses) and a greater resilience to developing clinical signs of infection (Buono et al. 2023; Mathewos et al. 2021). In terms of specific parasites, horses show a higher mean fecal egg count, indicating more severe infections compared to donkeys (Mathewos et al. 2022). Additionally, seroprevalence studies reveal that donkeys have a significantly higher rate of Toxoplasma gondii antibodies (92.16%) compared to horses (42.77%) (Kirbas 2022). These findings suggest that while both species share some parasitic challenges, donkeys may experience higher infection rates with less severe clinical manifestations, highlighting the need for tailored management strategies for each species.

In this chapter we will discuss parasitology, a very old and very complex science in all three species. Parasitology is a branch of biology that includes parasites, their hosts, and the relationship between them. In veterinary medicine, parasitology is of great significance due to the impact parasites have on animal health, productivity, and public health. Parasitology aids in diagnosing parasitic infections in animals. Understanding the life cycle of parasites allows veterinarians to develop effective treatment and prevention strategies. Parasites can cause various diseases in domestic and wild animals. Studies on parasites help in controlling and preventing these diseases, which is crucial for maintaining the health of livestock and pets. Parasitic infections can lead to reduced productivity in horses, donkeys and mules (sport and work performances), which has economic implications for the industry. Effective parasite control helps minimize these losses. Many parasites that infect animals can be zoonotic, meaning they can be transmitted from animals to humans (e.g., toxoplasmosis or echinococsis). Parasitology helps in identifying and controlling such infections, thus protecting public health. Proper nutrition, environmental management, and hygiene measures can reduce animals' exposure to parasites. Veterinarians use drugs to prevent and treat parasitic infections, but it is important to manage their use carefully to prevent the development of resistance. In the future, developing vaccines against certain parasites may significantly reduce the incidence of infections. Parasitology is a key component of veterinary science that contributes to the health of animals and humans, as well as economic prosperity.

The goals of this chapter are to provide an overview of the key parasites that parasitize horses, donkeys and mules, their impact on the hosts and various control methods. Additionally, the economic aspect of parasitosis, as well as advice for owners and veterinarians will be discussed.

2 Worldwide Distribution and Prevalence of Parasites

Parasites of any kind, as dependent organisms, have to follow their hosts' distribution, of course. There are numerous reports about prevalence in parasites in equids from all over the world. A recent Italian study performed on donkey's report that *Strongyles* were the most common parasites detected (84.9%), followed by *Dictyocaulus arnfieldi* (6.9%), *Oxyuris equi* (5.8%), *Parascaris* spp. (3.6%), *Anoplocephala* spp. (1.0%) and *Strongyloides westeri* (0.3%) (Buono et al. 2021). The same authors reported that coprocultures revealed an omnipresence of cyathostomins (100%), followed by *Strongylus vulgaris* (31.0%), *Poteriostomum* spp. (25.0%), *Triodontophorus* spp. (9.0%), *Strongylus edentatus* (7.0%) and *Strongylus equinus* (5.0%). On the other side of the world, in Brazil, a paper from 2024 showed similar results in horses: the parasite prevalence was 73.3%, with 72.1% present in the adult population and 80.6% in young horses. *Strongyles* had a prevalence of 72.1% and *Parascaris* spp. had a prevalence of 5.8% (Molento et al. 2024).

However, many situations are more complicated. *Cryptosporidium* spp. affect both horses and donkeys, but research in China, revealed that significantly higher *Cryptosporidium* infection rates were present in donkeys than in horses; 21.1% versus 1.4% (Li et al. 2019). Another study, from Iraq, confirms findings from many other studies and suggests that horses and donkeys in that area have roughly the same prevalence of *Cryptosporidium* spp. and all the major parasites, including *Strongylus vulgaris*, *Parascaris equorum* and *Oxyuris equi* (Wannas et al. 2012).

Epidemiology, as an integrated part of parasitology, is dynamic. Discoveries of emerging parasites in new geographic areas can occur, usually due to world commerce or world travel of people and animals, but sometimes the result of climate change. One recent example of this occurred in Ireland, where authors (Tinkler et al. 2024) have reported the first finding of *Besnoitia bennetti* in Irish donkeys. Other authors published the first report of infestation of a horse with *Setaria digitata* in Malaysia (Peng et al. 2019). Also, as an example, there is a first report of *Echinococcus equinus* in a mule in Turkey (Simsek and Cevik 2014).

Other research describes uncommon parasites in equines in which the horse, donkey or mule are accidental hosts. *Fasciola hepatica*, the common liver fluke, is a trematode parasite mostly associated with disease in sheep and cattle. Traditionally, liver flukes have been considered to be uncommon in horses but they have been described (Williams and Hodgkinson 2017), while donkeys are more compliant hosts (Quigley et al. 2020). However, there is serological evidence about the presence of *F. hepatica* in horses, donkeys and mules in Mexico with prevalence of 10.4%, 13.9% and 17.9% respectively (Villa-Mancera and Reynoso-Palomar 2020).

3 Major Parasites Infecting Horses, Donkeys and Mules

Parasites can be broadly divided in two large groups: external and internal. External parasites include mosquitoes, ticks, flies and lice. Internal parasites include several types of worms—large, small, and tapeworms, that infest the digestive track, as well as blood parasites like Babesia or piroplasmosis (Taylor et al. 2016).

3.1 External Parasites

External parasites affect the skin and can cause significant discomfort and health issues. Mosquitoes can transmit diseases like West Nile virus, which can affect all three equine species and humans. Ticks are known vectors of Lyme disease and anaplasmosis. Flies, particularly stable flies and horseflies can cause skin irritation and can also transmit diseases. Lice infestations are more common in the winter and can lead to itching, hair loss, and skin infections (Radostits et al. 2007).

3.1.1 Skin Nematodes

There are two families in the order *Nematodes* that affect skin: *Onchocercidae* and *Habronematidae*.

Family *Onchocercidae* has one genus *Onchocerca,* which has at least three species; *Onchocerca gutturosa, Onchocerca reticulata* and *Onchocerca cervicalis* (Scott 1988). These nematodes are filarial and thread-like. Adults live in the nuchal ligament or in flexor tendons and the suspensory ligament of the fetlock (Melhorn

2008). Donkeys in Africa are especially affected, with sporadic reports from Europe (Italy) (Papini et al. 2020).

Family *Habronematidae* has two genera; *Habronema* and *Draschia*. Most important species are *Habronema muscae*, *Habronema majus (Habronema microstoma)* and *Draschia megastoma* (Vasey 1981). Habronematidosis has a worldwide distribution. The muscid flies act as intermediate hosts and as a reservoir. Larva and adult stages parasitize the stomach of equids. However, the larvae can also be found on the skin, causing lesions known as "summer sores or jack sores", especially in warm weather, in the ocular and genital mucosa (Barlaam et al. 2020).

3.1.2 Ticks

Horses are susceptible to infestations by many species of ticks that feed on wild or domestic ungulates in different regions. Both soft (*Argasidae*) and hard ticks (*Ixodidae*) feed on horses and donkeys. *Ixodes* spp., *Dermacentor* spp., *Boophilus* spp., *Amblyomma* spp., and *Otobius megnini* (spinose ear tick) are most often identified (Sellon and Long 2007). In addition to feeding on blood, ticks are vectors to equine piroplasmosis, *Borrelia burgdorferi* and *Anaplasma phagocytophilum*, see Sect. 3.2.2.

3.1.3 Mites

Mites are the causative agent of mange. Equids can be affected by sarcoptic mange (*Sarcoptes scabei var. equi*), psoroptic mange (*Psoroptes equi, P. ovis, P. natalensis,* and *P. cuniculi*) and chorioptic mange (*Chorioptes bovis*). This and many other species of mites can affect multiple species, including humans. Mange can also be caused by demodex (*Demodex caballi, D. equi*), called demodectic mange. Clinically, we can observe irregular skin lesions, from erythematous areas to papules and crust formation, with severe itching and sometimes biting of affected areas, sometimes leading to decreased feed intake (Osman and Hanafy 2006).

3.1.4 Lice

Equids are hosts for two species of lice: the sucking louse, *Haematopinus asini,* and the biting louse, *Bovicola (Werneckiella) equi,* which is more common. Dermatological lesions, anemia and alopecia are common findings, together with secondary skin infections, reduced newborn birth weights and even abortions (Sorrell et al. 2010).

3.1.5 Mosquitoes

Mosquitoes are a large group of flying insects. The most important genera are *Culex, Aedes,* and *Anopheles.* Depending on the species, a film of water on leaf litter, maintained for the duration of larval and pupal periods, is sufficient. This means that they can breed almost anywhere in standing water (Foster and Walker 2002). Mosquitoes play a significant role in the transmission of various pathogens affecting equids, particularly through viral and parasitic infections. West Nile virus (WNV) is a flavivirus disease transmitted by mosquitoes, with neurological sign, encephalitis, ataxia, limb weakness, recumbency, muscle fasciculations and mortality nearly 50%

in clinically affected horses (Angenvoort et al. 2013). Venezuelan equine encephalitis virus (VEEV) is alphavirus, mainly affects different species of equids, humans, and wild animals, with a mosquito transmission (Guzmán-Terán et al. 2020). Eastern equine encephalitis (EEEV) is a related alphavirus to EEE. Both viral pathologies (EEE and VEE) can cause severe compromise of the central nervous system and death, they are emerging mostly in different countries of the Americas (Bonilla-Aldana et al. 2023). See neurology Chap. 8.

3.1.6 Biting Flies

The family *Diptera*, commonly known as tabanids, has more than 4434 species, currently allocated into 156 genera (Pape et al. 2011). They inhabit all continents except Antarctica. Males are nectarivores' as opposed to females which are hematophagous, seeking blood meals in mammals, birds, amphibians and reptiles (Foil and Hogsette 1994). Lesions on the skin result from the bites of tabanids or the horse flies (*Tabanus* spp., *Haematopota* spp., and *Chrysops* spp.), stable flies (*Stomoxys calcitrans*), and black flies (*Simulium* spp.). The bites result in crusted papules that usually last for a few days. Hypersensitivity reactions ranging from urticaria to eosinophilic (collagenolytic) granulomas or pruritic dermatoses have been associated with the bites of these insects (Von Tscharner et al. 2000). Tabanids are biological and mechanical vectors of a whole series of diseases: anthrax, tularemia and various forms of trypanosomiasis (Baldacchino et al. 2014).

3.1.7 Biting Midges

Tiny flies from genera *Culicoides* and *Leptoconops* have many names: midges, no-see-ums, punkies or sand flies. Despite their small size they cause big problems, one of which is equine insect bite hypersensitivity (IBH), commonly known as sweet itch or summer eczema. This condition is a recurrent skin disease due to an immune reaction to the bites of midges (Lomas and Robinson 2018). Additionally, the biting midge (*Culicoides imicola*) is the main vector of African Horse Sickness (AHS) in Africa (Ilchmann et al. 2009).

3.1.8 Hippobosca

The superfamily *Hippoboscoidea*, genera *Hippobosca* have seven species with the most prominent being *Hippobosca equina*. This species is commonly known as the forest fly, as well as flat or iron fly, due to the strong dorso-ventral flattening of its chitin exoskeleton (Sokół et al. 2015). This is an obligate hematophagous ectoparasite of many vertebrate species (Maślanko et al. 2022) including horses, donkeys and mules. The consequences are negative effects on the host's well-being, agitation and irritation, intensive tail lashing and frequent scraping against posts or tree trunks, as well as skin inflammation, loss of body weight, and a decrease of the quality of the host's skin, hair, and fleece (Romaniuk and Gad 2007) (Fig. 1). *H. equina* is a biological and mechanical transmitter of various pathogens between different animal species.

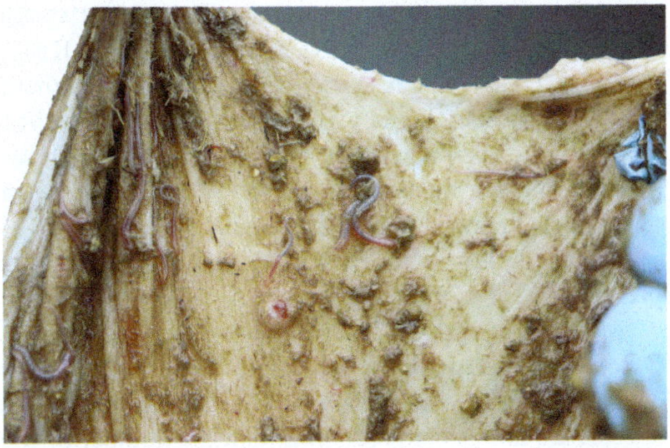

Fig. 1 Large and small strongyles photographed in the small intestium

3.1.9 Nonbiting Flies

The family *Muscidae* has over 100 genera. The genera *Musca* and *Hydrotaea,* are the most important from this group of ectoparasites. Muscid flies include significant blood-feeding parasites, vectors of disease agents, species that annoy humans and animals (Moon 2019). Significant species are known as the house fly (*Musca domestica*), bazaar fly (*M. sorbens*), bush fly (*M. vetustissima*), dung fly (*Scathophaga stercoraria*), stable fly (*Stomoxys calcitrans*) and false stable fly (*Muscina stabulans*).

Hydrotea flies (garbage flies) has seven known species, with at least one species in every biogeographic region (Huckett and Vockeroth 1987). The most important representatives are the black garbage fly (*Hydrotaea ignava*) and black dump fly (*Hydrotaea aenescens*).

3.1.10 Hypodermiasis (Warbles)

Warbles are caused by the larval stages of the heel flies, *Hypoderma bovis, H. lineatum* and *H. diana*, which is the deer warble. These are primarily parasites of ruminants and equids are dead-end hosts. If affected, large subcutaneous nodules can be clinically observed (Borges et al. 2019).

3.1.11 Accidental Parasites

Horses, donkeys and mules can be an accidental host of a large variety of ectoparasites in various parts of the world. Myiasis is caused by blow flies and flesh flies with the genera *Lucilia, Calliphora, Phormia, Chrysomyia,* and *Sarcophaga* (Catis and Mullen 2002). Equids are accidental hosts to these parasites.

3.2 Internal Parasites

Internal parasites are a major concern for equine health. Large strongyles (*Strongylus vulgaris*) and small strongyles (cyathostomins) are common in horses, donkeys, and mules, causing weight loss, diarrhea, and colic. Donkeys usually have no clinical signs, but overworked animals, with a poor condition display clinical signs (Matthews and Burden 2013; Tavassoli et al. 2015). Tapeworms (*Anoplocephala perfoliata*) can also cause colic and intestinal obstruction (Nielsen et al. 2014).

Roundworms (*Parascaris* spp.) primarily affect younger animals and can cause respiratory issues and poor growth. *Strongyloides westeri* affects foals, causing diarrhea and weight loss. Pinworms (*Oxyuris equi*) cause severe itching around the tail area (Reinemeyer and Nielsen 2013).

The taxonomy of internal parasites is quite large and challenging and will be presented next as a brief review.

3.2.1 Nematodes

The nematodes, roundworms or eelworms, are a diverse animal phylum. Nematode species are not easy to recognize and the number of species is the subject of debate, ranging from over 25,000 (Zhang 2013) to up to 100 million species (Floyd et al. 2002) with a conservative assessment of 1 million species (Blaxter 2016). There are roughly 2271 genera which are placed in 256 families (Anderson 2000), from which one third are parasites of vertebrates.

Gastrointestinal Nematodes

Dozens of species of similar nematodes are classified in the family *Strongylidae* and are commonly known as strongylid nematodes or strongyles. They are divided into two major groups or subfamilies *Strongylinae* (large strongyles, strongylins) and *Cyathostominae* (small strongyles, cyathostomins, cyathostomes) (Lichtenfels 1975).

Further, the subfamily *Strongylinae* has four genera: *Strongylus, Triodontophorus, Craterostomum,* and *Oesophagodontus.* Most important representatives are: *Strongylus vulgaris, Strongylus edentatus* and *Strongylus equinus.* Strongylosis has been reported from all parts of the world and almost affects more than 90% of the horse population (Nielsen et al. 2006). Clinically, infection with strongyles can cause mild clinical signs such as weight loss, loss of appetite, poor hair coat, intermittent diarrhea, lethargy, deterioration of condition, peripheral edema and disordered intestinal motility (Matthews and Morris 1995). While adult strongyles live in the cecum and large intestine, L4 and L5 stage larvae are responsible for arteritis, necrosis, and fibrosis of the cranial mesenteric artery and its branches (Kuzmina et al. 2012). As a result of migrations of larvae, thrombosis and embolism in blood vessels leads to infarction of the intestinal tract, severe colic and death (Marinković et al. 2009) (Figs. 1 and 2).

Cyathostominae (Cyathostomes, Cyathostomins, Small Strongyles, Trichonemes) has eight genera: *Cyathostomum, Cylicocyclus, Cylicodontophorus, Cylicostephanus, Coronocyclus, Parapoteriostomum, Poteriostomum* and *Petrovinema* (Lichtenfels et al. 1998). There are approximately 50 species of Cyathostomins. Most prominent species are: *Cyathostomum catinatum, Cylicocyclus nassatus, Cylicostephanus longibursatus, Cylicostephanus minutus, Cylicostephanus calicatus* and *Cyathostomum insigne* (Corning 2009). They are the most common and pathogenically significant parasite to affect horses around the

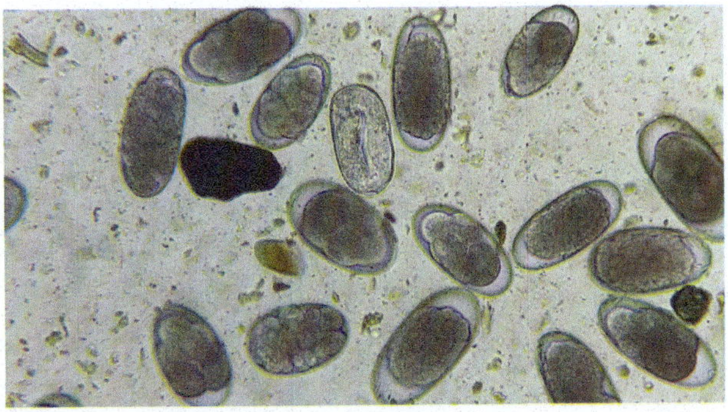

Fig. 2 *Strongylus* sp., egg

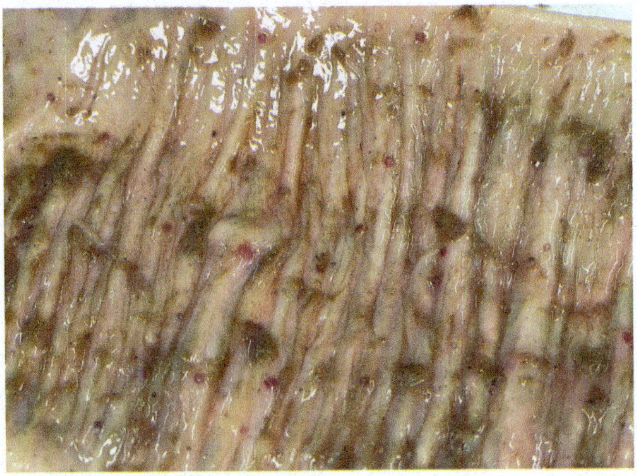

Fig. 3 *Cyathostominae* spp., larvae in the intestinal mucosa, visible as red nodules

world (Love et al. 1999). Adult worms in large numbers may cause lethargy, sudden weight loss, debilitation, and diarrhea but cyathostomin larval stages can cause serious damage to the intestinal mucosa when, at the encysted stage, tens of thousands of encysted larvae may literally cover the mucosa wall, severely damaging it and greatly reducing nutritional metabolism (Collobert-Laugier et al. 2002) (Fig. 3).

The family *Ascaridoidea* (roundworms) has one genus *Parascaris* with two species *Parascaris equorum* and *Parascaris univalens. Parascaris* spp. are large nematodes that predominantly infect foals and weanlings. The severity of the infection is more pronounced in donkeys than in horses, irrespective of their age (Getachew et al. 2008). The adult forms, in large numbers, can cause coughing, anorexia, lethal intestinal obstruction or rupture, and even death (Morsy et al. 2016), while larval forms can cause hepatitis, pneumonitis and associated respiratory disorders (Jabbar et al. 2014). We can distinguish the two species by no. of chromosomes. *P. univalens* has $2n = 2$ in contrast to *P. equorum* which has $2n = 4$. Recently, potential new species were collected from a donkey in China, described as *Parascaris* sp. with $2n = 6$ (Zhou et al. 2023) (Figs. 4 and 5).

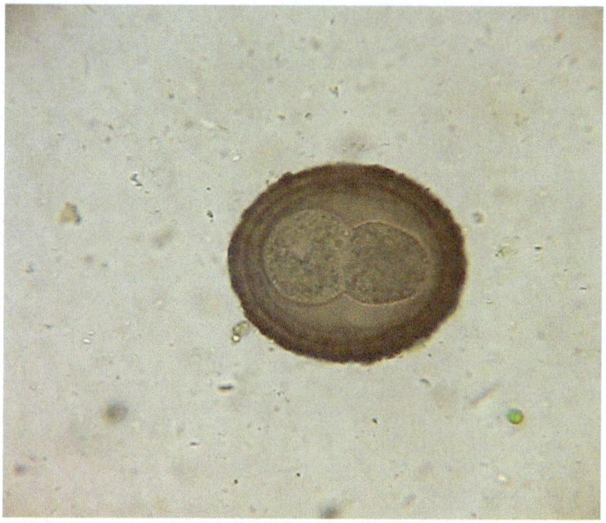

Fig. 4 *Parascaris* sp., egg

Fig. 5 *Parascaris* sp., adults

The order Rhabditida, family Strongylidae, genus Strongyloides, species *Strongyloides westeri,* called the threadworm parasite is mainly found in the small intestine of foals up to 4 months of age (Lyons 1994). Infested foals usually appear healthy (Miller et al. 2017). The infection route is lactogenic, with third stage larvae (L3) (Lyons et al. 1973). Also, L3s can be acquired from the environment through oral ingestion and transcutaneous invasion.

The order *Oxyurida,* genus *Oxyuris* has one species *Oxyuris equi,* called the pinworm. It has a worldwide distribution and an interesting life cycle: after fertilization, the males die. Then the female travels through the host's intestine to the anal opening and almost fully protrudes outside. The nematode's vulva begins to excrete a viscous greyish-white mass which consists of numerous eggs and mucus (Yevstafieva et al. 2020). Commonly, there is a presence of tail rubbing, damages the hair and causes regions of hair loss at the base of the tail or rump as well as excoriations and lesions in more intense cases (Reinemeyer and Nielsen 2014).

Lung Nematodes

The class *Secernentea*, family *Dictyocaulidae*, genus *Dictyocaulus* has one species: *Dictyocaulus arnfieldi*, the equine lungworm (Round 1976). *Dictyocaulus arnfieldi* is a lungworm affecting donkeys, mules and horses throughout the world. Donkeys and mules are considered to be the natural hosts of *D. arnfieldi* while this condition in horses is usually found in those that have been in the company of donkeys and mules (Rose and Hodgson 2000). Clinical manifestation includes infection of the lower respiratory tract, usually resulting in bronchitis or pneumonia and it can be observed in horses. In reservoir species (donkeys and mules), even in the face of prolonged patent infections they will not display any clinical sign (Round 1976), even if large numbers of parasites are accumulated in the lungs (Ibrahim 2017).

Nematode Infection of Neural and Somatic Tissues

The order *Rhabditida* and family *Paragrolaimidae* has eight species and the dominant is *Halicephalobus gingivalis*, previously known as *Micronema deletrix* or *Halicephalobus deletrix* (Onyiche and Okute 2018). *H. gingivalis* has been reported to infect humans and horses (Anderson et al. 1998) and donkeys (Schmitz and Chaffin 2004). Nematodes enter the tissues through wounds and are carried to other parts of the body through the hematogenous route in blood and lymph (Henneke et al. 2014). When it gets to the central nervous system (CNS), necrosis and inflammation occur, due to visceral migration (Kreuder et al. 1996). Other organs that may be affected include kidneys, oral and nasal cavities, lymph nodes, spinal cord, adrenal glands, heart, stomach, liver, ganglion and bone. The clinical manifestations depend on the affected organ or organs. The wide spread of the symptoms may include nasal tumor (Anderson and Bemrick 1965), osteomyelitis (Kreuder et al. 1996). granulomatous prepuce infection (Muller et al. 2008), granulomatous nephritis (Akagami et al. 2007), meningoencephalitis (Akagami et al. 2007), and disseminated disease (Ruggles et al. 1993). In cases with dominant CNS affection ataxia is common, (Bryant et al. 2006). Massive intracranial invasion in horses is acute and of short duration (Jubb and Huxtable 1993).

Nematode Infection of Ocular Tissues

The order *Rhabditida,* family *Thelaziidae,* genus *Thelazia* has one species of interest in equids: *Thelazia lacrymalis. Thelazia* (*Spirurida, Thelaziidae*) or eyeworms, cause ocular infestations in horses (Anderson 2000) with a limited no. of reports in donkeys (Mulwa 2019) and mules (Lyons et al. 2000). Localized on the surface of the eye, under the eyelids and the conjunctiva, on the lacrimal gland and in its excretory ducts, disease ranges from subclinical to clinical, with signs ranging from conjunctivitis, epiphora and photophobia to keratitis and ulcers; sometimes endophthalmitis and blindness may occur (Lyons and Drudge 1975). For its development, *Thelazia* requires an intermediate host from the genus M*usca*. (O'Hara and Kennedy 1991).

3.2.2 Protozoa

The phylum *Apicomplexa*, family *Piroplasmidae* has two genera; *Babesia* (*B. caballi*) and *Theileria* (*T. equi*—formerly known as *B. equi*). These piroplasms cause equine piroplasmosis, which affects all equids. The disease is described as equine malaria or horse tick fever. Obligatory intraerythrocytic organisms causes varying degrees of hemolytic anemia (de Waal and van Heerden 2004).

The order *Eucoccidiorida*, family *Sarcocystidae,* genus *Sarcocystis* has several species: *Sarcocystis neurona, S. falcatula* and *S. speeri. S. neurona* cause equine protozoal myeloencephalitis (EPM), which is a progressively debilitating disease affecting the CNS of horses, with opossums as the definitive hosts (Dubey 2001). *Sarcocystis neurona* is also serologically (Borges et al. 2017; Llano et al. 2021) and molecularly confirmed in donkeys and mules (Amairia et al. 2023). Currently, in donkeys, name *S. bertrami* is used for the sarcocysts diagnosed in the muscles of donkeys (Dubey et al. 2016; Passantino et al. 2019; Zhang et al. 2022).

3.3 Cestodes

The family *Anoplocephalidae* includes three species: *Anoplocephala perfoliata*, *Anoplocephala magna* and *Anoplocephaloides mamillana*. They infect the gastrointestinal (GI) tract of horses and donkeys. During their development they have a cysticercoid larval stage in the oribatid mites. *A. perfoliata* is currently considered as the most common and the most pathogenic of the equine tapeworm species worldwide (Lyons et al. 2000). They are associated with bowel irritation, intussusceptions and intestinal obstruction in horses (Gasser et al. 2005). Further, infestation with *Anoplocephala* sp. increases the risk of ileal impaction and spasmodic colic, particularly in horses chronically infected with many tapeworms (Traub-Dargatz et al. 2001). The presence of all three species is described also in donkeys (Getachew et al. 2010), usually without the described clinical signs. Regarding the mules, reports are on the level of *Anoplocephala* spp. (Takele and Nibret 2013) without further findings (Fig. 6).

Fig. 6 *Anoplocephala perfoliata*, adults in intestine, ileo-cecal valve

Fig. 7 *Gasterophilus* sp. larvae attached to the stomach mucosa

3.4 Bots

The family *Oestridae*, genus *Gasterophilus,* has nine species, and the most prominent are *G. flavipes* and *G. pecorum*. They are commonly known as bot flies, warble flies, heel flies or gad flies (Saini and Sankhala 2015). Flies of the family *Oestridae* are large robust flies, mostly bee- or wasp-like, without vibrissae (Marshall et al. 2017). *Gasterophilus* sp. are obligate parasites adapted to a larval life in the intestinal tract of equids (including horses, donkeys and zebras) (Colwell et al. 2006) These larvae cause gastrointestinal myiasis leading to gastrointestinal ulcerations, gut obstructions or volvulus and rectal prolapses (Hoseini et al. 2017). Clinical findings also include destruction of the gums, ulcerations, peritonitis, anemia, heavy debilitation and blockage of the gastrointestinal tract, and even death (Bezdekova et al. 2007) (Fig. 7).

4 Development Cycles

Given the large number of genera and species of parasites, the development cycle of only certain representatives of large groups of parasites will be presented here.

The life cycle of *Babesia caballi* (protozoa) involves complex interactions between the parasite and its tick vectors, primarily *Dermacentor* spp. and *Rhipicephalus* spp. In *Dermacentor* spp. *B. caballi* undergoes several developmental stages within the tick. Initially, small spherical bodies are formed in the gut, which develop into larger vermicular forms that can invade the tick's tissues, including the salivary glands, where they undergo multiple fissions to produce infectious sporozoites by day 4 of feeding (Holbrook et al. 1968; Blouin and Waal 1989). Transmission to equine hosts occurs when these sporozoites are injected during tick feeding. Notably, the transmission is limited to one generation of ticks in the absence of reinfection from a host, indicating a short-term reservoir potential for the parasite (Schwint et al. 2008). Additionally, *Rhipicephalus* spp. can also transmit *B. caballi*, albeit with a low parasitemia and a short patent period (De Waal and Potgieter 1987). This highlights the intricate life cycle and transmission dynamics of *B. caballi*, emphasizing the role of tick vectors in its propagation.

Some intestinal parasites also require an intermediate or transitional host, like *Anoplocephala perfoliata* which is a common tapeworm parasite primarily affecting horses. The definitive hosts are equids where the adult tapeworm resides in the intestine, particularly the ileocecal region. The life cycle begins when eggs are shed in the feces of infected horses, which are then ingested by oribatid mites, the intermediate hosts. Inside the mites, the eggs develop into cysticercoids, which are subsequently consumed by horses during grazing, thus completing the cycle (Gasser et al. 2005) (Figs. 8, 9 and 10).

The life cycle of *Dictyocaulus arnfieldi*, which is the primary parasitic lungworm (nematode) infecting donkeys begins when donkeys ingest infective third-stage larvae (L3) from contaminated pasture. These larvae are typically found on grass,

Fig. 8 *Anoplocephala perfoliata*, egg

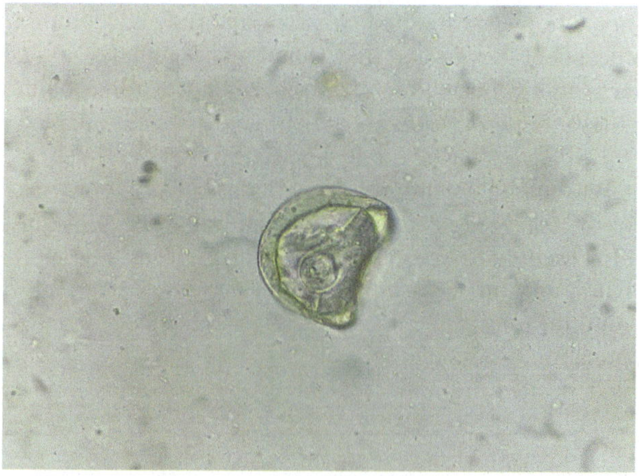

Fig. 9 *Anoplocephala perfoliata,* in situ

having migrated from fecal deposits. Once ingested, the larvae penetrate the intestinal wall and enter the lymphatic system, eventually reaching the bloodstream. They are then carried to the lungs, where they migrate to the bronchi and bronchioles. In the bronchi and bronchioles, the larvae develop into fourth-stage larvae (L4) and then into adult worms. This process takes approximately 3–4 weeks. The adult worms reside in the respiratory tract, where they reproduce. Female worms lay eggs, which are coughed up and swallowed by the host. These eggs then hatch into first-stage larvae (L1) in the intestines and are excreted in the feces. In the environment, the first-stage larvae (L1) molt to second-stage larvae (L2) and then to infective third-stage larvae (L3), completing the cycle. This development depends on environmental conditions, such as temperature and humidity (Smith and Jones 2020; Williams 2019; Taylor et al. 2016).

The life cycle of ectoparasites can be illustrated by that of *Hippobosca equina*. The female insects are macro-larviparous, as females keep the larva in their uterus until the end of the third instar; the three larval instars feed on the milk glands then pupation occurs rapidly after pre-pupae laying (Mehlhorn 2016). The larva (or pupa) is deposited in birds' cages and feathers or on the skin of a mammalian host (Halos et al. 2004). Adult insects increase in summer and autumn (Sokół and Michalski 2015) (Figs. 11 and 12).

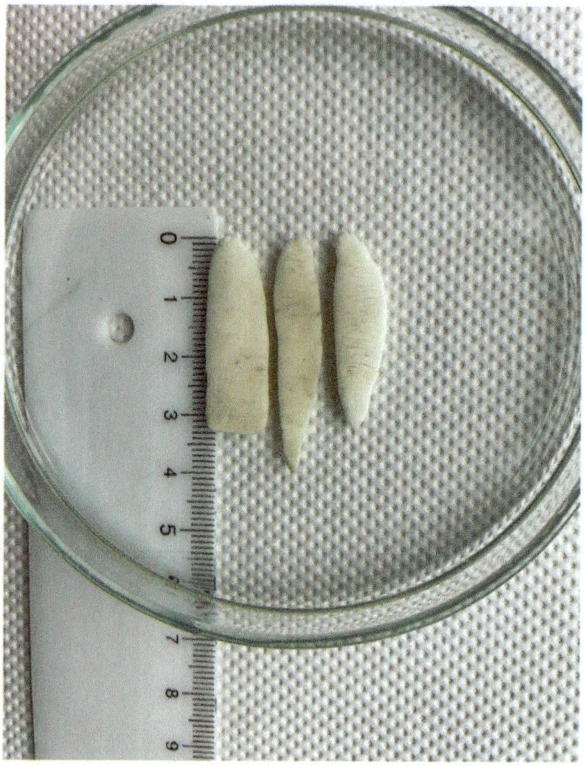

Fig. 10 *Anoplocephala perfoliata*, adult

5 Factors Which Affect the Survival and the Transmission of the Parasites in Horses, Donkeys and Mules

The survival and spread of parasites in horses and donkeys are influenced by several interrelated factors. Co-grazing between these species facilitates the transmission of helminths, as donkeys can act as reservoirs for parasites like *Dictyocaulus arnfieldi* and *Fasciola hepatica*, which can infect horses without overt signs developing in donkeys (Matthews and Burden 2013; Mas-Coma et al. 2020). Environmental conditions, particularly during warm and moist periods, enhance the viability of parasite eggs and larvae, leading to increased pasture contamination (Epe et al. 2001). Additionally, the overuse of anthelmintics has led to resistance, complicating control efforts and allowing parasites to thrive (Pavan et al. 2024). The social structure and population density of equids also play a role; lower densities may reduce the need for prophylactic treatments, while higher densities can exacerbate transmission risks through contamination (Epe et al. 2001). Thus, effective management strategies must consider these ecological and behavioral dynamics to mitigate parasite spread and ensure the health of both horses and donkeys.

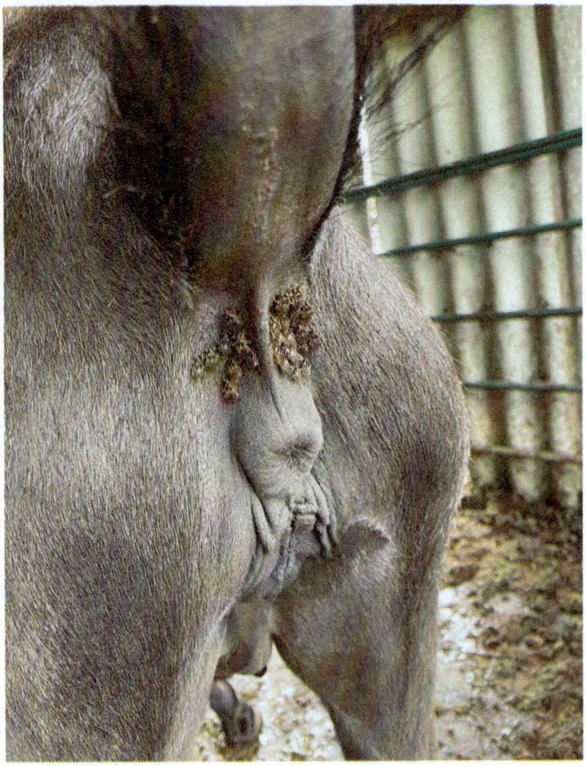

Fig. 11 *Hippobosca equina* on a donkey hair, adult

6 Diagnostics

6.1 Diagnostics of Endoparasites

Accurate diagnosis of parasitic infections is crucial for effective treatment and control. Methods such as fecal egg count (FEC) are commonly used to quantify the burden of gastrointestinal nematodes. Serological tests, including ELISA, are employed to detect specific antibodies against parasites like *Strongylus vulgaris* (Nielsen et al. 2015). Due to a large number of different parasitic species there are no developed tests specific for every species. Some of them are just recently developed, like the ELISA test for *Anoplocephala* spp. (Jürgenschellert et al. 2020). The method of polymerase chain reaction (PCR) is applied for parasite identification, for studying genetic diversity between phylum's, families and species of parasites, as well as for the diagnosis and epidemiological investigation of parasitic diseases (Trivizaki and Theodoropoulos 2018). The biggest advantage of PCR is its higher sensitivity in comparison to other classical indirect diagnostic methods. While these methods provide valuable information, they also have limitations, such as variability in egg shedding and cross-reactivity in serological assays.

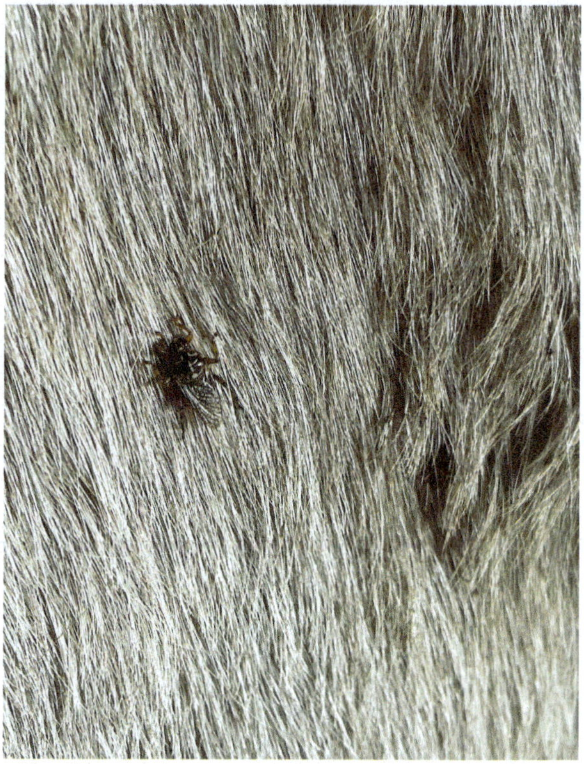

Fig. 12 *Hippobosca equina* under a donkey tail, adult

6.2 Diagnostics of Ectoparasites

On the other hand, the diagnosis of ectoparasite infections is a more straightforward process than that for endoparasites (Wells et al. 2018). Standard tests are microscopic analysis of scrapings of skin or whiskers. Additional tests like ELISA with greater sensitivity and specificity can be used (Ochs et al. 2001). In addition, novel molecular technologies are emerging for use in diagnosis and in developing novel diagnostics; for example, multiplexed real-time PCR assays, genotyping and gene clustering with pathway analysis (Wells et al. 2018).

7 Distribution

Many factors have an effect on the distribution of parasites, including geography, age and season, along with other factors that are not yet known. Some of the parasites have local distribution, some affect equines on the one continent and some have worldwide distribution even if all of them belong to the same genus. *Onchocerca gutturosa* affect skin in a form of cutaneous and ocular lesions in North America, Africa, Australia, and Europe.; *Onchocerca reticulata* appeared in Europe and Asia and *Onchocerca cervicalis* can be founded worldwide (Scott 1988). A survey of yearling horses in Australia revealed the highest prevalence of strongyle eggs (82%), while foals had the highest prevalence of *Parascaris* spp. (35%) (Abbas et al. 2023). In France, strongyle eggs were found in 97% of equids, with peaks in September and October, indicating a seasonal dynamic in parasite distribution (Merlin et al. 2023). While the studies indicate a clear pattern of parasite distribution influenced by age, season, and geography, the complexity of interactions among these factors necessitates ongoing research to develop effective management strategies.

8 Management

Management of external parasites includes regular grooming, use of insect repellents, and maintaining clean and dry living environments. In some cases, veterinary intervention is necessary to treat infestations and secondary infections (Bowman et al. 2021).

Effective management of internal parasites involves regular deworming programs based on fecal egg counts, good pasture management to reduce exposure, and proper sanitation practices. It is essential to tailor deworming protocols to the specific needs of horses, donkeys, and mules, as their susceptibility to different parasites and their responses to infections can vary (Kaplan and Nielsen 2010).

9 Therapy

The most common anthelmintic drugs used in equids are benzimidazoles (BZDs), tetrahydropyrimidines (THPs), macrocyclic lactones (MLs) and praziquantel (PZQ) (Gokbulut and McKellar 2018). Although mules and donkeys are often treated as horses from a therapeutic point of view, the studies on the effectiveness of antiparasitic drugs are limited. According to the reported studies, there is a difference in pharmacokinetics of medicines in horses and donkeys, and very limited information when it comes to mules. Only a few drugs are registered for use in donkeys (none exclusive). Therefore, additional research might result in a more optimized dose and protocol for both mules and donkeys (Bazzano et al. 2020; Buono et al. 2023).

In a class of benzimidazoles (BZDs), fenbendazole (FBZ) represents the most common BZDs used in horses. Fenbendazole is effective against large and small strongyles, ascarids, oxyurids and strongyloides species.

Pyrantel (PYR) is in the tetrahydropyrimidines class (THPs) and it is effective against large and small strongyles, ascarids, oxyurids and tapeworms, but only in twice the nematocidal dose (13.2 mg kg BW).

Macrocyclic lactones (MLs) are a class of anthelmintics of which ivermectin (IVM), moxidectin (MOX), doramectin (DRM) and eprinomectin (EPM) are the most commonly administered in equids. MLs are highly effective against both endo- (nematodes) and ectoparasites (mites, flies, lice).

Praziquantel (PZQ) is an isoquinoline used for controlling cestodes and trematodes in equids. It is usually used for the control of tapeworms, but it is reported that PZQ is poorly tolerated in donkeys, so using praziquantel at a dose of 13.2 mg kg BW would be safer in donkeys (Reinemeyer and Nielsen 2016; Gokbulut and McKellar 2018; Buono et al. 2023).

Ectoparasite treatment strategies include topical and systemic drug applications and treatment of the environment. Unfortunately, worldwide, the number of approved products for horses is limited, and even more so for donkeys and mules. However, drugs used in the control of ectoparasites are organophosphates, pyrethrin, pyrethroids, insect growth regulators (IGR), acid salt and biological control (Karasek et al. 2020). Drugs come in different pharmacological formulations: shampoo, powder, oil-based solution, spray, impregnated ear tags for collars etc. They can be administered orally, topically, by pour-on technique or in form subcutaneous injection.

Important considerations for owners is that we cannot use some drugs safely in donkeys and mules even though they may be widely used in other animals. For instance, amitraz, which is widely used in dogs and bees, applied orally, could lead to death and, when applied locally, could lead to a severe poisoning in both horses and donkeys (Duarte et al. 2003).

10 Drug Resistance

Parasitic resistance to drugs in equids is an emerging concern, as studies indicate a growing prevalence of resistance among equine parasites. Research highlights that the overuse and misuse of anthelmintics have contributed significantly to this issue, leading to reduced efficacy of commonly used treatments (Elghryani et al. 2024). Given the availability of deworming preparations, the owners usually do the parasite control themselves which leads to errors in therapy, use of incorrect doses, inadequate selection of medicine or formulation of medicines (Kukurić et al. 2024). Emphasize is on the need for strategic deworming practices to mitigate resistance development, suggesting that rotational use of different drug classes may be beneficial. Moreover, other studies (Elghryani et al. 2024) report that certain parasites, such as cyathostomins, have shown notable resistance to benzimidazole and pyrantel, which are frequently employed in equine parasitology. This resistance not only complicates treatment protocols but also raises concerns about the long-term sustainability of current anthelmintic strategies (Nielsen 2022). In light of these findings, it is crucial for equine practitioners to adopt integrated parasite management approaches, including fecal egg count monitoring and targeted treatments, to combat the rising threat of drug resistance in equine parasites (Voronkova and Usachev 2024).

Parasites in horses, particularly cyathostomins, develop resistance to anthelmintic drugs through several mechanisms, primarily driven by the extensive and repeated use of these medications. Studies indicate that resistance to commonly used drugs, such as benzimidazoles, pyrantel, and macrocyclic lactones like ivermectin, is increasingly prevalent across various equine populations globally (Elghryani et al. 2024; Merlin et al. 2024; Nielsen 2022). The genetic diversity and population structure of these parasites play a crucial role in resistance development. For instance, research has identified specific genetic candidates associated with pyrantel resistance, suggesting that selective pressure from drug use has led to the emergence of resistant strains (Sallé et al. 2023). Additionally, the shortening of egg reappearance periods (ERPs) post-treatment indicates a potential adaptation of parasites, allowing them to resume reproduction more quickly after anthelmintic administration (Nielsen 2022). Overall, the combination of genetic adaptation, selective pressure from drug use, and the lack of new anthelmintic classes has contributed to the growing challenge of drug resistance in equine parasites (Voronkova and Usachev 2024; Nielsen 2022).

Drug resistance in parasites affecting donkeys and mules is also a significant concern, particularly regarding strongyle infections. A study in Egypt found a high prevalence of strongyle species (73.02%) among donkeys, with evidence of ivermectin resistance, as the efficacy of this drug dropped to 91.6% after treatment (Hamed et al. 2019). Furthermore, in the UK, resistance to all three classes of anthelmintics has been documented in donkeys, complicating control strategies (Matthews and Burden 2013). The presence of drug-resistant parasites is exacerbated by co-grazing with horses, which can facilitate the spread of infections and resistance (Buono et al. 2023). Overall, the increasing prevalence of drug resistance highlights the need for careful management and selective treatment strategies to preserve the efficacy of available anthelmintics (Kuzmina et al. 2020).

One of the assessment tools to assess drug resistance is the Fecal Egg Count Reduction (FECR) test which is a critical tool for assessing the efficacy of anthelmintic treatments against gastrointestinal nematodes in livestock in general. Recent studies emphasize the importance of methodological rigor and the need for a genus-specific approach to enhance the accuracy of FECR results. The FECR test evaluates the reduction in fecal egg counts (FEC) pre- and post- treatment, with guidelines recommending a paired study design for improved accuracy (Kaplan et al. 2023). Depending on the size of the horse farm including at least 5–10 horses is necessary. Recommended cut-off values as general guidelines for strongyle nematodes are: benzimidazoles 95%, pyrantel 90%, ivermectin 95% and moxidectin 95%. If the farm average FECR falls below these values, anthelmintic resistance should be suspected (Sellon and Long 2007). Accordingly, there is a need for further education of breeders and veterinarians, on the importance of coprological examination, as a useful tool in assessing the severity of infection, need for therapy, environmental contamination and assessing the effectiveness of anthelmintics, which aims at their rational use and long-term successful application (Kukurić et al. 2022).

11 Therapeutic Alternatives

Given the undeniable problem with resistance to antiparasitic drugs, we are forced to look for alternatives. Since ancient times, there is knowledge that some plants have antiparasitic effects, but, of course, at that time, without scientific verification or any scientific explanation. Currently, there are numerous reports regarding the use of plants or plant extracts in antiparasitic treatment in equines. Australian authors (Payne et al. 2013) tested 37 species of plants against cyathostomin spp. eggs from which seven completely inhibited egg development (*Acacia baileyana, Acacia melanoxylon, Acacia podalyriifolia, Alectryon oleifolius, Duboisia hopwoodii, Eucalyptus gomphocephala* and *Santalum spicatum*). The list of plants used in this study is quite long including *Artemisia dracunculus*, *Eucalyptus camadulensis, Mentha pulegium, Zataria multiflora* and *Allium sativum* against *Parascaris equorum* larvae with a successful rate of 98% (Rakhshandehroo et al. 2017). The definitive list has yet to be concluded.

The list of bioactive compounds from plants is also long. It includes alkaloids, flavonoids, tannins, saponins, coumarins, terpenes, lignans, and amino acids. Important considerations in the war against antiparasitic resistance is that these compounds show various mechanisms of action against parasites, making it much harder for the parasite to develop a resistance (Elghandour et al. 2023).

In general, there is a great potential with different plants, different bioactive compounds and their combinations which require further investigation.

12 Economic Aspects of Equid Parasite Infections

Various studies have examined the economic implications of parasitic infections and they found that the consequences are profound, as they can cause reduced feed efficiency, weight loss, and lowered performance in working and sport horses (Ogbeina et al. 2022). Furthermore, the cost of antiparasitic treatments and preventive measures adds to the financial burden on horse owners and the equine industry as a whole. Since we know that the racehorse industry is worth about 300 billion dollars annually on the planetary scale or 115 billion dollars in the USA alone, economic aspects gain in importance.

In veterinary economics, the analysis of costs associated with equine health management is crucial. The direct costs, including veterinary care, medication, and preventive measures, must be weighed against the indirect costs, such as decreased productivity and potential losses due to premature death. Effective control of parasitic infections can lead to improved health outcomes for equines, resulting in economic benefits that outweigh the expenses involved in their management.

13 Guidelines for Owners and Veterinarians

Different national and international bodies have issued guidelines for owners and veterinarians. One of the most suitable is issued by AAEP (American Association of Equine Practitioners). To summarize those guidelines for internal parasites (AAEP 2024):

- Perform fecal egg count reduction tests (FECRT) annually to ensure that you are using effective dewormers.
- Recognize that no anthelmintic will eliminate all parasitic stages from a horse.
- Test fecal egg counts to stratify horses into low, medium, and high shedders to reduce pasture contamination.
- Deworm all horses at a baseline rate (once or twice a year) and target selected horses more often.
- Do not use fecal egg count to diagnose disease in horses.
- Discontinue deworming all horses with fixed intervals, and stop blindly rotating anthelmintic classes.

These guidelines are easily applied and are cost effective, but on the contrary most of them are not applied in practice (Scare et al. 2018).

14 Frequently Asked Q&A

This are common question that came from owners and the answers that is a product of veterinarians who work with equids:

1. What are the most common parasites in equids?
   The most common parasites in horses include:
   - Large strongyles (*Strongylus vulgaris*): These are particularly harmful and can cause colic and other severe health issues.
   - Small strongyles (cyathostomes): These are the most prevalent and can cause weight loss, diarrhea, and colic.
   - Roundworms (*Parascaris* spp.): Primarily affect foals and young horses, leading to respiratory issues, poor growth, and colic.
   - Tapeworms (*Anoplocephala* spp.): Can cause colic and intestinal blockages.
   - Pinworms (*Oxyuris equi*): Cause tail rubbing and irritation around the anus.
   - Bots (*Gasterophilus* spp.): Cause irritation in the stomach and intestines.
2. How can I recognize the signs of a parasitic infection in my horse?
   Signs of parasitic infection can vary but commonly include: weight loss or poor body condition, dull coat, diarrhea or loose stools, colic (abdominal pain), poor growth in young horses, respiratory issues such as coughing (mainly in young horses with roundworms), tail rubbing and irritation around the anus (often caused by pinworms), decreased performance and lethargy.
3. How often should I treat my horse for parasites?
   The frequency of treatment depends on several factors including the horse's age, environment, and overall health. Generally, it is recommended to deworm:
   - Foals: Every 2–3 months starting at 2 months of age.
   - Adult horses: Typically, 2–4 times a year, depending on the risk factors and results of fecal egg counts.
   - High-risk horses: More frequently based on veterinary advice and fecal egg counts.

4. Which deworming products are the most effective?
   Effective deworming products include:
   - Ivermectin: Effective against many types of parasites including large and small strongyles, bots, and pinworms.
   - Moxidectin: Similar to ivermectin but has a longer duration of action.
   - Fenbendazole: Effective against large strongyles, pinworms, and roundworms.
   - Pyrantel: Effective against large and small strongyles, pinworms, and tapeworms (when used at a double dose).
   - Praziquantel: Often combined with ivermectin or moxidectin to treat tapeworms.
5. Is there a risk of parasites developing resistance to deworming medications?
   Yes, there is a risk of parasites developing resistance, especially with overuse or incorrect use of deworming medications. To minimize resistance:
   - Use deworming medications strategically based on fecal egg count results.
   - Rotate between classes of dewormers, but not too frequently.
   - Only deworm when necessary and at the appropriate dose.
6. How should deworming products be used correctly (dosage, frequency, etc.)?
   To use deworming products correctly:
   - Dosage: Administer the correct dose based on your horse's weight. Underdosing can contribute to resistance.
   - Frequency: Follow veterinary advice and use fecal egg counts to determine the appropriate frequency.
   - Method: Ensure the horse swallows the entire dose. Some products come in paste form, which can be administered orally.
7. Is it necessary to analyze my horse's feces before treatment?
   Yes, performing fecal egg counts (FEC) is recommended before treatment to:
   - Determine the presence and type of parasites.
   - Assess the parasite load.
   - Help develop an effective deworming schedule.
   - Monitor the effectiveness of deworming treatments and check for resistance.
8. How can I prevent parasitic infections in the stable and on the pasture?
   Preventive measures include:
   - Regularly removing manure from pastures and stalls to reduce the parasite load.
   - Rotating pastures to prevent overgrazing and reduce parasite exposure.
   - Keeping feed and water sources clean and elevated off the ground.
   - Implementing quarantine protocols for new horses to prevent introducing new parasites.
9. What are the best practices for pasture management to reduce parasite exposure?
   Effective pasture management practices:
   - Pasture rotation: Move horses to different pastures to break the parasite lifecycle.
   - Manure removal: Regularly remove manure to reduce the parasite burden.

- Cross-grazing: Grazing other species, such as sheep or cattle, can help reduce the parasite load, as different species are affected by different parasites.
- Stocking density: Avoid overcrowding to reduce parasite transmission.

10. Are there natural or alternative methods for controlling parasites?

    Some natural or alternative methods include:
    - Diatomaceous earth: Claimed to help control internal parasites, though scientific evidence is limited.
    - Herbal remedies: Such as wormwood, garlic, and pumpkin seeds, though effectiveness and safety can vary.
    - Pasture management: Good pasture hygiene and rotation can significantly reduce the need for chemical dewormers.
    - Biological control: Using beneficial nematodes to reduce fly populations that can spread parasites.

However, it's important to consult with a veterinarian before relying solely on natural methods, as they may not be as effective as traditional dewormers.

15 Conclusions

Equine parasitology includes a large body of knowledge, including the biology of hosts and parasites, the interaction between hosts (donkeys, mules and horses) and parasites, and the pharmacokinetics of drugs that are poorly tested in donkeys and mules, in some cases even in horses. Due to a very large number of species of parasites affecting horses, donkeys and mules, understanding their life cycle is crucial for effective management and treatment of infections. Further, emerging resistance to common drugs is a great concern and the possibility of alternative therapies is developing. There is a need for further education of breeders and veterinarians in all aspects of parasitology in horses, donkeys and mules. Further studies on the pharmacokinetics of anthelminthics, especially in donkeys and mules, and diagnostics must be improved. Surveillance for resistance to common drugs and the development of alternative treatments is also imperative.

References

Abbas G, Ghafar A et al (2023) Understanding temporal and spatial distribution of intestinal nematodes of horses using faecal egg counts and DNA metabarcoding. Vet Parasitol. Available from: https://doi.org/10.1016/j.vetpar.2023.110094

Akagami M, Shibahara T, Yoshiga T, Tanaka N, Yaguchi Y, Onuki T, Kondo T, Yamanaka T, Kubo M (2007) Granulomatous nephritis and meningoencephalomyelitis caused by *Halicephalobus gingivalis* in a pony gelding. J Vet Med Sci 69:1187–1190

Amairia S, Jbeli M, Mrabet S, Mahjoubi Jebabli L, Gharbi M (2023) Molecular Prevalence of Sarcocystis spp. and Toxoplasma gondii in slaughtered equids in Northern Tunisia. J Equine Vet Sci 129:104894. https://doi.org/10.1016/j.jevs.2023.104894

American Association of Equine Practitioners Parasite control guidelines (2024) https://aaep.org/resource/internal-parasite-control-guidelines/

Anderson RC (2000) Nematode parasites of vertebrates. Their development and transmission, 2nd edn. CABI Publishing, Guilford, pp 428–430

Anderson RC, Linder KE, Peregrine AS (1998) Halicephalobus gingivalis (Stefanski, 1954) from a fatal infection in a horse in Ontario, Canada with comments on the validity of H. deletrix and a review of the genus. Parasite 5:255–261

Angenvoort J, Brault AC et al (2013) West Nile viral infection of equids. Vet Microbiol 167(1–2): 168–180. https://doi.org/10.1016/j.vetmic.2013.08.013. Epub 2013 Aug 28. PMID: 24035480; PMCID: PMC4581842

Anderson RV, Bemrick WJ (1965) Micronema deletrix, a saprophagous nematode inhabiting a nasal tumor of a horse. Proc. Helminthol. Soc. Wash., 32, pp. 74–75

Baldacchino F, Desquesnes M et al (2014) Tabanids: neglected subjects of research, but important vectors of disease agents! Infect Genet Evoltion 28:596–615. https://doi.org/10.1016/j.meegid.2014.03.029

Barlaam A, Traversa D, Papini R, Giangaspero A (2020) Habronematidosis in equids: current status, advances, future challenges. Front Vet Sci 7:358

Bazzano M, Di Salvo A, Diaferia M, Veronesi F, Galarini R, Paoletti F, Tesei B, McLean A, Veneziano V, Laus F (2020) Anthelmintic efficacy and pharmacokinetics of ivermectin paste after oral administration in mules infected by cyathostomins. Animals 10(6):934

Bezdekova B, Jahn P et al (2007) Pathomorphological study on gastroduodenal ulceration in horses: Localisation of lesions. Acta Vet Hung.;55:241–249. pmid:17555289

Blaxter M (2016) Imagining Sisyphus happy: DNA barcoding and the unnamed majority. Philos Trans R Soc London B 371 (1702):20150329. https://doi.org/10.1098/rstb.2015.0329. PMC 4971181.

Blouin EF & De Waal DT (1989) The fine structure of developmental stages of Babesia caballi in the salivary glands of Hyalomma truncatum. Onderstepoort J Vet Res 56(3):189–193

Bonilla-Aldana DK, Bonilla Carvajal CD, Moreno-Ramos E, Barboza JJ, Rodriguez-Morales AJ (2023) Mapping Eastern (EEE) and Venezuelan Equine Encephalitides (VEE) among Equines using geographical information systems, Colombia, 2008–2019. Viruses 15(3):707. https://doi.org/10.3390/v15030707

Borges F, Sybrecht GW, von Samson-Himmelstjerna G (2019) First reported case of Hypoderma diana Brauer, 1985 (Diptera: Oestridae)-associated myiasis in a horse in Germany. Equine Vet Educ 31(3):122–125. Available from: https://doi.org/10.1111/EVE.12810

Bowman DD, Lucio-Forster A, Janeczko S (2021) Internal Parasites. In Infectious Disease Management in Animal Shelters (eds L. Miller, S. Janeczko and K.F. Hurley)

Bryant UK, Lyons ET, Bain FT, Hong CB (2006) Halicephalobus gingivalis-associated meningoencephalitis in a thoroughbred foal. J Vet Diagn Invest 18:612–615

Buono F, Veronesi F et al (2021) Helminth infections in Italian donkeys: Strongylus vulgaris more common than Dictyocaulus arnfieldi. J Helminthol 95(e4):1–10. https://doi.org/10.1017/S0022149X20001017

Buono F, Veneziano V et al (2023) Horse and donkey parasitology: differences and analogies for a correct diagnostic and management of major helminth infections. Parasitology. https://doi.org/10.1017/S0031182023000525

Catis EP, Mullen GR (2002) Myiasis (Muscidae, Oestridae). In: Mullen G, Durden L (eds) Medical and veterinary entomology. Academic, Amsterdam

Collobert-Laugier C, Hoste H, Sevin C, Dorchies P (2002) Prevalence, abundance and site distribution of equine small strongyles in Normandy, France. Vet Parasitol 110:77–83

Colwell DD, Hall MJR et al (2006) In: Colwell DD, Hall MJR, Scholl PJ (eds) The oestrid flies: biology, host-parasite relationships, impact and management. CABI, Wallingford. https://doi.org/10.1079/9780851996844.0000

Corning S (2009) Equine cyathostomins: a review of biology, clinical significance and therapy. Parasites Vectors 2(Suppl 2):S1. https://doi.org/10.1186/1756-3305-2-S2-S1

Costa Marques Borges AM, Yeargan MR, Silva LG, Taques IIGG, Howe D, Aguiar DM (2017) Antibodies Against Sarcocystis neurona, Neospora spp., and Toxoplasma gondii in Horses and

Mules from the Northern Pantanal Wetland of Brazil. J Equine Vet Sci 56:19–25. https://doi.org/10.1016/j.jevs.2017.04.007
De Waal DT, Potgieter FT (1987) The transstadial transmission of Babesia caballi by Rhipicephalus evertsi evertsi. Onderstepoort J Vet Res
De Waal DT, van Heerden J (2004) Equine piroplasmosis. In: Coetzer JAW, Tustin RC (eds) Infectious diseases of livestock. Oxford University Press, Cape Town, pp 425–433
Duarte MD, Peixoto PV et al (2003) Intoxicações natural e experimental por amitraz em eqüídeos: aspectos clínicos Pesq. Vet Bras 23(3):105–118
Dubey JP (2001) Recent developments in the biology of Sarcocystis neurona and equine protozoal myeloencephalitis. J Vet Parasitol 15:91–102
Dubey JP, Van Wilpe E, Verma SK, Hilali M (2016) Ultrastructure of Sarcocystis bertrami sarcocysts from a naturally infected donkey (Equus asinus) from Egypt. Parasitology 143:18–23. https://doi.org/10.1017/S0031182015001432
Elghandour MMMY, Maggiolino A et al (2023) Moringa oleifera as a natural alternative for the control of gastrointestinal parasites in equines: a review. Plants 12:1921. https://doi.org/10.3390/plants12091921
Elghryani N, Lawlor A, McOwan T, Waal T (2024) Unravelling the effectiveness of anthelmintic treatments on equine strongyles on Irish farms. Animals. https://doi.org/10.3390/ani14131958
Epe C, Kings M et al (2001) The prevalence and transmission to exotic equids (equus quagga antiquorum, equus przewalskii, equus africanus) of intestinal nematodes in contaminated pasture in two wild animal parks. J Zoo Wildlife Med 32(2):209–216. https://doi.org/10.1638/1042-7260(2001)032[0209:TPATTE]2.0.CO;2
Floyd R, Abebe E et al (2002) Molecular barcodes for soil nematode identification. Mol Ecol 11(4): 839–850. Bibcode:2002MolEc. 11.839F. https://doi.org/10.1046/j.1365-294X.2002.01485.x. PMID 11972769. S2CID 12955921
Foil LD, Hogsette JA (1994) Biology and control of tabanids, stable flies and horn flies. Rev Sci Tech 13:1125–1158
Foster WA, Walker ED (2002) Mosquitoes (Culicidae). In: Mullen G, Durden L (eds) Medical and veterinary entomology. Academic, Amsterdam
Gasser RB, Williamson RMC et al (2005) Anoplocephala perfoliata of horses – significant scope for further research, improved diagnosis and control. Parasitology 131:1–13
Getachew AM, Innocent GT, Trawford AF, Feseha G, Reid SJW, Love S (2008) Equine parascarosis under the tropical weather conditions of Ethiopia: a coprological and postmortem study. Vet Rec 162:177–180. https://doi.org/10.1136/vr.162.6.177
Getachew M, Trawford A, Feseha G et al (2010) Gastrointestinal parasites of working donkeys of Ethiopia. Tropic Anim Health Prod 42:27–33. https://doi.org/10.1007/s11250-009-9381-0
Gokbulut C, McKellar QA (2018) Anthelmintic drugs used in equine species. Vet Parasitol 261:27–52
Guzmán-Terán C, Calderón-Rangel A, Rodriguez-Morales A, Mattar S (2020) Venezuelan equine encephalitis virus: The problem is not over for tropical America. Ann Clin Microbiol Antimicrob 19:19
Halos L, Jamal T et al (2004) Role of Hippoboscidae flies as potential vectors of Bartonella spp. infecting wild and domestic ruminants. J Appl Environ Microbiol 70(10):6302–6305. https://doi.org/10.1128/AEM.70.10.6302-6305.2004
Hamed MI, El-Allawy TA, Hassnein E (2019) Prevalence and anthelmintic resistance of strongyle infection of donkeys in El-Wadi El-Gadid, Egypt. J Adv Vet Res 9(4):144–150
Henneke C, Jespersen A et al (2014) The distribution pattern of Halicephalobus gingivalis in a horse is suggestive of a haematogenous spread of the nematode. Acta Vet Scand 56(1):56
Holbrook AA, Anthony DW, Johnson AJ, (1968) Observations on the development of Babesia caballi (Nuttall) in the tropical horse tick Dermacentor nitens Neumann. J Eukaryotic Microbiol. Available from: https://doi.org/10.1111/J.1550-7408.1968.TB02143.X

Hoseini SM, Zaheri BA et al (2017) Histopathological study of esophageal infection with Gasterophilus pecorum (Diptera: Oestridae) in Persian onager (Equus hemionus onager). J Arthropod-Borne Dis 11(3):441–445

Huckett HC, Vockeroth JC (1987) Muscidae. In: JF MA, Peterson BV, Shewell GE, Teskey HJ, Vockeroth JR, Wood DM (eds) Manual of Nearctic Diptera volume 2, monograph no. 28. Canada Communications Group, Ottawa, pp 1115–1131

Ibrahim N (2017) Equine lung worm: a systematic review. Global J Med Res G Vet Sci Vet Med 17(2) Version 1.0 Publisher: Global Journals Inc. (USA) Online ISSN: 2249-4618 & Print ISSN: 0975-5888

Ilchmann G, Clausen P-H et al (2009) Die Afrikanische Pferdepest – eine weitere Tierseuche ante portas? Pferde Spiegel 12:62–66. https://doi.org/10.1055/s-0029-1185887

Jabbar A, Littlewood DT et al (2014) The mitochondrial genome of Parascaris univalens–implications for a "forgotten" parasite. Parasites Vectors 7:428

Jubb KVF, Huxtable CR (1993) The nervous system in Jubb KVF, et al. (Eds.), Pathology of domestic animals (4.ed.), V.1, Academic, San Diego pp. 267–439

Jürgenschellert L, Krücken J et al (2020) Investigations on the occurrence of tapeworm infections in German horse populations with comparison of different antibody detection methods based on saliva and serum samples. Parasites Vectors, Available from: https://doi.org/10.1186/S13071-020-04318-5

Kaplan RM, Nielsen MK (2010) An evidence-based approach to equine parasite control: It ain't the 60s anymore. Equine Veterinary Education 22:306–316. https://doi.org/10.1111/j.2042-3292.2010.00084.x

Karasek I, Butler C, Baynes R, Werners A (2020) A review on the treatment and control of ectoparasite infestations in equids. J vet Pharmacol Therap 43:421–428. https://doi.org/10.1111/jvp.12874

Kirbas A (2022) Investigation of seroprevalence of toxoplasmosis in horses and donkeys in Muş Province of Turkey. J Hellenic Vet Med Soc, Available from: https://doi.org/10.12681/jhvms.25571

Kreuder C, Kirker-Head CA et al (1996) What is your diagnosis? (lameness caused by Micronema deletrix). J Am Vet Med Assoc 209(6):1070–1071

Kukurić T, Erdeljan M, Simin S et al (2022) The importance of coprological diagnostic for the rational use of anthelmintics in horses. Annual consultation of doctors of veterinary medicine of the Republic of Srpska (Bosnia and Herzegovina) scientific meeting with international participation, Trebinje. ISBN 978-99955-770-9-4, pp 175–178

Kukurić T, Erdeljan M, Simin S (2024) Significance of horse parasitosis and practical aspects of control, Symposium Health care and reproduction of ungulates, ungulates, poultry and carnivores, Belgrade. ISBN 978-86-900417-8-7, pp 149–156

Kuzmina TA, Lyons ET et al (2012) Fecundity of various species of strongylids (Nematoda: Strongylidae) parasites of domestic horses. Parasitol Res 111:2265–2271

Kuzmina TA, Zvegintsova NS et al (2020) Anthelmintic resistance in strongylids (Nematoda: Strongylidae) parasitizing wild and domestic equids in the Askania Nova Biosphere Reserve, Ukraine. Ann Parasitol. Available from: https://doi.org/10.17420/AP6601.237

Llano HAB, Soares RM, Acevedo-Gutierrez LY, Rodas JD, Polo G, Borges-Silva W, Jesus RF, Gondim LFP (2021) Seroepidemiology of sarcocystis neurona and Neospora spp. in horses, donkeys, and mules from Colombia. Acta Trop 220:105970. ISSN 0001-706X

Li F, Su J, Chahan B, Guo Q, Wang T, Yu Z, Guo Y, Li N, Feng Y, Xiao L (2019) Different distribution of Cryptosporidium species between horses and donkeys. Infect Genet Evol 75: 103954. ISSN 1567–1348, https://doi.org/10.1016/j.meegid.2019.103954

Lichtenfels JR (1975) Helminths of domestic equids: illustrated keys to genera and species with emphasis on North American forms. Proc Helm Soc Wash 42:1

Lichtenfels JR, Kharchenko VA, Krecek RC et al (1998) An annotated checklist by genus and species of 93 species level names for 51 recognized species of small strongyles (Nematoda: Strongyloidea: Cyathostominea) of horses, asses, and zebras of the world. Vet Parasitol 79:65

Lomas HR, Robinson PA (2018) A pilot qualitative investigation of stakeholders' experiences and opinions of equine insect bite hypersensitivity in england. Vet Sci 5:3. https://doi.org/10.3390/vetsci5010003
Love S, Murphy D, Mellor D (1999) Pathogenicity of cyathostomes infection. Vet Parasit 85:113–122
Lyons ET (1994) Vertical transmission of nematodes: Emphasis on Uncinaria lucasi in northern fur seals and Strongyloides westeri in equids. J Helmin Soc Wash 61:169–178
Lyons ET, Drudge JH (1975) Occurrence of the eyeworm, Thelazia lacrymalis, in horses in Kentucky. J Parasitol 61:1122–1124
Lyons ET, Drudge JH, Tolliver SC (1973) On the life cycle of Strongyloides westeri in the equine. J Parasitol 59(5):780–787
Lyons ET, Swerczek TW, Tolliver SC et al (2000) Prevalence of selected species of internal parasites in equids at necropsy in central Kentucky (1995–1999). Vet Parasitol 92:51
Marinković D, Sanja AK et al (2009) Morphological findings in the cranial mesenteric artery of horses with verminous arteritis. Acta Vet 59:231–241
Marshall SA, Kirk-Spriggs AH et al (2017) [Chapter] 12. key to Diptera families. In: Kirk-Spriggs AH, Sinclair BJ (eds) Manual of Afrotropical Diptera. Volume 1. Introductory chapters and keys to Diptera families. Suricata 4. South African National Biodiversity Institute, Pretoria, pp 267–355
Mas-Coma S, Buchon P et al (2020) Donkey fascioliasis within a one health control action: transmission capacity, field epidemiology, and reservoir role in a human hyperendemic area. Front Vet Sci. Available from https://doi.org/10.3389/FVETS.2020.591384
Maślanko W, Szwaj E et al (2022) Hippobosca equina L. (Hippoboscidae: Hippobosca)—An old enemy as an emerging threat in the Palearctic Zone. Int J Environ Res Public Health 19(24): 16978. https://doi.org/10.3390/ijerph192416978
Mathewos M, Girma D et al (2021) Prevalence of gastrointestinal helminthiasis in horses and donkeys of Hawassa District, Southern Ethiopia. Vet Med Int. Available from https://doi.org/10.1155/2021/6686688
Mathewos M, Teshome D, Fesseha H (2022) Study on Gastrointestinal Nematodes of Equines in and around Bekoji, South Eastern Ethiopia. J Parasitol Res. https://doi.org/10.1155/2022/8210160
Kaplan RM et al (2023) World Association for the Advancement of Veterinary Parasitology (W.A.A.V.P.) guideline for diagnosing anthelmintic resistance using the faecal egg count reduction test in ruminants, horses and swine. Vet Parasitol 318:109936. https://doi.org/10.1016/j.vetpar.2023.109936
Matthews JB (2008) An update on cyathostomins: Anthelmintic resistance and worm control. Equine Vet Educ 20:552–560
Matthews C, Burden FA (2013) Common helminth infections of donkeys and their control in temperate regions. Equine Vet Educ 25(9):461–467
Matthews AG, Morris JR (1995) Cyathostomiasis in horses. Vet Rec 136:52
Mehlhorn H (2016) Hippoboscidae. Encyclopedia of Parasitology, pp 1252–1252. https://doi.org/10.1007/978-3-662-43978-4_3942
Melhorn H (2008) Encyclopedia of Parasitology. Springer, Berlin/Heidelberg/New York
Merlin A, Ravinet N et al (2023) Prevalence and seasonal dynamic of gastrointestinal parasites in equids in France during two years. Preventive Veterinary Medicine, Available from. https://doi.org/10.1016/j.prevetmed.2023.106100
Merlin A, Larcher N, Carlos J (2024) The first report of triple anthelmintic resistance on a French Thoroughbred stud farm. Available from: https://doi.org/10.1016/j.ijpddr.2024.100528
Miller FL, Bellaw JL et al (2017) Strongyloides westeri worm and egg counts in naturally infected young horses. Vet Parasitol 248:1–3. ISSN 0304-4017. https://doi.org/10.1016/j.vetpar.2017.10.010

Molento MB, Pires LSA et al (2024) Prevalence and risk factors of gastrointestinal helminths infection in Brazilian horses: a retrospective study of a 12-year (2008–2019) diagnostic data. Res Vet Sci 173:105272. ISSN 0034-5288, https://doi.org/10.1016/j.rvsc.2024.105272

Moon RD (2019) Chapter 17 – Muscid Flies (Muscidae)In: Mullen GR, Durden LA (eds) Medical and veterinary entomology, 3rd edn. Academic, pp 345–368. ISBN 9780128140437. https://doi.org/10.1016/B978-0-12-814043-7.00017-0

Morsy K, Bashtar AR et al (2016) Description of two equine nematodes, Parascaris equorum Goeze 1782 and Habronema microstoma Schneider 1866 from the domestic horse Equus ferus caballus (Famisly: Equidae) in Egypt. Parasitol Res 115:4299–4306

Muller S, Grzybowski M, Sager H, Bornand V, Brehm WA (2008) Nodular granulomatous posthitis caused by *Halicephalobus* sp in a horse. Vet Dermatol 19(1):44–48

Mulwa, N. N. (2019). Occurrence and risk factors of endoparasites and associated lesions in donkeys in selected abattoirs in Kenya. PhD Thesis. University of Nairobi

Nielsen MK (2022) Anthelmintic resistance in equine nematodes: current status and emerging trends. Available from: https://doi.org/10.1016/j.ijpddr.2022.10.005

Nielsen MK, Monard J, Olsen SN (2006) Prescription -only anthelmintics- a questionnaire surveillance and control of equine strongyles in Denmark. Vet Parasitol 135:47–55

Nielsen MK, Pfister K, Von Samson-Himmelstjerna G (2014) Selective therapy in equine parasite control-application and limitations. Vet Parasitol 202:95–103

Nielsen MK, Vidyashankar AN, Bellaw J, Gravatte HS, Cao X, Rubinson EF, Reinemeyer CR (2015) Serum Strongylus vulgaris-specific antibody responses to anthelmintic treatment in naturally infected horses. Parasitol Res 114(2):445–451. https://doi.org/10.1007/s00436-014-4201-5

O'Hara JE, Kennedy MJ (1991) Development of the nematode Thelazia skrjabini (Nematoda: Thelazioidea) in experimentally infected face flies Musca autumnalis (Diptera: Muscidae). J Parasitol 77:417–425

Ochs H, Lonneux J et al (2001) Diagnosis of psoroptic sheep scab with an improved enzyme linked immunosorbent assay. Vet Parasitol 96:233e42

Ogbeina KE, Dogob AG, Oshadub DO, Edehc ER (2022) Gastrointestinal parasites of horses and their socioeconomic impact in Jos Plateau-Nigeria. Appl Vet Res. https://doi.org/10.31893/avr.20220

Onyiche TE, Okute TO (2018) Parasitic and zoonotic meningoencephalitis in humans and equids: current knowledge and the role of Halicephalobus gingivalis. Parasite Epidemiol Control 3(1): 36–42. ISSN 2405-6731. https://doi.org/10.1016/j.parepi.2017.12.002.

Osman SA, Hanafy A (2006) Clinical and therapeutic studies on mange in horses. Vet Parasitol 141: 191–195

Pape T, Blagoderov V, Mostovski MB (2011) Order Diptera Linnaeus, 1758. In: Zhang Z-Q (ed) Animal biodiversity: an outline of higher-level classification and survey of taxonomicrichness. Zootaxa 3148:222–229.

Papini RA, Lubas G, Sgorbini M (2020) Incidental detection of onchocerca microfilariae in donkeys (Equus asinus) in Italy: report of four cases. Front Vet Sci 7:569916

Passantino G, Lia RP, Latrofa S, Annoscia G, Šlapeta J, Otranto D, Rossi R, Zizzo N (2019) Sarcocystis bertrami in skeletal muscles of donkeys (Equus africanus asinus) from Southern Italy. Vet Parasitol Reg Stud Rep 16:100283. https://doi.org/10.1016/j.vprsr.2019.100283

Pavan ALB et al (2024) Trombos de Strongylus vulgaris aderidos na valva semilunar aórtica e aorta em equino. Pubvet. Available from: https://doi.org/10.31533/pubvet.v18n07e1623

Payne SE, Kotze AC, et al. (2013) Australian plants show anthelmintic activity toward equine cyathostomins in vitro. Vet Parasitol 196(1–2):153–160. ISSN 0304-4017. https://doi.org/10.1016/j.vetpar.2013.01.012

Peng TL, Armiladiana MM et al. (2019) First report of equine Setaria digitata (von Linstow 1906) infestation in Malaysia. Vet Parasitol Reg Stud Rep 17:100310. ISSN 2405-9390. https://doi.org/10.1016/j.vprsr.2019.100310

Quigley A, Sekiya M, Garcia-Campos A, Paz-Silva A, Howell A, Williams DJL, Mulcahy G (2020) Horses are susceptible to natural, but resistant to experimental, infection with the liver fluke, Fasciola hepatica. Vet Parasitol 281:109094. https://doi.org/10.1016/j.vetpar.2020.109094

Radostits OM, Gay CC, Hinchcliff KW, Constable PD (2007) Veterinary medicine: a text book of the diseases of cattle, horses, sheep, pigs and goats, 10th edn. Elsevier, London, pp 399–438

Rakhshandehroo E, Asadpour M et al (2017) The anthelmintic effects of five plant extracts on the viability of Parascaris equorum larvae 29(Issue4):219–224. https://doi.org/10.1111/eve.12676

Reinemeyer CR, Nielsen MK (2013) Handbook of equine parasite control (pp. ix+-210)

Reinemeyer CR, Nielsen MK (2014) Review of the biology and control of Oxyuris equi. Equine Vet Educ 26:584–591

Reinemeyer CR, Nielsen MK (2016) Control of helminth parasites in juvenile horses. Equine Vet Educ 29:225–232

Romaniuk K, Gad K (2007) Występowanie muchówki Hippobosca equina u koników polskich. Med Weter 63:1100–1101

Ruggles AJ, Beech J, Gillette DM, Midla LT, Reef VB, Freeman DE (1993) Disseminated Halicephalobus deletrix infection in a horse. J Am Vet Med Ass 203:550–552

Rose FR, Hodgson RDS (2000) Manual of Equine practice. 2nd Ed., USA, Saunders, Pp: 224

Round MC (1976) Lungworm infection (Dictyocaulus arnfieldi) of horses and donkeys. Vet Rec 99:393–395

Saini RK, Sankhala LN (2015) A case report of bovine hypodermosis. Int J Sci Environ Technol 4(3):725–726

Sallé G, Courtot E et al (2023) Spatio-temporal diversity and genetic architecture of pyrantel resistance in Cylicocyclus nassatus, the most abundant horse parasite. bioRxiv. Available from: https://doi.org/10.1101/2023.07.19.549683

Schwint N, Knowles D, Ueti M, Kappmeyer L, Scoles G (2008) Transmission of babesia caballi by dermacentor nitens (acari: ixodidae) is restricted to one generation in the absence of alimentary reinfection on a susceptible equine host. J Med Entomol 45(6):1152–1155

Scare JA, Steuer AE et al. (2018) Management practices associated with strongylid parasite prevalence on horse farms in rural counties of Kentucky. Vet Parasitol Reg Stud Rep 14:25–31. ISSN 2405-9390. https://doi.org/10.1016/j.vprsr.2018.08.002.

Schmitz DG, Chaffin MK (2004) What is your diagnosis? Halicephalobus gingivalis. J Am Vet Med Assoc 225:1667–1668

Scott DW (1988) Large animal dermatology. Saunders, Philadelphia

Sellon DC, Long MT. (2007) Equine infectious diseases / [edited by]– Second edition

Simsek S, Cevik A (2014) First detection and molecular characterization of Echinococcus equinus in a Mule in Turkey. Acta Parasit. 59:773–777. https://doi.org/10.2478/s11686-014-0308-1

Smith DG, Burden FA (2013) Practical Donkey and Mule Nutrition. In: Geor R, Coenen M, Harris P, editors. Equine Applied and Clinical Nutrition, Saunders. 304–316

Smith AB, Jones CD (2020) The impact of parasitic lungworms on equine health. J Vet Parasitol 45(3):234–245

Sokół R, Michalski M (2015) Occurrence of Hippobosca equina in Polish primitive horses during the grazing season. Ann Parasitol 61:118–122

Sokół R, Raś-Noryńska M, Michalczyk M, Raś A, Rapacz-Leonard A, Koziatek S (2015) Estimation of infection of internal parasites in horses from different type of farms, farms. Ann Parasitol 61(3):189–192

Sorrell MS, Fish RE, Taylo KH (2010) Pediculosis in two research ponies (Equus caballus). J Am Assoc Lab Anim Sci 49(4):487–490

Takele B, Nibret E (2013) Prevalence of gastrointestinal helminthes of donkeys and mules in and around Bahir Dar, Ethiopia. Ethiopian Vet J 17(1)

Tavassoli M, Yamchi JA, Hajipour N (2015) A survey on the prevalence of strongyles species in working donkeys in North-West of Iran. J Parasitic Dis 40:1210–1212

Taylor MA, Coop RL, Wall RL (2016) Veterinary parasitology, 4th edn. Wiley-Blackwell, Oxford

Tinkler et al (2024) First report of Besnoitia bennetti in Irish donkeys: an emerging parasitic disease in Europe. Irish Vet J 77:2. https://doi.org/10.1186/s13620-024-00263-2

Traub-Dargatz JL, Koral CA et al (2001) Estimate of the national incidence of colic and operation-level risk factors for colic among horses in the United States, spring 1998 to spring 1999. J Am Vet Med Assoc 219:67–71
Trivizaki S, Theodoropoulos G (2018) Application of polymerase chain reaction (PCR) in parasitology. J Hellenic Vet Med Soc 51(1):16–21. https://doi.org/10.12681/jhvms.15652
Vasey JR (1981) Equine cutaneous habronemiasis. Compend Contin Educ Pract Vet 3:S290–S300
Villa-Mancera A, Reynoso-Palomar A (2020) The prevalence and risk factors of liver fluke infection in Mexican horses, donkeys, and mules in tropical and temperate regions. Parasitol Res 119:3699–3703. https://doi.org/10.1007/s00436-020-06863-2
Von Tscharner C, Kunkle G, Yager J (2000) Stannard's illustrated equine dermatology notes. Vet Dermatol 11:163
Voronkova OB, Usachev II (2024) Comparative efficacy and anthelmintic resistance in therapy with ivermectin-type drugs in horses of herd and stable maintenance. Ветеринария, зоотехния и биотехнология. Available from: https://doi.org/10.36871/vet.zoo.bio.202404015
Wannas HY, Dawood KA, Gassem GA (2012) Prevalence of gastro-intestinal parasites in horses and donkeys in Al Diwaniyah Governorate. 11(1):148–155. Available from: 10.29079/VOL11ISS1ART183
Wells K, Gibson DI, Clark NJ, Ribas A, Morand S, McCallum HI (2018) Global spread of helminth parasites at the human–domestic animal–wildlife interface. Glob Change Biol 24:3254–3265. https://doi.org/10.1111/gcb.14064
Williams DJ, Hodgkinson J (2017). Fasciolosis in horses: A neglected, re-emerging disease. Equine Veterinary Education 29(4):202–204. https://doi.org/10.1111/eve.12521
Williams JC (2019) The epidemiology and control of Dictyocaulus spp. in domestic animals. Parasitol Res 118(2):567–574
Yevstafieva VO, Prykhodko YO et al (2020) Biological specifics of exogenous development of Oxyuris equi nematodes (Nematoda, Oxyuridae). Biosyst Divers 28(2):125–130. https://doi.org/10.15421/012017
Zhang Z (ed) (2013). Animal biodiversity: an update of classification and diversity in 2013. Animal biodiversity: An update of classification and diversity (Addenda 2013). Zootaxa 3703(1):5–11. https://doi.org/10.11646/zootaxa.3703.1.3
Zhang M, Wei K, Wu Z, Sun J, Hu J, Deng S, Tao J (2022) Morphological and molecular characterization of a Sarcocystis species infecting donkeys from China. Parasitol Res 121: 2917–2926. https://doi.org/10.1007/s00436-022-07616-2
Zhou M, Lu Y et al. (2023) Exploration of Parascaris species in three different Equus populations in China. Parasites Vectors 16:202. https://doi.org/10.1186/s13071-023-05768-3. https://aaep.org/resource/internal-parasite-control-guidelines/

Biomechanics in Donkeys and Mules

Rebecca S. V. Parkes and Syed. S. U. H. Bukhari

Abstract

This chapter explores the biomechanics of donkeys and mules, focusing on their locomotor patterns, anatomical adaptations, and interactions with saddlery, harness and terrain. While equine biomechanics is fairly well-studied in horses, research on donkeys and mules remains limited despite their global importance for draft and load carrying, particularly in low- and middle-income countries (LMICs). Donkeys are more efficient in their locomotion than horses, and gait characteristics can differ. Their different anatomical features and species-specific adaptations are likely affect movement and load-bearing capacity. The chapter also examines hoof conformation and the impact of saddlery and harness design, emphasizing the need for equipment that accommodates the distinct anatomy of donkeys and mules. By integrating existing equine research with emerging data on long-eared equids, this chapter advocates for targeted biomechanical studies to improve welfare, performance, and working conditions.

Keywords

Locomotor biomechanics · Conformation · Harness · Gait

R. S. V. Parkes (✉)
Hong Kong Jockey Club, Hong Kong SAR, China

Department of Veterinary Clinical Sciences, Jockey Club College of Veterinary Medicine and Life Sciences, City University of Hong Kong, Hong Kong SAR, China
e-mail: rebecca.sv.parkes@hkjc.org.hk

S. S. U. H. Bukhari
Department of Veterinary Clinical Sciences, Jockey Club College of Veterinary Medicine and Life Sciences, City University of Hong Kong, Hong Kong SAR, China

Centre for Animal Health and Welfare, Jockey Club College of Veterinary Medicine and Life Sciences, City University of Hong Kong, Hong Kong SAR, China

A. K. McLean et al. (eds.), *Donkey and Mule Medicine: Foundations, Breeds, and Management*, https://doi.org/10.1007/978-3-032-12726-6_22

1 Introduction

Locomotor biomechanics is the science of how animals move, and how they interact with the world around them. For equids, this includes the movement of the limbs and axial skeleton (spine and pelvis), hoof-surface interactions and the impact this has upon the body and limbs, and interactions between the tack, harness, rider and animal. Biomechanics is relatively well-studied in the horse but studies of donkeys and mules are in their infancy. This chapter will explore the principles of biomechanics in equids, what we know about donkeys and mules, and how we might apply some principles from horse research to their long-eared cousins.

2 Principles of Equine Biomechanics

Horses and donkeys share common ancestors, and although asses and zebras diverged from horses approximately 2.4 million years ago (Oakenfull and Clegg 1998), they have a similar musculoskeletal system and are well adapted for economical movement over long distances. Even the speed at which horses change gait can be explained by a need for efficiency; given the choice of when to change speed, horses will change from a walk to a trot and from a trot to a canter at the point that it expends relatively less energy to change gait (Taylor and Hoyt 1981). Donkeys are more efficient than horses, exhibiting lower energy expenditure than predicted for an animal with their body size, indicating further adaptations for efficient movement (Maloiy et al. 2009).

3 Donkey Gaits

Broadly speaking, donkeys seem to have the same gaits as horses. They walk, trot, canter, and gallop, and some breeds are gaited, for example, the Pêga donkey from Brazil. Mules can be bred to exhibit gaited characteristics, though the extent to which these are shown is likely to depend on the heritability of the gait in question.

One additional gait that warrants further exploration may be termed the 'amble', which has been reported in Andalusian donkeys (Navas and Delgado 2016) and in donkeys in Pakistan (Bukhari and Parkes 2022). Rather than exhibiting a true four-beat walk as horses do, many donkeys walk with a lateralised stride pattern nearing a pace rather than the true four-time walk which is usually observed in horses, Fig. 1. Donkeys moving in this way also have a shorter stride time, stance time and swing time (forelimbs only) in comparison to ponies walking at the same speed, but have similar swing phases in the hindlimbs (Bukhari and Parkes 2022).

This lack of understanding of donkey and mule gaits may have an impact on their veterinary care. Assessment of lameness in the donkey clinically is similar to assessing lameness in the horse (Thiemann and Poore 2019), but in many ways, it lags behind. Lameness scoring systems for the donkey are based on those for the horse (Khan et al. 2023), and although objective measures of lameness are now

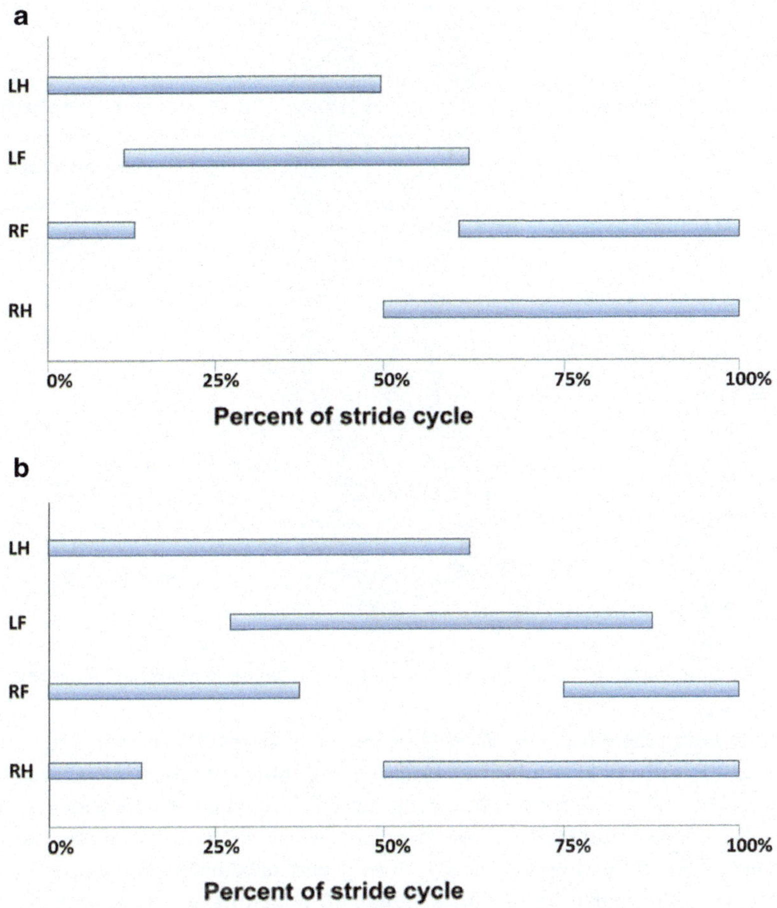

Fig. 1 Foot fall patterns of walking (**a**) Sperki breed donkeys and (**b**) horses

widely used to assess horses with several commercially available tools to support clinicians (e.g., Keegan et al. 2010; Lawin et al. 2023; Pfau et al. 2005), so far, no such studies have investigated objective lameness assessment for donkeys and mules.

The distal limb of equids is light, with reduced numbers of metacarpal/metatarsal bones. The superficial digital flexor tendon on the palmar aspect of the distal limb acts as a spring, storing energy when loaded and releasing it when unloaded (McGuigan and Wilson 2003), allowing for more efficient locomotion, (Fig. 2). A similar mechanism is seen in the *Biceps brachii* muscle of the horse (Wilson et al. 2003), allowing for reduced muscular effort during limb protraction. The nuchal ligament in the neck of the horse also stores up to 59% of the energy required to nod the head at the walk; this swinging motion is consistent with minimizing energy expenditure. The same movement pattern is seen in other species with mid-length

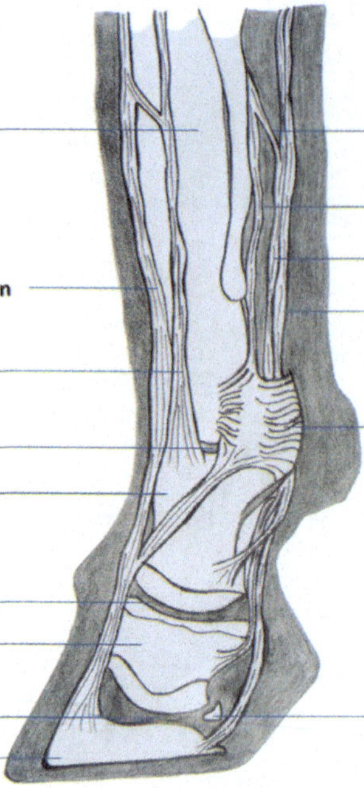

Fig. 2 Distal limb anatomy of the donkey showing joints, bones, tendons, and ligaments

necks, including the African wild ass (Loscher et al. 2016). It is reasonable to assume that given their physical similarities, donkeys and mules function in a similar way to horses and the African wild ass; however, there is almost no research focused purely on the locomotor biomechanics of these species, and there are some anatomical differences that could impact energy storing and efficiency. For example, in the donkey, an additional inferior check ligament, attaching to the superficial digital flexor tendon, appears to be present in at least some breeds such as miniature donkeys studied in Poland (Nazem and Sajjadian 2017).

During locomotion, energy is stored and released not only in the limbs but also in the back, which functions as an additional "spring" during canter and gallop in horses. The epaxial muscles and bones act like a "bow," while the sternum and abdominal muscles function as a "string." Contraction of the abdominal muscles, particularly the rectus abdominis, flexes the back by tensing the bow, a process also influenced by the retraction of the forelimbs or the protraction of the hind limbs. The fusion of the aponeuroses of the *gluteus medius* and *longissimus dorsi* muscles may contribute to this mechanism. Conversely, the back extends when the 'string' is tensed through the protraction of the forelimbs, retraction of the hind limbs, and the weight of the abdominal organs (van Weeren 2004). Unlike horses and ponies, donkeys have a flatter back profile cranially to caudally, with less prominent withers (Maśko et al. 2022) and are anecdotally reported to exhibit less vertical back movement during locomotion, though this has not been scientifically quantified. During load carrying, horses experience increased extension of the spine, particularly in the lumbosacral area, and when carrying a rider, the rider's posture impacts on spinal movement and limb kinematics (Clayton et al. 2023). It has been suggested that their reduced back movement and flatter back is why donkeys and mules make excellent pack animals. However, the impact of these structural differences on locomotion, load-carrying ability, and overall back function in donkeys remains unclear, highlighting the need for further research.

4 Foot Surface Interactions and Working in Different Situations

Interactions between the hoof and the ground influences musculoskeletal adaptation and injury risk in the horse, especially as horses lack dynamic adaptability seen in some other species. Both internal and external factors influence the mechanical interaction between the hoof and the ground, including hoof shape, shoeing, gait, surface properties such as hardness, and surface maintenance (Parkes and Witte 2015). Despite the fact donkeys and mules have a reputation for being well-adapted to working on difficult terrain, little is known about any adaptations to this work. We know that morphologically, donkeys have a more upright hoof than horses, and therefore donkeys should be trimmed accordingly; it is likely to be detrimental to trim donkey hooves to more closely resemble horse hooves, as lower heels increase force on the navicular bone in horses (Eliashar et al. 2004), and based on the anatomical similarity, we can assume that the impact would be similar in donkeys. Donkeys have a wider frog than horses, while mules fall somewhere in between the two (Phannithi et al. 2024); we know that, like horses, hoof asymmetry and conformational abnormalities are linked to lameness in donkeys (Khan et al. 2023). Donkeys have also been reported to bear weight in a five-point pattern, at each quarter, the heel and the toe, with some weight bearing on the sole at the toe, which is in contrast to horses bearing weight predominantly along the hoof wall (Thiemann and Poore 2019). What is not clear is how these factors might influence the ability of these species to work on difficult terrain, and there are no objective assessments of hoof-ground interactions in donkeys or mules.

Given that donkeys and mules frequently work on steep terrain, a fuller understanding of the impact of this on the animals would be beneficial. We know that in the horse, running on an incline shifts their weight onto the quarters, and on a decline, weight is shifted to the forelimbs (Parsons et al. 2008). The same is likely true in donkeys and mules, as we know that working equids are more prone to wither wounds with incorrectly fitted packs if going downhill (Sells et al. 2010).

5 Saddlery and Harness

The impact of tack and harnesses is a relatively new area in equine research. However, tack has been shown to have a significant impact on performance in sports horses, and a deeper understanding of the mechanics of equid-tack interactions has the potential to improve welfare. For example, studies assessing the impact of noseband pressure have led to modifications to FEI (Fédération Equestre Internationale) dressage rules and the implementation of formal, objective monitoring of noseband tightness in competition (FEI 2025). We know that for horses, too wide saddles increase pressure on the cranial spine and too narrow saddles increase pressure on the back caudally, and that saddle fit also impacts on spinal kinematics (MacKechnie-Guire et al. 2019). We also know that peak pressures under the girth correspond to the most common locations for 'girth galls', and that modification of girth shapes reduce this pressure (Murray et al. 2013). While it is relatively simple to apply these principles to donkey and mule harnessing and saddlery, the design of tack for these species would benefit hugely from species-specific research; however, at present, there is not even a benchmark for 'normal' back movement in the donkey or the mule.

We do know a little about appropriate harness and tack for donkeys based on objective data. A study by Rodrigues et al. (2021) evaluated seven different collar types used on working donkeys in Europe to understand their efficiency and impact on animal welfare. Using pressure mapping technology, the researchers assessed how each collar distributed pressure across the donkey's neck and shoulders during work. The findings indicated significant variations in pressure distribution among the collars, highlighting that certain designs, particularly those designed for horses, exerted uneven pressure, potentially leading to discomfort or injury. The study emphasizes the importance of selecting collars specifically designed for donkeys, as their anatomical structure differs from that of horses. Ill-fitted or poorly designed collars can cause inefficient power transfer and may increase injury risk. Therefore, it is crucial to choose equipment that accommodates these anatomical differences to ensure both efficiency and the well-being of the animal (refer to chapters 13 and 14).

Understanding how loading impacts animals is not the same as implementing changes to improve their welfare. While we know that stride parameters change in response to loading, helping us broadly estimate appropriate load limits, there is significant individual variation. This poses a challenge for workers in the field, who may struggle to assess the impact of loading on each animal accurately. Further research is required to understand the biomechanical differences between donkeys, mules, and horses, particularly in relation to the spine, limbs, and the effects of loading. A critical question remains: how much weight is appropriate for these animals to carry and/or pull safely?

Despite their importance for livelihoods, donkeys and mules have received far less research attention than horses, largely due to funding priorities. Even as mechanization increases, these animals continue to play vital roles in transport and labor. Research into the best tack options for horses has expanded in the last decade, particularly concerning saddles, nosebands, bits, and girths. However, little progress has been made in studying draught animals, leaving their equipment design largely based on traditional practices rather than scientific evidence. Additionally, while studies on load-bearing in horses often focus on the weight of human riders, working equids frequently carry a wide range of inert loads or human riders, and the way that these loads interact and place pressure on the back may differ. More research is needed to ensure their welfare and optimize their working conditions.

6 Conclusion

Donkeys and mules are frequently used for load carrying and draft work, both for leisure and as critical sources of labor in LMICs. Despite this, their locomotor biomechanics and interactions with loads and tack remain highly under studied in comparison to horses, and objective assessment of lameness in donkeys and mules has not yet been attempted. Biomechanical studies in horses are leading to objective assessments, alterations in saddlery and competition rules to improve welfare; while we can apply basic principles from horse biomechanics to donkeys and mules, anatomical differences and differences in use are likely to compromise these comparisons. As working equids continue to support livelihoods worldwide, targeted biomechanical research should become a priority.

Study Questions

1. **What Structure in the Equine Distal Limb Acts to Store Energy during Locomotion?**
   A. The navicular bone
   B. The fetlock joint
   C. The accessory ligament of the deep digital flexor tendon
   D. The superficial digital flexor tendon
2. **Donkey hoof anatomy differs from horse hoof anatomy. What is one way in which their anatomy differs?**
   A. More upright hoof-pastern axis*
   B. Mediolateral asymmetry is common
   C. The frog is narrower
   D. Donkeys bear weight mostly on the hoof wall
3. **What is a known consequence of poorly fitted packs on donkeys working downhill?**
   A. Increased hoof wear
   B. Reduced stride length
   C. **Wither wounds***
   D. Abdominal strain

References

Bukhari SSUH, Parkes RSV Quantification of basic stride parameters of walking donkeys international conference on equine exercise physiology, Upsalla, Sweden, June 2022

Clayton HM, MacKechnie-Guire R, Hobbs SJ (2023) Riders' effects on horses—biomechanical principles with examples from the literature. Animals 13:3854. https://doi.org/10.3390/ani13243854

Eliashar E, McGuigan MP, Wilson AM (2004) Relationship of foot conformation and force applied to the navicular bone of sound horses at the trot. Equine Vet J 36(5):431–435. https://doi.org/10.2746/0425164044868378

Fédération Équestre Internationale (2025) FEI dressage rules—changes for 2025. Available at: https://inside.fei.org/fei/disc/dressage/rules/changes-2025. Accessed 28 Feb 2025

Keegan KG, Dent EV, Wilson DA, Janicek J, Kramer J, Lacarrubba A, Walsh DM, Cassells MW, Esther TM, Schiltz P, Frees KE, Wilhite CL, Clark JM, Pollitt CC, Shaw R, Norris T (2010) Repeatability of subjective evaluation of lameness in horses. Equine Vet J 42(2):92–97. https://doi.org/10.2746/042516409X479568

Khan RZU, Rosanowski SM, Parkes RSV (2023) Hoof morphometry in a population of lame and nonlame working donkeys in Pakistan. Equine Vet J 55(3):435–445. https://doi.org/10.1111/evj.13861

Lawin FJ, Byström A, Roepstorff C, Rhodin M, Almlöf M, Silva M, Andersen PH, Kjellström H, Hernlund E (2023) Is markerless more or less? Comparing a smartphone computer vision method for equine lameness assessment to multi-camera motion capture. Animals (Basel) 13(3):390. https://doi.org/10.3390/ani13030390

Loscher DM, Meyer F, Kracht K, Nyakatura JA (2016) Timing of head movements is consistent with energy minimization in walking ungulates. Proceedings of the Royal Society B: Biological Sciences. Nov 30;283(1843):20161908. https://doi.org/10.1098/rspb.2016.1908

MacKechnie-Guire R, MacKechnie-Guire E, Fairfax V, Fisher D, Fisher M, Pfau T (2019) The effect of tree width on thoracolumbar and limb kinematics, saddle pressure distribution, and thoracolumbar dimensions in sports horses in trot and canter. Animals (Basel) 9(10):842. https://doi.org/10.3390/ani9100842. PMID: 31640213; PMCID: PMC6827167

Maloiy GMO, Rugangazi BM, Rowe MF (2009) Energy expenditure during level locomotion in large desert ungulates: the one-humped camel and the domestic donkey. J Zool 277:248–255

Maśko M, Wierzbicka M, Zdrojkowski Ł, Jasiński T, Sikorska U, Pawliński B, Domino M (2022) Comparison of donkey, pony, and horse dorsal profiles and head shapes using geometric Morphometrics. Animals (Basel) 12(7):931

McGuigan MP, Wilson AM (2003) The effect of gait and digital flexor muscle activation on limb compliance in the forelimb of the horse *Equus caballus*. J Exp Biol 206:1325–1336

Murray R, Guire R, Fisher M, Fairfax V (2013) Girth pressure measurements reveal high peak pressures that can be avoided using an alternative girth design that also results in increased limb protraction and flexion in the swing phase. Vet J 198(1):92–97

Navas FJ, Delgado JV (2016) Ambling and walk gaits preliminary characterization and therapeutic application in donkeys. In: X Iberian congress on animal genetical resources, p. 118. Sociedade Portuguesa de Recursos Genéticos Animais (SPREGA)/Sociedad Española para los Recursos Genéticos Animales (SERGA) Castelo Branco, Portugal

Nazem MN, Sajjadian SM (2017) Anatomical transverse magnetic resonance imaging study of ligaments in palmar surface of metacarpus in Miniature donkey: identification of a new ligament. Folia Morphol (Warsz) 76(1):110–116

Oakenfull EA, Clegg JB (1998) Phylogenetic relationships within the genus *Equus* and the evolution of alpha and theta globin genes. J Mol Evol 47(6):772–783. https://doi.org/10.1007/pl00006436

Parkes RS, Witte TH (2015) The foot-surface interaction and its impact on musculoskeletal adaptation and injury risk in the horse. Equine Vet J 47(5):519–525. https://doi.org/10.1111/evj.12420

Parsons KJ, Pfau T, Wilson AM (2008) High-speed gallop locomotion in the thoroughbred racehorse. I. The effect of incline on stride parameters. J Exp Biol 211(Pt 6):935–944. https://doi.org/10.1242/jeb.006650

Pfau T, Witte TH, Wilson AM (2005) A method for deriving displacement data during cyclical movement using an inertial sensor. J Exp Biol 208:2503–2514

Phannithi T, Laikul A, Pathomsakulwong W, Rungsri P, Apichaimongkonkun T, Watchrarat K, Cherdchutham W (2024) External hoof measurements of untrimmed and unshod mules in northern Thailand. Animals 14:1197

Rodrigues JB, Garrett C, Norris SL, Albuquerque F, Queijo L, Cooke F, Judge A (2021) Collar pressure mapping: an evaluation of seven collar types used on working donkeys in Europe. Vet Rec 189(8):e772. https://doi.org/10.1002/vetr.772

Sells PD, Pinchbeck G, Mezzane H, Ibourki J, Crane M (2010) Pack wounds of donkeys and mules in the Northern High Atlas and lowlands of Morocco. Equine Vet J 42(3):219–226

Taylor CR, Hoyt DF (1981) Gait and energetics of locomotion in horses. Nature 292:239–240

Thiemann AK, Poore LA (2019) Hoof disorders and farriery in the donkey. Vet Clin North Am Equine Pract 35:643–658

van Weeren PR (2004) Structure and biomechanical concept of the equine back. Pferdeheilkunde 20(4):341–348

Wilson AM, Watson JC, Lichtwark GA (2003) Biomechanics: a catapult action for rapid limb protraction. Nature 421(6918):35–36. https://doi.org/10.1038/421035a

Orthopedic Considerations in Donkeys and Mules

Tracy A. Turner

Abstract

Orthopedics is the branch of veterinary medicine concerned with the diagnosis, treatment, and prevention of diseases and injuries related to the musculoskeletal system, which includes bones, joints, ligaments, tendons, and muscles. In donkeys and mules, orthopedic issues can significantly affect their mobility, performance, and quality of life. Horse orthopedic issues are well documented in textbooks and peer reviewed manuscripts. Theoretically, since the species are similar, the same issues affecting horses can affect donkeys and mules, but that's not always true. Understanding the unique anatomical and physiological characteristics of donkeys and mules is crucial for effective orthopedic care, as their musculoskeletal system differs in several ways from that of horses.

This chapter explores the most common orthopedic conditions in donkeys and mules, diagnostic methods, treatment strategies, and preventive care practices to promote long-term musculoskeletal health.

Keywords

Musculoskeletal · Orthopedics · Ligaments · Tendons · Joints

T. A. Turner (✉)
American College of Veterinary Surgeons, American College of Veterinary Sports Medicine and Rehabilitation (equine), American Academy of Thermology, Turner Wilson Equine Consulting, Stillwater, MN, USA

A. K. McLean et al. (eds.), *Donkey and Mule Medicine: Foundations, Breeds, and Management*, https://doi.org/10.1007/978-3-032-12726-6_23

1 Anatomy of the Musculoskeletal System in Donkeys and Mules

The musculoskeletal system provides structure and enables movement. Donkeys and mules share similar musculoskeletal features with horses but exhibit unique traits that influence their orthopedic care. The skeletons of donkeys and mules are robust and well-adapted to weight-bearing tasks. Their limb bones, especially in donkeys, are denser and more compact than those of horses, allowing them to endure harsh terrain and heavy loads. Mules, as hybrids, inherit characteristics from both their horse and donkey parents, resulting in musculoskeletal traits that are distinct from either species.

Donkeys have lower withers and more prominent anterior manubria (the sternum protrudes further cranially), and the donkey pelvis is a different shape. (1) The shape of the donkey sacrum and the spacing and direction of the spines of the sacrum and coccygeal vertebrae differ in the donkey. Also, the location of the end of the spinal cord and the end of the dura differs from those of the horse. Injection of a local anesthetic or other analgesic into the epidural region necessitates knowledge of these differences. The sacrum of the horse is a relatively flat bone with five segments. The donkey's sacrum also has five segments, but often the first coccygeal vertebra is fused to the sacrum with occlusion of the sacro-coccygeal space. Further, the donkey sacrum is curved dorso-ventrally, directing the spinal canal ventrally in the caudal segments. The sacral spines point backwards and rapidly decrease in length caudally. The spinal canal slopes backward more steeply to end at the third coccygeal vertebra. The spinal cord ends at the second sacral vertebral segment, whereas the dural sheath extends back to the first or second coccygeal vertebra. The coccygeal vertebrae are much better developed in donkeys than in horses with the vertebral arches of the first three being complete and each having interarcuate ligaments. The first intercoccygeal space in the donkey is narrower than the second space. In horses, the spinal cord ends at the junction of the first and second sacral vertebral segments and the dural envelope at the fourth sacral segment. The vertebral arches of the second and third coccygeal vertebrae of horses are incomplete with the absence of any interarcuate ligaments. In the horse, epidural anesthesia is injected in the first inter-coccygeal space at an angle of about 45° from the horizontal. In the donkey, the injection site is in the second inter-coccygeal space, directing the needle at an angle of 30° from the horizontal. This site is more suitable because there is a wide interval between the second and third coccygeal vertebral spines, and the anesthetic can be injected forward under the second coccygeal spine. The needle can be introduced into the vertebral canal here because there are no large tail muscles, and the spines of the sacral segments and the coccygeal spines are more easily palpated in the donkey.

Donkeys have tighter ligaments and more upright joints compared to horses, which can influence their susceptibility to certain joint issues. Mules, being hybrids, may have a variable range of joint characteristics depending on their specific genetic makeup.

Donkeys and mules tend to have a higher proportion of slow-twitch muscle fibers, which contribute to their stamina and ability to work for extended periods (Li et al. 2022). Comparative Transcriptome Analysis of Slow-Twitch and Fast-Twitch Muscles in Dezhou Donkeys reports that the psoas major muscle (PM) of Dezhou donkeys had a fiber area percentage (FAP) of slow-twitch (Type I fibers) around 56%, which is significantly higher than in other muscles tested like the longissimus dorsi and biceps femori This muscle composition, while beneficial for endurance, requires special attention when managing orthopedic conditions related to muscle strain and overuse. Because donkeys and mules rely so heavily on slow-twitch endurance fibers, they can work longer with fewer outward signs of fatigue, but this same trait predisposes them to cumulative microtrauma, muscle strain, and orthopedic stress injuries if not carefully managed with rest, nutrition, and proper load distribution.

1.1 Why High Slow-Twitch Muscle Fiber Composition Can Lead to Strain/Overuse Issues

1. **Endurance vs. Recovery**
   - Slow-twitch (Type I) fibers are fatigue-resistant and excel at sustained, low-intensity activity (like walking long distances or carrying loads).
   - However, they are less powerful and produce less force per contraction compared to fast-twitch fibers. When donkeys or mules are asked to perform repetitive, prolonged work, these muscles stay engaged for hours with little rest, which can lead to chronic strain.
2. **High Workload, Low Relief**
   Donkeys and mules are often used for pack, draught, or repetitive agricultural labor in difficult terrain. Because their musculature is adapted for endurance, they can mask fatigue longer than horses. This means handlers may not notice early signs of overuse until micro-damage accumulates.
3. **Orthopedic Stress**
   - Slow-twitch fibers rely on oxidative metabolism and require a continuous oxygen supply. Prolonged activity can cause localized ischemia or oxidative stress, especially if tack, harnesses, or poor hoof balance alters biomechanics.
   - Over time, this can contribute to chronic muscle soreness, tendon strain, and even joint overload, since the musculoskeletal system must continuously stabilize under load.
4. **Limited Fast-Twitch Support**
   - Horses have a greater proportion of fast-twitch (Type II) fibers that can generate powerful bursts to assist when muscles are stressed. Donkeys and mules, with fewer Type II fibers, may lack this "backup power," leaving slow-twitch units to absorb the bulk of repetitive strain.

1.2 Hoof Conformation of Donkeys and Mules Effects on Soft Tissue

Hoof shape has a direct effect on how forces are distributed through the limb, which in turn can influence soft tissue stress and injury risk in donkeys and mules. Donkeys have small boxy hooves with thicker hoof walls and a more upright hoof angle.

The donkey hoof has a distinctive oval shape compared with the horse and generally has a dorsal hoof wall that is 5–10° more upright. In the healthy donkey foot, the frog is well developed, especially at the palmar aspect but the apex does not extend under the pedal bone as in the horse. The sole is thick and the wall thickness remains constant from toe to heel. The narrow conical shape of the hoof has allowed the donkey to be used on narrow trails and to be sure footed on difficult terrain.

Compared with the horse the mule's hoof has a distinctive upright dorsal wall angle and a broken forward hoof-pastern axis; the cartilages of the distal phalanx are much more developed towards the palmar/plantar parts of the hoof; the inclination degree of the hoof wall to the ground, at the toe, the quarters and heels is almost vertical; the sole has elevated moisture content in its deep layers and the horn tubules are more evident; the coronary dermis together with the large coronary groove are higher; the laminar dermis of the wall segment is less extended; the mule's hoof is smaller, longer and narrower.

1. **Donkey Hoof Shape vs. Horse Hoof**
   - Donkeys have a more upright hoof capsule, a straighter dorsal wall, a thicker sole, and a broader frog compared to horses.
   - Mules typically have hooves shaped somewhere between a donkey and a horse, but often narrower and tougher.

   Because of these differences:
   - A more upright hoof shifts load more vertically, reducing breakover leverage but increasing stress on deep flexor tendons and the navicular region if trimmed incorrectly.
   - If donkey hooves are trimmed to mimic horse hooves (lower heels, longer toe), this alters their natural biomechanics and can increase strain on the deep digital flexor tendon (DDFT) and navicular bone (Eliashar et al. 2010).
2. **Shock Absorption and Sole Loading**
   - Donkey soles are naturally thicker and tougher, adapted for arid, rocky terrain. This makes them less prone to bruising but also less flexible, which may reduce shock absorption.
   - Increased concussion on hard ground may then shift more stress to the suspensory ligament, superficial digital flexor tendon (SDFT), and distal interphalangeal joint.
3. **Frog and Weight Distribution**
   - Donkeys have a broader frog that participates more in weight-bearing. This helps spread forces across the hoof, protecting soft tissue structures.

  - If the frog is underdeveloped (e.g., due to poor trimming or environment), the hoof loses this support, concentrating stress on the laminar tissues and flexor tendons.
4. **Terrain and Hoof Mechanics**
  - Donkeys and mules often work on steep or uneven terrain, where hoof conformation determines how effectively forces are absorbed.
    - On uphill terrain, stress shifts more to the hind limb suspensory and flexor apparatus.
    - On downhill terrain, stress increases on forelimb tendons, ligaments, and shoulders ().

In summary, the natural upright, tough hoof of donkeys and mules is well adapted to arid, rocky terrain, but if improperly trimmed (to resemble horse hooves) or worked on hard/steep surfaces, this can increase strain on soft tissues such as tendons, ligaments, and the navicular apparatus, leading to higher injury risk.

2 Common Orthopedic Conditions

Donkeys and mules are prone to several orthopedic issues due to their body structure, environment, and workload. Some of the most common conditions are discussed below.

2.1 Laminitis

Laminitis is one of the most prevalent and serious orthopedic conditions in donkeys and mules. It involves inflammation of the laminae within the hoof, leading to severe pain, lameness, and, in chronic cases, hoof deformities.

Laminitis is the most serious disease of the equine hoof because it causes pathological changes in anatomy that lead to a devastating loss of function. The simplest definition of laminitis is: failure of the attachment between the distal phalanx (coffin bone) and the inner hoof wall. A horse has laminitis when the lamellae of the inner hoof wall, which normally suspend the distal phalanx from the inner surface of the hoof capsule, fail. Without the distal phalanx properly attached to the inside of the hoof, the weight of the horse and the forces of locomotion drive the bone down into the hoof capsule, shearing and damaging arteries and veins, crushing the corium of the sole and coronet, causing unrelenting pain and a characteristic lameness. Numerous causes of the disease have been defined but the leading cause remains gastrointestinal disturbances. Although the condition primarily affects the foot, the disease is a systemic disease which causes disturbances in most body systems. The disease can be arbitrarily divided into four phases: developmental, acute, chronic, and post chronic. But it must be remembered that these phases overlap.

The acute phase of laminitis which lasts from the onset of clinical foot pain and lameness at the trot, to the time when there is clinical (usually radiological) evidence of displacement of the distal phalanx within the hoof capsule. The signs of this phase include: "Bounding" digital pulse, warm feet, "camped in front" stance, and abnormal gait. The severity of the lameness relates to the severity of damage to the laminae. The severity of lameness as outlined by the Obel grading system can be helpful in determining the severity of laminar damage.

Laminitis can also affect donkeys, although the causes of the syndrome may be different between the species. Veterinary opinions vary on whether donkeys can develop the syndrome.

Horses develop laminitis because they are overweight and eat a diet high in carbohydrates. In the publication, *The Horse* Dr. Nicola Menzies-Gow, Diplomate of the European College of Equine Internal Medicine in England, states that donkeys affected by laminitis are younger than horses affected, and are not always overweight, don't eat too many carbohydrates, or have other health problems (Menzies-Gow et al. 2021).

Researchers found that nearly 50% of the donkeys studied developed at least one episode of laminitis compared to 34% of horses and ponies. Forty percent of the donkeys had at least one recurrent episode, and most of the donkeys had chronic, rather than acute, laminitis (Menzies-Gow et al. 2021). Laminitis involves inflammation of the structures in the hoof wall that hold the coffin bone in place. Inflammation of the laminae is extremely painful (Menzies-Gow et al. 2021; Lesté-Lasserre 2022). Cushing's disease in donkeys can contribute to the development of laminitis as it can in horses (Lesté-Lasserre 2022).

While most horses develop the disease in the spring due to ingesting spring grass, most cases in donkeys in England occurred in the winter from October to February when the animals were coming off pasture and into stalls (Menzies-Gow et al. 2021). It is believed laminitis may be related to the environment where the donkeys were walking on concrete in a barn, and the lameness was simply more noticeable on concrete versus out on the pasture.

In horses a common scale used to determine lameness is the Obel grading system from 1 to 4.

Obel grade 1: the equid exhibits constant shifting of its weight from leg to leg. The horse's gait is stiff and stilted at the trot but not the walk.

Obel grade 2: lameness is characterized by a stiff and stilted gait at the walk and trot. The horse is reluctant to rest full weight on a leg, but a leg can still be held up.

Obel grade 3: shows increasing degree of the stiff and stilted gait. It is nearly impossible to hold up one of the horse's legs. Permanent morphologic change occurs within 12 h after the onset of Obel grade 3 lameness.

Obel grade 4: is so severe that the horse spends much of its time recumbent.

The Obel scale can be used in donkeys and mules, but with caution—it may underestimate severity due to species differences in pain expression and locomotion.

Many clinicians supplement it with other tools (hoof inspection, digital pulse, radiographs, stance assessment).

Key Adaptations to Obel Scale for Donkeys and Mules

- Donkeys in particular may *mask pain* — standing quietly and appearing "normal" despite advanced pathology. Careful observation of stance, digital pulses, and hoof temperature are critical.
- Mules often fall between horse and donkey presentations; they may show more obvious lameness than donkeys, but subtler than horses.
- Chronic laminitis is far more common in donkeys than acute laminitis, so hoof morphology must always be considered in grading.

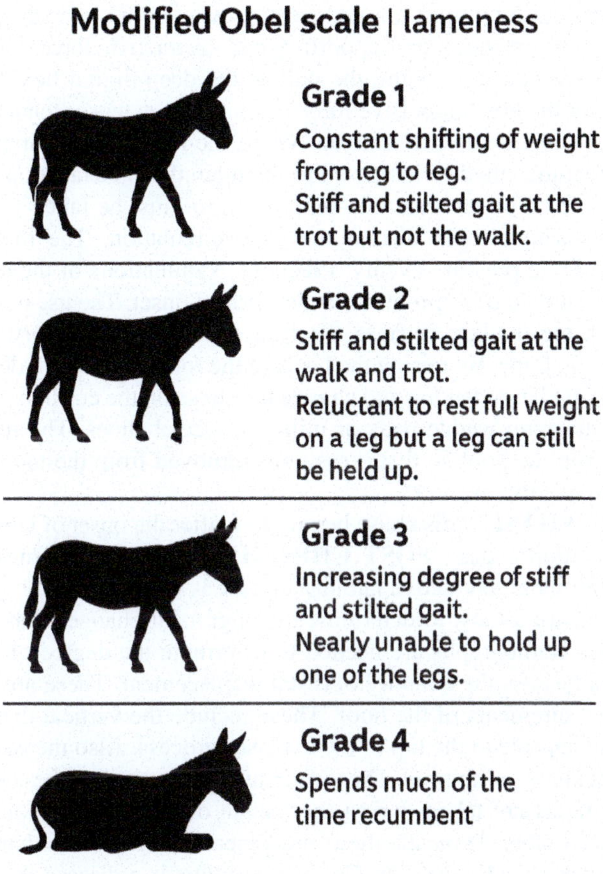

The lameness is due to pain and inflammation from ischemia to the secondary laminae. This phase is of short duration lasting only 48 h or when coffin bone displacement occurs.

Treatment is to prevent coffin bone rotation. This is achieved by continuing vasodilator and antithrombotic therapy. Anti-inflammatory therapy is important to reduce laminar swelling, ease pain and therefore indirectly aid the circulation. Non-steroidal anti-inflammatory drugs (NSAIDs) are among the most effective drugs. Phenylbutazone dosage starts at 4.4 mg/kg twice daily. After the initial effect, dosage should be reduced to the minimum effective dose. This is in order to reduce the possibility of adverse side effects. Flunixin meglumine is the other most commonly used NSAID. A dose of 1.1 mg/kg is administered either intramuscularly or intravenously twice daily. In addition to its anti-inflammatory effects, Flunixin has beneficial anti-endotoxin properties. Dimethyl sulfoxide (DMSO) is a potent anti-inflammatory that is administered at 1 gm/kg intravenously as a 20% solution. The treatment can be repeated daily or every other day. Some clinicians soak the equid's feet in DMSO. Corticosteroids are absolutely contraindicated. They potentiate the effects of catecholamines and can exacerbate the disease.

Care of the foot is of utmost importance to prevent coffin bone displacement. The foot is in equilibrium between supportive and distractive forces. The distractive forces include the equid's weight, the pull of the deep digital flexor tendon and a shearing force at the toe. Supportive forces include the extensor tendon, the sole and frog, and most importantly the laminae. When the laminae are damaged, the other supportive structures must be increased to maintain the skeletal support. Packing of the frog and sole helps increase this support. Care must be taken not to apply too much pressure which could disrupt the solar circulation. The material must be applied to distribute pressure evenly. Frequent examinations of the hoof are necessary to prevent or recognize problems at the earliest onset. The use of the heart bar is less helpful in the donkey because the frog does not extend forward enough to support the coffin bone. By raising the heels of the foot 12–20° will decrease the pull of the deep digital flexor tendon and change the stress on the coffin bone. This can be achieved by attaching wedge pads or utilizing special shoes. The more traditional method is to trim the hoof so that pressure is removed from the toe so that support can be applied evenly.

CHRONIC PHASE: This phase begins 48 h after the onset of lameness or when coffin joint rotation occurs. **POST CHRONIC PHASE** is lameness due to third phalanx displacement, but active laminar disease is over.

Radiographs are a must to determine the alignment changes that have occurred within the foot. Radiographs are needed to determine the degree of rotation of the coffin bone and assess the quantity of distal displacement. These are both achieved by making measurements of the hoof. These include the vertical distance from the top of the hoof capsule to the top of the extensor process. Also measuring the dorsal hoof wall thickness is helpful. These measurements tend to be variable but the sooner radiographs are taken during the course of disease the more accurate one can assess the problem. Typically the vertical measurement is less than 5 mm and the hoof wall thickness is 18–20 mm. Chronic laminitis is a disease that is principally treated mechanically in part due to the difficulty of early detection. Once the disease has been detected through signalment or diagnostic imaging, significant mechanical damage has occurred to the supportive tissues of the hoof. An important aspect of mechanical management involves the stabilization of these structures to reduce further mechanical damage, to minimize inflammation and to control pain *(additional information on shoeing can be found in detail in the podiatry and farrier chapters)*.

Although the relative importance of different management practices remains controversial, the goals of these practices are generally based on mediating the detrimental biomechanical principles that are responsible for the exacerbation of this disease. These principles include at least the following:

Deep digital flexor tendon (DDFT) loading. This tendon serves to connect the flexor digitorum profundus muscle situated in the upper limb to the third phalanx within the hoof. Part of the function of this tendon/muscle group is to assist in flexing the joints of the pastern. During laminitis, the stress on this tendon can exacerbate the rotation of the third phalanx.

Breakover. Bending forces on the front of the hoof wall, particularly during breakover in the posterior aspect of the stance phase of the stride, can exacerbate damage to the hoof wall and supporting lamina during laminitis.

Skeletal loading on the distal phalanx. As third phalanx rotation occurs during laminitis, the extensor process of the third phalanx rotates dorsally. This change in the axial alignment of the bone can intensify the mechanical forces of the skeletal load that cause rotation.

Solar compression. As the third phalanx rotates, the dorsal aspect of the solar margin of the third phalanx can begin to compress the vasculature of the sole, particularly at the toe of the hoof. If sufficient trauma occurs, solar bruising and/or necrosis can occur.

Treatment is to maintain or reestablish the coffin bone/hoof wall alignment, which is achieved by shoeing to support the skeletal column. Shoeing approaches include at least the following *(additional information on shoeing can be found in the Podiatry and Farrier chapters)*:

1. *Modified toes.* Shoes designed to lessen the effect of the DDFT load and to minimize break over stresses of the dorsal hoof wall usually feature some type of toe modification. This may include a rocker toe, a square toe, or a rolled toe, or a shoe that is "set back" under the toe of the hoof.
2. *Sole protection or relief.* Shoes designed for this purpose will prevent direct, non-uniform pressure on the sole. These shoes may feature a concave sole surface, a wide web, a "treatment plate" or pad, or the shoes may be placed on backwards.
3. *Wedges.* To mediate the adverse effect of the DDFT, wedges may be incorporated in the shoe to elevate the heels of the hoof. In some cases the wedge may be reversed to elevate the toe with the intent of mediating the adverse effect of skeletal loading and restoring normal alignment of the third phalanx after rotation has occurred.
4. *Sole support.* Some shoes may feature some type of sole packing or frog support to provide support to the ventral aspect of the third phalanx. The features of this shoe may include a sole packing of a variety-of-hardnesses that covers all or a portion of the sole. These shoes may also feature a "heart bar" that supports the

frog but some treatments that work well for horses may not work the same on a donkey or mule considering the differences in hoof conformation (e.g. heart bar shoe).
5. *Wall support.* The hoof wall of the laminitic equine will become weakened with the increasing loss of internal structural support as the disease progresses. To address this situation a bar shoe may be used to aid the structural integrity of the

Shoes that have been commonly used for donkeys and mules are reverse shoes with frog packing, rail shoes, and modifications of the above make sure and work with your veterinarian, podiatrist and farrier as a team to come up with the best solution for your donkey or mule's condition. Selection should be based on the support needs of the foot, secondary pathology within the foot, and the skill levels of the farrier.

Hoof wall resection is a technique that removes or thins the hoof wall down to the level of the sensitive laminae. It is used to expose necrotic tissue, provide drainage of infected laminae, and relieve pressure on the coronary band to allow proper hoof wall growth and to achieve better stabilization of the bone with less impairment to the circulation.

Tendon releasing procedures to either decrease or relieve the pull of the deep flexor tendon on the coffin bone. Inferior check desmotomy reduces deep flexor tendon pull but may not provide enough relief. Deep flexor tenotomy eliminates the deep flexor pull. It can be performed either standing or recumbent. The location of the cut can be either at the mid metacarpal level or at the pastern.

PROGNOSIS is dependent on numerous factors but the most important are:

A. Severity of damage
B. Duration of signs
C. Response to treatment

3 Developmental Orthopedic Disease (DOD)

Donkeys and mules are subject to the same growth-related issues as horses, but they occur less commonly. Genetics, in utero nutrition and positioning may play a role. In horses, the push for fast growth is a prime factor in causing these issues. Donkeys and mules are not under these pressures and therefore far fewer DOD problems occur.

DOD can be loosely defined as any musculoskeletal problem that occurs in the young animal as a result of growth and maturation. Invariably it is related to the process of endochondral ossification and the maturation of cartilage to bone in the developing skeletal system. In the horse there are 4 clinical manifestations; physitis, osteochondrosis, angular limb deformity, and flexural limb deformity. Etiological factors involved include trauma, excessive concussion, illness, conformation, genetic predisposition, and nutrition.

DOD in animals under 4 months of age is due to inadequate intake of a balanced ration, nursing from a mare who is an excellent milker, or some other cause of a rapid growth rate in the foal. Nutritional factors related to DOD in animals under 8 weeks of age include unbalanced mineral intake of either major or trace minerals, or excessive quantities of mare's milk or unbalanced mineral in the milk. DOD. in weanlings and yearlings is usually due to a ration incorrectly balanced, or a rapid growth rate, or excess body weight.

An example of an unbalanced diet would be a mature horse diet that is fed to any horse under 2 years of age. Feeding 128% or more of energy requirements will produce developmental orthopedic disease. In donkeys and mules were are still learning about DOD and know it has been reported and seen in larger donkey breeds such as the Mammoth. A key to a healthy diet for donkeys and mules like with horses is forage first, providing more fiber versus non-soluble carbohydrates, adequate vitamins and minerals and not over supplementing any one nutrient to cause an imbalance. Diets with excessive protein are usually NOT a problem.. Horses should be fed as a percentage of their body weight, but this differs for both donkeys and mules (please refer to *Donkey Nutrition and Mule Nutrition chapters*). Management factors that affect DOD. are ignoring the pregnant mare, feeding fast growers the same as the slow growers, and feeding an unbalanced ration. If you do not account for the nutrients in forage or limit the forage and feed >10 lbs of grain/day you may cause DOD Confining the growing horse and allowing it access to free choice grain, or feeding adequate protein and energy with inadequate minerals are other methods of causing DOD and again we have little research in donkeys and mules but we can use horses as a suggested model on how this occurs in young growing equine. DOD can also occur if the developing bones are traumatized due to being overweight or due to excessive work, such as lunging a young horse >10 min/day and similar conditions are likely found in donkeys and mules that may start to work carrying heavy loads or work for long periods of time. D.O.D. can be prevented by closely monitoring the rate of growth and adjusting the feed intake to maintain an appropriate growth rate.

Osteochondrosis is a local or generalized failure of endochondral ossification affecting epiphyseal and/or metaphyseal cartilage. In the horse there are 2 clinical syndromes: osteochondrosis dessicans (OCD) and subchondral cyst-like lesions (bone cysts). The same conditions can be seen in donkeys and mules but are less documented. One study has looked at anatomical development in a young mule foal and a few in donkeys (Nocera et al. 2023). Normal endochondral ossification occurs in the bones of the extremities, vertebral column, pelvis, and base of the skull. A hyaline cartilage matrix is replaced by bone. There are several steps in normal endochondral ossification. Chondrocytes mature and enlarge. The intracellular matrix thins, and the mature chondrocyte produces alkaline phosphatase which causes calcification of the intercellular matrix. Calcification of the matrix causes chondrocytes to die due to lack of adequate nutrition. Capillaries invade the region bringing osteoblasts and the osteoblasts produce bone which matures to lamellar bone.

Osteochondrosis occurs when cartilage cells proliferate normally but maturation and differentiation is abnormal. Failure of the cartilage matrix to calcify or failure of vascular penetration and osseous replacement results in thickened cartilage. Cartilage necrosis develops in deeper layer fissures due to lack of nutrition and a cartilage flap (OCD) formation occurs. Another scenario is the thickened cartilage persists without fissuring and becomes surrounded by subchondral bone as adjacent cartilage continues the process of endochondral ossification. The retained cartilage dies resulting in a subchondral cyst-like lesion.

Conservative therapy for osteochondrosis consists of rest and limited exercise. The equid's feed intake should be decreased to decrease total energy in the diet. The diet should be analyzed for mineral imbalances. Medications such as hyaluronic acid administered intraarticularly (IA) or polysulfated glycosaminoglycans (PSGAGs) administered intramuscularly (IM) or IA are useful adjunctive therapy. Surgical therapy requires either arthroscopy or arthrotomy to debride the lesion. Arthroscopy allows better inspection of the joint. The objective of surgery is to remove cartilage flaps and osteochondral fragments, leave sharp edges around cartilage defects, and debride the lesion down to bleeding bone. Curettage (forage) of cystic lesions and leaving sharp cartilage edges, removing all debris within the cyst and exposing bleeding bone is the treatment for cyst-like lesions. Frequently surgery needs to be combined with conservative therapy. In many cases, surgery significantly increases the prognosis for an athletic animal.

Flexural deformities are developmental problems affecting tendons and ligaments. Flaccidity or weakness of the flexor tendons in newborn foals is not uncommon. It is due to musculotendinous weakness due systemic illness or lack of exercise or excessively long toes. Clinically, these foals walk on the caudal aspect of the foot and the toe of the hoof "rocks-up". Generally, the condition occurs in rear feet or all four legs. Treatment is to trim the hooves so that the foal has a flat weight bearing surface. Exercise is encouraged (swimming is excellent). A more severe presentation is digital hyperextension deformity in foals. The etiology is unknown, but the clinical signs are apparent immediately after birth, characterized by extreme dorsiflexion of the phalangeal joints. Severely affected animals will bear weight on the ventral surface of the proximal sesamoid bones. Treatment consists of supporting the hoof with heel extensions. Bandaging methods are not satisfactory as they cause further weakness. Light weight fetlock supporting braces may be of some benefit. There is a report of surgical shortening of the flexor tendons; however, the results are poor. The prognosis for these animals is poor to guarded.

Congenital flexural deformities are usually due to uterine malpositioning, genetic defects (such as trisomy), toxic and infectious insults during pregnancy, and recently congenital hypothyroidism has been identified as a cause. Clinical signs consist of either fetlock flexion deformity (fetlock knuckling; most common) or DDF contraction (club foot). Generally, the deep and superficial flexor tendons are involved. Congenital flexion deformities of the carpus also occur. Individual tendon involvement is difficult to define in these cases. Contraction of the carpal fascia is an important component and may prevent any correction. Correction in some foals is spontaneous, but others may need treatment. Oxytetracycline (3 grams IV, up to 2 injections 48 h apart) has been used very effectively to relax the contracture in horses but refer to pharmacology chapter for appropriate doses for mules and donkeys. In addition, splints and casts can be used for fetlock and coffin joint deformities, but they require careful monitoring. If the limb cannot be straightened in extension with splints, the prognosis is poor. Surgery, either inferior and/or superior check desmotomy or superficial flexor, may be required to straighten the leg (s) in some cases.

Acquired flexural deformities are frequently related to pain. Any pain in the limb will initiate the flexor withdrawal reflex and this in turn may lead to a flexural deformity. Faulty nutritional management constitutes the most important cause of contracted tendons. Overfeeding and excess energy protein in the diet leads to epiphysitis or osteochondrosis. Imbalanced feeding such as calcium/phosphorus imbalances or deficiencies in copper, vitamin E etc. have resulted in flexural deformities. One study in Portugal evaluating Vit E and Se concentrations in pregnant Mirandes donkeys found that many foals that were born with orthopedic conditions were from dams with low Vit E and Se (Queresma et al. 2021). These imbalances may induce tendon contracture directly or indirectly via epiphysitis and osteochondrosis. Anatomical changes that occur are ill-defined. Some researchers think that bone growth exceeds tendon lengthening but this has not been proven. Deep digital tendon contracture (coffin joint flexural deformity) presents with the hoof having a raised heel or "club foot" appearance. The deep digital flexor tendon is very tense in these cases. Two stages of contracture are documented: stage 1 is when the dorsal surface of the hoof wall is less than 90° and Stage 2 is when the dorsal surface is greater than 90°. Treatment consists of making necessary dietary changes (grass hay and ration balancing). Additionally, both exercise to stretch the tendon and complete rest to relax the muscle have been advocated. Likewise, hoof trimming to lower the heel, if possible, is performed but maintenance of comfortable heel contact is imperative. A toe extension is usually helpful. Phenylbutazone or other NSAIDs can be used for analgesia. Inferior check desmotomy has shown good results for stage 1 but deep flexor tenotomy is usually necessary for stage 2. Tenotomy may be performed above or below fetlock. Some cases do not respond due to joint capsule contracture.

Superficial digital contracture (fetlock flexural deformity) presents as knuckling at the fetlock with the hoof in normal alignment. Tendon involvement is complex and variable. In most cases, contraction of both the SDF and DDF tendons is present, and the DDF may even be primary. Careful palpation in the flexed and extended positions will help elucidate the relative involvement of the tendons. Treatment consists of dietary changes to decrease the caloric intake. Exercise should be encouraged to stretch the tendons. Raised heel shoes with toe extensions have been shown to be useful. NSAIDs can be used for analgesia. Surgery is not as successful and the surgery should be determined by carefully assessing the structures involved. Superior check desmotomy lengthens the SDF, while inferior check desmotomy lengthens the DDF. Combinations of superior and inferior check desmotomy are usually necessary. Superficial digital flexor tenotomy or suspensory ligament desmotomy may be required to straighten the leg and combinations of surgery are often necessary.

Rupture of the common digital extensor tendon can occur spontaneously but they rarely cause a problem, however contracted tendons may occur secondarily. Clinical signs consist of mild carpal flexion with normally extended fetlocks. The most characteristic feature is swelling over the craniolateral aspect of the carpus. Recovery is often spontaneous but casts or splints may be helpful to keep the carpus straight and help prevent tendon contracture. Suturing of the tendon ends is not indicated because the prognosis is good without surgical intervention.

Angular limb deformities are defined as any medial or lateral deviation of joints from their normal axis. This primarily involves the carpus, tarsus, or fetlocks. There are 4 causes: 1. tendon or ligament laxity (joint instability), 2. compressional physitis, 3. Malformed or displaced carpal or tarsal bones, and 4. Fractures of the cuboidal bones. Radiographs are needed to establish the cause and location of the deviation, as well as to determine the prognosis. Conservative treatment and management consist of corrective trimming to keep the hooves level and splinting or casts applied to the concave side of leg. Surgical techniques include periosteal stripping/release, which is performed above the slow growing physis. The release allows the slow growing side to "catch up". Physeal bridging procedures are performed across the rapid growing physis to slow growth and allow the opposite side to "catch up". Nutritional management to decrease energy and to balance the ration is recommended. Exercise should be controlled until the leg straightens.

Compressional physitis, or "epiphysitis," is defined as structural changes in the physis due to compression which may result in growth defects and/or pain. Different parts of the physis may grow at different rates. Physitis most commonly involves the distal radius, the distal tibia, distal cannon bones, and proximal first and second phalanges. Two age groups are commonly affected: young foals with pre-existing angular limb deformities and weanlings or yearlings with straight legs who may be lame. They have 'hour-glass' knees and ankles and are usually on high caloric or high supplemented diets. The cause of physitis is both nutritional and mechanical with compression of the growth plate as the final common pathway. Treatment consists of managing the diet (usually seen in animals on high energy diets may also involve Ca, P, Zn, and/or Cu imbalances) and the judicious use of analgesics and/or anti-inflammatories.

4 Capsulitis/Synovitis

Capsulitis and synovitis are inflammatory conditions affecting the joints in equine species, often causing pain, swelling, and reduced mobility. Capsulitis refers to inflammation of the joint capsule, which encases and protects the joint, while synovitis involves inflammation of the synovial membrane, responsible for producing synovial fluid that lubricates and nourishes joint structures. These conditions are commonly seen equidae where repetitive stress or trauma to the joints can lead to overuse injuries. Clinical signs may include heat, tenderness, and joint effusion, which can impair the horse's ability to move freely and perform activities. Treatment typically involves anti-inflammatory medications, rest, and in some cases, joint injections or physical therapy to reduce inflammation and promote healing. Early diagnosis and intervention are crucial to preventing chronic damage and ensuring long-term joint health in affected horses. If inflammation is not controlled, can lead to degenerative joint disease.

5 Osteoarthritis (Degenerative Joint Disease)

Degenerative joint disease (DJD), also known as osteoarthritis, is a common condition in equines that affects the cartilage within joints. DJD in equidae can arise from both intrinsic and extrinsic factors, often combining to exacerbate the progression of the disease This disease is characterized by the gradual breakdown of cartilage, leading to pain, inflammation, and decreased mobility. DJD can occur in any joint but is most frequently observed in the knees, hocks, and fetlocks of equidae. The condition is often a result of repetitive stress and trauma to the joints, which can be exacerbated by factors such as age, conformation, and activity level. Symptoms of DJD in horses include lameness, stiffness, and swelling in the affected joints.

Early diagnosis and management are crucial to slow the progression of the disease and improve the quality of life for the affected horse. An accurate diagnosis of DJD involves a combination of clinical evaluation and diagnostic imaging. Physical examination including observing gait abnormalities, palpating joints for swelling, and assessing pain responses. Clinical signs are lameness, joint swelling, stiffness, and a reduced range of motion. Radiography (X-rays) reveal bone remodeling, joint space narrowing, and osteophytes (bone spurs). Ultrasound imaging of soft tissue structures, including ligaments and tendons, to evaluate secondary damage is important. Arthroscopy involves the direct visualization of the joint interior for detailed assessment and, in some cases, therapeutic intervention. Synovial fluid analysis tests for inflammatory markers and infection within the joint.

While DJD is not curable, early intervention and appropriate management can significantly improve the quality of life for affected equidae. Treatment strategies aim to alleviate pain, reduce inflammation, and preserve joint function. Non-steroidal anti-inflammatory drugs (NSAIDs): Medications such as phenylbutazone or flunixin meglumine provide pain relief and reduce inflammation. Intra-articular injections: Corticosteroids and hyaluronic acid or regenerative therapies can be injected directly into the joint to decrease inflammation and improve lubrication. Considering there may be a link to DMS (Donkey Metabolic Disease) and laminitis always consult with your veterinarian on treatment. Before administering corticosteroids to a donkey or mule, a veterinarian should take the following precautions:

- **Assess baseline risk**: Your vet should thoroughly evaluate the animal's weight, body condition, and overall health.
- **Screen for metabolic disease**: Blood tests for insulin dysregulation and PPID should be performed, especially if the animal is overweight or older.
- **Use the right drug**: Choose the least potent and shortest-acting corticosteroid necessary to treat the condition.
- **Consider alternative therapies**: In cases where the risk of laminitis is deemed too high, your vet may opt for alternative anti-inflammatory medications.
- **Monitor closely**: After treatment, the animal should be monitored for signs of lameness or other symptoms of laminitis.

Disease-modifying agents such as polysulfated glycosaminoglycans (PSGAGs) or bisphosphonates may slow cartilage degradation and improve joint health. Controlled, low-impact exercise can help maintain mobility without exacerbating joint damage. Weight management to maintain a healthy body weight reduces stress on joints. Environmental adaptations such as providing soft bedding, smooth surfaces, and appropriate shoeing to minimize joint strain.

In severe cases, surgical intervention may be warranted. Common procedures include Joint arthrodesis which stabilizes a painful joint by fusing its bones. Debridement of damaged cartilage or bone fragments via arthroscopy is the most common surgery.

Prevention plays a crucial role in minimizing the risk of DJD in equidae. Key strategies include: Proper training: Gradual conditioning to strengthen muscles and minimize joint stress. Regular hoof care: Correct trimming and shoeing to ensure balanced weight distribution. Routine veterinary check-ups: Early detection of joint abnormalities or injuries. Nutritional support: Adequate intake of omega-3 fatty acids, glucosamine, and chondroitin to promote joint health.

Degenerative joint disease is a significant challenge in equine care, impacting performance, comfort, and longevity. Understanding its causes, symptoms, and treatment options is essential for horse owners, trainers, and veterinarians. Through a combination of medical interventions, lifestyle adjustments, and preventive measures, equidae with DJD can experience improved quality of life and prolonged utility in their roles—be it as athletes, companions, or working animals.

6 Tendon and Ligament Injuries

Tendon and ligament injuries are among the most common and challenging issues faced. These injuries can limit an equine's movement and, in severe cases, jeopardize its career or life. Tendons connect muscle to bone, while ligaments connect bone to bone, playing a critical role in the stability and movement of a horse's musculoskeletal system. Understanding the causes, symptoms, diagnosis, treatment, and prevention of these injuries is essential for maintaining equine health and performance.

Tendons are strong, fibrous tissues that transmit the force generated by muscles to bones, enabling movement. In horses, the superficial digital flexor tendon (SDFT) and deep digital flexor tendon (DDFT) are the most commonly affected. These structures are vital for absorbing the shock and strain of high-speed movement and bearing the horse's weight during activity.

Ligaments stabilize joints by connecting bone to bone. Key ligaments in horses include the suspensory ligament, which supports the fetlock joint, and the collateral ligaments that stabilize various joints, such as the knee and hock. Ligament injuries can range from minor strains to complete ruptures, significantly affecting the horse's mobility and performance.

Tendon and ligament injuries in equine can occur due to acute trauma or repetitive strain. Understanding the common causes helps in early detection and prevention. Overloading or excessive strain from high-intensity activities like running and jumping can lead to small tears or ruptures. Poor conformation or less-than-ideal limb alignment or body structure are more prone to soft tissue injuries. Sudden trauma such as tripping, slipping, or collisions during exercise can cause acute injuries. Fatigue, when muscles are tired, tendons and ligaments are forced to absorb more strain, increasing the risk of injury. Finally, age-related degeneration: older equidae may experience cumulative wear and tear, making their tendons and ligaments more susceptible to injury.

The signs of tendon and ligament injuries can vary depending on the severity and location of the damage. Common symptoms include: Swelling and heat in the affected area. There maybe lameness or reluctance to bear weight on the injured limb. Often there is visible thickening or bulging of the tendon or ligament. There is usually pain upon palpation of the affected structure. This leads to reduced performance or unwillingness to work. Early recognition of these signs is critical to prevent further damage and to initiate appropriate treatment promptly.

Accurate diagnosis is essential to determine the extent of the injury and to develop an effective treatment plan. Diagnostic tools often include a thorough physical examination, including palpation of the tendons and ligaments, as the first step. The veterinarian assesses swelling, heat, and pain response.

Ultrasound imaging is the most commonly used diagnostic tool for tendon and ligament injuries. It provides detailed images of the soft tissues, allowing the veterinarian to identify tears, fiber disruption, or other abnormalities.

Magnetic Resonance Imaging (MRI) is used for detailed imaging of complex cases, especially for injuries that cannot be fully visualized with ultrasound. It is particularly useful for diagnosing suspensory ligament injuries.

Thermographic imaging detects heat patterns in the horse's limb, which can indicate inflammation and injury. While not a standalone diagnostic tool, it is useful for initial assessments.

Treating tendon and ligament injuries requires a combination of rest, therapeutic interventions, and careful management to ensure proper healing and prevent re-injury. Rest is crucial to allow the damaged structures to heal. Initially, equidae are confined to a stall with minimal movement. Controlled exercise is gradually introduced to promote proper tissue remodeling and prevent stiffness.

Applying cold packs or ice reduces inflammation and alleviates pain in the acute phase of the injury.

Non-steroidal anti-inflammatory drugs (NSAIDs) are often prescribed to reduce swelling and discomfort. In some cases, corticosteroid injections may be used with caution.

Extracorporeal shockwave therapy stimulates tissue repair and reduces pain, making it a valuable tool in treating soft tissue injuries.

Platelet-Rich Plasma (PRP) and Stem Cell Therapy are regenerative therapies that are growing in popularity. PRP involves injecting concentrated platelets from the horse's blood into the injured area to promote healing, while stem cell therapy uses cells that can differentiate into tendon or ligament tissue.

Specialized bandages or support boots may be used to immobilize the affected area and reduce strain during the healing process.

Rehabilitation plays a crucial role in recovery. A well-structured program allows gradual reintroduction of exercise while minimizing the risk of re-injury. It typically involves walking exercises to build strength and flexibility. Hydrotherapy is used to reduce weight-bearing stress during movement. Regular monitoring through ultrasound or physical exams is important to assess healing progress.

Preventing tendon and ligament injuries involves a combination of good management practices, proper conditioning, and regular veterinary care. Warm-up and cool-down periods are important periods to help reduce strain on soft tissues. Avoiding hard, slippery, or uneven surfaces reduces the risk of injury. A balanced nutrition provides a diet rich in essential nutrients to support overall tissue health.

Routine farriery to establish proper trimming and shoeing optimize limb optimizes limb alignment and reduces stress. Regular veterinary checks lead to early detection of minor issues and can prevent severe injuries.

Tendon and ligament injuries in equines pose significant challenges but can be successfully managed with prompt diagnosis, appropriate treatment, and a well-planned rehabilitation program. By recognizing the importance of prevention and understanding the anatomy and function of these critical structures, horse owners and veterinarians can ensure the longevity and well-being of equine athletes. Investing in proper care and management not only enhances performance but also strengthens the bond between horses and their caretakers.

7 Volar Annular Ligament Constriction

The volar annular ligament, a vital structure in the anatomy of equidae, plays an essential role in the stability and efficiency of the musculoskeletal system, particularly in the limbs. Its constriction, a relatively uncommon but significant condition, can lead to profound implications for the mobility and health of equidae.

The volar annular ligament is a fibrous band located at the palmar or volar aspect of the limb, encircling structures that maintain the integrity of flexor tendons. It is part of a complex system that ensures proper function and smooth gliding of tendons as they pass over joints. In equidae, this ligament is particularly important for supporting the weight-bearing capacity of their limbs and enabling effective locomotion.

Constriction of the volar annular ligament may arise due to several factors. (1) Trauma to the limb can induce fibrosis or scarring, leading to constriction of the ligament. (2) Chronic inflammation due to repetitive strain or conditions like tendonitis can result in swelling and eventual thickening of the ligament. (3) Congenital malformations may predispose certain equidae to ligament constriction. (4) As equidae age, natural wear-and-tear can affect the ligaments, leading to structural changes and constriction.

Constriction of the volar annular ligament manifests through varied symptoms. Early detection often relies on observing changes in the animal's gait or behavior. Key signs include Swelling or bulging at the affected site, typically in the palmar region. There maybe lameness of the affected limb and reluctance to bear weight. There will be pain and tenderness upon palpation of the ligament area. The constriction causes reduced range of motion in the joints beneath the ligament. There is often compensatory lameness affecting other limbs due to altered movement patterns.

A thorough physical examination, paying close attention to the affected limb's appearance, palpation findings, and range of motion. The presence of pain, swelling, and thickening of the annular ligament often raises suspicion. Ultrasonic imaging is the gold standard for diagnosing annular ligament constriction. It allows veterinarians to visualize ligament thickening, tendon involvement, and associated soft tissue changes. Doppler ultrasonography may further assess blood flow and detect inflammatory foci. Radiography, while not as definitive as ultrasonography, helps rule out concurrent conditions such as fractures, arthritis, or bone abnormalities in the fetlock region. In complicated or ambiguous cases, magnetic resonance imaging (MRI) or computed tomography (CT) may be employed to provide detailed insights into soft tissue and tendon pathology.

Treatment for annular ligament constriction in equidae is multifaceted and depends on the severity of the condition. The goals are to reduce pain, restore normal tendon function, and prevent recurrence. In mild cases, non-surgical treatment options may suffice. Restricting the horse's activity can reduce strain on the affected limb and allow inflammation to subside. Non-steroidal anti-inflammatory drugs (NSAIDs), such as phenylbutazone or flunixin meglumine, can alleviate pain and swelling. Applying ice packs or cold compresses to the fetlock region can help minimize inflammation during the acute phase or reduce pain temporarily for rehabilitative exercises. Compression bandages may provide support and reduce swelling. However, conservative management often provides only temporary relief, especially in cases involving significant ligament thickening or tendon impingement.

For more severe or refractory cases, surgical treatment is often necessary. Annular ligament desmotomy, involves the precise cutting or resection of the constrictive ligament to release pressure on the underlying tendons. It is typically performed under general anesthesia or local anesthesia with a standing horse. Techniques include open desmotomy, a direct incision is made over the affected area to access and transect the ligament. Endoscopic desmotomy is a minimally invasive approach using an arthroscope or tenoscope, which reduces postoperative pain and promotes faster recovery. Postoperative care is critical to ensure successful outcomes. Equine should undergo a structured rehabilitation program, including controlled exercise and physiotherapy.

Several adjunctive therapies can enhance recovery and prevent recurrence. Physiotherapy consisting of regular stretching and strengthening exercises improve ligament elasticity and tendon movement. Platelet-Rich Plasma (PRP) Therapy injection into the affected area can promote healing by delivering growth factors. Extracorporeal shockwave therapy stimulates tissue regeneration and reduces inflammation.

The prognosis for horses with annular ligament constriction depends on several factors, including the severity of the condition, the underlying cause, and the timeliness of treatment. Equidae treated surgically generally have a favorable prognosis, with many returning to their previous levels of activity. Early intervention and appropriate postoperative care are crucial for achieving optimal outcomes.

Constriction of the annular ligament in equidae is a condition that requires a nuanced approach to diagnosis and treatment. Advances in imaging and surgical techniques have significantly improved outcomes for affected horses. By understanding the underlying causes and implementing preventive measures, equine caretakers can help ensure the long-term health and mobility of these animals. With proper care, many horses can overcome this condition and continue leading active, fulfilling lives.

8 Rupture of the Peroneus Tertius

The rupture of the peroneus tertius in equidae is a rare but notable musculoskeletal injury. The peroneus tertius is a critical component of the reciprocal apparatus in the hind limb, which ensures synchronized movement of the stifle and hock joints. Damage to this structure significantly impacts a equine's mobility and functionality, often requiring careful diagnosis and management to ensure optimal recovery.

The peroneus tertius is a dense, fibrous structure located in the horse's hind limb, originating from the distal femur and inserting into the third metatarsal bone and the fourth tarsal bone. It is a key element of the reciprocal apparatus, which couples the movement of the stifle and hock joints during locomotion. When the stifle joint flexes, the peroneus tertius ensures that the hock joint flexes simultaneously, allowing for efficient and synchronized movement. This mechanism is vital for galloping, jumping, and other high-mobility actions characteristic of equine behavior.

The rupture of the peroneus tertius is most often caused by excessive force being applied to the hind limb. This injury can result from a sudden hyperextension of the hind limb during activities such as jumping or galloping. Trauma from slipping or falling on uneven terrain can cause the injury. Such stress leads to the tearing or complete rupture of the structure, disrupting the reciprocal motion of the stifle and hock joints. In some cases, the rupture occurs alongside other injuries, such as fractures or damage to surrounding ligaments and tendons, which can complicate diagnosis and recovery.

The rupture of the peroneus tertius presents specific clinical signs that make it relatively distinguishable from other hind limb injuries. The inability of the hock to flex when the stifle is flexed. This can be seen when the horse's limb extends backward abnormally during motion. There is usually visible swelling or pain localized around the site of the injury. The animal will exhibit a reluctance to move or an altered gait, often described as awkward and uncoordinated. In severe cases, the animal may completely avoid placing weight on the affected limb, leading to secondary issues such as muscle atrophy or laminitis in the contralateral limb due to uneven weight-bearing.

Proper diagnosis of a peroneus tertius rupture involves a combination of clinical examination and imaging techniques. Veterinarians typically perform the following steps including physical examination which includes observing the horse's gait and manually manipulating the limb to assess joint movement and detect abnormalities, specifically the pathognomonic movement, stifle flexion with hock extension. Ultrasound imaging to identify tears or disruptions in the peroneus tertius structure. Radiography may be used to rule out concurrent injuries, such as fractures, which could complicate the condition. Early and accurate diagnosis is essential to prevent further damage and to determine the most effective course of treatment.

Treatment for a ruptured peroneus tertius is primarily conservative, focusing on rest and rehabilitation. The key steps in managing this injury include immobilization, The affected limb must be kept immobilized to allow the injured fibers of the peroneus tertius to heal. This is typically achieved through stall rest but on occasion may involve the use of supportive bandaging or splints.

After an initial rest period, controlled and gradual reintroduction of exercise is necessary to rebuild strength and flexibility in the limb. Activities such as hand-walking or light groundwork may be prescribed. Medications such as non-steroidal anti-inflammatory drugs (NSAIDs) are often used to manage pain and reduce inflammation. Physical therapy techniques like massage, cold therapy, and laser treatments may be employed to stimulate blood flow and promote healing in the injured area. Follow-up evaluations, including repeated ultrasounds, are crucial to track the progress of healing and adjust the rehabilitation plan as needed.

The prognosis for a horse with a ruptured peroneus tertius varies depending on the severity of the injury and the timeliness of intervention. In many cases, horses can return to normal activity levels with proper care and rehabilitation. However, complications such as incomplete healing or the development of scar tissue may limit the horse's performance, particularly in high-demand disciplines like show jumping or racing.

It is worth noting that early identification and adherence to a veterinarian's recovery plan significantly enhance the likelihood of a favorable outcome.

The rupture of the peroneus tertius in equidae, while rare, is a significant injury that requires prompt attention and careful management. Understanding the anatomical function, causes, and treatment options is vital for horse owners, trainers, and veterinarians alike. By implementing preventive measures and adhering to a structured recovery plan, affected horses have a strong chance of regaining their mobility and returning to their routines. As with any injury, early intervention remains the cornerstone of successful outcomes in equine care.

9 Rupture of the Gastrocnemius

The gastrocnemius muscle, located in the hind limb of equids, plays a critical role in weight-bearing and locomotion by providing propulsion and support during movement. A rupture of this muscle, though relatively rare, can have profound implications for the affected animal, leading to severe lameness and significant functional impairment.

The gastrocnemius muscle is a key component of the equine hind limb, forming part of the Achilles tendon and extending from the distal femur to the calcaneus. It comprises two muscle bellies, the medial and lateral heads, which converge into a common tendon. The primary function of the gastrocnemius is to extend the hock joint and flex the stifle joint, facilitating efficient locomotion. In equids, the gastrocnemius works synergistically with other muscles, including the superficial digital flexor, to enable powerful strides and explosive movements.

Rupture of the gastrocnemius muscle in equids can result from a variety of causes, often linked to overexertion or trauma. Common etiologies include traumatic injury such as a sudden impact or fall can overstretch the muscle, leading to partial or complete rupture. Lacerations may also lead to rupture of the gastrocnemius muscle. Overexertion during intense physical activity may strain the gastrocnemius beyond its physiological limits. Pre-existing conditions or degenerative changes in the tendon or muscle, possibly due to age or chronic overuse, can predispose an equid to rupture. Nutritional deficiencies causing imbalances in calcium or other essential nutrients may weaken the musculoskeletal system, increasing vulnerability to injury.

The clinical signs of a gastrocnemius rupture in equids are often dramatic and may include acute non-weight-bearing lameness, characterized by an inability to extend the hock joint. A dropped hock is a hallmark of complete rupture, the hock appears to drop closer to the ground due to the loss of tensile support from the gastrocnemius. Swelling and pain of the area around the ruptured muscle is noted. In cases of partial rupture, the horse may exhibit a stiff or guarded gait.

An accurate diagnosis of a gastrocnemius rupture is essential for effective management. Typically employ a combination of clinical examination and advanced imaging techniques to confirm the condition. The initial assessment involves observing the horse's stance and gait, palpating the affected limb, and identifying signs of pain or swelling. The "dropped hock" posture is a strong indicator of complete rupture. Ultrasonography is the diagnostic tool of choice for evaluating soft tissue injuries. It allows visualization of the gastrocnemius muscle and tendon, helping to identify the extent and location of the rupture. Radiography, while less commonly used, may be helpful in ruling out concurrent fractures or calcaneal injuries. Magnetic Resonance Imaging (MRI) in complex cases can provide detailed images of the muscle and surrounding structures, offering insights into the severity of the injury.

The treatment of gastrocnemius rupture in equids depends on the severity of the injury and the intended use of the animal. Management strategies include both conservative and surgical approaches. For partial ruptures or cases where surgery is not feasible, conservative treatment may be employed. This typically involves confinement to stall rest to minimize movement and allow healing. Bandages or splints may be used to stabilize the limb and reduce strain on the muscle. Non-steroidal anti-inflammatory drugs (NSAIDs) help manage pain and swelling. Gradual reintroduction of physical activity, including controlled walking, helps restore function over time.

In cases of complete rupture or when conservative management fails, surgical repair may be necessary. The procedure typically involves reattaching the torn ends of the muscle or tendon and reinforcing the repair with sutures or grafts. Postoperative care includes strict rest, physical therapy, and regular monitoring.

The prognosis for gastrocnemius rupture in equids varies based on the severity of the injury and the timeliness of treatment. Partial ruptures often have a favorable outcome with appropriate management, allowing the horse to return to normal activity. However, complete ruptures may result in long-term functional deficits.

Rupture of the gastrocnemius in equids is a serious condition that requires prompt diagnosis and tailored management. While the injury can be debilitating, advancements in veterinary care and rehabilitation strategies have improved outcomes for affected horses. Through a combination of preventative measures and effective treatment, the equine industry can continue to safeguard the health and well-being of these remarkable animals.

10 Fractures

Fractures in donkeys and mules, while less common due to their sturdy bone structure, do occur, particularly in the limbs. Fractures require immediate veterinary attention. Fractures in equidae, represent a significant challenge in veterinary medicine. These injuries, when left untreated, can lead to severe pain, lameness, and even death. However, advances in diagnostic tools, surgical techniques, and rehabilitation protocols have greatly improved outcomes for these animals.

Fractures in equidae can be classified based on their location, pattern, and severity. Simple fractures are clean breaks where the bone pieces remain aligned. Compound fractures are breaks where the bone pierces penetrate through the skin, increasing the risk of infection. Comminuted fractures occur when the bone is shattered into multiple fragments, complicating treatment. Stress fractures result from micro-damage caused by repetitive strain resulting in a crack in the bone. Avulsion fractures occur when a ligament or tendon pulls off a piece of bone.

Equidae are most susceptible to fractures in weight-bearing bones due to their large size and the forces exerted during movement. The most commonly affected areas include: the cannon bone (metacarpal/metatarsal III), the sesamoid bones within the fetlock joint, the pelvis, the humerus and radius in the forelimb and the tibia and femur in the hindlimb.

Fractures can occur due to a variety of reasons. Falls, kicks from other horses, or collisions during exercise or transport may cause fractures of any bone. Overuse, especially on hard surfaces, can lead to stress fractures. Weak bone quality from underlying metabolic disorders, malnutrition, or age-related degeneration can increase susceptibility.

Fractures often present with acute and severe lameness. There is swelling or deformity at the site of injury. Reluctance to bear weight on the affected limb is very common. There may be visible wounds or protrusions in the case of compound fractures. In milder cases one may only note localized heat and sensitivity. In some cases, stress fractures might present subtle signs, such as mild lameness that worsens with activity.

An accurate requires a thorough physical examination including palpation to identify swelling, sensitivity, or deformity. Radiography is the primary imaging tool for identifying fractures. Ultrasonic imaging is useful for soft-tissue injuries associated with fractures but can identify disruption of the boney cortex in areas that are difficult to radiograph. Computed tomography (CT): provides detailed images for complex fractures, particularly in areas like the skull or pelvis. CT can also be helpful in the planning of fracture reconstruction. Bone scintigraphy (Nuclear Imaging) is effective in detecting stress fractures not apparent on X-rays.

For minor, stable fractures, especially stress fractures, conservative treatment may suffice. This includes stall rest, restricting movement to allow natural healing. Bandages, splints, or casts to immobilize the affected area. Finally, anti-inflammatory medication to manage pain and swelling.

Complex or severe fractures often require surgical repair. Techniques include internal fixation using screws, plates, or pins to stabilize the bone. External fixation uses frames and rods applied outside the body to hold bones in place. Bone grafting may be necessary to enhance healing especially in cases of significant bone loss. Advances in equine surgery, including minimally invasive techniques, have greatly improved success rates.

In cases where treatment is not feasible or the animal's quality of life is severely compromised, euthanasia may be the most humane option. This is particularly true for catastrophic fractures or those with a poor prognosis.

Rehabilitation is critical for successful recovery. Key components include controlled exercise with a gradual increase in activity to restore strength and mobility. Physical therapy utilizing techniques such as hydrotherapy, laser therapy, or massage to promote healing and overall body comfort. Dietary support is extremely important, adequate nutrition, including calcium and vitamin D, supports bone repair.

Regular follow-ups with a veterinarian, including imaging studies, ensure that the fracture is healing correctly and complications are addressed promptly. Despite best efforts, fractures in equidae can lead to complications, such as: delayed or incomplete healing, infections, particularly in compound fractures, joint stiffness or arthritis if the fracture involves a joint and psychological stress from prolonged confinement.

Fractures in equidae, while challenging, are no longer the insurmountable issue they once were. With prompt diagnosis, advances in treatment, and dedicated rehabilitation, many affected animals can return to a high quality of life. This underscores the importance of a multidisciplinary approach, involving veterinarians, trainers, and owners, to ensure the best outcomes for these magnificent creatures.

11 Conclusion

Orthopedic health is critical to the overall well-being and productivity of donkeys and mules. By understanding the unique musculoskeletal features of these animals, veterinarians, handlers, and owners can better diagnose, treat, and prevent common orthopedic conditions. Regular monitoring, appropriate care, and prompt intervention are key to maintaining mobility and preventing long-term disability in donkeys and mules.

Questions

1. **Question 1:** Which anatomical difference requires a change in the epidural injection site for donkeys compared to horses?
   A. Donkeys have longer necks
   B. The sacral spines are more backward-curved and the spinal cord ends at S2
   C. Donkeys have shorter legs
   D. Donkeys have fewer vertebrae.
   Answer: B
2. **Question 2:** Why are donkeys and mules more susceptible to cumulative muscle strain despite their endurance capabilities?
   A. They have a higher proportion of fast-twitch fibers
   B. They have fewer muscles overall
   C. Their slow-twitch fibers stay engaged for long periods, masking fatigue
   D. They cannot sweat efficiently
   Answer: C
3. **Question 3:** Which feature of donkey hooves contributes to increased stress on the deep digital flexor tendon if trimmed like a horse's hoof?
   A. Smaller sole area
   B. More upright dorsal wall and thicker sole
   C. Softer frog
   D. Longer toes than horses
   Answer: B
4. **Question 4:** How does the frog of donkeys and mules help protect the hoof?
   A. It absorbs shock and distributes weight across the hoof
   B. It strengthens the coffin bone
   C. It provides a cosmetic appearance only
   D. It reduces blood flow to the laminae.
   Answer: A
5. **Question 5:** Compared to horses, donkeys affected by laminitis are more likely to be:
   A. Older and overweight
   B. Younger and not necessarily overweight
   C. Affected only in spring
   D. Immune to recurrent episodes
   Answer: B
6. **Question 6:** Which of the following is a primary goal in the treatment of laminitis in donkeys and mules?
   A. Increase hoof growth rate only
   B. Prevent or correct coffin bone rotation and relieve stress on supportive structures

C. Reduce blood flow to the hoof
D. Use corticosteroids to reduce inflammation.
Answer: B

7. **Question 7:** Developmental Orthopedic Disease (DOD) in donkeys and mules occurs:
   A. More frequently than in horses due to rapid growth
   B. Less frequently than in horses, often related to genetics, nutrition, or trauma
   C. Exclusively in adult animals
   D. Only in mules
   Answer: B

References

Eliashar E, Li Y, Ma Q, Shi X, Yuan W, Liu G, Wang C (2010) Comparative transcriptome analysis of slow-twitch and fast-twitch muscles in Dezhou donkeys. Genes (Basel). https://www.ncbi.nlm.nih.gov/pmc/articles/PMC9498731/

Lesté-Lasserre C (2022) Study: laminitis might affect younger, less overfed donkeys. The Horse. https://thehorse.com/1112410/study-laminitis-might-affect-younger-less-overfed-donkeys/

Li Y, Ma Q, Shi X, Yuan W, Liu G, Wang C (2022) Comparative transcriptome analysis of slow-twitch and fast-twitch muscles in dezhou donkeys. Genes 13(9):1610. https://doi.org/10.3390/genes13091610

Menzies-Gow NJ, Wakeel F, Little H, Buil J, Rickards K (2021) Cross-sectional study to identify the prevalence of and factors associated with laminitis in UK donkeys. Equine Vet J. https://doi.org/10.1111/evj.13505

Nocera I, Sorvillo B, Sgorbini M, Aliboni B, Citi S (2023) Radiographic appearance of the fore digit and carpal joint in the mule foal from birth to 3 months of age. Animals 13(15):2417. https://doi.org/10.3390/ani13152417

Parsons KJ, Pfau T, Wilson AM (2008) High-speed gallop locomotion in the thoroughbred racehorse. I. The effect of incline on stride parameters. J Exp Biol 211(6):935–944

Quaresma M, Marín C, Bacellar D, Nóvoa M, Navas FJ, McLean A (2021) Selenium and vitamin E concentrations in Miranda jennies and foals (Equus asinus) in Northeast Portugal. Animals 11(6):1772. https://doi.org/10.3390/ani11061772

Sells PD, Pinchbeck G, Mezzane H, Ibourki J, Crane M (2010) Pack wounds of donkeys and mules in the northern high atlas and lowlands of Morocco. Equine Vet J 42(3):219–226. https://doi.org/10.2746/042516409X478532

Podiatry: Hoof Care and Hoof Related Disease of the Donkey and Mule

Sammy L. Pittman

Abstract

Hoof diseases represent a major welfare concern and productivity challenge for donkey and mule populations worldwide. These often stoic animals, historically used as beasts of burden, are prone to hoof-related conditions such as laminitis, white line disease, abscesses, keratomas, navicular syndrome, and flexor contractures. Despite their rugged appearance and ability to mask pain, donkeys and mules frequently present with advanced hoof pathology due to neglect, inadequate farriery, and lack of early veterinary intervention. Anatomical distinctions—including a more upright hoof conformation, thicker sole, and distinct frog position—require tailored management approaches, particularly in response to environmental challenges such as moisture, poor footing, and nutritional imbalances.

This chapter provides a comprehensive overview of donkey and mule hoof anatomy, biomechanics, and disease management, emphasizing the role of radiographic and venographic imaging in diagnosing subtle or internal lesions. Biomechanical imbalances, such as altered hoof-pastern axis and mediolateral deviations, often contribute to chronic lameness and structural failure. A four-point trimming approach, modeled after the natural wear pattern of the donkey and mule hoof, is advocated to reduce leverage while maintaining sole depth and functional loading zones.

Preventive care—including routine farriery, proper environmental hygiene, and nutritional regulation—is essential to improving long-term hoof health. In low-income regions where these animals remain vital to livelihoods, the high prevalence of hoof-related lameness further underscores the need for education, accessible veterinary services, and structured hoof care programs. By integrating clinical experience with peer-reviewed literature, this chapter aims to provide

S. L. Pittman (✉)
Innovative Podiatry, Collinsville, TX, USA

A. K. McLean et al. (eds.), *Donkey and Mule Medicine: Foundations, Breeds, and Management*, https://doi.org/10.1007/978-3-032-12726-6_24

veterinarians, farriers, and caretakers with practical tools and strategies to improve outcomes and reduce suffering in these often underserved equids.

Keywords

Podiatry · Mules · Donkeys · Navicular · Keratoma · Abscesses · Navicular disease · White line · Flexor contractures

1 Introduction

The prevalence of hoof diseases in donkey populations globally is a significant concern, with hoof-related conditions being one of the leading causes of morbidity and mortality. Despite their often-resilient appearance, donkeys and mules are particularly vulnerable to hoof disorders, which can severely impact their quality of life, productivity, and overall welfare. Many of these conditions are preventable with improved management practices, regular hoof care, and a deeper understanding of the unique anatomy and biomechanics of donkey hooves.

Donkey hooves differ anatomically from those of horses, with a more upright and oval shape, a thicker sole, and a well-developed but differently positioned frog. These structural differences not only affect the way donkeys move but also how their hooves bear weight and respond to environmental conditions. The biomechanics of the distal limb in donkeys and mules, including the way forces are distributed through the hoof capsule, play a crucial role in maintaining hoof health. Any disruption in these biomechanics, whether due to poor hoof care, inappropriate diet, or environmental factors, can lead to a range of hoof diseases, including laminitis, white line disease, and hoof abscesses.

Radiographic studies have shown that hoof diseases in donkeys are often associated with specific biomechanical issues, such as abnormal hoof conformation, lateromedial imbalances, and alterations in the hoof-pastern axis. These conditions can be exacerbated by environmental factors such as wet, muddy ground or poor-quality bedding, which weaken the hoof structure and increase the risk of infection. Radiographs are essential tools in diagnosing these conditions, providing detailed insights into the internal structure of the hoof and revealing issues that may not be visible during a physical examination.

The incidence of hoof-related diseases in donkeys is alarmingly high across various regions and environments. Studies have shown that in some populations, up to 37% of donkeys suffer from foot lameness, with conditions like hoof abscesses, white line disease, and laminitis being the most common (Amene et al. 2019; Bukhari and Ali 2023; Zhao and Li 2022; Bonelli et al. 2019; Tesfaye et al. 2010). These diseases are not only a major cause of pain and suffering for the animals but also lead to significant economic losses in communities that rely on donkeys for work and transport. In many cases, the detection of lameness in donkeys is delayed due to their stoic nature, which masks the severity of their discomfort until the condition has progressed significantly.

Globally, the impact of hoof-related diseases on donkey populations is profound. In developing countries, where donkeys are often a critical component of rural livelihoods, hoof diseases can lead to decreased productivity, increased veterinary costs, and even loss of the animal, which can have devastating effects on the owners. Moreover, the high prevalence of these conditions highlights the need for better education and resources for donkey and mule owners, as well as more widespread adoption of preventive measures.

Effective hoof care for donkeys and mules relies on close collaboration between veterinarians, farriers, and owners. Regular trimming, proper nutrition, and maintaining a clean, dry environment are key strategies in preventing hoof diseases. Additionally, understanding the specific anatomical and biomechanical characteristics enables professionals to provide more targeted and effective care, reducing the incidence of hoof-related diseases and improving the overall welfare of these animals.

This chapter will explore the various aspects of donkey and mule hoof care, including the anatomy of their hooves, the biomechanics of the distal limb, and the use of radiographs to aid in management. By addressing these issues, we aim to provide a comprehensive guide to improving hoof health and reducing the burden of these conditions on donkey populations worldwide.

1.1 Distal Limb Anatomy: Comparing Donkeys, Mules, and Horses

The distal limb anatomy of donkeys, mules, and horses, share fundamental similarities but exhibit relevant differences that have significant implications for hoof care, biomechanics, and disease management. Understanding these distinctions is crucial for veterinarians and farriers to provide optimal care and prevent hoof-related issues in these animals (refer to Fig. 1) (Misk and Hifny 1983).

1.2 Common Anatomical Structures

The distal limb of all three species—donkeys, mules, and horses—consists of several key structures that work together to support the animal's weight, absorb shock, and facilitate movement. These structures include (refer to Fig. 2):

1. **Bones**: The distal limb in all three species includes the cannon bone (metacarpal or metatarsal III), the two proximal sesamoid bones, the long and short pastern bones (proximal and middle phalanges), the coffin bone (distal phalanx), and the navicular bone (distal sesamoid).
2. **Joints**: The fetlock joint (metacarpophalangeal or metatarsophalangeal joint), pastern joint (proximal interphalangeal joint), and coffin joint (distal interphalangeal joint) are present in all species and allow for the flexion and extension necessary for movement.

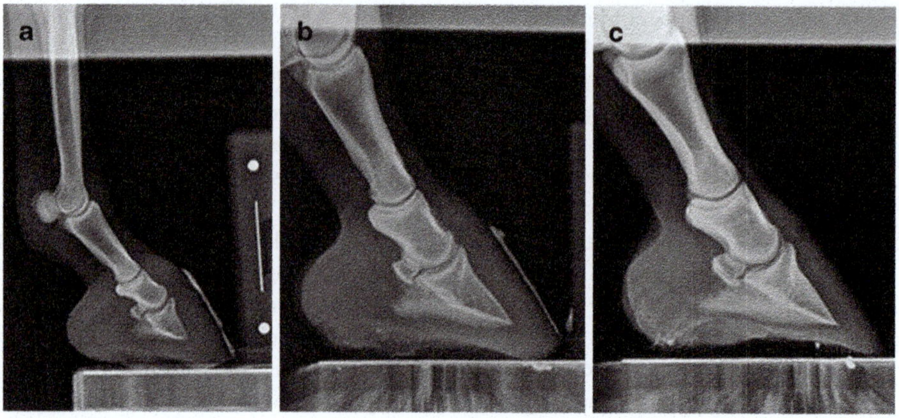

Fig. 1 Comparison of radiographic anatomy of the donkey (A), mule (B) and horse (C). Note the greater hoof mass to coffin bone ratio in the donkey compared to the horse. Hoof walls and digital cushion are much more robust. Mules fall somewhere in between and have characteristic of both the horse and the mule

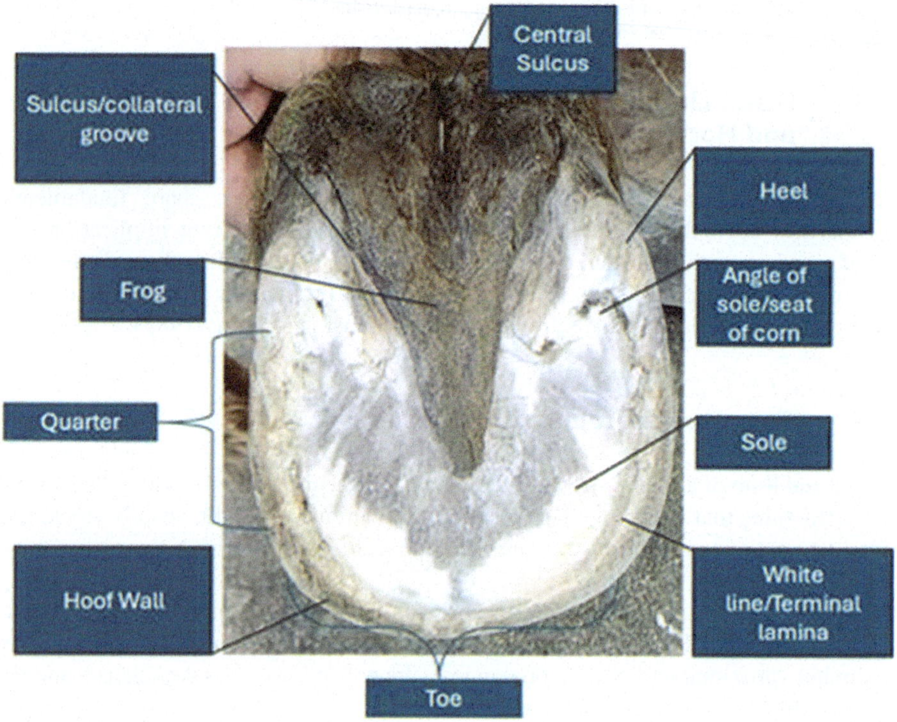

Fig. 2 Solar anatomy is named the same for all 3 species but varies in proportions

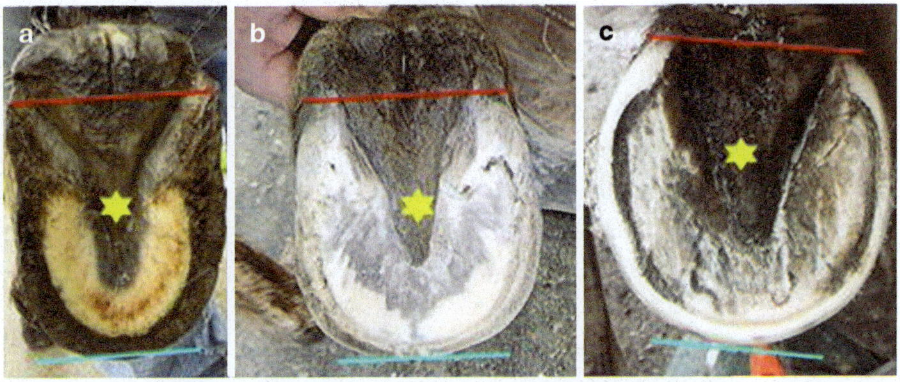

Fig. 3 Donkeys (**a** far left) and mules (**b**, middle) have a more robust frog that extends past the point of the heels compared to horses (**c** far right). The yellow star indicates the estimated center of rotation of the coffin joint. Note the amount of hoof from the star to the red line or point of heels and the turquoise line at the toe

3. **Tendons and Ligaments**: The superficial and deep digital flexor tendons, the suspensory ligament, and the extensor tendons are critical in supporting the limb and controlling movement. The deep digital flexor tendon (DDFT) is particularly important as it attaches to the solar surface of the coffin bone, playing a significant role in the biomechanics of the hoof.
4. **Hoof Capsule**: The hoof capsule, composed of the hoof wall, sole, frog, and laminae, protects the internal structures and bears the animal's weight. It also plays a crucial role in shock absorption and traction (refer to Fig. 3).

1.3 Differences in Distal Limb Anatomy

While the overall structure is similar across donkeys, mules, and horses, there are important anatomical differences that influence hoof care and susceptibility to certain hoof conditions (refer to Fig. 4). It is also important to note the differences in proportions of the coffin bone to the overall hoof mass when compared to their equine counterparts. Donkeys tend towards a larger mass of horn and corium relative to the coffin bone size. This also affects the biomechanical considerations as the center of rotation (COR) of the coffin joint is located more dorsally, creating a more innate balanced system when considering toe length to heel support around the COR. Mules can have similar ratio depending on whether they exhibit more donkey or horse type hoof characteristics (Phannithi et al. 2023).

1. **Hoof Shape and Structure**:
   - **Donkeys**: Donkey hooves are more oval to U shaped and upright compared to horse hooves. The dorsal hoof wall is more vertical, and the frog is more robust, especially at the palmar aspect. The sole is thicker, and the hoof wall thickness is relatively uniform from toe to heel. The heels typically terminate well before the widest part of the frog.

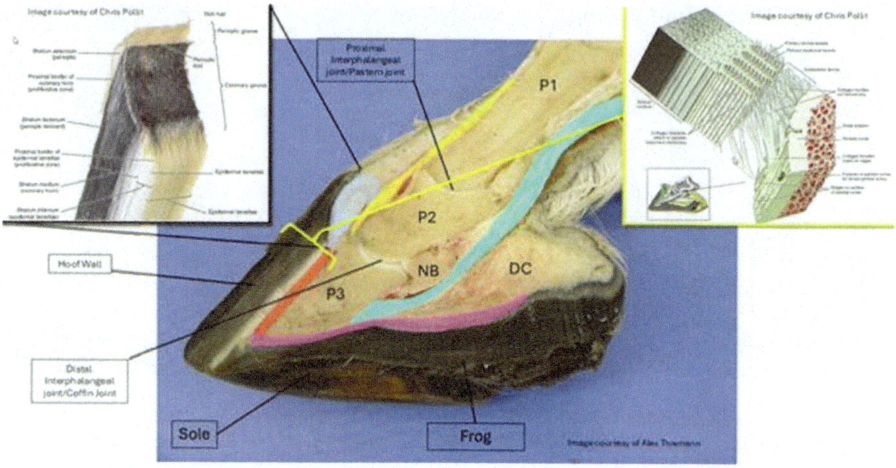

Fig. 4 Gross and microanatomy is very similar in all 3 species with different overall proportions. P1-first phalanx or long pastern bone, P2- second phalanx or short pastern bone, P3- third phalanx or coffin bone, DC-digital cushion, deep digital flexor tendon highlighted in turquoise, common digital extensor tendon highlighted in yellow, coronary corium highlighted in light blue, lamellar corium highlighted in red and solar corium highlighted in pink

- **Mules**: Mules typically inherit a combination of hoof characteristics from both their donkey and horse parents. Their hooves are often narrower and more upright than those of horses,

2. **Bone Structure**:
   - **Coffin Bone (Distal Phalanx)**: The coffin bone in donkeys tends to be more compact and angled more upright than in horses, contributing to the overall upright hoof conformation. In mules, the coffin bone usually mirrors the donkey's structure more closely, but with some variation depending on the mule's lineage.
   - **Navicular Bone**: The navicular bone in donkeys is smaller and less prone to diseases such as navicular syndrome, which is more common in horses due to their lower and longer hooves that increase stress on this bone.
3. **Soft Tissue Composition**:
   - **Digital Cushion**: The digital cushion in donkeys is generally more fibrous and less fatty compared to horses, providing more resilience and shock absorption in rough terrains. Mules tend to have a digital cushion that combines the characteristics of both species, offering durability and shock absorption.
   - **Frog and Ungual Cartilages**: The frog in donkeys is larger and more central to the hoof's weight-bearing function compared to horses. This difference is less pronounced in mules, but they still tend to have a more substantial frog than horses.
   - **Vascular Corium**: The vascular corium, the tissue layer that supplies blood to the hoof and supports growth and health of the hoof wall, sole, and frog, is

more substantial in donkeys compared to horses. This increased vascularity is crucial for donkeys' hoof resilience, especially in their natural habitats, which often involve hot, dry terrains. The robust vascular corium enhances the donkey's ability to maintain healthy hooves despite these challenging conditions and dissipate heat and concussion.

1.4 Routine Hoof Care in Donkeys and Mules

Routine hoof care is the cornerstone of preventing most hoof pathologies encountered in donkeys and mules. Unlike their equine counterparts, donkeys and mules are often kept in environments or under management systems that delay or limit regular farriery, leading to complications such as overgrowth, imbalanced load distribution, white line disease, abscesses, and lamellar structural failure (refer to Fig. 5).

The author advocates a consistent trimming protocol modeled after the natural wear patterns of free-roaming donkeys and mules, which often inhabit dry, abrasive terrain. This approach centers on a four-point trim that focuses on reducing leverage at the toe and preserving a healthy solar depth while maintaining the hoof's four principal loading pillars—two at the toe quarters and two at the heel buttresses. This trim minimizes mechanical strain on the lamellae, reduces abnormal tension on the DDFT, and promotes even load distribution.

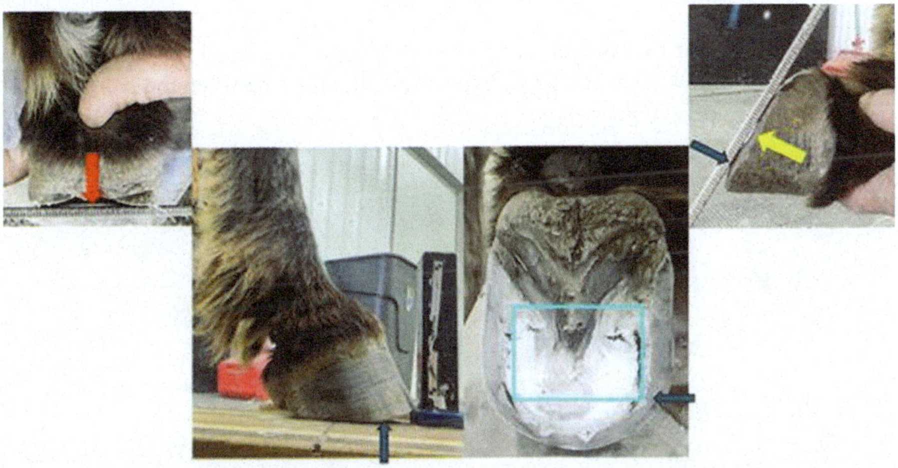

Fig. 5 Demonstrates the characteristics of a four point style trim that reduces leverage at the toe and aids in keeping the white line/terminal lamina and wall from becoming stretched and packed with debris. Blue arrows indicate the pillars at the toe. Yellow arrow indicates the relief placed in the quarters. Red arrow demonstrates the sole relief created by the trim as to not directly load the sole in this region. The turquoise box is created by drawing a line from the angle of the sole towards the toe. Where this intersects the white line/terminal lamina is a good guideline and reference for the toe pillar that is left for primary weight bearing zone along with the heel. This is a very helpful management trim when accompanied by the appropriate environment

Routine trimming should occur every 6–8 weeks, even for animals not in work. This frequency supports optimal hoof mechanics and helps avoid pathological loading patterns. Combined with a clean, dry environment and balanced nutrition tailored to prevent metabolic dysfunctions, this proactive maintenance plan is vital in sustaining long-term hoof health.

The distal limb anatomy of donkeys, mules, and horses, while sharing many similarities, exhibits important differences that impact their hoof care and overall health. Understanding these differences is essential for effective management, particularly in preventing and treating hoof diseases. By tailoring hoof care practices to the specific anatomical and functional and environmental needs of each species, veterinarians and farriers can help ensure the long-term soundness and well-being of these animals.

1.5 Radiographic Evaluation Techniques and Parameters

To effectively evaluate critical hoof parameters, radiographs must be taken with precision and consistency. Hoof Care-friendly radiographs focus not only on the bony structures but also on capturing the mechanical properties of the hoof, which are crucial for designing effective therapeutic shoeing programs. The following steps help ensure high-quality, consistent radiographs (refer to Figs. 6 and 7) (Redden 2003; Redden and Bras 2017).

1. **Hoof Placement on Blocks**: Both hooves should be placed on blocks that are designed to allow the X-ray beam to penetrate the hoof between the palmar rim of

Fig. 6 Appropriate positioning for podiatry style hoof radiographs

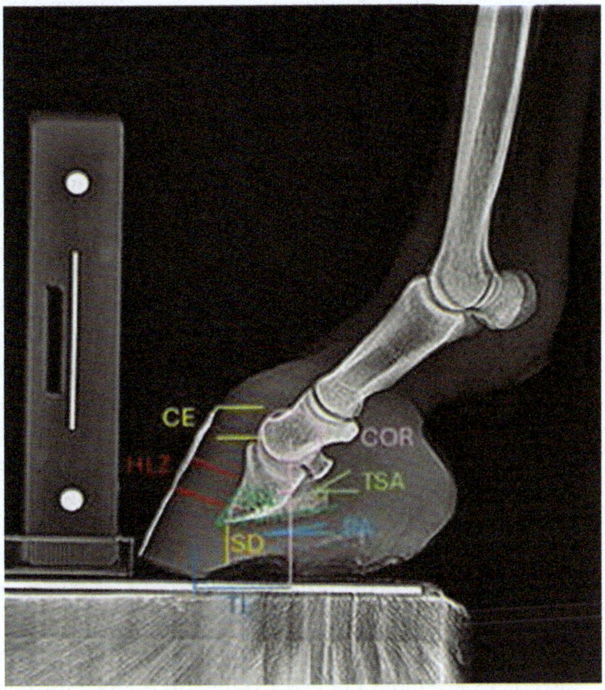

Fig. 7 Radiographic parameters that are helpful in managing hoof related problems. CE-coronary band to extensor process distance, HLZ-horn-lamellar zone, SD-sole depth, TL-toe lever, PA-palmar angle, TSA-tendon surface angle, COR-center of rotation of the coffin joint

the coffin bone and the ground surface. The blocks should align with the equid's natural stance, accommodating any toe-in or toe-out conformation.

2. **Beam Orientation**: The X-ray beam should be centered on the area of interest, particularly focusing on the palmar angle. A low beam orientation is necessary to capture accurate measurements of the hoof's mechanical properties.
3. **Use of Radiographic Paste**: Applying radiographic paste along the dorsal hoof wall allows for accurate measurement of parameters such as the coronary band to extensor process distance and the horn-lamellar zones. This paste should extend from the last hair follicle down to the entire length of the hoof capsule.
4. **Calibration and Consistency**: Consistent focal film distance (typically 24 to 28 inches) and the use of a calibration instrument in the radiographic field help ensure that measurements are accurate and comparable over time. Calibration accounts for magnification and provides reliable data for assessing hoof changes.
5. **Soft Tissue Detail**: Achieving the right balance of soft tissue detail in the radiograph is essential. This includes visualizing the deep flexor tendon and the dermo-epidermal junction.

1.6 Radiographic Measurements and their Implications

Several measurements are critical when evaluating the hoof's biomechanics and designing therapeutic interventions:

- **Coronary Band to Extensor Process (CE)**: This distance, measured from the top of the hoof capsule to the extensor process of the coffin bone, can indicate shifts in hoof position due to conditions like laminitis.
- **Horn-Lamellar Zone (HLZ)**: The thickness of the horn-lamellar zone, particularly around the extensor process and the apex of the coffin bone, helps identify displacements and rotations within the hoof. Changes in this zone can be early indicators of laminitis, white line disease, and clubbed feet.
- **Sole Depth (SD)**: The sole depth, measured from the tip of the coffin bone to the ground surface of the sole, is a crucial indicator of hoof health. Depth of sole can range from 10 to 18 mm depending on the size of the donkey or mule. A general reference is the proximal/upper HLZ measurement and sole depth under the tip of the coffin bone should be similar in a healthy scenario. Thin soles indicate a compromised hoof that requires intervention to increase hoof mass and protect internal structures.
- **Digital Breakover (DB) and Toe Lever (TL)**: These measurements assess the lever arm created by the hoof shape, shoeing and coffin bone characteristics. Longer toe levers increase strain on the lamellar bond, DDFT, navicular apparatus and solar corium under tip of and are often associated with lameness, making it important to manage this parameter through shoeing and or trimming adjustments.
- **Bone Angle (BA)**: The bone angle of the coffin bone, per the author's findings, averages around 55 degrees (range of 45–60) for fronts and 50 degrees (range of 45 to 55) in hinds, and dictates the shape and character of the hoof. Variations of the bone angle will create different innate biomechanical properties specific to each hoof. Considering this parameter in your evaluations will help make decisions on management. For example, a coffin bone with a steep bone angle will create a greater toe angle, higher heel profile and an increased likelihood of a broken forward axis in the distal interphalangeal joint. Conversely a lower coffin bone angle (30-40dg) will create lower profile heel and longer toe lever as a rule.
- **Tendon Surface Angle (TSA)**: This angle is measured on the distal aspect of the flexor surface of the navicular bone is important for assessing the engagement of the flexor tendon. Adjustments to this angle can relieve pressure on the navicular apparatus and improve hoof function.
- **Palmar Angle (PA)**: The palmar angle is the angle formed between the solar surface of the coffin bone (distal phalanx) and a level ground surface. Measuring this angle is crucial as it helps monitor and evaluate the tension on the deep digital flexor tendon (DDFT) and the distribution of load across the hoof. A typical healthy palmar angle in donkeys and mules ranges between 3° and 10°, although this can vary depending on the individual animal's hoof conformation and health. Measuring the palmar angle accurately can be challenging, especially in hooves with significant bone remodeling. The measurement is typically taken from the wings of the coffin bone relative to a level ground surface or an embedded wire in a block used during radiography. Using the wings of the coffin bone offers the most consistent and reliable measurement, allowing for an accurate evaluation. In general, lowering the palmar angle increases tendon tension; while raising it

decreases tension. However, the palmar angle should be evaluated in the context of other factors, such as sole depth, clinical examination findings, and digital alignment.

2 Biomechanics of the Distal Limb

Understanding the biomechanics of the distal limb in donkeys and mules is essential for proper hoof care and disease prevention. The distal limb biomechanics of these animals involve the interplay of several anatomical structures, including tendons, bones, and the hoof capsule, which work together to manage and distribute the forces exerted on the limb. These forces impact how the hooves grow, wear, and respond to environmental conditions, affecting the animal's overall health and mobility.

2.1 Suspension and Support Structures

At the core of distal limb biomechanics is the suspension mechanism provided by the anatomy including deep digital flexor tendon (DDFT) and the lamellar bond. This tendon originates from the deep flexor muscle bellies and runs down the palmar (or plantar) aspect of the limb, over the navicular bone, and attaches to the semi-lunar crest on the solar aspect of the coffin bone. The DDFT, along with the lamellar network that attaches the coffin bone to the hoof wall, forms a suspension system that supports coffin bone inside the hoof capsule. This system is crucial in maintaining hoof stability and managing the distribution of forces within the hoof capsule.

The frog, sole, ungual cartilages, and digital cushion also play vital roles as support structures, helping to absorb and distribute load across the hoof. These structures work in tandem with the suspension system to ensure that the forces generated during movement are evenly distributed, preventing excessive pressure on any single part of the hoof.

2.2 Load Distribution and Vascular Effects

The way in which load is distributed across the hoof can have significant implications for hoof health. When weight is placed on the limb, it causes vascular compression within the hoof, moving blood away from the loaded areas to those that are unloaded. This redistribution is generally harmless if the compression is temporary and evenly distributed. However, chronic or uneven loading can lead to reduced nutrient flow to the affected areas, resulting in issues such as poor hoof growth, bone remodeling, and soft tissue damage.

For example, in donkeys and mules with high-grade club feet or flexor contractures, the tight suspension of the DDFT transmits a greater portion of the load to the toe, leading to vascular compression and decreased growth of the sole under the apex of the coffin bone and dorsal hoof wall. Conversely, in donkeys and mules with low heels and long toes, the reduced tension in the DDFT allows for more load at the heels, often leading to heel crushing and difficulty growing adequate heel height.

3 Pathology and Treatment Consideration

3.1 Laminitis/Lamellar Structural Failure

Donkeys and mules present unique challenges in the diagnosis and treatment of lamellar structural failure (LSF), a condition more traditionally known as laminitis (refer to Fig. 8). Due to anatomical and metabolic distinctions from horses, appropriate adaptation of mechanical, medical, and diagnostic strategies is essential (refer to Fig. 9).

3.2 Anatomical Considerations

Donkeys and mules possess a relatively smaller and more upright distal phalanx (coffin bone) in relation to the overall hoof capsule size when compared to horses. Additionally, their vascular corium is notably more robust, supporting more

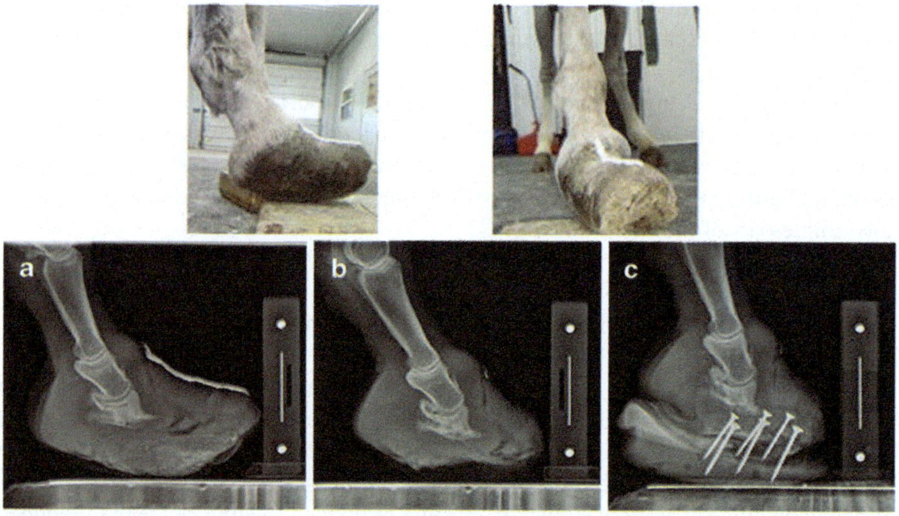

Fig. 8 Chronic lamellar structural failure/laminitis pre trim (**a**), post trim (**b**), and post shoeing (**c**). Some cases benefit from the improved mechanics of the trim alone others will benefit from the addition of a shoe that provides load transition to the heels and reduces leverage. Image C shows a fully rockered wedge plastic clog

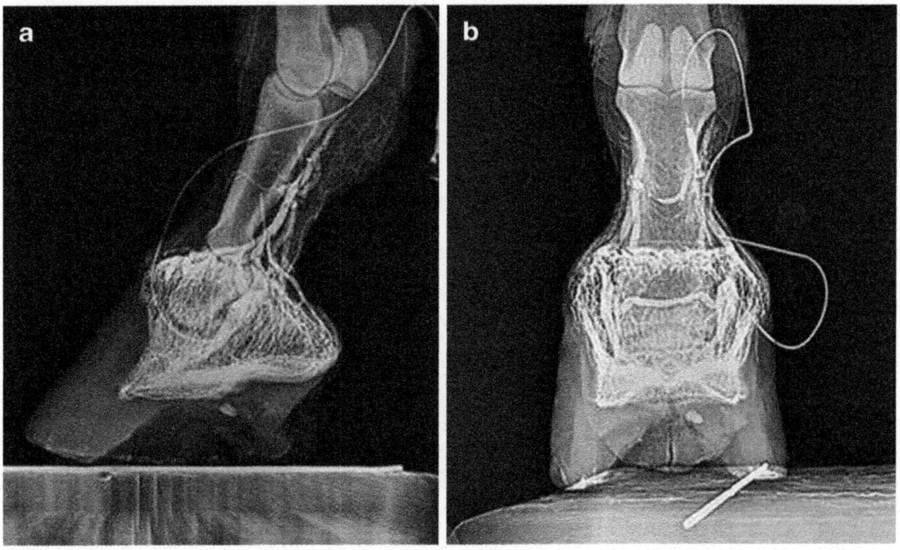

Fig. 9 Lateral venogram image (**a**) and dorsopalmar venogram image (**b**). Venograms highlight the soft tissue and vascular corium that isn't visible in plain radiographs. This can be very helpful in managing and investigating the degree of lamellar structural failure

effective temperature regulation, load dissipation, and tissue recovery. This vascular difference may help explain why donkeys often display a greater capacity for healing, even in cases of chronic LSF.

3.3 Most Common Causes

Obesity-induced insulin dysregulation, resulting in hyperinsulinemia, is the most frequent cause of LSF in donkeys and mules. Additional triggers include abrupt dietary changes, stress, corticosteroid use, and underlying endocrinopathies such as pituitary pars intermedia dysfunction (PPID). Cornielus citation corticosteroid use risk in laminitis, De lat hyperinsulinema reference.

3.4 Diagnostic Approach

As in horses, diagnosis relies heavily on integration of the physical examination, radiographs, and digital venograms (Nocera et al. 2022). Donkeys tend to be stoic and may not show overt pain, making reliance on imaging more important. Lateral and dorsopalmar radiographs should be acquired with a podiatry-style approach, often utilizing a wooden block and radiopaque paste on the dorsal wall. The palmar angle (PA), coronary extensor distance (CE), sole depth (SD), and horn-lamellar zone (HL) are all evaluated.

Digital venograms remain a critical tool for assessing internal vascular architecture, particularly the circumflex vessels, coronary papillae, and solar plexus (Baldwin and pollit 2010); Nocera et al. 2022; Rucker 2017).

Mechanical Management

Mechanical treatment decisions must be scaled to match the level of structural compromise. Rockered wedge shoes, static wedges (such as modified Ultimate Wedge platforms or clogs), or deep flexor tenotomy are selected based on displacement seen on imaging and venographic changes (Hood 1999; Stewart 2018; Taylor and Hood 2003; O'Grady and Parks 2008; Parks and O'Grady 2003).

Repeat venographic imaging every 7–30 days allows close monitoring and escalation or reduction of mechanical therapy. Sole depth of 15–18 mm and even wall growth patterns are key milestones before down-regulating support. If mechanical response is inadequate, tenotomy may be warranted.

Medical and Dietary Management

Management of insulin dysregulation and concurrent pain is as important as mechanical correction. Dict must focus on minimizing non-structural carbohydrates (NSCs), primarily through grass hay (≤10% NSC) and eliminating grain and sugary treats. In high-risk animals or those recovering from LSF, soaking hay can further reduce carbohydrate content.

Recommended medications include:

- NSAIDs such as phenylbutazone or flunixin meglumine to manage acute pain and inflammation.
- Pentoxifylline to improve microvascular perfusion.
- Metformin and SGLT2 Inhibitors (Ertugliflozin, bexagliflozin) as insulin regulators especially in donkeys with metabolic syndrome.
- Levothyroxine sodium (Thyro-L) may help increase insulin receptor sensitivity and promote weight loss in refractory metabolic cases.
- Pergolide mesylate (Prascend) for confirmed or suspected PPID.

Regular monitoring of insulin, ACTH, and leptin levels should be incorporated into case management. Weight loss through dietary restriction and increased movement, when feasible, remains foundational. Body condition scores should be recorded and monitored to maintain donkeys in the ideal range of 4–5/9.

4 Conclusion

Successful management of lamellar structural failure in donkeys and mules relies on early identification of metabolic dysfunction, precise diagnostic imaging, and scaled mechanical therapy. Due to their anatomical advantages and more substantial vascular corium, donkeys may respond favorably when provided with timely and tailored care. Radiographs, venograms, and clinical markers should be repeated frequently to evaluate response. When paired with focused medical and dietary management, these tools help restore hoof function and minimize long-term structural compromise.

4.1 Abscesses

Hoof abscesses are arguably the most common cause of acute, severe lameness in donkeys and mules. While the clinical presentation often mimics more serious pathologies such as fractures or laminitis, most hoof abscesses follow a predictable course and resolve with appropriate care. Their frequency in donkeys and mules can be attributed to prolonged exposure to wet or unsanitary environments, as well as suboptimal hoof care practices which predispose the foot to infection. Understanding the pathogenesis, diagnostic approach, and treatment strategies is essential to proper case management.

4.2 Etiology and Predisposing Factors

In donkeys and mules, hoof abscesses are most caused by breakdown of the terminal lamina (white line) allowing bacterial infiltration into the solar or sub mural corium. This breakdown may result from excessive moisture, poor hygiene, or chronic mechanical stress. Other frequent routes of entry include direct puncture wounds from nails or sharp debris and propagation through vertical hoof wall cracks. Abscess formation is essentially the body's inflammatory response to bacterial invasion in a confined, non-compliant structure.

White line stretching, common in animals with metabolic dysfunction and lamellar structural failure (LSF), compromises the natural barrier of the hoof. These compromised areas create pathways for foreign debris and pathogens to access sensitive internal tissues. In donkeys and mules affected by LSF, recurrent abscesses may result from repeated trauma or continued white line separation, highlighting the need to treat both the infection and its mechanical cause.

4.3 Diagnosis

The hallmark of a hoof abscess is acute onset, non-weight-bearing lameness of a singular limb, often with a bounding digital pulse and distal limb heat. Diagnosis is facilitated by careful application of hoof testers and exploration with a probe or blunt pick. In many cases, careful inspection of the white line or solar surface will reveal a defect, fissure, or soft spot suggestive of a draining tract.

Direct puncture wounds require special attention and more advanced diagnostics. If the puncture occurs near the midline frog, sulci, or over areas such as the navicular bursa or coffin joint, radiographs and potentially advanced imaging are warranted to assess deeper structures and determine involvement of the coffin bone, navicular bone, or associated synovial structures. Infection involving these deeper tissues necessitates more aggressive and systemic treatment and is best managed in collaboration with a veterinarian.

4.4 Treatment

The primary treatment goal in managing hoof abscesses is drainage. In most cases associated with moisture and terminal lamina breakdown, abscesses can be vented by carefully following the tract using a blunt probe, bent horseshoe nail, or narrow loop knife. Over-paring or debriding the solar surface is discouraged, as this can create further instability or predispose the area to secondary infection. When feasible, drainage via the hoof wall, rather than the sole, simplifies ongoing care and reduces contamination from ground contact.

In cases where a drainage tract is not easily found, warm water soaks using Animalintex or similar poultice pads are highly effective. This helps soften the surrounding hoof capsule, relieve pressure, and facilitate natural drainage through the coronary band or sole. Bandaging the foot after drainage, or using a protective boot or shoe, helps to keep the area clean and free of new debris, which is essential to avoid recurrence.

If the abscess originated from a puncture wound, more aggressive measures may be warranted. This may include systemic antibiotics, regional limb perfusion, radiographs, hoof wall resection, specialized shoeing to offload the defect, or even surgical debridement. These cases carry a more guarded prognosis and should be managed with veterinary oversight.

Abscesses secondary to lamellar structural failure require a dual-pronged treatment approach. Addressing the abscess alone without managing the underlying LSF leads to recurrence and persistent lameness. As discussed in the earlier sections of this chapter, appropriate radiographic and venographic assessment paired with mechanical offloading is essential for long-term success in these cases.

4.5 Conclusion

Hoof abscesses in donkeys and mules are a frequent and often painful condition that requires attentive management. Prevention through regular farriery, environmental hygiene, and metabolic health is ideal. However, when abscesses do occur, prompt identification and appropriate venting, combined with protection of the defect and management of underlying structural or metabolic causes, provide a high rate of resolution. Practitioners must remain vigilant for cases involving deeper structures and should not hesitate to escalate care when anatomical involvement beyond the corium is suspected.

4.6 Navicular Syndrome in Donkeys and Mules

Navicular syndrome, while commonly associated with horses, is notably less prevalent in donkeys and mules. However, mules, being hybrids of horses and donkeys, inherit conformation and biomechanical traits from both species, making them more susceptible to navicular-related pathology than donkeys but still less so than horses.

In donkeys, navicular syndrome is rare due to their naturally upright hoof conformation, robust frog structures, and a more substantial vascular corium. These anatomical advantages help in mitigating the biomechanical stresses that commonly predispose horses to navicular disease. Donkeys generally possess a higher dorsal hoof wall angle and a shorter toe-lever, reducing the strain on the deep digital flexor tendon (DDFT) and navicular apparatus.

Mules, however, present a more complex picture. Their hoof conformation often blends donkey uprightness with the flatter, longer-toe characteristics of horses. This combination can predispose some mules to soft tissue overload within the navicular apparatus, particularly in animals with inherited horse-like hoof morphologies and biomechanical inefficiencies.

The pathophysiology in mules mirrors that of horses: chronic overload of the DDFT, impar ligament strain, and resultant inflammation and degeneration of the navicular bone's soft tissue attachments. Key contributing factors include long toe-low heel conformation, crushed heels, and poor mediolateral balance.

Diagnostically, the approach parallels equine protocols. A thorough clinical examination focusing on hoof tester sensitivity in the heel region, gait analysis for toe-first landings, and diagnostic analgesia (e.g., palmar digital nerve blocks) are foundational. Radiographs remain the primary imaging modality, with dorsopalmar and lateral views essential for evaluating the navicular bone and overall hoof conformation. Although magnetic resonance imaging (MRI) offers superior soft tissue detail, it is often impractical due to cost and availability, particularly in donkeys and mules.

Mechanical management remains the cornerstone of treatment. As in horses, reducing the toe lever and raising the palmar angle with shoeing mechanics and or trimming are a primary intervention, effectively decreasing DDFT tension and load on the navicular bone. Correcting mediolateral imbalances and restoring appropriate palmar angle further alleviates stress on the navicular apparatus.

Medical management is adjunctive, with non-steroidal anti-inflammatory drugs (NSAIDs) utilized for pain control, and adjunctive therapies like bisphosphonates, coffin joint injections, or navicular bursa injections considered in refractory cases. Importantly, any treatment plan must address the underlying biomechanical faults to prevent recurrence.

Consistent monitoring through serial radiographs is critical, ensuring that shoeing modifications maintain proper alignment and load distribution. Rehabilitation should proceed cautiously, with gradual increases in activity levels to avoid exacerbating the condition.

In conclusion, while navicular syndrome is less common in donkeys and moderately more so in mules, its presence warrants a comprehensive diagnostic and management approach. Understanding species-specific anatomical and biomechanical nuances is essential in optimizing outcomes for these unique equids.

4.7 White Line Disease in Donkeys and Mules

White line disease, also referred to as onychomycosis, is an opportunistic fungal and bacterial infection of the inner hoof wall, particularly targeting the non-pigmented stratum medium—the so-called "white line." Though less of a misnomer than in the past, confusion remains in terminology. The true white line is not actually white and refers to the terminal laminae, which appear yellow to brown. The stratum medium is the region that is invaded by opportunistic organisms (Thiemann 2019).

In donkeys and mules, white line disease is a common and often underdiagnosed cause of lameness, particularly in moist environments and when ideal hoof care management is lacking. Donkeys appear more susceptible due to a combination of environmental conditions and mechanical predispositions, such as excess toe leverage or poorly managed wall defects.

The disease typically gains entry through structural weaknesses such as stretched terminal lamina, vertical wall cracks, or previous abscess tracts. Fungi and bacteria exploit these defects, infiltrating the stratum medium and, if unchecked, progressively undermining the integrity of the hoof wall. Most unilateral cases in horses, donkeys and mules involve club feet or subtly upright conformation patterns—mechanical contributors that increase tension on the white line.

Clinically, the disease is characterized by a loss of wall integrity at the solar surface. The affected area may appear as a separation between the wall and sole and tapping may produce a hollow sound. The sole may migrate forward, displacing the terminal laminae proximally and anteriorly. Dirt, debris, and sand are often packed into the crevice, visibly contributing to the lesion's progression.

Radiographically, white line disease presents as a lucent, irregular air pocket along the horn side of the horn-lamellar (HL) zone. This is distinct from laminitis, which involves the lamellar side and shows more uniform radiographic displacement of the coffin bone. The lesion may extend proximally toward the apex of P3, with irregular, radiolucent borders often superimposed on bone. These differences in appearance help distinguish between white line disease and lamellar structural failure (LSF) despite similar clinical signs.

The author has found that donkeys and mules respond best to a four-point style trim that reduces toe leverage and mechanical strain on the white line for maintenance trimming. Routine hoof cleaning is critical to reduce the risk of reinfection or debris impaction. Shoeing strategies that reduce deep digital flexor tendon (DDFT) tension and leverage on the toe are crucial. Rockered shoes, clogs, and other mechanical offloading devices have been instrumental in successful outcomes. These tools help reduce further tearing and allow for horn regrowth while unloading compressed and undermined areas of the hoof.

In more extensive cases, mechanical stabilization using adhesives or casting techniques may be required. Stabilizing the hoof during the regrowth phase ensures continued structural integrity and prevents additional mechanical separation. Once all active infection is cleared through appropriate topical therapy and debridement, the author prefers to seal the debrided area with a high-concentration copper sulfate–impregnated polymethacrylate product such as Equilox or Hoof MD. This method helps prevent further infection while allowing the defect to grow out with proper support.

The clinical approach should begin with accurate identification of the affected region using radiographs and probing. Debridement of the affected wall is critical to expose the infected horn and allow for direct topical therapy. Care must be taken to remove only the compromised horn while preserving healthy structures. Post-debridement care may include stabilization with adhesives or hoof casting, and application of antifungal and antibacterial agents to the exposed site.

Return to work or training should only be considered once the lesion has fully grown out, the hoof wall is stable, and the mechanical and environmental contributors have been addressed. With early recognition, appropriate mechanical offloading, and strict hygiene, donkeys and mules can recover effectively and resume their roles with minimal long-term consequences.

4.8 Keratomas in Donkeys and Mules

Keratomas are benign, space-occupying masses composed of keratin that develop within the hoof capsule. While relatively uncommon across all equids, their occurrence in donkeys and mules may be underdiagnosed due to species-specific anatomical and behavioral factors. Donkeys, for instance, possess more upright, cylindrical hooves and often exhibit subtle signs of discomfort, making early detection challenging.

4.9 Pathophysiology and Etiology

Keratomas typically arise from abnormal proliferation of keratin-producing epithelial cells, often following chronic irritation, trauma, or infection. In donkeys, recent studies have identified the presence of keratinopathogenic fungi and Treponema spp. spirochetes within keratoma lesions, suggesting a potential microbial component in their development. The specific causes are not well understood, but potential factors include repeat trauma and foreign body entrapment.

4.10 Clinical Presentation

Affected animals may present with:

- Recurrent hoof abscesses, particularly in the same location.
- Varying degrees of lameness, often correlating with the size and location of the keratoma.
- Hoof wall deformities, including bulges or deviations in the white line.
- Sole pain, especially upon pressure or during hoof testing.

Radiographic imaging often reveals a radiolucent area adjacent to the distal phalanx, indicative of the keratoma's presence.

4.11 Diagnosis

Accurate diagnosis involves a combination of:

- Clinical examination, assessing signs of lameness and hoof deformities.
- Radiography, identifying characteristic radiolucent and radiopaqe zones within the hoof capsule.
- Histopathology, confirming the keratoma's benign nature and ruling out other pathologies.

Advanced imaging modalities, such as venograms, MRI or CT scans, may offer enhanced visualization in complex cases.

4.12 Treatment and Prognosis

The primary treatment for keratomas is surgical excision, aiming to remove the mass entirely and alleviate pressure on surrounding structures. Post-operative care includes:

- Protective shoeing to support the hoof during regrowth.
- Regular monitoring to ensure proper healing and early detection of potential recurrence.
- Antimicrobial therapy if secondary infections are present.

Prognosis is generally favorable, with many animals returning to their previous level of activity post-recovery. However, incomplete removal can lead to recurrence, necessitating further intervention.

4.13 Occurrence in Donkeys and Mules

While comprehensive data on keratoma prevalence in donkeys and mules is limited, a study involving 167 donkeys with keratomas highlighted the condition's presence in this species. Given the diagnostic challenges and subtle clinical signs in donkeys and mules, it's plausible that keratomas are underreported in these animals.

4.14 Conclusion

Keratomas, though benign, can significantly impact the hoof health and overall well-being of donkeys and mules. Early recognition, accurate diagnosis, and appropriate surgical intervention are key to successful outcomes. Regular hoof care and monitoring are essential, especially in animals with a history of hoof trauma or chronic infections.

4.15 Flexor Tendon Contractures and Laxity in Donkeys and Mules

Flexural limb deformities, encompassing both tendon contractures and laxity, are significant orthopedic concerns in donkeys and mules. These conditions can arise congenitally or be acquired due to factors such as malnutrition, overwork, or inadequate hoof care. In neglected cases, it's not uncommon to observe a combination of contracture in one limb and laxity in the contralateral limb, complicating diagnosis and treatment (refer to Fig. 10).

Pathophysiology

- **Flexor Contractures**: These involve a disproportionate length of the distal long bones to the length of deep digital flexor tendon (DDFT), superficial digital flexor tendon (SDFT), or both. This results in over-flexion of joints, leading to an upright or knuckled stance. In severe cases, the animal may walk on the dorsal aspect of the hoof or fetlock.
- **Flexor Laxity**: Characterized by insufficient tension in the flexor tendons, leading to hyperextension of joints. Affected animals may exhibit dropped fetlocks or walk on the palmar/plantar aspect of the pastern.

Clinical Presentation

- **Contractures**:
  - Upright or knuckled posture.
  - Difficulty in placing the heel on the ground.
  - Resistance to limb extension.
- **Laxity**:
  - Dropped fetlocks.
  - Hyperextension of joints.
  - Difficulty in maintaining a normal stance.

Diagnosis

Accurate diagnosis involves:

- **Clinical Examination**: Observing gait, stance, and limb conformation.
- **Radiography**: Assessing bone alignment and joint angles.
- **Ultrasonography**: Evaluating tendon integrity and detecting any lesions.

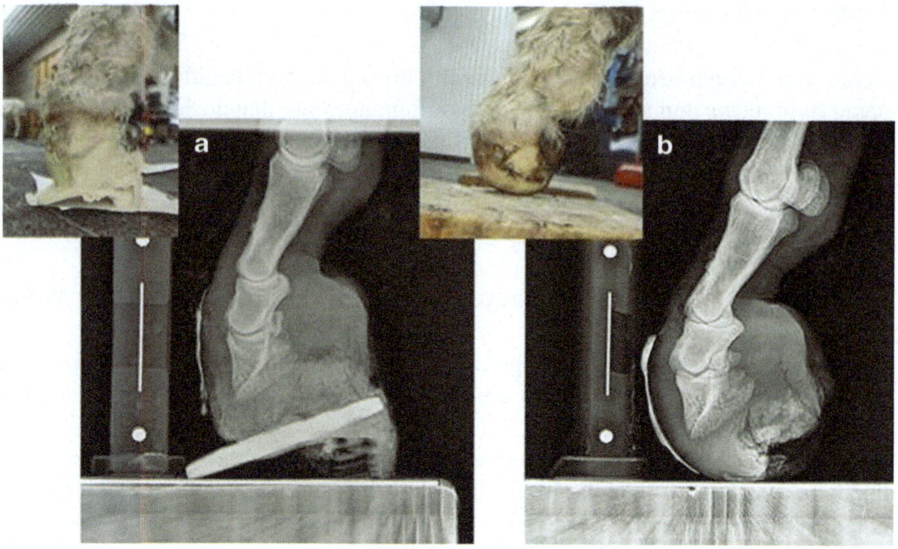

Fig. 10 This is a deep flexor contracture in a donkey with pre shoeing radiograph at (**b**) and post trim and extended toe, wedged heel shoe at (**a**). This shoeing will offer palliative care and improve the ability to walk and stand. However, most of the upper-level cases require surgery to cut the deep flexor tendon combined with this shoeing and casting for initial management. A gradual weaning down of the angle is recommended over 7–10 days and dependent on patient comfort

Management Strategies

- **Conservative Treatment**:
  - **Physical Therapy**: Regular stretching exercises to improve tendon flexibility. **Corrective Trimming and Shoeing**: Flexor laxity generally requires reducing the toe leverage and moving the point of heel contact as far palmar as possible, which may necessitate heel extensions based on the specific circumstances. Flexor contracture often involves a combination of trimming down the heel while replacing it with a wedge and adding a toe extension to provide mechanical assistance and improve the donkey or mule's ability to stand.
  - **Nutritional Support**: Ensuring adequate intake of essential nutrients to support musculoskeletal health.
- **Surgical Intervention**:
  - **Desmotomy**: Surgical release of the check ligaments to alleviate tension in contracted tendons.
  - **Tenotomy**: Cutting of the tendon in severe cases to allow for realignment and improved function.

Post-operative care includes controlled exercise, continued corrective trimming, shoeing, possibly casting and monitoring for signs of recurrence or complications (Kramer 2006).

Prognosis
With early detection and appropriate management, the prognosis for donkeys and mules with flexural deformities is generally favorable. However, chronic or severe cases may result in permanent lameness or reduced functionality, emphasizing the importance of prompt intervention.

4.16 Conclusion

Effective hoof care in donkeys and mules requires a thorough understanding of anatomy, biomechanics, and the environmental and metabolic contributors unique to these species. Whether managing white line disease, lamellar structural failure, navicular changes, or flexor contracture syndromes, successful outcomes rely on accurate diagnosis, mechanical insight, and diligent follow-through.

Routine hoof trimming using a four-point model, consistent environmental management to reduce moisture and debris accumulation, and nutritional support targeting insulin regulation and metabolic stability are foundational to long-term soundness. Practitioners must remain vigilant to subtle changes in gait, hoof conformation, and response to load—early intervention often makes the difference between full recovery and chronic pathology.

With a structured approach rooted in both experience and evolving science, the donkey and mule hoof can be managed effectively, supporting their welfare, comfort, and continued usefulness across a range of disciplines and environments.

Test Questions

1. What is the most common cause of acute severe lameness in donkeys and mules?
2. What environmental condition most commonly contributes to hoof abscesses in donkeys?
3. What anatomical hoof structure is often compromised in cases of white line isease?
4. Which diagnostic tool is essential to identify internal hoof pathologies like abscesses or WLD?
5. What is the preferred method of drainage for most hoof abscesses?
6. How can soaking in Animalintex poultice pads assist abscess treatment?
7. What kind of mechanical trimming is recommended for donkeys and mules by the author?
8. Which four pillars are preserved in the author's four-point trimming style?
9. What metabolic condition commonly contributes to lamellar structural failure in donkeys?
10. What are the two main components of the hoof wall affected in white line disease?
11. What soft tissue tendon structures can be involved in flexor contractures?

12. What imaging technique is recommended when assessing puncture wounds?
13. Which shoeing methods are often employed for WLD management?
14. What is the purpose of venograms in evaluating hoof health?
15. What component can be applied to debrided WLD defects for antimicrobial protection?
16. What is a keratoma, and how might it present radiographically?
17. What microbial organisms have been found in keratoma lesions in donkeys?
18. What kind of trimming mimics the natural wear patterns in donkeys and mules?
19. What is a common presentation of a donkey with laminitis on physical exam?
20. Which hormone-related condition is treated with pergolide in donkeys?
21. Why is identifying the mechanical cause of LSF essential for abscess management?
22. What shoeing style is used to reduce DDFT tension in laminitic cases?
23. What are key diagnostic signs of WLD visible on radiographs?
24. Why is the laminar corium significant in evaluating hoof vascular integrity?
25. What is emphasized in the conclusion as crucial for long-term hoof health?

Answer Key

1. Hoof abscesses
2. Wet or unsanitary environments
3. Stratum medium and terminal lamina
4. Radiographs
5. Venting through the hoof wall or following the tract conservatively
6. Softens the horn and promotes drainage to the coronary band
7. Four-point style trim
8. Toe pillars, heel pillars
9. Equine Metabolic Syndrome (EMS)
10. Stratum medium (white zone) and stratum internum
11. Deep digital flexor tendon and superficial digital flexor tendon
12. Radiographs
13. Rockered shoes, clogs, adhesives or casting
14. To assess vascular supply and compression
15. Copper sulfate impregnated polymethacrylate (e.g., Equilox)
16. A benign mass that appears as a circular radiolucent defect
17. Keratinopathogenic fungi and Treponema spp. spirochetes
18. Four-point style trimming
19. Bounding digital pulses, reluctance to move, heat in the hoof
20. Pituitary Pars Intermedia Dysfunction (PPID)
21. To prevent recurrence by reducing mechanical stress
22. Rockered shoes or clogs
23. Lucent zones, irregular borders at the stratum medium
24. It reflects perfusion and inflammation related to hoof health
25. Routine care, nutrition, and environmental management

References

Amene AS, Fekadu B, Molla D, Sheferaw B (2019) The prevalence of foot related problems in working donkeys and its implication on the livelihood of donkey owners in Hawassa City, Southern Ethiopia. Int J Livestock Product 10(3):86–93. https://doi.org/10.5897/IJLP2018.0527

Baldwin GI, Pollitt CC (2010) Progression of venographic changes after experimentally induced laminitis. Vet Clin N Am Equine Pract 26(2):339–359

Bonelli F, Rota A, Sgorbini M et al (2019) Prevalence and risk factors for lameness in dairy donkeys in Italy. Livest Sci 223:27–32. https://doi.org/10.1016/j.livsci.2019.03.008

Bukhari SSUH, Ali MM (2023) Investigation of foot pain, lameness and gait abnormality in draught donkeys in Pakistan. Vet Med Sci 9(1):204–213. https://doi.org/10.1002/vms3.1017

Cornelisse CJ, Robinson NE (2013) Glucocorticoid therapy and the risk of equine laminitis. Equine Vet Educ 25:39–46

De Laat MA, McGowan CM, Sillence MN (2010) Equine laminitis: induced by 48 h hyperinsulinaemia in Standardbred horses. Equine Vet J 42(2):129–135

Hood DM (1999) The mechanisms and consequences of structural failure of the foot. Vet Clin N Am Equine Pract 15(2):437–461

Karikoski NP, McGowan CM, Singer ER, Asplin KE, Sillence MN (2014) Pathology of natural cases of equine endocrinopathic laminitis associated with hyperinsulinemia. Vet Pathol 51(3): 731–740

Kramer J (2006) Deep digital flexor tenotomy. In: Manual of equine field surgery. Saunders, pp 98–102

Misk N, Hifny A (1983) Anatomy of the hoof in donkeys. Assiut Vet Med J 10:1–10

Nocera I, Aliboni B, Ben David L, Gracia-Calvo LA, Sgorbini M, Citi S (2022) Radiographic and venographic appearance of healthy and laminitic feet in Amiata donkeys. Front Vet Sci 9: 923486. https://doi.org/10.3389/fvets.2022.923486

O'Grady SE, Parks AH (2008) Farriery options for acute and chronic laminitis. In: Proceedings of the AAEP convention, San Diego, CA

Parks AH, O'Grady SE (2003) Chronic laminitis: current treatment strategies. Vet Clin N Am Equine Pract 19(2):393–416

Phannithi T, Laikul A, Pathomsakulwong W, Rungsri P, Apichaimongkonkun T, Watchrarat K, Cherdchutham W (2023) External hoof measurements of untrimmed and unshod mules in Northern Thailand. Animals 13(5):871. https://doi.org/10.3390/ani13050871

Pollitt CC (2015) The illustrated horse's foot. Saunders

Redden RF (1997) Shoeing the laminitic horse. In: Proceedings of the American Association of Equine Practitioners, AAEP

Redden RF (2003) Clinical and radiographic examination of the equine foot. In: Proceedings of the international equine laminitis conference

Redden RF, Bras RJ (2017) How to obtain and use radiographic information to enhance therapeutic shoeing and trimming in horses. In: Proceedings of the American Association of Equine Practitioners, San Antonio, TX

Rucker A (2017) The digital venogram. In: Equine Laminitis. Wiley, pp 240–251

Stewart ML (2018) How to construct and apply atraumatic therapeutic shoes to treat acute or chronic laminitis in the horse. In: Proceedings of the American Association of Equine Practitioners

Taylor D, Hood DM, Wagner IP (2003) Short-term effect of therapeutic shoeing on severity of lameness in horses with chronic laminitis. J Equine Vet Sci 23(3):113–118

Tesfaye A, Mengistu A, Regassa A (2010) Assessment of working donkeys' hoof conditions and management practices in eastern Ethiopia. Trop Anim Health Prod 42(6):1241–1245. https://doi.org/10.1007/s11250-010-9555-5

Thiemann AK, Poore LA (2019) Hoof disorders and farriery in the donkey. Vet Clin N Am Equine Pract 35(3):589–608. https://doi.org/10.1016/j.cveq.2019.07.008

Zhao Y, Li X (2022) Evaluation of the welfare status of donkeys on smallholder farms in China. Animals 12(14):1792. https://doi.org/10.3390/ani12141792

Farrier Practices: Maintaining Donkey and Mule Hooves

Erek Grohs CF

Abstract

The design and structure of donkey and mule feet is dramatically different from that of a horse. Donkeys have a much more upright foot, with deeper sulci around the frog, and outward flares of the point of heel. Due to the thickness of the hoof structure, and the upright conformation, they do not possess the same shock absorbing capabilities that a horse's foot needs. Mules, being a hybrid of the donkey and horse, have a hoof structure that may be more similar to a donkey or a horse. Mules and donkeys may prove to be challenging to work with and require additional restraints such as blind folds and nose twitches along with patience. Special attention to your positioning relative to the animal, and how you hold their legs and feet is important for your safety and the animal's comfort. Unfortunately, most farrier schools do not spend much time focusing on this group of equids and there is not much literature available. I will share my personal experience trimming and shoeing both mules and donkeys here in my own practice in Oregon and in Latin America with the Equitarian Initiative.

Keywords

Farrier · Donkey hooves · Mule hooves · Equitarian Initiative

1 Introduction

As a farrier who works with a variety of equids, and across disciplines I have experienced that often the hoof trimming and shoeing of donkeys and mules has been inadequate due to a lack of understanding of their needs. Unfortunately, most

E. Grohs CF (✉)
Northwest HoofCare, Myrtle Creek, OR, USA
e-mail: erek@northwesthoofcare.com

A. K. McLean et al. (eds.), *Donkey and Mule Medicine: Foundations, Breeds, and Management*, https://doi.org/10.1007/978-3-032-12726-6_25

farrier schools do not spend much time focusing on this group of equids and there is not much literature available. In this chapter, I will relate my experience working with donkeys and mules as a hoof care provider, as well as my work with Equitarian Initiative in rural underdeveloped parts of Latin America, to describe the proper management of their hooves and some of the common hoof diseases in these equids.

The design and structure of donkey and mule feet is dramatically different from that of a horse. Donkeys have a much more upright foot, with deeper sulci around the frog, and outward flares of the point of heel. They also generally have a thicker sole, frog, and hoof wall than that of a horse. This design is suited to their natural desert habitat of rocky, uneven terrain, allowing for more precise foot placement, traction, and resistance to abrasion. While donkeys can certainly move with speed, they are more prone to casually walking along as compared to the horse which has a propensity to gallop long distances. Due to the thickness of the hoof structure, and the upright conformation, they do not possess the same shock absorbing capabilities that a horse's foot needs.

Mules, being a hybrid of the donkey and horse, have a hoof structure that may be more similar to a donkey or a horse. Like donkeys, mules tend to have a thicker sole and hoof wall, and a straighter heel, compared to horses. When trimming or shoeing these animals, it is important to consider their unique anatomy, as typical practices for horse foot care will not be suitable.

The handling of mules and donkeys can be difficult for someone used to working with horses due to the differences in their behavior. They are often suspicious by nature, and it requires more effort to earn their trust. They also have more flexibility and range of motion with their hind legs, and as such can be more dangerous. Sometimes it may be necessary to use aides such as blindfolds, twitches, or foot ropes to safely handle the animal (See Fig. 1). Special attention to your positioning relative to the animal, and how you hold their legs and feet is important for your safety and the animal's comfort.

2 Restraints and Handling in the Field

Most farriers and veterinarians learn good horsemanship skills that can be applied to all equids, however, there are a few things that apply specifically to donkeys and mules which will be helpful in performing proper hoof maintenance (see Behavior Chapter).

First off, donkeys and mules tend to be more wary of new people and experiences than horses.

When approaching, I find that confidently directing yourself to the front quarter is more effective than head on. This approach also keeps you out of striking zone from a front foot. "Disciplining" a donkey or mule is rarely effective. If they pull away, kick, etc… it is best to back off some and release pressure, and then re-approach calmly. With horses you may be able to demand compliance, however, with donkeys and mules you need to earn their trust. In my experience, once they trust you, they never forget.

Fig. 1 Utilizing a rope to manage the hind foot of a mule

Begin by gently touching the equid's shoulder, and slowly working your hand down its side. Be cautious as you work your way to the hind end, as the hind legs are perfectly capable of swinging forward and out if the animal feels uncomfortable. Special attention to body language is important. In my experience, while a horse may sometimes react suddenly without much warning, a donkey or mule generally gives some indication about their feelings and intentions. When they seem to be relaxed you can work your hand down the side of a front leg. If the animal becomes agitated, back up some and then repeat the process. Once you can reach the cannon bone without agitation, gently wrap your hand around the limb and ask for the foot (See Fig. 2). I find that applying pressure to the tendons as we may with a horse, is counterproductive. Once they have willingly lifted the foot, position it between your legs as you would a horse and begin your trim. If you're working with a smaller donkey, it may be easier to hold the foot with one hand and use your tools with the other.

The same applies to the hind end. Carefully work your way over the rump and down the side. Often, they will become agitated and sometimes kick out when you get close to the hock. If this occurs, give them some release by backing off, and then try again. Once you can touch the cannon bone, they should willingly give up the foot (See Fig. 3).

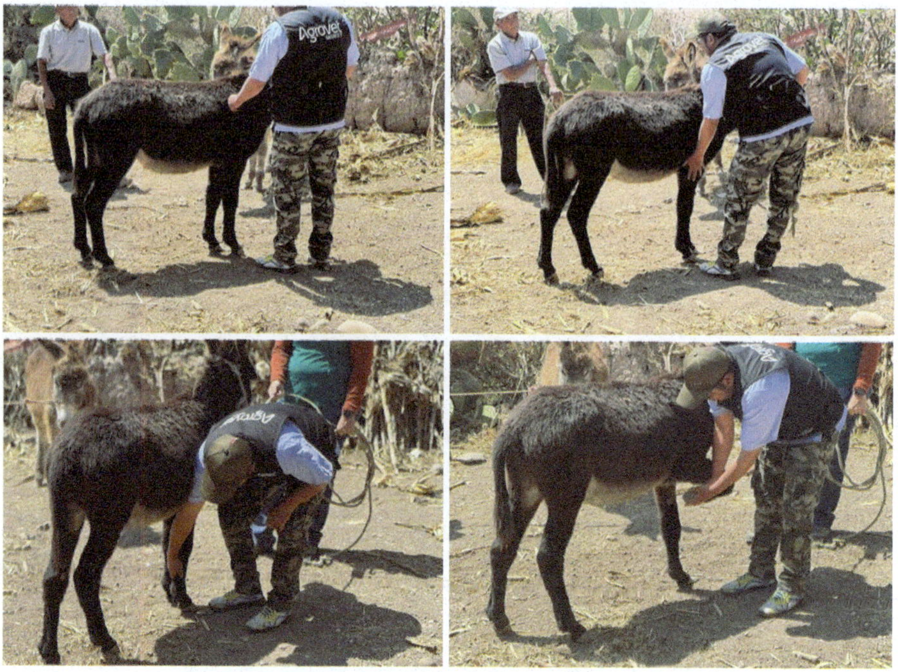

Fig. 2 Demonstration of approaching a donkey and asking for its front foot

Sometimes while holding the cannon bone, they will jerk their leg forward and back a few times before settling. Hold onto the leg until they've settled and then they should be ready for the trim. With the hind leg on smaller animals, I will often get on my knees with my arm over the hock and around the medial aspect of the limb, and the lower limb cradled in my lap (See Fig. 4). This position keeps you relatively safe and keeps them comfortable. Small concessions to your comfort to aid theirs will usually result in more compliance from the donkey or mule.

3 The Trim

Donkeys and mules require the same frequency of trimming as a horse. On average, they should be trimmed every 6–8 weeks. While their hooves tend to be more sturdy and less prone to distortion, the excess leverage caused by overgrowth still greatly affects the tendons, ligaments, and joints. Regular trimming is also important to assess and address any disease or defects that may occur.

As stated previously, the donkey and mule foot anatomy is different from the horse (see Illustration 1). Most of the time, we consider the angle of the dorsal hoof wall relative to the palmar aspect of the foot set at 55° to be an appropriate angle to shoot for on a horse hoof, whereas a donkey's ideal angle is closer to 60–65°. They

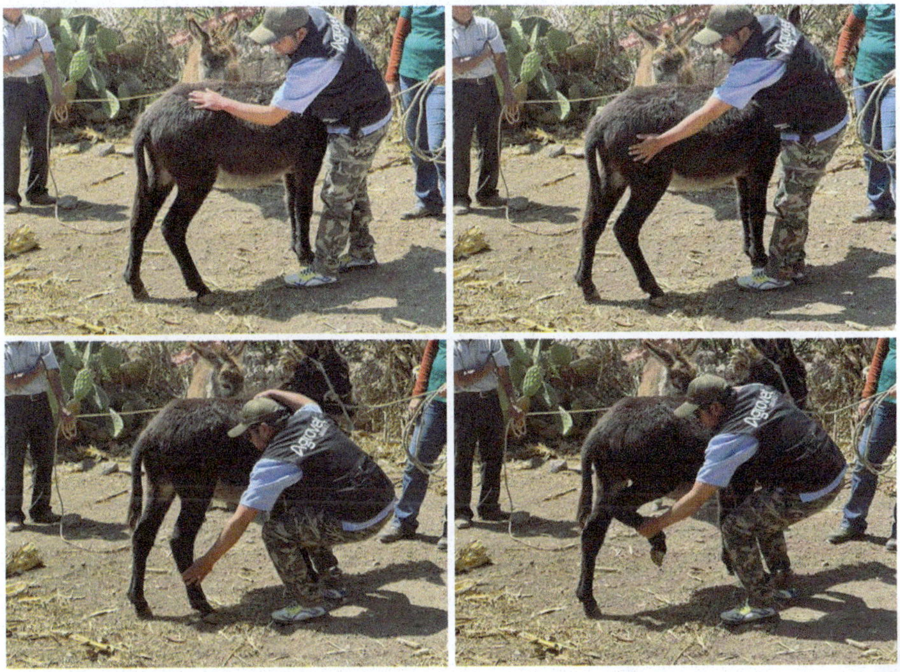

Fig. 3 Demonstration of approaching a donkey and asking for its hind foot)

Fig. 4 Sometimes you have to adapt your position to keep the animal comfortable

also have a much deeper and wider frog. The shape of a donkey foot is also different. While a horse's front foot is generally somewhat circular and hind feet somewhat spade shaped, all four feet of a donkey should resemble a U, with a slight outward flare at the points of the heels. The conformation of a mule foot can vary widely from horse-like to donkey-like but tends to more closely resemble a donkey. For the sake

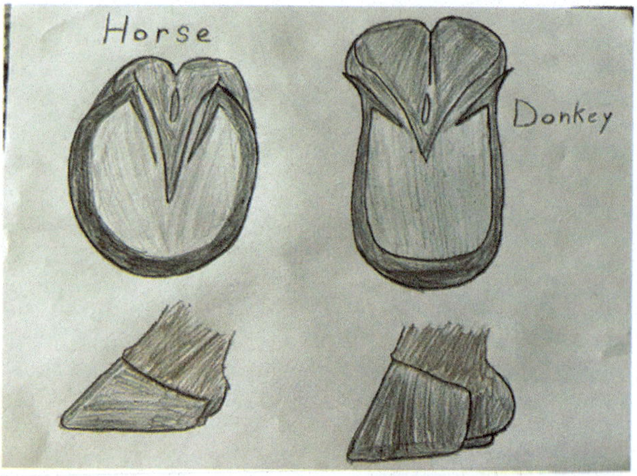

Illustration 1 Difference in shape; Horse Vs. Donkey foot

of this chapter, I will focus on trimming a donkey. When trimming mules, you will need to make decisions based upon the structure you encounter.

I begin my trim by picking out any loose dirt from the sole and commissures. If the frog appears infected or there are loose bits, conservatively trim away to healthy solid material. Ideally, there will still be contact between the frog and the ground when the trim is completed.

Next, I will use my knife to exfoliate any excess sole until I find the solid border between sole and wall. Donkeys and mules tend to "stack" on more sole than horses, so you may need to go deeper than accustomed to with a horse.

To begin my trimming of the wall, I will start my nipper run at the heel. I identify my starting point by observing where the heel growth deviates from straight. It can be helpful to use a sharpie to mark each heel and then draw a line following the border of wall and sole. Use your nippers to trim flat from one heel to the other. Next, use your rasp to insure a flat level ground-bearing surface. To check your trim, hold the donkey's leg by the cannon bone and sight down the back side. For proper medial lateral balance, the solar surface of the foot should be perpendicular to the limb, with no high spots. After establishing your balance, use the fine side of your rasp to round the perimeter of the hoof. Removing the sharp edge caused by the initial trim will help to prevent breakage of the edge of the wall (See Fig. 5).

Once you've trimmed the bottom of the foot, let the donkey stand on it and observe the angle and shape. Most of the time, donkeys do not require any top dressing (rasping of the wall). However, any flares aside from the points of the heel should be filed down. If there appears to be excess length at the toe, use your nippers to square it off.

4 Shoeing

Donkeys rarely require shoes. Their hoof walls and soles are quite durable, and the shape and size of the foot does not lend itself easily to applying a shoe. Mules however, depending on the integrity of the foot, the work they are asked to perform, and the terrain they are working on, may need shoes for comfort and traction.

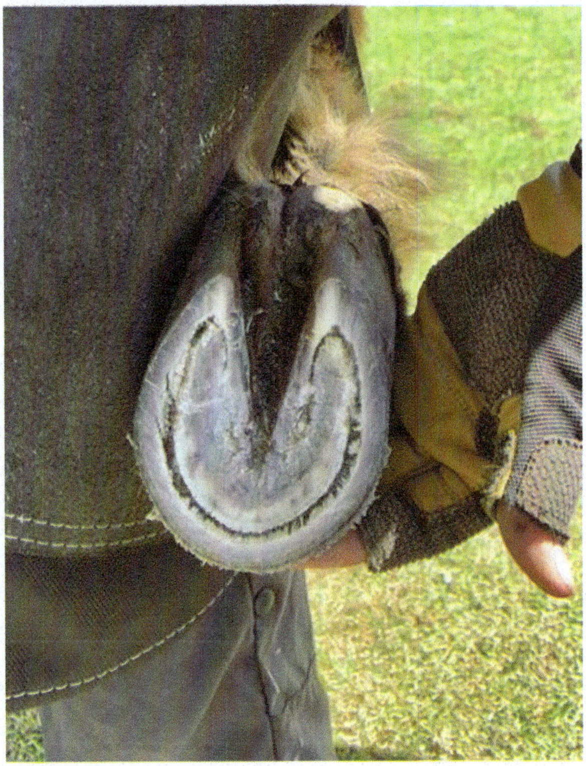

Fig. 5 Freshly trimmed mule foot with contracted heels and sulci thrush

Shoe selection for mules is important. Mule shoes differ from horseshoes in their shape. A mule shoe should have straight branches, and a blunter toe than a horseshoe. Often a plain mule keg shoe is sufficient, but if more traction is necessary, they come with heel and toe caulks, or you can add traction by braising on borrium, or applying stud pins.

To appropriately place a mule shoe, it should be perimeter fit. Take care to avoid excess heel length on the shoe, as this may be caught by another foot, root, rock etc., causing the loss of the shoe. Also note that the heel of the shoe should not extend past the heel of the foot and cover up the commissures and frog, as this may result in rocks being stuck between the shoe and foot causing discomfort and potentially abscessation.

Nail placement can be a challenge for those accustomed to working with horses. A mule's hoof walls tend to be straighter, and as such you will not need to pitch your nails much. The nails should exit the hoof wall approximately 1/3 of the way up and be clenched securely. On horses I prefer to only use the first three nail holes on each branch, avoiding driving any nails behind the widest part of the foot. With mules, especially those used for backcountry packing, I will use four per side in an effort to more securely hold the shoe in place.

5 Disease

While generally hardy, mules and donkeys encounter many of the same hoof issues that horses do. The most common issues that I encounter are thrush, white line disease, and abscesses. Often, I find that donkey and mule feet do not get the same attention as their horse counterparts. Frequent picking and cleaning of the foot is essential to prevent and treat these issues. Donkeys and mules can also be prone to metabolic issues that lead to laminitis (Refer to nutrition and endocrine chapters).

Abscesses form when the sensitive structures beneath the horny sole become inflamed and infected. This can occur when a rock or other foreign material gets lodged in the foot, a nail is driven too deep, or when walking on sharp gravel with soft or over-trimmed soles. When lameness is noted, careful examination of the foot is necessary to determine the location of the abscess. If wearing shoes, carefully remove the shoe. When you've located the abscess, carefully use your knife to create a tract for the abscess to drain. Once you have relieved the pressure, you can pack the area with a mixture of sugar and iodine, and wrap it, changing the wrap and packing every 2–4 days until the hole is closed and clinical signs have alleviated. If the abscess cannot be accurately located, or if drainage requires excessive removal of material, soak the foot in warm water with Epsom salts for 20 30 min, and then proceed with the packing and wrap. If the abscess has not resolved after one week, it may be prudent to use a radiograph to determine the size, location, and severity of infection, and then perform a more invasive removal of the infected tissue. If this is the case, attention should be paid to the sterility of your instruments, and proper bandaging of the foot afterwards (See Fig. 6).

Thrush is a fungal infection, usually affecting the frog and sole. It can be identified by a strong odor and sometimes the presence of dark "goo" in the commissures of the frog. This can cause a great deal of irritation and discomfort for the animal. The best treatment is prevention! Meaning, maintaining a clean foot

Fig. 6 Heel abscesses in Costa Rica. After resection of necrotic tissue, the opening was flushed with betadine, and then Beeswax was melted in to seal the wound

Fig. 7 Overgrown mule foot with white line separation and thrush

and sanitary environment. Mud and manure constantly packed in the foot will create a perfect environment for this fungus to grow. If the foot becomes infected, frequent cleaning and application of any antifungal medication will quickly address the problem.

White line disease occurs when there is a separation between the sole and wall. This is most commonly the result of allowing the hooves to grow too long (See Fig. 7). As the hoof grows away from the sole, pressure from standing on it or walking pulls the wall away from the connective structures. Once there is an opening, anaerobic bacteria start to grow and eat away at the healthy material deeper inside. This can ultimately result in a total loss of the hoof wall if not treated.

Once the foot has been trimmed, it is necessary to continue removing walls around the infected area until no more disease is found. Being an anaerobic infection, allowing air to access the area should prevent any further issues (See Fig. 8). However, it will be necessary to keep the affected area clean and dry until new growth has fully come in.

Fig. 8 Trimmed donkey foot exposing white line pocket

6 Conclusion

Working with donkeys and mules one must be patient, consider anatomical differences in the hoof anatomy related to the animal's conformation. Consider diseases found in horses to be similar or the same in mule and donkey hooves such as thrush, white line and any boney changes within the hoof capsule may occur. When working with both mules and donkeys in some cases the mules may be more relaxed with a familiar person or owner. Consider the restraint and handling techniques mentioned in this chapter along with my personal experience trimming and shoeing donkey and mule hooves. Hopefully, my personal experiences and tips will help you in the field with mule and donkey patients/clients (See Fig. 9).

Questions

1) Name two anatomical differences in donkey hooves compared to horse hooves
2) Name two anatomical differences in mule hooves compared to horse hooves
3) Name a disease that affects the white line of both donkeys and mules
4) Describe the differences in mule shoes compared to horseshoes

Fig. 9 One of my donkey cases is from Peru

Answers

1) Two key anatomical differences between donkey and horse hooves are:
   1. **Hoof Shape and Structure**: Donkey hooves are more upright, cylindrical, and have a more oval or boxy shape compared to the more rounded and angled hooves of horses. The donkey hoof is naturally steeper and narrower, adapted for dry, arid terrain (See Fig. 10).
   2. **Frog and Sole Anatomy**: Donkeys have a more reduced frog and a thicker, tougher sole than horses. The frog is often less developed and recessed, and the sole tends to be more concave and resilient to rocky surfaces.
2) Two anatomical differences in mule hooves compared to horse hooves are:
   1. **Hoof Shape**: Mule hooves are generally more upright, oval-shaped, and narrow compared to the rounder and broader hooves of horses (refer to Fig. 11). This reflects the influence of the donkey parent, which contributes to a more efficient hoof for rugged terrain.

      Hoof Wall Density: Mules tend to have denser and tougher hoof walls than horses, making them more resilient and less prone to chipping or cracking, especially in dry or rocky environments.

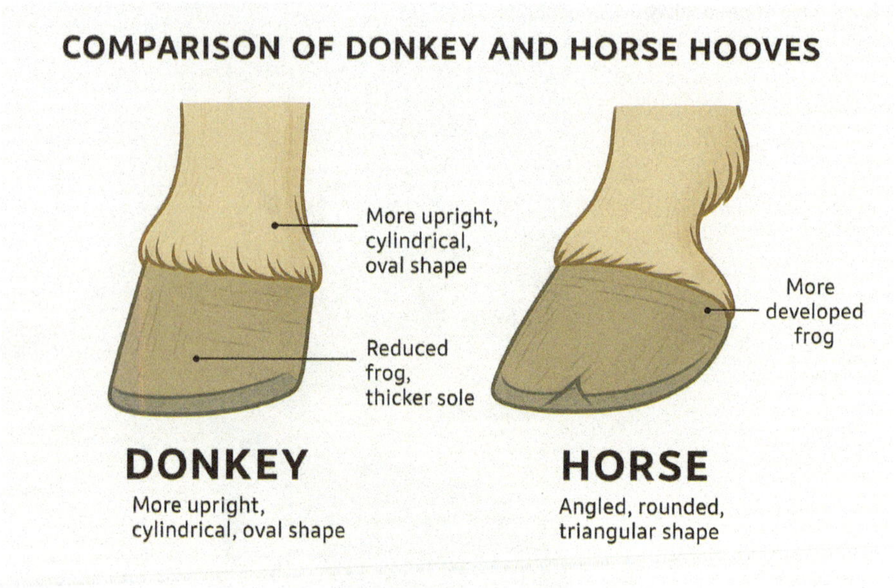

Fig. 10 Comparing the donkey and horse hoof (AI generated image)

Fig. 11 Comparing a mule hoof to a horse hoof

3) **White Line Disease**
4) Mule shoes differ from horseshoes in several important ways due to the unique anatomy and biomechanics of mule hooves:

1. **Shape and Size**: Mule shoes are typically narrower and more oval-shaped compared to the rounder and wider horse shoes. This reflects the mule's more upright, boxier hoof shape inherited from donkeys.
2. **Heel Fit**: Mule shoes usually have a longer heel extension to accommodate the more upright and contracted heel of a mule's hoof. This helps provide better support and prevents underrun heels, a common issue in mules.
3. **Material and Thickness**: Mule shoes are often made slightly lighter and thinner than horse shoes of comparable size, as mules tend to have harder hooves and are more efficient movers, requiring less shock absorption.
4. **Nailing Pattern**: The nail holes in mule shoes are positioned differently to match the tighter white line and stronger hoof wall typical of mules. This helps avoid injury and improves shoe retention.

The manufacturer's authorised representative in the EU is Springer Nature Customer Service Centre GmbH, Europaplatz 3, 69115 Heidelberg, Germany. If you have any concerns regarding our products, please contact ProductSafety@springernature.com

Printed and bound by CPI Group (UK) Ltd, Croydon, CR0 4YY
05/10/2026
02228035-0006